Geology and Paleontology of
Seymour Island, Antarctic Peninsula

View of Seymour Island looking northeastward from the Cretaceous terrane of the López de Bertodano Formation toward the Eocene rocks of the La Meseta Formation. In the midground, the intervening Cross Valley Formation is exposed in parts of the broad valley of the same name. Photo by Rodney M. Feldmann, 1984.

The Geological Society of America
Memoir 169

Geology and Paleontology of Seymour Island, Antarctic Peninsula

Edited by

Rodney M. Feldmann
Department of Geology
Kent State University
Kent, Ohio 44242

Michael O. Woodburne
Department of Earth Sciences
University of California at Riverside
Riverside, California 92521

1988

Published by The Geological Society of America, Inc.
3300 Penrose Place, P.O. Box 9140, Boulder, Colorado 80301

GSA Books Science Editor Campbell Craddock

Printed in U.S.A.

Library of Congress Cataloging-in-Publication Data

Geology and paleontology of Seymour Island, Antarctic
  peninsula.

  (Memoir / Geological Society of America ; 169)
  Includes bibliographies and index.
  1. Geology—Antarctic regions—Seymour Island.
2. Paleontology—Antarctic regions—Seymour Island.
I. Feldmann, Rodney M.   II. Woodburne, Michael O.
III. Series: Memoir (Geological Society of America) ;
169.
QE350.G45   1988   559.8'9   88-11013
ISBN 0-8137-1169-X

10 9 8 7 6 5 4 3 2

Contents

Foreword

This volume is the outgrowth of several field seasons of work on Seymour Island, Antarctica. The book was originally conceived on the return voyage from the island in January 1984, and, subsequent to that, the editors solicited potential articles from all individuals who had been engaged in paleontological and geological research in the area. The results of that effort are embraced within the articles that follow. The treatment is not comprehensive; rather, the papers represent the present state of our understanding of the geology and paleontology of Seymour Island within the limits of interest and ability of the authors.

Publication of these contributions in a single volume has a number of clear advantages, perhaps the most important of which is that access to a diverse range of studies will be simplified for future workers. Also, collective publication reflects the major importance of Seymour Island as a research site for the interpretation of the geological history of the Antarctic Peninsula.

As the work progressed, it became clear that the volume could be strengthened by exchange of ideas between authors and discussion of tentative results of the research. To that end, several of the papers were presented under the auspices of a symposium entitled, "Contributions of polar research to the earth sciences," convened by David L. Elliot and Rodney M. Feldmann at the meeting of the North-Central Section, Geological Society of America, at Kent State University, Kent, Ohio, on April 24–25, 1986. Fourteen of the twenty-two authors were able to attend the meetings and points of common interest were discussed at length. Among other results of that meeting, a generally uniform stratigraphic nomenclature was accepted, the areas of potential differences in interpretation were discussed, problem areas for future research were defined, and matters of style, format and deadline were resolved.

As with nearly all research in the Antarctic region, the primary source of support has been the Division of Polar Programs of the National Science Foundation. Grants from that agency to Rosemary A. Askin, David Elliot, Rodney M. Feldmann, Michael O. Woodburne, and William J. Zinsmeister supported a combination of field work, laboratory study, and publication without which this research could not have been conducted.

To assure that the individual contributions meet the standards of quality demanded by the Geological Society of America, technical reviewers were selected from among earth scientists who had previously been careful, authoritative reviewers and who were not involved in the research on Seymour Island. We gratefully acknowledge the efforts, in this regard, of the following: John Anderson, Don Baird, John Barron, Gale Bishop, David Bottjer, Jon Branstrator, John Buckeridge, John Carter, Richard Cifelli, G. A. Cooper, Murray Copeland, Maureen Downey, Jim Doyle, Lucy Edwards, Robert Frey, Joseph Ghiold, Andrew Gombos, Jr., Alan Graham, Richard Grant, Daniel Habib, Carl Hansen, Donald Hattin, R. A. Hewitt, George Jeletzky, Mark Leckie, Larry Marshall, Ursula Marvin, Richard Minnich, William Newman, Donald Palmer, Robert Pease, George Pemberton, LouElla Saul, Samuel Savin, Frederick Schram, Rudolf Schuster, William Simpson, William

Sliter, Norman Sohl, Lewis Stover, Sherman Suter, Neil Wells, Austin Williams, Jack Wolfe, and Ellis Yochelson. Finally, we acknowledge the efforts of Campbell Craddock, books science editor of the Geological Society of America, who provided valuable information to the editors during the preparation of the manuscripts.

Preface

Seymour Island is a unique locality in Antarctica, and ranks as one of the more important localities in the Southern Hemisphere for the study of Upper Cretaceous and Paleogene strata. The chapter in this volume by William J. Zinsmeister provides an excellent introduction to the early geologic exploration of the island and the realization of the great potential that lies in Cretaceous- and Tertiary-age faunas. After the Swedish South Polar Expedition in the early years of this century, more than 40 years passed before the island was visited again, this time by members of the Falkland Islands Dependencies Survey, now the British Antarctic Survey. Because of uncertainties in ice conditions between James Ross Island, Seymour Island, and Snow Hill Island, that visit and others by British geologists in the 1950s were rather brief. Nevertheless, new fossil specimens were collected and important stratigraphic relations between the Tertiary and Cretaceous rocks on Seymour were established, though never published. The published reports deal mainly with paleontology; Spath (1953) and Howarth (1958) both described ammonite faunas, Ball (1960) documented rotularids, and Marples (1953) described fossil penguin remains.

The establishment of the Argentine station Marambio on the island in 1969 initiated the modern phase of geologic and paleontologic studies. U.S. participation began in the austral summer of 1974–1975, when an American team of four joined geologists of the Argentine Antarctic Institute in a cooperative program. There have been four subsequent U.S. expeditions, funded by the National Science Foundation as part of the U.S. Antarctic Research Program; the most recent was during the 1986–1987 season.

Seymour Island lacks a permanent ice cap. At the height of summer, the snow remaining is largely confined to dirt-covered accumulations in steep-sided gulleys and ravines. Exposure of the rock sequence is excellent, though not perfect; most of the sediments are poorly consolidated and therefore erode easily due to disaggregation caused by permafrost formation and summertime melting. The contrast between the inland dip slopes and cuestas and the cliff sections—for example, along the southern coast—is striking. Unfortunately, the cliff sections are virtually inaccessible and rather dangerous due to frequent rockfalls.

The sedimentary sequence exposed on the island comprises a thick section of Campanian, Maastrichtian, Paleocene, and upper Eocene beds. They form the only marine sequence of this age that crops out in Antarctica. The beds span the Cretaceous-Tertiary boundary in a sequence of fine-grained clastic sediments that appears to be relatively complete, though possibly condensed in part, for this important paleontologic time interval. The precise position of this boundary has yet to be established, assuming such is possible in a continuous clastic sequence. The biota contained in both the Cretaceous and Paleogene beds is unusually rich and diverse (Feldmann, 1984); in many groups it is exceptionally preserved, providing the paleontologist with a wealth of unusual material for study. It is probably fair to predict that both sequences will continue to yield a rich record of the high-latitude Southern Hemisphere biota rivaled only by that from New Zealand and southeastern Australia.

A quick review of the contents of this memoir will show the breadth of studies conducted. It should be noted that the more obvious gaps, such as the lack of any contribution on the ammonites, are not gaps due to the lack of research efforts. Some results have already been published, for instance, by Macellari (1984, 1986), Zinsmeister (1979, 1982, 1984), Woodburne and Zinsmeister (1982, 1984), and Grande and Eastman (1986).

The chapters herein document some of the richness and diversity of the biota and the important faunal similarities with Patagonia, as well as with New Zealand and Australia. Further, the remarkable fauna of echinoderms (Blake and Zinsmeister, 1979) and decapod crustaceans (Feldmann and Zinsmeister, 1984), together with the Eocene molluscs, formed the basis for arguing that shallow-water Antarctic environments may have formed centers for the origin of a variety of Recent taxa, taxa that currently inhabit subtropical environments ranging from the deep oceans to the shelf seas (Zinsmeister and Feldmann, 1984). This example of heterochroneity in the high southern latitudes supports similar arguments for northern high-latitude terrestrial floras (Hickey et al., 1983), and suggests that polar regions may be more important in biologic evolution than previously thought.

These chapters bring together in one volume much of the paleontology and geology of the island, and thereby reinforce the view that these strata are indeed unusual and present a unique opportunity for research. The papers do not provide final solutions to all the geologic problems of the island, which is to be expected at this stage of investigation. Since submission and review of the chapters for this memoir, another field season has been conducted—during the 1986–1987 austral summer. The results of that effort demonstrated that much has yet to be learned and that many fossil discoveries are yet to be made. Seymour Island is surely destined to be recognized as one of the more important localities for Cretaceous and Paleogene paleontological research in the Southern Hemisphere, if not the world at large.

David H. Elliot
Byrd Polar Research Center
The Ohio State University

REFERENCES CITED

BALL, H. W. 1960. Upper Cretaceous Decapoda and Serpulidae from James Ross Island, Graham Land. Falkland Islands Dependencies Survey Scientific Report, 30 p.

BLAKE, D. B., AND W. J. ZINSMEISTER. 1979. Two early Cenozoic sea stars (Class Asteroidea) from Seymour Island, Antarctic Peninsula. Journal of Paleontology, 53:1145–1154.

FELDMANN, R. M. 1984. Seymour Island yields a rich fossil harvest. Geotimes, 29:16–18.

——— , AND W. J. ZINSMEISTER. 1984. New fossil crabs (Decapoda: Brachyura) from the La Meseta Formation (Eocene) of Antarctica: paleogeographic and biogeographic implications. Journal of Paleontology, 58:1046–1061.

GRANDE, L., AND J. T. EASTMAN. 1986. A review of Antarctic ichthyofaunas in the light of new fossil discoveries. Paleontology, 29:113–137.

HICKEY, L. J., R. M. WEST, M. R. DAWSON, AND D. K. CHOI. 1983. Arctic terrestrial biota: paleomagnetic evidence of age disparity with mid-northern latitudes during the Late Cretaceous and early Tertiary. Science, 221:1153–1156.

HOWARTH, M. K. 1958. Upper Jurassic and Cretaceous ammonite faunas of Alexander Island and Graham Land. Falkland Islands Dependency Survey Scientific Report, 16 p.

MACELLARI, C. E. 1984. Revision of serpulids of the genus *Rotularia* (Annelida) at Seymour Island (Antarctic Peninsula) and their value in stratigraphy. Journal of Paleontology, 58:1098–1116.

——— . 1986. Late Campanian–Maastrichtian ammonite fauna from Seymour Island (Antarctic Peninsula). Journal of Paleontology, v. 60/Paleontological Society Memoir 18, 55 p.

MARPLES, B. J. 1953. Fossil penguins from the mid-Tertiary of Seymour Islands Falkland Island Dependencies Survey Scientific Report, 15 p.

SPATH, L. F. 1953. The Upper Cretaceous cephalopod fauna from Graham Land. Falkland Islands Dependencies Survey Scientific Report, 66 p.

WOODBURNE, M. O., AND ZINSMEISTER, W. J. 1982. The first land mammals from Antarctica. Science, 218:284–286.

——— . 1984. The first land mammal from Antarctica and its biogeographic implications. Journal of Paleontology, 58:913–948.

ZINSMEISTER, W. J. 1979. Biogeographic significance of the Late Mesozoic and Early Tertiary molluscan faunas of Seymour Island (Antarctic Peninsula) to the final breakup of Gondwanaland, p. 349–355. *In* J. Gray and H. J. Boucott (eds.), Historical Biogeography, Plate Tectonics, and the Changing Environment. Oregon State University Press, Corvallis.

——— . 1982. Late Cretaceous—early Tertiary molluscan biogeography of the southern circum-Pacific. Journal of Paleontology, 56:84–102.

——— . 1984. Late Eocene bivalves (Mollusca) from the La Meseta Formation collected during the 1974–75 joint Argentine-American Expedition to Seymour Island, Antarctic Peninsula. Journal of Paleontology, 58:1497–1527.

——— , AND R. M. FELDMANN. 1984. Cenozoic high latitude heterochroneity of Southern Hemisphere marine faunas. Science, 224:281–283.

Geological Society of America
Memoir 169
1988

Early geological exploration of Seymour Island, Antarctica

William J. Zinsmeister
Department of Earth and Atmospheric Sciences, Purdue University, West Lafayette, Indiana 47907

ABSTRACT

During the latter part of the 19th century, the waters off the coast of the Antarctic Peninsula began to be visited by ships searching for new whaling grounds. During the austral summer of 1892–1893, two of these early whaling expeditions visited the Seymour Island region on the northeast tip of the peninsula. The Norwegian ship *Jason*, commanded by Captain C. A. Larsen, reached Antarctica in November 1892; after sailing south along the east side of the peninsula as far as the sea ice would allow, the crew returned to Seymour Island in mid-November and landed on the east side of the island in search of seals. While ashore, Captain Larsen collected the first fossils from the continent of Antarctica. Although he never published any account of his discovery, we know many of the particulars of Larsen's discovery from Dr. Donald, one of two naturalists assigned to the Dundee Whaling Expedition that arrived in the area shortly after the discovery.

In 1901, explorer/scientist Otto Nordenskjöld organized the Swedish South Polar Expedition (1901–1903) to explore the Seymour Island region. Nordenskjöld's expedition was forced, due to the loss of his ship, the *Antarctic,* to spend 2½ years in Antarctica. During his forced stay, he explored Seymour Island and the surrounding area and made a number of important scientific discoveries. Although the Swedish South Polar Expedition was one of the most scientifically successful expeditions in the annals of Antarctic exploration, Nordenskjöld did not succeed, due to chance and circumstances beyond his control, in exploiting the truly remarkable fossil deposits that are present on Seymour Island.

INTRODUCTION

During the past decade, geological exploration on Seymour Island (Fig. 1) has revealed that this small, desolate island, located off the northeast tip of the Antarctic Peninsula, contains one of the most important records of Late Cretaceous and Early Tertiary life in the Southern Hemisphere (Fig. 2). Although the geological community is only now recognizing the importance of Seymour Island, the early explorer/scientist Otto Nordenskjöld (1905, p. 252) clearly knew the importance of these deposits:

The collection of fossils we have brought home will be the thread which will gradually lead to discoveries enabling us to form a picture of the chief features of the nature of the antarctic regions, from the Jurassic period down to our own times . . . where, maybe, many animals and plants were first developed that afterwards found their way as far as to northern lands.

Nordenskjöld's prophetic words have been borne out by the spectacular paleontologic discoveries during the last ten years.

These discoveries have not only provided new insights into the geologic history of Antarctica, they have also provided answers to questions about life in the Southern Hemisphere that have puzzled naturalists since Charles Darwin's voyage on HMS *Beagle.*

One of the puzzling questions about early exploration of Seymour Island was the failure of the paleontological community for nearly 70 years to fully recognize the scientific potential of these deposits. Although Nordenskjöld made a number of important fossil finds, his collections gave no hint of the magnitude of the scientific treasures that lay waiting on the island. His collections were viewed only as providing information concerning the presence of life on Antarctica prior to the onset of glacial conditions. It was not recognized that the fossil deposits on Seymour Island would provide answers to many fundamental questions concerning the origin and evolution of life on the southern continents, as well as insight into the climatic history of the Southern

1

Figure 1. Oblique view of northern half of Seymour Island. 1 = Cross Valley; 2 = plateau at north end of Island; 3 = Cape Wiman; 4 = Larsen Cove.

Hemisphere. We shall see that circumstances during the Swedish South Polar Expedition, combined with chance, prevented Nordenskjöld from fully exploiting the fossil deposits on Seymour Island.

Although the first fossils to be described from Antarctica were from Seymour Island, the first report of fossil remains from Antarctica was nearly 80 years earlier by James Eights (1833), a member of the First American Expedition to Antarctica in 1830. While visiting one of the islands in the South Shetland group near the northwest tip of the Antarctic Peninsula, Eights (1833, p. 64) reported seeing ". . . a fragment of carbonized wood imbedded in the conglomerate. It was in a vertical position, about two and half feet in length and four inches in diameter, its color is black, exhibiting a fine ligneous structure, the concentric circles are distinctly visible. . . ."

Although Eights did not mention which island he landed on, recent discoveries of wood and plant fossils on King George Island suggest that Eights probably found the fossil log there (Fig. 2). He made no mention of collecting the fossil, and the absence of any subsequent description indicates he probably left it where he found it.

CAPTAIN C. A. LARSEN AND THE DISCOVERY OF THE FIRST FOSSILS FROM ANTARCTICA

The first documented collection of fossils from Antarctica was made by the Norwegian whaling captain C. A. Larsen (Fig. 3) on Seymour Island during the austral summer of 1892–1893. Confusion exists concerning the site where the collection was made and who actually collected the fossils. Sharman

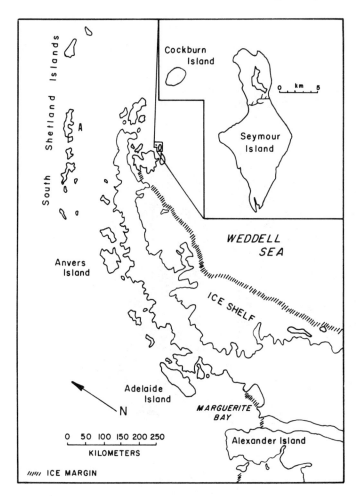

Figure 2. Index map of northern part of the Antarctic Peninsula and Seymour Island. A = King George Island, the probable location where Eights reported seeing a petrified log.

the early accounts of whaling activities of the Dundee expedition provides an answer to the question of who actually collected the fossils, these authors have provided only indirect evidence about the exact location of the discovery.

During the late 19th century, it was becoming clear that the whaling grounds in the Northern Hemisphere were rapidly being depleted, and whalers began to look elsewhere for new whaling grounds. In his epic voyage to Antarctica from 1839 to 1843, Sir James Clark Ross reported encountering many whales in the waters around Antarctica. The prospects of finding a new and profitable whaling ground along the coast of Antarctica prompted two separate whaling expeditions—a Norwegian expedition on the *Jason* commanded by C. A. Larsen, a second from Dundee consisting of four ships (the *Active, Balaena, Diana,* and *Polar Star*) to explore the northern part of the Antarctic Peninsula in 1892. Both expeditions were disastrous financially, but Larsen's discovery of the first fossils from Antarctica opened a new chapter in exploration of the continent.

Larsen arrived in the vicinity of Seymour Island in early November 1892. As far as is known, Larsen's ship was the first to visit this region since Ross's voyage to this part of Antarctica nearly 50 years earlier. Although Larsen did not publish an account of his voyage, we know from a later account he wrote (Larsen, 1894) that he initially sailed south of Seymour Island, along what is now known as the Larsen Ice Shelf, as far south as the ice would permit, before returning north. On December 4, 1892, Larsen and a small party from the ship went ashore on the east coast of Seymour Island to look for fur seals. It was while they were ashore that they made their momentous discovery.

Dr. C. W. Donald's account of the voyage of the *Active* has provided some of the details of Larsen's discovery. The Royal Geographical Society of Edinburgh had convinced the Dundee sponsors of the expedition to include two surgeons who were also to act as naturalists on two of the expedition's four ships, Dr. Donald on the *Active* and Dr. James Bruce on the *Balaena*. Although these men believed they would have a free hand to carry out scientific observations during the course of the whaling season, the captains of the two ships viewed them as nuisances; as a consequence, their scientific activities were severely limited. Both the naturalists met with endless frustrations during the austral summer of 1892–1893. A passage by Murdoch (1894, p. 251) clearly reflects this:

We have been close enough to the land to enable us to distinguish the colouring of the rocks. . . . It is almost unbearable to see the land so close and yet have no means to land on it. We feel tempted to jump overboard and swim. All our boats hang idle on davits, yet we are not allowed one to land with. . . . Captain Larsen of the *Jason* has landed, and tells us he found beds of fossils on the beach. . . ."

Many details concerning the location and events that occurred during the voyages of the *Active* and the *Balaena* are vague or nonexistent. For example, Bruce (1896) gave virtually a day-to-day account of the voyage of the *Balaena* through December 24, but thereafter noted only occasionally the dates for

and Newton (1894), in their description of the collection from Seymour Island made no mention of Larsen, stating the fossils were "obtained by Dr. Donald." The only locality data concerning the fossils was a reference to Seymour Island in the title of the paper. No further references to Seymour Island or to the circumstances leading to the discovery were mentioned in their short paper. In the title of a subsequent paper describing additional fossils from Seymour Island, Sharman and Newton (1894) referred to the fossils as having been collected by both "Dr. Donald and Captain Larsen." Two questions thus arise: Who actually found the fossils, and what was the location of the discovery?

Although Sharman and Newton made note in their second paper of Dr. Donald's account of Larsen's discovery, no mention was made of Donald's role. Thus they continued to imply that he took an active part in the discovery. Although an examination of

Figure 3. Members of the Swedish South Polar Expedition on their departure from Gothenburg, Sweden. (From Nordenskjöld, 1905.)

events that occurred during the remaining part of the cruise, even though he listed daily temperatures and bathymetric soundings for the entire cruise. These omissions and vague accounts may be attributed to the secrecy that surrounded many of the early whaling expeditions. Since the objective of these expeditions was to locate new whaling grounds, captains and owners frequently considered any discoveries or geographic data privileged information. This may explain why Bruce in his early reports never mentioned the *Balaena*'s encounter with the *Jason*.

We do know from an account given by Murdoch (1894, p. 253), in a book titled *From Edinburgh to the Antarctic. An artist's notes and sketches during the Dundee Antarctic Expedition of 1892–1893,* that the *Balaena* met the *Jason* on December 28 and that Bruce went aboard the *Jason* to care for a sick seaman: "In the evening, Larsen came aboard the *Balaena* followed later by the 'Steersman' or first mate and . . . had a jolly good evening in the cabin, smoking and yarning."

Although Murdoch stated that Larsen boarded the *Balaena* on the 28th, his earlier passage expressing the frustration of not being allowed ashore was written on December 27, suggesting

that the *Balaena* must have had some contact prior to the 28th. When this earlier meeting occurred is not known. The puzzling question is why Bruce made no mention of the meeting or the fossil discovery. His only meniton of Larsen was when he referred to Donald's description of Larsen's discovery.

In 1930, Aagaard (1930, p. 84) provided a fascinating account of Larsen's evening visit on December 28. The *Jason* had been at sea for several months, and many of the supplies, including tobacco, had been exhausted:

As there was a terrible lack of tobacco on board [the *Jason*] . . . the demand for it was urgent, which Bruce took advantage of and traded most of the fossils Sorell and Hansen and the other crewmembers had found on Seymour Island. 'What were fossils good for when you have Navy cut and juicy quids.' The few fossils left over after Dr. Bruce's visit were acquired by Donald a few days later against chewing tobacco, while Larsen, the clever man that he was, let a little bit of his brandy supply go in exchange for tobacoo for his pipe, but his fossils he took home to Norway where they ended up at the University of Oslo.

I have not been able to determine whether the fossils Donald gave to Sir Archibald Geikie upon his return to Edinburgh consisted of only the fossils that he obtained, or whether they also included those acquired by Bruce.

Of the early accounts by Donald and Bruce, the only published record of the encounter of the ships of the Dundee expedition with Larsen and the *Jason* was the one given by Donald. On January 24, 1893, Donald described the encounter of the *Active* with the *Jason* off the east coast of Seymour Island, and the subsequent evening he spent with Larsen. It was during the evening of the 24th that Donald learned of the fossils on Seymour Island. Donald (1893b, p. 438) provided the following account of Larsen's discovery of the fossils on Seymour Island:

Captain Larsen, of the Norwegian ship *Jason*, landed on the island to the north of Snow Hill, which may be called Seymour Island [James Ross 50 years earlier originally referred to the island as Cape Seymour], and obtained there a number of fossils. . . . Captain Larsen reports that he found no traces of vegetation there, the surface being formed of volcanic debris and (a) number of these fossils.

In a subsequent, more complete description of the voyage, Donald (1896, p. 638) elaborated slightly on Larsen's discovery ". . . [Larsen] landed on Seymour Island. There he got a number of fossil shells and pieces of fossil wood, and a few round stones with a concentric arrangement of layers—to all appearance an old lava bomb." In the following, somewhat confusing, paragraph, Donald stated, "Though we saw any quantity of these fossils and ball, we saw no signs of vegetation." Although the use of "we" in the passage would seem to imply that Donald actually visited Seymour Island, it is clear from the tracks of the *Active* and *Balaena* (Bruce, 1896) that the *Active* remained some distance from Seymour Island. Since Donald (1896) gave detailed accounts of other landings he made on islands north of Seymour Island, it would be very surprising, considering the importance of fossils, had he landed on Seymour Island and made no mention of it.

The only information concerning the location of Larsen's discovery comes from Donald's oblique reference to volcanic debris and round concretions. From geologic work done on the island during the last 12 years, volcanic erratics are now known to be scattered all over the island. Likewise, Donald's reference to "round balls with concentric arrangement of layers (concretions)" provides little guidance, since concretions are common throughout both the Cretaceous and Tertiary deposits on the island. The only clues about Larsen's locality, based on Donald's accounts, come from the fossils themselves. All the specimens are poorly preserved and are abraded. Sharman and Newton (1894, p. 707) noted that the fossils were "much denuded, having apparently been long exposed to the weather." Since Larsen never sailed along the west side of the island (Nordenskjöld, 1905), the landing must have been somewhere along its east coast.

From my work on the Tertiary deposits on Seymour Island (Zinsmeister, 1977, 1982), we know that several fossiliferous outcrops occur at the base of the sea cliffs about a mile north of Cross Valley (Fig. 4), and the molluscs from these localities are the same species that Sharman and Newton (1894) described.

The fossils at these outcrops occur in massive, calcareous conglomeratic lenses that are rapidly being eroded from the outcrop by tides and sea ice. Because of the hard, calcareous nature of the matrix, the fossils are very resistant to abrasion and are frequently transported along the beach a kilometer or more from the source by longshore currents and ice. Further evidence that these fossiliferous outcrops at the base of the sea cliffs were the source of Larsen's fossils comes from an account by Aagaard (1930, p. 70) of Larsen's landing: "On the night of December 4th, Larsen went ashore with some of his crew among them Th. Sorlle and Karl Hansen to look for seals. . . . The island consisted mostly of clay, 'cement' and shell-sand and there were hundreds of petrified objects of which Larsen and his crew took some along to the ship. . . ."

A single specimen of the Cretaceous bivalve (*Lahilla*) was also collected during Larsen's first visit to Seymour Island, but was not described until 1898 by Sharman and Newton. This specimen apparently was misplaced and was not made available to Sharman and Newton for study until sometime after the publication of their first paper.

Since no Cretaceous rocks occur along the beach where Larsen made his landing, it would appear that he left the beach and ventured inland. However, I believe that Larsen probably did not go inland, and the specimen of *Lahilla* was found on the beach. My opinion is based on the fact that where *Lahilla* occurs inland, there are large numbers of well-preserved specimens of *Lahilla* with many other invertebrates. It would be surprising if Larsen had gone inland and picked up only a single poorly preserved specimen and none of the other Cretaceous fossils that are so abundant in the upper part of the Cretaceous rocks just south of Cross Valley. Examination of Larsen's specimen reveals that it is also extremely abraded, which suggests that it, too, was found on the beach by chance. It would not be too surprising to find an occasional Cretaceous fossil on the beach, since the upper part of the Cretaceous sequence is exposed in sea cliffs several kilometers south of Cross Valley. Although Sharman and Newton were not aware of the existence of Cretaceous beds on Seymour Island at the time of their publication, and because the other fossils are of obvious Tertiary age, they concluded that this species of *Lahilla* was also of Tertiary age.

Larsen (1894) indirectly provided support for the location of the beach locality in his account of his second landing on Seymour Island on November 18th of the following year. From his account in the *Geographical Journal*, it is clear that his landing was on the "middle part of the island," which would be just south of Cross Valley. Larsen landed just before 6 p.m. with two boats. Upon his arrival, he sent a mate in one boat north along the coast to a "small bay (now known as Larsen Cove) for reconnoitering" and two other men inland, but he did not provide any account of where these two parties went or if they found anything. The third party, consisting of Larsen and two other men, headed inland in a westerly direction for about a "quarter of a Norwegian mile from the shore" to a point about 300 feet above sea level. He described the topography of this part of the island as, "hilly and intersected

Figure 4. East coast of Seymour Island. Probable location at base of seacliff (1) where Larsen made his fossil discovery in 1892. The low hills on the left center of photo (2) is where Larsen encountered the fossil wood during his second visit to Seymour Island in 1893. Nordenskjöld is believed to have collected his leaf fossils from a concretionary horizon (3) just below the crest of the flat-topped hill in center of photo. (From Nordenskjöld, 1905.)

by deep valleys, some of the hills are conical, and consist of sand, small gravel, and cement: here and there is some petrified wood." The only fossils he collected from this area were several pieces of wood, ". . . which looked as if they were of deciduous trees; the bark and branches, as also the year-ring, were seen in the logs."

Larsen's description of hills and the occurrence of petrified wood clearly indicates that he was just south of Cross Valley. The topography of this part of Seymour Island is characterized by small conical hills composed of fine sandstones with occasional gravels. Although fossilized wood may be found throughout the Cretaceous and Tertiary rocks on Seymour Island, it generally occurs as the odd fragment associated with marine fossils. I think that it is safe to assume that if there were any marine invertebrates associated with the "petrified wood," Larsen would have mentioned their occurrence and collected them. The only place on the island where petrified wood occurs in any abundance is the region immediately south of the head of Cross Valley. I have en-

countered numerous fossil logs, some in this region as much as 1 m in diameter and 10 m in length. No marine fossils have been found in direct association with fossil wood in this area of the island.

After gathering a small collection of fossil wood, Larsen's party turned south and headed toward an unnamed cape (now known as Penguin Point), with a penguin rookery, approximately 4 km away. Larsen's party reached the rookery by early evening and then returned because of darkness. He had planned to "visit the spot where we had been during the first year, but could not do so on the account of darkness setting in" while they were still "several miles" from it. The only fossiliferous outcrops of Tertiary age within such a distance of the penguin rookery are those at the base of the sea cliffs just north of Cross Valley. Because of this reference to his desire to return to the place where they had collected the fossils during the previous year, which was several miles away from the rookery, we can be fairly confident that the source of Larsen's fossils was the beach just north of Cross Valley (Fig. 4).

Although Larsen was able to collect only a few poorly preserved fossils during his two brief visits to Seymour Island, this marked the first time any fossils had been collected from Antarctica, and led Sharman and Newton (1898, p. 61) to remark,

the number of forms known (from Antarctica) is at present few, yet these give the promise of a large field of fossiliferous rocks as yet unexplored, . . . and it is of extreme interest to find that in this distant southern area Lower Tertiary deposits occur with forms of life so very similar to those with which we are familiar in our own northern country.

It is apparent that these prophetic words prompted Dr. Otto Nordenskjöld to select the Seymour Island area as the primary region to be explored by the Swedish South Polar Expedition, an expedition that proved to be one of the most scientifically successful expeditions in the annals of Antarctic exploration.

SWEDISH SOUTH POLAR EXPEDITION, 1901–1903

Otto Nordenskjöld came from a long line of explorers and naturalists. His uncle Baron Adolf Erik Nordenskiold* was the first to navigate the Northeast Passage on the *Vega* in 1878–1880. The literal translation of Nordenskjöld is "northern shield," or "protector of the north," a very appropriate name for a polar explorer. The exploits of his famous uncle undoubtedly influenced Otto's decision to select a career of polar exploration. His first expedition was to the Magellan region of southern South America in 1895–1897. He subsequently went to Alaska in 1898 and then to East Greenland in 1900.

Nordenskjöld planned the Swedish South Polar Expedition with the express purpose of investigating the fossiliferous deposits of Seymour Island on the northeastern tip of the Antarctic Peninsula and exploring the east coast of the peninsula as far south as possible. Nordenskjöld financed the expedition privately. The debts incurred by the expedition were to plague him for the rest of his life. The expedition left Gothenburg, Sweden, on October 16, 1901. Because of Larsen's experience along the east side of the Antarctic Peninsula, Nordenskjöld selected him to command the expedition's ship *Antarctic.* Nordenskjöld's main objective was to establish a wintering camp as far south as possible, but if no suitable site was found, his plans called for a winter camp on Seymour Island. At the end of the season, the *Antarctic* was to return to southern South America and carry out a scientific survey of the region during the austral winter. At the beginning of the following summer, the *Antarctic* was to return to Seymour Island and pick up Nordenskjöld's party.

On the morning of November 16, 1901, the *Antarctic* reached the northern point of Seymour Island. Although the pack ice was broken up, it was not until the following morning that they were able to sail an additional 9 km south and reach a

*The use of "j" and "i" in -skjöld and -skiold reflects the differences between Finnish and Swedish spellings of Nordenskjöld.

position off Penguin Point. On the morning of the 17th, Nordenskjöld took two boats and went ashore at Penguin Point to establish a depot in the middle of the penguin rookery in case they were forced to retreat from a more southern camp. Little did he realize that the penguins would become the chief source of food for his party for the next two years. One of the ironies of the expedition was that Nordenskjöld's first landing on Seymour Island was at one of the few places on the island that is virtually unfossiliferous. He had expected to find the rich fossiliferous beds that he believed existed on the island. He wrote (1905, p. 56–57), "I was bent to the highest point of expectation. But, . . . my hopes were thoroughly disappointed. . . . It was not possible for me to know then that just that part of the island was very poorest, . . . the impressions I gained during that landing were decisive of my determination not to make Seymour Island our wintering-station." Several months later, on his first visit to Seymour Island after establishing his wintering station just south of Seymour Island on Snow Hill Island, he discovered the presence of vast numbers of fossils on Seymour Island and realized his mistake.

After leaving Seymour Island, the expedition continued south along the coast, but heavy sea ice conditions at about 66°S forced them to turn back. For the next couple of weeks, the *Antarctic* sailed eastward and carried out the first oceanographic survey of the western part of the Weddell Sea. On February 12, 1902, the expedition sailed into Admiralty Sound just west of Seymour Island. Nordenskjöld believed that establishing his winter station somewhere along the coast of Admiralty Sound would afford protection against the winter storms.

Late in the day on the 16th, they approached the north end of Snow Hill Island, and after a brief visit to the island, Nordenskjöld decided to establish his wintering quarters at the head of a small bay on the west side of the island. His decision was influenced by the discovery of a number of ammonites in concretions on the beach of the bay. This discovery on Snow Hill Island on February 12 was the first indication of the presence of Mesozoic rocks in Antarctica. By February 17, the hut was finished and occupied by the wintering party, which consisted of Nordenskjöld, Gosta Bodman, Jose M. Sobral, Ole Jonassen, Eric Ekelof, and Gustof Akerlund (Fig. 5). The location of the hut—in a broad amphitheater with a high ridge directly south behind the hut—appeared to be the ideal location for maximum protection against the coming winter storms (Fig. 6). Unfortunately, rather than protecting the winter station, the amphitheater actually amplified the south winds into gales, a fact that became apparent when the roof of the hut was nearly blown away during the first storm. Nordenskjöld was forced to reinforce the hut by propping two massive beams on the hut's lee side. The almost continuous strong southern gales were to greatly affect the field work during the following winter. In his narrative of the expeditions, Nordenskjöld felt that no other polar explorer up to that time had ever encountered such violent wind conditions.

Nordenskjöld made his first sledge trip to Seymour Island on February 27. The purpose of the trip was to check the depot at Penguin Point and to conduct a brief survey of the southern part

Figure 5. Nordenskjöld's wintering party in front of Hut on Snow Hill Island. Front row: Bodman, Nordenskjöld, Sobral; back row: Jonassen, Ekelof, Akerlund. (From Nordenskjöld, 1905.)

of the island. To reach Seymour Island, the party had to cross a narrow strait (Admiralty Inlet) between Snow Hill and Seymour Islands. Although the strait is less than 2 km wide, crossing with sledges required the presence of firm sea ice. Because of the short length of the days in late February, the party managed to travel only a relatively short distance, but they were able to cross the strait. The party established a camp for the night at the head of a small bay at the south end of Seymour Island. From this campsite, the party could go to Penguin Point, either along the coast or by an overland route. The coastal route required traveling along a narrow beach that was exposed only during low tide, and depending on the prevailing winds, could be impassable because of the jamming of sea ice against the sea cliffs. If the conditions were right, the coastal route would be the fastest. The overland route, on the other hand, required traveling over a number of steep hills; if the temperatures were above freezing, overland travel could be very difficult because of the number of soft, muddy creek bottoms that had to be crossed. During this first trip to the penguin rookery, Nordenskjöld chose the overland route. Most of the later

trips were made either along the beach or by boat because of the need to reach the penguin rookery quickly.

On their way to the rookery, Nordenskjöld was "astonished" by the abundance of well-preserved fossils they encountered (Fig. 7). This discovery led him to make the following entry in his notebook "Seymour Island is most undoubtedly a wonderful land and it is decidedly unfortunate that we have not chosen it as the site of our wintering station." After determining the state of the depot at the rookery, the party returned to the previous night's camp site. Upon reaching the south end of the island, Nordenskjöld decided to continue on and reached his hut on Snow Hill Island after dark.

Nordenskjöld's party covered approximately 25 km from the camp at the south end of the island to the rookery and back to the hut on Snow Hill Island during this first journey to Seymour Island—a long day, even under the best travel conditions. From my own experience on Seymour Island, such a trip of 25 km would have been long and arduous, and would have left little time to do any geology, except to pick up the occasional fossil.

Figure 6. Nordenskjöld's Hut on Snow Hill Island. Note prominent basaltic dikes on hill behind hut. (From Nordenskjöld, 1905.)

Although a few fossils were collected during this brief excursion, no locality data were recorded by Nordenskjöld. This first trip seems to have established a pattern for future work on Seymour Island: short trips with a minimal amount of preparation. These trips contrast sharply to the longer sledging expeditions to the south and to James Ross Island, which were carefully planned. The closeness of Seymour Island to the wintering hut may have led Nordenskjöld to believe that work on Seymour Island could be done anytime and did not require extensive preparation. Unfortunately, the purpose of most of the later trips to Seymour Island was to obtain food from the penguin rookery and what geology was done was as an afterthought. Consequently, the remarkable nature of the fossil deposits of Seymour Island were never fully recognized.

Nordenskjöld did not return to Seymour Island until the following spring. On November 27, he sent Bodman, Ekelof, and Jonassen on a sledge journey to Seymour Island to collect penguin eggs at Penguin Point and then to Cockburn Island. Nordenskjöld accompanied them as far as the south end of Seymour Island and then returned to the hut. The sledge party traveled along the west side of Seymour Island as far as the prominent point (Cape Bodman) that was visible from the hut and then turned east and went directly to the penguin rookery. After collecting a number of penguin eggs, they proceeded directly to Cockburn Island and camped at a small penguin rookery on the northwest side of the island—the only place James Ross landed during his cruise along the east side of the Antarctic Peninsula 50 years earlier. Although there is no record of the exact route of the sledge party, it is most likely they traveled on the sea ice to Cape Bodman. After rounding Cape Bodman, they probably continued eastward on the sea ice to a point near the mouth of Cross Valley and then went overland to Cross Valley.

The presence of high sea cliffs along Cape Bodman and a narrow beach that is exposed only at low tide lends support to the idea of a sea-ice route to Penguin Point. As they traveled toward Penguin Point from the mouth of Cross Valley, they passed through the upper part of the Cretaceous deposits that are abundantly fossiliferous. It was during this part of the journey that they collected some "beautiful ammonites" for Nordenskjöld. Unfortunately, the only samples they collected when they reached Cockburn Island were a few pieces of "volcanic tuffs," and they did not recognize that most of Cockburn Island was composed of Cretaceous sediments, capped only by a thin layer of basalts. None of the members of this party were trained scientists, and the cone-shaped nature (Fig. 8) of the island led them to believe that the entire island was of volcanic origin. Consequently, Nordenskjöld decided not to return to Cockburn Island and to concentrate his efforts elsewhere. It was not until Gunnar Andersson visited Cockburn Island a year and a half later that the true nature of the island was recognized and the presence of a thick Cretaceous sequence, Tertiary beds distinct from those on Seymour Island, and the Plio-Pleistocene "Pecten conglomerate" were discovered.

An interesting historical sidelight to Andersson's discoveries on Cockburn Island is the comment made nearly 50 years earlier by Sir James Hooker. During the epic voyage of Sir James Ross to Antarctica during the early 1840s, Ross, together with Captain Crozier and Hooker, made a brief visit to Cockburn Island to take possession of the region for the British Empire on January 6, 1843. Ross had two surgeon/naturalists with him during the voyage (James Hooker and Robert McCormick). Hooker was in charge of making botanical collections and McCormick had been designated the expedition geologist. Unfortunately, McCormick was not a member of the shore party. (Ross had a policy during the voyage that one of the surgeons had to remain on board anytime the other went ashore, and it was Hooker's turn to go ashore.) If McCormick had gone ashore with Ross, evidence of past life in Antarctica might have been discovered 50 years earlier.

As the shore party neared Cockburn Island, Hooker noted:

On approaching Cockburn Island, the cliffs above were seen to be belted with yellow, which, as it were, streams down to the ocean among the rocky debris. The colour was too pale to be caused by iron ochre, which it otherwise resembles. (in Turrill, 1953, p. 138).

Figure 7. Cretaceous invertebrate fossils (left, *Pachydiscus ultimus;* right, *Pinna anderssoni*) discovered during the Swedish South Polar Expedition. (From Nordenskjöld, 1905.)

He attributed the color to be the result of the surface being covered with lichens. The yellow color that Hooker observed in the cliffs and in the gullies undoubtedly could be attributed to Cretaceous sediments that underlie the basaltic plateau.

The beautifully preserved ammonites collected by Bodman's group during their trek across Seymour Island prompted Nordenskjöld to make his third journey with the objective of exploring the Cross Valley region of the island. On December 2, Nordenskjöld, together with Jonassen and Akerlund, set off for Seymour Island. They proceeded directly to the depot at Penguin Point and then to Cross Valley, to the vicinity where Larsen discovered his petrified wood. It was here that Nordenskjöld made what he considered to be the most important of his discoveries—numerous impressions of leaves (Fig. 9a). This discovery clearly proved to Nordenskjöld that Antarctica had not always been covered with ice, but had been warm enough in the past to support an abundant and diverse flora.

Earlier discoveries of plant fossils in the Arctic, together with Nordenskjöld's finds on Seymour Island, convinced him that the climates of the polar regions of the earth had changed considerably. On the following day, he continued along the east coast of the island toward a prominent high plateau at the north end of the island (Fig. 1). Approximately 1 km north of the head of Cross Valley, the party left the beach and proceeded along a prominent terrace at the base of the plateau. As they made their way along the terrace, they came upon a small, sandy hill with many large, well-preserved bones. It was not until they had returned to Sweden that it was recognized that these bones represented remains of several types of penguins, some of which were as tall as 1.5 m (Wiman, 1905). In addition, several large whale bones and some invertebrates were collected from the surface of the terrace. Later in the day, Nordenskjöld decided to return to his leaf locality and collect additional material. He wanted to stay longer on Seymour Island, but decided to return to Snow Hill Island because he had not brought any equipment for collecting small, delicate leaf fossils. He was also expecting the return of the *Antarctic* soon and was anxious to begin preparations to depart. At that moment, he was not aware that the *Antarctic* was caught in the ice and would be lost, or that he would be forced to remain in Antarctica for nearly two more years. Although his third trip to Seymour Island lasted just two days, Nordenskjöld was able to make a number of important discoveries, the most important of

Figure 8. Cockburn Island with James Ross Island in the background. Photo taken from Cross Valley.

which were the leaf-bearing deposit and the find of vertebrate bones. He also recognized that the Cretaceous deposits on Seymour Island were distinctly different from those on Snow Hill Island.

Nordenskjöld decided to return to Seymour Island in late December to continue his geological survey, but upon reaching the Admiralty Inlet, he discovered that the sea ice was gone and the only way to reach Seymour Island was by boat. He did not try to reach Seymour Island again until the middle part of January 1903, when he made a one-day trip to the island by boat. Nordenskjöld made no mention of doing any geological work on this trip. He probably went directly to the rookery to collect penguin meat for the coming winter; by now he realized that there would be no relief ship and they were going to have to spend a second winter in Antarctica.

On February 6, he returned once again to the rookery to obtain as many penguins as possible for the coming winter. On the following day, Nordenskjöld and Ekelof climbed the high plateau at the north end of the island to survey the ice conditions

in Erebus and Terror Gulf and to try to see if the *Antarctic* was in the vicinity. Once again, their track on this trip took them along the east side of the island. Although no mention was made of any fossil collecting, it appears from the locality data of some of the fossils described by Wilckens (1911) that a few Tertiary molluscs were collected during this trip. They returned to the rookery and killed approximately 400 penguins before returning to the hut on the 15th. No further attempt was made to return to Seymour Island until the following spring. The remainder of the fall of 1903 was spent obtaining seal meat for the rapidly approaching winter.

After an uneventful winter, Nordenskjöld decided to mount a sledging trip around James Ross Island to survey the geology and explore the large sound that separated James Ross Island from the mainland. It was during this journey that one of the most extraordinary encounters in the annals of Antarctic exploration occurred.

On October 16, as Nordenskjöld's sledging party reached the northeastern end of James Ross Island, they ran into Gunnar

W. J. Zinsmeister

Figure 9. A, Tertiary leaf fossils collected by Nordenskjöld from Seymour Island. B, Jurassic plants collected by Andersson from Hope Bay. (From Nordenskjöld, 1905.)

Figure 10. A, Sketch of Andersson's field part at the time of their meeting with Nordenskjöld near Vega Island. B, Andersson's party in front of Nordenskjöld's hut. (From Nordenskjöld, 1905.)

Andersson's party which was heading south from Hope Bay toward Snow Hill Island. When Nordenskjöld first spotted Andersson's party several miles away, he initially thought that he was about to make the first meeting of indigenous natives of Antarctica. He recorded (1905, p. 308) his reaction:

...What is it I at last see before me? Two men, black as soot from top to toe; men with black clothes, black faces and high black caps, and with their eyes hidden by peculiar wooden frames, which were attached to the face that they reminded one of black silk masks with pierced pieces of wood for the eyes. Never before have I seen such a mixture of civilization and the extremest degree of barbarousness . . . to what race of men these creatures belong (See Fig. 10).

Nordenskjöld was astounded when he realized that they were part of the *Antarctic* party; he had not dreamed that they were members of his own expedition. Andersson told Nordenskjöld of their extraordinary adventures of being stranded during the preceding fall.

When the *Antarctic* returned from South America during the spring of 1902 to relieve the wintering party on Snow Hill Island, the ship encountered extremely heavy sea-ice conditions at the north end of Erebus and Terror Gulf. At this point, the ship was close enough for them to see the top of Cockburn Island, but it was impossible for them to proceed any farther south. After sailing to the north for a week or so, the *Antarctic* made another attempt to reach Snow Hill Island, but was again stopped by the pack. At this point, Andersson decided to sledge south to Snow Hill Island to meet Nordenskjöld. Before heading south, they established a depot at Hope Bay on the mainland west of Joinville Island. The original plan called for Andersson to reach Snow Hill Island and inform Nordenskjöld of the difficulties the *Antarctic* was having. If the *Antarctic* did not reach Snow Hill Island by February 10, both parties would sledge north to Hope Bay and wait to be picked up by *Antarctic*.

After leaving the *Antarctic,* Andersson and his party, consisting of S. Druse and T. Grunden, headed south toward Snow Hill Island. It soon became apparent, however, that because of the summer melting and breakup of the pack along the edge of Sydney Herbert Sound, they were not going to be able to reach Snow Hill Island. Andersson decided to return to the depot and await the return of the ship, unaware that during their absence the *Antarctic* had been crushed by the pack and had sunk. By mid-February, Andersson's party began to make preparations in the event they would be forced to winter over; on March 10, realizing that the *Antarctic* was not going to return, the men started to build a stone hut for the coming winter. The depot that they had established had insufficient food supplies to last the winter, and they were forced to kill several hundred penguins and seals for both food and fuel. Because Andersson had taken only the most essential equipment for the sledge journey, they had no scientific instruments, and consequently were not able to conduct any scientific activity during the winter except for a short trip to Mount Flora near Hope Bay, where they discovered a spectacular Late Jurassic to Early Cretaceous flora (Fig. 9). Andersson's

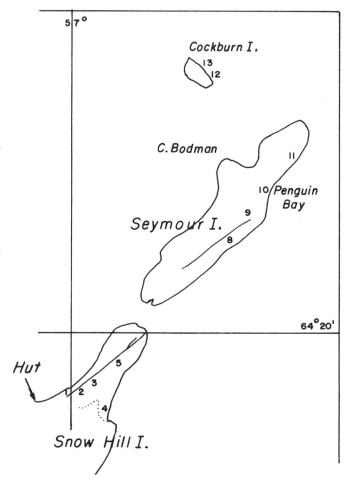

Figure 11. Nordenskjöld's locality map showing the location of major fossil localities on Seymour, Snow Hill, and Cockburn Islands. Many of the fossils collected by Nordenskjöld's party were never plotted on his map. Map redrawn from original. (From Nordenskjöld, 1905.)

enforced stay during the winter of 1903 represents the first time in the annals of Antarctic exploration that a party was forced to survive a winter by living off the land. The following spring, Andersson headed south where he met Nordenskjöld's party as it approached James Ross Island.

After about a week's rest at the hut, Andersson (Fig. 10) decided to visit Cockburn Island and investigate the reported volcanic rocks on the island. A three-man party, consisting of Andersson, Bodman, and Jonassen, left on October 21. During their short visit to Cockburn Island, they recognized the true nature of the island. The basalts formed only a resistant layer on top of the island; the remainder of the island was composed of Cretaceous and Tertiary sediments. They also noted the presence of a Plio-Pleistocene terrace deposit on the top of the basalts, which they referred to as the "Pecten Conglomerate" (Andersson, 1906).

Figure 12. Tertiary fossiliferous outcrop on west side of plateau at north end of Seymour Island.

Upon being informed of Andersson's discoveries, Nordenskjöld immediately decided to mount another expedition to Seymour Island and left on October 26. As in the past, they went directly to the penguin rookery and established a temporary base camp there. For the next four days, each member of the party went in separate directions along the eastern side of the island, studying the geology and making collections of fossils. On October 30, Nordenskjöld and Andersson returned to the bone locality on the terrace on the east side of the island and collected additional material. While Andersson and Sobral remained on Seymour Island, Nordenskjöld returned to the hut and made preparations for the return of the *Antarctic*. On November 6, Nordenskjöld returned once again to Seymour Island, but on his way he met Andersson and Sobral, who were returned to Snow Hill Island. While preparing a meal the night before, Andersson had burned his hand, which required medical attention. This ended their last effort to study the geology of Seymour Island.

On November 7, Larsen and two seamen arrived in a small boat from Paulet Island; it was only then that Nordenskjöld

learned of the fate of the *Antarctic*. Fortunately, Nordenskjöld did not have time to think about having to spend a third winter because the Argentine rescue ship *Uruguay,* commanded by Captain Irizar, arrived the following day. By November 10, Nordenskjöld's party, equipment, and samples had been taken aboard the *Uruguay,* and except for two brief stops to pick up Andersson's Mount Flora samples and the remainder of the crew of the *Antarctic* on Paulet Island, they proceeded directly to South America.

EPILOGUE

An analysis of the fossils collected by the Swedish South Polar Expedition, together with our recently acquired knowledge of the geology of Seymour Island, answers the question of why Nordenskjöld found only a relatively small number of species on an island that contains one of the most abundant and diverse Cretaceous and Tertiary faunas and floras in the Southern Hemisphere. Although Nordenskjöld and members of his party made a number of trips to Seymour Island, except for the final 10-day trip, all were short, lasting no more than a few days. Also, most of the forays to Seymour Island were devoted to collecting food for survival during the winter, with little attention given to scientific endeavors.

Except for the two trips along the west side of the island to Cape Bodman during November and December 1902, Nordenskjöld always traveled along the east side of the island to Penguin Point. Accounts of these trips to Penguin Point never stated whether they traveled along the beach or across the island, except for the first trip on which Nordenskjöld recognized his mistake of not establishing the winter station on Seymour Island. It is reasonable to assume that most of these trips were along the beach because it was the quickest route to the rookery. Based on our extensive work on Seymour Island during the last 10 years, we know that the least fossiliferous region of Seymour Island is on the eastern side—from the small bay at the south end of the island to Penguin Point, and to the plateau at the north. The coastal exposures along the east side of the island, from the south end to the high plateau at the north end, are characterized by coarse-grained unfossilerous glauconitic sands. Except for the uppermost part of the Cretaceous sequence just south of Penguin Point, these glauconitic sands have only a few poorly preserved fossils.

With the exception of Nordenskjöld's Cretaceous localities 8 and 9 (Fig. 11), the Cretaceous fossils that Wilckens (1910) described consist of the odd specimens that various members of the expedition collected as they crossed the island. All of Nordenskjöld's Tertiary localities are on the east side of the plateau. Although the Tertiary section exposed on the east side of the plateau contains a number of fairly fossiliferous horizons, they cannot compare in diversity and abundance with those that exist on the west side. If Nordenskjöld had ventured down the west side of the northern plateau, he would have encountered the vast shell banks that characterize most of the western side of the island (Fig. 12). Whereas Nordenskjöld obtained only 26 species of invertebrates from his localities on the east side of the plateau, approximately 200 species of invertebrates and remains of land mammals have been found in the shell banks on the western slopes of the plateau. It is not uncommon on the west side of the plateau to find the surfaces of many of the hills literally paved with fossils.

Although Nordenskjöld recognized the potential of Seymour Island in deciphering the history of Antarctica and made a number of important paleontological discoveries, he failed—due to circumstances and chance—to realize the full scientific potential of Seymour Island. It would be another 70 years before the scientific community recognized the importance of this small, desolate island at the northeast tip of the Antarctic Peninsula.

ACKNOWLEDGMENTS

I extend my deep appreciation to Dr. Fred Goldberg of Ledigo, Sweden, for translating several passages from Aagaard's book *Fangst og forskning i Sydishavet,* and for preparing copies of several of Nordenskjöld's photographs. I am particularly grateful to Peter J. Anderson, Institute of Polar Studies, The Ohio State University, for Murdock's book *From Edinburgh to the Antarctic.* I also thank the Division of Polar Programs, National Science Foundation, for the years of support of my geological program on Seymour Island (namely grants DPP-7421509, -7721585, and -7920215).

REFERENCES CITED

AAGAARD, B. 1930. Fangst og forskning i Sydishavet, Oslo, 354 p.

ANDERSSON, J. G. 1904. Swedish Antarctic expedition. III. The scientific operations on board the *Antarctic* in the summer of 1902–1903. IV. The sledge-exepdition from the *Antarctic.* Geographic Journal of London, 23:215–220.

BRUCE, W. S. 1896. Cruise of the *Balaena* and the *Active* in the Antarctic Seas, 1892–1893. Pt. 1. The *Balaena.* Geographical Journal of London, 7:502–521.

DONALD, C. W. 1893. On the Antarctic Expedition of 1892–1893. British Association for the Advancement of Science, Report.

——. 1896. Cruise of the *Balaena* and the *Active* in the Antarctic seas, 1892–1893. Geographical Journal of London, 6:625–643.

EIGHTS, J. 1833. Description of a new crustaceous animal and on the shores of the South Shetland Islands, with remarks on their natural history. Transactions of the Albany Institute, 2(1):53–69.

LARSEN, C. A. 1894. The voyage of the *Jason* to the Antarctic regions. Geographical Journal of London, 10(4):333–344.

MURDOCK, W.G.B. 1894. From Edinburgh to the Antarctic: An Artist's Sketches during the Dundee Antarctic Expedition. Green and Company, London, 364 p.

NORDENSKJÖLD, O. 1905. Antarctica, or Two Years Amongst the Ice of the South Pole, (English edition) Macmillan Co., New York, 608 p.

SHARMAN, G., AND E. T. NEWTON. 1894. Notes on some fossils from Seymour Island, in the Antarctic regions, obtained by Dr. Donald. Transactions of the Royal Society of Edinburgh, 37, part 3(30):707–709.

——. 1898. Notes on some additional fossils collected on Seymour Island, Graham's Land, by Dr. Donald and Captain Larsen. Proceedings of the Royal Society of Edinburgh, 22(1):58–61.

TURRILL, W. B. 1953. Pioneer Plant Geography: The Phytogeographical Researches of Sir Joseph Dalton Hooker. The Hague, Martinus Nijhoff, 325 p.

WILCKENS, O. 1910. Die anneliden, bivalven und gastropoden der antarktischen Kreideformation. Wissenschaftliche Ergebnisse der Schwedischen Sudpolar-expedition, 1901–1903, 3(13):1–132.

——. 1911. Die mollusken der antarktischen Tertiarformation. Wissenschaftliche Ergebnisse der Schwedischen Sudpolar-expedition, 1901–1903, 3(13): 1–62.

WIMAN, C. 1905. Uber die altertiaren vertebraten der Seymour-Insel. Wissenschaftliche Erbgebnisse der Schwedischen Sudpolar-expedition, 1901–1903, 3(1):1–37.

ZINSMEISTER, W. J. 1977. Note on a new occurrence of the Southern Hemisphere aporrhaid *Struthioptera* Finlay and Marwick on Seymour Island, Antarctica. Journal of Paleontology, 51(2):399–404.

——. 1982. Late Cretaceous–Early Tertiary molluscan biogeography of the southern circum-Pacific. Journal of Paleontology, 56(1):84–102.

MANUSCRIPT ACCEPTED BY THE SOCIETY SEPTEMBER 1, 1987

Geological Society of America
Memoir 169
1988

Topographic map of Seymour Island

Henry H. Brecher and Robert W. Tope
Byrd Polar Research Center, The Ohio State University, Columbus, Ohio 43210

ABSTRACT

A topographic map of Seymour Island was compiled at a scale of 1:20,000 with a 10-m contour interval. One-color Ozalid copies were produced for field use. Small-format reconnaissance aerial photographs were used for the compilation, scale control was obtained from a *Landsat* return beam vidicon image, and leveling was on shorelines. The limitations on precision, accuracy, and reliability arising from use of these unconventional methods are discussed. The map is published here in a three-color version.

INTRODUCTION

After paleontologists and geologists had worked on Seymour Island for several seasons without a topographic map for use in the field and as a base for geologic mapping, the first author was asked to investigate the possibility of producing such a map from existing materials. The only readily available photographic coverage was 1:100,000-scale aerial photography taken by the British Antarctic Survey (BAS) in January 1979 with a modified and calibrated Vinten 70-mm reconnaissance camera. A single recoverable ground control point, a Doppler satellite position at Base Vicecomodoro Marambio, established by the U.S. Geological Survey in 1976, existed (Fig. 1). A surveyed height above sea level for this point was also provided by BAS and attributed to the Instituto Antártico Argentino.

Although it would be necessary to use some unconventional methods to overcome the problem of insufficient ground control, and the limited metric quality of the photographs would have to be accepted, it appeared that a map at a reasonable scale and with a useful contour interval could be compiled. The original product was a Mylar transparency at 1:20,000 scale with a 10-m contour interval, reproduced as black-line Ozalid copies for use in field work. The map is published here in a cartographically more satisfactory form (Fig. 2, in pocket inside back cover). The cartographic drafting is by the second author.

CONTROL

The control points necessary to scale, orient, and level the individual stereoscopic models were obtained by conventional aerial block triangulation of the photographs to be plotted. The "ground control" for the triangulation was derived from a *Land-sat 3* return beam vidicon (RBV) image (30350-12063-D, Path 230, Row 105, Subscene D, scale 1:500,000, 18 February 1979) and from use of the fact that all shorelines are at sea level. The orientation of the RBV image was obtained by overlaying it on a published map at the same scale (British Antarctic Survey, 1979). The estimated error in azimuth due to this procedure, about 4 min of arc, is negligible, and the resultant errors in latitude and longitude have no effect on relative positions on the map. For scaling, 16 reasonably uniformly distributed coastal points were selected and matched binocularly on the RBV image and the aerial photographs. Preliminary adjustments indicated that their precision fell into two distinctly different groups; they were weighted accordingly in the final adjustment. The nominal scale of the RBV image was accepted as errorless. For leveling, 28 points on the shorelines were held to sea-level elevation, and 32 points on floating ice well offshore were assigned elevations of 2 m and were lightly weighted (Fig. 1).

MAP PROJECTION

The map is on the SOM (Space Oblique Mercator) projection using the WGS60 ellipsoid (semi-major axis 6,378,165 m; flattening 1/298.3) because it is derived from the *Landsat 3* RBV image, which is cast on this projection. The projection is conformal and is "for practical purposes distortion free" with "scale differences of no more than 1:10,000" (U.S. Geological Survey, 1979, p. AD-3). A local rectangular coordinate system has been added for convenience with northing parallel to the 56°50′W meridian and arbitrary origin near the lower left corner of the map.

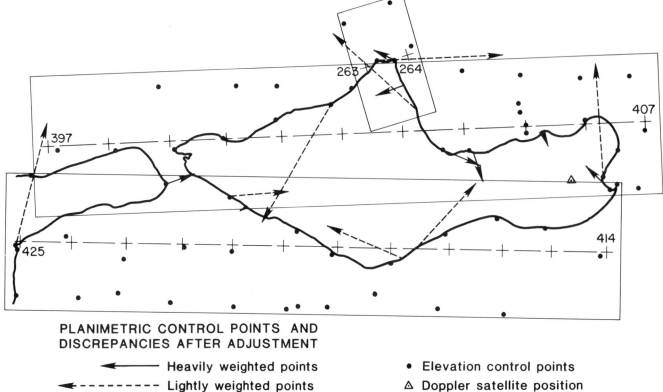

PLANIMETRIC CONTROL POINTS AND
DISCREPANCIES AFTER ADJUSTMENT

←———— Heavily weighted points   • Elevation control points

←- - - - - - - Lightly weighted points   △ Doppler satellite position

Discrepancy scale

Photographic coverage

Figure 1. Photographic coverage and control points.

PLOTTING

In order to accommodate the unusual lens distortion and focal length of the photography, restitution was carried out on analytical plotters. Difficulty in obtaining good relative orientations, first experienced during the aerial trangulation, was also evident during plotting. Substantial model deformation was also present, which was easily detected because almost all models contained long stretches of shoreline. The maximum departure from flatness was generally 5 to 8 m; in one or two models it amounted to more than 15 m. Even though this was undesirable, it was necessary to use points on floating ice for relative orientations. This is not believed to be the cause of the model deformations but it cannot definitely be excluded. All attempts to eliminate this deformation failed, and ultimately it had to be accepted. Elevations are therefore appreciably and unpredictably in error, and some mismatch between adjoining models was evident.

The planimetry, consisting almost entirely of stream beds, was plotted at a scale of 1:20,000, which was considered conven-

ient for field use. Since the density of contours in many places turned out to be greater than had been foreseen, because of regions of unexpectedly steep terrain, contours were plotted at a scale of 1:10,000 and reduced photographically to the 1:20,000 publication scale. Although the 10-m contour interval was not justified by the map accuracy, this interval was selected to permit reasonable depiction of the terrain. No attempt was made to differentiate between dry and active stream beds. Both banks were plotted when streams were wide enough to do so. Although an attempt was made to plot streams down to a given width only, in doubtful cases, streams were plotted rather than omitted.

PRECISION, ACCURACY, RELIABILITY

The block aerial triangulation yielded a photo coordinate standard deviation of 19.8 μm. This is about twice the error expected from the repeatability of measurements but is considered acceptable and represents errors of about 3 m in plan and 5 m in elevation in the terrain. Root-mean-square coordinate errors from the aerial triangulation were 21.7 m in easting, 20.3 m in northing, and 2.9 m in elevation. Discrepancies between "ground control" positions derived from the RBV image, and adjusted positions from the aerial triangulation block adjustment at the individual control points are shown in Figure 1. Clearly, these are due to the combination of errors arising from the measurements of the aerial photography and those of the RBV image. Measurement precision on the RBV image is equivalent to about 7 m in position in the terrain. It is likely that most of the disagreement is due to the uncertainty of identification of features in the RBV image since "ground resolution is estimated to be between 26 and 40 m . . ." (U.S. Geological Survey, 1979, p. AE-2). This, therefore, appears to be the limiting factor in the procedure.

It is hoped that the fairly large number of reasonably well-distributed points used yields a better overall scale result than reliance on a unique solution, i.e., two points only. Furthermore, the redundant points allow some estimate of the quality of the result to be attempted. The sea-level elevations assigned to the shore points were used as the elevation datum in preference to the single elevation at the Marambio VOR building. Possible effects of tides were considered. Tides at the time of photography were found to be negligibly small. The adjusted elevation of the ground surface at the VOR building was 4.2 m higher than the surveyed elevation given for the plaque in the entrance. The model deformation mentioned cannot be further quantified. It thus appears that precision of relative positions and elevations of points "near" each other is a few meters and therefore within the precision of measurements made on the map. Accuracy is much less easily estimated, but it is believed that errors are not likely to exceed 30 m in position and 20 m in elevation. Additional details are given in Brecher (1984).

ACKNOWLEDGMENTS

William Zinsmeister took the initiative to have a map of Seymour Island produced. He and David Elliot secured the necessary funding and provided contacts with BAS. Lloyd Herd, chief of the Aerial Engineering Section of the Ohio Department of Transportation, made available much instrument time free of charge. Henderson Aerial Surveys made an essential instrument available without charge and did the photocopying to produce various map transparencies. Larry Mumford gave valuable help with this aspect of the work. Free time on a plotter and stereocomparator of The Ohio State University Department of Geodetic Science and Surveying was provided through the courtesy of Dean Merchant and Armin Gruen. Mostafa Madani provided his aerial triangulation block adjustment program and gave much help with its use. Michael Thomson of BAS was most helpful in providing copies of the aerial photographs and related information. Caroline Clarke of the BAS section, Directorate of Overseas Surveys, made extraordinary efforts to track down ground control. Jane Ferigno of the U.S. Geological Survey's EROS program helped locate a suitable *Landsat 3* RBV image. Carlos Macellari provided much "user input" during the map compilation. The Ohio State University Instruction and Research Computer Center provided computer time free of charge. The work was supported in part by Division of Polar Programs, National Science Foundation Grant DPP-8020096A02.

REFERENCES

BRECHER, H.H. 1984. Landsat 3 RBV imagery as scale control for a topographic map of Seymour Island, Antarctica from non-metric aerial photographs. Surveying and Mapping, 44(3):253–258.

BRITISH ANTARCTIC SURVEY 1979. Northern Graham Land and South Shetland Islands. British Antarctic Territory, geological map, scale 1:500,000, series BAS 500G, sheet 2, edition 1.

U.S. GEOLOGICAL SURVEY 1979. Landsat Data Users Handbook, 3rd rev. ed. Sioux Falls, SD.

MANUSCRIPT ACCEPTED BY THE SOCIETY SEPTEMBER 1, 1987

Geological Society of America
Memoir 169
1988

Techniques used in collecting fossil vertebrates on the Antarctic Peninsula

Dan S. Chaney
Department of Paleobiology, National Museum of Natural History, Smithsonian Institution, Washington, D.C. 20560

ABSTRACT

Fossil vertebrates were collected on Seymour Island, Antarctica, using traditional field methods with minor modifications. It was found that freezing of plaster bandages prior to setting is not detrimental to their strength. The "cap" of the field jacket not only must cover the exposed surface but also must be secured to the jacket by taking several turns around the entire block to hold the two parts of the jacket securely together. Water-catalyzed polyurethane-impregnated fiberglass bandages were tested and found to protect the specimen in transit well, although they were difficult to apply to small- and medium-sized blocks. Polyethylene glycol is useful for consolidating fragile specimens in weather conditions typical of Antarctica.

INTRODUCTION

Materials and procedures required to remove a fossil from the ground and ensure its safe arrival at the laboratory vary, depending on the enclosing matrix, the specimen's state of preservation, and working conditions. Low temperatures, wind-chill effects, and permafrost, encountered in Antarctica, are usually not important considerations, as most fossil localities are accessible during a relatively warm, dry season. However, localities with weather conditions similar to those of Antarctica exist on other continents not only at high latitudes but also at high altitudes. These conditions must be overcome or compensated for if fossils are to be collected. The most common methods employed to collect vertebrate fossils are surface-prospecting, screen-washing, and excavation. This chapter discusses the effects of low temperatures on traditional collecting procedures and suggests some useful variations and materials to alleviate some of the adverse effects of the cold.

On Seymour Island, Antarctic Peninsula, most specimens have been collected simply by gathering them from the surface. Matrix from the "Mammal Site" in Breakwind Canyon, where fossil marsupial specimens were discovered in 1982 (Woodburne and Zinsmeister, 1982), was screen-washed in the 1984–1985 field season. Mosquito net bags, found useful for the recovery of fossils in Nebraska and Wyoming (Grady, 1978), proved to be effective for wet-washing matrix on Seymour Island (Case et al., this volume). Specimens that are large or fragile, or both, must be excavated, and usually require consolidation and encasement in a protective jacket prior to removal from the ground.

Previous paleontological expeditions to Antarctica used a variety of materials for fossil vertebrate collection. Polyethylene glycol (PEG), however, was not one of the materials employed. Jensen used a portable shed, which could be placed over specimens for local climate control, at Coalsack Bluff in 1969. In the rare instances when this became necessary, Jensen melted beeswax with a propane torch and poured it onto fragile specimens (James Jensen, personal communication, 1984). In 1977–1978 (Shackleton-MacGregor Glacier region) and 1985–1986 (Beardmore Glacier area), a Scott tent was erected for climate control at productive sites, and specimens were collected using either plaster medical bandages, paper, aluminum foil, or a combination thereof. Specimens were wrapped in plastic bubble wrap or burlap for final shipment out of Antarctica (Sherri DeFauw, personal communication, 1986). Consolidants were not used in the field on either the 1977–1978 or the 1985–1986 expedition. On Seymour Island, small shelters—constructed of aluminum foil, backpacks, locality stakes, or a rock box—were found adequate to protect the specimen and a small white-gas stove (used in the field to melt snow and wax) from wind and snow.

CONSOLIDATION

A specimen not consolidated in the field may be damaged during shipment to the laboratory. Frozen fossils may appear to be sturdy while held together by ground ice, but irreparable damage may result if they are not properly treated when thawing occurs en route to the laboratory. Polyvinyl acetate, as with any alcohol- or acetone-soluble plastic, when applied to frozen specimens will remain on the surface rather than penetrate to consolidate the specimen. Lack of penetration, due to the presence of water or ice, is indicated by the formation of a white, gummy film that is weak and will allow the specimen to deteriorate as it thaws. Since all the specimens in the Antarctic are frozen, an alternate method needed to be developed.

It was known that waxes and wax-like substances are useful in cold climates: they can be melted and poured onto the specimens, where they solidify into thick, rigid coatings. When the specimen thaws, the wax remains solid, holding the specimen together. Beeswax and carnauba wax, at one time the waxes most often used in the preparation of vertebrate fossils (Rixon, 1976), have several undesirable properties, including drying and cracking over time, incompatibility with other materials, and irreversibility. As natural waxes dry, they lose the strength to hold the specimen together. Other more desirable materials will not adhere to wax, and the natural waxes are difficult, if not impossible, to remove completely. Once natural wax has been applied to a specimen, wax must be used for any further repairs.

An alternative to natural wax is Carbowax 3350 (Union Carbide Corp.), a polyethylene glycol that is water-soluble and melts upon heating to 58°C. PEG can be removed from the specimen more easily than natural wax. The water-soluble properties of PEG may result in the absorption of moisture as the specimen thaws and thus further impregnation of the specimen may be achieved. Application of PEG to the specimen after it has been trenched is a simple task. The PEG is melted and poured onto the specimen in sufficient quantities to maintain the integrity of the block. Should more than one coat be required, a short wait between applications may be necessary to allow the PEG to solidify. Carbowax shrinks as it solidifies; thus it is useful to apply heat locally to the specimen, after application, to prevent curling and aid penetration. Local heating is easily accomplished with a propane torch. Conditions permitting, the specimen may then be encased in a plaster jacket.

FIELD JACKET

The material from which field jackets for frozen or near-frozen specimens may be made represents another problem. In the laboratory, to test the applicability of plaster-impregnated bandages and plaster of Paris for use in field jackets in Antarctica, freezing water (0°C) was mixed with plaster bandages and plaster of Paris that had been cooled to 0°C or below. The wet bandages and plaster were placed in a freezer overnight, then removed and allowed to warm to room temperature (23°C). The previously frozen materials were compared with a standard plaster bandage and with plaster of Paris that had been mixed with tap water and allowed to cure at room temperature.

Both the test and control sets of bandages and plaster of Paris set, but the frozen plaster of Paris had voids where ice crystals had grown. This condition is the result of the plaster of Paris having been poured into a thick block. The cloth and the thin layers of bandages that were applied to the surface of a cold cube prevented the growth of large crystals. The plaster of Paris was also slightly chalky and came off on the finger when rubbed. This did not appear to be the case with the medical bandages, as they contain adhesive agents that bond the plaster to the fabric, thus reducing chalkiness (Johnson & Johnson, 1959).

Plaster must be applied at temperatures above the freezing point of water to allow proper mixing and should, if possible, be maintained at higher temperatures (20°C+) until set. After setting, plaster requires time to dry before it reaches its greatest strength. In damp weather, plaster in field jackets will set but not dry. An alternate material, without the drawbacks of plaster, had to be found. Scotchflex, Conformable Casting Tape (3M Corp.), a water-catalyzed polyurethane resin and fiberglass bandage, had been used to collect a fossil whale from a spring in Florida (Anonymous, Technology and Conservation Magazine, 1982). The fiberglass bandages set quickly, thus eliminating the problem of freezing. They also attain their greatest strength when submerged in water, thus alleviating the need for further drying.

In the 1983–1984 field season, these water-catalyzed polyurethane resin and fiberglass bandages were used on a highly fractured fossil whale vertebra found in the unconsolidated sands of the late Eocene La Meseta Formation (Fig. 1). PEG was applied to the loose portions of the specimen. The specimen was then trenched and more PEG applied where required. Subsequently, fiberglass bandages were applied; the PEG served as a separator between the specimen and the field jacket. The fiberglass bandages were found to be too stiff to conform well to the irregular contours of the block. They were also extremely sticky and adhered to gloved hands more readily than to the block—a property that could lead to specimen damage, as extensive contact with the block may be required to apply the bandages properly. Nevertheless, with two people working on a larger block, fiberglass bandages could be useful in the Antarctic for specimen collecting.

During the 1984–1985 field season, plaster-impregnated medical bandages were used to collect part of a plesiosaur from the upper Cretaceous Lopez de Bertodano Formation. After the application of PEG, wet paper was applied to serve as a separator between the matrix and the plaster bandages, all of which froze within a minute after being applied. The large blocks required additional strengthening braces. Since no wood was immediately available, plastic locality strikes were used (Fig. 2). Because weather conditions were deteriorating, the frozen blocks were transported back to camp without being capped. The following morning, the blocks, still frozen, were capped in the usual manner. Although the caps appeared to be securely in place, the newly applied plaster bandages did not laminate well to the frozen jackets. The cap on one block was partially torn away during shipment of the specimen to the laboratory, resulting in some damage to the specimen.

Field jackets, particularly for small blocks, may be constructed from other materials. For example, a thick layer of PEG-

Figure 1.1. Whale vertebra as found in the late Eocene La Meseta Formation, Seymour Island, Antarctic Peninsula. Note the fragmentary nature of the specimen and the friable matrix. The scale, a film can, is 5 cm long. 1.2, Polyethylene glycol has been applied to the specimen, which has been trenched for application of the field jacket. Water, in which bandages will be wetted, is heating on small gas stove. 1.3, Fiberglass bandages have been applied to the block in preparation for removal from the ground. 1.4, The block ready for the application of the cap. The cap will complete the protective field jacket. Rock box, used to carry supplies, is being used as a wind break.

Figure 2. Collection of a Portion of a plesiosaur skeleton from the upper Cretaceous Lopez de Bertodano Formation, Seymour Island. Plaster medical bandages are being applied to hold reinforcing plastic splints in place.

impregnated cheese cloth or several layers of aluminum foil can be wrapped tightly around the block.

I did not have the opportunity to work on any of the blocks mentioned above because they went to institutions other than my own for preparation. William Dailey (personal communication, 1984) informed me that the whale vertebra encased in the fiberglass bandages was prepared out in as good a condition as the specimen allowed. Brian Small (personal communication, 1985) explained that the plesiosaur material, encased in the plaster jacket, did not fare as well; this was due in part to damage during shipping, as well as to the problem of poor lamination between jacket and cap. A better procedure for capping frozen blocks is to use several turns of the capping bandages completely around the block to hold the cap secure in place.

CONCLUSIONS

The properties of polyethylene glycol make it useful for the consolidation of specimens under weather conditions encountered in Antarctica. With due care and special considerations for freezing of medical bandages, they may be used as field jackets in subfreezing temperatures. Problems will arise in using plaster of Paris in temperatures slightly above freezing, or on damp or rainy days when the plaster will neither freeze nor dry. Under the latter conditions, polyurethane and fiberglass bandages are probably preferable for use on larger blocks. For smaller blocks, layers of PEG-impregnated cheesecloth wrapping, or even several layers of aluminum foil formed tightly to the block will suffice as a good field jacket.

ACKNOWLEDGMENTS

The Department of Paleobiology, National Museum of Natural History, allowed me to work in Antarctica during the austral summers of 1983–1984 and 1984–1985. My involvement in this project would not have been possible without invitations from M. O. Woodburne and W. J. Zinsmeister to accompany them to Seymour Island. Brian Small and Scott Spesshardt of Texas Tech University, and Tim Horner, presently at Ohio State University, helped collect the plesiosaur. This project was supported by National Science Foundation Grants DPP-8215493 and DPP-8521368. Earlier drafts of this manuscript have been read by Robert Emry, Arnold Lewis, and Clayton Ray, Department of Paleobiology, National Museum of Natural History, Smithsonian Institution; Sherri DeFauw, Department of Biological Sciences, Wayne State University; Rodney Feldmann, Geology Department, Kent State University; and Michael Woodburne, Department of Earth Sciences, University of California at Riverside. The final manuscript was reviewed by Donald Baird, Museum of Natural History, Princeton University, and William Simpson, Department of Geology, Field Museum of Natural History. I appreciate all of their suggestions, which improved the manuscript. The 3M Corporation provided the fiberglass bandages used in 1983–1984 at no charge.

REFERENCES CITED

TECHNOLOGY AND CONSERVATION MAGAZINE. 1982. Technology trends: a mighty cast. 1982. Technology and Conservation Magazine, 7(4):5-6.

GRADY, F. vH, 1979. Some new approaches to microvertebrate collecting and processing. The Chiseler, 1(1), 1978. *Reprinted in* Newsletter of the Geological Curators Group, 2(7):439-442.

JOHNSON & JOHNSON. Technical aspects of plaster of Paris. 4 p.

RIXON, A. E., 1976. Fossil Animal Remains: Their Preparation and Conservation. Athlone Press, University of London, London, 304 p.

WOODBURNE, M. O., AND W. J. ZINSMEISTER. 1982. Fossil land mammal from Antarctica. Science, 218:284-286.

MANUSCRIPT ACCEPTED BY THE SOCIETY SEPTEMBER 1, 1987

Geological Society of America
Memoir 169
1988

Stratigraphy, sedimentology, and paleoecology of Upper Cretaceous/Paleocene shelf-deltaic sediments of Seymour Island

Carlos E. Macellari
Earth Sciences and Resources Institute, University of South Carolina, Columbia, South Carolina 29208

ABSTRACT

Richly fossiliferous Upper Cretaceous–Paleocene beds exposed on Seymour Island provide an excellent opportunity for a combined sedimentological-paleoecological analysis of a high-latitude shallow marine sequence. The depositional paleoenvironments of this sequence were interpreted on the basis of the stratigraphy, sedimentological analysis (including grain-size analysis) of the sediments, and auto- and synecological evaluation of the fauna present.

The sequence is composed of two formations. The underlying López de Bertodano Formation (upper Campanian to Paleocene) consists of 1,190 m of gray to tan, friable, sandy, muddy siltstone, and is subdivided into 10 informal units. The lower six units (informally named the *Rotularia* Units) contain a depauperate macrofauna and are dominated by the annelid *Rotularia*. They were deposited in a shallow marine environment, near a delta or estuary. Units 7 through 10 include an abundant macrofauna (Molluscan Units). Units 7 through 9 are interpreted as progressively deeper water deposits, with Units 7 and 8 representing middle shelf facies, and Unit 9, the outer shelf facies. Macrofauna in the most offshore portion of Unit 9 is characterized by an epifaunal suspension-feeding bivalve asssemblage dominated by *Pycnodonte* cf. *P. vesiculosa*, and an increased percentage of cosmopolitan ammonites. Regressive facies appear 30 m below the inferred Cretaceous/Tertiary boundary (contact between Units 9 and 10). Unit 10 was probably deposited in a middle shelf to inner shelf environment.

The overlying Sobral Formation (Paleocene) follows unconformably, and is composed of as much as 255 m of maroon, well-laminated silts at the base, followed by cleaner sandstones that become more glauconitic and crossbedded toward the top. The Sobral is not very fossiliferous and represents the filling of the basin by the progradation of a deltaic system. Units 1 and 2 are interpreted as pro-delta facies followed by clean sands of a coastal barrier (Unit 3). The uppermost Sobral (Units 4 and 5) contains the delta top facies, mostly representing the lateral accretion of distributary channels.

Three unconformity-bounded depositional sequences are recognized in the studied section. The lower sequence extends from Units 1 to 9 of the López de Bertodano Formation (upper Campanian to upper Maastrichtian). The intermediate sequence is restricted to Units 9 and 10 of the same formation (upper Maastrichtian to lower Paleocene); the upper sequence is represented by the Sobral Formation (Paleocene). Uncertainties in the dating of these sequences make their comparison with worldwide sea-level fluctuations premature.

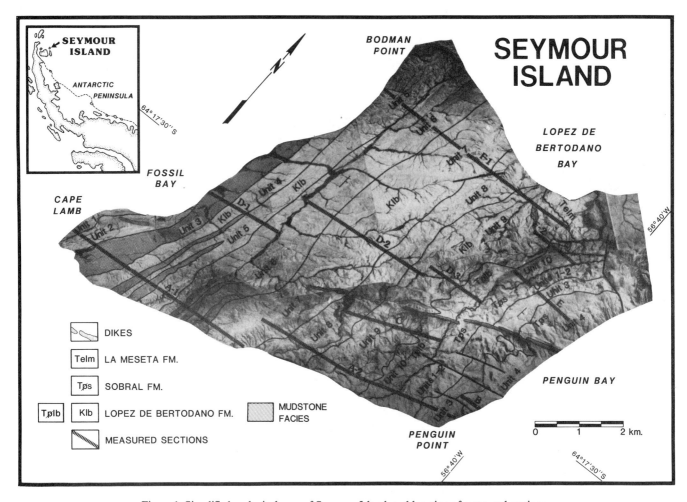

Figure 1. Simplified geological map of Seymour Island, and location of measured sections.

INTRODUCTION

Seymour Island (Isla Vicecomodoro Marambio in the Argentine literature), located on the northeastern tip of the Antarctic Peninsula at latitude 64°15′S and longitude 56°45′W (Fig. 1), contains a thick and richly fossiliferous shallow marine sequence. The Cretaceous to Paleocene strata of the island have been previously described in several reports (Andersson, 1906; Bibby, 1966; Rinaldi et al., 1978; Elliot and Trautman, 1982; Zinsmeister, 1982; Huber, 1984; Macellari, 1984a; Palamarczuk et al., 1984); however, no detailed analysis of the stratigraphy or depositional environments of the sequence as a whole has been published to date.

The monotonous lithology (composed mostly of muddy sandy silt) and the paucity of sedimentary structures do not provide enough information for detailed analysis of the depositional environments of the sequence. In contrast, the abundant and well-preserved macrofauna preserved in these sediments display marked variations in composition and abundance through time, thus helping to distinguish variations in the depositional environments otherwise impossible to distinguish with the lithological data alone.

This study concentrates on the stratigraphy, sedimentology, and depositional environments of the López de Bertodano and Sobral Formations cropping out in the southern part of Seymour Island. The diverse and abundant macrofauna, as well as the excellent exposures of these rocks, provide the basis for a combined sedimentological-paleoecological analysis of the sequence. This work is based on data collected during two field seasons (1982, 1983-1984), detailed photointerpretation of the exposures, and subsequent sedimentological and macropaleontological laboratory studies of the samples collected.

STRATIGRAPHY

Six sections were measured through the Cretaceous–Lower Tertiary sequence of Seymour Island (Figs. 1, 2). This homoclinal sequence dips 8 to 10° to the east, and is divided into the López

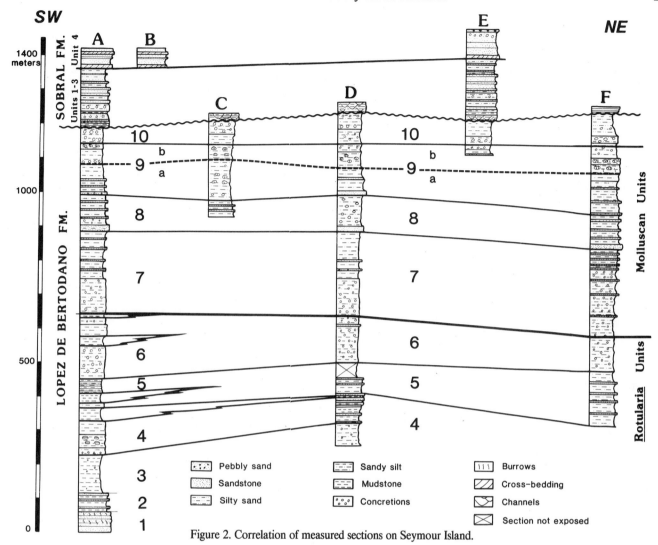

Figure 2. Correlation of measured sections on Seymour Island.

de Bertodano and Sobral Formations. The two formations are subdivided into several informal units having a continuous morphologic expression. These units were distinguished by photointerpretation, on the basis of prominent and continuous stratal surfaces, and usually contain a distinctive lithology. This subdivision provides a valuable tool for correlation between localities and different sections in the lithologically monotonous sequence of Seymour Island.

An erosional unconformity marks the contact between the López de Bertodano and Sobral Formations. The unconformable nature of this contact is reflected in the erosion of a portion of the underlying López de Bertodano Formation, the sporadic occurrence of a lag conglomerate at the base of the Sobral Formation (Macellari, 1984a), and in a jump in the vitrinite reflectance values across this contact (Palamarczuk et al., 1984).

The "datum" for correlation between different measured sections is the last ammonite-bearing bed, which is immediately overlain by a laterally continuous glauconitic unit where the Cre-

taceous/Tertiary boundary has been placed (Askin, 1984; Huber, 1984, 1987; Macellari, 1984a, 1985b; but see Huber, 1987). The "datum" is parallel to the boundary of the upper units of the López de Bertodano Formation.

López de Bertodano Formation

This formation, defined by Rinaldi et al. (1978), is composed of friable sandy siltstones with intercalations of more indurated calcareous horizons. The maximum thickness was found in Section A (1,190 m), and the minimum thickness in Section F (920 m) (Fig. 2). This change in thickness is due to both the preservation of older strata and to a greater rate of sedimentation in the southwest portion of the island. Grain size and mud percentages remain fairly constant in the unit, but there is an increase upward in glauconite as well as in the volcanic component of the sand fraction. The López de Bertodano Formation is of Late Campanian to Paleocene age, consisting mostly of Maastrichtian-

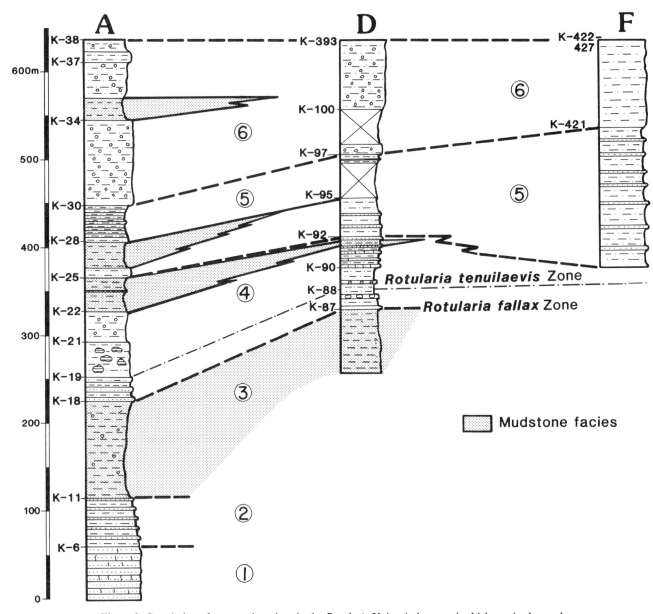

Figure 3. Correlation of measured sections in the *Rotularia* Units. A decrease in thickness is observed from Sections A through F. Symbols are the same as in Figure 2. Numbers on the left of the columns indicate localities marking the boundary of the units.

age strata (Askin, this volume; Huber et al., 1983; Huber, 1984, this volume; Macellari, 1984a, 1986).

The poorly indurated strata of the López de Bertodano Formation has been subdivided into 10 units based on differences in lithologic, faunal, and physiographic characteristics (Fig. 1). Recognition of these units in the field, however, is not always possible. The macrofauna is abundant and allows the clustering of the 10 units into two informal, larger units. Units 1 through 6 are characterized by the conspicuous presence of the annelid worm tube *Rotularia* and a poorly preserved depauperate macrofauna. Also characteristic of most of these beds is the presence of an agglutinated foraminiferal fauna (Huber, 1984). Units 7 through 10 (Molluscan Units) bear an abundant and diverse bivalve, gastropod, and ammonite macrofauna.

Rotularia *units (Units 1-6).* The lowermost beds on Seymour Island form a distinctive lithologic sequence composed of friable silty mudstone and sandy siltstone predominantly gray in color, usually strongly bioturbated, and interbedded with fine-grained gray sandstone (Figs. 3, 4.1,2). Concretions are sparse but increase in abundance toward the top of the unit (Fig. 3). These beds continue laterally into Snow Hill Island.

The maximum thickness of the *Rotularia* Units occurs in Section A (635 m) on the southwest portion of the island. The *Rotularia* Units show a decrease in thickness toward the northeast, as the gray mudstone units pinch out. Another lateral difference is the northeasterly decrease in the amount of fine sand, in such a way that in Section F the sandstone beds form discrete packages among a predominantly clayey, sandy siltstone sequence (Fig. 4.1). Fossils, though very sparse, are better preserved in this direction. The lithologic units identified are:

Unit 1, only observed in Section A, has a total thickness of 59 m and is composed of well-lithified, moderately well-bedded, fine-grained, well-sorted, angular, greenish to grayish sandstone in beds 20 cm to 1 m thick, with irregular tops and bottoms. The sandstone beds are separated by thin siltstone-shale intercalations. Round to subround concretions as much as 30 cm in diameter that have a yellowish patina are found in this unit. The sandstones have a calcareous cement, and bioturbation is very common. At the base of the section (Station K-1), there are common, small (ca. 2 mm in diameter), randomly oriented tubes with the appearance of feeding traces of polychaete worms. Horizontal tubes are more common toward the top of this unit (Fig. 5.1). Fossils are very sparse and poorly preserved. They include *Rotularia (Austrorotularia) fallax* (rare), *Gunnarites bhavaniformis*, *Oistotrigonia* sp., Gastropoda sp. indet., and fossil wood. All macrofossils are preserved as internal molds. Morphologically, this unit presents a prominent cuesta topography.

Unit 2 is found only in Section A and has a thickness of 56 m. It is composed of greenish gray to tan, very fine-grained sandstone (usually 50 cm thick) with 12 m intervals of friable siltstone. Laminations are preserved in some portions of the clayey siltstone. As is also the case in Unit 1, fine-grained sandstone beds are grain-supported and contain little matrix. Fossils are sparse, preserved only as internal molds. These include *Nordenskjoldia* sp., unidentified bivalves, ammonites, corals, and fossil wood. This unit gives rise to a terrane of gently undulating low relief, with more resistant sandstone beds.

Unit 3 is found in Sections A and D and has a maximum thickness of 110 m. It consists of friable, sandy, muddy siltstone, massive in appearance with colors ranging from gray to olive green. A grain-size analysis of this unit yielded 72 percent of mud (silt plus clay) (26 percent clay). Rounded, calcareous, reddish concretions (10 to 15 cm in diameter) are common. The only fossil found here is *R. (Austrorotularia) fallax*, which is moderately abundant. This unit forms a low-relief badlands topography (Fig. 4.2).

The base of Unit 4 is defined by a break in topography, and is easily delineated in aerial photographs (Fig. 4.2). The lithology of the unit is composed of gray sandy siltstone with small round concretions, interrupted by more resistant layers of 0.3 to 0.5 m thick, gray, moderately bedded, calcareous sandstone. The fine-grained sandstone beds generally show a basal, erosional contact, sometimes contain mud intraclasts, and commonly have small, solitary corals exposed on the bedding planes. Some levels are strongly bioturbated. Intervals of muddy siltstone, similar to those of Unit 3, are intercalated, but disappear rapidly in a northeasterly direction. The most conspicuous feature of Unit 4 is the presence of large (as much as 1.5 m in diameter), rounded, calcareous

TABLE 1. MACROFAUNA FOUND IN UNIT 5 OF THE LOPEZ de BERTODANO FORMATION*

Annelida R.(Austrorotularia) tenuilaevis (A,F)	**Gastropoda** Eunaticina arctowskiana (F) Perissoptera sp. (D,F) Cryptorhytis philippiana (F) Taioma charcotianus (F)
Bivalvia Panopea clausa (A,F) Pinna sp. (F)	
	Other Corals (D,F) Echinoderm spines (A,F)
Ammonoidea Maorites tuberculatus (F)	Fossil wood (F) Marine reptiles (F)

*Letters in parentheses indicate the section in which the fossil was found.

concretionary horizons that preserve the original bedding planes of the rock. With the exception of *Rotularia,* fossils are poorly preserved and include corals, *R. (Austrorotularia) tenuilaevis,* unidentified ammonites and gastropods, the decapod *Hoploparia stokesi,* abundant echinoderm spines, vertebrae of marine reptiles, and fossil wood. A size analysis of a representative sample of this unit yielded 64 percent mud (18 percent clay) (Sample K-22).

Unit 5 can be traced laterally in aerial photographs from one end of the island to the other. However, a facies change takes place along strike. Section A contains massive, gray, clayey, sandy silt beds that pinch out rapidly to the northeast. These muddy units have concave bases and flat tops in aerial view. Above these are found well-laminated (beds 1 to 1.5 cm) greenish gray clays, silts, and sandstones that are occasionally flaser-bedded (Fig. 4.3). Intercalations of well-lithified sandstone as much as 3 m thick are also observed. These usually contain small, solitary corals on the bedding planes. In Section F and in the lower part of Section D, the dominant lithology is 20 to 25-m-thick packages of massive, gray, friable sandy siltstone with intercalations of well-bedded, fine-grained, calcareous sandstone that are 20 to 25 m apart and approximately 0.2 to 0.4 m thick. Strongly bioturbated concretionary levels are common. A *Rotularia*–solitary coral–echinoderm faunal association generally dominates, with fossils better preserved and more diverse in Section F. Fossils found in this unit are listed in Table 1. A grain-size analysis of Sample K-28a indicated 74 percent mud, with 21 percent clay. The thickness of this unit ranges from 82 m (Section A) to 96 m (Section D) to 160 m (Section F).

Unit 6 is uniform in its distribution throughout the island, with the exception of one level of massive, gray mudstone present only in Section A. The dominant lithology of Unit 6 is a dark, apparently massive, sandy siltstone with abundant red-weathering concretions, between 10 and 30 cm in diameter, sometimes containing calcareous tubes 5 to 8 mm in diameter and 10 to 15 cm long. Minor, fine-grained, calcareous sandstone intercalations are

Figure 4. 1; Strata of Unit 5 of the López de Bertodano Formation in the vicinity of Cape Bodman. Lithology is composed of apparently massive gray sandy silt with sporadic intercalations of more indurated fine, calcareous sandstone. Cockburn Island can be observed in the background. 2; Panorama looking north, showing the contact between mudstones of Unit 3 and the more resistant beds of Unit 4. 3; Contact between well-bedded, brownish, clayey, sandy silt, and massive, gray, sandy silt (below). Unipod is 1.5 m; Station K-28, Unit 5.

also observed. Grain-size analysis of Unit 6 indicates a mud percentage ranging from 71.5 to 97 percent, with as much as 21.5 percent clay. Fossils are sparse, but are somewhat more abundant than in the underlying unit. The exception is *Rotularia,* which is extremely abundant at several levels. The most common association is that of *Rotularia* and echinoderms; a list of fossils found in this unit is shown in Table 2.

Molluscan units (Units 7–10). A thick sedimentary sequence of generally monotonous, sandy siltstone that contains a very abundant and well-preserved molluscan fauna overlies the *Rotularia* Units. These units were deposited with relatively constant thickness throughout the island (Fig. 6).

Unit 7 has an almost uniform thickness of 250 m. The dominant lithology is a dark gray, friable, bioturbated silty sandstone with minor fine-grained sandstone intercalations. Flaser bedding is present in some intervals. The lower 130 m of the unit in Section A carries abundant small, round concretions. This concretionary interval is much thicker in Section F, where concretions are large and irregular at the base, but become progressively smaller up-secion. Intercalations of thin (approximately 40 to 50 cm), gray, well-bedded, calcareous sandstone are most abundant in Section A. Fossils are very abundant and well preserved, particularly at the base of the unit. Bivalve molluscs generally are articulated, and many are preserved in life position, indicating no, or very little, postmortem reworking of the material. Fossils found in this unit are listed in Table 3. In addition, fossil wood, echinoderm spines, shark vertebrae, corals, marine reptile bones, and the decapod *Callianassa meridionalis* are present. The mud content of Unit 7 ranges from 67 percent (clay 17 percent; Station K-45) to 68 percent (clay 12 percent; Station K-38).

Unit 8 has a thickness ranging from 100 to 110 m. The base of the unit consists of 6 to 15 m of massive, relatively well-cemented, gray, silty sandstone with beds approximately 1 m thick (Fig. 5.2). In Section D, the same beds exhibit some minor clay layers and evidence of strong bioturbation. Horizontal and oblique burrows are common. Above this, there are better bedded sandy silts that are intercalated with massive, gray, silty sandstone. Concretionary horizons are more common toward the top of the unit. Glauconitic intervals, 40 to 50 cm thick, are present in Section A. Unit 8 has a larger mean grain size and sand percentage (41 to 63 percent sand) than the adjacent units. Solitary corals are abundant. Molluscs are well preserved, but are relatively rare, usually small, and not very diverse (Table 4). In contrast, the foraminiferal fauna shows a peak in diversity at this level (Huber, 1984, this volume).

Unit 9 displays an increase in the glauconite content of the sediment, as well as a very abundant and diverse macrofauna. The upper half of the Unit 9 (9b) includes a lower interval with very abundant macrofauna (*Pycnodonte* beds), followed by an upper interval with a less diverse molluscan assemblage.

Unit 9 is composed of gray to tan sandy silt, and is well bedded at the base and more massive toward the top. Concretions become abundant upward, some of them as much as 0.4 to 0.5 m in diameter. Numerous elongated concretions (which probably

TABLE 2. MACROFAUNA FOUND IN UNIT 6 OF THE LOPEZ de BERTODANO FORMATION*

Annelida	**Gastropoda**
R. (Austrorotularia)	*Cassidaria mirabilis* (A,D)
tenuilaevis (A,D,F)	*Cryptorhytis philippiana* (F)
R. (Rotularia) shackletoni (F)	*Eunaticina arctowskiana* (A)
	Taioma charcotianus (D)
Bivalvia	
Pycnodonte seymourianus (A)	
Dozyia dryqalskiana (D)	**Other**
Nucula suboblonga (F)	Corals (A,D)
Cucullaea antarctica (F)	Echinoderm spines (F)
Oistotrigonia sp. (D,F)	Marine reptiles (A)
Panopea clausa (D)	
Thracia sp. (A)	
Ammonoidea	
Diplomoceras lambi (A,D)	
Maorites seymourianus (F)	

*Letters in parentheses indicate the section in which the fossil was found.

represent burrows of this shape) are found at the top. The thickness of this unit up to the *Pycnodonte* beds ranges from 70 to 120 m. Grain-size analysis in Unit 9 shows a mud percentage ranging from 52 to 64 percent.

Fossils found in Unit 9a are included in Table 5. In addition, the following were found: echinoderm spines, marine reptile bones, shark vertebrae, the decapods *Hoploparia stokesi* and *Callianassa meridionalis,* Brachiopoda sp. indet., corals, and fossil wood, usually bored by pholadid bivalves. The macrofossils are well preserved and are commonly found in situ with articulated valves. Large concretions containing numerous juvenile ammonites, particularly *Maorites densicostatus,* and to a lesser extent *Grossouvrites gemmatus,* are also common here (Fig. 5.3).

The *Pycnodonte* beds do not have well-defined boundaries. They extend approximately from the upper part of the *Pachydiscus riccardi* to the *P. ultimus* zone (Macellari, 1985a), with a total thickness of approximately 40 to 50 m. Characteristics of these beds include the very abundant bivalve *Pycnodonte* cf. *P. vesiculosa* (Fig. 5.4) and extremely abundant bioturbation that is preserved in large (as much as 1 m) concretionary horizons (Fig. 5.5). The lithology consists of massive, gray, sandy siltstones and glauconitic beds, with small irregular concretions. Fossils found in the *Pycnodonte* beds are included in Table 6. In addition, the following are also present: echinoderm spines, marine reptile bones, *Hoploparia stokesi* (very abundant in Section F), corals, crinoid fragments, bored fossil wood, and coalified wood.

The uppermost part of the *Pycnodonte* beds coincides approximately with the last record of *Pachydiscus ultimus*. Very abundant bioturbation similar to *Thalassinoides* and a diverse and abundant macrofauna are found here. Bivalves in this uppermost interval consist mostly of epifaunal suspension feeders. At

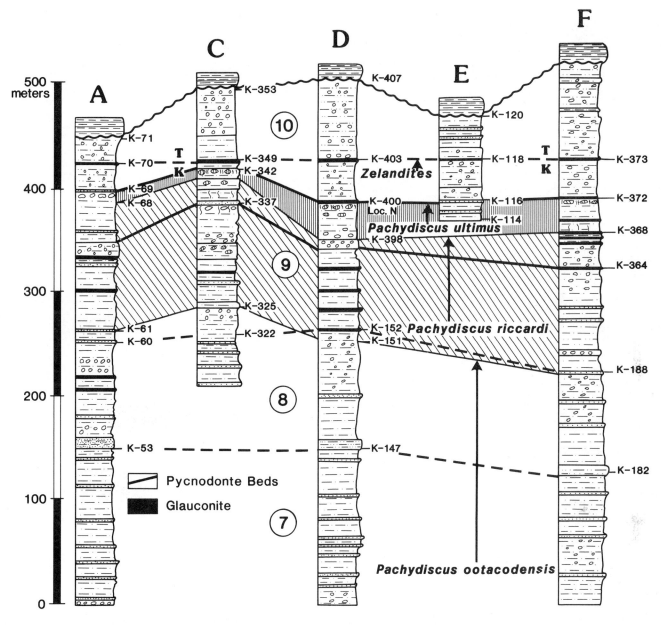

Figure 6 (above). Diagram showing correlation of sections measured in the Molluscan Units. See Figure 2 for an explanation of the symbols. Numbers to the right of the columns indicate localities marking the boundary of the units.

Figure 5. 1; Horizontal burrows in sandstone of Unit 1 of the López de Bertodano Formation (Station K-5). 2; Thick-bedded, silty sandstone forming the base of Unit 8 (Station K-182). 3; Calcareous concretion with very abundant juvenile specimens of *Maorites densicostatus;* Station K-377, Unit 9. 4; *Pycnodonte* beds. Note the large number of specimens of *Pycnodonte* cf. *P. vesiculosa* on the surface, as well as abundant bioturbated concretions (Station K-399). 5; Strongly bioturbated concretions. Note conspicuous tubes extending from a more expanded chamber. Top of *Pycnodonte* Beds, Station st. 7. Notebook is 30 cm long. 6; Base of Unit 1 of the Sobral Formation, showing flaser bedding and irregular channels with a higher sand content (lighter areas). Soft-sediment deformation is observed at the base of the channel (Station K-122). 7; *Skolithos* in an apparently massive, medium-grained sandstone (Unit 3 of the Sobral Formation). 8; Cross-bedded glauconitic sandstone at the base of Unit 4 of the Sobral Formation.

TABLE 3. MACROFAUNA FOUND IN UNIT 7 OF THE LOPEZ de BERTODANO FORMATION

Annelida
R. (Rotularia) shackletoni
R. (Austrorotularia) tenuilaevis
R. (Austrorotularia) zinsmeisteri

Cephalopoda
Maorites seymourianus (A,D,F)
Maorites wedellensis (A,D,F)
Kitchinites darwini (A,D,F)
Diplomoceras lambi (A,D,F)
Grossouvrites gemmatus (A,D,F)
Pachydiscus ootacodensis (D)
Eutrephoceras simile (A,D,F)

Gastropoda
Perissoptera sp. (A,D,F)
Cassidaria mirabilis (A,D)
Cryptorhytis philippiana (ab)(A,D,F)
Eunaticina arctowskiana (ab)(A,D,F)
Amberleya spinigera (ab) (A,D,F)
Pleurotomaria larseniana (A,D,F)
Taioma charcotianus (F)
Cerithium sp.

Bivalvia

Nucula suboblonga (A,D,F)
Linotrigonia pygoscelium (A,D,F)
Laevitrigonia regina (A,D,F)
Nordenskjoldia nordenskjoldi (D,F)
Austrocucullaea oliveroi (A)
Cucullaea antarctica (A,D,F)
Pinna anderssoni (A,D,F)
Acesta snowhillensis (F)
Seymourtula antarctica (F)
Dozyia drygalskiana (A)

Lucina scotti (D)
Lahilla larseni (F)
Thyasira townsendi (A)
Thracia sp. (A)
Solemya rossiana (A)
Panopea clausa (A,D,F)
Goniomya hyriiformis (A,D,F)
Pycnodonte cf. vesiculosa (A,D)
Pycnodonte seymourianus (A,D,F)

*Letters in parentheses indicate the section in which the fossil was found; **ab** = abundant.

TABLE 4. MACROFAUNA FOUND IN UNIT 8 OF THE LOPEZ de BERTODANO FORMATION*

Annelida
R. (Austrorotularia) sp.
R. (Rotularia) shackletoni

Ammonoidea
Grossouvrites gemmatus (A,D)
Maorites densicostatus (C,D)
Kitchinites laurae (D)
Pachydiscus riccardi (D)
Anagaudryceras seymouriense (D)

Bivalvia
Nucula suboblonga (A)
Cucullaea antarctica (C,D,F)
Pulvinites antarctica (D)
Laevitrigonia regina (C,D,F)
Linotrigonia pygoscelium (C,D)
Lahilla larseni (A)
Panopea clausa (A,F)
Goniomya hyriiformis

Gastropoda
Perissoptera nordenskjoldi (A,C,D,F)
Amberleya spinigera (A,C,D)
Eunaticina arctowskiana (F)
Cassidaria mirabilis (C,D,F)
Cryptorhytis philippiana (C)
Taioma charcotianus (D)

Scaphopods

Other
Corals
Echinoderm spines
Fossil wood

*Letters in parentheses indicate the section in which the fossil was found.

TABLE 5. MACROFAUNA FOUND IN UNIT 9 OF THE LOPEZ de BERTODANO FORMATION*

Annelida
R. (Rotularia) shackletoni (A,C,F)
R. (Austrorotularia) zinsmeisteri (A)
R. (Austrorotularia) sp. (C,F)
Ditrupa sp.

Cephalopoda
Maorites densicostatus (A,C,D,F)
Grossouvrites gemmatus (A,C,D,F)
Diplomoceras lambi (A,C,D,F)
Kitchinites laurae (C,D,F)
Pachydiscus riccardi (A,C,D,F)
Anagaudryceras seymouriense (C,D)
Pseudophyllites loryi (F)
Eutrephoceras simile (D,F)

Gastropoda
Perissoptera nordenskjoldi (A,C,D,F)
Cassidaria mirabilis (C,D,F)
Cryptorhytis philippiana (D,F)
Amberleya spinigera (C,D,F)
Pleurotomaria larseniana (A,D,F)
Taioma charcotianus (C,D,F)
Cerithium sp. (D)
Eunaticina arctowskiana (F)

Other
Solitary corals (ab)

Bivalvia

Linotrigonia pygoscelium (D)
Laevitrigonia regina (C,D,F)
Cucullaea antarctica (C,D,F)
Pinna anderssoni (C,D,F)
Pulvinites antarctica (C)
Phleopteria sp. (D)
Entolium sp. A (D,F)
Entolium sp. B
Acesta n. sp. (D)
Seymourtula antarctica (D,F)
Dozyia drygalskiana (d)

Lucina scotti (D)
Lahilla larseni (C,D,F)
Cyclorisma n. sp. (C,D)
Thyasira townsendi (D)
Solemya rossiana (D)
Panopea clausa (D,F)
Goniomya hyriiformis (D)
Pycnodonte cf. vesiculosa (D,F)
Surobula n. sp. (D,F)

*Letters in parentheses indicate the section in which the fossil was found; **ab** = abundant.

one locality (Locality N), a level with intraclasts and reworked macrofauna was found within these upper beds (Fig. 6). The *Pycnodonte* beds are followed by 25 to 30 m of massive, gray, sandy siltstone with small, round to irregular concretions. In previous works this interval was recognized only in Section D (Macellari, 1984a, 1986). However, subsequent detailed field mapping by P. Sadler and B. Huber demonstrated that the contact between Units 9 and 10 was misplaced in Sections A, E, and F, and in all of them an interval 25 to 30 m thick was found between the last record of *Pachydicus ultimus* and the base of Unit 10. In Section C this uppermost interval appears to be much thinner, but this could be attributed to measurement errors caused by a more complicated topography.

The contact between Units 9 and 10 is placed at a laterally continuous, 40-cm-thick glauconitic horizon that coincides with the inferred Cretaceous/Tertiary boundary (Askin, this volume, b; Macellari, 1984a, 1985; Huber, this volume; Sadler, this volume). Beneath this glauconite are found the last ammonites, represented by the species *Zelandites varuna, Kitchinites* sp., *Maorites densicostatus,* and *Diplomoceras lambi.*

TABLE 6. MACROFAUNA FOUND IN UNIT 9b (*PYCNODONTE* BEDS) OF THE LOPEZ de BERTODANO FORMATION*

Annelida
R. (Rotularia) shackletoni
R. (Austrorotularia) sp.
  (A,C,D,E,F)
Ditrupa sp.

Gastropoda
Perissoptera nordenskjoldi (D,F)
Cassidaria mirabilis (D)
Cryptorhytis philippiana (C,D,F)
Amberleya spinigera (C,D,F)
Taioma charcotianus (C,D,F)
Cerithium sp.
Eunaticina arctowskiana (C)

Cephalopoda
Maorites densicostatus
  (A,C,D,F)
Grossouvrites gemmatus
  (C,D,F)
Diplomoceras lambi (C,D.F)
Kitchinites sp. (D)
Pachydiscus riccardi (C,D)
Pachydiscus ultimus
  (A,C,D,E,F)
Anagaudryceras
  seymouriense (D,F)
Zelandites varuna (D)
Pseudophyllites loryi (C)
Eutrephoceras simile (C,D)

Bivalvia

Laevitrigonia regina (C,D,F)	Lahilla larseni (C,D,F)
Cucullaea antarctica (C,D)	Cyclorisma n. sp. (C)
Pinna anderssoni (C,D,F)	Panopea clausa (C,D) (C,D)
Pulvinites antarctica (C)	Goniomya hyriiformis (C,D)
Acesta n. sp. (C,D,F)	Pycnodonte cf. vesiculosa
Seymourtula antarctica (C,D)	(A,C,D,E,F)

*Letters in parentheses indicate the section in which the fossil was found.

TABLE 7. MACROFAUNA FOUND IN UNIT 10 OF THE LOPEZ de BERTODANO FORMATION*

Annelida
R. (Austrorotularia) sp.†

Bivalvia
Nucula suboblonga (C,D)
Cucullaea ellioti (A,C,D,E,F)
Lahilla larseni (A,C,D,E,F)

Gastropods
Perissoptera nordenskjoldi
  (A,C,D,F)
Pleurotomaria larseniana (A)
Austrosphaera patagonica (C)

Other
Echinoderms spines
Bored fossil wood (**ab**)

*Letters in parentheses indicate the section in which the fossil was found. **ab** = abundant.
†Present only at the base of the unit.

The thickness of the overlying Unit 10 is directly controlled by the degree of erosion prior to the deposition of the Sobral Formation. Unit 10 is composed of green to greenish brown, sandy siltstone that becomes progressively more sandy upward. Large, round concretions are usually found at the top of the unit, as well as concretionary horizons and irregular concretions. Glauconite is common, particularly in Section A. The impoverished macrofauna found in Unit 10 is restricted to a few species of bivalves and gastropods (Table 7). The most typical element is the venerid bivalve *Lahilla larseni,* which occurs in large concentrations, commonly associated with the winged gastropod *Perissoptera nordenskjoldi.* In addition, bored fossil wood is particularly abundant throughout this interval.

Sobral Formation

These beds are approximately equivalent to the "couches a *Lahilla luisa* Wilckens sans ammonites" of Kilian and Reboul (1909). They were included in the "Older Seymour Island Beds" of Andersson (1906) and in the "Snow Hill Island Series" of Bibby (1966), together with the López de Bertodano Formation. Rinaldi et al. (1978) considered these beds an independent lithostratigraphic unit; this interpretation was followed by Rinaldi (1982), Zinsmeister (1982), and subsequent workers. The Sobral Formation is assigned a lower Paleocene (Danian) age based on

dinoflagellates (Pallamarczuk et al., 1984), siliceous microfossils (Harwood, this volume), and foraminifera (Huber, this volume). In previous works, I have included the upper part of the Sobral Formation in the Cross Valley Formation (Macellari, 1984a, 1986; Macellari and Zinsmeister, 1983). However, recent mapping in the area indicates that the type section of the Cross Valley Formation is separated by an erosional unconformity from the upper Sobral Formation (Sadler, this volume). Most likely, this unconformity involves only a minor time gap, and the Cross Valley Formation represents the culmination of the shallowing trend initiated in the Sobral Formation (cf. also Sadler, this volume). The maximum measured thickness of the Sobral Formation is 255 m.

The Sobral Formation was recently divided into five informal units on the basis of detailed mapping (Sadler, this volume). The lower contact of the Sobral Formation is defined at the base of a very distinctive and laterally continuous 15- to 20-m-thick, dark brown, well-laminated to flaser-bedded, silty mudstone that contains numerous channels of varying thicknesses (Figs. 5.6, 7, 8). The channels are usually 1 to 3 m wide, almost coalesce at certain intervals, and are cut sometimes by large vertical burrows. Much larger scale channels (on the order of 50 to 100 m) are also observed at the base of the sequence, cutting regionally into the underlying López de Bertodano Formation. This unit has been informally called the "brown chocolate layer," and it makes a distinctive break in the topography that is easily traceable in aerial photographs (Unit 1). Shark teeth are locally abundant at the base of these beds. At Section A, the basal layer is more glauconitic and contains large, round concretions as much as 1 m in diameter and a basal lag conglomerate with material reworked from the López de Bertodano Formation. These beds are followed by a light gray, massive, silty sandstone (Unit 2).

A 0.5- to 1-m-thick, white tuff horizon is observed at approximately 40 m above the base of the formation (Fig. 8). The upper part of Unit 2 contains some glauconitic intervals and sandstone

Figure 7. Base of the Sobral Formation. BC = brown chocolate layer, note the presence of small channels; G = glauconite-rich bed (Stations K-120-122).

beds rich in pumice fragments and glass shards. Fossils are very sparse and not very diverse, being relatively more abundant in Section A than in Section E (Table 8). The invertebrates of Units 1 and 2 are similar to those of Unit 10 of the López de Bertodano Formation, with the exception of a more abundant and diverse gastropod fauna.

Unit 3 is approximately 95 m thick and consists of very friable, gray to light green, silty sandstone, with intercalations (usually 0.5 to 2 m thick) of massive to moderately bedded sandstone. Levels with distinctive red horizons are found toward the top; other yellowish (limonitic?) levels are particularly common in Section E.

Petrographically, sandstones from Unit 3 are almost devoid of matrix, and much more quartz-rich than those of the underlying units. Grain-size analysis shows a marked increase in the sand percentage of the rock. Toward the top of the unit, particularly in Section E, oblique and vertical burrows (*Skolithos*) are abundant (Fig. 5.7). This unit has a typical "soft" relief and forms a low topography. Unbored fossil wood is very common. With the exception of poorly preserved specimens of the bivalves *Australoneilo gracilis* and *Cucullaea ellioti*, the macrofauna is almost completely absent.

Units 4 and 5 have a maximum thickness of 76 m, measured in Section E. The dominant lithology is a greenish to dark brown, friable sand with large fragments of fossil wood, some of which are carbonaceous. Spherical concretions with a yellow patina are intercalated with more consolidated sandstone layers approxi-

mately 1 m thick. Yellowish and yellow-orange layers, probably limonitic, become more abundant upward.

The base of Unit 4 is placed at the first conspicuous cross-bedded interval found in the sequence (Fig. 5.8). This cross-bedded unit forms a prominent break in topography (Fig. 1). Similar beds are found in the northern tip of the island (not shown in Fig. 1) in the Cape Wiman area. These lower beds are 0.6 to 1 m thick, tabular as well as trough cross-bedded, and composed of medium-grained glauconitic sandstone. Foreset laminae are 0.5 to 1.5 cm thick and dip between 38° and 8°, with a mean value of 23° (S = 7°). Cross-bedding measurements in the southeastern part of the island (n = 63) indicate paleocurrent directions toward the ENE (N66°), whereas in the Cape Wiman area, paleocurrents flowed almost due east (N88°) (Fig. 9).

There is a noticeable upward increase in grain size in Unit 4, and the highest beds on Section E are composed of a pebbly sandstone in which the pebbles are predominantly volcanic (Unit 5). Numerous randomly oriented burrows having the appearance of roots are present at this higher level.

GRAIN-SIZE ANALYSIS

The unconsolidated nature of most of the Cretaceous/Paleocene sequence of Seymour Island provides a good opportunity for grain-size analysis of these sediments. Forty-one samples were analyzed to determine sand content, mean grain size, sorting, and general trends in the grain size populations in order to better

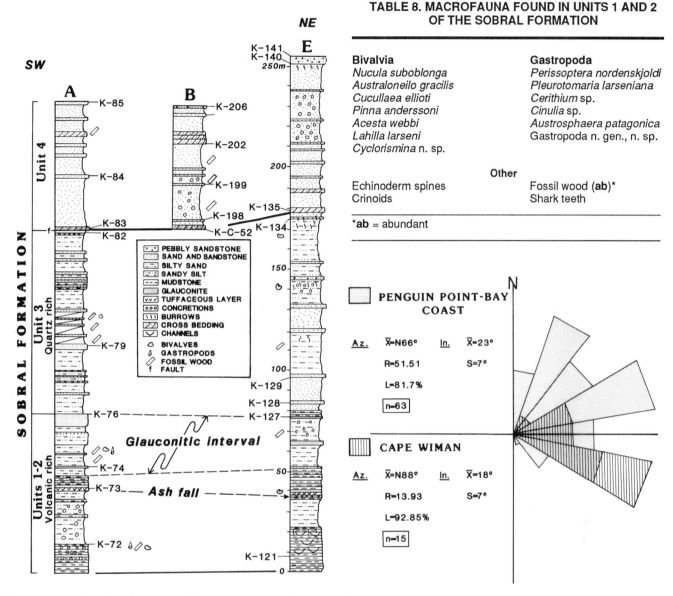

Figure 8. Correlation of sections measured in the Sobral Formation.

TABLE 8. MACROFAUNA FOUND IN UNITS 1 AND 2 OF THE SOBRAL FORMATION

Bivalvia	Gastropoda
Nucula suboblonga	*Perissoptera nordenskjoldi*
Australoneilo gracilis	*Pleurotomaria larseniana*
Cucullaea ellioti	*Cerithium* sp.
Pinna anderssoni	*Cinulia* sp.
Acesta webbi	*Austrosphaera patagonica*
Lahilla larseni	Gastropoda n. gen., n. sp.
Cyclorismina n. sp.	

	Other	
Echinoderm spines		Fossil wood (**ab**)*
Crinoids		Shark teeth

*****ab** = abundant

Figure 9. Cross-bedding measurements in the basal sand of Unit 4 of the Sobral Formation.

understand the depositional environments of these sediments and the associated energy levels.

The data on grain size were obtained through the conventional methods of dry sieving for particles coarser than 63 microns and pipetting for finer fractions (see Macellari, 1984a, for a detailed description of the methodology used).

Results

Most samples from the López de Bertodano Formation are sandy silts to clayey silts, with occasional silty sands (Fig. 10).

Clay percentages are relatively constant, with values around 17 percent. The mud content (clay plus silt) is more variable, ranging from a maximum of 98 percent (Sample K-33) to 37 percent (Sample K-147) (Figs. 10, 11). An upward trend of decreasing mud percentages, as well as of increasing grain size, is observed from the López de Bertodano to the Sobral Formations. In both sections analyzed, an increase in sand occurs at Units 8 (Samples K-54, K-147, K-149), and 10 (Samples K-71 and K-119).

A marked change in the sand content occurs within the Sobral Formation. Unit 1 is similar in sand content (35 percent) to the López de Bertodano Formation, but increases sharply in

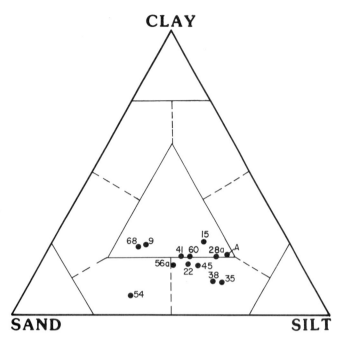

CLAY

SAND   SILT

Figure 10. Diagram showing percentages of clay, silt, and sand in some of the López de Bertodano samples (diagram after Shepard, 1954).

Unit 3 to approximately 85 percent. In the upper portion of the Sobral Formation, samples are almost exclusively composed of sand and pebbly sands.

Grain size

Mean grain size and sorting were calculated for all the samples analyzed, using the moment method (e.g., Krumbein and Pettijohn, 1938) and a programmable pocket calculator. For samples involving analysis of silt and clay sizes, the total mean grain size and total sorting were calculated. However, in the samples in which only the total amount of mud was determined, the mean grain size and sorting of only the sand fraction was calculated. In either case, the sorting curve is essentially a mirror image of the mean grain-size curve. The results are similar to those for the central Oregon continental shelf (Carey, 1972), where a decrease in grain size in an offshore direction was mirrored by an increase in the sorting coefficient. The trend observed here (i.e., negative correlation between grain size and sorting) could be explained in terms of more uniform flow regimes associated with the finer grained sediments (better sorted) and the more variable flow regimes associated with the coarser grained sediments (more poorly sorted). Alternatively, this decrease in sorting in the coarse-grained sediments may be a reflection of mixing from different sources under higher energy conditions.

Mean grain size and trends in grain size up-section, obtained by sieving the unconsolidated samples, are very similar to trends observed in thin sections. The sands of the López de Bertodano

Formation, when considered without the mud, are very well sorted (mean S: 0.31 Φ). In Unit 3 of the Sobral Formation, sands are coarser grained and well sorted (mean 2: 0.39 Φ), but the sorting decreases in the upper part of Unit 4 (S: 1.2 Φ).

Sorting and mean grain size remain approximately constant throughout the López de Bertodano and Sobral Formations. Departures from the general trend are observed in Units 1, 8, and 10 of the López de Bertodano Formation. The increase in the mean grain size at Sample K-54 (Unit 8) occurs because the total sand content increases, not because of a real increase in grain size. In fact, mean sand size is almost constant throughout the formation (Fig. 11).

The almost constant mean grain size of the sand fraction, regardless of important variations in the mud content of the samples, suggests two important inferences: (1) the process of mud deposition did not affect the sand fraction; and (2) a uniform fine sand population was available throughout the deposition of these sediments, or, alternatively, there was uniform flow regime during sand deposition. These observations imply that variation in the mud content is a reflection of the effectiveness of the winnowing of the fine materials; most likely, the sand input remained relatively constant. Alternatively, the variation in sand content can be a reflection of the dominance at a given time of one of two sources, one providing the mud and the other fine sand.

Grain-size curves

Histograms from the López de Bertodano Formation are unimodal and strongly positively skewed (long tail of fine-grained particles), with a mode in the very fine sand or coarse silt. A very different distribution is found in some sands from Unit 4 of the Sobral Formation (i.e., Sample K-140) in which no well-defined mode is observed and the distribution is more widely spread (poorly sorted).

Clastic sediments have been interpreted as mixtures of three or less log-normally distributed populations (i.e., Moss, 1962, 1963). These different populations, when plotted on log-probability paper, form identifiable straight lines (i.e., Visher, 1969). Grains transported by three main types of processes are usually resolvable as identifiable straight segments of a log-normal distribution: (1) traction or surface creep population, consisting of the coarser grained material; (2) saltation population, consisting of the moving bed layer or traction carpet; and (3) suspension population, consisting of the finer grained material (Visher, 1969). However, some doubt exists regarding the exact correlation of these populations with the proposed transport mechanisms (i.e., Middleton, 1976). Christiansen et al. (1984) concluded that the grain-size populations are better described as log-hyperbolic distributions rather than a mixture of log-normal distributions. In addition, Sengupta (1979) found that a mixture of different straight segments on a log-probability plot need not result from different modes of transportation; instead, the segments may be related to particular flow conditions or types of source material. The grain-size distribution is related to local

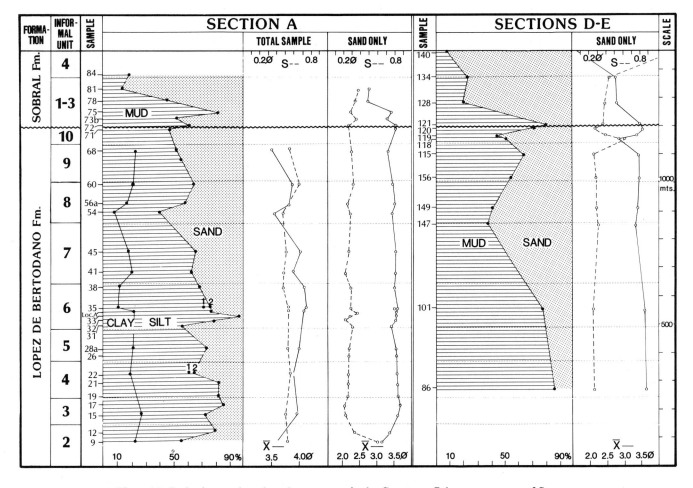

Figure 11. Grain-size trends and sand percentages in the Cretaceous-Paleocene sequence of Seymour
Island (Sections A and D-E): 1 and 2 are different values obtained in a duplicate analysis of the same
sample. X = mean grain size; S = sorting.

depositional conditions, but as pointed out by Taira and Scholle
(1979), it does not necessarily reflect a particular environment.

Analysis of the sediments of Seymour Island on log-
probability plots yields important information on their transport
mechanisms. When used with other environmental parameters,
these data are an important aid in the interpretation of the transport and depositional conditions of these sediments.

Some representative log-probability plots from the Creta-
ceous/Paleocene sequence of Seymour Island are shown in
Figure 12. In general, three types of graphs are observed here.
The first type of curve, which was consistently found in the López
de Bertodano Formation, is portrayed in Figure 12, 1–3. In
general, two well-defined populations are apparent: a very well-
sorted saltation population, and a moderately to poorly sorted
suspension population. The truncation of the saltation population
takes place at approximately 4.5 Φ. This value is an indication of
the amount of turbulence of the environment (Visher, 1969), a
truncation value that indicates fairly quiet deposition from sus-

pension. This also indicates the presence of either a nearby source
of mud (i.e., major river outflow) or a relatively enclosed or quiet
environment that prevented bypassing of the fine-grained sediments to the deeper portions of the basin.

The second type of curve, found in Units 3 and 4 of the
Sobral Formation, has three populations (Fig. 12.4; i.e., K-84).
The coarse-grained fraction (as much a 1.5 percent) represents a
moderately sorted traction population. The break between this
fraction and the saltation population occurs around a diameter of
2 Φ. This break, which commonly occurs at 2 Φ, has been
interpreted as the transition in size above which inertial forces
produce rolling or sliding of particles rather than saltation (i.e.,
Visher, 1969). The saltation population is well sorted (steep
slope), whereas the suspension population (which accounts for as
much as 35 percent of the sample) is moderately sorted (K-81,
K-84; Fig. 12.4). Truncation of these last two populations occurs
at 3 Φ, indicating relatively high turbulence. The suspension population is composed of fine sand and finer materials, and the data

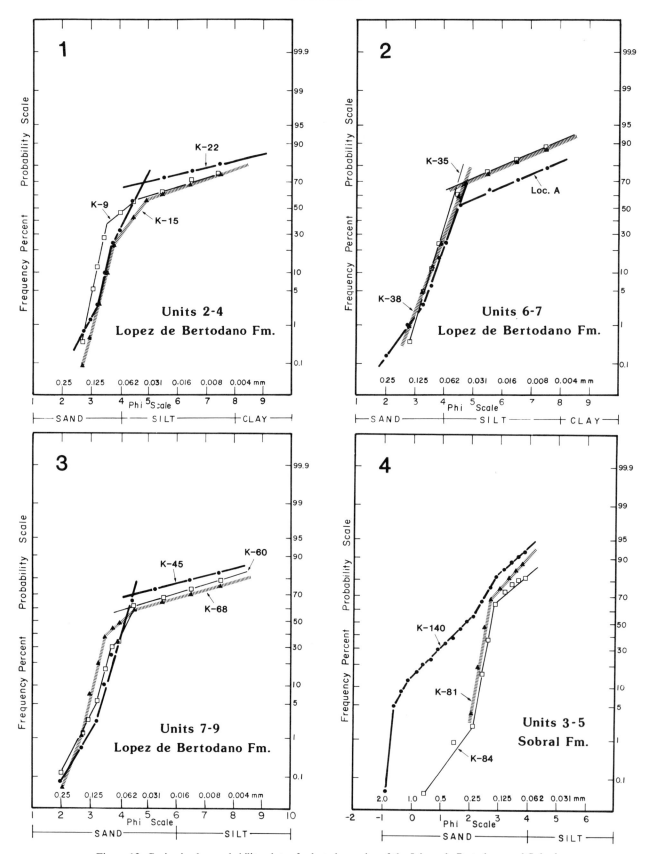

Figure 12. Grain-size log-probability plots of selected samples of the López de Bertodano and Sobral Formations.

points do not align very well into a straight line, suggesting mixing between saltating and suspension populations. This curve is typical of a high-energy environment such as is found in a very shallow marine setting near the surf zone. Sands deposited near the surf zone are usually clean, mainly because the mud is taken away in suspension. The presence of a large suspension population in sands from Unit 4 of the Sobral Formation may be related to a large amount of fine materials being supplied to this area from a nearby source.

The third type of curve, found only in the uppermost Sobral Formation, shows very poor sorting (Locality K-140, Fig. 12.4). A poorly sorted surface creep population includes approximately 40 percent of the distribution, and the break is observed at 2 Φ. The saltation population is better sorted, accounting for approximately 40 percent of the distribution, whereas the suspension population accounts for 20 percent of the distribution. Truncation occurs at 3 Φ. The high concentration of a poorly sorted surface creep population suggests dumping of this fraction. Although the identification of environments on the basis of grain-size distributions is very difficult and not always reliable, this curve is very similar to that found by Saitta (1968, *in* Visher, 1969) in a Pennsylvanian delta distributary sand, which in turn is similar to those sands of modern tidal channels (Visher, 1969). A similar grain-size distribution was observed by Glaister and Nelson (1974) for distributary channel deposits of a deltaic system. Dumping of a poorly sorted coarse fraction could take place in an estuary where a large tidal range originates reversing bottom currents. Alternatively, decreased flow velocity at the mouth of distributary channels would also result in the dumping of the coarse fraction.

In conclusion, the dominant type of grain-size distribution found in the López de Bertodano Formation indicates the presence of a semi-enclosed or quiet environment (obviating the winnowing of fine particles), and the presence of a nearby source of mud. The general increase in grain size and the change in the grain-size distribution up-section, together with other information, indicate a regressive event (possibly because of the progradation of a nearby delta), changing the environment from shelf to nearshore to possibly deltaic in character.

PALEOECOLOGY AND DEPOSITIONAL ENVIRONMENTS

The López de Bertodano and Sobral Formations represent marine deposition extending from late Campanian to Paleocene time. Sedimentation took place in relatively quiet conditions, with a strong terrigenous source derived from the nearby high terrain of the Antarctic Peninsula. Several major fluvial systems may have contributed the large amounts of silt and clay characteristic of this sequence.

Because the lithology is relatively homogenous throughout, studies of the macrofaunal assemblages, as well as the autoecological aspects of the various taxa, are of great assistance in the interpretation of the depositional environments represented in this sequence. Fossils provide information on the environment in which they lived by their distribution, by their abundance and diversity as a total assemblage, and by their particular mode of life and the ecological requirements of individual species.

The following two sections concentrate on the paleoecological aspects of the macrofauna—as pertaining to individual species, and as concerning assemblages of species or fossil associations.

Mode of life of macrofossils

*Annelids (*Rotularia*).* There are very few paleoecological studies on the mode of life of these organisms. I have suggested, based on the evidence from Seymour Island, that *Rotularia* shows a preference for sediments with a high mud content, and that it may have inhabited with its tube pointing upward, probably extending above the sediment-water interface (Macellari, 1984b). Recent observations on James Ross Island, Antarctica, tend to confirm this mode of life (E. Olivero, personal communication, 1986).

Bivalves. The paleoecological requirements of the bivalves from Seymour Island are relatively easy to evaluate because many taxa found in this sequence have Recent representatives of the same genus. Moreover, there is an extensive literature covering the ecology/paleoecology of Recent and fossil bivalves. Additionally, functional morphology analysis of this group allows for interpretation of the mode of life in many extinct taxa. The stratigraphic distribution of bivalves, with their feeding preferences, the vertical variations in diversity, and the relative abundance of each trophic group are presented in Figure 13 (see Wilckens, 1910, and Zinsmeister and Macellari, this volume, for a discussion of the taxonomy of this group).

Nucula is a moderately active, shallow burrower that uses its long labial palps to gather food from the sediment (deposit feeder). *Nucula* is associated with very fine-grained substrates deposited in quiet conditions. Depth does not seem to be a controlling factor in its distribution. For example, *Nucula proxima* (Say) is found in organic-rich, muddy, medium sand at depths of 0.5 to 3.0 m below mean low tide near Cape Cod, Massachusetts (Stanley, 1970). The same species, however, is a typical intermediate shelf constituent (22 to 72 m) of the northern Gulf of Mexico (Parker, 1956, 1960). Other species of *Nucula* are found in much deeper settings (as much as 1,640 m) off the northern Oregon coast (Pereyra and Alton, 1972).

The new genus *Australoneilo* (Zinsmeister and Macellari, this volume) is very similar to the Recent *Malletia*. *Malletia* is a labial palp deposit feeder, possibly of high mobility as inferred from its compressed shape (Rhoads et al., 1972). It is usually found in mobile sediments at variable depths, ranging from shallow settings to abyssal depths (Nicol, 1966–1967).

Nordenskjoldia and *Austrocucullaea* have no Recent representatives. *Nordenskjoldia* has been considered an epibyssate suspension feeder (Freneix, 1981). In the case of *Austrocucullaea,* the well-rounded venter, crenulated inner margin, and low elonga-

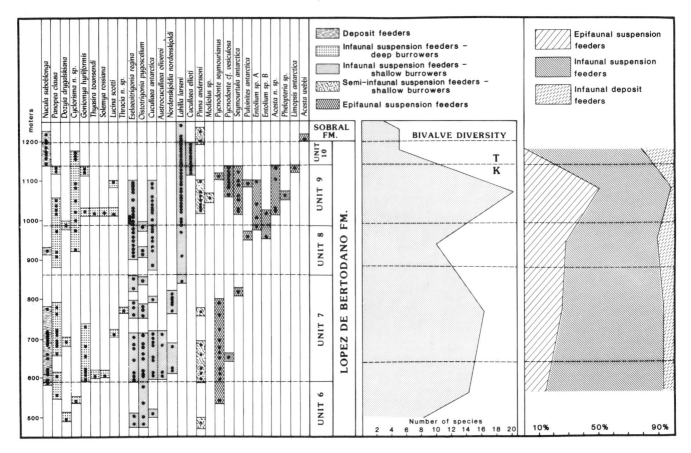

Figure 13. Stratigraphic distribution, variations in diversity, and variations in feeding types of bivalves from the López de Bertodano and Sobral Formations.

tion (L/H, <1.35) are more consistent with semi-infauna, endobyssate, or sessile life position (cf. Stanley, 1970; Thomas, 1978). *Nordenskjoldia,* however, is more elongated (L/H, >1.35), and some specimens have a flat venter; thus an epibyssate mode of life is very possible for *Nordenskjoldia.*

Cucullaea is a shallow burrower, infaunal suspension feeder. According to Thomas (1978), most arcoids live in periodically unstable environments, although the infauna arcoids (i.e., *Cucullaea*) are excluded from constantly mobile substrates by their weak ligaments and lack of siphons, factors which preclude both deep and rapid burrowing.

Solemya, found on Seymour Island, has several Recent representatives. For example, the suspension feeder *Solemya velum* Say is a moderately rapid burrower that forms a U-shaped burrow as much as 6 to 7 cm deep. The preferred environment of this species is organic-rich, muddy, medium sands at water depths of 2 to 4 m, although it occasionally is found living intertidally in fine sand (Stanley, 1970). Other modern species (e.g., *Solemya agassizi* Dall) are found to depths of 1,920 m off the northern Oregon coast (Pereyra and Alton, 1972), and Parker (1964) included the species *S. valvulus* Carpenter in his bathyal-abyssal

assemblage. *Solemya* inhabits cohesive, very fine sand or mud, but does not seem to be particularly controlled by depth.

Modiolus sp. is very rare in the Seymour Island assemblage. Recent species of *Modiolus* are semi-infaunal suspension feeders, usually living in shallow depths (cf. Stanley, 1970).

Pinna is a typical endobyssate genus usually living in sandy or mud-sand substrates. Its apex is deeply buried and it has a mass of byssal threads that help individuals attach to small stones or fragments of shells. The posterior margin projects approximately 2 to 3 cm above the surface of the sediment (Abbott, 1974). Recent species of *Pinna* are usually restricted to shallow-water settings, as much as 30 to 40 m deep. According to Keen (1971), *Pinna* is usually found on mud banks on the Pacific coast.

Phelopteria is a Cretaceous genus characterized by its long posterior wing, very similar to the Recent *Pteria.* Rhoads et al. (1972), by comparison with *Pteria,* considered that *Phelopteria* was a suspension feeder that lived attached on the surface by means of a byssus. Observations on Seymour Island show that *Phelopteria* lived gregariously and attached to hard objects—in particular, in this case, to a large shell of the gastropod *Amberleya spinigera.*

Pulvinites is an extinct suspension-feeding genus that probably lived on the sediment surface attached by a byssus. The smooth pectinid *Entolium* was probably an epifaunal, free-swimming suspension feeder (cf. Warren and Speden, 1977).

Acesta is a sessile, epifaunal suspension feeder. Recent species inhabit the substrate of shallow to very deep waters; for example, *Acesta mori* Hertlein was found to depths as much as 1,400 m (Abbott, 1974).

Seymourtula (Zinsmeister and Macellari, this volume) was probably an epibyssate suspension feeder that lived attached to hard substrate, similar to the Recent *Limatula.* On Seymour Island, *Seymourtula* was found at several localities attached to ammonite shells. Recent species of *Limatula* are common in deep-water habitats (cf. Abbott, 1974).

Eselaevitrigonia and *Oistotrigonia* were probably suspension feeders that lived infaunally, with the truncated posterior end lying parallel to, and level with, the surface of the substratum (cf. Stanley, 1977). Stanley (1977) also concluded that most trigoniids were restricted to shallow-water facies.

Lucina is an infaunal suspension feeder with a long siphon and low mobility (Stanley, 1970). This genus is presently represented in shallow subtidal settings (Stanley, 1970), but it is also common in the middle shelf (24 to 80 m) of the northern Gulf Coast of Mexico on sandy substrate (Parker, 1960), and in deeper waters off the Gulf of California (Parker, 1964).

Thyasira lives erect, deeply buried in the substrate, and is a suspension feeder of low mobility. The presence of a long inhalant tube allows individuals to burrow deeply in the sediment and enables them to withstand conditions deleterious to other groups. For this reason, *Thyasira* is able to populate environments characterized by a limited food supply, and, in some cases, oxygen-poor, hydrogen sulfide–rich waters (Kauffman, 1969). Recent *Thyasira* species are most diverse and abundant in the outer sublittoral zone of continental shelves (70 to 200 m), but they are also known from shallow waters of the inner shelf to depths of more than 2,000 m (Kauffman, 1969). *Dozyia* may have been an infaunal, siphonate suspension feeder, as deduced from its morphology.

Lahilla is inferred to be a shallow, infaunal, suspension feeder that, in life position, was oriented with its posterior margin parallel to the sediment-water interface. In the Cretaceous of the Magallanes Basin (southern South America), it is typically found in shallow-water settings. Specimens found on Seymour Island are relatively large (as much as 10 cm) and are very abundant—in many cases, the dominant species.

Cyclorisma was probably an infaunal siphonate suspension feeder. Considering the large pallial sinus observed in *Cyclorisma* from Seymour Island, it possibly was a deep burrower.

Thracia is represented on Seymour Island by only a few examples. Recent species live buried in mud at a depth of about 12 cm, with the commissural plane horizontal and the left valve upward. *Thracia* is a suspension feeder found at depths ranging from 20 to more than 2,000 m.

The presence of the large anterior and posterior gapes in *Goniomya* indicates that this was a deep burrower that accommodated a strong siphon. Freneix (1981) regarded this genus as a suspension feeder of low mobility.

Panopea is a suspension feeder that normally lives deeply buried in the substrate and feeds through long, prominent siphons. *Panopea clausa* was observed in situ on several occasions, in an almost vertical position. Recent species are restricted to fine-grained sediments and quiet environments (mud flats), with most species recorded in shallow waters extending to depths of 50 m.

Pycnodonte is an epifaunal suspension feeder common to many Late Cretaceous molluscan assemblages. The two forms found on Seymour Island, however, may represent two different strategies for adapting to a soft substrate. The flat *Pycnodonte seymourensis* usually had a small attachment surface and overcame the problem of living free on soft substrate by means of the "snowshoe strategy" (Jablonski and Bottjer, 1983). In this strategy, free-living adults assume a broad, flat form, thus distributing the weight of the organism over a large surface area. The grypheate-shaped *Pycnodonte* cf. *P. vesiculosa* adapted to the substrate by means of the "iceberg strategy" (Jablonski and Bottjer, 1983). In this case, a free-living adult expands one of its valves ventrally so that the individual is supported by the denser sediment at depth, but still maintains contact with the sediment-water interface. According to Jablonski and Bottjer (1983), *Pycnodonte* of the Campanian-Maastrichtian coastal plains of the United States is a ubiquitous component of subtidal assemblages, but appears in greater abundance in offshore shelf habitats, i.e., the Turonian Greenhorn Limestone of Kansas, which represents deposition of predominantly pelagic carbonate muds under low-energy conditions in relatively deep water (Bottjer et al., 1978). Sohl (1977) noted that *Pycnodonte* specimens from the Navesink Formation of New Jersey (early Maastrichtian), a strongly bioturbated, clayey, glauconitic sand, are more strongly convex than those occurring in the more nearshore environments of the overlying and underlying formations.

Gastropods. *Pleurotomaria* is an epifaunal browser that inhabits soft substrates (Warren and Speden, 1977). In two localities on Seymour Island, *Amberleya* was found inside the living chamber of large specimens of the ammonite *Maorites,* and thus it could be inferred that it was an epifaunal carnivore or scavenger. *Eunaticina,* like most naticids, was probably an epifaunal carnivore. *Perissoptera* was a deposit feeder; however, there is some debate as to whether *Perissoptera* lived infaunally or epifaunally (cf. Popenoe, 1983). The Recent *Aphorrais pespelicani* (Lamarck) spends part of its time buried beneath the sea-bottom, and probably shorter intervals epifaunally (Popenoe, 1983). *Taioma* was possibly an epifaunal carnivore, whereas *Cerithium,* well known in the Recent, is an epifaunal grazer that lives predominantly in very shallow waters. However, more offshore species of *Cerithium* are also known.

Ammonites. Since this is an extinct group, ecological inferences about their habitats and feeding preferences depend on functional morphological analysis and comparisons with the living *Nautilus.* According to Kennedy and Cobban (1976), the majority of ammonites exploited the low levels of the food pyramid, eating phyto- and zooplankton (first-order carnivores). Other ammonites were benthic, probably vegetarian browsers. Still another group was carnivorous and scavenger, such as the Recent *Nautilus,* but probably lacked the ability to capture large, active prey.

Of particular interest is the uncoiled heteromorph *Diplomoceras.* This recurved, U-shaped (ancylocone) genus was obviously not an active swimmer. According to Klinger (1981, among others), *Diplomoceras* was a planktonic floater; however, it is not known if the animal was restricted to a habitat close to the substrate or if it had the ability to migrate vertically through the water column. According to Klinger (1981), the U-shaped morphology also indicates an adaptation to a soft substrate, floating close to the sediment-water interface, yet with the respiratory system held clear from the sediment.

The superfamily Lytocerataceae (which includes *Anagaudryceras, Vertebrites, Zelandites, Pseudophyllites*) occurs more frequently in offshore, deeper water facies than other ammonite groups (Scott, 1940; Ward and Signor, 1983). This observation is consistent with the cosmopolitan distribution observed for many of these genera. Kennedy and Cobban (1976) warned, however, that the presence of Lytocerataceae in shallow-water deposits would indicate that they were not confined to deep-water environments, and they concluded that "offshore" rather than "deep water" may have been the milieu of this group.

The ornamentation of the ammonites may have served as protection from predators (Ward, 1981). One specimen of *Maorites densicostatus* found on Seymour Island has several circular orifices, distributed in a broad curve, in the region where the shell is crushed. This may have been produced by a predator (mosasaur?) in the manner described by Kauffman and Kesling (1960). The more compressed *Pachydiscus* and smoother *Maorites* found up-section, in the López de Bertodano Formation, probably were better adapted for swimming than those more coarsely ornamented and evolute species found down-section (Macellari, 1986).

Large concentrations of juvenile specimens, such as those observed in Unit 9 of the López de Bertodano Formation, are common in portions of the Cretaceous western interior of the United States at localities hundreds of kilometers from the paleo-strandline (Kennedy and Cobban, 1976). Juveniles apparently concentrated farther offshore and then moved to more nearshore environments upon attaining mature stages.

Paleoecology

Although the total macrofauna of the López de Bertodano and Sobral Formations on Seymour Island include more than 90 taxa, its diversity is quite low compared with other Maastrichtian localities of the world, i.e., the Fox Hill Formation (Waage, 1968; Speden, 1970) and the Ripley Formation (Wade, 1926; Sohl, 1960, 1964) of North America. This decreased diversity is much more apparent in the case of the gastropod fauna. The low macro-

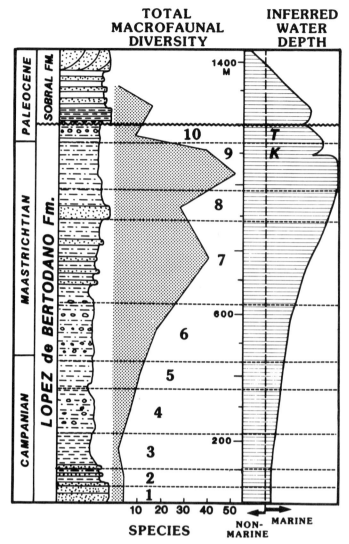

Figure 14. Vertical trends in total macrofaunal diversity of the López de Bertodano and Sobral Formations, and inferred variations in relative water depth.

faunal diversity in the López de Bertodano Formation is comparable with other assemblages found in the Late Cretaceous of New Zealand, New Caledonia, and southern South America (Weddellian Province of Zinsmeister, 1979). Thus, it may reflect a biogeographical or latitudinal control in the distribution of the taxa.

Macrofaunal diversity increases steadily through the López de Bertodano Formation, with a maximum in the middle of Unit 9, followed by a rapid drop after the Cretaceous/Tertiary boundary (Fig. 14). As suggested by the review of macrofaunal habitats in the previous section, the distribution of many organisms is more dependent on the type of substrate than on the water depth per se. Because water depth has a direct influence on the energy of the environment, and thus on the type of substrate available, an indirect relationship does exist between organisms and depth of the water column. Changes in depth will cause changes not only in the substrate, but also in very important factors such as temperature, salinity, and oxygen concentration of the bottom water. These parameters are important in restricting the distribution of species and in controlling the total diversity of a given fauna.

On Seymour Island, changes in diversity are generally independent of the lithology, which is fairly homogeneous throughout. Several authors (i.e., Purdy, 1964; Franz, 1976; Boucot, 1981) have pointed to the intimate dependence of benthic organisms on sediment type. Purdy (1964) showed a positive correlation between number of deposit-feeding species and the amount of silt and clay in a given sample. This relationship may occur because a decrease in water turbulence allows organic detritus and clay particles with adsorbed organic materials to settle out in quieter, deeper water settings. Carey (1972) also noted a steady increase in deposit feeding and epifaunal organisms, together with an increase in the biomass, moving offshore on the central Oregon continental shelf.

The observed general increase in diversity in the macrofauna could be related to an increase in the depth of the basin through time. Maximum diversity is generally accepted to occur in more stable environments (e.g., Heckel, 1972; Valentine, 1973). Valentine (1973) also drew attention to the importance of resource level and spatial heterogeneity as primary controlling factors of total diversity. Because shallow environments are more susceptible to disturbances, they should sustain a less diverse fauna than their deeper counterparts, all other factors being equal. Spatial heterogeneity, as observed in the sediment, does not seem to have varied significantly through time.

The lowermost units of the López de Bertodano Formation (Units 1 through 4) contain a depauperate macrofauna. In part, this may be a preservation problem, because most specimens are preserved as internal molds (with the exception of *Rotularia*). Foraminiferal faunas are dominated by agglutinated forms, although diverse and abundant calcareous faunas are present at some intervals (Huber, 1984, this volume). A study of the palynomorphs from these units suggests that samples from Units 1 and 2 contain a diverse nonmarine and less diverse marine palynomorph assemblage, probably indicative of a shallow marine environment (Askin, this volume). Samples from Unit 3 also indicate a nearshore marine setting. One sample, however, yielded abundant small acanthomorph acritarchs, which could be indicative of brackish conditions (Askin, this volume).

Evidently, the *Rotularia* facies were deleterious to most macro- and microfaunal groups. Palynomorphs, abundant organic fibers, and fragments of unbored fossil wood indicate proximity to a coast. Low macrofaunal diversity may be the result of brackish conditions, low oxygen content, or high turbidity of the water that precluded colonization by suspension feeders. The presence of abundant bioturbation and the lack of lamination indicate extensive burrowing uncharacteristic of low oxygen environments. This faunal assemblage may have been affected by inter-

mittent exposure to brackish conditions in a shallow marine environment, as might occur near the mouth of an estuary or delta. Very small, solitary corals settled during episodes of diminished deltaic or estuarine influence. The presence of stromatolites at the top of Unit 4, usually restricted to very thin horizons (ca. 5 cm) is a further indication of a shallow-water setting (cf. Gasparini et al., 1987).

Faunas in Units 5 and 6 are also dominated by the annelid *Rotularia*. A gradual increase in faunal diversity is observed upsection. Better defined marine conditions are indicated by the presence of levels with ammonites (*Maorites tuberculatus* and *Diplomoceras lambi*). The typical association for this interval is *Rotularia*, solitary corals, and echinoderm spines (Fig. 15). The common presence of mosasaur remains provides additional evidence for proximity to the coast (Gasparini and del Valle, 1984). Most bivalves found in and above Units 5 and 6 have the two valves articulated and are commonly still in a life position, indicating a low degree of disturbance on the bottom. Units 6, 7, and 8 are characterized by infaunal suspension-feeding bivalves (Fig. 13). This mode of feeding is typical of Late Cretaceous nearshore bivalve assemblages (Jablonski et al., 1983).

Unit 7 has a relatively diverse macrofauna, probably indicating open marine conditions. The deposit feeder *Nucula* is common, but the macrofauna is dominated by infaunal suspension feeders; among these are several deep burrowers, such as *Panopea, Goniomya*, and *Tracia*, usually found in quiet environments. Diversity decreases in Unit 8, which is dominated by shallow burrowing, infaunal suspension-feeding bivalves.

The change in the bivalve assemblage between Units 7 and 8 is related to the more sand-rich substrate of Unit 8. Most of the species of Unit 8 probably had the ability to burrow rapidly in order to adapt to shifting substrate conditions. *Rotularia*, which has a preference for mud-rich substrates (Macellari, 1984b), is almost completely absent in this unit.

The most diverse and abundant fauna of the López de Bertodano Formation occurs in Unit 9 (Fig. 16). Particularly interesting is the appearance of numerous suspension-feeding epifaunal bivalves, such as the ostreid *Pycnodonte* cf. *P. vesiculosa*, which dominates in the upper part of Unit 9. *Pycnodonte*-dominated paleocommunities are well known and widespread in Late Cretaceous offshore marine assemblages on soft substrate (Sohl and Koch, 1982; Jablonski and Bottjer, 1983; Jablonksi et al., 1983). Other epifaunal taxa found in the upper part of Unit 9 (*Pycnodonte* Beds) are *Seymourtula, Acesta, Phelopteria, Pulvinites,* and *Entolium*. Sedimentation conditions had to have been very quiet. One specimen of *Seymourtula* was found attached to an ammonite, but with its two valves dislodged and slightly displaced. The presence of the loose valve on the surface of the ammonite implies very weak local currents. Irregular tubes similar to *Planolites* and to *Thalassinodes* are very common in beds approximately coinciding with the last record of *Pachydiscus ultimus* (Fig. 6).

Other faunal data suggest that Unit 9 is the most offshore environment of the Seymour Island sequence. First is the occurrence of horizons containing very abundant, small, juvenile am-

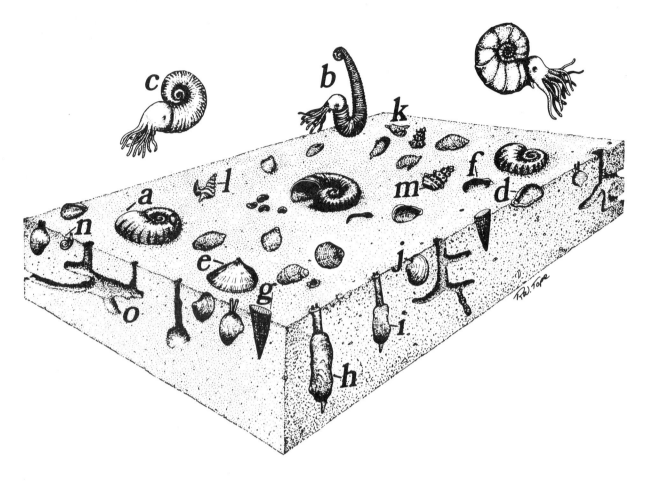

Figure 15. *Rotularia* assemblage, dominant in the lower part of the López de Bertodano Formation. a = *Rotularia;* b = echinoderms.

monites; as discussed previously, such horizons are common in offshore facies of the Western Interior of the United States. Second is the appearance of several cosmopolitan genera of Lytocerataceae (*Anagaudryceras, Vertebrites, Pseudophyllites*), some of which attained very large sizes. These are generally considered as offshore or deep-water indicators. Third is the appearance of more involute, smoother, and streamlined species of *Pachydiscus* and *Maorites,* indicating better adaptation for swimming and, possibly, more open-water conditions (Macellari, 1986). It is difficult to assess the depth at which these sediments were deposited, but it was probably less than 200 m.

The upper 30 m of Unit 9 contain a much less diverse macrofauna, and again the dominant group of bivalves is composed by infaunal suspension-feeding taxa (Fig. 13). The ammonite fauna in this interval is dominated by endemic species, and only in the very last meter does the cosmopolitan genus *Zelandites* occur. These macrofaunal changes could be interpreted in terms of changes in sedimentation rate (i.e., increased sedimentation in the uppermost interval of Unit 9) or in terms of bathymetric changes (i.e., rapid drop in sea level after the *Pachydiscus ultimus* Zone).

Faunal diversity decreases in Unit 10 of the López de Bertodano Formation. Abundant large fragments of fossil wood suggest proximity to a coast. Typical elements of Unit 10 are the infaunal suspension feeder bivalve *Lahilla larseni* and the deposit feeders *Perissoptera nordenskjoldi* and *Nucula suboblonga.* Some of the changes in diversity, however, may be related to the pervasive extinction event associated with the Cretaceous/Tertiary boundary. The overall macrofaunal change suggests, however, a shallower water setting than that of the preceding Unit 9. Unit 10 contains a zone of poorly preserved calcareous foraminifera (dissolution zone of Huber, this volume). The abundance of *Lahilla* and *Perissoptera* in both Units 9 and 10, and the presence of bored fossil wood in the upper levels of Unit 9, suggest that no large facies change occurred across the Cretaceous/Tertiary boundary on Seymour Island.

Units 1 and 2 of the Sobral Formation contain a fauna similar to that of Unit 10 of the López de Bertodano Formation,

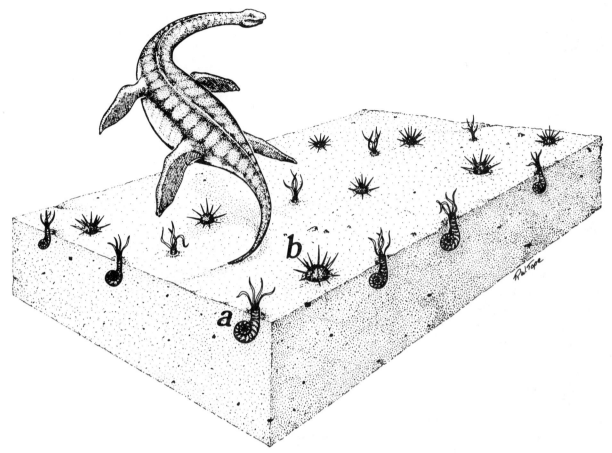

Figure 16. Macrofaunal assemblage found in the *Pycnodonte* Beds. a = *Anagaudryceras;* b = *Diplomoceras;* c = *Grossouvrites;* d = *Pycnodonte* cf. *vesiculosa;* e = *Acesta;* f = *Entolium;* g = *Pinna;* h = *Goniomya;* i = Panopea; j = *Lahilla;* k = *Phelopteria;* l = *Perissoptera;* m = *Amberleya;* n = *Rotularia* sp.; o = *Thalassinoides. Pachydiscus ultimus* is also characteristic of this assemblage.

but a marked decrease in the abundance of specimens is observed. Fossil wood is very abundant, but logs are devoid of the encrusting bivalves typically found in fossil wood in the López de Bertodano Formation. This means that the wood underwent little transport in sea water, suggesting proximity to the coast. Vertical burrows (*Skolithos*), usually indicative of shore or very shallow-water environments (Seilacher, 1967), provide additional evidence of a shallow-water setting.

Fossil invertebrates are absent in Units 3 through 5, with the exception of a poorly preserved, yet unstudied fauna found approximately 80 m above the base of Unit 4 in the Cape Wiman area. Fragments of fossil wood, some carbonaceous, are very common, and the presence of structures similar to rootlets, as well as thin coal laminae, suggest that at least a portion of these sediments was deposited above sea level.

Environmental restoration

Most sediments exposed on Seymour Island are younger than those present in other localities of the James Ross Basin. As a result, reconstruction of the paleogeographic setting of the Seymour Island sequence is only tentative, since there are no laterally equivalent age beds to provide a three-dimensional perspective. However, an evaluation of all the available sedimentological, paleontological, stratigraphical, and petrographical data does provide definite constraints for the reconstruction of the paleoenvironments of these deposits.

The abundant macrofauna indicates a marine environment for most of the studied sequences, and the presence of abundant mud is indicative of a major source of fine-grained material, probably an estuary or delta actively introducing sediments to the basin. An additional constraint is the necessity of a relatively quiet environment (low-wave energy) that would prevent the bypassing of sediments to the outer shelf or slope. This interpretation is supported by the near-absence of traction structures (i.e., cross-bedding and widespread sand sheets) expected in the nearshore portion of an open shelf environment.

López de Bertodano Formation. An important feature of the *Rotularia* Units is the rapid decrease in thickness in a northerly direction. This occurs because several massive, strongly bioturbated mudstone beds pinch out in this direction. This lateral variability suggests the presence of two major sources: one, distributed over the entire island, provided the sandy siltstones and cleaner sandstones, and the other, mostly affecting the southwestern portion of the island, provided the source of mud.

Rivers are the main source of mud on the shelf environment (Drake, 1976). However, a large percentage of this mud is trapped in estuaries and coastal wetlands (McCave, 1972; Meade, 1972; Meade et al., 1975; Drake, 1976), particularly in areas where there is restricted wave activity. Clay particles transported in fresh water tend to flocculate when entering the ocean, but this process does not seem to be important in trapping fine sediments in estuaries (Drake, 1976). As demonstrated by petrographical data, sediments from the López de Bertodano Formation were derived from the Antarctic Peninsula (Macellari, 1984a). It is envisioned that a major river outflow was located south of Seymour Island, influencing mostly the southern portion of the island. This estuary opened into a very shallow marine setting and shed large amoutns of fine material into the basin. Mudstone facies are interpreted as nearshore mud tongues that developed parallel to the coast where a large percentage of the fine material from the nearby fluvial system became trapped (Fig. 17). Occasional periods of more brackish conditions may have limited the ability of most macrofauna to settle in this region and resulted in an almost monospecific abundant "community" of *Rotularia*. However, O^{18} values found in tests of calcareous benthic foraminifera do not show fluctuations in salinity (E. Barrera, personal communication, 1986).

Other units of the *Rotularia* facies were deposited in a semi-protected shallow shelf environment, not exposed to pervasive wave action. Periodic storms or times of increased current activity and decreased fine sediment input resulted in the winnowing of the fine materials, forming relatively clean sand surfaces on which solitary corals settled. Periods of even stronger current activity resulted in the formation of mud intraclasts and well-sorted pellets. The presence of flaser bedding implied that both sand and mud were available and that periods of current activity alternated with periods of deposition from suspension (Reineck and Singh, 1975). Flaser bedding is characteristic of tidally influenced environments (Reineck and Singh, 1975), but is also found in the marine delta front (Coleman and Gagliano, 1965). Thick units of well-bedded, brownish, muddy, sandy siltstone (Fig. 4.3) are also observed in the *Rotularia* facies. These well-bedded units are usually devoid of macrofauna and bioturbation, but in some cases (i.e., Station K-28), the microfauna (foraminifera) is abundant, diverse (Huber, 1984, this volume), and similar to that of the massive gray siltstones. Similar well-bedded sediments are common in the delta front platform of the Mississippian Delta (Scruton, 1960; Shepard, 1960), where they also sustain an impoverished macrofauna (Parker, 1956).

Sediments observed in the northern part of the island are generally similar to those in the south. The scarcity of fauna and the dominance of *Rotularia* again indicate a restricted environment. However, a horizon containing a moderately abundant and

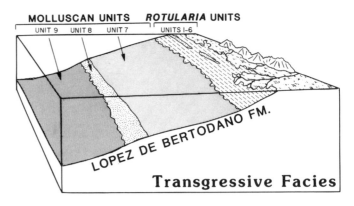

Figure 17. Interpreted paleoenvironmental setting of Units 1 to 9 of the López de Bertodano Formation.

well-preserved fauna, including several specimens of the ammonite *Maorites tuberculatus,* was found at Station G-83 (Unit 5, near Cape Bodman) and forms a laterally continuous horizon throughout the island. This horizon approximately coincides with a peak in foraminiferal diversity (Station 165; Huber, 1984), indicating the establishment of more open marine conditions.

Variations in mud content of the López de Bertodano Formation are an indication of the effectiveness of winnowing of the fine materials, or, alternatively, an indication of variations in input to the basin. The sparsely intercalated and laterally continuous, grain-supported sandstone beds in the predominantly muddy sequence can be compared to cheniers. According to Elliot (1978), cheniers form in conditions of low wave energy, low tidal range, effective longshore currents, and a variable supply of predominantly fine-grained sediments. Periods of high sediment supply produce poorly sorted mudflats that prograde seaward because longshore currents are unable to sort the sediments. During periods of diminished sediment supply, longshore currents and waves erode and winnow the mudflat sediments into chenier ridges. Although these sandy layers were not generally exposed— as suggested by the widespread occurrence of echinoderm spines, solitary corals, and the foraminiferal evidence—they may have formed in very shallow settings, as indicated by the localized presence of thinly bedded, thin (<5 cm) stromatolites (cf. Gasparini et al., 1987).

A middle shelf setting is proposed for Unit 7, one of the Molluscan Units (Fig. 17). An increase in water depth from the *Rotularia* to the Molluscan Units is suggested by the more laterally continuous beds and facies observed in the latter, as well as by the presence of a much more diverse and abundant macrofauna. Sandstones within this interval have a large percentage of matrix, indicating that the winnowing process was not as effective as in the *Rotularia* Units. Sedimentation took place in quiet conditions, as shown by the truncation values (ca. 4.5 Φ) found between the saltation and suspension populations in the grain-size analysis,

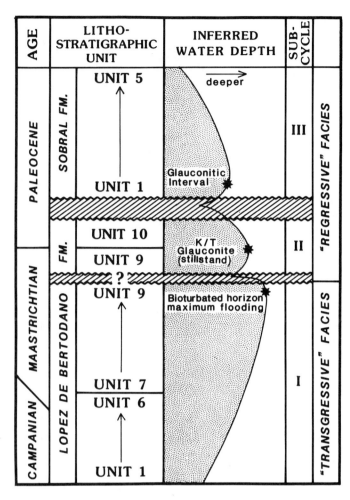

AGE	LITHO-STRATIGRAPHIC UNIT		INFERRED WATER DEPTH	SUB-CYCLE	
PALEOCENE	SOBRAL FM.	UNIT 5	deeper →	III	"REGRESSIVE" FACIES
		UNIT 1	Glauconitic Interval		
MAASTRICHTIAN	FM.	UNIT 10	K/T Glauconite (stillstand)	II	
		UNIT 9			
		?			
	LOPEZ DE BERTODANO	UNIT 9	Bioturbated horizon maximum flooding		"TRANSGRESSIVE" FACIES
		UNIT 7		I	
CAMPANIAN		UNIT 6			
		UNIT 1			

Figure 18. Interpreted depositional sequences of the López de Bertodano and Sobral Formations on Seymour Island. See text for comments and explanation.

and the presence in many horizons of bivalves in life position. Most sedimentation took place by quiet deposition from suspension, although a well-sorted saltation population is present.

The more sandy sediments of Unit 8 can be interpreted as representing shelf/slope break facies. Stanley and Wear (1978) have noted that sediments in the immediate vicinity of the shelf break are commonly coarser than those of the adjacent continental shelf. The shelf break is a zone of almost continuous resuspension, with only a minimum accumulation of fines (Stanley et al., 1983). Turbulence in this region is usually caused by shelf edge currents, upwelling, and possibly, breaking internal waves (Southard and Cacchione, 1972). Infaunal biomass is generally greatly decreased in this region, but sand-sized planktonic foraminifera may show a relative increase (Blake and Doyle, 1983). A peak in foraminiferal diversity was found in Sections D and F, at the base of Unit 8, but no size sorting was noted (Huber, 1984). The interpretation of Unit 8 as a shelf break facies would suggest a

deepening of the basin. However, these sediments may represent a time of increased energy conditions (or reduced mud input) in a shelf environment. This last explanation is more in accord with the general shallow setting of the López de Bertodano Formation.

Unit 9 represents the deepest, or most offshore facies of the López de Bertodano Formation and the peak of the Maastrichtian transgression on Seymour Island. The most important sedimentological change observed here is the appearance of widespread glauconite beds. Glauconites are common in the most offshore shelf environments of the Late Cretaceous Atlantic Coastal Plains of the United States (i.e., Jablonski and Bottjer, 1983). In Recent environments, abundant (as much as 98 percent of the coarse fraction) glauconite occurs at the outer edge of the Oregon shelf (Kulm et al., 1975), whereas Hein et al. (1974) found that glauconite is concentrated on slopes between the shelf edge and canyon floors off the California coast. In general, this authigenic mineral forms at depths ranging from 100 to 500 m, and the extensive formation of glauconite suggests low terrigenous sedimentation rates (cf. Odin and Matter, 1981).

Glauconite is also very abundant in Unit 10 of the López de Bertodano Formation and in the Sobral Formation, both of which are clearly shallower water deposits. In this upper portion of the sequence, the occurrence of glauconite could be associated with the appearnce of a volcanic signal (i.e., increase in the plagioclase/K-feldspar ratio, increase in the percentage of "clean" quartz, and increase in volcanic fragments); as a result, the presence of glauconite in these sediments may be also controlled by the availability of a suitable substrate or chemical conditions necessary for its development.

A more cohesive bottom was probably present during deposition of the upper level of Unit 9, allowing the development and preservation of large and abundant trace fossils (burrows). It is tentatively proposed here that the bioturbated horizon and intraclasts found at the top of the *Pachydiscus ultimus* Zone represent a small hiatus in sedimentation. This would explain the presence of these sediments, as well as the change in macrofauna observed in the overlying strata. A new cycle of sedimentation probably started immediately above this bioturbated horizon. The reappearance of cosmopolitan, smooth ammonites (*Zelandites*) at the top of this interval may indicate a new deepening of the basin. This interpretation will have to be confirmed with further detailed mapping of the 50 m below the Cretaceous/Tertiary glauconite. Particularly important is to establish the lateral continuity of the bioturbated horizon and the distribution of reworked fauna and detailed macrofaunal trends within the recognized subunits. The Cretaceous/Tertiary boundary glauconite represents chemical sedimentation at a time of reduced clastic input into the basin. The glauconite represents a condensed interval, possibly coiniciding with a high stand of sea level (Figs. 14, 18).

Unit 10 is interpreted as an inner shelf deposit (Fig. 19). A slight increase in grain size, the presence of abundant fossil wood, and a fauna similar to that of the overlying Sobral Formation suggest a shallower water setting than that interpreted for the preceding unit.

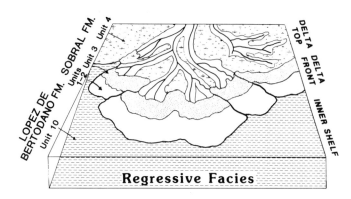

Figure 19. Interpreted depositional environments of the upper portion of the López de Bertodano and the Sobral Formations.

Sobral Formation

The unconformity observed at the base of the Sobral Formation indicates the initiation of a new depositional cycle. The Sobral Formation is interpreted as the prograding facies of a delta complex. The well-bedded sediments of Unit 1 are similar to the Recent pro-delta slope facies of the modern Niger Delta (Allen, 1970); similar bedding types have been reported by Coleman and Gagliano (1965). Units 1 and 2 were probably deposited in relatively quiet conditions with a high sediment supply, as inferred from the abundant mud matrix, the lack of high-energy sedimentary structures in most horizons, and the near-absence of bioturbation. The presence of a glauconite-rich level at the top of Unit 2 (Figs. 7, 8) may represent the maximum flooding achieved during this new depositional cycle.

The more sandy Unit 3 is characterized by well-sorted sands with sand grains more rounded than in either higher or lower units. These sediments have clearly been deposited above wave base, in a relatively high-energy environment such as the foreshore of a beach. The occurrence of *Skolithos* also suggests nearshore deposition. The presence of organic matter fragments and very abundant wood fragments indicates the proximity to the delta front. This environment has a Recent analogue in the lower portion of the coastal barrier sand of the Niger Delta (Allen, 1970; Oomkens, 1974).

Units 4 and 5 of the Sobral Formation represent the final progradation of a deltaic sequence, displaying the transition from marine to nonmarine facies (Fig. 19). The glauconitic grains found here, however, have been transported, as demonstrated by the uniformity of grain size of the glauconites and the other constituents, and their well-rounded edges. Unimodal tabular and trough cross-bedding, mostly dipping basinward (east-northeast), indicate dispersal of these sediments from the Antarctic Peninsula. This laterally continuous unit was formed by the lateral migration of distributary channels, thereby developing a broad delta

front sheet sand. This interpretation coincides with that of Trautman (1976) for the Cross Valley Formation. The basal layer of Unit 4 is overlain by relatively clean, well-sorted sands with rounded grains, indicating the persistence of deposition near the surf zone. Coarser grained, poorly sorted pebbly sandstones at the top of the sequence examined have a grain-size distribution comparable to that found in interdistributary channels of a delta complex (Glaister and Nelson, 1974). Well-preserved leaf fossils and minor coal laminae are concentrated in the interdistributary areas of the delta top. Portions of the Units 4 and 5 were probably deposited under non-marine conditions, but episodes of marine influence are apparent, as shown by a horizon containing a poorly preserved molluscan fauna.

CONCLUSIONS

Cretaceous to Paleocene strata of Seymour Island record a major transgressive/regressive cycle composed of possibly three lower order cycles (Figs. 14, 18). Macrofauna provide invaluable assistance in recognizing the facies succession in a predominantly featureless, fine-grained sequence.

Units 1 through 6 of the López de Bertodano Formation were deposited in a very shallow delta/estuary–influenced environment. Mud facies may have developed in subaqueous mud lenses, whereas sporadic intercalations of clean sand formed during periods of more intense winnowing or decreased fluvial input. Units 7, 8 and 9 were deposited in progressively deeper water conditions, with Units 7 and 8 forming under middle shelf conditions, and Unit 9 in outer shelf conditions. Unit 9 contains the most offshore facies of the studied sequence. It is difficult to evaluate the absolute depth at which these sediments were deposited, but microfaunal (Huber, 1984, this volume) and macrofaunal studies suggest that maximum depths did not exceed 150 to 200 m. The relative terms offshore and onshore are more appropriate for depicting most changes observed in this sequence.

Possibly a second subcycle of sedimentation was initiated in the late Maastrichtian (top of Unit 9) and was continued into the early Paleocene (Unit 10). The Cretaceous/Tertiary boundary coincides with a continuous glauconitic horizon and represents a condensed interval. Unit 10 was probably deposited in a middle to inner shelf environment.

A third depositional subcycle started during the early Paleocene with the sedimentation of the Sobral Formation. This formation formed as the basin was filled by the progradation of a deltaic system.

Units 1 and 2 are interpreted as pro-delta facies, followed by clean sands deposited as a coastal barrier (Unit 3). Units 4 and 5 contain the delta top facies, including distributary channels and interdistributary marshes. Fluctuations in the extent of delta progradation produced an alternation of nonmarine and very shallow marine facies in this unit.

It is difficult at this point to make specific correlations between the interpreted Maastrichtian/Paleocene depositional cycles of Seymour Island and worldwide sea-level fluctuations. This is due to the yet-uncertain precise dating of several of the intervals, and also to the tentative nature of some of the interpreted cycles.

ACKNOWLEDGMENTS

This work is based on a portion of a Ph.D. dissertation submitted to the Department of Geology and Mineralogy, Ohio State University. I thank W. J. Zinsmeister (Purdue University) and D. Elliot and P. Anderson (Institute of Polar Studies) for their support and encouragement throughout this project. Different portions of this paper were reviewed by L. Krissek, D. H. Elliot, P. N. Webb, and B. Huber (Ohio State University), W. J. Zinsmeister (Purdue University), and N. H. Sohl (U.S. National Museum of Natural History). I have benefited by the participation of B. Huber in the field work. I thank C. Heaton (University of South Carolina), and R. Topes (Ohio State University) for drafting several of the figures. This work was supported by National Science Foundation Grants DPP-8213985 (field work and laboratory) and EAR-8418145 (preparation of the manuscript).

REFERENCES

ABBOTT, R. T. 1974. American Seashells, 2nd ed. Van Nostrand-Reinhold Co., New York, 663 p.

ALLEN, J.R.L. 1970. Sediments of the modern Niger Delta: a summary and review. *In* J. P. Morgan and R. H. Shaver (eds.), Deltaic Sedimentation, Modern and Ancient. Society of Economic Paleontologists and Mineralogists Special Publications, 15:138–151.

ANDERSSON, G. L. 1906. On the geology of Graham Land. University of Uppsala Geological Institute Bulletin, 7:19–71.

ASKIN, R. A. 1984. Palynological investigations of the James Ross Island basin and Robertson Island, Antarctic Peninsula. Antarctic Journal of the U.S., 19(5):6–7.

BIBBY, J. S. 1966. The stratigraphy of part of north-east Graham Land and the James Ross Island Group. British Antarctic Survey Scientific Report 53, 37 p.

BLAKE, N.J., AND L. J. DOYLE. 1983. Infaunal-sediment relationships at the shelf-slope break. *In* D. J. Stanley and G. T. Moore (eds.), The Shelfbreak: Critical Interface on Continental Margins. Society of Economic Paleontologists and Mineralogists Special Publications, 33:381–389.

BOTTJER, D. J., C. ROBERTS, AND D. E. HATTIN. 1978. Stratigraphic and ecologic significance of *Pycnodonte kansasense*, a new Lower Turonian oyster from the Greenhorn Limestone of Kansas. Journal of Paleontology, 52(6):1208–1218.

BOUCOT, A. J. 1981. Principles of Benthic Marine Paleoecology. Academic Press, New York, 463 p.

CAREY, A. G., Jr. 1972. Ecological observations on the benthic invertebrates from the central Oregon continental shelf, p. 422–443. *In* A. T. Pruter and D. L. Alverson (eds.), The Columbia River Estuary and Adjacent Ocean Waters: Bioenvironmental Studies. University of Washington Press, Seattle.

CHRISTIANSEN, C. P. BLAESILD, AND K. DALSGAARD. 1984. Re-interpreting "segmented" grain-size curves. Geological Magazine, 121(1):47–51.

COLEMAN, J. M. AND S. M. GAGLIANO. 1965. Sedimentary structures: Mississippi River deltaic plain. *In* G. V. Middleton (ed.), Primary Sedimentary Structures and Their Hydrodynamic Interpretation. Society of Economic Paleontologists and Mineralogists Special Publications, 12:133–148.

DRAKE, D. E. 1976. Suspended sediment transport and mud deposition on continental shelves, p. 127–158. *In* D. J. Stanley and D.J.P. Swift (eds.), Marine Sediment Transport and Environmental Management. John Wiley, New York.

ELLIOT, D. H., AND T. A. TRAUTMAN. 1982. Lower Tertiary strata on Seymour Island, Antarctic Peninsula, p. 287–297. *In* C. Craddock (ed.), Antarctic Geoscience. University of Wisconsin Press, Madison.

ELLIOT, T. 1978. Clastic shorelines. *In* H. G. Reading (ed.), Sedimentary Environments and Facies, p. 143–177. Elsevier, New York.

FRANZ, D. 1976. Benthic molluscan assemblages in relation to sediment gradients in northeastern Long Island Sound, Connecticut. Malacology, 15:377–399.

FRENEIX, S. 1981. Faunes de bivalves du Sénonien de Nouvelle Calédonie: Analyses paléobiogéographique, biostratigraphique, paléoécologique. Annales Paléontologie (Invert.), 67(1):13–32.

GASPARINI, Z., AND R. DEL VALLE. 1984. Mosasaurios (Reptilia, Sauria) Cretácicos, en el continente Antártico. IX Congreso Geológico Argentino, S. C. Bariloche, Actas, 4:423–431.

——, ——, AND R. GOÑI. 1987. Un Elasmosaurido (Reptilia, Plesiosauria) del Cretácico Superior de la Antártida. Contribuciones Instituto Antártico Argentino (in press).

GLAISTER, R. P., AND H. W. NELSON. 1974. Grain-size distributions, an aid in facies identification. Canadian Petroleum Geologists Bulletin, 22(3):203–240.

HECKEL, P. H. 1972. Recognition of ancient shallow marine environments. *In* J. K. Rigby and W. K. Hamblin (eds.), Recognition of Ancient Sedimentary Environments. Society of Economic Paleontologists and Mineralogists Special Publications, 16:226–286.

HEIN, J. R., A. O. ALLWARDT, AND G. B. GIGGS. 1974. The occurrence of glauconite in Monterrey Bay, California: diversity, origins, and sedimentary environmental significance. Journal of Sedimentary Petrology, 44(2):562–571.

HUBER, B. T. 1984. Late Cretaceous foraminiferal biostratigraphy, paleoecology, and paleobiogeography of the James Ross Island region, Antarctic Peninsula. Unpublished M.Sc. thesis, The Ohio State University, Columbus, 246 p.

——. 1987. The location of the Cretaceous-Tertiary contact on Seymour Island, Antarctic Peninsula. Antarctic Journal of the U.S., 20(5) (in press).

——, S. M. HARWOOD, AND P. N. WEBB. 1983. Upper Cretaceous microfossil biostratigraphy of Seymour Island, Antarctic Peninsula. Antarctic Journal of the U.S., Annual Review, 18(5):72–74.

JABLONSKI, D., AND D. J. BOTTJER. 1983. Soft-bottom epifaunal suspension-feeding assemblages in the Late Cretaceous, p. 747–812. *In* M.J.S. Tevesz and P. L. McCall (eds.), Biotic Interactions in Recent and Fossil Benthic Communities. Plenum Press, New York.

JABLONSKI, D., J. J. SEPKOSKI, JR., D. J. BOTTJER, AND P. M. SHEEHAN. 1983. Onshore-offshore patterns in the evolution of Phanerozoic shelf communities. Science, 222(4628):1123–1124.

KAUFFMAN, E. G. 1969. Bivalvia: form, function, and evolution, p. N129–N205. *In* R. C. Moore (ed.), Treatise on Intertebrate Paleontology: Part N, Mollusca 6 [1 of 3]. University of Kansas Press, Lawrence.

——, AND R. V. KESLING. 1960. An Upper Cretaceous ammonite bitten by a mosasaur. University of Michigan Museum of Paleontology Contributions, 15:193–248.

KEEN, A. M. 1971. Sea Shells of Tropical West America: Marine Mollusks from Baja California to Peru, 2nd ed. Stanford University Press, Stanford, CA, 1064 p.

KENNEDY, W. J., AND W. A. COBBAN. 1976. Aspects of Ammonite Biology, Biogeography, and Biostratigraphy. Special Papers in Palaeontology 17, 94 p.

KILIAN, W. AND P. REBOUL. 1909. Les Céphalopodes néocretacées des Îles Seymour et Snow Hill. Wissenschaftliche Ergebnisse der Schwedischen Südpolar-Expedition 1901–1903, Stockholm, 3(6):1–75.

KLINGER, H. C. 1981. Speculations on buoyancy control and ecology in some heteromorph ammonites, p. 337–355. In M. R. House and J. R. Senior (eds.), The Ammonoidea [The Systematics Association Special Volume 18]. Academic Press, New York.

KRUMBEIN, W. C. AND F. J. PETTIJOHN. 1938. Manual of Sedimentary Petrography. Appleton-Century, New York, 549 p.

KULM, L. D., ET AL. 1975. Oregon continental shelf sedimentation: interrelationships of facies distribution and sedimentary processes. Journal of Geology, 83:145–176.

MACELLARI, C. E. 1984a. Late Cretaceous stratigraphy, sedimentology, and macropaleontology of Seymour Island, Antarctic Peninsula. Unpublished Ph.D. dissertation, The Ohio State University, Columbus, 599 p.

——. 1984b. Revision of serpulids of the genus Rotularia (Annelida) at Seymour Island (Antarctic Peninsula) and their value in stratigraphy. Journal of Paleontology, 58(4):1098–1116.

——. 1985a. Paleobiogeografía y edad de la fauna de Maorites-Gunnarites (Ammonoidea) del Cretácico Superior de la Antártida y Patagonia. Ameghiniana (Buenos Aires), 21(2-4):223–242.

——. 1985b. El límite Cretácico/Terciario en la Península Antártica y en el Sur de Sudamérica: Evidencias Macropaleontológicas. VI Congreso Latinoamericano de Geologia, Bogotá, I:266–278.

——. 1986. Late Campanian-Maastrichtian ammonites from Seymour Island, Antarctic Peninsula. Journal of Paleontology Memoir 18, Vol. 60, suppl. to N. 2: 1–55.

——, AND W. J. ZINSMEISTER. 1983. Sedimentology and macropaleontology of the Upper Cretaceous to Paleocene sequence of Seymour Island. Antarctic Journal of the U.S., Annual Review, 18(5):69–71.

McCAVE, I. N. 1972. Transport and escape of fine-grained sediment from shelf areas, p. 225–248. In D.J.P. Swift, D. B. Duane, and O. H. Pilkey (eds.), Shelf Sediment Transport: Process and Pattern. Dowden, Hutchinson and Ross, Stroudsburg, PA.

MEADE, R. H. 1972. Transport and deposition of sediments in estuaries. In B. W. Nelson (ed.), Environmental Framework of Coastal Plain Estuaries. Geological Society of America Memoir, 133:91–120.

——, P. L. SACHS, F. T. MANHEIM, J. C. HATHAWAY, AND D. W. SPENCER. 1975. Sources of suspended matter in waters of the Middle Atlantic Bight. Journal of Sedimentary Petrology, 45:171–188.

MIDDLETON, G. V. 1976. Hydraulic interpretation of sand size distributions. Journal of Geology, 84:405–426.

MOSS, A. J. 1962. The physical nature of common sandy and pebbly deposits, Part I. American Journal of Science, 260:337–373.

——. 1963. The physical nature of common sandy and pebbly deposits, Part II. American Journal of Science, 261:297–343.

NICOL, D. 1966–1967. Descriptions, ecology, and geographic distribution of some Antarctic pelecypods. Bulletin of American Paleontology, 51(231):1–88.

ODIN, G. S., AND A. MATTER. 1981. De glauconiarum origine. Sedimentology, 28:611–641.

OOMKENS, E. 1974. Lithofacies relations in the late Quaternary Niger Delta complex. Sedimentology, 21:195–222.

PALAMARCZUK, S., ET AL. 1984. Las Formaciones López de Bertodano y Sobral en la Isla Vicecomodoro Marambio, Antártida. IX Congreso Geológico Argentino, S.C. Bariloche, Actas, 1:399–419.

PARKER, R. H. 1956. Macro-invertebrate assemblages as indicators of sedimentary environments in the east Mississippi Delta region. American Association of Petroleum Geologists Bulletin, 40 (2):295–376.

——. 1960. Ecology and distributional patterns of marine macro-invertebrates, northern Gulf of Mexico, p. 302–337. In F. P. Shepard, F. B. Phleger, and T. H. Van Andel (eds.), Recent Sediments, Northwest Gulf of Mexico. American Association of Petroleum Geologists, Tulsa.

——. 1964. Zoogeography and ecology of macro-invertebrates of the Gulf of California and the continental slope of western Mexico. In Marine Geology of the Gulf of California—A Symposium. American Association of Petroleum Geologists Memoir, 3:331–376.

PEREYRA, W. T., AND M. S. ALTON. 1972. Distribution and relative abundance of

invertebrates off the northern Oregon coast, p. 444–474. In A. T. Pruter and D. L. Alverson (eds.), The Columbia River Estuary and Adjacent Ocean Waters: Bioenvironmental Studies. University of Washington Press, Seattle.

POPENOE, W. P. 1983. Cretaceous Aporrhaidae from California: Aporrhainae and Arrhoginae. Journal of Paleontology, 57(4):742–765.

PURDY, E. G. 1964. Sediments as substrates, p. 238–271. In J. Imbrie and N. Newell (eds.), Approaches to Paleoecology. John Wiley & Sons, New York.

REINECK, H. E. AND I. B. SINGH. 1975. Depositional Sedimentary Environments. Springer-Verlag, Berlin, 439 p.

RHOADS, D. C., SPEDEN, I. G., AND WAAGE, K. M. 1972. Trophic group analysis of Upper Cretaceous (Maestrichtian) bivalve assemblages from South Dakota. American Association of Petroleum Geologists Bulletin, 56(6):1100–1113.

RINALDI, C. A. 1982. The Upper Cretaceous in the James Ross Island Group, p. 281–286. In C. Craddock (ed.), Antarctic Geoscience. The University of Wisconsin Press, Madison.

——, ET AL. 1978. Geología de la Isla Vicecomodoro Marambio. Contribuciones Instituto Antártico Argentino, 217:1–37.

SCOTT, G. 1940. Paleoecological factors controlling distribution and mode of life of Cretaceous ammonoids in the Texas area. American Association of Petroleum Geologists Bulletin, 24(7):1164–1203.

SCRUTON, P. C. 1960. Delta building and the deltaic sequence, p. 82–102. In F. P. Shepard, F. B. Phleger, and T. H. Van Andel (eds.), Recent Sediments, Northwest Gulf of Mexico. American Association of Petroleum Geologists, Tulsa.

SEILACHER, A. 1967. Bathymetry of trace fossils. Marine Geology, 5:413–428.

SENGUPTA, S. 1979. Grain-size distribution of suspended load in relation to bed materials and flow velocity. Sedimentology, 26:63–82.

SHEPARD, F. P. 1954. Nomenclature based on sand-silt-clay ratios. Journal of Sedimentary Petrology, 24:151–158.

——. 1960. Mississippi Delta: marginal environments and growth, p. 56–81. In F. P. Shepard, F. B. Phleger, and T. H. Van Andel (eds.), Recent Sediments, Northwest Gulf of Mexico. American Association of Petroleum Geologists, Tulsa.

SOHL, N. F. 1960. Archaeogastropoda, Mesogastropoda, and Stratigraphy of the Ripley, Owl Creek, and Prairie Bluff Formations, p. 1–151. U.S. Geological Survey Professional Paper 331-A.

——. 1964. Neogastropoda, Ophisthobranchia, and Basommatrophora from the Ripley, Owl Creek, and Prairie Bluff Formations, p. 153–334. U.S. Geological Survey Professional Paper 331-B.

——. 1977. Benthic marine molluscan associations from the Upper Cretaceous of New Jersey and Delaware. In J. P. Owens, N. F. Sohl, and J. P. Minard (eds.), A Field Guide to Cretaceous and Lower Tertiary Beds of the Raritan and Salisbury Embayments: New Jersey, Delaware, and Maryland. American Association of Petroleum Geologists/Society of Economic Paleontologists and Mineralogists Joint Convention, Washington, D.C.

——, AND C. F. KOCH. 1982. Substrate preference among some Late Cretaceous shallow-water benthic mollusks. Geological Society of America Abstracts with Programs, 14:84.

SOUTHARD, J. B., AND D. A. CACCHIONE. 1972. Experiments on bottom sediment movement by breaking internal waves, p. 83–97. In D.J.P. Swift, D. B. Duane, and O. H. Pilkey (eds.), Shelf Sediment Transport: Process and Pattern. Dowden, Hutchinson and Ross, Stroudsburg, PA.

SPEDEN, I. G. 1970. The type Fox Hills Formation, Cretaceous (Maestrichtian), South Dakota: Part 2, Systematics of the Bivalvia. Peabody Museum of Natural History Bulletin, 33:1–222.

STANLEY, D. J. AND C. M. WEAR. 1978. The "mud-line": an erosion-deposition boundary on the upper continental slope. Marine Geology, 28:M19–M29.

——, S. K. ADDY, AND E. W. BEHRENS. 1983. The mudline: variability of its position relative to shelfbreak. In D. J. Stanley and G. T. Moore (eds.), The Shelfbreak: Critical Interface on Continental Margins. Society of Economic Paleontologists and Mineralogists Special Publications, 33:279–298.

STANLEY, S. M. 1970. Relation of shell form to life habits in the Bivalvia. Geological Society of America Memoir, 125:1–296.

——. 1977. Coadaptation in the Trigoniidae: a remarkable family of burrowing bivalves. Palaeontology, 20(4):869–899.

TAIRA, A., AND P. A. SCHOLLE. 1979. Discrimination of depositional environments using settling tube data. Journal of Sedimentary Petrology, 49(3):787–800.

THOMAS, R.D.K. 1978. Shell form and the ecological range of living and extinct Arcoida. Paleobiology, 4(2):181–194.

TRAUTMAN, T. A. 1976. Stratigraphy and petrology of Tertiary clastic sediments, Seymour Island. Unpublished M.Sc. thesis, The Ohio State University, Columbus, 170 p.

VALENTINE, J. W. 1973. Evolutionary paleoecology of the marine biosphere. Prentice-Hall, Englewood Cliffs, NJ, 511 p.

VISHER, G. S. 1969. Grain size distributions and depositional processes. Journal of Sedimentary Petrology, 39(3):1074–1106.

WAAGE, K. M. 1968. The Type Fox Hills Formation, Cretaceous (Maestrichtian), South Dakota: Part 1, Stratigraphy and paleoenvironments. Peabody Museum of Natural History Bulletin 27, 171 p.

WADE, B. 1926. The Fauna of the Ripley Formation on Coon Creek, Tennessee. U.S. Geological Survey Professional Paper 137, 272 p.

WARD, P. 1981. Shell sculpture as a defensive adaptation in ammonoids. Paleobiology, 7(1):96–100.

——, AND P. W. SIGNOR III. 1983. Evolutionary tempo in Jurassic and Cretaceous ammonites. Paleobiology, 9(2):183–198.

WARREN, G., AND I. SPEDEN. 1977. The Piripauan and Haumurian stratotypes (Mata Series, Upper Cretaceous) and correlative sequences in the Haumuri Bluff district, South Marlborough. New Zealand Geological Survey Bulletin, 92:1–59.

WILCKENS, O. 1910. Die Anneliden, Bivalven und Gastropoden der Antarktischen Kreideformation. Wissenschaftliche Ergebnisse der Schwedischen Südpolar-Expedition 1901–1903, Stockholm, 3(12):1–132.

ZINSMEISTER, W. J. 1979. Biogeographic significance of the late Mesozoic and early Tertiary molluscan faunas of Seymour Island (Antarctic Peninsula) to the final breakup of Gondwanaland, p. 349–355. *In* J. Gray and A. Boucot (eds.), Historical Biogeography, Plate Tectonics and the Changing Environment. Proceedings of the 37th Annual Biology Colloquium and Selected Papers. Oregon State University Press, Corvallis.

——. 1982. Review of the Upper Cretaceous–Lower Tertiary sequence on Seymour Island, Antarctica. Journal of the Geological Society, London, 139(6):779–786.

MANUSCRIPT ACCEPTED BY THE SOCIETY SEPTEMBER 1, 1987
INSTITUTE OF POLAR STUDIES CONTRIBUTION NO. 559

Geological Society of America
Memoir 169
1988

Upper Cretaceous and lower Paleocene diatom and silicoflagellate biostratigraphy of Seymour Island, eastern Antarctic Peninsula

David M. Harwood

Byrd Polar Research Center and Department of Geology and Mineralogy, The Ohio State University, Columbus, Ohio 43210

ABSTRACT

Diverse, siliceous microfossil assemblages—including marine diatoms, silicoflagellates, ebridians, endoskeletal dinoflagellates, chrysophyte cysts (archaeomonads), radiolarians, and sponge spicules—were recovered from Seymour Island. Their stratigraphic occurrence is documented from the ~1,400-m-thick section of the López de Bertodano (upper Campanian into lower Paleocene) and Sobral (lower Paleocene) Formations. These units consist of detrital silt and fine sand deposited in a quiet shelf environment. The following new diatoms are proposed: *Coscinodiscus sparsus, Gladius antarcticus, Gladius antarcticus* f. *alta, Hemiaulus huberi, Hemiaulus seymouriensis, Pterotheca minor, Pterotheca trojana,* and *Wittia macellarii.*

The microfossil assemblages compare well with floras of similar age from diatomites in the Ural Mountain region of the Soviet Union and from Deep Sea Drilling Project (DSDP) Hole 275 in the southwest Pacific. The present work documents the fourth occurrence of Campanian/Maastrichtian, and the first of Danian diatoms and silicoflagellates from the Southern Hemisphere. High sedimentation rates have preserved a detailed record of Late Cretaceous and early Paleocene siliceous microfossil evolution and extinction.

Silicoflagellate assemblages show an abrupt composition change from *Lyramula*-dominated to *Corbisema*-dominated floras a few meters above a resistant glauconitic sandstone within the upper López de Bertodano Formation. The abundance of diatom resting spores increases from ~5 percent in the uppermost Maastrichtian to ~35 percent a short distance above this glauconite and continues at these high values through the lower Paleocene. These and other fossil data suggest that the Cretaceous/Tertiary (K-T) boundary is situated a few meters above the glauconitic sandstone. Unlike most other known K/T sections, clastic sedimentation was continuous across the boundary. Most Cretaceous diatom species (as much as 84 percent) continue into Tertiary beds. This suggests that the extinction event responsible for devastating the other major groups of Cretaceous plankton did not affect the diatoms; resting spore formation may have aided their survival.

GEOLOGIC SETTING OF SEYMOUR ISLAND

Seymour Island (Isla Vicecomodoro Marambio) (64°15′S latitude; 56°45′W longitude) is a small ice-free island adjacent to the northeast tip of the Antarctic Peninsula (Fig. 1). Fossiliferous sediments on Seymour Island and on other islands of the James Ross Basin represent the only extensive exposures of Upper Cretaceous and Lower Tertiary marine strata between South America and New Zealand and thus are important for future paleobiogeographic and paleontologic reference. Zinsmeister (1982, 1986) provided a general overview of the history, geology, paleontology, and scientific potentials of Seymour Island.

Seymour Island contains a thick and fossiliferous shallow marine sequence of upper Campanian through Eocene fine-grained sediments. These sediments were deposited under relatively quiet-water conditions on the continental shelf, adjacent to

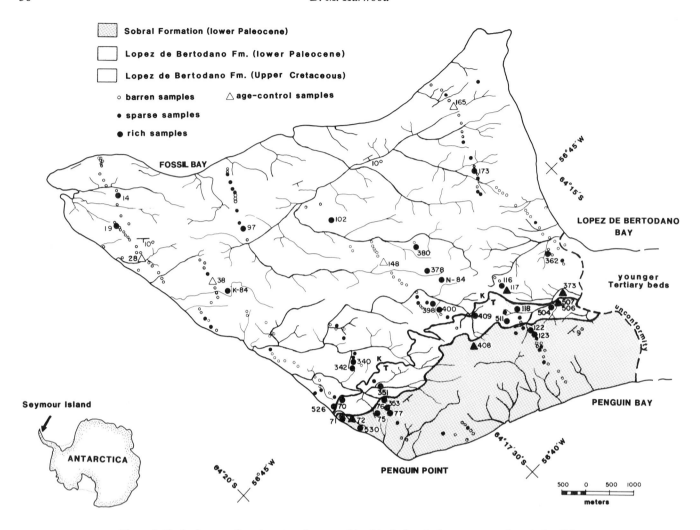

Figure 1. Geologic map of southwestern Seymour Island including drainage systems. Sample localities that yielded rich siliceous microfossil assemblages (data included in Table 4) are numbered and indicated by large solid circles or triangles. Small solid circles = samples containing traces of siliceous microfossils; open circles = barren samples; open triangles = samples barren of siliceous microfossils but which provide age control from the occurrence of calcareous nannoplankton. Sublinear array of circles reflects sample-traverse and location of measured sections. Cretaceous/Tertiary boundary is indicated at the contact between shaded and unshaded regions.

a major source of volcanic and continental detritus from mountains in the Antarctic Peninsula.

Upper Cretaceous and Paleocene sediments on Seymour Island show only minor homoclinal tilting and appear to be unaffected by thermal or burial metamorphism. Pliocene diabase dikes cut the Cretaceous and Paleocene sequence but show only local metamorphic effects (Zinsmeister, 1982; Huber, 1984; Macellari, 1984a).

STRATIGRAPHY

Upper Cretaceous and lower Paleocene strata on Seymour Island are part of a gently northward dipping (as much as 10°)

homoclinal sequence (Fig. 1) composed of unconsolidated, well-exposed (Macellari, Fig. 4, this volume), fine-grained clastic sediments. These sediments constitute the upper Marambio Group of Rinaldi (1982), which is divided into three units; two of these, the López de Bertodano Formation and the overlying Sobral Formation, are discussed briefly below. More detailed stratigraphic descriptions for these sediments on Seymour Island and in the James Ross Island region are included in Andersson (1906), Bibby (1966), Rinaldi et al. (1978), Olivero (1981), del Valle and others (1982), Macellari and Huber (1982), Rinaldi (1982), Zinsmeister (1982), Palamarczuk and others (1984), Huber (1984; this volume), Macellari (1984a; this volume), and Sadler (this volume).

A wide variety of fossil groups are present in the Upper Cretaceous and lower Paleocene sediments on Seymour Island, including dinoflagellate cysts, acritarchs, nonmarine palynomorphs, and spores from fresh-water algae (Askin, this volume); plant debris, wood, ammonites, bivalves, gastropods, brachiopods, corals, echinoids, and serpulid worms (Macellari, 1984a,b, 1986, this volume; Zinsmeister and Macellari, 1983; this volume); decapod crustraceans (Tschudy and Feldmann, this volume); marine reptiles, fish and shark remains (Chatterjee and Zinsmeister, 1982); Foraminifera (Huber, 1984, 1987b, this volume); calcareous nannoplankton (Huber et al., 1983); silicoflagellates, diatoms, endoskeletal dinoflagellates, chrysophyte cysts (archaeomonads), ebridians, radiolarians, and sponge spicules (this chapter).

López de Bertodano Formation

A maximum thickness of 1,190 m of the López de Bertodano Formation crops out on Seymour Island (Macellari and Huber, 1982). This unit is monotonous, showing little lithologic variation, and consists of friable, gray, unconsolidated, sandy siltstone that is massive to well bedded (Fig. 2). Glauconite is abundant in the upper López de Bertodano Formation and in the overlying Sobral Formation. Rounded, irregular, calcareous concretions and concretionary horizons are present throughout. Macrofossils (ammonites, bivalves, gastropods, echinoids, corals, and fossil wood) are common, except in the lower 300 m.

The López de Bertodano Formation was deposited in quiet water on a stable shelf, as indicated by sedimentological and paleontological data (Huber, this volume; Macellari, 1984a, this volume), which include the absence of slope instability and traction current structures, a high mud content, and recovery of in situ macrofauna. Macellari (this volume) suggested the lower 600 m (*Rotularia* facies) was deposited in shallow, turbid water under the influence of a nearby delta or estuary. Diverse assemblages of siliceous and calcareous microfossils between 200 and 300 m (Huber, Fig. 3, this volume), however, suggest open-marine conditions for this interval. The remaining upper portion of this formation was deposited in progressively deeper water from middle to outer shelf depths that probably did not exceed the shelf/slope break (Macellari, this volume).

Sobral Formation

The contact between the underlying López de Bertodano and the Sobral Formation is erosional; basal channeling is present in several sections. A maximum thickness of 255 m of the Sobral Formation crops out on Seymour Island. This unit consists of well-laminated silts at the base, followed by cleaner sandstones that become more glauconitic, coarser grained, and cross-bedded at the top (Macellari, this volume) (Fig. 2). Bedding and sedimentary structures are well defined in the Sobral Formation, with more pronounced lateral variations and discontinuous beds than in the underlying unit. Round, irregular calcareous concretions

and concretionary horizons are frequent. Macrofossils are less abundant and less diverse than in the López de Bertodano Formation (Macellari, this volume).

The lower portion of the Sobral Formation was initially deposited in a delta-front environment, below wave base. With continued filling of the basin by deltaic progradation, the depositional environment progressed through coastal barrier sands to a delta top facies in the upper Sobral (Macellari, this volume).

DATING OF THE LOPEZ DE BERTODANO AND SOBRAL FORMATIONS ON SEYMOUR ISLAND

The López de Bertodano Formation ("Older Seymour Beds" of Andersson, 1906) was, on the basis of ammonite evidence, considered to be Albian to Campanian by Kilian and Reboul (1909); Campanian by Howarth (1958, 1966), Rinaldi and others (1978), and Olivero (1981); upper Campanian to possible Maastrichtian by Spath (1953); and upper Campanian with the overlying Sobral Formation being of Maastrichtian or Danian age by Rinaldi (1982) and Palamarczuk and others (1984), respectively. The micropaleontologic investigations of Huber and others (1983); Huber (1984, 1987b, this volume); Askin (this volume); and the present report, have refined the dating of this sequence. The López de Bertodano Formation is now considered to include upper Campanian through lower Danian sediments, and to contain a more or less continuous, although probably condensed, section across the Cretaceous/Tertiary boundary. The overlying Sobral Formation is considered to be lower Paleocene.

The lower 400 m of the López de Bertodano Formation on Seymour Island are upper Campanian. The Campanian/Maastrichtian boundary is approximated in the interval between 350 to 415 m from calcareous nannoplankton-based age extrapolation. The co-occurrence of calcareous nannoplankton *Nephrolithus frequens* Gorka (mid- to upper Maastrichtian according to Wind, 1979, 1983), and *N. corystus* Wind (upper Campanian to mid-Maastrichtian according to Wind, 1979, 1983) in Sample SI-38 (Fig. 1) suggests a middle Maastrichtian age for this sample. The occurrence of *N. corystus* and the absence of *N. frequens* in Sample SI-28 (Fig. 1) suggests a Campanian to mid-Maastrichtian age at this level. The occurrence of *N. frequens* and the absence of *N. corystus* in Samples SI-38 and SI-148, 300 and 500 m below the Cretaceous/Tertiary boundary (Fig. 1) suggests that a thick sequence of upper Maastrichtian sediments is present on Seymour Island. These and other calcareous nannoplankton are illustrated, and the age control they provide is further discussed in Huber and others (1983).

The Cretaceous/Tertiary boundary is placed within a glauconite-rich interval approximately 1.5 meters above a resistant glauconitic sandstone (Askin, Fig. 2, this volume), and is situated 50 to 90 m below the Sobral/López de Bertodano unconformable contact (Figs. 1, 2). While there is no apparent lithologic change associated with the K/T transition, besides a possible nondepositional or decreased sedimentation event repre-

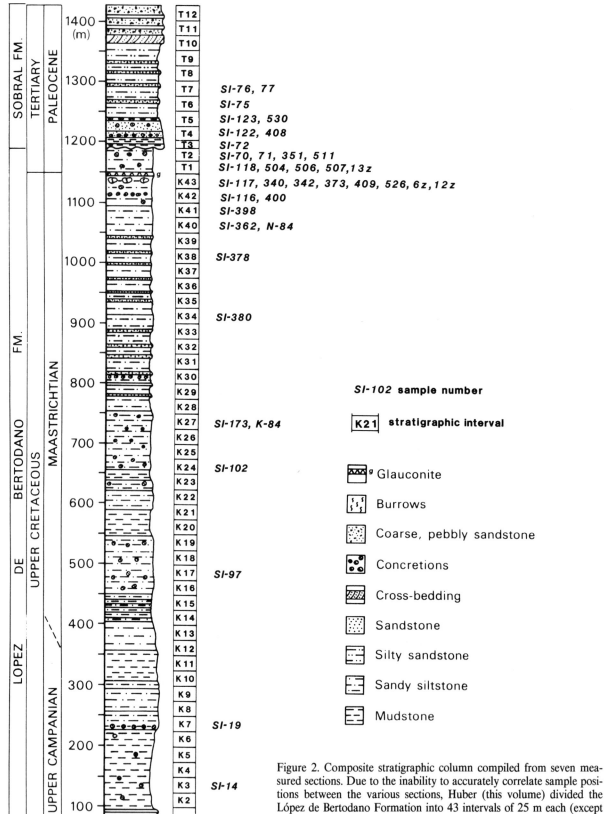

SI-102 sample number

K21 stratigraphic interval

▵▵ᵍ Glauconite

Burrows

Coarse, pebbly sandstone

Concretions

Cross-bedding

Sandstone

Silty sandstone

Sandy siltstone

Mudstone

Figure 2. Composite stratigraphic column compiled from seven measured sections. Due to the inability to accurately correlate sample positions between the various sections, Huber (this volume) divided the López de Bertodano Formation into 43 intervals of 25 m each (except for the lowest and highest four intervals). The overlying Tertiary section is similarly divided here into 14 intervals. Only the samples that yielded siliceous microfossil assemblages (Table 4) are listed here. Figure modified from Huber (this volume).

sented by the glauconitic bed, the macro- and microfauna and -flora show an abrupt transition.

Several difficulties exist in identifying and correlating the K/T boundary and the glauconite bed laterally (Huber, 1987a). These are particularly evident on the southeast corner of the island, where the outstanding problems include (1) multiple and discontinuous glauconitic beds; (2) incomplete exposure; (3) poor sample data limited by dip-slope exposure of critical intervals; (4) faulting and the potential for repeated sections; (5) lag deposits of ammonites on the surface of dip slopes; (6) anomalous ammonite occurrence (recognized by Huber in 1985 and by others in 1987) above a resistant glauconite bed; and (7) the presence of Zone 3 (R. A. Askin, personal communication, 1985) palynomorphs (mid-Maastrichtian) in the above ammonites from what should be uppermost Zone 4 (uppermost Maastrichtian) sediments. While the above factors pose stratigraphic problems in this area of Seymour Island, the biostratigraphic succession to the north, afforded by molluscs, dinoflagellates, siliceous microfossils, and foraminifera, results in the reliable and consistent definition of the Cretaceous/Tertiary boundary.

The top of the Cretaceous sequence is marked by the highest occurrence of Cretaceous silicoflagellates *Lyramula* spp., *Vallacerta tumidula,* and *Corbisema geometrica;* the diatoms *Gladius antarcticus* and *G. pacificus;* the planktonic foraminifera *Heterohelix globulosa, Globigerinelloides multispinatus,* and *Hedbergella monmouthensis* (Huber, this volume); Zone 4 palynomorphs (Askin, this volume), ammonites, and other Cretaceous molluscs (Zinsmeister and Macellari, 1983; Macellari, 1984a; this volume; Macellari and Zinsmeister, 1983), all of which have their highest occurrence at or a few meters above the resistant glauconite bed (Figure 2).

The presence of Danian sediments in the López de Bertodano Formation above the resistant glauconite bed is best indicated by the appearance of characteristic Danian dinoflagellate cysts of zone 5 (Askin, Fig. 2, this volume). The first demonstrably lower Paleocene silicoflagellate assemblage of the *Corbisema hastata* Zone of Bukry and Foster (1974) occurs ~35 m (Sample SI-511) above the resistant "K/T glauconite," within the uppermost López de Bertodano Formation (Figs. 2, 7). The lowest occurrence of *Corbisema aspera,* 11 m (Sample SI-507) above the glauconite bed, may represent the lowest Paleocene identified by silicoflagellates. Too little is known of the biostratigraphic range of *C. aspera*; it is only known from three other Paleocene sites and one Eocene site (Schulz, 1928), and has not been reported in Cretaceous sediments.

A diverse lower Paleocene silicoflagellate assemblage (13 taxa), including an ebridian, occurs 40 m above the base of the Sobral (Sample SI-408) and includes the Paleocene silicoflagellates *Corbisema aspera, Corbisema hastata, Corbisema hastata* f. *miranda, Corbisema hastata* f. *Globulata, Corbisema inermis,* and *Corbisema apiculata* f. *minor* (Fig. 7). Comparison with Gleser's (1966) description of lower Paleocene silicoflagellate assemblages in the eastern Ural Mountains and Middle Volga area, USSR, as well as the absence of Cretaceous genera, suggest that the Sobral

Formation and the upper portion of the López de Bertodano Formation, above the glauconitic bed, are lower Paleocene.

It may also be possible to identify lowest Paleocene sediments on Seymour Island by the consistent increase in the abundance of diatom resting spores above the glauconitic bed (Fig. 8). If this increase is repeated in other K/T sections, the lowest Paleocene can be identified on Seymour Island, by siliceous microfossils, within ~5 m above the glauconitic bed.

Wide spacing of samples yielding siliceous microfossils and the requisite use of carbonate concretions and material from within macrofossils to recover diatoms and silicoflagellates are the principal factors that prevent these microfossils from precisely identifying the K/T boundary at this time. Silicoflagellates are able to, and diatoms have high potential to mark the position of the K/T boundary on Seymour Island. More sample material is needed from the interiors of cemented macrofossils and concretions from numerous measured sections across this interval.

The Danian foraminifera *Globastica daubjergensis* occurs in the uppermost López de Bertodano Formation, 1 m below the base of the Sobral/López de Bertodano contact and near the base of the Sobral Formation (Huber, this volume). It is unlikely that any of the Sobral Formation is upper Paleocene. Characteristic upper Paleocene diatoms, including *Hemiaulus incurvus* Schibkova, *H. periterus* Fenner, and *H. inequilaterus* Gombos, and silicoflagellates *Naviculopsis* spp. and *Crassicorbisema* spp., which make up a substantial part of known upper Paleocene siliceous microfossil assemblages (Mukhina, 1974; Gombos, 1977; Ling, 1981; Fenner, 1985; Perch-Nielsen, 1985), are not present as high as 250 m above the base of the Sobral Formation.

A carbonate dissolution facies, which either prevented deposition or dissolved all calcareous microfossils and etched the macrofauna, is suggested in the interval immediately above the K/T glauconite and up to a few meters below the contact between the López de Bertodano and Sobral Formations (Huber, 1987a; this volume). This dissolution interval may be related to the hiatus and residual boundary clays present in carbonate K/T sections around the globe. In clastic sequences such as Seymour Island, where sedimentation was largely independent of ocean chemistry, the record of extinction/evolution events is apparently preserved through continued clastic sedimentation.

METHODS OF STUDY

Samples used in this study (Figs. 1, 2; Table 1) were collected by B. Huber and C. Macellari during the Ohio State University Institute of Polar Studies' 1982 and 1984 expeditions to Seymour Island (Macellari and Huber, 1982; Huber, 1984, this issue; Macellari, 1984). Geographic and stratigraphic location of samples listed in Table 1 can be found in the above references. Samples were taken from pits dug into the permafrost layer and from carbonate concretions and internal casts of macrofossils adjacent to the collecting sites. Samples were not collected from dip slopes with limited stratigraphic control, nor from displaced concretions and/or ammonites in residual lag deposits where

TABLE 1. SAMPLES CONTAINING SILICEOUS MICROFOSSILS

Upper Cretaceous			Lower Paleocene	
SI-14*†	SI-102*†	SI-342*	SI-70*	SI-204†
SI-19*	SI-116*	SI-362†	SI-71*	SI-350*
SI-37	SI-117*	SI-372*	SI-72*	SI-351*
SI-47†	SI-163†	SI-373*	SI-75*	SI-353
SI-56b	SI-170†	SI-378*	SI-76*	SI-405*
SI-67	SI-171†	SI-380	SI-77	SI-408*
SI-69a	SI-173*†	SI-397*	SI-79†	SI-504*
SI-87†	SI-175†	SI-398*	SI-83	SI-506*
SI-88†	SI-178†	SI-400†	SI-118	SI-507
SI-89†	SI-187†	SI-409	SI-119*	SI-507.5*
SI-89†	SI-194†	SI-501*	SI-121*	SI-511
SI-90†	SI-319	SI-526*	SI-122*	SI-513
SI-95†	SI-325†	SI-N-84*	SI-123	SI-521.5*
SI-96	SI-331	SI-K-84*	SI-126	SI-530
SI-97*	SI-340		SI-127	SI-540
			SI-130	SI-541
			SI-137	

*Good siliceous microfossil assemblages with assemblage data presented in Table 4.
†Pyritization and iron-staining of siliceous microfossils.

diatom and silicoflagellate assemblages were recovered in some samples (Figs. 1 through 3; Table 1). Apparently, early cementation in calcareous concretions and within the shells of macrofossils protected the siliceous microfossils from diagenetic effects that removed them from the surrounding unconsolidated sediments. The benefit of sampling calcareous concretions and macrofossils when studying early Cenozoic and Cretaceous siliceous microfossils, or when the effects of diagenesis are evident, was noted by Bramlette (1946) in the Miocene Monterey Formation of California (Lohman, 1960); in the Aptian Doncaster Member of the Rolling Downs Group in Queensland, Australia (Harper, 1977; unpublished report); and in a summary by Blome and Albert (1985).

Approximately 5 cc of material from calcareous concretions and cemented interiors of macrofossils was placed in a 400-ml beaker, covered with equal volumes of concentrated HCl and 30 percent H_2O_2, and gently heated in a hot water bath. When the bubbling reaction stopped, the sample was washed repeatedly with distilled water and centrifuged at 1,500 rpm for 4 min until a neutral pH was reached. A combination of settling and sieving was applied to concentrate the rare diatoms from the silt-rich sediments. The sample was then washed into a 600-ml beaker, sufficiently mixed (by addition of a Calgon solution, stirring, and brief ultrasonic treatment) and allowed to settle for 30 sec through a 10-cm column of water. The suspension was then siphoned through a >25-μm sieve. The residue remaining in the beaker was agitated, broken with a stirring rod, and resuspended in a 10-cm column of water and allowed to settle for an additional 30 sec. The suspended material was filtered through the same >25-μm sieve. This process of settling was repeated until no material remained in suspension after 30 sec. The 30-sec settling time removed much of the abundant sand- and silt-sized clastic material. Material that passed through the >25-μm micrometer sieve was collected and allowed to settle for 12 hr through a 10-cm column of distilled water and Calgon solution. Clay-sized material that remained in suspension after 12 hr was decanted. This process was repeated several times for clay-rich samples. Without the use of sieves, the high diatom diversity and large number of productive samples would not have been achieved.

Preliminary examination of each sample was performed on a slide of the <25-μm fraction. This size fraction was chosen for preliminary examination to determine if diatoms were present in the sample. To determine if diatoms were present, 10 traverses of a 22- × 40-mm cover glass were done on this slide at 600× magnification. If diatoms were noted, an average of 8 slides (six of the >25 and two of the <25 fractions) were mounted with Permount to decrease the optical relief of the common quartz grains, and thoroughly examined. Microfossil-rich samples were later mounted with Hyrax for identification and photographic documentation.

Abundance data were collected from unsieved material on only those samples that yielded relatively rich and diverse microfossil assemblages; presence/absence data was collected on all others from sieved material (Table 4). Bias due to poor preserva-

stratigraphic position could not be assured. Sample positions were located on seven measured sections that were correlated by air photo interpretation. A composite section was drawn from these seven sections by Huber (this volume), who divided the composite section into 25-m intervals (Fig. 2); his notation will be followed here and slightly modified where appropriate. Some biostratigraphic detail is lost by these summary intervals, but this method is warranted here because it enables correlation of the several sections and combination of the sparse diatom-bearing samples. High rates of sedimentation, between 50 to 400 m/m.y. (Sadler, this volume), suggest that the 20- and 25-m intervals do not individually represent a substantial amount of time.

All of the samples examined for diatoms (including barren samples) are indicated in Figure 1. Sample spacing averaged 15 m in each section, with an average 3-m sample spacing in composite section. There were 260 samples examined in this study; siliceous microfossil remains were recovered in 79 of these (Figs. 1, 2; Table 1). Of these 79 samples, 38 yielded siliceous microfossil assemblages of sufficient number and adequate preservation for detailed study and stratigraphic documentation.

Unconsolidated material was processed in the early phase of this research for calcareous nannoplankton and diatoms, and the Seymour Island sequence appeared to be barren of siliceous microfossils. Diatoms were first noted by C. Macellari in thin sections of calcareous concretions. Standard siliceous microfossil preparation techniques with HCl and H_2O_2 were performed on the calcareous concretions and internal casts of macrofossils. Rich

TABLE 2. SILICEOUS MICROFOSSIL DIVERSITY ON SEYMOUR ISLAND

	Lower Paleocene Genera/Species	Upper Campanian, Maastrichtian Genera/Species
Diatoms	43/161	40/167
Silicoflagellates	3/14	5/13
Ebridians	1/1	----
Endoskeletal Dinoflagellates	1/1	2/2

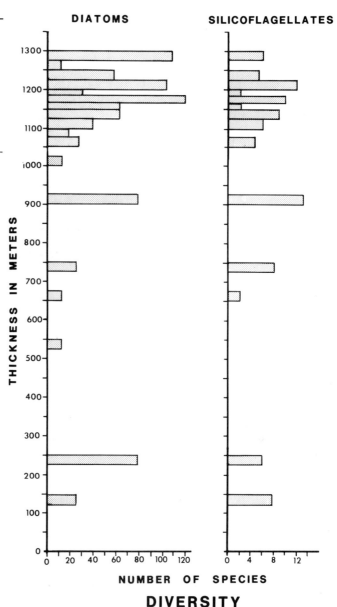

Figure 3. Diatom and silicoflagellate diversity by stratigraphic interval. Values reflect total number of species and varieties encountered in all samples from a particular interval. High diversity and consistent microfossil recovery above 1,000 m reflect increased volcanic ash content that aided the preservation of siliceous microfossils. A rough correlation exists between microfossil-bearing samples and intervals with calcareous concretions and concretionary horizons (Fig. 2).

tion was too great in many samples to yield valid quantitative results. Abundance data in percent of the total siliceous microfossils assemblage is presented in Table 4, for values greater than 2 percent; lower values are indicated with an "X" representing the presence of a particular taxon.

SILICEOUS MICROFOSSIL ASSEMBLAGES

Upper Cretaceous and lower Paleocene siliceous microfossils from Seymour Island were first reported in Huber and others (1983). Subsequent discussion and descriptions of these assemblages are by Martinez-Macchiavello in Palamarczuk and others (1984) and by Harwood (1985, 1986). The work herein documents the stratigraphic distribution of diatoms (192 species), silicoflagellates (27 species and varieties), ebridians (1 species), endoskeletal dinoflagellates (2 species), and the general occurrence of chrysophyte cysts (archaeomonads), radiolarians, and sponge spicules throughout the ~1,400-m composite section of the López de Bertodano and Sobral Formations (Tables 2, 4; Figs. 9-24).

Diversity values for individual Seymour Island siliceous microfossil assemblages are given in Figure 3; combined diversity values for the lower Paleocene and upper Campanian/Maastrichtian appear in Table 2. Diversity fluctuations reflect both preservational biases and variance in original productivity. Intervals of high diatom diversity (Fig. 3) correspond to intervals of high dinoflagellate diversity (Askin, this volume), indicating perhaps greater marine influence and/or less turbidity at these intervals.

Microfossil abundance, preservation, diversity, and occurrence were considerably greater in the Tertiary (lower Sobral and uppermost López de Bertodano Formation) than in either the Cretaceous portion of the López de Bertodano Formation or the upper Sobral. The enhanced preservation and higher abundance in this interval corresponds with an increase in content of volcanic ash to 20 percent (Macellari, 1984a; this volume) of the sand-size fraction in the lower 100 m of the Sobral Formation and in the Paleocene section of the López de Bertodano Forma-

TABLE 3. COMPOSITION OF CRETACEOUS AND TERTIARY
DIATOM ASSEMBLAGES ON SEYMOUR ISLAND IN PERCENT

Upper Cretaceous (147 species)*		Danian (161 species)†	
Hemiaulus	18.0	*Hemiaulus*	21.0
Pterotheca	10.0	*Pterotheca*	12.0
Stephanopyxis	11.5	*Stephanopyxis*	10.5
Trinacria	9.0	*Trinacria*	6.0
Sceptroneis/Incisoria	5.5	*Sceptroneis/Incisoria*	3.0
Triceratium	3.5	*Triceratium*	5.5
Pseudopyxilla	3.5	*Pseudopyxilla*	4.0
Ancanthodiscus	3.0	*Ancanthodiscus*	3.0

Note: The structure and composition of Upper Cretaceous and Danian diatom assemblages are similar. This similarity suggests no major reorganization of diatom assemblages from the Cretaceous to the Paleocene.

*These eight genera represent 63.5 percent of the total assemblage of 40 Cretaceous diatom genera.

†These eight genera represent 65 percent of the total assemblage of 43 Danian diatom genera.

tion, continuing several meters below the K/T glauconite in the Cretaceous. A correlation between diatom and diatomite occurrence and volcanism was first noted by Ehrenberg (1846) and reaffirmed by many later workers (Taliaferro, 1933). While the diatoms are living, volcanic ash provides a source of soluble silica, perhaps increasing productivity if other nutrients are also available. The ash buffers the diatom frustule against dissolution by increasing the dissolved silica content in the water column and sediment pore waters.

Diatom assemblage composition at the generic level varies little throughout the ~1,400-m sequence. The same eight diatom genera make up more than 60 percent of the total assemblages for the Cretaceous and the lower Paleocene (Table 3). Structure of the diatom assemblages, in terms of ranking the genera by number of species, changes very little from Cretaceous to Tertiary.

PALEOENVIRONMENTAL INTERPRETATION

Diatom assemblage composition indicates deposition on the continental shelf under normal marine conditions. Diatom resting spores, which compose a large portion of the Seymour Island assemblages (Tables 3, 6; Figs. 17, 18), are indicative of shallow-water, shelfal conditions (Ross and Sims, 1973; Strelnikova, 1974; Jousé, 1978). Kitchell and others (1986) have shown that in the Late Cretaceous, polar diatoms were well adapted to the seasonal stresses of changing light, temperature, and nutrients. This was indicated by the alternating abundance of vegetative diatom cells and resting spores in laminated diatomaceous sedi-

ments from the Arctic Ocean. The recovery of aggregates of predominantly resting spores in several samples from Seymour Island also indicates a strong seasonality there. The general paucity of benthic diatom species, which are abundant in some Cretaceous and Paleocene diatomaceous deposits (Moreno Shale and Western Siberian/Northern Ural deposits), suggests deposition below the euphotic zone. Because the depth of the euphotic zone varies with season, latitude (incident angle of the sun to the water surface), and the amount of particulate matter (terrigenous or biogenic) in the water column, depth estimates based on the absence of benthic diatoms are often difficult. The low number of benthic diatoms on Seymour Island indicates that deposition was either in waters greater than 75 m, or possibly shallower if the water column contained a high volume of suspended material. The nature of the sediment, fine-grained silts and sands, and macrofossil paleoecology (Macellari, 1984a, b; this volume), suggests a high sediment input and turbid conditions.

The suggestion of turbid conditions may be supported by the environmental interpretation offered by Tappan (1962) to explain pyritized siliceous microfossils. Tappan interprets the pyritized casts of radiolarians in Cretaceous sediment from Alaska to reflect rapid burial in turbid, inner, sublittoral environments. Similar facies are interpreted for pyritic diatoms and radiolarians by Given and Wall (1971) and Wall (1975). Thus, the common occurrence of pyritized and iron-stained siliceous microfossils is in agreement with the depositional environment suggested by Macellari (this volume): influenced by a nearby deltaic environment. Just how close Seymour Island was to the delta at a particular time is unknown.

SILICEOUS MICROFOSSIL BIOSTRATIGRAPHY

Existing reference material and literature on Upper Cretaceous and lower Paleocene siliceous microfossils

The biostratigraphic record of diatoms and silicoflagellates from the Maastrichtian through Danian time is poorly known. The works of Gleser (1966), Hajós and Stradner (1975), Strelnikova (1974), and Barron (1985) represent the only treatment of diatoms of this age from a stratigraphic framework. Other reports on Upper Cretaceous and lower Paleocene diatoms are largely descriptive.

The Upper Cretaceous diatom and silicoflagellate record includes minor assemblages from the Turonian, Santonian, and major deposits from the Campanian and Maastrichtian. Turonian diatoms are known from Europe. Müeller (1912) described a questionable centric diatom, *Actinoclava frankei,* from Rilmerich, Westfalia. Schulz (1928) and Weisner (1936) reported pyritized diatom remains from a Turonian claystone in Czechoslavakia, and Deflandre (1941) reported several semi-dissolved, partly pyritized diatoms from France. Santonian diatoms are known in the eastern slopes of the northern Urals from the works of Strelnikova (1968) in the Leplya River basin in the Western Siberian Lowland from Voronkov (1959).

The Santonian and lower Campanian mark the start of widespread deposition of rich and diverse deposits of siliceous microfossils. The most continuous diatomaceous sequences of Upper Cretaceous sediment are known from outcrops and drillholes in the Western Siberian Lowlands and eastern slopes of the northern Ural Mountains. Diatomite deposits as much as 500 m thick and ranging from Santonian to upper Paleocene occur there. These deposits are described by Strelnikova (1974, 1975) and Gleser (1966). Russian researchers have pioneered Cretaceous and Paleocene diatom studies and applied their results to geological interpretation for more than 50 yr. Diatoms from deposits in the Western Siberian Lowlands and the eastern slopes of the northern Urals are best documented by Strelnikova (1974). Additional reports of diatoms and silicoflagellates from these deposits are found in the works of Jousé (1948, 1949a, b, 1951b, 1955), Krotov (1959), Voronkov (1959), Voszhennikova (1960), Krotov and Schibkova (1961), Schibkova (1961), Vekshina (1961a, b), Gleser (1966), and Strelnikova (1964, 1965a, 1966a, b, 1971, 1974). Gleser (1959) reviewed the silicoflagellate family Vallacertaceae and gave original descriptions of all silicoflagellate species in Santonian to Campanian deposits of the eastern slopes of the Urals.

According to Strelnikova (1974, 1975), diatom floras from Western Siberia have several similarities: (1) to Campanian floras known from the Ulyanovsk Province (Simbirsk is the old name) in the Middle Volga area, with lower Campanian deposits near Penza (Jousé, 1949b; Strelnikova, 1974, 1975) and upper Campanian deposits near the town of Inza (Strelnikova, 1975); (2) to upper Senonian diatom floras known from Kunashir Island of the Lesser Kuril Island group in the western Bering Sea (Jousé,

1951b, 1968; Strelnikova, 1974); and (3) to Campanian diatom floras recovered from pore channels of hexactinellid sponges from Gdansk (Danzig), Poland (Schulz, 1935). Silicoflagellates are also known from the "Danzig sponges" and were described in Schulz (1928) and Gemeinhardt (1930). Zittel (1876) described Senonian silicoflagellates from Westphalia.

Other Campanian deposits with diatoms and silicoflagellates include Campanian (Maastrichtian?) deposits from Alpha Ridge in the Arctic Ocean (Bukry, 1981a, 1985; Barron, 1985; Kitchell, 1980; Kitchell and Clark, 1982; Kitchell et al., 1986; Ling et al., 1973); reworked Cretaceous diatoms in glacial deposits from Banks Island in the Canadian Arctic (Vincent et al., 1983); upper Campanian and possibly Maastrichtian diatomaceous sediments from the Campbell Plateau in the southwest Pacific at DSDP Site 275 (Hajós and Stradner, 1975; Perch-Nielsen, 1975a, b; and Bukry, 1975a); Campanian-Maastrichtian diatomaceous sediments in the Indian Ocean at DSDP Site 216 (Bukry, 1974; Gresham, 1985, 1986); and Upper Cretaceous diatoms and silicoflagellates from the Tonga Trench (Balance et al., 1986). Silicoflagellate assemblages from numerous Upper Cretaceous to lower Miocene DSDP sites were investigated by Bukry and Foster (1974). Silicoflagellates are also known from the Maastrichtian at Stevns Klint (Perch-Nielsen, 1985).

Several Campanian/Maastrichtian diatom occurrences are reported from North America. Fenner (1982) reported Maastrichtian diatoms from dredged sediments in Emperor Canyon off the New Jersey coast. Other reports include pyritized diatoms from Alberta (Given and Wall, 1971; Wall, 1975), from the Peedee Formation in South Carolina (Abbott, 1978), and from the Pierre Shale in Wyoming (Bergstresser and Krebs, 1983). Upper Cretaceous silicoflagellates from Wyoming were documented by Klement (1963).

The best studied Maastrichtian diatomaceous deposit is the Moreno Shale from California. Diatoms were discovered by Anderson and Pack in 1915 and later described by Hanna (1927a, 1934), Lefébure and Chenevière (1939), Long, Fuge and Smith (1946), Barker and Meakin (1946, 1949), and Abbott and Harper (1982). Silicoflagellates were described by Hanna (1928a), Deflandre (1950), Mandra (1960, 1968), Cornell (1974), and Ling (1972). Archaeomonads (Chrysomonad statocysts) were examined by Rampi (1940). Other Maastrichtian diatomaceous deposits include those reported by Weidemann (1964) from the foothills of the Swiss Alps, and by Strelnikova (1975) from the Apuku River basin in the Koryakskaya Range, western Bering Sea coast, northeast of the Kamchatka Peninsula, USSR. Seymour Island represents the fourth known and the highest latitude occurrence of Upper Cretaceous siliceous microfossils from the Southern Hemisphere: the others were from the Tonga Trench and from DSDP Holes 275 and 216.

Rich lower Paleocene diatomaceous deposits were previously known only from two regions in Russia, in the Middle Volga area and from the eastern slopes of the northern Ural Mountains. The diatoms and silicoflagellates from deposits in the Urals described by Jousé (1951a), Krotov (1957a, b), Krotov

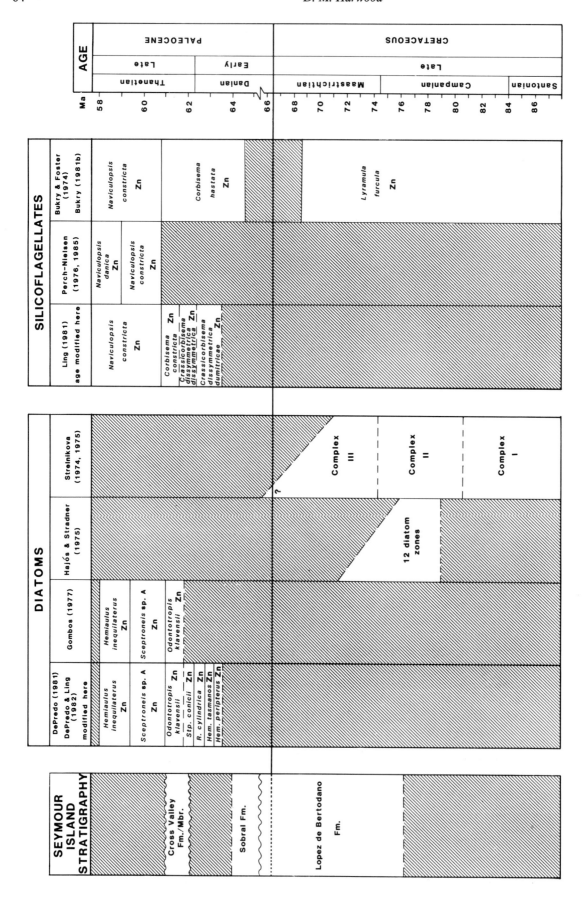

Figure 4. Existing Cretaceous and Paleocene biostratigraphic division of diatom- and silicoflagellate-bearing sequences with a relative correlation of the Seymour Island strata to these zonal schemes. Ages assigned to DSDP Hole 208 sequences are adjusted here to slightly younger intervals than suggested by Ling (1981) and DePredo (1980).

and Schibkova (1961), Gleser (1962, 1966), Gleser and others (1974), and Strelnikova (1966b) are older than the deposits in the Volga region. The Volga deposits were first discovered at Bekle-mischewo near the town of Karsun (west of Ulyanovsk) by Pacht in 1854. Diatoms from this deposit were described by Weisse (1854) and Ehrenberg (1855) and also studied by Eulenstein and Weissflog, with some illustrations published between 1876 and 1879 in Schmidt and others (1874–1959). All reports of diatoms from Simbirsk (Ulyanovsk) prior to 1884 were from this deposit. In 1884 Lahusen discovered a new diatomaceous deposit containing richer material from Archangelsk-Kurojedowo (also near Karsun). These diatoms were described by Witt (1885). The individuality of these two deposits was later smeared, and all deposits were labeled "Simbirsk" (Ulyanovsk). Names of other lower Paleocene localities that appear in the literature include Barysh, Carlovo, Inza, Kissatib, Kusnetzk, Onianino (Ananino), Penza, and Sysran. Descriptions of diatoms from these lower Paleocene deposits can be found in Schmidt and others (1874–1959), Grunow (1884), Pantocsek (1889, 1892), Jousé (1949b), and Krotov and Schibkova (1959). A history of the study of these deposits is presented in Burke and Woodward (1963–1974). Other lower Paleocene diatom occurrences were reported in Chenevière (1934) and Rogozin (1913).

A considerable amount of work has focused on Paleocene siliceous sediments recovered from DSDP Hole 208 in the southwest Pacific (Bukry, 1973; Dumitrica, 1973a, b; DePredo, 1981; DePredo and Ling, 1981; Ling, 1981; Gresham, 1985, 1986). Some of these authors believe that most of the lower Paleocene is represented at this site. A quite different lower Paleocene diatom assemblage from Seymour Island argues against the recovery of lower Paleocene sediments from DSDP Hole 208; they are probably no older than uppermost upper Paleocene. The absence of lower Danian Foraminifera from DSDP Hole 208 (Webb, 1975) supports this interpretation.

Recent reviews of Cretaceous and Paleogene diatom (Fenner, 1985) and silicoflagellate (Perch-Nielsen, 1985) biostratigraphy provide descriptions, illustrations, and range data of important biostratigraphic species. The work of Gleser (1966) on silicoflagellates from the USSR should also be consulted; diatoms occurring with the silicoflagellates are listed in this work. Biostratigraphic zones proposed for Cretaceous and Paleocene siliceous microfossils are related to each other and the Seymour Island sequence in Figure 4.

López de Bertodano and Sobral Formations

In this report, the biostratigraphic ranges of 223 siliceous microfossil species and varieties are documented in 38 samples from the upper Campanian through lower Danian (Table 4). The biostratigraphic ranges of more common diatoms are given in Figure 5. Poor preservation, and in many cases low abundance, limit the amount of detailed biostratigraphic information available from the Seymour Island sequence. True first and last occurrences are probably not reflected in the collected data. Instead,

these datums are concentrated at samples containing diverse, well-preserved assemblages (Fig. 6). For these reasons, division of the sequence into new biostratigraphic zones is not attempted. Species included in Figure 5 are those which would be useful in a zonal scheme.

Silicoflagellate distribution data are presented in Figure 7 and Table 4. The Cretaceous/Tertiary boundary interval is recognized with siliceous microfossils on Seymour Island by the replacement of *Lyramula/Vallacerta*-dominated silicoflagellate assemblages (*Lyramula furcula* Zone of Bukry and Foster, 1974) with a *Corbisema*-dominated assemblage (*Corbisema hastata* Zone of Bukry and Foster, 1974) (Fig. 7). Cretaceous silicoflagellate genera *Lyramula* and *Vallacerta* and diatom genus *Gladius* have their highest occurrence a few meters above the resistant glauconite in samples SI-Z-6 and SI-Z-12 from the cemented interiors of ammonites. A 35-m gap exists between the highest *L. curcula* (Cretaceous) and the lowest *C. hastata* (Paleocene); the marker species for the above silicoflagellate zones. This gap contains an assemblage similar to lower Paleocene floras from the Ural Mountains reported in Gleser (1966), where small, heavily silicified species such as *Corbisema apiculata* f. *minor* are common. Similar small, heavily silicified species, and *C. hastata* f. *miranda,* first appear near the top of the López de Bertodano Formation with the first occurrence of *Corbisema hastata.* Lowermost Paleocene assemblages similar to this have not been reported in the literature of the Deep Sea Drilling Project and are known only from Russian deposits (Gleser, 1966). However, the replacement of Cretaceous genera *Lyramula* and *Vallacerta* by Paleocene assemblages rich in *Corbisema* species is well documented by DSDP data (Bukry, 1981b). The potential for the poorly known silicoflagellate species *Corbisema aspera* to mark the lowest Paleocene on Seymour Island was discussed earlier (Fig. 7). The increase in silicoflagellate diversity through the lower Paleocene, as noted by Gleser (1966), is also reflected in the Seymour Island sequence. Low-diversity silicoflagellate assemblages consisting of two species (including common *Corbisema aspera*) in interval T1 just above the "K/T glauconitic interval" are replaced upsection by assemblages with 7 species in interval T2 and by assemblages with 14 species in interval T3 (Fig. 7).

Diatom distribution data through the Upper Cretaceous and lower Paleocene sequence on Seymour Island (Fig. 6) clearly illustrate the effects of the few samples with well-preserved, diverse assemblages on diatom ranges. Plateaus in Figure 6 at samples SI-19, SI-380, SI-511, SI-408, SI-76, and SI-77 reflect the high number of species that have a lowest or highest appearance in these samples. Most of the species occurring in the Cretaceous continue into the Tertiary, and appear to be unaffected by the Cretaceous/Tertiary boundary event, which devastated several major planktonic microfossil groups (Table 5; Fig. 6).

Cross Valley and La Meseta Formations

A preliminary examination of siliceous microfossils from several samples from the Cross Valley and La Meseta Formations

TABLE 4. STRATIGRAPHIC DISTRIBUTION AND PERCENT ABUNDANCE OF SILICEOUS MICROFOSSILS FROM SEYMOUR ISLAND

DIATOMS

AGE	FORMATION	SAMPLES	Acanthodiscus antarcticus	Acanthodiscus ornatus	Acanthodiscus schmidtii	Acanthodiscus vulcaniformis	Actinoptychus packi	Actinoptychus punctulatus	Actinoptychus taffi	Anaulus acutus?	Anaulus incisus	Anaulus sibericus	Anaulus sp. A.	Anaulus weyprechtii	Arachnoidiscus simbirskianus	Aulacodiscus simbirskianus	Aulacodiscus sp. cf. breviprocessus	Aulacodiscus sp. cf. nigrescens	Aulacodiscus spp. fragments	Aulacodiscus? sp. A.	Auliscus sp. A.	Auliscus sp. cf. priscus	Biddulphia cretacea	Biddulphia fistulosa	Biddulphia sparsipunctata	Chasea bicornis	Chasea proshkinae-lavrenkoae	Cladogramma sp. cf. jordani	Cladogramma sp. cf. simplex	Coscinodiscus morenoensis	Coscinodiscus solidus	Coscinodiscus sparsus	Drepanotheca bivittata	Endictya lunyacsekii	Epithelion spp.	Eunotogramma enorme	Eunotogramma gibbosa	Eunotogramma polymorpha
DANIAN	SOBRAL	SI-77	X	.	.	.	X	.	.	.	.	.	.	.	.	.	.	.	.	.	.	X	6	.	.	.	.	.	.	.	3	.	.	.	.	.	.	.
		SI-76	X	X	X	2	X	.	X	X	.	.	.	X	X	X	X	.	.	X	.	X	X	3	X	.	X	.	X	.	2	.	.	.	.	.	.	X
		SI-75	.	.	.	.	.	.	.	.	.	.	.	.	.	.	.	.	.	.	.	X	X	.	.	.	.	.	X	.	.	.	.	.	.	.	.	.
		SI-530	X	X	3	4	.	.	.	.	X	.	X	.	.	.	3	.	.	.	.	2	.	X	.	2	.	.	X	.	3	.	.	.	.	.	.	.
		SI-123	.	.	.	X	X	.	.	2	.	.	.	.	.	.	.	X	.	.	.	.	3	1	.	.	.	.	.	.	.	.	.	.	.	X	X	.
		SI-408	.	.	X	X	X	X	X	.	X	.	.	X	.	.	.	.	X	.	.	X	5	.	.	.	.	.	X	.	.	.	.	.	.	.	.	X
		SI-122	X	.	.	X	.	.	.	.	.	.	.	.	.	.	.	.	.	.	.	X	X	.	.	X	.	.	X	.	.	.	.	.	.	.	.	.
		SI-72	X	.	X	X	.	.	.	.	.	.	.	.	.	.	.	.	.	.	.	X	X	.	.	.	.	.	.	.	.	.	.	.	.	.	.	.
	LOPEZ DE BERTODANO	SI-511	.	X	7	5	X	.	X	X	.	X	.	X	.	.	X	X	.	X	.	X	X	.	.	5	.	.	.	.	X	5	.	X	X	.	.	X
		SI-71	.	.	X	X	X	.	.	.	.	.	.	X	.	.	.	.	.	.	.	.	X	.	.	.	.	.	X	.	.	.	.	.	.	.	.	.
		SI-70	.	.	X	1	.	.	X	.	.	.	X	.	.	.	.	.	.	.	.	X	2	X	.	X	.	.	2	.	.	.	X	.	.	.	.	.
		SI-351	.	.	3	3	.	.	.	.	.	.	.	.	.	.	.	.	.	.	.	.	X	.	.	X	.	.	3	.	.	.	.	.	.	.	.	.
		SI-118	.	.	.	X	.	.	.	.	.	.	.	.	.	.	.	.	.	.	.	.	X	.	.	X	.	.	2	.	.	.	.	.	.	.	.	.
		SI-507	X	.	8	X	X	.	.	.	.	.	.	.	.	.	X	.	.	.	.	.	3	6	.	3	.	X	.	X	X	.	.	.	.	.	.	.
		SI-506	2	.	1	3	.	.	.	.	.	.	.	.	.	.	X	.	.	.	.	.	.	.	.	.	.	.	.	.	3	.	.	.	.	.	.	.
		SI-504	X	.	2	1	.	.	.	.	.	.	.	.	.	.	.	.	.	.	.	X	.	.	.	2	.	.	2	.	.	.	.	.	.	.	.	.
		SI-13z	.	.	.	7	.	.	.	.	.	.	.	.	.	.	.	.	.	.	.	.	.	.	.	.	.	.	.	.	.	.	.	.	.	.	.	.
		SI-526	.	.	.	.	.	.	.	.	.	.	.	.	.	.	.	.	.	.	.	.	.	.	.	.	.	.	.	.	.	.	.	.	.	.	X	.
MAASTRICHTIAN		SI-12z	.	.	.	X	.	.	.	.	.	.	.	.	.	.	.	.	.	.	.	.	.	.	.	X	.	.	.	.	.	.	.	.	.	.	.	.
		SI-6z	.	.	.	X	.	.	.	.	.	.	.	.	X	.	.	.	.	X	.	X	X	.	.	X	.	.	.	.	X	.	.	.	.	.	.	.
		SI-373	.	X	.	.	.	.	.	.	.	.	.	.	X	X	.	.	.	.	.	.	.	.	.	1	.	.	.	.	X	.	.	.	.	.	.	.
		SI-409	.	.	.	X	.	.	.	.	.	.	.	.	.	.	.	.	2	.	.	.	.	.	.	.	.	.	2	.	.	.	2	.	X	X	.
		SI-117	.	.	2	1	.	.	.	.	.	.	.	.	.	.	.	.	.	.	.	.	.	.	.	.	.	.	.	.	.	.	2	.	.	.	.	.
		SI-342	X	5	.	X	.	.	X	.	.	.	.	.	.	.	.	.	.	.	.	.	.	.	.	X	.	.	X	.	.	.	X	.	.	.	.	.
		SI-340	.	.	X	.	.	.	.	.	.	.	.	.	.	.	.	.	X	.	.	.	.	.	.	X	.	.	5	.	.	.	.	.	.	.	.	.
		SI-400	.	.	.	.	.	.	X	.	.	.	.	.	.	.	4	.	.	.	.	.	.	.	.	.	.	.	.	.	.	.	X	.	.	.	.	.
		SI-116	2	X	.	X	.	X	.	X	.	.	.	.	.	.	X	2	.	.	.	.	.	.	.	.	.	.	X	.	.	.	X	.	.	.	.	.
		SI-398	X	.	.	X	.	.	.	.	.	.	.	.	.	.	.	.	.	.	.	.	.	.	.	.	.	.	X	.	.	.	.	.	.	.	.	.
		SI-362	5	.	.	X	.	.	.	.	.	.	.	.	.	.	.	.	.	.	.	.	.	.	.	.	.	.	.	.	2	.	.	.	.	.	.	.
		SI-N84	X	.	X	X	.	.	.	.	.	.	.	.	.	.	.	.	.	.	.	.	.	.	.	.	.	.	.	.	.	.	X	.	.	.	.	.
		SI-378	.	.	4	.	.	.	.	.	.	.	.	.	.	.	.	.	.	.	.	.	.	.	.	.	.	.	X	.	.	.	.	.	.	.	.	.
		SI-380	.	.	X	X	X	.	X	.	.	.	.	.	.	.	X	.	X	.	.	.	.	.	.	X	.	.	X	.	.	.	X	.	.	.	.	.
		SI-K84	3	4	.	X	.	.	.	.	.	.	.	.	.	.	.	.	X	.	.	.	.	.	.	X	.	.	2	.	.	.	.	.	.	.	.	.
		SI-173	.	.	.	.	.	X	.	.	.	.	.	.	.	.	.	.	.	.	.	.	.	.	.	.	.	.	X	X	.	.	.	.	.	.	.	.
		SI-102	.	.	X	X	.	.	.	.	.	.	.	.	.	.	.	.	.	.	.	.	.	.	.	.	.	.	.	.	.	.	.	.	.	.	.	.
		SI-97	X	.	.	.	.	.	.	.	.	.	.	.	.	.	.	.	.	.	.	.	.	.	.	X	.	.	.	.	.	.	.	.	.	.	.	.
CAMP.		SI-19	X	.	.	X	X	.	X	.	X	.	4	.	.	.	.	.	X	.	.	.	.	.	.	X	X	.	.	X	3	.	.	.	.	.	X	.	X
		SI-14	.	.	.	.	X	.	X	.	X	.	.	.	.	.	.	.	X	.	.	.	.	.	.	X	.	.	.	.	X	.	.	.	.	.	.	.

TABLE 4. STRATIGRAPHIC DISTRIBUTION AND PERCENT ABUNDANCE OF SILICEOUS MICROFOSSILS FROM SEYMOUR ISLAND (continued)

DIATOMS

Eunotogramma producta v. recta	Eunotogramma weissei	Gladius antarcticus	Gladius antarcticus f. alta	Gladius pacificus	Gladius speciosus group	Gn. et sp. indet. Jousé, 1951	Goniothecium odontella	Helminthopsis wornardti	Hemiaulus ambiguus	Hemiaulus andrewsi	Hemiaulus assymetricus	Hemiaulus bipons	Hemiaulus curvatulus	Hemiaulus danicus	Hemiaulus echinulatus	Hemiaulus elegans	Hemiaulus gleseri	Hemiaulus hostilis group	Hemiaulus huberi	Hemiaulus incisus	Hemiaulus includens	Hemiaulus kittonii	Hemiaulus kondai	Hemiaulus polymorphus v. frigida	Hemiaulus polymorphus v. morsianus	Hemiaulus praelegans	Hemiaulus proshkinae-lavrenkoae	Hemiaulus rossicus	Hemiaulus seymouriensis	Hemiaulus sibericus	Hemiaulus simplex ?	Hemiaulus sp. A.	Hemiaulus sp. B.	Hemiaulus sp. C.	Hemiaulus sp. D.	Hemiaulus sp. E.	Hemiaulus sp. F.	Hemiaulus sp. G.	Hemiaulus sp. H.	SAMPLES
.	.	.	.	.	.	.	.	.	.	.	.	X	.	.	.	.	.	.	.	.	.	.	.	.	.	.	.	.	2	.	.	.	.	.	.	.	.	.	.	SI-77
.	.	.	.	.	.	.	.	.	.	X	.	X	.	X	.	X	.	.	X	.	.	X	X	.	.	X	2	.	X	.	X	.	.	X	.	.	X	X	X	SI-76
.	.	.	.	.	.	.	.	.	.	.	.	.	.	.	.	.	.	.	.	.	.	.	.	.	.	.	.	.	.	.	.	.	.	.	.	.	.	.	.	SI-75
.	.	.	.	.	.	.	.	.	.	.	.	.	.	.	.	.	.	X	.	.	.	.	X	.	.	.	X	.	.	.	.	.	.	.	.	.	.	X	.	SI-530
.	.	.	.	.	.	3	.	.	.	.	.	.	.	.	.	.	.	.	.	.	.	.	.	.	.	.	1	.	.	.	X	.	.	.	.	X	.	.	.	SI-123
.	.	.	.	X	X	.	.	X	.	1	X	X	.	X	.	X	.	X	.	X	.	X	X	X	X	.	X	.	.	.	.	.	.	X	.	.	.	X	X	SI-408
.	.	.	.	.	.	.	.	.	.	.	.	X	.	.	.	.	.	X	.	.	.	.	.	.	.	.	X	.	.	.	.	.	.	X	.	.	.	X	.	SI-122
.	.	.	.	.	.	.	.	.	.	.	.	.	.	.	.	.	.	X	.	.	.	.	.	.	.	.	X	.	.	.	.	.	.	.	.	.	.	.	.	SI-72
.	X	.	.	.	.	.	.	.	X	.	.	.	X	.	X	.	.	2	2	.	X	X	3	X	.	X	.	X	.	X	.	.	.	.	X	.	.	.	X	SI-511
.	X	.	.	.	.	.	.	.	.	.	.	.	.	.	.	.	.	.	.	.	X	.	.	.	.	.	X	.	.	.	.	.	.	.	.	.	.	.	.	SI-71
.	X	.	.	.	.	2	.	.	.	.	.	X	X	.	.	.	.	.	.	.	X	.	.	.	.	.	X	.	.	.	.	.	.	.	.	.	.	.	.	SI-70
.	.	.	.	.	.	.	.	.	.	.	.	.	.	.	.	.	.	.	.	.	.	.	.	.	.	.	.	.	.	.	.	.	.	.	.	.	.	.	.	SI-351
.	.	.	.	.	.	.	.	.	.	.	.	.	X	X	.	.	.	.	.	.	X	.	.	.	.	.	X	.	.	.	.	.	.	.	.	.	.	X	.	SI-118
.	.	.	.	.	.	.	.	.	.	.	.	.	.	.	.	.	.	.	.	.	.	X	.	.	.	.	X	.	.	X	.	.	.	.	.	.	.	.	.	SI-507
.	X	.	.	.	.	.	.	.	X	.	.	.	.	.	.	.	.	.	.	.	.	.	.	.	.	.	X	.	.	.	.	.	.	.	.	.	.	X	.	SI-506
.	.	.	.	.	.	4	.	.	.	.	.	.	.	.	.	.	.	.	.	.	.	.	.	.	.	.	.	.	.	.	.	.	.	.	.	.	.	.	.	SI-13z
.	.	.	.	.	.	.	.	.	.	.	.	.	.	.	.	.	.	.	.	.	.	.	.	.	.	.	X	.	.	.	.	.	.	.	.	.	.	.	.	SI-526
.	.	X	.	X	.	X	.	.	.	.	.	.	.	.	.	.	.	.	.	.	.	.	.	.	.	.	.	.	.	.	.	.	.	.	.	.	.	.	.	SI-12z
.	.	X	?	X	.	.	.	.	X	.	.	.	.	.	.	.	.	.	.	.	.	X	.	.	.	.	X	.	.	.	.	.	.	.	.	.	.	.	.	SI-6z
.	4	X	5	.	.	.	.	.	.	2	.	.	.	.	.	.	.	.	.	.	.	.	2	.	.	.	.	.	.	.	.	2	.	.	.	.	.	.	.	SI-373
.	X	.	X	.	.	.	.	.	.	.	.	.	.	.	.	.	.	.	.	.	.	.	X	.	.	.	.	.	.	.	.	.	.	.	.	.	.	.	.	SI-409
.	.	.	.	.	.	.	.	2	.	.	.	.	.	.	.	.	.	.	.	.	.	.	.	.	.	.	.	.	.	.	.	.	.	.	.	.	.	.	.	SI-117
.	.	.	.	X	.	.	.	.	.	.	.	.	.	.	.	.	.	.	.	.	.	.	.	.	.	.	.	3	.	.	.	.	.	.	.	.	.	.	.	SI-342
.	.	.	.	X	.	.	.	.	.	.	.	.	.	.	.	.	.	.	.	.	.	.	.	.	.	.	.	8	.	.	.	.	.	.	.	.	.	.	.	SI-340
.	X	.	.	.	.	.	.	.	.	.	.	.	.	.	.	.	.	.	.	.	.	.	.	.	.	.	.	.	.	.	.	.	.	.	.	.	.	.	.	SI-400
.	.	.	.	.	.	.	.	.	.	.	.	.	.	.	.	.	.	.	.	.	.	.	.	.	.	.	.	.	.	.	.	.	.	.	.	.	.	.	.	SI-116
.	.	.	.	.	.	.	.	.	.	.	.	.	.	.	.	.	.	.	.	.	.	.	.	.	.	.	.	.	.	.	.	.	.	.	.	.	.	.	.	SI-398
.	.	4	.	.	.	.	.	.	.	.	.	.	.	.	.	.	.	.	.	.	.	.	.	.	.	.	.	.	.	.	.	.	.	.	.	.	.	.	.	SI-362
.	.	X	.	X	.	.	.	.	.	.	.	.	.	.	.	.	.	.	.	.	.	.	X	.	.	.	.	.	.	.	.	.	.	.	.	.	.	.	.	SI-N84
.	.	.	.	.	.	.	.	.	.	.	.	.	.	.	.	.	.	.	.	.	.	.	X	.	.	.	.	.	.	.	.	.	.	.	.	.	.	.	.	SI-378
.	.	3	2	X	.	.	X	.	.	X	.	.	X	.	X	.	X	X	.	.	X	.	X	.	.	.	.	.	7	X	.	.	.	.	.	X	.	.	X	SI-380
.	.	.	.	.	.	.	.	.	.	.	.	.	.	.	.	.	.	.	.	.	.	.	.	X	X	.	X	.	.	.	.	.	.	.	.	.	.	.	.	SI-K84
.	.	.	.	.	.	.	.	.	.	.	.	.	.	.	.	.	.	.	.	.	.	.	.	.	.	.	.	.	.	.	.	.	.	.	.	.	.	.	.	SI-173
.	.	.	.	.	.	.	.	.	.	.	.	.	.	.	.	.	.	.	.	.	.	.	.	.	.	.	X	.	.	.	.	.	.	.	.	.	.	.	.	SI-102
.	.	.	.	.	.	.	.	X	.	.	.	.	.	.	.	.	.	.	.	.	.	.	.	.	.	.	.	.	.	.	.	.	.	.	.	.	.	.	.	SI-97
2	X	.	.	X	X	.	X	X	.	.	.	3	.	.	X	X	.	X	X	.	X	X	X	X	X	.	4	.	X	.	.	6	X	.	.	.	.	.	.	SI-19
.	.	.	.	.	.	.	.	X	.	.	.	X	.	.	.	.	.	.	.	.	X	.	.	.	.	.	X	.	.	.	.	.	X	.	.	.	.	.	.	SI-14

TABLE 4. STRATIGRAPHIC DISTRIBUTION AND PERCENT ABUNDANCE OF SILICEOUS MICROFOSSILS FROM SEYMOUR ISLAND (continued)

DIATOMS

AGE	FORMATION	SAMPLES	Hemiaulus sp. I.	Hemiaulus sp. J.	Hemiaulus sp. K.	Hemiaulus sp. L.	Hemiaulus sp. M.	Hemiaulus sp. cf. tumidicornis	Hemiaulus sporialis	Horodiscus anastomosans	Huttonia antiqua	Hyalodiscus russicus	Incisoria inordinata	Incisoria lanceolata	Incisoria paleoceanica v. fuscina	Incisoria punctata	Incisoria sp. A.	Kentrodiscus aculeatus	Kentrodiscus fossilis	Kittonia sp.	Melosira ? patera	Melosira vetustissima	Odontotropis cristata	Odontotropis sp. Hajós	Omphalotheca sp. cf. jutlandica	Paralia sulcata group	Peponia barbadensis	Podosira simpla	Pseudopodosira westii	Pseudopyxilla aculeata	Pseudopyxilla americana	Pseudopyxilla dubia	Pseudopyxilla hungarica	Pseudopyxilla rossica	Pseudopyxilla sp. Strelnikova	Pseudostictodiscus picus	Pterotheca aculeifera
DANIAN	SOBRAL	SI-77	·	·	·	·	·	·	·	·	·	X	·	·	X	·	·	2	X	·	X	·	·	·	·	X	·	·	·	·	·	·	·	X	X	·	·
		SI-76	·	·	X	X	·	X	X	X	·	X	·	·	·	·	·	X	X	·	X	X	·	·	·	X	·	·	·	·	·	·	X	X	X	X	X
		SI-75	·	·	·	·	·	·	·	·	·	·	·	·	·	·	·	·	·	·	·	·	·	·	¹	·	X	X	·	·	·	·	·	·	·	·	·
		SI-530	·	·	·	·	X	2	·	·	·	·	·	·	·	·	·	·	·	·	X	·	·	·	·	X	·	·	·	X	·	X	·	·	·	·	·
		SI-123	·	·	·	·	·	·	·	·	·	·	·	·	·	·	·	·	·	·	X	·	X	·	·	X	·	·	·	·	·	·	·	·	·	·	·
		SI-408	·	·	X	·	X	·	X	·	X	·	·	X	X	·	X	X	X	X	·	X	X	X	·	X	·	·	·	·	·	·	X	X	·	X	2
		SI-122	·	·	·	·	X	·	X	·	·	·	·	·	·	·	·	·	·	·	X	X	·	X	·	X	·	·	·	·	·	·	X	·	·	·	X
		SI-72	·	·	·	·	X	·	·	·	·	·	·	·	·	·	·	·	·	·	X	·	·	·	·	X	·	·	·	·	·	·	X	X	·	·	·
	LOPEZ DE BERTODANO	SI-511	X	X	X	·	·	·	X	·	·	·	·	·	X	X	·	2	X	·	X	·	·	·	·	X	·	·	·	X	X	·	X	X	X	X	X
		SI-71	·	·	·	·	·	·	X	·	·	·	·	·	·	·	·	·	·	·	·	·	·	·	·	·	·	·	·	·	·	·	·	·	·	X	·
		SI-70	·	·	·	·	X	·	X	·	·	·	·	·	·	·	·	·	X	X	·	X	·	·	·	X	·	X	·	·	·	·	X	·	·	·	·
		SI-351	·	·	·	·	·	·	·	·	·	·	·	·	·	·	·	·	·	·	·	·	·	·	·	·	·	·	·	·	·	·	·	·	·	·	·
		SI-118	·	·	·	·	X	·	·	·	·	·	·	·	·	·	·	·	X	·	·	·	·	·	·	·	·	·	·	·	·	·	X	·	·	·	·
		SI-507	X	·	·	·	·	X	·	·	·	·	·	·	·	·	·	X	·	·	·	·	·	·	·	X	·	·	·	X	·	·	X	X	·	·	·
MAASTRICHTIAN		SI-506	·	·	·	·	·	·	·	·	·	·	·	·	·	·	·	·	·	·	·	·	·	·	·	·	·	·	·	·	·	·	·	·	·	·	·
		SI-504	·	·	·	·	·	·	·	·	·	·	·	·	·	·	·	·	·	X	·	X	·	·	·	·	·	·	·	2	·	·	X	·	·	·	·
		SI-13z	·	·	·	·	·	·	·	·	·	·	·	·	·	·	·	·	·	7	·	2	·	·	·	·	·	·	·	·	·	1	·	·	·	·	·
		SI-526	·	·	·	·	·	·	·	·	·	·	·	·	·	·	·	·	·	X	·	·	·	·	·	·	·	·	·	·	·	·	·	·	·	·	·
		SI-12z	·	·	·	·	·	·	·	·	·	·	·	·	·	·	·	·	·	·	·	·	·	·	·	·	·	·	·	·	·	·	·	·	·	·	·
		SI-6z	·	·	·	·	·	·	·	·	·	·	·	·	·	·	·	·	·	·	·	·	·	·	·	·	·	·	·	·	·	·	·	·	·	·	·
		SI-373	·	·	·	2	·	·	·	·	·	·	·	X	·	·	·	·	·	·	·	·	·	·	·	·	·	·	·	·	·	·	·	·	·	·	·
		SI-409	·	·	·	·	·	X	·	·	·	·	·	·	·	·	·	·	X	·	·	·	·	·	·	·	·	·	·	X	·	·	·	·	·	·	·
		SI-117	·	·	·	·	·	·	·	·	·	·	·	·	·	·	·	·	·	2	·	·	·	·	·	·	·	·	·	·	·	2	·	·	·	·	·
		SI-342	·	·	·	·	·	·	·	·	·	·	·	·	·	·	·	·	·	·	·	·	·	·	·	·	·	·	·	·	·	·	·	·	·	·	·
		SI-340	·	X	·	·	·	·	·	·	·	·	·	·	·	·	·	·	·	·	·	·	·	·	·	·	·	·	·	·	·	·	·	X	·	·	·
		SI-400	·	·	·	·	4	·	·	·	·	·	·	·	·	·	·	·	·	·	·	·	·	·	·	·	·	·	·	·	·	·	·	X	·	·	·
		SI-116	·	·	·	·	·	·	·	·	·	·	·	·	·	·	·	·	·	·	·	·	·	·	·	·	·	·	·	·	·	·	·	·	·	·	·
		SI-398	·	·	·	·	·	·	·	·	·	·	·	·	·	·	·	·	·	·	·	·	·	·	·	·	·	·	·	·	·	·	·	·	·	·	·
		SI-362	·	·	·	·	·	·	·	·	·	·	·	2	·	·	·	·	·	2	·	·	·	·	·	·	·	·	·	·	·	·	·	·	·	·	·
		SI-N84	·	·	·	·	·	·	·	·	·	·	·	·	·	·	·	·	·	X	·	·	·	·	·	·	·	·	·	·	·	·	·	·	·	·	·
		SI-378	·	·	·	·	·	X	·	·	·	·	·	·	·	·	·	·	·	7	·	·	·	·	·	·	·	·	·	·	·	·	·	·	·	·	·
		SI-380	·	X	·	X	·	X	·	X	X	·	X	·	X	X	·	·	·	X	·	·	·	·	·	X	·	·	·	X	·	·	X	·	X	·	·
		SI-K84	·	·	·	·	·	·	·	·	·	·	·	·	·	·	·	X	·	·	·	·	·	·	·	·	·	·	·	·	·	·	·	·	·	·	·
		SI-173	·	·	·	·	·	·	·	·	·	·	·	·	·	·	·	·	·	·	·	·	·	·	·	·	·	·	·	·	·	·	·	·	X	·	·
		SI-102	·	·	·	·	·	·	·	·	·	·	·	·	·	·	·	·	·	·	·	·	·	·	·	·	·	·	·	·	·	·	·	·	·	·	·
		SI-97	·	·	·	·	·	X	·	·	·	·	·	·	·	·	·	·	·	·	·	·	·	·	·	·	·	·	·	·	·	·	·	X	·	X	·
CAMP.		SI-19	·	·	·	·	·	·	·	·	·	·	4	3	·	X	2	X	·	·	·	X	·	·	·	·	·	·	·	X	X	X	·	·	X	X	X
		SI-14	·	·	·	·	·	·	·	·	·	·	·	·	X	X	·	·	·	·	·	·	·	·	·	X	·	·	·	·	·	·	·	·	·	·	·

TABLE 4. STRATIGRAPHIC DISTRIBUTION AND PERCENT ABUNDANCE OF SILICEOUS MICROFOSSILS FROM SEYMOUR ISLAND (continued)

DIATOMS

P. carinifera	P. carinifera v. curvirostris	P. carinifera v. tenuis	P. clavata	P. costata	P. cretacea	P. crucifera	P. danica	P. evermanni	P. infundibulum	P. kittoniana	P. major	P. minor	P. pokrovskajae	P. pokrovskajae var.	P. sp. A.	P. sp. B.	P. spada	P. subulata	P. trojana	P. uralica	Pyrgodiscus sp.	Pyrgodiscus? triangulatus	Pyxidicula minuta	Rhizosolenia cretacea	Rhizosolenia sp. A.	Rhizosolenia sp. B.	Rutilaria? sp.	Sceptroneis grunowii	Sceptroneis praecaducea	Sceptroneis sp. A.	Sceptroneis? sp. B.	Skeletonema penicillis	Skeletonema polychaetum	Skeletonema sp. A.	Skeletonema subantarctica	Sphynctolethus monstrosus	Stellarima steinyi	Stephanopyxis broschii	Stephanopyxis delectabilis	SAMPLES
1	2											2	3	6		X	X					2	2	X										X						SI-77
X			X							X	X	X		5	2	X	X	X	X		X		X	2	X						X				X			A		X	SI-76
	X																						X																	SI-75
5									X			X	7	3							2	2	4	X					X	X				X						X	SI-530
1						X		2	X	X		8	5								X	3	3							X										X	SI-123
X			X				X	X	X	X		8	X	X	X	X	X		X		8	5	2	X		X	2	X				2	X	X	X	X				SI-408
X	X		X				X			X			X				X	X	X		X	X	X				X				X									SI-122
											X				X			X				X		X																SI-72
2				X								2	5	X	X	X	X	X		X	X	X	X	4	4	X			X				3						2	SI-511
										X			X														X				X									SI-71
2	X								X				X		4	3					X			X	2							X							X	SI-70
6							2					1		X							6																			SI-351
2		X				X		X	X		2	2			X					X		X	4				X				X								X	SI-118
3						3				5	X									2	X	X	X		X															SI-507
5										6	3																													SI-506
X					X				X		8	X	X							4	X																		2	SI-504
			X						1											X																			SI-13z
									X	X																													SI-526
									X																														SI-12z
	X			X				X	X							X	X		X		X	X		X							X								SI-6z
																																							SI-373
X									X	X										4		X																	SI-409
2																				2	X				X														SI-117
																				2	X				X														SI-342
																					X																		SI-340
6								8	4											X	2		X																SI-400
							X																		X														SI-116
																				X																			SI-398
3																				X	2																		SI-362
																				X	X																		SI-N84
																				X																			SI-378
			X								X		X		X		X		X		X	X	X		X		X							X	2	X	SI-380
X			X								2	3																											SI-K84
X			X							X											X																	SI-173
									X	X	X																												SI-102
									X												X																		SI-97
	X			2	X				X			X		3	8	X			X	X	X	X		X							X							X	SI-19
		X				X			X					X	X							X	X																SI-14

TABLE 4. STRATIGRAPHIC DISTRIBUTION AND PERCENT ABUNDANCE OF SILICEOUS MICROFOSSILS FROM SEYMOUR ISLAND (continued)

DIATOMS

AGE	FORMATION	SAMPLES	Stephanopyxis edita	Stephanopyxis hannai	Stephanopyxis lavrenkoi	Stephanopyxis marginata	Stephanopyxis maxima	Stephanopyxis simonseni	Stephanopyxis sp. A	Stephanopyxis sp. B	Stephanopyxis sp. C	Stephanopyxis sp. D	Stephanopyxis sp. cf. grunowii	Stephanopyxis superba	Stephanopyxis turris	Stephanopyxis turris v. cylindrus	Stephanopyxis uralensis	Stephanopyxis weyprechtii	Thalassiosiropsis wittiana	Triceratium cellulosum	Triceratium edgardi	Triceratium flos	Triceratium flos var. intermedia	Triceratium indefinitum	Triceratium nobile	Triceratium sentum	Triceratium simplicissimum	Triceratium solenoceros	Triceratium sp. A	Trinacria acutangulum	Trinacria aries	Trinacria excavata	Trinacria grevillei	Trinacria heibergii v. rostratum	Trinacria insipiens	Trinacria pileolus	Trinacria sp. A
DANIAN	SOBRAL	SI-77	.	7	X	.	.	4	.	8	.	.	.	.	8	3	.	X	.	.	.	.	.	.	.	.	.	2	X	X	.	3	.	.	.	.	X	.
DANIAN	SOBRAL	SI-76	X	4	.	.	X	4	.	X	.	.	.	4	5	X	4	X	.	.	.	X	X	X	2	.	X	X	5	5	2	X	X	X	X	X	.
DANIAN	SOBRAL	SI-75	.	.	.	.	.	X	.	.	.	.	.	.	X	.	.	.	.	.	.	.	.	.	.	.	.	X	.	.	.	.	.	.	.	X	.
DANIAN	SOBRAL	SI-530	X	2	.	.	X	X	.	2	.	.	.	X	2	7	.	X	.	4	X	.	.	.	2	.	.	X	.	5	X	6	.	.	.	X	.
DANIAN	SOBRAL	SI-123	.	.	.	.	3	X	.	2	.	.	.	2	.	8	.	.	.	.	X	.	.	.	.	.	.	.	.	X	.	.	.	.	.	X	.
DANIAN	SOBRAL	SI-408	.	X	X	.	X	X	.	.	.	.	.	2	3	.	2	X	X	.	.	.	.	.	.	.	X	X	X	.	X	.	.	X	X	.	.
DANIAN	SOBRAL	SI-122	.	.	.	.	.	X	.	.	.	.	.	X	X	.	X	.	X	.	.	.	.	.	X	.	.	X	.	X	X	X	.	X	.	.	.
DANIAN	SOBRAL	SI-72	.	.	X	.	.	X	.	.	.	.	.	X	X	X	.	X	.	.	.	.	.	X	.	.	X	X	X	X	.	X	.	.	.	.	.
DANIAN	SOBRAL	SI-511	X	2	X	.	.	2	.	2	.	.	X	2	1	X	3	X	X	X	.	.	.	X	X	X	X	.	X	X	X	X	2	.	.	X	.
DANIAN	LOPEZ DE BERTODANO	SI-71	X	.	.	.	.	.	.	.	.	.	.	.	X	.	.	.	X	.	.	.	.	.	.	.	.	.	.	X	.	.	.	.	.	X	.
DANIAN	LOPEZ DE BERTODANO	SI-70	1	4	X	.	X	1	.	5	X	.	.	4	X	.	5	.	.	X	.	.	.	X	.	.	.	X	.	.	.	X	.	X	.	X	.
DANIAN	LOPEZ DE BERTODANO	SI-351	2	.	.	.	.	2	.	6	5	.	.	X	6	.	2	.	.	.	.	.	.	2	.	.	.	.	.	5	1	.	.	.	.	.	.
DANIAN	LOPEZ DE BERTODANO	SI-118	3	9	2	.	.	2	X	1	.	.	.	3	3	.	3	.	.	.	.	X	.	.	.	.	.	X	.	.	.	X	.	.	.	.	.
DANIAN	LOPEZ DE BERTODANO	SI-507	1	3	2	.	2	2	X	4	X	6	.	2	6	X	4	.	X	.	X	.	.	.	.	.	.	X	.	.	3	X	X	.	X	.	3
DANIAN	LOPEZ DE BERTODANO	SI-506	2	1	.	.	3	3	.	1	.	1	.	.	X	.	.	.	.	.	.	.	.	.	.	.	.	.	.	3	.	.	.	.	.	.	.
DANIAN	LOPEZ DE BERTODANO	SI-504	8	1	.	X	X	8	X	1	.	.	X	7	4	.	X	.	X	.	.	.	.	X	.	.	.	X	X	.	.	.	.	.	.	X	.
DANIAN	LOPEZ DE BERTODANO	SI-13z	X	3	8	.	.	2	.	2	.	.	.	.	1	.	.	.	.	.	.	.	.	.	.	.	.	.	.	.	.	.	.	.	.	.	.
MAASTRICHTIAN	LOPEZ DE BERTODANO	SI-526	.	X	.	.	.	6	2	.	.	.	8	2	.	.	.	.	.	.	.	.	.	.	.	3	.	.	.	.	.	.	.	.	.	X	.
MAASTRICHTIAN	LOPEZ DE BERTODANO	SI-12z	.	X	.	.	.	.	.	X	X	X	.	.	.	.	.	.	.	.	.	.	.	.	X	.	.	.	.	X	.	.	.	.	.	X	.
MAASTRICHTIAN	LOPEZ DE BERTODANO	SI-6z	X	.	X	X	.	X	.	.	.	.	.	X	.	.	.	.	.	.	.	.	.	.	.	.	.	.	X	.	X	.	.	.	.	X	.
MAASTRICHTIAN	LOPEZ DE BERTODANO	SI-373	.	.	.	3	X	8	.	.	.	X	2	.	.	.	.	.	.	2	.	.	X	.	2	.	.	X	X	.	.	.	.	.	3	.	.
MAASTRICHTIAN	LOPEZ DE BERTODANO	SI-409	5	6	X	.	3	5	3	2	.	3	1	2	3	.	.	.	.	.	.	.	2	.	3	.	X	.	3	.	X	.	2	.	.	.	.
MAASTRICHTIAN	LOPEZ DE BERTODANO	SI-117	.	.	4	.	4	.	.	.	.	4	6	.	.	.	.	.	.	.	.	.	.	.	.	.	.	8	.	.	.	.	.	.	.	.	.
MAASTRICHTIAN	LOPEZ DE BERTODANO	SI-342	.	6	6	3	.	5	5	X	.	.	6	5	6	.	3	.	.	.	5	.	.	4	X	X	X	.	X	7	5	.	.	.	3	.	.
MAASTRICHTIAN	LOPEZ DE BERTODANO	SI-340	.	1	5	2	.	2	2	X	.	.	3	6	1	.	X	.	.	.	.	.	3	.	3	.	.	3	3	.	.	.	.	.	.	.	.
MAASTRICHTIAN	LOPEZ DE BERTODANO	SI-400	.	.	8	.	.	6	.	.	.	.	2	3	6	.	.	.	.	.	.	.	.	8	.	.	.	.	.	4	2	.	.	.	.	6	.
MAASTRICHTIAN	LOPEZ DE BERTODANO	SI-116	.	2	3	2	.	2	X	.	2	.	2	2	4	.	6	.	.	.	.	.	.	X	.	.	.	2	.	X	.	X	2	X	.	.	.
MAASTRICHTIAN	LOPEZ DE BERTODANO	SI-398	.	8	8	.	.	X	X	X	.	.	3	2	.	2	.	.	.	.	.	.	X	.	X	.	.	X	X	.	.	X	.	.	.	.	.
MAASTRICHTIAN	LOPEZ DE BERTODANO	SI-362	.	1	9	.	.	X	4	.	X	.	1	6	.	2	.	.	.	.	.	.	X	.	.	.	.	X	.	.	.	.	.	.	.	.	.
MAASTRICHTIAN	LOPEZ DE BERTODANO	SI-N84	.	X	X	.	.	X	X	.	.	.	X	X	.	.	.	X	.	.	.	.	.	.	.	.	.	.	.	.	.	.	.	.	.	.	.
MAASTRICHTIAN	LOPEZ DE BERTODANO	SI-378	.	X	X	.	.	X	.	.	.	.	3	X	.	X	.	.	.	.	.	.	.	.	.	.	.	.	.	.	.	.	.	.	.	.	.
MAASTRICHTIAN	LOPEZ DE BERTODANO	SI-380	.	X	X	X	X	X	X	.	.	X	.	2	.	.	X	2	.	.	.	.	.	X	X	X	.	.	.	X	5	X	.	.	X	4	.
MAASTRICHTIAN	LOPEZ DE BERTODANO	SI-K84	.	X	.	.	X	X	.	.	.	.	8	.	X	.	.	.	.	.	.	.	.	1	X	.	.	.	.	X	.	.	.	.	.	4	.
MAASTRICHTIAN	LOPEZ DE BERTODANO	SI-173	.	.	.	.	X	.	X	.	.	.	X	X	.	X	.	.	.	.	.	.	.	.	X	.	.	.	.	.	X	.	.	.	.	.	.
MAASTRICHTIAN	LOPEZ DE BERTODANO	SI-102	.	.	.	.	X	.	.	.	.	.	X	X	.	.	.	.	.	.	.	.	.	.	.	.	.	.	.	X	.	.	.	.	.	.	.
MAASTRICHTIAN	LOPEZ DE BERTODANO	SI-97	.	.	.	.	X	.	.	.	.	.	X	.	X	.	.	.	.	.	.	.	.	.	.	.	.	.	.	X	.	.	.	.	.	.	.
MAASTRICHTIAN	LOPEZ DE BERTODANO	SI-19	.	.	.	.	X	.	.	.	.	.	3	.	X	X	6	.	.	.	.	.	2	X	.	.	.	.	.	X	X	X	X	.	.	X	2
CAMP.	LOPEZ DE BERTODANO	SI-14	.	.	.	.	X	.	.	.	.	.	X	.	.	X	.	.	.	X	.	.	.	X	.	.	.	.	.	X	.	.	.	.	.	.	.

TABLE 4. STRATIGRAPHIC DISTRIBUTION AND PERCENT ABUNDANCE OF SILICEOUS MICROFOSSILS FROM SEYMOUR ISLAND (continued)

DIATOMS						OTHER						SILICOFLAGELLATES																												SAMPLES
Trinacria sp. B.	Trinacria sp. C.	Trinacria swastika	Trinacria tristictia	Wittia macellarii	Xanthiopyxis granti	Archaeomonadaceae	archaeomonad sp. A.	Actiniscus pentasterias	Carduifolia onoporides	Ammodochium danicum ?	radiolarians	Corbisema apiculata	Corbisema apiculata f. minor	Corbisema aspera	Corbisema geometrica	Corbisema hastata	Corbisema hastata f. miranda	Corbisema hastata v. globulata	Corbisema hastata? var.	Corbisema inermis var.	Corbisema inermis var.	Corbisema lateradiata	Corbisema sp. cf. archangelskiana	Corbisema sp. variant	Cornua sp. A.	Cornua trifurcata	Dictyocha sp. cf. navicula f. minor	Dictyocha staurodon f. minor	Genus et species indet. A.	Lyramula deflandrei	Lyramula furcula	Lyramula minor	Lyramula simplex	Lyramula sp. cf. furcula	Silicoflagellate genus A.	Silicoflagellate genus B. sp.1	Silicoflagellate genus B. sp.2	Vallacerta siderea	Vallacerta tumidula	
.	.	.	.	.	2	.	X	.	.	.	X	X	.	.	.	.	.	.	.	.	.	.	.	.	.	.	.	.	.	.	.	.	.	.	.	.	.	.	.	SI-77
.	.	.	.	.	X	X	X	.	.	X	.	X	.	X	.	X	.	X	.	.	.	.	X	.	.	.	.	.	.	.	.	.	.	.	.	.	.	.	.	SI-76
.	.	.	.	.	X	.	.	.	.	.	.	.	.	.	.	.	.	.	.	.	.	.	.	.	.	.	.	.	.	.	.	.	.	.	.	.	.	.	.	SI-75
X	.	.	X	X	.	.	.	.	2	.	X	2	X	X	.	X	.	X	.	X	.	.	.	.	.	.	.	.	.	.	.	.	.	.	.	.	.	.	.	SI-530
.	.	.	.	X	.	X	.	.	.	X	X	.	.	.	.	.	.	.	.	.	.	.	.	.	.	.	.	.	.	.	.	.	.	.	.	.	.	.	.	SI-123
X	.	.	.	X	.	X	.	.	.	X	X	X	2	X	.	X	X	X	X	X	.	X	.	X	X	.	X	X	.	.	.	.	.	.	.	.	.	.	.	SI-408
.	.	.	.	.	X	.	.	.	.	X	X	.	.	.	.	.	.	.	.	X	.	X	.	.	.	.	.	.	.	.	.	.	.	.	.	.	.	.	.	SI-122
X	.	.	X	X	X	.	.	.	2	.	X	2	X	.	.	X	X	X	.	X	X	.	X	.	.	.	.	.	.	.	.	.	.	.	.	.	.	.	.	SI-72
.	.	.	.	.	.	.	.	.	2	.	X	2	X	X	X	.	.	.	.	.	.	.	.	.	.	.	.	.	.	.	.	.	.	.	.	.	.	.	.	SI-511
.	.	.	X	X	.	.	.	.	.	X	.	X	.	X	.	.	.	.	.	.	.	.	.	.	.	.	.	.	.	.	.	.	.	.	.	.	.	.	.	SI-71
.	.	.	X	X	.	.	.	.	.	.	X	.	.	.	.	.	.	.	.	X	.	.	X	.	.	.	.	.	.	.	.	.	.	.	.	.	.	.	.	SI-70
.	.	.	.	.	.	.	.	.	.	.	X	X	.	.	.	.	.	.	.	.	.	.	.	.	.	.	.	.	.	.	.	.	.	.	.	.	.	.	.	SI-351
.	.	.	X	X	.	.	.	.	.	.	.	.	.	.	.	.	.	.	.	.	.	.	.	.	.	.	.	.	.	.	.	.	.	.	.	.	.	.	.	SI-118
.	.	X	.	.	X	2	.	X	.	.	2	.	3	.	.	.	.	.	.	.	.	.	.	.	.	.	.	.	.	.	.	.	.	.	.	.	.	.	.	SI-507
.	.	.	.	.	X	.	.	X	.	.	X	.	.	.	.	.	.	.	.	.	.	.	.	.	.	.	.	.	.	.	.	.	.	.	.	.	.	.	.	SI-506
.	.	.	.	.	X	.	.	X	.	.	X	.	.	.	.	.	.	.	.	.	.	.	.	.	.	.	.	.	.	.	.	.	.	.	.	.	.	.	.	SI-504
.	.	X	.	.	.	.	.	X	.	.	X	.	.	.	.	X	.	.	.	X	.	X	.	.	.	.	.	.	.	.	.	.	X	.	.	.	.	.	4	SI-13z
.	.	.	.	.	X	.	.	.	.	X	X	.	.	.	.	X	.	.	.	X	.	X	.	.	.	.	1	4	6	X	.	.	.	.	.	.	.	.	4	SI-526
.	.	.	.	.	.	.	.	.	.	.	.	.	.	.	.	.	.	.	.	.	.	.	.	.	.	.	.	.	X	.	.	.	.	.	.	.	.	.	.	SI-12z
.	.	.	.	.	.	.	.	.	.	.	X	.	.	.	.	.	.	.	.	X	.	.	.	.	.	.	X	X	X	.	.	.	.	.	.	.	.	X	.	SI-6z
.	.	X	.	.	.	.	X	.	.	X	.	X	.	.	.	X	.	.	.	.	.	.	.	.	.	.	.	.	2	.	.	.	.	.	.	.	.	.	.	SI-373
.	.	.	.	.	2	.	X	2	.	X	.	.	.	.	.	.	.	X	.	2	.	.	.	.	.	.	.	.	2	X	.	.	.	.	.	.	.	.	.	SI-409
.	.	.	.	.	.	.	.	.	.	.	X	.	.	.	.	.	.	.	.	X	.	.	.	.	.	.	.	.	2	8	1	2	X	X	.	.	.	.	.	SI-117
.	.	2	7	.	.	.	.	X	.	X	.	.	.	.	.	.	.	.	.	X	.	.	.	.	.	.	.	.	.	X	.	.	.	.	.	.	.	.	.	SI-342
.	.	.	X	.	.	.	.	5	.	X	.	.	.	3	.	.	.	.	.	2	.	.	.	.	.	.	.	.	1	3	.	.	.	.	.	.	.	.	X	SI-340
.	.	.	X	.	.	.	4	.	2	.	.	.	.	.	.	.	.	.	.	.	.	.	.	.	.	.	.	.	2	X	X	X	.	.	.	.	.	X	.	SI-400
.	.	.	.	2	.	X	.	.	.	.	X	.	.	.	.	.	.	.	.	.	.	.	.	.	.	.	.	.	2	.	.	.	.	.	.	.	.	.	.	SI-116
.	.	.	.	.	.	.	.	X	.	.	.	.	.	.	.	.	.	.	.	.	.	.	.	.	.	.	.	.	.	.	.	.	.	.	.	.	.	.	.	SI-398
.	.	.	.	.	.	.	.	X	.	.	.	.	3	.	.	.	.	.	.	.	.	.	.	.	.	.	.	.	2	.	.	.	.	.	.	.	.	.	X	SI-362
.	.	.	.	.	.	.	.	X	.	.	.	.	.	.	.	.	.	.	.	.	.	.	.	.	.	.	.	.	X	.	.	.	.	.	.	.	.	.	X	SI-N84
.	.	X	.	.	.	X	.	.	.	.	X	.	.	.	.	.	.	.	.	.	.	.	.	.	.	.	.	.	X	.	.	.	.	.	.	.	.	.	2	SI-378
X	X	3	.	.	.	.	.	.	3	.	X	X	.	.	2	.	.	.	.	X	.	4	.	.	.	.	X	4	7	X	4	.	.	.	.	.	X	2	.	SI-380
.	.	.	.	.	.	.	X	.	3	.	.	.	.	.	.	.	.	7	.	.	.	.	.	.	.	.	5	2	4	.	.	.	.	X	4	5	.	.	.	SI-K84
.	.	.	.	.	.	.	.	X	.	.	.	.	.	.	.	.	.	.	.	.	.	.	.	.	.	.	.	.	.	.	.	.	.	.	.	.	.	.	X	SI-173
.	.	.	.	.	.	.	.	X	.	.	.	.	.	.	.	.	.	.	.	.	.	.	.	.	.	.	X	X	.	.	.	.	.	.	.	.	.	.	.	SI-102
X	.	.	.	.	X	X	.	X	X	.	X	.	.	.	X	.	.	.	.	.	.	.	.	.	.	.	4	8	X	.	.	.	.	.	.	.	X	X	.	SI-97
.	.	.	.	.	.	.	.	X	.	X	.	.	X	.	.	.	.	.	.	X	X	.	.	.	.	.	X	X	X	.	.	.	.	.	.	.	.	X	.	SI-19
.	.	.	.	.	.	.	.	.	.	.	.	.	.	.	.	.	.	.	.	.	.	.	.	.	.	.	.	.	.	.	.	.	.	.	.	.	.	.	X	SI-14

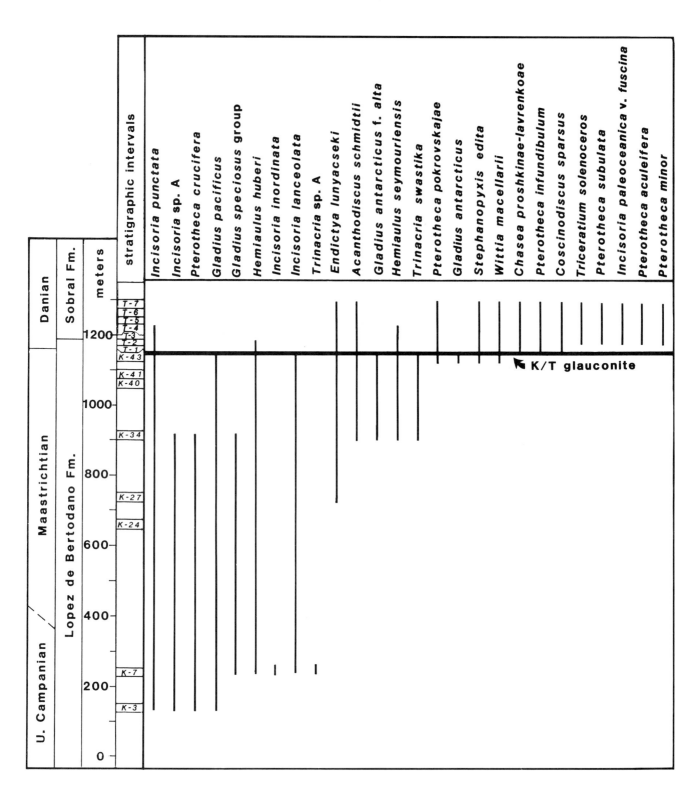

Figure 5. Biostratigraphic occurrence chart of select diatom species from Seymour Island. Diatoms datums may be able to identify the Cretaceous/Tertiary boundary but because this is the only known section with a diatom record across the K/T boundary, it is unknown which species are reliable markers of this horizon.

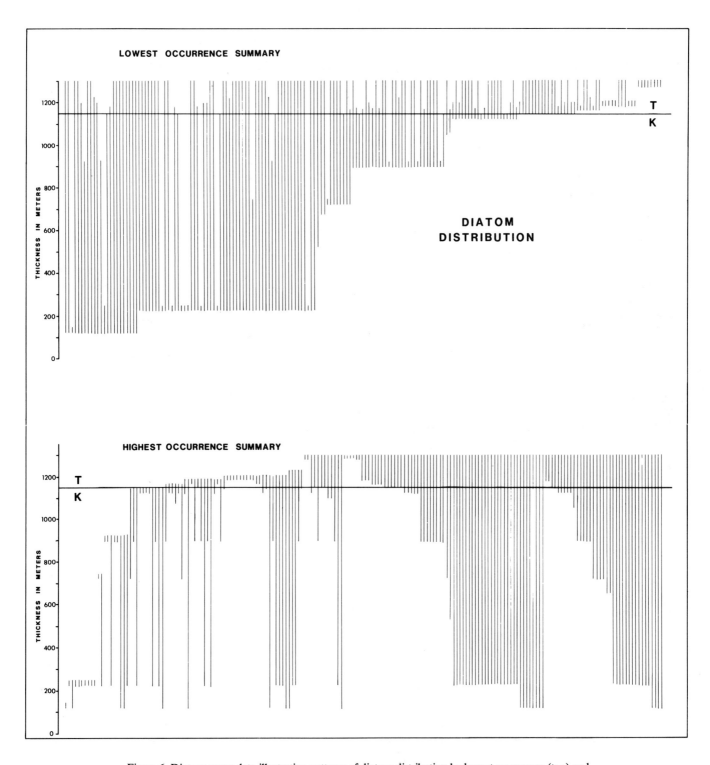

Figure 6. Diatom range data illustrating patterns of diatom distribution by lowest occurrence (top) and highest occurrence (bottom). Plateaus reflect many new occurrences due to well-preserved assemblages at these levels. A large number of Cretaceous species with a lowest occurrence in the Cretaceous continue into the Tertiary (top), while only a limited number of diatom species (31) are restricted to the Cretaceous (far left). Of these 31, only 11 disappear in the uppermost Maastrichtian; the others have their highest appearance lower in the section.

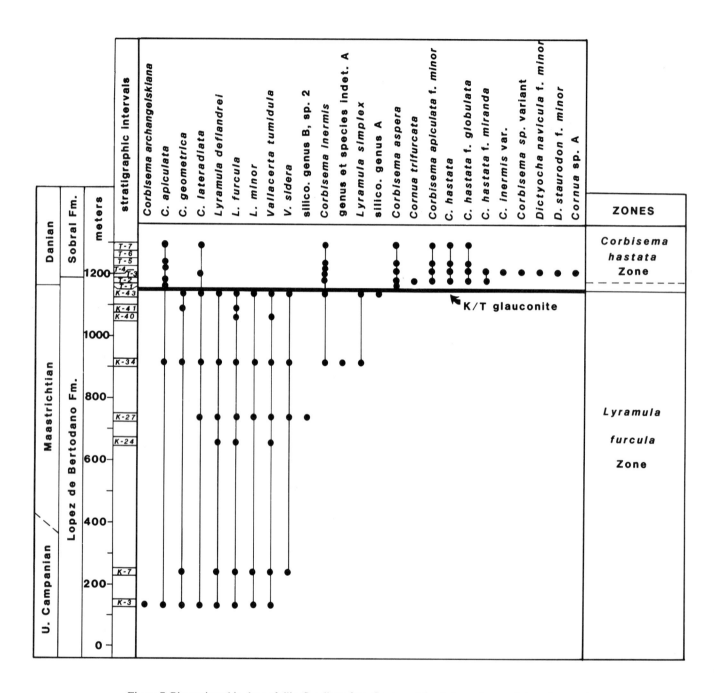

Figure 7. Biostratigraphic chart of silicoflagellates from Seymour Island. A nearly complete replacement of the Cretaceous *Lyramula*-dominated assemblage by a Paleocene *Corbisema*-dominated assemblage identifies the K/T transition on Seymour Island.

TABLE 5. DIATOM EVENTS SURROUNDING THE
CRETACEOUS/TERTIARY BOUNDARY

Seymour Island Diatom Data Alone	Seymour Island Diatom Data With Species Ranges Extended Above and Below the K/T Boundary by Information From the Literature
27% of the lower Paleocene species have first occurrences there. (44 of 161 species)	17% of the lower Paleocene species have first occurrences there. (18 of 166 species)
79% of the Upper Cretaceous species continue into the Paleocene. (116 of 147 species)	84% of the Upper Cretaceous species continue into the Paleocene. (138 of 184 species)
21% of the Upper Cretaceous species have last occurrences there. (31 of 147 species)	16% of the Upper Cretaceous species have last occurrences there. (26 of 164 species)

provides the following age data for these sediments. Sample 13-18 from the top of Section 13, measured in the Cross Valley Formation by Elliot and Trautman (1982), yielded the following silicoflagellates: *Crassicorbisema disymmetrica disymmetrica, Corbisema naviculoides, C. constricta, C. cuspis,* and others. According to the biostratigraphy of Ling (1981), this assemblage is upper Paleocene. The overlying La Meseta Formation is dated here as upper lower Eocene based on the presence of diatoms *Coscinodiscus bulliens, Craspedodiscus molleri, Melosira architecturalis, Pseudorutilaria monile, Pyxilla gracilis;* and by the absence of diatom genera *Asterolampra, Brightwellia,* and *Rylandsia,* which first appeared in the middle Eocene (Gombos, 1982). These diatoms were recovered from Sample 13-19 (Elliot and Trautman, 1982), collected by W. J. Zinsmeister from the lower third of the La Meseta Formation, which occurs as an outlier on the crest of a ridge within Cross Valley (W. J. Zinsmeister, personal communication, 1985).

RESPONSE TO THE CRETACEOUS/
TERTIARY EVENT

This study is the first to record siliceous microfossil assemblages across a K/T boundary sequence. Seymour Island offers many advantages over other K/T sections in evaluating the effects of the Cretaceous/Tertiary extinction event. These advantages include abundant, diverse, and well-preserved fossils from a variety of macro- and microfossil groups; high sediment accumulation rates ranging from 50 to 400 m/m.y. (Sadler, this volume); good exposure; stratigraphic continuity across the K/T boundary within the López de Bertodano Formation; and a noncarbonate sequence where siliceous and organic-walled microfossils, rather than calcareous nannoplankton, are the dominant primary producers.

The replacement of Cretaceous silicoflagellates by Tertiary species is abrupt (Bukry, 1981a) (Fig. 7). Gleser (1966) reported that *Lyramula furcula* and *Vallacerta simplex* are ". . . represented in Paleocene deposits by negligible amounts of isolated skeletons, and thus have no practical significance." Considering the large volume of Cretaceous diatomites in the eastern slopes of the northern Urals and Western Siberian Lowlands, some reworking of older forms into younger sediments would be expected. From the Seymour Island data it appears that silicoflagellates experienced an abrupt extinction at the end of the Cretaceous (Fig. 7), and a gradual recovery in the early Paleocene, as indicated by the upsection increase in diversity, was also noted by Gleser (1966).

Numerous estimates of the extinction of diatoms from the Cretaceous into the Tertiary have been proposed. These range from no extinction to 69 percent extinction of Cretaceous genera (Russell, 1979; Emiliani, et al., 1981; Thierstein, 1982). These estimates were reviewed by A. M. Gombos, Jr. (written communication, 1984). Extinction estimates at the species level include: 87 percent extinction of Cretaceous diatom species from the Moreno Shale (Wornardt, 1972); 50 percent extinction of Cretaceous species from the USSR (Strelnikova, 1975); and 79 percent extinction of Cretaceous species from DSDP Hole 275 (Weaver and Gombos, 1981). These values do not reflect extinction at a K/T boundary, only the absence of many Cretaceous species from younger Tertiary deposits.

Because no data existed for lowest Paleocene diatom occurrences prior to the present study, the ranges of species into the earliest Tertiary was unknown and led to high extinction values in the above works. Percentage data on Seymour Island diatom species that are restricted to the Cretaceous, cross the K/T boundary, and first appear in the lower Paleocene are calculated from Table 4 and Fig. 6 and are shown in Table 5. Seymour Island diatom data indicate that more than 79 percent of the

TABLE 6. LIST OF DIATOM RESTING SPORES FROM SEYMOUR ISLAND

Acanthodiscus antarcticus	*Odontotropis* sp. Hajos	*Pterotheca crucifera*
Acanthodiscus ornatus	*Omphalotheca* sp. cf. *jutlandica*	*Pterotheca danica*
Acanthodiscus schmidtii	*Podosira simpla*	*Pterotheca evermanni*
Acanthodiscus vulcaniformis	*Pseudopodosira westii*	*Pterotheca infundibulim*
Anaulus acutus	*Pseudopyxilla aculeata*	*Pterotheca kittoniana*
Anaulus sibericus	*Pseudopyxilla americana*	*Pterotheca major*
Anaulus sp. A.	*Pseudopyxilla dubia*	*Pterotheca minor*
Biddulphia cretacea	*Pseudopyxilla hungarica*	*Pterotheca pikrovskajae*
Chasea bicornis	*Pseudopyxilla rossica*	*Pterotheca pikrovskajae* var.
Chasea proshkinae-lavrenkoae	*Pseudopyxilla* sp. Strelnikova	*Pterotheca* sp. A.
Cladogramma sp. cf. *jordani*	*Pseudostictodscus picus*	*Pterotheca* sp. B.
Cladogramma sp. cf. *simplex*	*Pterotheca aculeifera*	*Pterotheca spada*
Epithelion spp.	*Pterotheca carinifera*	*Pterotheca subulata*
Goniothecium odontella	*Pterotheca carinifera* v. *curvirostris*	*Pterotheca trojana*
Hemiaulus hostilis group	*Pterotheca carinifera* v. *tenuis*	*Pterotheca uralica*
Hemiaulus sporialis	*Pterotheca clavata*	*Pyrgodiscus* sp.
Kentrodiscus aculeata	*Pterotheca costata*	*Pyrgodiscus? triangulatus*
Kentrodiscus fossilis	*Pterotheca cretacea*	*Skeletonema penicillis*
Odontotropis cristata		

species that occur in the upper Campanian/Maastrichtian have a continued range into the lower Paleocene (Table 5, Fig. 6).

This survivorship rate is increased when biostratigraphic data from the literature are combined with Seymour Island data. Due to assemblage biasing from inconsistent microfossil occurrence and preservation on Seymour Island and the potential for diachronous highest and lowest occurrences around the globe, the ranges of some Seymour Island diatom species are probably incomplete. For some species, older or younger occurrences are indicated in the published literature. Table 5B presents recalculated survival and extinction percentage values, taking into account ranges extended above and below the K/T boundary. From this exercise, it appears that as much as 84 percent of the Upper Cretaceous species continue across the K/T boundary into the Tertiary. It should be emphasized that the fact that 16 percent (26 species) of Cretaceous species do not occur in Tertiary sediments does not imply that they all disappeared at the K/T boundary. In fact, only 10 of the 26 species in this category have their highest occurrence within the uppermost Maastrichtian stratigraphic interval K-43 (Fig. 6). This raises the estimate of diatom survivorship across the K/T boundary to above 90 percent.

The devastation of numerous pelagic microfossil groups at the end of the Cretaceous (Thierstein, 1982) did not appear to differentiate between phytoplankton and zooplankton, nor did it select between groups with calcareous and siliceous skeletons; the silicoflagellates and calcareous nannoplankton suffered a similar fate to that of the planktonic foraminifera. Why then did the diatoms survive the K/T event? One possible explanation of the high diatom survivorship across the K/T may be a result of the diatom's ability to form resting spores in response to environ-

mental stress, resulting from the relatively long and dark high-latitude winters (Milne and McKay, 1982). Diatom resting spores are common in the Seymour Island sequence (Table 6; Figs. 8, 17, 18), and increase in abundance from ~7 percent below to ~35 percent of the total diatom assemblage above the K/T boundary (Fig. 8). This increase in spores following the "K/T event" was predicted by Kitchell and others (1986), whereas the meroplanktonic life history involving resting spores ". . . may have been differentially enhanced during the Late Cretaceous mass extinction regime." With this survival mechanism already built into their life cycle (Hargraves and French, 1983; French and Hargraves, 1980; Garrison, 1984), the diatoms may have endured the terminal Cretaceous extinction event in an encysted stage, reappearing in the early Paleocene when favorable conditions returned (Harwood, 1985,1986).

Diatom assemblages immediately above the "K/T glauconite" on Seymour Island are rich but of low diversity and are dominated by *Stephanopyxis* spp. (Table 4). Low diatom and silicoflagellate diversity in the earliest Tertiary can be explained by stressed environmental conditions following the extinction event. Askin (this volume) also reported low diversity/high dominance in dinoflagellate assemblages from the interval between the glauconite bed and the base of the Sobral Formation, i.e., the dissolution facies of Huber (1987b). Of additional interest is the dominance of pyritized and Fe-stained diatoms and silicoflagellate assemblages in the Upper Cretaceous and the general absence of this type of preservation in the lower Paleocene on Seymour Island (Table 1). While this diagenetic effect may be due to an increase of volcanic material in the lower Paleocene (Macellari, 1984a; this volume), it may also reflect chemical changes in the depositional environment (Huber, 1986).

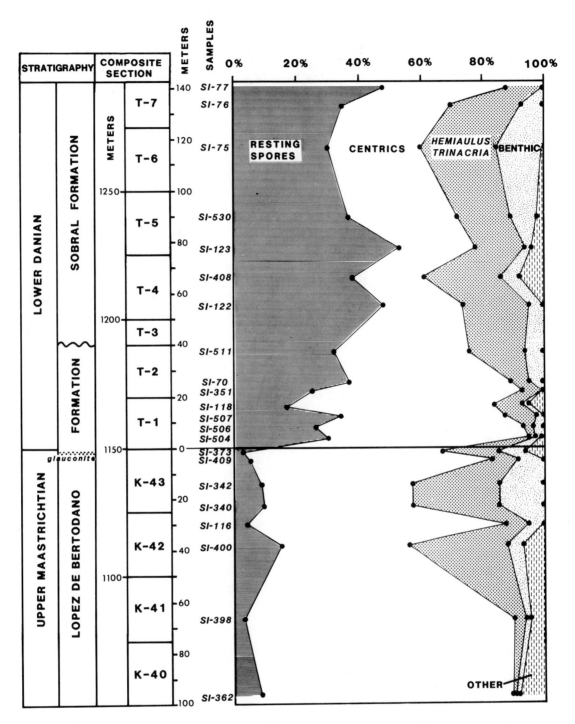

Figure 8. Diatom abundance data surrounding the Cretaceous/Tertiary boundary on Seymour Island. Diatom assemblages are divided into five general groups: (1) resting spores—see list of species in Table 6; (2) centrics, including species of Suborders Coscinodiscineae and Rhizosoleniineae; (3) benthics, including species of the Family Heliopeltaceae and Subfamilies Stictodiscoideae and Eupodiscoideae; (4) diatom species of the genera *Hemiaulus* and *Trinacria;* and (5) other diatom species not included in the above categories (classification follows Simonsen, 1979). Diatom resting spores show a marked increase in abundance a few meters above the "K/T glauconite" and may provide a means to distinguish Upper Cretaceous from lower Paleocene sediments on Seymour Island and perhaps elsewhere.

Information provided by the distribution of a variety of microfossil and macrofossil groups, such as preserved on Seymour Island, should provide insight into the nature of the "Terminal Cretaceous Event." In light of the role resting spores apparently played in the diatom's survival, arguments against the "K/T darkened sky hypothesis"—which maintain that, because diatoms survived, the world was never dark for an extended period—may not be valid. Because of their position at the base of the food chain, diatoms, like calcareous nannoplankton, have a potential causal link to the fate of other organisms in higher trophic levels (Milne and McKay, 1982). The survival of diatoms across this major boundary may shed light on, and limit our speculation about, the nature of the K/T event. Upper Cretaceous diatom-rich sediment sequences where the biotic response to the K/T event is recorded in areas of high siliceous productivity should be identified and studied in order to determine whether a trophic collapse, through loss of calcareous nannoplankton, contributed to the extinction of other organisms. If a high rate of diatom productivity continued through the K/T event in fertile, upwelling environments, for example, the macrofaunal response to the K/T event may be quite different from that in carbonate environments where calcareous nannoplankton were the dominant producers and extinction levels were high. Association with these siliceous "extinction refugia" environments may help to explain the survivorship of some Cretaceous marine species into the Tertiary.

CRETACEOUS PENNATE DIATOMS

Relatively diverse assemblages of pennate diatoms including the genera *Incisoria* and *Sceptroneis,* were recovered from the upper Campanian through Danian sequence on Seymour Island (Figs. 17.1 through 17.14), in the past, primarily because of limited data on Cretaceous diatoms, workers have discounted the presence of pennate diatoms in pre-Tertiary rocks (Schrader, 1969; Tappan and Loeblich, 1973). Although Strelnikova (1974, 1975) reported rapheless pennates of the genus *Sceptroneis* in the upper Campanian, and Hajós and Stradner (1975) recovered several species of *Sceptroneis, Incisoria, Tubularia,* and other

pennates in the Campanian, review articles on fossil marine diatoms (Burckle, 1978; Schrader and Schuette, 1981; and Tappan, 1980) maintain that pennate diatoms first appeared in the Early Tertiary. Schrader and Schuette (1981) recognized the work of Hajós and Stradner (1975), but suggested that the pennates are contaminants from younger sediment. The fact that pennate diatom assemblages are also known from Seymour Island, the Ural Mountain region in Russia (Strelnikova, 1974), and the Moreno Shale in California (Abbott and Harper, 1982), and that DSDP Hole 275 recovered only Upper Cretaceous sediment covered with a thin veneer of Quaternary foram ooze, indicate that pennate diatoms must have evolved by late Campanian time.

ACKNOWLEDGMENTS

I thank B. T. Huber, C. E. Macellari, and W. J. Zinsmeister for supplying sample material from the López de Bertodano and Sobral Formations, and for helpful discussions. A. M. Gombos, Jr., provided assistance and guidance through the initial phase of this study. B. T. Huber, A. M. Gombos, J. A. Barron, and V. Porguen reviewed this manuscript. J. Fenner and A. M. Gombos provided insight and discussion on diatom identifications; D. Bukry provided comments regarding silicoflagellate identifications. H. Harper and J. A. Barron are thanked for sharing diatom slides from the Maastrichtian Moreno Shale and Campanian Arctic cores, respectively, and the Deep Sea Drilling Project for supplying comparative study material from DSDP Holes 275, 216, and 208. V. Porguen provided helpful discussion regarding diatom deposits in the Soviet Union. M. Grant, M. Karrer, M. Marks, and M. Renz assisted with sample preparation. K. Doddroe typed the manuscript.

This research was supported by National Science Foundation Grants DSDP-8020096 and DPP-82113985 (to W. J. Zinsmeister and D. H. Elliot) for field sampling, and in part by Grant DPP-8214174 (to P.-N. Webb) for laboratory and microscope analysis. Support was also provided by a 1984 Research Grant from the Geological Society of America and by a 1985 Grant-in-Aid from the American Association of Petroleum Geologists (to D.M.H.).

SYSTEMATIC PALEONTOLOGY

Diatoms

Genus *ACANTHODISCUS* Pantocsek, 1892

Acanthodiscus antarcticus Hajós & Stradner, 1975
Figure 9.1
Acanthodiscus antarcticus Hajós & Stradner, 1975, p. 934, Plate 13, Figures 1-6.

Acanthodiscus ornatus Hajós & Stradner, 1975
Figure 9.2, 3
Acanthodiscus ornatus Hajós & Stradner, 1975, p. 394, Plate 13, Figures 7, 9, 10; Plate 28, Figure 4.

Acanthodiscus schmidtii Jousé, 1951a
Figure 9.4, 5
Acanthodiscus schmidtii Jousé, 1951a, Plate 1, Figure 8a,b.
[Unnamed form] from Simbirsk in Schmidt et al., 1874–1859, Plate 139, Figure 13.
Discussion. A variety of forms are included here. Most resemble those illustrated but others have fewer and smaller punctae.

Acanthodiscus sp.
Figure 9.6

Ancanthodiscus vulcaniformis Jousé 1951a
Figures 10.8-12
Acanthodiscus vulcaniformis Jousé, 1951a, Plate 3, Figures 1a,b,c
Chasea ornata Hajós & Stradner, 1975, p. 928, Plate 5, Figures 4, 5; Plate 27, Figure 4.

Genus *ACTINOPTYCHUS* Ehrenberg, 1841

Actinoptychus packi Hanna, 1927a
Figure 9.7, 8
Actinoptychus packi Hanna, 1927a, p. 12, Plate 1, Figures 1-3; Hajós & Stradner, 1975, p. 928, Plate 5, Figures 23-24; Plate 29, Figures 1-4, Plate 30, Figures 1-4.

Actinoptychus punctulatus Pantocsek, 1886
no illustration
Actinoptychus punctulatus Pantocsek, 1886, Bd. 1, p. 64, Plate 8, Figure 60; Kanaya, 1957, p. 97, Plate 7, Figure 16; Strelnikova, 1974, p. 69, Plate 14, Figure 7a,b.

Actinoptychus taffi Hanna, 1927a
Figure 9.9
Actinoptychus taffi Hanna, 1927a, p. 13, Plate 1, Figure 4.

Genus *ANAULUS* Ehrenberg, 1844

Anaulus acutus? Brun, 1896
Figure 9.10, 11
Anaulus acutus Brun, 1896, p. 231, Plate 20, Figures 15-18.
Anaulus subantarcticus Hajós in Hajós & Stradner, 1975, p. 935, Plate 13, Figure 26.

Genus and species indet. Fenner, 1977, Plate 33, Figure 8.
Also see.—Anaulus birostratus Grunow in Grove & Sturt, 1887, Plate 5, Figure 25; Van Heurck, 1885, Plate 103, Figures 1-3.

Anaulus incisus Hajós & Stradner, 1975
no illustration
Anaulus incisus Hajós & Stradner, 1975, p. 935, Figure 18; Plate 13, Figure 23.

Anaulus sibericus Strelnikova, 1974
Figure 9.12-14
Anaulus sibericus Strelnikova, 1974, p. 106, Plate 52, Figures 1-5.
Discussion. The dominance of *A. sibericus* and *Hemiaulus elegans* in resting spore and vegetative cell diatom assemblages, respectively, from the Arctic Ocean (Kitchell et al., 1986) suggests that *A. sibericus* is a resting spore of *H. elegans.*

Anaulus weyprechtii Grunow, 1884
Figure 9.15
Anaulus weyprechtii Grunow, 1884, p. 58, Plate 2, Figure 18; Hustedt, 1930, p. 894, Figure 538; Krotov & Schibkova, 1959, Plate 5, Figures 11, 12; Proshkina-Lavrenko, 1949, p. 212, Plate 99, Figure 19.

Anaulus sp. A - Figure 9.16, 17

Genus *ARACHNOIDISCUS* Deane ex Pritchard, 1852

Arachnoidiscus simbirskianus Pantocsek, 1892
Figure 9.18, 19
Arachnoidiscus simbirskianus Pantocsek, 1892, Plate 15, Figure 223.
Arachnoidiscus ehrenbergii f. *paleoceanicus* Jousé, 1951a, p. 30, Figures 1, 2; Gleser et al., 1974, Plate 16, Figure 10.

Genus *AULACODISCUS* Ehrenberg, 1844

Aulacodiscus simbirskianus Pantocsek, 1889
no illustration
Aulacodiscus simbirskianus Pantocsek, 1889, Plate 18, Figure 289.

Aulacodiscus sp. cf. *A. breviprocessus* Strelnikova, 1974
Figure 12.3
cf. *Aulacodiscus breviprocessus* Strelnikova, 1974, p. 76, Plate 25, Figure 1-4.

Aulacodiscus sp. cf. *A. nigrescens* Pantocsek, 1889
Figure 12.2
cf. *Aulacodiscus nigrescens* Pantocsek, 1889, Plate 26, Figure 394.

Aulacodiscus? sp. A
Figure 12.4, 5
Discussion. In his review of this paper, Victor Porguen questioned whether this form is an *Aulacodiscus,* and noted a greater affinity with the genus *Anthemodiscus* described by Barker and Meakin (1943, p. 252, Plate 38, Fig. 5).

Genus *AULISCUS* Ehrenberg, 1843

Auliscus sp. cf. *A. priscus* Long, Fuge & Smith, 1946
Figure 9.20
cf. *Auliscus priscus* Long, et al., 1946, p. 99, Plate 14, Figure 4.

Auliscus sp. A—Figure 9.21

Genus *BIDDULPHIA* Gray, 1921 emend. Van Heurck, 1885

Biddulphia cretacea Hajós & Stradner, 1975
Figure 12.20, 21
Biddulphia cretacea Hajós & Stradner, 1975, p. 930, Plate 5, Figures 6, 7; Plate 34, Figures 1, 2.
Discussion.—This is probably a resting spore of *Hemiaulus* sp. G, as the two were found attached.

Biddulphia fistulosa Pantocsek, 1892
Figure 12.22
Biddulphia fistulosa Pantocsek, 1892, Plate 18, Figure 273.

Biddulphia sparsipunctata Hajós, 1975
no illustration
Biddulphia sparsipunctata Hajós in Hajós & Stradner, 1975, p. 930, Plate 5, Figures 8, 9.

Genus *CHASEA* Hanna, 1934
Discussion.—Species of this genus are probably resting spores of *Hemiaulus* and *Cerataulina*. *Chasea proshkinae-lavrenkoae* may be related to the spore genus *Syringidium* Ehrenberg, as it bears resemblance to *S. bicorne* Ehrenberg (see illustrations in Hasle and Syvertsen, 1980, and Hasle and Sims, 1985). Also refer to *Ceratulina bergonii* Peragallo in Gleser et al., 1974, Plate 9, Figures 13, 14, and Hustedt, 1930, p. 869, Figure 517, for a spore similar in structure to *C. proskinae-lavrenkoae*.

Chasea bicornis Hanna, 1934
Figure 10.1, 2
Chasea bicornis Hanna, 1934, p. 354, Plate 48, Figures 12–16.

Chasea proshkinae-lavrenkoae new combination
Figure 10.3–7
Discussion.—This form is similar to resting spores of *Hemiaulus proshkinae-lavrenkoae* Jousé, 1951, Plate 3, Figures 9a–c. The spores of this species are transferred to spore genus *Chasea.* Although this form is similar to *Chasea bicornis, C. bicornis* is smaller and lacks the characteristic second elevated "bubble" located in the center of the valve, as illustrated here and in Jousé, 1951a. Perhaps this species is better placed in *Syringidium* or *Cerataulina* (see above).

Genus *CLADOGRAMMA* Ehrenberg, 1854

Cladogramma sp. cf. *C. jordani* Hanna, 1927a
no illustration
cf. *Cladogramma jordani* Hanna, 1927a, p. 16, Plate 2, Figure 1; Hajós & Stradner, 1975, p. 928, Plate 4, Figure 9–11.
Discussion.—This form may be the hypovalve of *Pterotheca aculeifera,* as indicated in Strelnikova, 1974, Plate 57.

Cladogramma sp. cf. *C. simplex* Hajós & Stradner, 1975
Figure 10.13, 14
cf. *Cladogramma simplex* Hajós & Stradner, 1975, p. 928, Plate 4, Figures 7, 8; Plate 28, Figure 5.

Genus *COSCINODISCUS* Ehrenberg, 1838

Coscinodiscus morenoensis Hanna, 1927a
Figure 10.15, 16
Coscinodiscus morenoensis Hanna, 1927a, p. 18, Plate 2, Figures 3, 4; Hajós & Stradner, 1975, p. 927, Plate 3, Figures 6, 7; Plate 4, Figure 2; Plate 24, Figures 1–3; Plate 25, Figures 1–5.

Conscinodiscus solidus Strelnikova, 1971
no illustration
Coscinodiscus solidus Strelnikova, 1971, p. 44, Plate 1, Figure 8; Strelnikova, 1974, p. 63, Plate 10, Figures 1–5.

Coscinodiscus sparsus new species
Figure 11.11, 12
Description.—Valve is circular, convex, and covered with irregularly spaced, sparse areolae that extend onto the broad valve margin (as much as one-fifth valve radius). Areolae are largest in valve center, decreasing toward the margin. Tangential arrangement of areolae often visible in valve center is interrupted by numerous, irregular hyaline areas covering between 5 and 40 percent of valve face.
Holotype. Figure 11.11, USNM 416128.
Type locality.—Upper López de Bertodano Formation, Sample SI-511 (concretion), Seymour Island, northeast Antarctic Peninsula.
Type level.—Lower Paleocene.

Genus *DREPANOTHECA* Schrader, 1969

Drepanotheca bivittata (Grunow & Pantocsek in Pantocsek, 1886) Schrader, 1969
no illustration
Eunotogramma? bivittata Grunow & Pantocsek in Pantocsek, 1886, p. 48, Plate 26, Figure 247; Grove & Sturt, 1887, p. 77, Plate 6, Figure 24.
Eunotogramma bivittata Grunow & Pantocsek (type 2), Brun, 1894, Plate 5, Figure 2–3.
Eunotogramma marginopucntatum Long et al., 1946, p. 106, Plate 16, Fig. 14.
Drepanotheca bivittata (Grunow & Pantocsek) Schrader, 1969, p. 123, Plate 38, Figure 9.

Genus *ENDICTYA* Ehrenberg, 1845

Endictya lunyacsekii Pantocsek, 1889
Figure 19.15, 16
Endictya lunyacsekii Pantocsek, 1889, p. 114; Pantocsek, 1892, Plate 10, Figure 168.

Genus *EPITHELION* Pantocsek, 1892

Epithelion sp.—Figure 22.1

Genus *EUNOTOGRAMMA* Weisse, 1854

Eunotogramma enorme Krotov, 1959
Figure 21.22
Eunotogramma enorme Krotov, 1959, p. 111, Figure 11.

Eunotogramma gibbosa Strelnikova, 1965a
Figure 22.3
Eunotogramma gibbosa Strelnikova, 1965a, p. 36, Plate 2.
Eunotogramma sp. Long et al., 1946, p. 106, Plate 16, Figure 4.
Eunotogramma enorme Krotov, Strelnikova, 1974, p. 108, Plate 52, Figures 16–19.

Eunotogramma polymorpha Strelnikova, 1965a
Figure 10.17, 18
Eunotogramma polymorpha Strelnikova, 1965a, p. 36, Plate 1, Figures 7, 8; Strelnikova, 1974, p. 107, Plate 52, Figures 8–13.
Also see.—*E. productum* and *E. variable* in Fenner, 1977; Van Heurck, 1885; and Gleser et al.,1974.

Eunotogramma producta var. **recta** Long, Fuge & Smith, 1946
Figure 10.19
Eunotogramma productum var. *rectum* Long et al., 1946, p. 106, Plate 16, Figures 10, 11.

Eunotogramma weissei Ehrenberg, 1855
Figure 10.20
Eunotogramma weissei Ehrenberg, Grunow, 1884, p. 59, Plate 2, Figures 21–22; Schmidt et al., 1874–1959, Plate 144, Figures 38, 42, 43, Jousé, 1955, Plate 5, Figure 7; Gleser et al., 1974, Plate 15, Figure 6.

Genus *GLADIUS* Forti & Schulz, 1932

Discussion.—A new morphologic group within the genus *Gladius* is recognized in the Upper Cretaceous material from Seymour Island. Unlike most *Gladius* species with valves extended along the pervalvar axis, this new group—represented by *Gladius antarcticus* n. sp. and *G. antarcticus* forma *alta* n. sp., n. form—possesses short pervalvar axes with most of the frustule extended in only two dimensions. Comparison of the photographs in Figure 11 to those of Strelnikova (1966a, 1974) clearly illustrates the relationship between the three-dimensional *Gladius* species (*G. speciosus* group and *G. pacificus*) and the largely two-dimensional *G. antarcticus* n. sp. Several species closely related to *G. antarcticus* n. sp. are known from the Upper Cretaceous Moreno Shale in California, including *Craspedodiscus morenoensis* and *Coscinodiscus morenoensis* Hanna var. in Long et al. (1946), *Pomphodiscus morenoensis* (Long, Fuge & Smith) Barker & Meakin (1946), *Craspedodiscus morenoensis* in Strelnikova (1974), and perhaps *Coscinodiscus centroaculeatus* Strelnikova (1971, 1974) and *Benetorus fantasmus* Hanna (1927a). The description given for the fine structure of *Pomphodiscus moenoensis* by Barker & Meakin (1946), involving very fine puncta in quincunx arrangement beneath the hexagonal cellules and forming a marginal ring, applies well for most *Gladius* species, as can be seen in Figure 11. *P. morenoensis* differs from *G. antarcticus* n. sp. only in the shape, asymmetry, and fine structure of the central connecting apparatus. It is clear that *Pomphodiscus morenoensis* Baker & Meakin and *Coscinodis-*

cus morenoensis Hanna var. Long, Fugre & Smith belong with the genus *Gladius.* An extensive study of *Benetorus fantasmus* Hanna and *Coscinodiscus centroaculeatus* Strelnikova utilizing scanning electron microscopy is needed to clarify their relationship within this group.

Gladius antarcticus new species
Figure 11.1, 2
Coscinodiscus morenoensis Hanna, variations of Long et al., 1946, p. 104, Plate 17, Figure 3. *Not Coscinodiscus morenoensis* Hanna, 1927a, p. 18, Plate 2, Figures 3, 4.
Description.—Frustule heterovalvar, ranging in size from 110 to 130 μm in diameter. Valve flat from margin to one-fourth of the radius from center where either subcentral elevated ring (Fig. 11.1) or central circular elevation (30 to 35 μm in diameter) (Fig 11.2) rises between 3 to 7 μm from valve plain. Complex connecting structure is present both within the subcentral elevated ring and on the central elevated region. The connecting structure consists of central labiate process surrounded by a finely reticulate pad consisting of fine punctae. These punctae radiate in an irregular anastamose pattern from the central process to two-thirds of the radius of the central area. Ring of small areolae (8 in 10 μm) divides connecting pad from elevated ring and planar portion of the valve. Subhexagonal areolae (4 to 5 in 10 μm) arranged in radial pattern decrease in size only slightly toward central area. Beneath the hexagonal framework a siliceous layer consisting of fine punctae (20 punctae in 10 μm) arranged in quincunx pattern extends from the central connecting pad to the margin where narrow ring of fine punctae surrounds the valve.
Discussion.—The elevated central area most likely fits within the elevated ring of an adjacent valve as a method of chain formation. Similar connecting structures are known from other *Gladius* species, as illustrated by Strelknikova (1971, 1974). Long et al. (1946) apparently did not know where to place this species and lumped it as a variation of *C. morenoensis.*
Holotype.—Figure 11.2 USNM 416131.
Paratype.—Figure 11.1, USNM 416132.
Type locality.—Upper López de Bertodano Formation, Sample 373 (concretion) Seymour Island, northeast Antarctic Peninsula.
Type level.—Uppermost Maastrichtian, Upper Cretaceous.

Gladius antarcticus f. **alta** new species, new form
Figures 11.3–5, 7–10; 12.25
Description.—The Frustule is heterovalvar, ranging in size from 60 to 85 μm m in diameter. One valve is flat, with a subcentral elevated ring one-third of the radius from center. The other valve is flat, from the margin to one-third of the radius from the center, where the central portion of the valve is greatly elevated (10 to 80 μm from the valve base), bending upward to form a narrowing cylinder, and constricting to the narrowest diameter just below the apex of the elevation where it widens to a flat platform 22 to 32 μm in diameter. Central areas on both valves possess the complex connecting structure described above for *G. antarcticus* and a similar fine structure of punctae beneath the subhexagonal areolae.
Discussion.—This form is separated from *G. antarcticus* by the much greater height of the central elevation, by the smaller diameter, and by the stratigraphic distance between the two forms. *G. antarcticus* f. *alta* is from the lower upper Maastrichtian to uppermost Maastrichtian. This form is apparently transitional between the planar upper Maastrichtian *G. antarcticus* n. sp. and high elevation *G. speciosus* group and the majority of other *Gladius* species.
Holotype.—Figure 11.4, USNM 416129.
*Paratype.*Figure 11.10, USNM 416130.
Type locality. Upper López de Bertodano Formation, Sample 380 (concretion), Seymour Island, northeast Antarctic Peninsula.
Type level.—Lower upper Maastrichtian, Upper Cretaceous.

Gladius pacificus Hajós & Stradner, 1975
Figures 10.23, 11.6?
Gladius pacificus Hajós and Stradner, 1975, p. 933, Plate 11, Figures 1,
2; Plate 26, Figure 2; Barron, 1985, p. 141, Plate 10.2, Figure 15.

Gladius speciosus Schulz, 1935 group
Figure 10.24
Gladius speciosus Schulz, 1935, p. 391, Plate 1, Figures 6–8; Hajós and
Stradner, 1975, p. 933, Plate 11, Figures 5, 6; Plate 26, Figure 3; Jousé,
1955, p. 76, Figure 4; Strelnikova, 1966a; Strelnikova, 1974, p. 104,
105, Plates 49, 51; Barron, 1985, p. 141, Plate 10.2, Figure 14.

Genus ***GONIOTHECIUM*** Ehrenberg, 1841

Goniothecium odontella Ehrenberg, 1844
Figure 10.21,22
Goniothecium odontella Ehrenberg, Jousé, 1951b, p. 60, Plate 5, Figures
1–7; Hajós & Stradner, 1975, p. 935, Plate 10, Figures 10, 11; Strelni-
kova, 1974, p. 116, Plate 55, Figures 1–12; Plate 56, Figures 1–5;
Barron, 1985, p. 141, Plate 10.2, Figure 13.

Genus ***HELMINTHOPSIS*** Van Heurck, 1892

Helminthopsis wornardti Hajós 1975
no illustration
Helminthopsis wornardti Hajós in Hajós & Stradner, 1975, p. 935, Plate
13, Figure 24.

Genus ***HEMIAULUS*** Ehrenberg, 1844
Discussion.—The Genus *Notiostyrax* DePredo & Ling, 1981, was recently described from Paleocene sediments in DSDP Hole 208. The
described forms are broken horns of a *Hemiaulus* species.

Hemiaulus ambiguus Grunow, 1884
Figure 15.2,3
Hemiaulus ambiguus Grunow, 1884, p. 10, Plate 2, Figures 25, 26;
Hustedt, 1930, p. 876, Figure 520.

Hemiaulus andrewsi Hajós, 1975
no illustration
Hemiaulus andrewsi Hajós in Hajós & Stradnew, 1975, p. 931, Plate 7,
Figure 8.

Hemiaulus assymetricus Jousé, 1951b
Figure 13.3
Hemiaulus assymetricus Jousé, 1951b, p. 52, Plate 111, Figure 2; Strel-
nikova, 1974, p. 97, Plate 47, Figures 1–6; Gleser et al., 1974, Plate 9,
Figure 4.
Note.—Difficulty was experienced in distinguishing small forms of this
species from *H. echinulatus* Jousé.

Hemiaulus bipons (Ehrenberg) Grunow in Van Heurck, 1882
Figures 13.5–11
Hemiaulus bipons (Ehrenberg) Grunow in Van Heurck, 1882, Plate 103,
Figures 6–9; Wolle, 1890, Plate 64, Figures 26, 28, 29.
Hemiaulus bifrons (Ehrenberg) Grunow, misspelling by Van Heurck,
1896, p. 456, Figure 182.
Hemiaulus hungaricus Pantocsek, 1886, p. 48, Plate 29, Figure 291;

Pantocsek, 1892, Plate 19, Figures 284, 286.
Hemiaulus ambiguus Grunow, Gleser et al., 1974, Plate 19, Figure 11.
Not H. ambiguus Grunow, 1884.

Hemiaulus curvatulus Strelnikova, 1971
Figure 13.12
Hemiaulus curvatulus Strelnikova, 1971, p. 49, Plate 1, Figures 12, 13;
Strelnikova 1974, p. 96, Plate 47, Figures 14–16; Hajós & Stradner,
1975, p. 931, Plate 6, Figure 8; Schrader & Fenner, 1976, p. 983, Plate
43, Figs. 10–11.

Hemiaulus danicus Grunow, 1878
Figure 13.16,17
Hemiaulus danicus Grunow, Grunow, 1884, p. 13 (65), Plate 2(B),
Figure 40; Schmidt et al. 1874–1959, Plate 143, Figure 43; Hustedt,
1930, p. 877, Figure 521; Hajós & Stradner, 1975, p. 391, Plate 5,
Figures 5, 6; Strelnikova, 1974, p. 100, Plate 43, Figure 19.

Hemiaulus echinulatus Jousé, 1949b
Figure 13.4
Hemiaulus echinulatus Jousé, 1949b, p. 186, Plate 72, Figure 5; Jousé,
1951b, p. 53, Plate 3, Figure 3a–c; Strelnikova, 1974, p. 100, Plate 46,
Figures 15–23; Hajós & Stradner, 1975, p. 931, Plate 5, Figures 21, 22.

Hemiaulus elegans (Heiberg, 1863) Grunow, 1884
Figure 13.18
Corinna elegans Heiberg, 1863, p. 53, Plate 3, Figures 1–5; Schmidt et
al., 1874–1959, Plate 143, Figures 54–55, Plate 144, Figures 2, 4.
Hemiaulus elegans (Heiberg) Grunow, 1884, p. 66, Plate 2(B),
Figure 51; Witt, 1885, p. 162, Plate 9, Figure 16; Pantocsek, 1889, Bd. 2,
p. 82; Hustedt, 1930, p. 881, Figure 527; Gleser et al., 1974, Plate 19,
Figure 6; Ross et al., 1977, p. 186–187, Plate 7, Figures 45–47.
Hemiaulus elegans var. *intermedius* Grunow, 1884, p. 67, Plate 2(B),
Figure 52.
Hemiaulus sp. Hajós & Stradner, 1975, Plate 34, Figure 6.

Hemiaulus gleseri Hajós, 1975
Figure 12.26
Hemiaulus gleseri Hajós in Hajós & Stradner, 1975, p. 931, Plate 5,
Figure 20; Plate 7, Figures 6, 7; Barron, 1985, p. 141, Plate 10.2, Figures
1, 2.

Hemiaulus hostilis Heiberg, 1863 group
Figures 14.1–5
Hemiaulus hostilis Heiberg, 1863, p. 48, Plate 1, Figure 11; Kitton, 1870,
p. 101, Plate 11, Figures 7–11; Schmidt et al., 1874–1959, Plate 143,
Figure 42; Plate 144, Figure 1; Grunow, 1884, p. 63, Plate 2(B), Figures
32, 33; Strelkinova, 1974, p. 99, Plate 47, Figures 10–13.
Discussion.—All varieties of *H. hostilis* are grouped here. The forms
from Seymour Island most closely resemble those forms illustrated in
Kitton, 1870, and *H. hostilis* f. *minuta* Grunow, 1884, p. 63, Plate 2,
Figure 32; Schmidt et al., 1874–1959, Plate 144, Figure 1 (although the
horns are longer on Seymour Island specimens) and *H. hostilis* var.
polaris Grunow sensu Krotov & Schibkova, 1959, Plate 4, Figures 4, 5.

Hemiaulus huberi new species
Figure 13.21–25
Description.—Valve circular, flat, with steep valve margin; two horns,
densely punctate, located 180° apart, rise from an elevated part of the
valve mantle; hyaline spine, one-half as long as the horn, at tip of horn;
faint keel runs along outside the horn from tip to valve base; areolae
often elongate, arranged in curved lines beginning at valve margin and

converging toward a process located at a distance of one-half the radius at a position equidistant from each horn and oriented in figured specimens at 9 o'clock; hyaline ridge at crest of valve margin surrounds valve face. Resting spore (*Acanthodiscus*) still attached to parent valve in Figure 13.25.

Discussion.—This diatom represents one of four known "circular" species of the genus *Hemiaulus*. Others are *Hemiaulus pacificus* (Hajós) Gombos, known from the lower Oligocene (see Gombos & Ciesielski, 1983, p. 602), plus *Hemiaulus rostratus* Pantocsek, 1892, Plate 37, Figures 521, 523, and *Hemiaulus seymourensis* n. sp. The radial arrangement of areolae in *H. pacificus* and the converging arrangement of areolae to an off-center process in *H. huberi* distinguish tese two taxa. *H. huberi* and *H. seymourensis* differ by subrectangular and finer areolae and by converging horns in *H. seymouriensis*. Named for Brian T. Huber, who, along with Carlos Macellari, mapped and described the Upper Cretaceous and lower Paleocene marine sequence on Seymour Island, and collected samples used in this study.

Holotype.—Figure 13.22, USNM 416133.
Paratye.—Figure 13.24, USNM 416134.
Type locality.—López de Bertodano Formation, Sample SI-19 (concretion), Seymour Island, northeast Antarctic Peninsula.
Type level.—Upper Campanian, Upper Cretaceous.

Hemiaulus incisus Hajós, 1976
Figure 14.10,11
Hemiaulus incisus Hajós, 1976, p. 829, Plate 23, Figures 4–9; Gombos, 1977, p. 594, Plate 15, Figure 3; Fenner, 1977, p. 521, Plate 25, Figures 6, 8; Gombos & Ciesielski, 1983, p. 602, Plate 20, Figure 6.

Hemiaulus includens (Henrenberg, 1855) Grunow, 1844
no illustration
Biddulphia includens Ehrenberg, 1855, p. 301.
Hemiaulus capitatus Greville, 1865, p. 54, Plate 6, Figure 24.
Hemiaulus includens (Ehrenberg) Grunow, 1884, p. 64, Plate 2(B), Figures 36–38; Strelnikova, 1974, p. 101, Plate 48, Figures 1–6.

Hemiaulus kittonii Grunow, 1884
Figure 13.15
Hemiaulus sp. Kitton 1870–1871, Plate 14, Figure 11.
Hemiaulus kittonii Grunow, 1884, p. 61; Schmidt et al., 1874–1959, Plate 142, Figures 2–11; Strelnikova, 1974, p. 96, Plate 42, Figures 92–94.
Hemiaulus speciosus Jousé, 1951b, p. 55, Plate 3, Figure 5.
Hemiaulus altus Hajós in Hajós & Stradner, 1975, p. 931, Plate 5, Figure 17–19; Gombos, 1977, p. 594, Plate 20, Figures 3, 4.

Hemiaulus kondai Hajós, 1975
Figure 13.28
Hemiaulus kondai Hajós in Hajós & Stradner, 1975, p. 932, Plate 6, Figures 10, 11.

Hemiaulus polymorphus var. *frigida* from Mors, Grunow, 1884
Figure 14.7
Hemiaulus polymorphus var. *frigida* von Mors Grunow, 1884, Plate 2(B), Figures 47, 48.
Hemiaulus polymorphus var. *frigida* Grunow, Gleser et al., 1974, Plate 15, Figure 9; Plate 19, Figure 1.

Hemiaulus polymorphus var. *frigida* from Franz Joseph, Grunow, 1884
Figure 13.29
Hemiaulus polymorphus var. *frigida* von Franz Joseph, Grunow, 1884, p. 14(66), Plate 2(B), Figure 49.

Hemiaulus polymorphus var. *morsianus* Grunow, 1884
Figure 14.12
Hemiaulus polymorphus var. *morsianus* Grunow, 1884, p. 14, Plate 5, Figure 53.
Note.—The figured specimen may actually be a longer form of *H. sibericus* (see Fig. 14.14).

Hemiaulus praelegans Jousé
Figure 13.19,20
Hemiaulus pra-elegans Jousé 1951b, p. 53, Plate 3, Figures 4a,b.
Hemiaulus prae-elegans Jousé, misspelled in Krotov, 1959, p. 107, Figure 2, and in Hajós & Stradner, 1975, p. 932, Plate 6, Figures 12, 14. The specimens illustrated by Krotov, 1959, could be spineless forms of *H. danicus* (see Fig. 13.16) from the center of a *H. danicus* chain.

Hemiaulus proshkinae-lavrenkoae Jousé, 1951a
Figure 10.7
Hemiaulus proshkinae-lavrenkoae Jousé, 1951a, p. 37, Plate 3, Figures 9a,b,c.
Discussion.—The figured specimen has a *Chasea proshkinae-lavrenkoae* n. comb. spore still attached.

Hemiaulus rossicus Pantocsek, 1889
Figures 14.18–21; 15.4,5
Hemiaulus antarcticus Ehrenberg sensu Weisse, 1854, p. 242, Plate 1, Figures 18a–f; Witt, 1885, Plate 6, Figs. 1,2; Schmidt et al., 1874–1959, Plate 144, Figures 28–35; DePredo, 1981, p. 94, Plate 2, Figure 9.
Hemiaulus polycystinorum var. ? *simbirskiana* Grunow, 1884, p. 65, Plate B (2), Figures 44, 45.
Hemiaulus polycystinorum var. *brevicornis* Jousé 1951b, p. 54, Plate 4, Figure 1a–d.
Hemiaulus polycystinorum Ehrenberg, Hajós & Stradner, 1975, p. 931, Plate 6, Figures 4–7; Plate 34, Figure 3; Plate 35, Figures 2–5.
Hemiaulus rossicus Pantocsek, 1889, p. 84; Strelnikova, 1974, p. 102, Plate 43, Figures 1–18; Gleser et al., 1974, Plate 9, Figure 6; Ross et al., 1977, p. 189, Plate 8, Figures 48–52.
Discussion.—Weisse misidentified this diatom as *Hemiaulus antarcticus* Ehrenberg, now known as *Eucampia antarctica* (Castracane) Mangin.

Hemiaulus seymourensis new species
Figure 12.23,24
Description.—Valve weakly silicified, circular to ellipsoidal, flat with steep valve margin. Tips of two horns converge upward toward valve center (out of focus in Fig. 12.23, broken in Fig. 12.24). Valve surface covered with subrectangular areolae in pattern radiating from a submarginal labiate process.
Holotype.—Figure 12.23, USNM 416135.
Type locality.—López de Bertodano Formation, Sample SI-380 (concretion), Seymour Island, northeast Antarctic Peninsula.
Type level.—Lower upper Maastrichtian, Upper Cretaceous.

Hemiaulus sibericus Grunow, 1884
Figure 14.13,14
Hemiaulus sibericus Grunow, 1884, p. 12, Plate 2(B), Figures 34–35; Schmidt et al., 1874–1959, Plate 118, Figure 8; Jousé, 1955, p. 86, Plate 2, Figure 9.

Hemiaulus simplex? Brun, 1894
Figure 14.16, 17
Hemiaulus simplex Brun, 1894, p. 77, Plate 6, Figures 8, 9.

Hemiaulus sporialis Strelnikova, 1971
Figure 15.7–10
Hemiaulus sporialis Strelnikova, 1971, p. 48, Plate 3, Figures 1–10;
Strelnikova, 1974, p. 95, Plate 41, Figures 1–10.

Hemiaulus sp. cf. *H. tumidicornis* Strelnikova, 1971
Figure 14.9,25
cf. *Hemiaulus tumidicornis* Strelnikova, 1971, p. 49, Plate 1, Figures
14–16; Strelnikova, 1974, p. 102, Plate 47, Figures 17–25; Barron, 1985,
p. 141, Plate 10.2, Figures 5–6, 12?

Hemiaulus sp. A—Figure 13.26,27

Hemiaulus sp. B—Figure 14.22

Hemiaulus sp. C—Figure 15.14–17

Hemiaulus sp. D—Figure 14.23,24

Hemiaulus sp. E—Figure 9.22

Hemiaulus sp. F
Figure 13.13,14
Description.—No whole frustules of this diatom encountered. Elongate
horn increasing in size toward terminus, where two to four spines are
present; sparse punctae on horn, narrow keel on outside edge of horn.
Punctae become more numerous on valve at base of horn.
Discussion.—The horn structure is similar to an unnamed form, Spring.
(Grundl.), sp.n.?, in Schmidt et al., 1874–1959, Plate 143, Figures 45,
46. The present form differs in that transverse sulci are present.

Hemiaulus sp. G—Figure 14.6

Hemiaulus sp. H
Figure 14.8
Discussion.—This diatom bears some similarity to *Hemiaulus originalis*
Greville, 1865, p. 29, Plate 3, Figure 9; and *Hemiaulus lobatus* Greville,
1865, p. 29, Plate 3, Figure 9.

Hemiaulus sp. I—Figure 21.23

Hemiaulus sp. J—Figure 15.6

Hemiaulus sp. K—Figure 15.1

Hemiaulus sp. L—Figures 14.15, 15.12, 13

Hemiaulus sp. M—Figure 12.16

Genus *HORODISCUS* Hanna, 1927a

Horodiscus anastomosans (Pantocsek, 1892) new combination
Figure 16.1
Coscinodiscus anastomosans Pantocsek, 1892, Plate 28, Figure 411.
Horodiscus rugosus Hajós in Hajós & Stradner, 1975, p. 925, Plate 1,
Figures 7, 8.
Discussion.—The transfer out of *Coscinodiscus* to *Horodiscus* is based
on the absence of areolae in this species and on the presence of distinct
radial costae.

Genus *HUTTONIA* Grove & Sturt, 1887

Huttonia antiqua Hajós & Stradner, 1975
Figure 12.17
Huttonia antiqua Hajós & Stradner, 1975, p. 929, Plate 5, Figures 13,
14; Plate 31, Figures 1–3.

Genus *HYALODISCUS* Ehrenberg, 1845

Hyalodiscus russicus Pantocsek, 1889
Figure 16.2
Hyalodiscus russicus Pantocsek, 1892, Plate 5, Figure 72.

Genus *INCISORIA* Hajós, 1975

Incisoria inordinata Hajós, 1975
Figure 17.3
Incisoria inordinata Hajós in Hajós & Stradner, 1975, p. 937, Plate 13,
Figures 20, 21.

Incisoria lanceolata Hajós & Stradner, 1975
Figures 12.18, 19; 17.4
Incisoria lanceolata Hajós & Stradner, 1975, p. 937, Plate 13, Figures
22, 25; Plate 36, Figure 5.

Incisoria paleoceanica var. *fuscina* (Schibkova, 1959), n. comb.
Figure 17.5, 6
Grunowiella palaeocenica var. *fuscina* Schibkova in Krotov & Schib-
kova, 1959, p. 128, Plate 5, Figure 14.

Incisoria punctata Hajós & Stradner, 1975
Figure 17.1, 2
Incisoria punctata Hajós & Stradner, 1975, p. 937, Plate 13, Figures 15,
16; Plate 36, Figure 6.

Incisoria sp. A—Figure 17.7, 8

Genus *KENTRODISCUS,* Pantocsek, 1889

Kentrodiscus aculeatus (Hajós, 1975) new combination
Figure 17.15
Pterotheca aculeata Hajós in Hajós & Stradner, 1975, p. 933, Plate 12,
Figures 10, 11.

Kentrodiscus fossilis Pantocsek, 1889
Figure 17.16–17.18
Kentrodiscus fossilis Pantocsek, 1889, p. 75 (mislabeled as *K.*
(*Dicladia*?) *russicus* in Plate 23, Figure 350); Jousé, 1949c, vol. 2, p.
205, Plate 75, Figure 11; Krotov & Schibkova, 1959, p. 126, Plate 5,
Figures 5–7.
Kentordiscus hungaricus Van Heurck, 1896, p. 430, Figure 150
(mislabeled).

Genus *KITTONIA* Grove & Sturt, 1887

Kittonia sp. (fragment)—Figure 13.1

Genus *MELOSIRA* Agardh, 1824

Melosira? patera Long, Fuge & Smith, 1946
Figure 16.7
Melosira patera Long et al., 1946, p. 109, Plate 17, Fig. 18; Hajós &
Stradner, 1975, p. 924, Plate 1, Figures 1, 2.
Also see.—*Ethmodiscus carinatus* Pantocsek, 1892, Plate 29, Figure
430, and *Ethmodiscus russicus* Pantocsek, 1892, Plate 34, Figures 484,
488.

Melosira vetutissima Hajós & Stradner, 1975
Figure 16.10
Melosira vetutissima Hajós & Stradner, 1975, p. 924, Plate 1, Figures
9–12; Plate 18, Figures 1, 2.

Genus *ODONTOTROPIS* Grunow, 1884

Odontotropis cristata (Grunow, 1882) Grunow, 1884
no illustration
Biddulphia? cristata (Grunow in Van Heurek, 1882, Plate 102, Figure 4.
Odontotropis cristata (Grunow) Grunow, 1884, p. 7, Plate 5(E), Figure
58; Hustedt, 1930, p. 857, Figure 511.

Odontotropis sp. Hajós, 1975
no illustration
Odontotropis sp. Hajós in Hajós & Stradner, 1975, p. 930, Figure 13.

Genus *OMPHALOTHECA* Ehrenberg, 1854

Omphalotheca sp. cf. *O. jutlandica* Grunow in Van Heurck, 1883
Figure 17.30, 31
cf. *Omphalotheca? jutlandica* Grunow in Van Heurck, 1883, Plate 83,
Figure 12.

Genus *PARALIA* Heiberg, 1863

Paralia sulcata group (Ehrenberg) Cleve 1873
Figure 18.6, 9
Melosira sulcata (Ehrenberg) Kützing, 1844, p. 55, Plate 2, Figure 7;
Schmidt et al., 1874–1959, Plate 178, Figures 1, 2, 7–24, 38, 39; Plate
183, Figure 10; Strelnikova, 1974, p. 49, Plate 1, Figures 9–14.
Paralia sulcata (Ehrenberg) Cleve, 1873, p. 7.

Genus *PEPONIA* Greville, 1863

Peponia barbadensis Greville, 1863
Figure 17.19
Peponia barbadensis Greville, 1863, p. 76, Plate 4, Figure 25; Schmidt et
al., 1874 1959, Plate 144, Figures 48, 49; Schrader & Fenner, 1976,
p. 992, Plate 36, Figures 10, 13.

Genus *PODOSIRA* Ehrenberg, 1840

Podosira simpla Jousé, 1949b
no illustration
Podosira simpla Jousé in Proschkina-Lavrenko, 1949b, p. 32, Plate 6,
Figure 6.
Pseudopodosira simplex Strelnikova, 1974, p. 51, 52, Plate 2,
Figures 10, 11.

Genus *PSEUDOPODSIRA* Jousé 1949

Pseudopodosira westii (W. Smith) Sheshukova-Poretzkaya & Gleser,
1964
Figure 17.20
Pseudopodosira westii (W. Smith) Sheshukova-Poretskaya & Gleser,
1964, Plate 1, Figures 3, 4; Hajós & Stradner, 1975, p. 924, Plate 1,
Figures 3, 4; Plate 19, Figures 1, 2, 4.
Also see.—*Hyalodiscus russicus* Pantocsek, 1892, Plate 5, Figure 72.

Genus *PSEUDOPYXILLA* Forti, 1909

Pseudopyxilla aculeata Jousé, 1951b
Figure 17.21 22
Pseudopyxilla aculeata Jousé, 1951b, p. 60, Plate 4, Figure 7; emend.
Jousé & Krotov in Krotov & Schibkova, 1959, p. 126, Plate 5, Figures
3, 4.
Pterotheca spada Tempère & Brun sensu Strelnikova, 1974, Plate 56,
Figures 10, 11a,b, *not* Figure 9.
Note.—The funnel-shaped spine-bearing process is extended to a greater
degree in upper Campanian forms (Fig. 17.21), and in Strelnikova
(1974) than in Danian forms (Fig. 17.22).

Pseudopyxilla americana (Ehrenberg) Forti, 1909
no illustration
Rhozosolenia americana Ehrenberg, 1843; Ehrenberg, 1854, Plate 33/3,
Figure 20; Plate 33/17, Figure 14.
Pyxilla americana (Ehrenberg) Grunow in Van Heurck, 1883, Plate 83
bis, Figures 1–2.
Pseudopyxilla americana (Ehrenberg) Forti, 1909, p. 28, 30, Plate 1,
Figures 6, 7; Strelnikova, 1974, p. 112, Plate 54, Figures 1–15.
Not *Pseudopyxilla americana* in Hajós & Stradner, 1975, p. 933, Plate
12, Figure 3.

Pseudopyxilla dubia (Grunow in Van Heurck, 1883) Forti, 1909
Figure 17.23, 24
Pyxilla dubia Grunow in Van Heurck, 1883, Plate 83, Figures 7, 8; Plate
83 bis, Figure 12.
Pseudopyxilla dubia (Grunow in Van Heurck) Forti, 1909, p. 12, Plate 1,
Figure 22; Fenner, 1977, p. 526, Plate 14, Figure 9, Plate 17, Figures
1–6; Gombos & Ciesielski, 1983, p. 603

Pseudopyxilla hungarica (Pantocsek, 1892) Forti, 1909
Figure 17.26, 27
Pyxilla hungarica Pantocsek, 1892, Plate 26, Figure 392.
Pseudopyxilla hungarica (Pantocsek) Forti, 1909, p. 28.
Pseudopyxilla americana (Ehrenberg) Forti, Hajós & Stradner, 1975,
p. 933, Plate 12, Figure 3.

Pseudopyxilla sp. Strelnikova, 1974
Figure 17.25
Pseudopyxilla sp. Strelnikova, 1974, Plate 54, Figure 16.

Pseudopyxilla rossica (Pantocsek, 1892) Forti, 1909
Figure 17.28, 29
Pyxilla rossica (russica) Pantocsek, 1892, Plate 19, Figure 277.
Pseudopyxilla rossica (Pantocsek) Forti, 1909, p. 14, Plate 1, Figure 13.

Genus *PSEUDOSTICTODISCUS* Grunow 1882 in Schmidt et al.,
1874–1959

Pseudostictodiscus picus Hanna, 1927a
Figure 22.2
Pseudostictodiscus picus Hanna, 1927a, p. 28, Plate 3, Figures 1–4.

Genus *PTEROTHECA* Grunow in Van Heurck, 1883

Pterotheca aculeifera (Grunow in Van Heurck, 1883)
Van Heurck 1896
Figure 18.3, 4
Pyxilla? aculeifera Grunow in Van Heurck, 1883, Plate 83, Figures 13, 14.
Pterotheca aculeifera (Grunow) Van Heurck, 1896, p. 430, Figure 151; Fenner, 1977, p. 527, Plate 17, Figures 8–21; Jousé, 1951b, p. 59, Plate 4, Figures 5a,b; Strelnikova, 1974, p. 114, Plate 57, Figures 5–11; *not* Figures 1–4, 12–22.

Pterotheca carinifera (Grunow in Van Heurck, 1883) Forti, 1909
Figure 18.6
Pyxilla? carinifera Grunow in Van Heurck, 1883, Plate 83, Figures 5, 6.
Pterotheca carnifera (Grunow) Forti, 1909, p. 13; Hanna, 1927b, p. 119, Plate 20, Figures 9, 10; Jousé, 1951a, p. 38, Plate 4, Figure 3.

Pterotheca carinifera var. *curvirostris* Jousé, 1955
Figure 18.7
Pterotheca carinifera var. *curvirostris* Jousé, 1955, p. 90, Plate 2, Figure 7.

Pterotheca carinifera var. *tenuis* Jousé, 1951a
Figure 18.8
Pterotheca carinifera var. *tenuis* Jousé, 1951a, p. 40, Plate 4, Figure 4.

Pterotheca clavata Strelnikova, 1974
Figure 17.32
Pterotheca clavata Strelnikova, 1974, p. 115, Plate 58, Figure 35; Dzinoridze et al., 1976, Plate 9, Figure 5.

Pterotheca costata Schibkova, 1959
no illustration
Pterotheca costata Schibkova in Krotov & Schibkova, 1959, p. 127, Plate 5, Figure 9.

Pterotheca cretacea Hajós & Stradner, 1975
Figure 18.9–11
Pterotheca cretacea Hajós & Stradner, 1975, p. 934, Plate 12, Figures 16–18, 21; Plate 26, Figure 1.

Pterotheca crucifera Hanna, 1927a
Figure 18.5
Pterotheca crucifera Hanna, 1927a, p. 30, Plate 4, Figure 5; Hajós & Stradner, 1975, p. 934, Plate 12, Figures 8, 9, 22; Plate 27, Figure 7; Plate 28, Figure 3.

Pterotheca sp. cf. *P. crucifera*
Figure 12.14, 15
cf. *Pterotheca crucifera* Hanna, 1927a, p. 30, Plate 4, Figure 5.

Pterotheca danica (Grunow in Van Heurck, 1883) Forti, 1909
Figure 18.12
Stephanogonia danica Grunow in Van Heurck, 1883, Plate 83 bis, Figures 7, 8.
Pterotheca danica (Grunow) Forti, 1909, p. 13; Hanna, 1927b, p. 119, Plate 20, Figure 11; Gombos & Ciesielski 1983, p. 603, Plate 13, Figures 1–3, 9.

Pterotheca evermanni Hanna, 1927a
Figure 18.13, 14
Pterotheca evermanni Hanna, 1927a, p. 31, Plate 4, Figure 6; Strelnikova, 1974, p. 112, Plate 56, Figures 12–15; Gleser et al., 1974, Plate 12, Figure 4.

Pterotheca infundibulum Krotov, 1959
Figure 18.15
Pterotheca infundibulum Krotov in Krotov & Schibkova, 1959, p. 127, Plate 5, Figure 10.

Pterotheca kittoniana Grunow in Van Heurck, 1883
Figure 18.1, 2
Pyxilla? kittoniana Grunow in Van Heurck, 1883, Plate 83, Figures 10, 11.
Pterotheca (Pyxilla?) kittoniana Grunow in Van Heurck, 1883, Plate 83, bis, Figures 9, 10.
Pterotheca aculeifera (Grunow), Jousé, 1951b, p. 59, Plate 4, Figures 5a, b; Strelnikova, 1974, Plate 57, Figures 1–4, 12–22, not Figures 5–11.

Pterotheca major Jousé, 1955
Figure 18.16
Pterotheca major Jousé, 1955, p. 101, Plate 6, Figure 2, and text Figure 1; Gombos & Ciesielski, 1983, p. 603, Plate 13, Figures 6–8.

Pterotheca minor new species
Figure 12.12,13
Description.—Epitheca (70 × 15 μm) hollow, cylindrical as much as one-half the valve height where one edge remains straight while rest of valve tapers and bends to meet this edge, resulting in an off-center conical top. Heavily silicified at base and in upper third of valve. The cylindrical portion is free of sculpture, whereas the conical portion bears weakly developed longitudinal ridges.
Holotype.—Figure 12.12, USNM 416136.
Paratype.—Figure 12.13, USNM 416137.
Type locality.—Upper López de Bertodano Formation, Sample SI-511 (concretion), Seymour Island, northeast Antarctic Peninsula.
Type level.—Danian, Paleocene.

Ptertotheca pokrovskajae Jousé, 1951b
Figure 18.19–23
Pterotheca pokrovskajae Jousé, 1951b, p. 58, Figure 2.
Discussion.—The presence of pores over the entire height of the valve and the curved nature of the process seem to characterize this species.

Jousé illustrates only the conical portion of the valve. The cylindrical portion bears abundant pores that render it fragile and easily broken. The presence of the pores on the conical valve seems unique among *Pterotheca.* If these pores are in some way environmentally controlled, then this species and the varieties may fit into groups mentioned previously, i.e., *P. danica, P. carinifera,* and *P. carnifera* var. *curvirostris.*

Several forms present in the Seymour Island material differ from *P. pokrovskajae* s. str. in that the elongate process is straight and not curved. Note that the pores increase in density toward the right and are most abundant to the left of the lateral siliceous flanges. Jousé's hand-drawn figure shows a random scatter of the pores.

Pterotheca pokrovskajae Jousé var.
Figure 12.9,10
Discussion.—This diatom has the same overall appearance of *P. pokrovskajae* s. str. but does not possess the characteristic pores of this species. Perhaps this diatom is a curved variety of *P. carinifera.*

Pterotheca spada Brun & Tempère, 1889
Figure 18.24,25
Pterotheca spada Brun & Tempère, 1889, p. 50, Plate 1, Figure 7; Dzinoridze et al., 1976, Plate 9, Figure 7; Strelnikova, 1974, p. 113, Plate 56, Figure 9, not Figures 10, 11; Gombos & Ciesielski, 1983, p. 603, Plate 13, Figures 4, 5.

Pterotheca subulata Grunow in Van Heurck, 1883
Figure 18.26
Pterotheca (Pyxilla??) *subulata* Grunow in Van Heurck, 1883, Plate 83 bis, Figure 6.
Pyxilla subulata (Grunow) Wolle, 1890, Plate 65, Figure 19.

Pterotheca trojana new species
Figure 12.11
Description.—Epitheca small (12×3 μm), cylindrical, weakly silicified, slightly curved. Terminus rounded and possessing a short blunt spine. Two transverse siliceous ribs, one near the terminus and the other one-third length down the valve, constrict the cylinder, giving the upper portion a bulbous shape.
Holotype.—Figure 12.11, USNM 416138.
Type locality.—Upper López de Bertodano Formation, Sample 380 (concretion), Seymour Island, northeast Antarctic Peninsula.
Type level.—Lower upper Maastrichtian, Upper Cretaceous.

Pterotheca uralica Jousé, 1949
Figure 19.1
Pterotheca uralica Jousé, 1949a, p. 68, Plate 1, Figure 7; Fenner, 1977, p. 527, Plate 17, Figure 33; Strelnikova, 1974, Plate 57, Figures 27–30a,b.

Pterotheca sp. A—Figure 19.2

Pterotheca sp. B—Figure 19.3,4

Genus *PYRGODISCUS* Kotton in Cleve, 1885

Pyrgodiscus sp.—Figure 19.5
Discussion.—This is similar to *Pyrgodiscus* from Simbirsk, illustrated in Schmidt et al., 1874–1959, Plate 100, Figure 14.

Pyrgodiscus? triangulatus Hajós & Stradner, 1975
Figure 19.6,7
Pyrgodiscus triangulatus Hajós & Stradner, 1975, p. 928, Plate 18, Figures 5, 6; Figure 11a,b.
Discussion.—This does not fit well in *Pyrgodiscus;* it is either the hypovalve of a *Pterotheca* or the epivalve with the spine broken off.

Genus *PYXIDICULA* Ehrenberg, 1838

Pyxidicula minuta Grunow, 1884
Figure 19.11,12
Pyxidicula minuta Grunow, 1884, p. 92, Plate 5(E), Figure 6; Hajós & Stradner, 1975, p. 925, Plate 1, Figures 16–18.

Genus *RHIZOSOLENIA* Ehrenberg, 1841

Rhizosolenia cretacea Hajós & Stradner, 1975
Figure 19.8
Rhizosolenia cretacea Hajós & Stradner, 1975, p. 929, Plate 7, Figure 7; Plate 31, Figures 4–6; Barron, 1985, p. 141, Plate 10.3, Figure 1.

Rhizosolenia sp. A—Figure 19.9

Rhizosolenia? sp. B—Figure 12.6–8

Genus *RUTILARIA* Greville, 1863

Rutilaria? sp. Figure 19.10

Genus *SCEPTRONEIS* Enrenberg, 1844

Sceptroneis grunowii Anissimova, 1938
Figure 17.9,10
Sceptroneis grunowii Anissimova, Hajós & Stradner, 1975, p. 998, Plate 22, Figures 26–28; Plate 23, Figure 8; Plate 25, Figures 7, 9; Schrader & Fenner, 1976, p. 998, Plate 22, Figures 26–28; Plate 23, Figure 8, Plate 25, Figures 7, 9.

Sceptroneis praecaducea Hajós & Stradner, 1975
Figure 17.11
Sceptroneis praecaducea Hajós & Stradner, 1975, p. 936, Plate 13, Figures 13, 14; Plate 36, Figures 1–4; Schrader & Fenner, 1976, p. 999, Plate 23, Figures 9, 21; Plate 22, Figure 36; Plate 25, Figure 13.

Sceptroneis sp. A—Figure 17.12

Sceptroneis? sp. B—Figure 17.13,14

Genus *SKELETONEMA* Greville, 1865

Skeletonema penicillis Grunow in Van Heurck, 1883
Figure 18.17,18
Skeletonema? penicillus Grunow in Van Heurck, 1883, Plate 83 ter., Figure 6.
Hemiaulus spore in Schmidt et al., 1874–1959, Plate 142, Figure 1.
Pterotheca cornuta Schibkova in Krotov & Schibkova, 1959, p. 126, Plate 5, Figure 8.

Skeletonema polychaetum Strelnikova, 1971
Figure 16.14,15
Skeletonema polychaetum Strelnikova, 1971, p. 42; Strelnikova, 1974, p. 54, Plate 3, Figures 3–7; Barron, 1985, p. 141, Plate 10, Figures 2–4.

Skeletonema subantarctica Hajós, 1975
Figure 16.16
Skeletonema subantarctica Hajós in Hajós & Stradner, 1975, p. 925, Plate 2, Figure 1.

Skeletonema sp. A—Figure 16.13

Genus **SPHYNCTOLETHUS** Hanna, 1927a

Sphynctolethus sp. cf. *S. monstrosus* Hanna, 1927a
no illustration
cf. *Sphynctolethus monstrosus* Hanna, 1927a, p. 32, Plate 4, Figures 7, 8.

Genus **STELLARIMA** Hasle & Sims, 1986

Stellarima steinyi (Hanna, 1927a) Hasle & Sims, 1986
Figure 20.3,4
Coscinodiscus steinyi Hanna, 1927a, p. 19, Plate 2, Figures 5, 6.
Symbolophora steinyi (Hanna) Nikolaev, 1983, p. 1125, Plate 2, Figures 8, 9.
Stellarima steinyi (Hanna) Hasle & Sims, 1986, p. 111.

Genus **STEPHANOPYXIS** Ehrenberg, 1844

Stephanopyxis broschii Grunow, 1884
Figure 15.11
Stephanopyxis broschii Grunow, 1884, p. 38, Plate 5(E), Figures 26–28; Strelnikova, 1974, p. 57, Plate 5, Figures 3a,b–6.

Stephanopyxis delicatabilis Pantocsek, 1889
Figure 19.13,14
Stephanopyxis delictabilis Pantocsek, 1889, Plate 23, Figures 408, 409.
Stephanopyxis antiqua Jousé, 1951b, p. 46, Plate 1, Figures 3a,b; Gleser et al., 1974, Plate 9, Figure 9; Fenner, 1985, p. 738, Figure 14.13, not *S. antiqua* Pantocsek, 1892, p. 96, Plate 19, Figure 280.

Stephanopyxis edita Jousé, 1955
Figure 19.18,19
Stephanopyxis edita Jousé, 1955, p. 85, Plate 1, Figure 7.

Stephanopyxis sp. cf. *S. grunowii* Grove & Sturt, 1888
no illustration
cf. *Stephanopyxis grunowii* Grove & Sturt, 1888 in Schmidt et al., 1874–1959, Plate 130, Figures 1–4; Gombos, 1977, p. 597, Plate 28, Figures 3–5; Plate 31, Figures 3–5; Plate 31, Figures 1, 2, 7; Plate 32, Figures 1–3.

Stephanopyxis hannai Hajós, 1975
Figure 19.20,21
Stephanopyxis hannai Hajós in Hajós & Stradner, 1975, p. 925, Plate 2, Figures 9, 10, text Fig. 5a,b.

Stephanopyxis lavrenkoi Jousé, 1949
Figures 19.22,23
Stephanopyxis lavrenkoi Jousé in Proschkina-Lavrenko, 1949, p. 40, Plate 10, Figure 9; Strelnikova, 1974, p. 60, Plate 7, Figures 7a,b.

Stephanopyxis marginata Grunow, 1884
Figure 12.1
Stephanopyxis marginata Grunow, 1884, p. 38(90), Plate 5(E), Figure 17; Hustedt, 1930, p. 309, Figure 148.
Stephanopyxis ferox Greville, Witt, 1886, Plate 6, Figure 16.

Stephanopyxis maxima Pantocsek, 1892
Figure 20.5
Stephanopyxis maxima Pantocsek, 1892, Plate 26, Figure 384.

Stephanopyxis simonseni Hajós, 1975
Figure 19.24
Stephanopyxis simonseni Hajós in Hajós & Stradner, 1975, p. 926, Plate 2, Figures 7, 8.

Stephanopyxis superba (Greville, 1861) Grunow, 1884
Figure 19.25
Cresswellia superba Greville, 1861, p. 68, Plate 8, Figures 3–5.
Stephanopyxis superba (Greville) Grunow, 1884, p. 39; Schmidt et al., 1874–1959, Plate 123, Figures 3–8; Hajós, 1976, p. 926, Plate 2, Figures 11, 12; Gombos, 1977, p. 597, Plate 29, Figures 1–4.

Stephanopyxis turris (Greville & Arnott, 1857) Ralfs in Pritchard, 1861
Figure 19.26, 27
Stephanopyxis turris (Greville & Arnott) Ralfs in Pritchard, 1861, p. 826, Plate 5, Figure 74; Grunow in Van Heurck, 1883, Plate 83 ter., Figure 12; Hustedt, 1930, p. 304, Figure 104.

Stephanopyxis turris var. **cylindrus** Grunow, 1884
Figure 19.28
Stephanopyxis turris var. *cylindrus* Grunow, 1884, p. 35, Plate 5(E), Figures 9, 14.

Stephanopyxis uralensis Strelnikova, 1971
Figure 19.17
Stephanopyxis uralensis Strelnikova, 1971, p. 42, Plate 1, Figures 6, 7; Strelnikova, 1974, p. 58, Plate 7, Figures 1–4; Gleser et al., 1974, Plate 11, Figure 11.

Stephanopyxis weyprechtii (Grunow, 1884) Hajós, 1975
no illustration
Pyxidicula weyprechtii Grunow, 1884, p. 92, Plate 5(E), Figure 5.
Stephanopyxis weyprechtii (Grunow) Hajós in Hajós & Stradner, 1975, p. 926, Plate 22, Figures 1, 2, 5.

Stephanopyxis sp. A—Figure 20.1

Stephanopyxis sp. B—Figure 20.2

Stephanopyxis sp. C—Figure 19.29,30

Stephanopyxis sp. D—Figure 11.13

Genus *THALASSIOSIROPSIS* Hasle & Syvertsen, 1985

Thalassiosiropsis wittiana (Pantocsek, 1889) Hasle, 1985
Figure 20.6
Coscinodiscus wittianus Pantocsek, 1889, p. 119; Gleser et al., 1974,
Plate 8, Figure 8; Barron, 1985, p. 141, Plate 10.1, Figure 5.
Coscinodiscus lineatus sensu Weiss, Stelnikova, 1974, p. 62, Plate 9,
Figures 3–12; Hajós & Stradner, 1975, p. 927, Plate 3, Figures 1–3; Plate
38, Figure 1.
Thalassiosiropsis wittiana (Pantocsek) Hasle in Hasle & Syvertsen, 1985,
p. 89, Plates 1–5.

Genus *TRICERATIUM* Ehrenberg, 1839

Triceratium cellulosum Greville, 1861
no illustration
Triceratium cellulosum Greville, 1861, p. 44, Plate 4, Figure 14; Schmidt
et al., 1874–1959, Plate 95, Figures 28–32; Fenner, 1985, p. 439, 440,
Figures 13.1–13.3
Triceratium cellulosum var. *simbirskiana* Witt, 1886, p. 168–169, Plate
12, Figures 8–10.

Triceratium edgardi Hajós, 1975
Figure 21.25
Triceratium edgardi Hajós in Hajós & Stradner, 1975, p. 929, Plate 8,
Figures 13, 14.

Triceratium flos Ehrenberg, 1855
no illustration
Triceratium flos Ehrenberg, Witt, 1885, p. 33, Plate 9, Figure 13; Hus-
tedt, 1930, p. 827, Fig. 489; Krotov & Schibkova, 1959, p. 116, Plate 2,
Figure 1; Strelnikova, 1974, p. 87, 88, Plate 31, Figures 15, 16.
Biddulphia flos (Ehrenberg) Grunow, 1884, p. 7, Plate 5(E), Figure 59;
Plate 2, Figure 19.

Triceratium flos var. **intermedia** Schmidt, 1886
Figure 20.10
Triceratium flos var. *intermedia* Schmidt et al., 1874–1959, Plate 95,
Figures 33, 34.

Triceratium indefinitum (Jousé, 1951b) Strelnikova, 1974
Figures 20.11, 21.4–7
Trinacria indefinita Jousé, 1951b, p. 50, Plate 2, Figures 5a,b; Gleser et
al., 1974, Plate 9, Figure 5.
Triceratium indefinitum (Jousé) Strelnikova, 1974, p. 82, Plate 30, Fig-
ures 1–29; Plate 31, Figures 1–6.

Triceratium nobile Witt, 1885
Figure 20.12
Triceratium nobile Witt, 1885, p. 34, Plate 10, Figure 3; Plate 12,
Figures 4, 7; Schmidt et al., 1874–1959, Plate 111, Figures 26–29; Plate
150, Figure 25; Jousé 1951a, Plate 3, Figure 4; Hajós & Stradner, 1975,
p. 930, Plate 9, Figure 3.

Triceratium sentum Witt in Schmidt, 1890
Figure 21.1–3
Triceratium sentum Witt in Schmidt et al., 1874–1959, Plate 150, Fig-
urcs 2–6.
Trigonium sentum (Witt) Hustedt in Schmidt et al., 1874–1959, Plate
467, Figures 25–26.

Triceratium simplicissimum Witt, 1885
Figure 20.13
Triceratium simplicissimum Witt, 1885, p. 34, Plate 8, Figures 7–9;
Schmidt et al., 1874–1959, Plate 150, Figure 16.
Discussion.—This species resembles a small *Trinacria excavata.*
Note.—Compare this with *T. mesoleum* Grunow in Van Heurck, 1883,
Plate 113, Figure 14; Jousé, 1951a, p. 31, Plate 3, Figure 5.

Triceratium solenoceros Ehrenberg, 1844
Figure 20.7, 8
Triceratium solenoceros Ehrenberg, 1844, p. 273; Ralfs in Pritchard,
1861, p. 856; Schmidt et al., 1874–1959, Plate 77, Figure 21; Plate 96,
Figure 11; Pantocsek, 1886, Plate 28, Figure 286; Wolle, 1880, Plate
101, Figure 3.
Triceratium sp. Schmidt et al., 1874–1959, Plate 96, Figure 9.

Triceratium sp. A—Figure 20.9

Genus *TRINACRIA* Heiberg, 1863

Trinacria acutangulum (Strelnikova, 1974) n. comb.
Figure 21.8–10, 12
Triceratium acutangulum Strelnikova, 1974, p. 83, Plate 32, Fig-
ures 1–10.
Triceratium idoneum Pantocsek sensu Hajós & Stradner, 1975, p. 930,
Plate 9, Figure 6.
Note.—Compare this with *Trinacria* sp. B, Figure 20.14, 15.

Trinacria aries Schmidt, 1886
Figure 21.14
Trinacria aries Schmidt et al., 1874–1959, Plate 96, Figures 14–17;
Plate 150, Figures 14, 15; Hanna, 1927a, p. 36, Plate 5, Figures 1–2.

Trinacria excavata Heiberg, 1863
Figure 21.15, 16
Trinacria excavata Heiberg, 1863, p. 51, Plate 4, Figure 9; Hustedt,
1930, p. 887, 888, Figure 352; Gombos, 1977, p. 599, Plate 37, Figure 6;
Gombos & Ciesielksi, 1983, p. 605, Plate 17, Figure 8; Fenner, 1977,
p. 535.

Trinacria grevillei Witt, 1885
Figure 21.17
Trinacria grevillei Witt, 1885, p. 35, Plate 12, Figure 11; Schmidt et al.,
1874–1959, Plate 96, Figure 32; Plate 110, Figure 6.

Trinacria heibergii var. **rostratum** Jousé, 1951a
Figure 21.18
Trinacria heibergii var. *rostratum* Jousé, 1951a, Plate 3, Figures 1a,b.

Trinacria insipiens Witt, 1885
Figure 21.19
Trinacria insipiens Witt, 1885, p. 36, Plate 10, Figure 1; Plate 11,
Figures 5, 7, 11; Plate 12, Figure 2; Hanna, 1927b, p. 123, Plate 21,
Figure 7.

Trinacria pileolus (Ehrenberg, 1844) Grunow, 1884
Figure 21.20
Trinacria pileolus Grunow, 1884, p. 68, Plate 2(B), Figures 59, 60;
Schmidt et al., 1874–1959, Plate 97, Figures 11–14; Plate 111, Figures
16, 17; Hajós & Stradner, 1975, p.932, Plate 9, Figures 7, 8.

Trinacria swastika (Long, Fuge & Smith, 1946) new combination
Figure 21.11
Triceratium swastika Long et al., 1946, p. 114, Plate 17, Figure 8.
Tortilaria briggeri Barker & Meakin, 1948, p. 234, Plate 28, Figures 5, 6.

Trinacria tristictia Hanna, 1927a
Figure 21.21
Trinacria tristictia Hanna, 1927a, p. 38, Plate 5, Figures 11, 12; Proschkina-Lavrenko, 1949, p. 194, Plate 96, Figure 5; Hajós &: Stradner, 1975, p. 932, Plate 9, Figures 5, 6.

Trinacria sp. A—Figure 21.24

Trinacria sp. B—Figure 20.14, 15
Note.—Compare with *Trinacria acutangulum* (Fig. 21.8–10) and with *Trinacria anissimovii* Jousé, 1949a, p. 66, Figures 1a,b; Hajós & Stradner, 1975, p. 932, Plate 10, Figure 2.

Trinacria sp. C—Figure 21.13

Genus *WITTIA* Pantocsek, 1889

Wittia macellarii new species
Figure 16.3–6
Description.—Valve discoid, undulating with elevated ring often present at half the radius; valve covered with closely spaced coarse radial costae that extend from broad hyaline margin to central hyaline ring; fine puncta; irregularly arranged between costae; as much as eight marginal processes equally spaced around margin; situated in the hyaline margin with a triangular extension of the marginal hyaline ring extending short distance toward valve center.
Discussion.—This diatom falls within the genus *Wittia* with the presence of the characteristic radiant striae in the median portion of the valve and the presence of a central hyaline ring (see Van Heurck, 1896, p. 499). Other features, however, suggest that it may be a new genus, distinct from *Wittia*. This species is similar to genus *Lepidodiscus,* yet it lacks the marginal crenulations and central structure. The presence of processes, as many as six, equally spaced around the margin, suggests a relationship with the genus *Aulacodiscus*. Simonsen (1972) suggested there are close ties between these genera. It is named after Carlos Macellari, who, along with B. Huber, mapped and described the Upper Cretaceous and lower Paleocene sequence on Seymour Island, and collected the samples used in this study.
Holotype.—Figure 16. 4, USNM 416139.
Type locality.—Sobral Formation, Sample SI-76 (concretion), Seymour Island, northeast Antarctic Peninsula.
Type level.—Danian, Paleocene.

Genus *XANTHIOPYXIS* Ehrenberg, 1844

Xanthipyxis granti Hanna, 1927a
Figure 16.11, 12
Xanthiopxyis granti Hanna, 1927a, p. 39, Plate 5, Figures 13, 14; Hanna, 1934, p. 355, Plate 48, Figures 10, 11; Hajós & Stradner, 1975, p. 927, Plate 4, Figures 16, 17; Plate 26, Figures 4, 5; Plate 35, Figure 7.

Genus and sp. indet. Jousé, 1951b
Figure 13.2
Genus et sp. indet. Jousé, 1951b, p. 62, Plate 4, Figures 9a,b.

Silicoflagellates

Genus *CORBISEMA* Hanna, 1928

Corbisema apiculata (Lemmermann, 1901) Hanna, 1931
Figure 22.4, 5
Dictyocha triancantha Ehrenberg var. *apiculata* Lemmermann, 1901, p. 259, Plate 10, Figure 19, 20. Schulz, 1928, p. 247–249, Figures 27, 28, 29a, 73; Gemeinhardt, 1930, p. 41, Figure 20; Deflandre, 1932b, Figure 32; Gleser, 1966, p. 245, Plate 6, Figure 5.
Corbisema apiculata (Lemmermann) Hanna, 1931, p. 198, Plate D, Figure 2; Frenguelli, 1940; Figure 2h; Perch-Nielsen, 1975b, p. 685, Plate 2, Figures 15, 16, 19; Plate 3, Figures 19, 20, 24; Plate 15, Figures 1, 2.

Corbisema apiculata f. ***minor*** (Schultz, 1928) n. comb.
Figure 22.6–9
Dictyocha triacantha var. *apiculata* f. *minor* Schulz, 1928, p. 249, Figure 29a, (not Fig. 29b); Gemeinhardt, 1930, p. 42, Figure 31; Gleser, 1966, p. 245, Plate 6, Figures 2–4; Plate 31, Figures 4–6.
Corbisema minor (Schulz) Perch-Nielsen, 1975b, p. 686, Plate 3, Figure 7 (not Figure 120.

Corbisema sp. cf. ***C. archangelskiana*** (Schult, 1928) Frenguelli, 1940
Figure 22.12
Dictyocha triacantha var. *archangelskiana* Schulz, 1928, p. 250, 281, Figs. 33a–c, 77, 78?; Gemeinhardt, 1930, p. 45, 46, Figure 37a,b?.
Dictyocha archangelskiana (Schulz) Gleser, 1966, p. 232–233, Figure 5.3, Plate 8, Figures 6, 7.
Corbisema archangelskiana (Schulz) Frenguelli, 1940, Figure 12a; Ling, 1972, p. 152–153, Plate 23, Figure 18; Bukry, 1974, Plate 1, Figure 1; Ciesielski, 1975, p. 654, Plate 3, Figures 1, 2; Bukry, 1975b, Plate 3, Figure 4.

Corbisema aspera (Schulz, 1928) n. comb.
Figure 22.10, 11
Dictyocha triacantha var. *apiculata* f. *aspera* Schulz, 1928, p. 249, 282, Figure 28.

Corbisema geometrica Hanna, 1928a
Figure 22.13, 14
Corbisema geometrica Hanna, 1928a, p. 261, Plate 14, Figures 1, 2; Deflandre, 1950, p. 133, 134, 136, 137, 139; Gleser, 1966, p. 253, Plate 9, Figure 7; Mandra, 1968, p. 248, Figures 3–5; Cornell, 1974, p. 4, Plate 1, Figures 1, 5, 9, 14, 16; Bukry & Foster, 1974, Figure 1b; Perch-Nielsen, 1975b, p. 685, Plate 2, Figures 1–6, 8; Hajós & Stradner, 1975, p. 938, Plate 15, Figure 1.

Corbisema hastata (Lemmermann, 1901) Bukry, 1973
Figure 22.15, 17, 18
Dictyocha triacantha var. *hastata* Lemmermann, 1901, p. 259, Plate 10, Figures 16, 19; Schulz, 1928, p. 249, Figures 31a,b; Gemeinhardt, 1930, p. 43, Figures 35a–c; Deflandre, 1932a, Figure 28; Gleser, 1966, p. 231, Plate 6, Figure 6; Plate 7, Figures 3, 5; Plate 31, Figure 9.
Corbisema hastata (Lemmermann) Frenguelli, 1940, Figure 12c; Bukry, 1973, p. 892, Plate 1, Figure 1; Bukry & Foster, 1974, Figure 1c.
Corbisema hastata hastata (Lemmermann) Bukry, 1975a, p. 853, Plate 1, Figure 9; Bukry, 1976, p. 892, Plate 4, Figures 9–16.

Corbisema hastata var. ***globulata*** Bukry, 1976
Figure 22.16, 19
Corbisema triacantha var. *hastata* Lemmermann, Gleser, 1966, p. 248, Plate 7, Figure 1.

Corbisema hastata globulata Bukry, 1976, p. 892, Plate 4, Figures 1–8; Bukry, 1977, p. 831, Plate 1, Figure 2; Bukry, 1978a, p. 815, Plate 1, Figures 3–5; Bukry, 1978b, p. 784.

Corbisema hastata f. *miranda* Bukry, 1984
Figure 23.2
Dichtyocha triacantha var. *apiculata* f. *minor* Schulz, 1928, p. 249, Figure 29b (not Figure 29a).
Corbisema hastata minor (Schulz) Bukry, 1975a, p. 854, Plate 1, Figure 10; Bukry, 1978a, Plate 1, Figures 8, 9.
Corbisema hastata miranda Bukry in Barron et al., 1984, p. 150.

Corbisema hastata? var.—Figure 23.1

Corbisema inermis Dumitrica, 1973b
Figure 23.3
Dictyocha triacantha var. *inermis* Lemmermann, 1901, p. 259, Plate 10, Figure 21; Schulz, 1928, p. 248, Figures 30a,b; Gemeinhardt, 1930, p. 43, Figure 43.
Dictyocha triacantha var. *inermis* f. *inermis* Gleser, 1966, p. 230, Plate 8, Figures 1, 2; Plate 32, Figure 1.
Corbisema inermis (Lemmermann) Dumitrica, 1973b, p. 845–846, Plate 12, Figures 7–9.
Corbisema inermis inermis (Lemmermann) Bukry, 1976, p. 892, Plate 5, Figures 1–3.

Corbisema inermis Dumitrica var.—Figure 24.9

Corbisema lateradiata (Schulz, 1928) Perch-Nielsen, 1975b
Figure 24.10–12
Dictyocha triacantha var. *apiculata* f. *late-radiata* Schulz, 1928, p. 281, Figure 73; Gemeinhardt, 1930, p. 43, Figure 32.
Corbisema geometrica var. *apiculata* Jousé, 1949a, p. 78, Plate 2, Figure 5; Jousé, 1951b, p. 63, Plate 6, Figure 2; Cornell, 1974, p. 42, Plate 1, Figures 8, 12, 13, 17; Hajós & Stradner, 1975, p. 938, Plate 15, Figures 2, 3, 5.
Corbisema geometrica Hanna, Ling, 1972, p. 154, Plate 24, Figure 2; McPherson & Ling, 1973, p. 476, Plate 1, Figures 1–4.
Corbisema geometrica lateradiata (Schulz) Bukry, 1975a, p. 853, Plate 1, Figure 8.
Corbisema lateradiata (Schulz) Perch-Nielsen, 1975b, p. 686, Plate 2, Figures 7, 10, 11, 17, 18.

Corbisema sp. variant
Figure 23.5,6
Discussion.—This species is similar to *Cornua* sp. A (below and *C. trifurcata,* yet only one corner of the basal ring is open. There are converging extensions of the basal ring and well-developed "feet" at the extensions of the basal ring. Also note that portions of the skeleton near the apices and at the end of the open portion of the ring are not hollow.

Genus *CORNUA* Schulz, 1928

Cornua sp. A
Figure 23.4,7
Discussion.—This species is similar to *Cornua trifurcata* Schultz in that the basal ring is incomplete. This silicoflagellate and the above variant of *Corbisema* sp. may represent (as suggested by Gleser, 1966) an evolutionary transition from *Lyramula* to *Corbisema,* with the progressive closure of the basal ring.

Cornua trifurcata Schulz, 1928
no illustration
Cornua trifurcata Schulz, 1928, p. 285, Figure 83a–c; Gleser, 1966, p. 239, 240, Plate 3, Figures 3–5.

Genus *DICTYOCHA* Ehrenberg, 1837

Dictyocha sp. cf. *D. navicula* var. *minor* Schulz, 1928
Figure 23.8
cf. *Dictyocha navicula* v. *minor* Schulz, 1928, p. 246, Figure 22.

Dictyocha staurodon f. *minor* Schulz, 1928
Figure 23.9
Dictyocha staurodon f. *minor* Schulz, 1928, p. 251, Figure 34d,e; Gemeinhardt, 1930, p. 46, Figure 38d.

Genus *LYRAMULA* Hanna, 1928a

Lyramula deflandrei Perch-Nielsen & Edwards, 1975
Figure 23.13,14
Lyramula deflandrei Perch-Nielsen & Edwards, in Perch-Nielsen, 1975b, p. 688, Plate 9, Figures 8–17; Hajós & Stradner, 1975, p. 938, Plate 16, Figures 1, 2.

Lyramula furcula Hanna, 1928a
Figure 23.15
Lyramula furcula Hanna, 1928a, p. 262, Plate 41, Figures 4, 5; Deflandre, 1950 p. 61, Figures 163, 167–169; Perch-Nielsen, 1975b, p. 688, Plate 9, Figures 18–21, 26, 27; Hajós & Stradner, 1975, p. 938, Plate 16, Figures 7, 8; Plate 40, Figure 2.

Lyramula sp. cf. *L. furcula* Hanna—Figure 23.18

Lyramula minor (Deflandre, 1940) Deflandre, 1950
Figure 23.165
Lyramula furcula var. *minor* Deflandre, 1940, p. 509, Figures 7–10; Hajós & Stradner, 1975, p. 938, Plate 16, Figures 4–6, Plate 40, Figure 1.
Lyramula minor (Deflandre) Deflandre, 1950, p. 62, Figures 170–173; Perch-Nielsen, 1975b, p. 688, Plate 9, Figure 17.

Lyramula simplex Hanna, 1928a
Figure 23.17
Lyramula simplex Hanna, 1928a, p. 262, Plate 41, Figure 6; Deflandre, 1950, p. 61, Figures 164, 165; Gleser, 1966, p. 221, Plate 2, Figures 1, 2, 4; Perch-Nielsen, 1975b, p. 688, Plate 9, Figures 24, 25; Hajós & Stradner, 1975, p. 938, Plate 17, Figure 1.

Genus *VALLACERTA* Hanna, 1928a

Vallacerta siderea (Schulz, 1928) Bukry, 1981a
Figure 24.4–7
Dictyocha siderea Schulz, 1928, p. 284, Figures 81a,b; Gemeinhardt, 1930, p. 56, Figure 47.
Dictyocha siderea var. *quadrata* Schulz, 1928, p. 284–285, Figures 82a,b, *not Vallacerta quadrata* Hajós in Hajós & Stradner, 1975, p. 939, Plate 16, Figure 9, which possesses a central dome.

Vallacerta hortoni Hanna (in part) in Gleser, 1966, p. 235.
Vallacerta siderea (Schulz) Ling, McPherson & Clark, 1973, p. 360–361, Figures 2a–e; McPherson & Ling, 1973, p. 478, Plate 2, Figures 1–2, not a valid transfer according to Bukry (1981a).
Vallacerta siderea (Schulz) Bukry, 1981a, p. 62.
Discussion.—The majority of specimens of *V. siderea* from Seymour Island are of the five-spine form.

Vallacerta tumidula Gleser, 1959
Figure 23.19, 20

Vallacerta tumidula Gleser, 1959, p. 107, Figures 4, 5; Gleser, 1966, p. 236, Plate 1, Figures 3–6; Hajós & Stradner, 1975, p. 939, Plate 17, Figures 4, 5.

Silicoflagellate genus A
Figure 23.10, 11

Discussion.—This form of silicoflagellate, similar to *Distephanus polyactis* (Ehrenberg) is unknown from the Cretaceous. While placement of this form in the genus *Paradictyocha* makes sense morphologically, the forms are probably homeomorphic because the phylogenetic history and younger geologic record of *Paradictyocha* suggests the two forms are unrelated (see Dumitrica, 1973b, for description and discussion). A silicoflagellate similar to the figured upper Maastrichtian specimen is recorded in Gemeinhardt, 1930, p. 69, Figures 58d, 59c from Mors, Denmark. A new genus is not proposed here because only one specimen was recovered.

Silicoflagellate genus B, species 1
Figure 23.12

Discussion.—The illustrated specimen is from DSDP Hole 275. Only two similar forms were encountered in Seymour Island material (Fig. 24.8). These differ by possession of an apical ring. It appears that a group of silicoflagellates with a general *Mesocena/Paradictyocha* structure was living during the upper Campanian and Maastrichtian. The blunt spines on the basal ring, plus the presence of basal pikes at and between the basal ring spines and the convexity of the basal ring between the apical spines, separate this genus from the above genus A (D. Bukry, personal communications, 1986). Perhaps this form is ancestral to genus A. These forms indicate the need for further discussion of Cretaceous and early Cenozoic silicoflagellate phylogeny. Did they give rise to Cenozoic distephanids, or did they disappear with *Lyramula* and *Vallacerta* at the K/T boundary?

Silicoflagellate genus B, species 2—Figure 24.8

Genus & species indet. A—Figure 24.1

Ebridians

Genus *AMMODOCHIUM* Hovasse, 1932c

***Ammodochium danicum*?** Deflandre, 1932b
Figure 23.21, 22

Ammodochium danicum Deflandre, 1932b, p. 305–307, Figures 14–23; Deflandre, 1951, p. 53, Figures 13, 14.
Discussion.—The specimens illustrated here represent partial specimens, including only the median tripod.

Endoskeletal dinoflagellates

Genus *ACTINISCUS* Ehrenberg, 1854

Actiniscus sp. cf. ***A. pentasterias*** Ehrenberg, 1840
Figure 24.3

cf. *Actiniscus pentasterias* Ehrenberg, Perch-Nielsen, 1975a, p. 882, Plate 10, Figures 2–10, 16.

Genus *CARDUIFOLIA* Hovasse, 1932a,c
See discussion of *Carduifolia* in Dumitrica (1973a).

Carduifolia onoporides Hovasse, 1932a ex Hovasse 1932c
Figure 23.23, 24

Carduifolia onoporides Hovasse, 1932a, p. 126, Figures 9a,b,c; Frenguelli, 1940, p. 87, Figure 25d;
Carduifolia sp. Frenguelli, 1940, p. 87, Figure 25e.
Carduifolia avipes Hovasse, 1943, p. 280, Figure 5.
Carduifolia cf. *C. onoporides* Hovasse, Perch-Nielsen, 1975a, p. 883, Plate 10, Figures 26–30.

Chrysophyte cysts (Archaeomonads)

Archaeomonad sp. A—Figure 24.2

→

Figure 9. Photomicrographs of diatoms from Seymour Island. Bar scales = 10 μm at magnifications of ×560, ×900, and ×1,400 (from left to right). 1, *Acanthodiscus antarcticus* Hajos & Stradner, SI-19, ×900. 2, 3, *Acanthodiscus ornatus* Hajos & Stradner, SI-76, ×560. 4, 5, *Acanthodiscus schmidtii* Jouse. 4, SI-76, ×900; 5, SI-76, ×560. 6, *Acanthodiscus* sp., SI-19, ×560. 7, 8, *Actinoptychus packi* Hanna, SI-76, ×900. 9, *Actinoptychus taffi* Hanna, SI-408, ×900. 10, 11, *Anaulus acutus?* Brun, SI-19, ×900. 12, 14, *Anaulus sibericus* Strelnikova. 12, SI-19, ×900; 14, SI-408, ×900. 13, *Anaulus sibericus?* Strelnikova, SI-19, ×900. 15, *Anaulus weyprechtii* (Grunow, SI-70, ×900. 16, 17, *Anaulus* sp. A, SI-19, ×1,400. 18, 19, *Arachnoidiscus simbirskianus* Pantocsek, 18. SI-76, ×560; 19, SI-408, ×900. 20, *Auliscus* sp. cf. *A. priscus* Long, Fuge & Smith, SI-408, ×1,400. 21, *Auliscus* sp. A, SI-76, ×560. 22, *Hemiaulus* sp. E., SI-123, ×1,400.

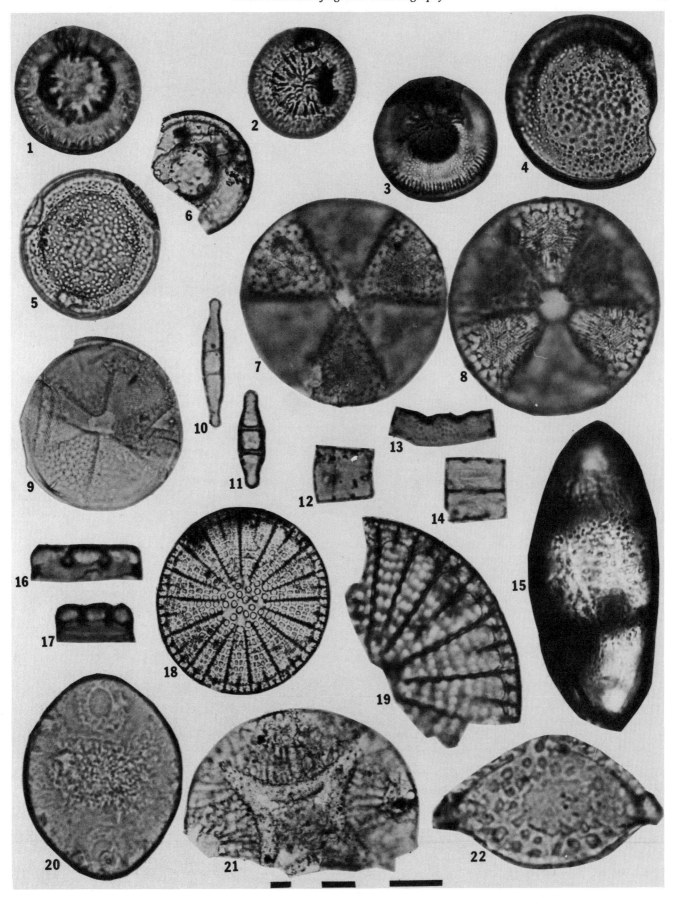

D. M. Harwood

Figure 10. Photomicrographs of diatoms from Seymour Island. Bar scales = 10 μm at magnifications of ×560, ×900, and ×1,400 (from left to right). 1, 2 *Chasea bicornis* Hanna. 1, SI-77, ×1,400; 2, SI-19, ×1,400. 3-7, *Chasea proshkinae-lavrenkoae* n. comb. 3, 4 SI-76, ×900; 5-7, SI-408, ×900; 7, with vegetative cell of *Hemiaulus proshkinae-lavrenkoae* Jousé attached. 8-12, *Acanthodiscus vulcaniformis* Jousé. 8, 9, SI-76, ×900; 10, 11, SI-70, ×900; 12, SI-19, ×900. 13, 14, *Cladogramma* sp. cf. *C. simlex* Hajós & Stradner, SI-19, ×900. 15, 16, *Coscinodiscus morenoensis* Hanna. 15, SI-380, ×1,400; 16, SI-19, ×900. 17, 18, *Eunotogramma polymorpha* Strelnikova, SI-76, ×1,400. 19, *Eunotogramma producta* var. *recta,* Long, Fuge & Smith, SI-19, ×1,400. 20, *Eunotogramma weissei* Ehrenberg, SI-19, ×900. 21, 22, *Goniothecium odontella* Ehrenberg. 21, SI-117, ×900; 22, SI-70, ×900. 23, *Gladius pacificus* Hajós & Stradner, SI-19, ×900. 24, *Gladius speciosus* Schulz, SI-19, ×900.

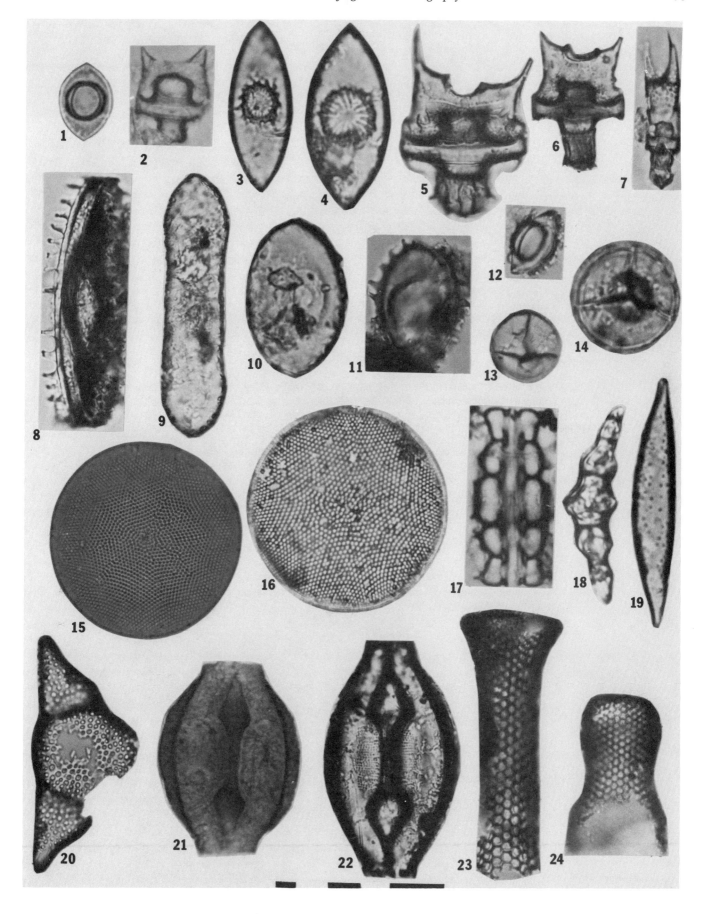

Figure 11. Photomicrographs of diatom genus *Gladius* and others from Seymour Island. Bar scales represent 10 μm at magnifications of ×560, ×900, and ×1,400 (from left to right). 1, 2, *Gladius antarcticus* n. sp., SI-373, ×900. 1, Paratype; 2, holotype. 3-5, 7-10, *Gladius antarcticus* f. *alta* n. sp., n. form. 3, Central connecting structure, SI-380, ×1,400; 4, holotype in two focal planes, SI-380, ×1,400; 5, 9, SI-380, ×1,400; 7, SI-373, ×900; 8, photograph through diameter of central elevation showing nature of the central process, SI-373, ×1,400; 10, paratype, SI-380, ×900. 6, *Gladius pacificus?* Hajós, SI-380, ×1,400. 11, 12, *Coscinodiscus sparsus* n. sp., holotypes SI-511, ×560. 11, Holotype; 12, paratype. 13, *Stephanopyxis* sp. D, SI-77, ×560. 14, *Gladius antarcticus* f. *alta?,* SI-380, ×900.

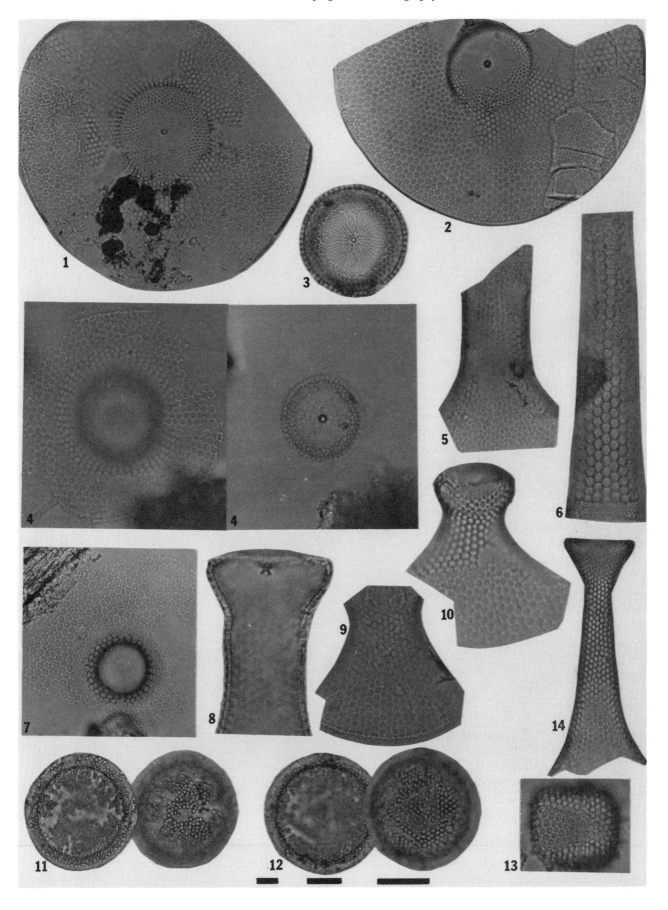

Figure 12. Photomicrographs of diatoms from Seymour Island. Bar scales = 10 μm at magnifications of ×560, ×900, and ×1,400 (from left to right). 1, *Stephanopyxis marginata* Grunow, SI-342, ×560. 2, *Aulacodiscus* sp. cf. *A. nigrescens* Pantocsek, SI-380, ×560. 3, *Aulacodiscus* sp. cf. *A. breviprocessus* Strelnikova, SI-373, ×560. 4, 5, *Aulacodiscus?* sp. A, SI-380, ×1,400. 6-8, *Rhizosolenia?* sp. B. 6, SI-511, ×560; 7, SI-507, ×900; 8, SI-507, ×560. 9, 10, *Pterotheca pokrovskja* Jousé var., SI-70, ×560. 11, *Pterotheca trojana* n. sp., holotype, SI-380, ×1,400. 12, 13, *Pterotheca minor* n. sp., SI-511, ×560. 12, Holotype; 13, paratype. 14, 15, *Pterotheca* sp. cf. *P. crucifera.* 14, SI-380, ×900; 15, SI-511, ×900. 16, *Hemiaulus* sp. M, SI-373, ×900. 17, *Huttonia antiqua* Hajós & Stradner, SI-380, ×1,400. 18, 19, *Incisoria lanceolata* Hajós & Stradner. 18, SI-19, ×900; 19, SI-380, ×1,400. 20, 21, *Biddulphia cretacea* Hajós & Stradner. 20, SI-511, ×560; 21, Si-511, ×900, 22, *Biddulphia fistulosa* Pantocsek, SI-511, ×900. 23, 24, *Hemiaulus seymouriensis* n. sp. 23, holotype, SI-380, ×900; 24, broken and distorted valve, SI-380, ×1,400. 25, *Gladius antarcticus* f. *alta* n. sp., n. form, fragment, SI-380, ×1,400. 26, *Hemiaulus gleseri* Hajós, SI-380, ×560.

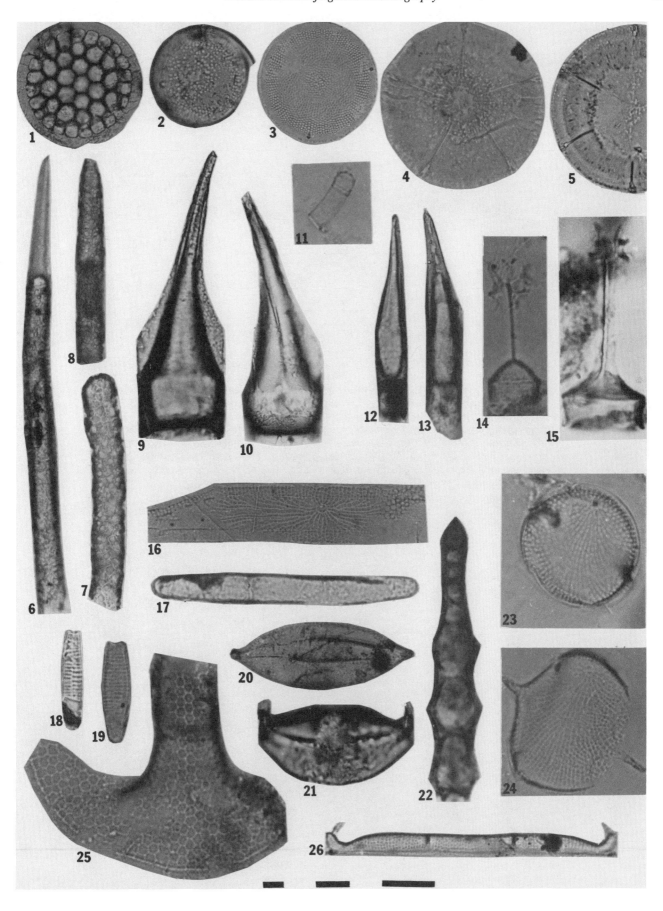

Figure 13. Photomicrographs of diatom genus *Hemiaulus* and others from Seymour Island. Bar scales = 10 μm at magnifications of ×560, ×900, and ×1,400 (from left to right). 1, *Kittonia* sp., SI-408, ×560. 2, Genus and species indeterminant Jousé, SI-408, ×900. 3, *Hemiaulus assymetricus* Jousé, SI-19, ×1,400. 4, *Hemiaulus echinulatus* Jousé, SI-19, ×1,400. 5-11, *Hemiaulus bipons* (Ehrenberg) Grunow. 5-9, 11, SI-408, ×900; 10, SI-19, ×1,400. 12, *Hemiaulus curvatulus* Strelnikova, SI-408, ×900. 13, 14, *Hemiaulus* sp. F, SI-408, ×900. 15, *Hamiaulus kittonii* Grunow, SI-408, ×900. 16, 17, *Hemiaulus danicus* Grunow, SI-408, ×900. 18, *Hemiaulus elegans* (Heiberg) Grunow, SI-19, ×1,400. 19, 20, *Hemiaulus praelegans (Jousé)* SI-19, ×900. 21-25, *Hemiaulus huberi* n. sp., SI-19, ×900. 26, 27, *Hemiaulus* sp. A, SI-19, ×1,400. 28, *Hemiaulus kondai* Hajós, SI-76, ×560. 29, *Hemiaulus polymorphus* var. *frigida* from Franz Joseph, Grunow, SI-19, ×900.

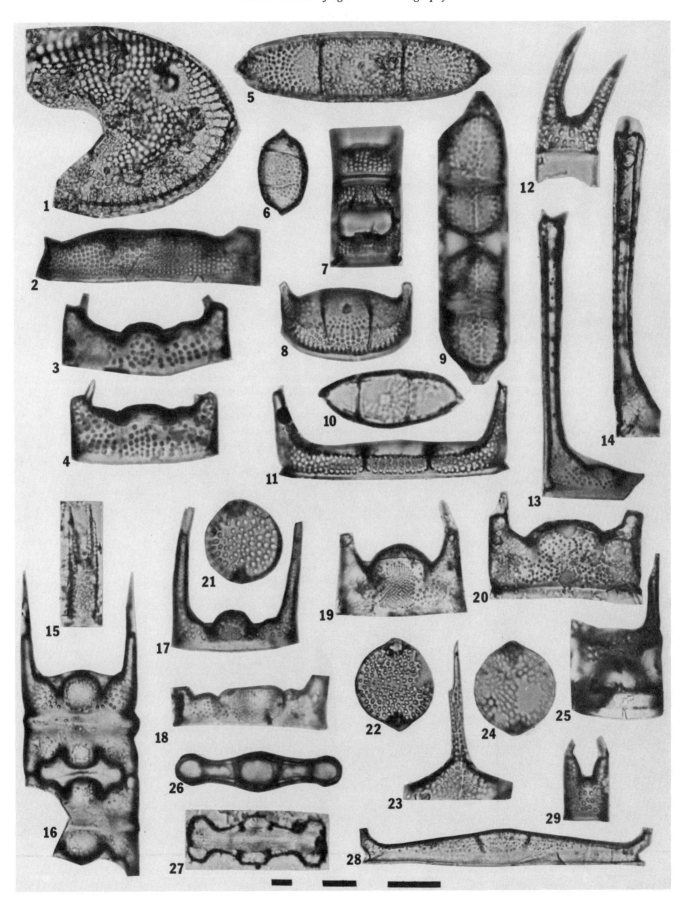

Figure 14. Photomicrographs of diatom genus *Hemiaulus* from Seymour Island. Bar scales = 10 μm at magnifications ×560, ×900, and ×1,400 (from left to right). 1-5, *Hemiaulus hostilis* Heiberg group. 1-2, SI-76, ×1,400; 3-5, SI-408, ×900. 6, *Hemiaulus* sp. G, SI-408, ×900. 7, *Hemiaulus polymorphus* var. *frigida* from Mors. Grunow, SI-408, ×900. 8, *Hemiaulus* sp. H, SI-76, ×900. 9, 25, *Hemiaulus* sp. cf. *H. tumidicornis* Strelnikova. 9, SI-76, ×1,400; 25, SI-408, ×900. 10, 11, *Hemiaulus incisus* Hajós. 10, SI-408, ×900; 11, SI-408, ×1,400. 12, *Hemiaulus polymorphus* var. *morsianus* Grunow, SI-19, ×900. 13, 14, *Hemiaulus sibericus* Grunow. 13, SI-408, ×900; 14, SI-19, ×900. 15, *Hemiaulus* sp. L, SI-76, ×900. 16, 17, *Hemiaulus simplex?* Brun. 16, SI-76, ×900; 17, SI-76, ×1,400. 18-21, *Hemiaulus rossicus* Pantocsek. 18, 19, SI-19, ×900; 20, 21, SI-76, ×900. 22, *Hemiaulus* sp. B, SI-408, ×900. 23, 24, *Hemiaulus* sp. D, SI-76, ×900.

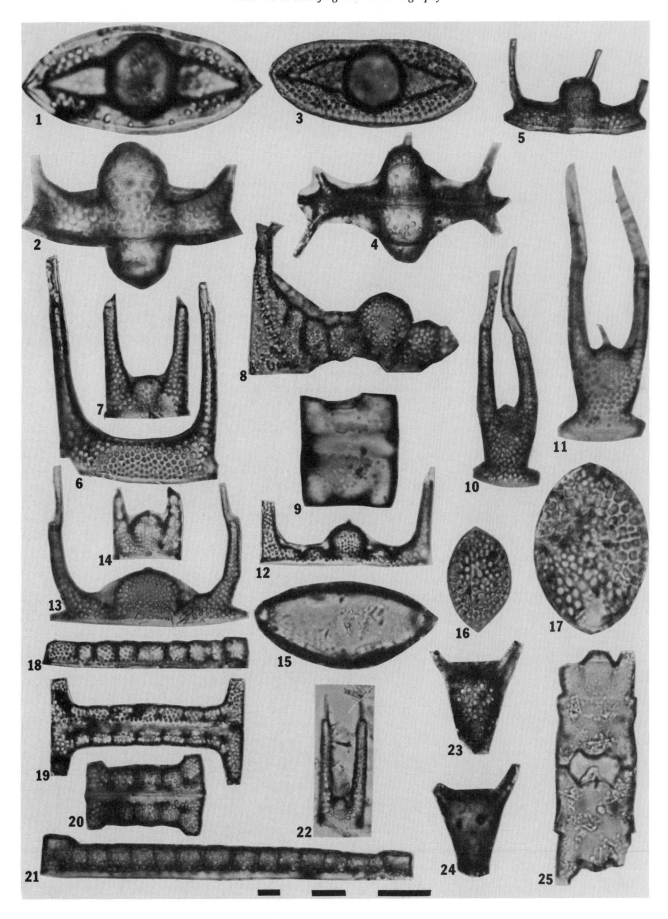

Figure 15. Photomicrographs of diatom genus *Hemiaulus* from Seymour Island. Bar scales = 10 μm at magnifications of ×560, ×900, and ×1,400 (from left to right). 1, *Hemiaulus* sp. K, SI-408, ×1,400. 2, 3, *Hemiaulus ambiguus* Grunow, SI-380, ×900. 4, 5, *Hemiaulus rossicus* Pantocsek, in a phase of encystment or division, SI-380, ×900. 6, *Hemiaulus* sp. J, SI-340, ×560. 7-10, *Hemiaulus sporialis* Strelnikova. 7, 8, Specimens with other vegetative and spore values, SI-380, ×900; 9, SI-380, ×900; 10, SI-380, ×560. 11, *Stephanopxis broschii* Grunow sensu Strelnikova, SI-511, ×1,400. 12, 13, *Hemiaulus* sp. L, SI-380, ×560. 14-17, *Hemiaulus* sp. C, SI-511, ×560.

Figure 16. Photomicrograph of diatoms from Seymour Island. Bar scales = 10 μm at magnifications of ×560, ×900, and ×1,400 (from left to right). 1. *Horodiscus anastomosans* (Pantocsek) n. comb., SI-76, ×560. 2. *Hyalodiscus russicus* Pantocsek, SI-408, ×900. 3-6, *Wittia macellarii* n. sp. 3, 4, 6, SI-76, ×900; 4, holotype; 5, detail of margin and nature of marginal processes, SI-76, ×1,400. 7, *Melosira? patera* Long, Fuge & Smith, SI-76, ×900. 8, 9, *Paralia sulcata* (Ehrenberg) Kützing. 8, SI-76, ×900; 9, SI-19, ×900. 10, *Melosira vetustissima* Hajós & Stradner, SI-76, ×1,400. 11, 12, *Xanthiopyxis grantii* Hanna, SI-76, ×560. 13, *Skeletonema* sp. A, SI-408, ×1,400. 14, 15, *Skeletonema polychaetum* Strelnikova. 14, SI-408, ×1,400; 15, SI-77, ×1,400. 16, *Skeletonema subantarctica* Hajós, SI-408, ×900.

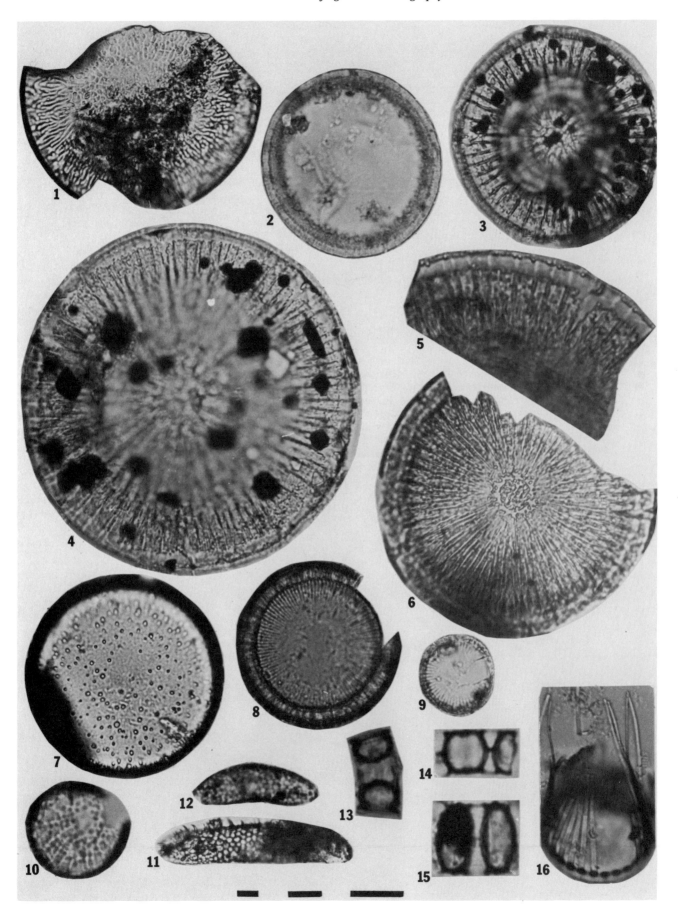

Figure 17. Photomicrographs of pennate diatoms (1 through 14) and diatom resting spores (15 through 32) from Seymour Island. Bar scales = 10 μm at magnifications of ×560, ×900, and ×1,400 (from left to right). 1, 2, *Incisoria punctata* Hajós & Stradner, SI-408, ×900. 3, *Incisoria inordinata* Hajós, SI-19, ×900. 4, *Incisoria lanceolata* Hajós & Stradner, SI-19, ×900. 5, 6, *Incisoria paleoceanica* var. *fuscina* (Schibkova) n. comb. 5, SI-408, ×1,400; 6, SI-408, ×900. 7, 8, *Incisoria* sp. A. 7, SI-19, ×1,400; 8, SI-19, ×900. 9, 10, *Sceptroneis grunowii* Anissimova, SI-408, ×900. 11, *Sceptroneis praecaducea* Hajós & Stradner, SI-76, ×1,400. 12, *Sceptroneis* sp. A, SI-19, ×900. 13, 14, *Sceptroneis?* sp. B. 13, SI-19, ×900; 14, SI-118, ×900. 15, *Kentrodiscus aculeatus* (Hajós) n. comb., SI-19, ×900. 16-18, *Kentrodiscus fossilis* Pantocsek. 16, SI-405, ×900; 17, 18, SI-408, ×900. 19, *Peponia barbadensis* Greville, SI-19, ×900. 20, *Pseudopodosira westii* (W. Smith) Sheshukova & Gleser, SI-72, ×900. 21, 22, *Pseudopyxilla aculeata* Jousé. 21, SI-408, ×900; 22, SI-19, ×900. 23, 24, *Pseudopyxilla dubia* (Grunow in Van Heurck) Forti. 23, SI-76, ×1,400; 24, SI-19, ×1,400. 25, *Pseudpyxilla* sp. Strelnikova, SI-76, ×1,400. 26, 27, *Pseudopyxilla hungarica* (Pantocsek) Forti, SI-408, ×900. 28, 29, *Pseudopyxilla rossica* (Pantocsek) Forti. 28, SI-76, ×1,400; 29, SI-408, ×900. 30, 31, *Omphalotheca* sp cf. *O. jutlandica* Grunow in Van Heurck. 30, SI-19, ×1,400; 31, SI-408, ×900. 32, *Pterotheca clavata* Strelnikova, SI-408, ×1,400.

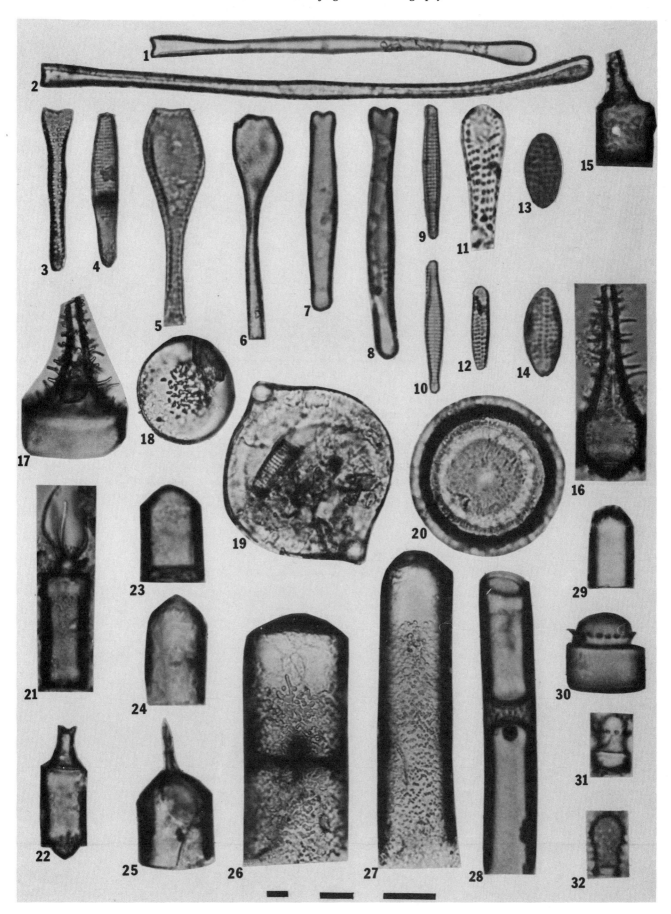

Figure 18. Photomicrographs of diatom resting spores from Seymour Island. Bar scales = 10 μm at magnifications of ×560, ×900, and ×1,400 (from left to right). 1, 2, *Pterotheca kittoniana* Grunow in Van Heurck, SI-76, ×1,400. 3, 4, *Pterotheca aculeifera* (Grunow in Van Heurck) Van Heurck, SI-408, ×900. 5, *Pterotheca crucifera* Hanna, SI-19, ×900. 6, *Pterotheca carnifera* (Grunow in Van Heurck) Forti, SI-118, ×900. 7, *Pterotheca carnifera* var. *curvirostris* Jousé, SI-19, ×1,400. 8, *Pterotheca carnifera* var. *Tenuis* Jousé, pyritized diatom, SI-14, ×1,400. 9-11, *Pterotheca cretacea* Hajós & Stradner, SI-19, ×900. 12, *Pterotheca danica* (Grunow) Forti, SI-76, ×560. 13, 14, *Pterotheca evermani* Hanna, SI-408, ×1,400. 15, *Pterotheca infundibulum* Krotov, SI-123, ×900. 16, *Pterotheca major* Jousé, SI-70, ×900. 17, 18, *Skeletonema penicillis* Grunow in Van Heurck, 17, SI-76, ×2,000; 18, SI-76, ×1,400. 19-23, *Pterotheca pokrovskajae* group Jousé. 19, SI-76, ×1,400; 20, SI-76, ×900; 21, 23, SI-408, ×900; 22, SI-76, ×560. 24, 25, *Pterotheca spada* Brun et Tempère, SI-408, ×900. 26, *Pterotheca subulata* Grunow in Van Heurck, SI-408, ×900.

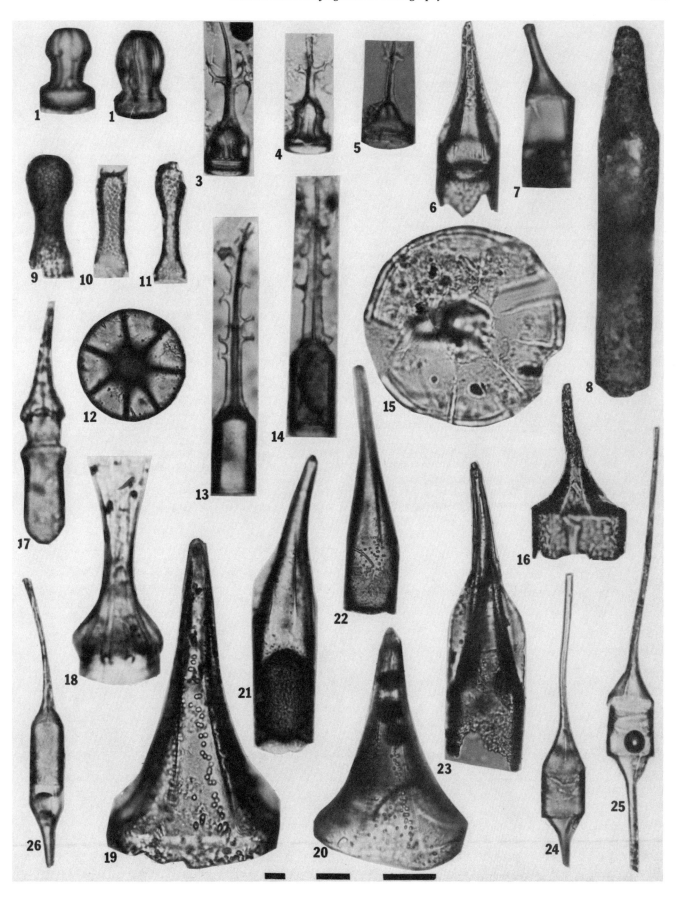

Figure 19. Photomicrographs of diatoms from Seymour Island. Bar scales = 10 μm at magnifications of ×560, ×900, and ×1,400 (from left to right). 1, *Pterotheca uralica* Jousé, SI-19, ×1,400. 2, *Pterotheca* sp. A, SI-76, ×560. 3, 4, *Pterotheca* sp. B. 3, SI-76, ×560; 4, SI-408, ×900. 5, *Pyrgodiscus* sp., SI-118, ×560. 6, 7, *Pyrgodiscus? triangulatus* Hajós & Stradner, SI-408, ×1,400. 8, *Rhizosolenia cretacea* Hajós & Stradner, SI-408, ×900. 9, *Rhizosolenia* sp. A, SI-408, ×900. 10, *Rutilaria?* sp., SI-408, ×1,400. 11, 12, *Pyxidicula minuta* Grunow. *11, SI-19,* ×900; 12, SI-97, ×900. 13, 14, *Stephanopyxis delectabilis* Pantocsek, SI-19, ×1,400. 15, 16, *Endictya lunyacsekii* Pantocsek. 15, SI-73, ×560; 16, SI-70, ×900. 17, *Stephanopyxis uralensis* Strelnikova, SI-76, ×560. 18, 19, *Stephanopyxis edita* Jousé. 18, SI-119, ×560; 19, SI-76, 400. 20, 21, *Stephanopyxis hannai* Hajós. 20, SI-76, ×560; 21, SI-118, ×560. 22, 23, *Stephanopyxis lavrenkoi* Jousé, SI-118, ×560. 24, *Stephanopyxis simonseni* Hajós, SI-76, ×560. 25, *Stephanopyxis superba* (Greville) Grunow, SI-76, ×560. 26, 27, *Stephanopyxis turris* (Greville et Arnott) Ralfs. 26, SI-97, ×900; 27, SI-19, ×560. 28, *Stephanopyxis turris* var. *cylindrus* Grunow, SI-76, ×900. 29, 30, *Stephanopyxis* sp. C, SI-116, ×900.

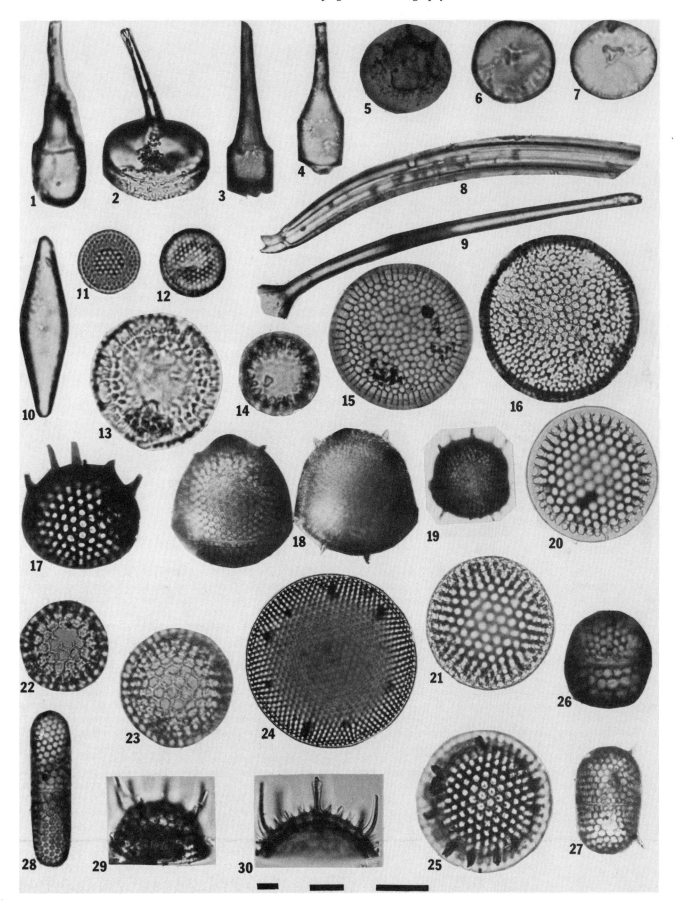

Figure 20. Photomicrographs of diatom genera *Stephanopyxis, Triceratium,* and others from Seymour Island. Bar scales = 10 μm at magnifications of ×560, ×900, and ×1,400 (from left to right). 1, *Stephanopyxis* sp. A, SI-118, ×560. 2, *Stephanopyxis* sp. B, SI-118, ×560. 3, 4, *Stellarima steinyi* (Hanna) Hasle and Sims, SI-76, ×560. 5, *Stephanopyxis maxima* Pantocsek, SI-373, ×560. 6, *Thalassiosiropsis wittiana* (Pantocsek) Hasle, SI-19, ×560. 7, 8, *Triceratium solenoceros* Ehrenberg. 7, SI-70, ×560; 8, SI-76, ×900. 9, *Triceratium* sp. A, SI-76, ×560. 10, *Triceratium flos* var. *intermedia* Schmidt, SI-76, ×900. 11, *Triceratium indefinitum* (Jousé) Strelnikova, SI-19, ×900. 12, *Triceratium nobile* Witt, SI-70, ×900. 13, *Triceratium simplicissium* Witt, SI-408, ×900. 14, 15, *Trinacria* sp. B. 14, SI-19, ×900; 15, SI-380, ×560.

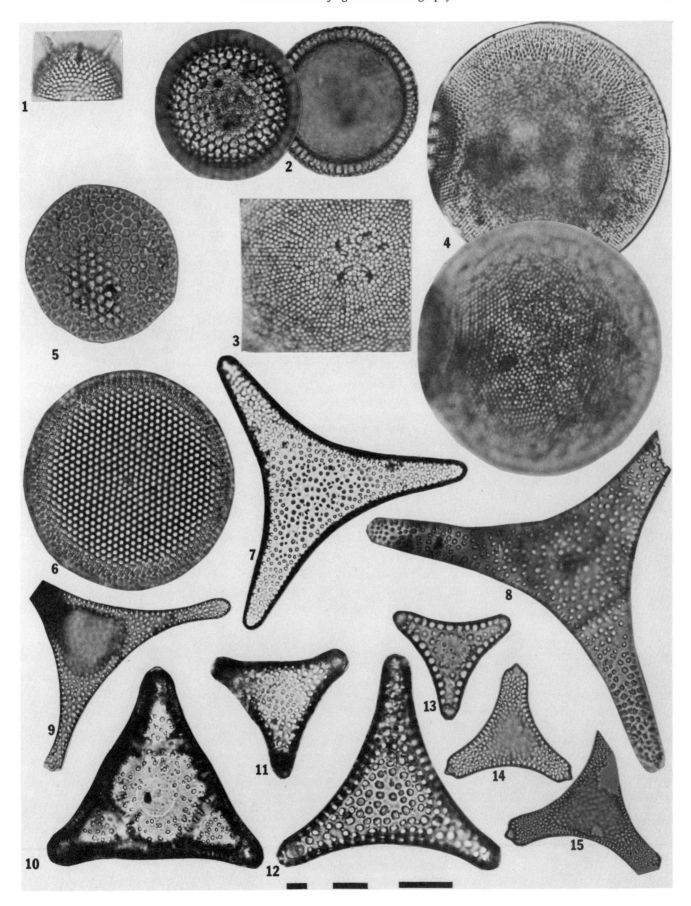

Figure 21. Photomicrographs of diatom genus *Trinacria* and others from Seymour Island. Bar scales = 10 μm at magnifications of ×560, ×900, and ×1,400 (from left to right). 1-3, *Triceratium sentum* Witt in Schmidt. 1, 2, SI-76, ×900; 3, SI-408, ×900. 4-7, *Triceratium indefinitum* (Jousé) Strelnikova. 4, SI-19, ×1,400; 5, 7, SI-408, ×900; 6, SI-19, ×900. 8, 10, *Trinacria acutangulum* (Strelnikova) n. comb., SI-76, ×560. 11, *Trinacria swastika* (Long, Fuge & Smith) n. comb., SI-380, ×560. 12, *Trinacria acutangulum* (Strelnikova) var. SI-76, ×900. 13, *Trinacria* sp. C, SI-380, ×560. 14, *Trinacria aries* Schmidt, SI-76, ×900. 15, 16, *Trinacria excavata* Heiberg. 15, SI-76, ×900; 16, SI-408, ×900. 17, *Trinacria grevillei* Witt, SI-76, ×360, 18, *Trinacria heibergii* var. *rostratum* Jousé, SI-76, ×560. 19, *Trinacria insipiens* Witt, SI-76, ×900. 20, *Trinacria pileolus* (Ehrenberg) Grunow, SI-118, ×900. 21, *Trinacria tristictia* Hanna, SI-342, ×560. 22, *Eunotogramma enorme* Krotov, SI-409, ×560. 23, *Hemiaulus* sp. I, SI-408, ×900. 24, *Trinacria* sp. A, SI-19, ×900. 25, *Trinacria edgardi* Hajós, SI-342, ×560.

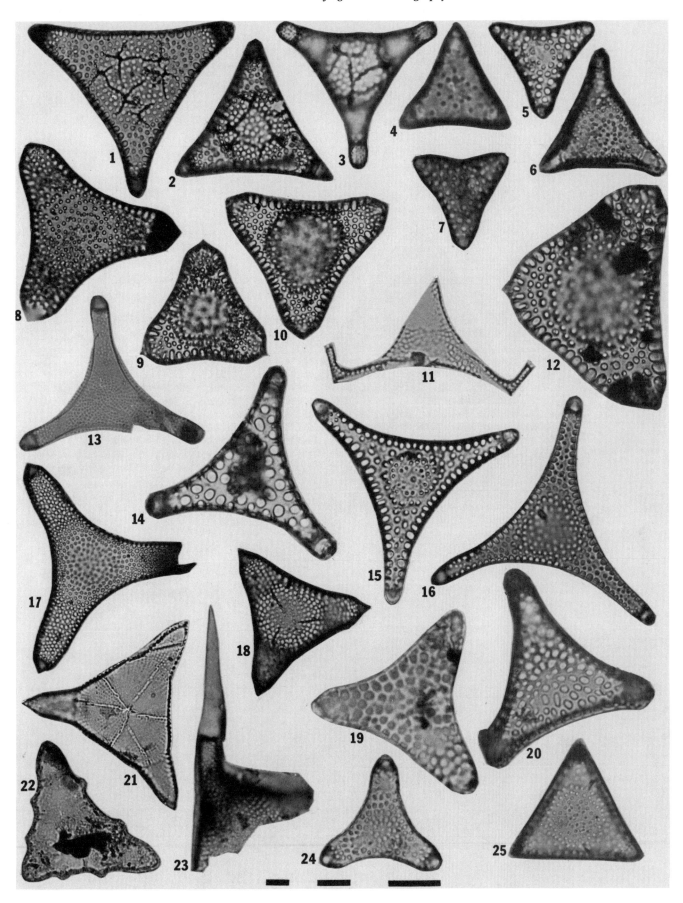

Figure 22. Photomicrographs of diatoms and silicoflagellates from Seymour Island. Bar scales = 10 μm at magnifications of ×560, ×900, and ×1,400 (from left to right). 1, *Epithelion* sp., SI-19, ×1,400. 2, *Pseudostictodiscus picus* Hanna, SI-408, ×900. 3, *Eunotogramma gibbosa* Strelnikova, SI-70, ×900. 4, 5, *Corbisema apiculata* (Lemmermann), SI-408, ×900. 6-9, *Corbisema apiculata* f. *minor* (Schulz) n. comb., SI-408, ×560. 10, 11, *Corbisema aspera* (Schulz) n. comb., SI-408, ×560. 12, *Corbisema* sp. cf. *C. archangelskiana* (Schulz) pyritized, SI-14, ×560. 13, 14, *Corbisema geometrica* Hanna. 13, SI-117, ×900; 14, SI-117, ×560. 15, 17, 18, *Corbisema hastata* (Lemmermann). 15, SI-408, ×900; 17, SI-408, ×560; 18, SI-76, ×560. 16, 19, *Corbisema hastata* var. *globulata* Bukry, SI-408, ×900.

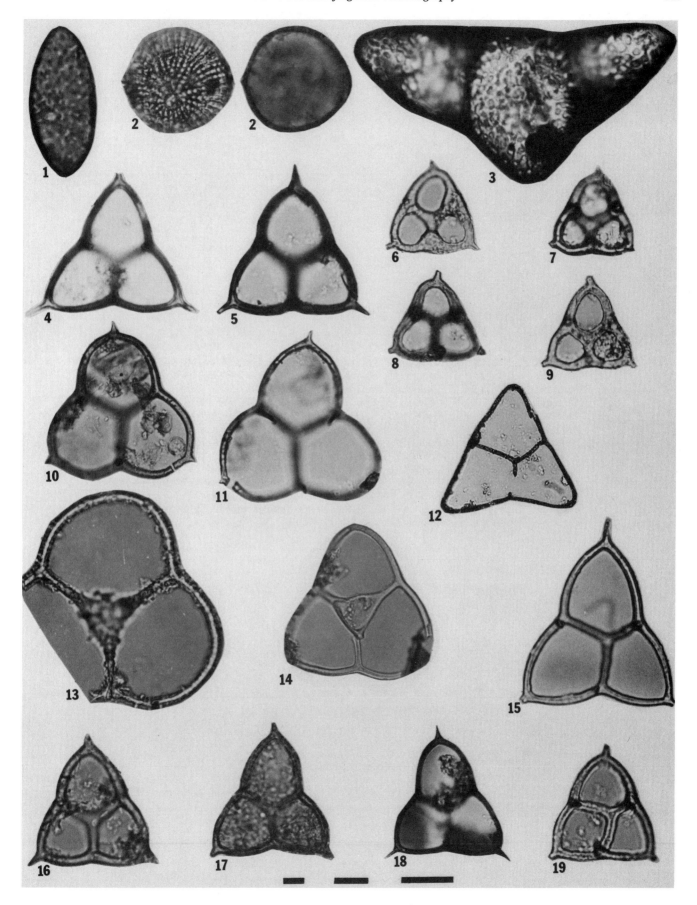

Figure 23. Photomicrographs of silicoflagellates and other siliceous microfossils from Seymour Island. Bar scales = 10 μm at magnifications of ×560, ×900, and ×1,400 (from left to right). 1, *Corbisema hastata* var., SI-480, ×560. 2, *Corbisema hastata* f. *miranda* Bukry, SI-408, ×560. 3, *Corbisema inermis* Dumitrica, SI-408, ×560. 4, 7, *Cornua* sp. A. 4, SI-408, ×900; 7, SI-408, ×560. 5, 6, *Corbisema* sp. variant. 5, SI-408, ×1,400; 6, SI-408, ×560. 8, *Dictyocha* sp. cf. *D. navicula* v. *minor* Schulz, SI-408, ×900. 9, *Dictyocha staurodon* f. *minor* Schula, SI-408, ×560. 10, 11, Silicoflagellate genus A. 10, SI-117, ×900; 11, SI-117, ×560. 12, Silicoflagellate genus B, species 1, DSDP Hole 275, Core 2-5, 103-105 cm, ×560. 13, 14, *Lyramula deflandrei* Perch-Nielsen & Edwards. 13, SI-117, ×900; 14, SI-14, ×310. 15, *Lyramula furcula,* Hanna, SI-14, ×560. 16, *Lyramula minor* (Deflandre) Deflandre, SI-117, ×900. 17, *Lyramula simplex* Hanna, SI-400, ×560. 18, *Lyramula* sp. cf. *L. furcula* Hanna, SI-117, ×560. 19, 20, *Vallacerta tumidula* Gleser. 19, SI-19, ×900; 20, SI-117, ×560. 21, 22, *Ammodochium danicum?* Deflandre (ebridian). 21, SI-408, ×900; 22, SI-408, ×1,400 (median tripod). 23, 24, *Carduifolia onoporides* Hovasse (endoskeletal dinoflagellate), SI-408, ×560.

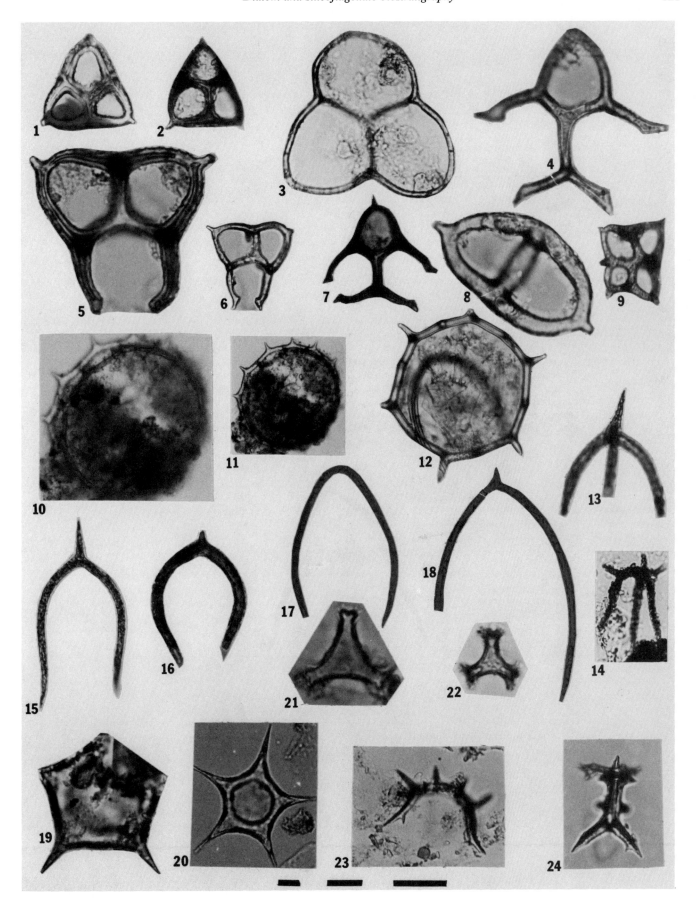

Figure 24. Photomicrographs of silicoflagellates and other siliceous microfossils from Seymour Island. Bar scales = 10 μm at magnifications of ×560, ×900, and ×1,400 (from left to right). 1, Genus et species indet. A, SI-380, ×560. 2, Archaeomonad sp. A, SI-400, ×900. 3, *Actiniscus* sp. cf. *A. pentasterias* Ehrenberg, SI-19, ×1,400. 4-7, *Vallacerta siderea* (Schulz), SI-380, ×560. 8, Silicoflagellate genus B, specis 2, SI-K-84, ×560. 9, *Corbisema inermis* Dumitricia var., SI-511, ×560. 10-12, *Corbisema latera-diata* (Schultz) Perch-Nielsen. 10, 11, SI-380, ×900; 12, SI-K-84, ×900.

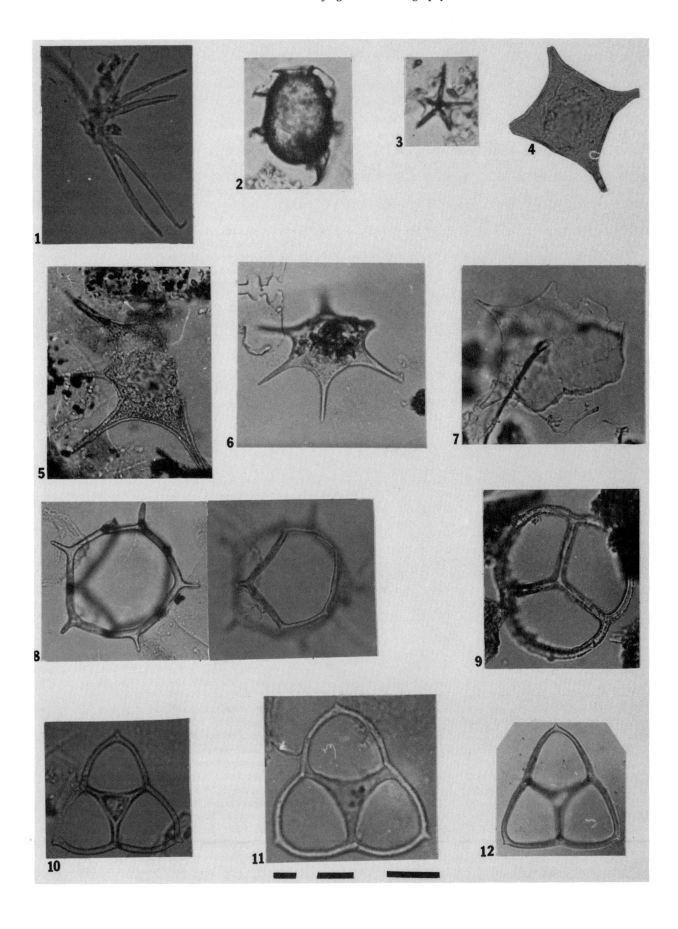

124   *D. M. Harwood*

REFERENCES

ABBOTT, W. H. 1978, Cretaceous diatoms from the Peedee Formation of South Carolina. Geologic Notes, 22(2):105–108.

——, AND H. E. HARPER. 1982. A biostratigraphic revisit of the Cretaceous Moreno Shale of California. Seventh International Symposium on Living and Fossil Diatoms, Philadelphia (abstract).

ANDERSSON, J. G. On the geology of Graham Land. Geological Institute, University of Uppsala, Bulletin, 7:19–71.

ANISIMOVA, N. 1938, Diatomovyye nizhnetretichnykh otlozhenii Sredney Volgi [Diatoms from the Lower Tertiary deposits in the Middle Volga Region]. Vsesoyu znyi Nauchnoissledovatel'skii Geologichneskii Institut (VSEGI), Leningrad.

BALANCE, P. F., ET AL. 1986. Exotic rocks on the landward wall of the Tonga Trench—Debris accepted from a downgoing seamount. Abstract, 12th International Sedimentological Congress, August, 1986, Canberra, Australia, p. 19.

BARKER, I. W., AND S. H. MEAKIN. 1943. New genera and species of diatoms from Russia. Journal of the Quekett Microscopical Club, Series 4, 1:251–255.

——. 1946. New diatoms from the Moreno Shale. Journal of the Quekett Microscopical Club, Series 4, 2:143–144.

——. 1948. New and rare diatoms. Journal of the Quekett Microscopical Club, Series 4, 2:233–235.

——. 1949. New and rare diatoms. Journal of the Quekett Microscopical Club, Series 4, 3:301–303.

BARRON, J. A. 1985. Diatom biostratigraphy of the CEASAR 6 core, Alpha Ridge, p. 137–148. In H. R. Jackson, P. J. Mudie, and S. M. Blasco, (eds.), Initial Geological Report on CESAR: The Canadian Expedition to Study the Alpha Ridge, Arctic Ocean. Geological Survey of Canada, Ottawa, Paper 84-22.

——, D. BUKRY, AND R. Z. POORE. 1984. Correlation of the middle Eocene Kellogg Shale of northern California. Micropaleontology, 30(2):138–170.

BERGSTRESSER, T. J., AND W. N. KREBS. 1983. Late Cretaceous (Campanian-Maastrichtian) diatoms from the Pierre Shale, Wyoming, Colorado, and Kansas. Journal of Paleontology, 57(5):883–891.

BIBBY, J. S. 1966. The stratigraphy of part of north-east Graham Land and the James Ross Island Group. British Antarctic Survey Scientific Report, 53:1–37.

BLOME, C. D., AND N. R. ALBERT. 1985. Carbonate concretions: an ideal sedimentary host for microfossils. Geology, 12(3):212–215.

BRAMLETTE, M. N. 1946. The Monterey Formation of California and the origin of its siliceous rocks. United States Geological Survey Professional Paper 212, 57 p.

BRUN, J. 1893. Notes sur quelques espèces nouvelles. Le Diatomiste 1:173–177.

——. 1894. Espèces nouvelles. Le Diatomiste, 2:72–78; 86–88.

——. 1896. Diatomées Miocènes. Le Diatomiste, 2:229–247.

——, AND J. Tempère. 1889. Diatomées fossiles du Japon. Mémoires de la Société de Physique et d'Histoire Naturelle de Genève, 30(9)1–75.

BUKRY, D. 1973. Coccolith and silicoflagellate stratigraphy, Tasman Sea and southwestern Pacific Ocean, Deep Sea Drilling Project Leg 21, p. 885–893. In R. E. Burns and J. E. Andrews, et al. (eds.), Initial Reports of the Deep Sea Drilling Project, 21. U.S. Government Printing Office, Washington, D.C.

——. 1974. Coccolith and silicoflagellate stratigraphy, eastern Indian Ocean, Deep Sea Drilling Project Leg 22, p. 601–607. In C. C. van der Borch, J. G. Sclater, et al. (eds.), Initial Reports of the Deep Sea Drilling Project, 22. U.S. Government Printing Office, Washington, D.C.

——. 1975a. Silicoflagellate and coccolith stratigraphy, Deep Sea Drilling Project Leg 29, p. 845–872. In D. E. Hayes, L. A. Frakes, et al. (eds.), Initial Reports Deep Sea Drilling Project, 29. U.S. Government Printing Office, Washington, D.C.

——. 1975b. Coccolith and silicoflagellate stratigraphy near Antarctica, Deep Sea Drilling Project, Leg 28, p. 825–839. in D. E. Hayes, L. A. Frakes, et al.

(eds.), Initial Reports of the Deep Sea Drilling Project, 28. U.S. Government Printing Office, Washington, D.C.

——. 1976. Cenozoic silicoflagellate and coccolith stratigraphy, south Atlantic Ocean, Deep Sea Drilling Project Leg 36, p. 885–917. In P. F. Barker, I.W.D. Dalziel, et al. (eds.), Initial Reports of the Deep Sea Drilling Project, 36. U.S. Government Printing Office, Washington, D.C.

——. 1977. Coccolith and silicoflagellate stratigraphy, south Atlantic Ocean, Deep Sea Drilling Project Leg 39, p. 825–839. In P. R. Supko, K. Perch-Nielsen, et al. (eds.), Initial Reports of the Deep Sea Drilling Project, 39. U.S. Government Printing Office, Washington, D.C.

——. 1978a. Cenozoic coccolith, silicoflagellate, and diatom stratigraphy, Deep Sea Drilling Project Leg 44, p. 807–863. In W. E. Benson, R. E. Sheridan, et al. (eds.), Initial Reports of the Deep Sea Drilling Project, 44. U.S. Government Printing Office, Washington, D.C.

——. 1978b. Cenozoic silicoflagellate and coccolith stratigraphy, northwestern Atlantic Ocean, Deep Sea Drilling Project Leg 43, p. 775–805. In W. E. Benson, R. E. Sheridan, et al. (eds.), Initial Reports of the Deep Sea Drilling Project, 44. U.S. Government Printing Office, Washington, D.C.

——, 1981a. Cretaceous Arctic silicoflagellates. Geo-Marine Letters, 1:57–63.

——. 1981b. Synthesis of silicoflagellate stratigraphy for Maastrichtian to Quaternary marine sediment, p. 433–444. In J. E. Warme, R. G. Douglas, and E. L. Winterer (eds.), The Deep Sea Drilling Program: A Decade of Progress. Society of Economic Paleontologists and Mineralogists Special Publication 32.

——. 1985. Correlation of Late Cretaceous Arctic silicoflagellates from Alpha Ridge, p. 125–135. In H. R. Jackson, P. J. Mudie, and S. M. Blasco (eds.), Initial Geological Report on CESAR: The Canadian Expedition to Study the Alpha Ridge, Arctic Ocean. Geological Survey of Canada, Ottawa, Paper 84-22.

——, AND J. H. FOSTER. 1974. Silicoflagellate zonation of Upper Cretaceous to lower Miocene deep-sea sediment: United States Geological Survey Research Journal, 2:303–310.

BURCKLE, L. H. 1978. Marine diatoms, p. 245–266. In B. U. Haq, and A. Boersma (eds.), Introduction to Marine Micropaleontology. Elsevier, New York.

BURKE, J. F., AND WOODWARD, J. B. 1963–1974. A review of the genus *Aulacodiscus*. Staten Island Institute of Arts and Sciences, New York, 360 p., 30 plates.

CHATTERJEE, S., AND W. J. ZINSMEISTER. 1982. Late Cretaceous marine vertebrates from Seymour Island, Antarctic Peninsula. Antarctic Journal of the United States, 17(5):66.

CHENEVIÈRE, E. 1934. Sur un dépôt fossile marin à diatomées suité à Kamischev (Russie centrale). Bulletin de la Société Francais de Microscopie, 3(3):33–36.

CIESIELSKI, P. F. 1975, Biostratigraphy and paleoecology of Neogene and Oligocene silicoflagellates from cores recovered during Antarctic Leg 28, p. 625–692. In D. E. Hayes, L. A. Frakes, et al. (eds.), Initial Reports of the Deep Sea Drilling Project, 28. U.S. Government Printing Office, Washington, D.C.

CORNELL, W. C. 1974. Maastrichtian silicoflagellates of the Great Valley, California. Geoscience and Man, 1:37–44.

DEFLANDRE, G. 1932a. Sur la systématique des Silicoflagellés. Société Botanique de France, Bulletin, 29:494–506.

——. 1932b. Remarques sur quelques ébriacées. Société Zoologique de France Bulletin, 57:302–315.

——. 1941. Sur la présence de diatomées dans certains silex creaux Turoniens et sur un nouveau mode de fossilisation de ces organismes. Comptes Rendus de L'Académie des Sciences, 213–878–880.

——. 1950. Contribution à l'étude des silicoflagellidés actuels et fossiles. Microscopie. 2:72–108, 117–142, 191–210.

——. 1951. Recherches sur les Ebridiens. Paléobiologie. Évolution. Systématique. Bulletin biologique de la France et de la Belgique, 85:1–84.

DEL VALLE, R. A., N. H. FOURCADE, AND F. A. MEDINA. 1982. The stratigraphy of Cape Lamb and The Naze, Vega and James Ross Islands, Antarctica, p. 275–280. *In* C. Craddock (ed.), Antarctic Geoscience. University of Wisconsin Press, Madison.

DEPREDO, C. A. 1981. High latitude Paleocene diatoms of the Southern Oceans. Unpublished M.S. thesis, Northern Illinois University, 244 p.

——, AND H. Y. LING. 1981. Early to early-Middle Paleocene diatom zonation. Antarctic Journal of the United States, 16(5):124–126.

DUMITRICA, P., 1973a. Cenozoic endoskeletal dinoflagellates in southwestern Pacific sediments cored during Leg 21 of the Deep Sea Drilling Project, p. 819–835. *In* R. E. Burns, J. E. Andrews, et al. (eds.), Initial Reports Deep Sea Drilling Project, 21. U.S. Government Printing Office, Washington, D.C.

——. 1973b. Paleocene, Late Oligocene and post-Oligocene silicoflagellates in southwestern Pacific sediments cored on DSDP Leg 21, p. 837–883. *In* R. E. Burns and J. E. Andrews, et al. (eds.), Initial Reports of the Deep Sea Drilling Project, 21. U.S. Government Printing Office, Washington, D.C.

DZINORDINZ, R. N., ET AL. 1979. Diatom and radiolarian Cenozoic stratigraphy, Norwegian Basin, Deep Sea Drilling Project Leg 38, p. 289–427. *In* M. Talwani, G. Udintsev, et al. (eds.), Initial Reports of the Deep Sea Drilling Project, Supplement to Volumes 38–41. U.S. Government Printing Office, Washington, D.C.

EHRENBERG, C. G. 1830. Beiträge zur Kenntniss der Organisation der Infusorien und iher geographischen Verbreitung, besonders in Sibirien, p. 1–88. Abhandlunger der Königlichen Akademie der Wissenschaften zu Berlin.

——. 1844. Mitteilung über zwei neue Lager von Gebrigsmassen aus Infusorien als Meeres-Absatz in Nord-Amerika und eine Vergleichung derselben mit den organischen Kreidegebilden in Europa und Afrika, p. 57–97. Berichte über Zur Beknntmachung geeigneten Verhandlungen der Königlichen Akademie der Wissenschaften zu Berlin.

——. 1846. On the remains of infusorial animalcules in volcanic rocks. Geological Society of London Quarterly Journal, 1846:73–91.

——. 1854. Mikrogeologie. Das Erden und Felsen schaffende Wirken des unsichtbar kleinen selbstandigen Lebens auf der Erde. Leopold Voss, Leipzig, 374 p., 40 plates.

——. 1855. Über ein europäisches marines Polygastern-Lager und über verlarvte Polythalamien in den marinen Polygastern Tripeln von Virginien und Simbirsk. Bericht der Köngilichen preuss, p. 292–305. Akademie der Wissenschaften zu Berlin.

ELLIOT, D. H., AND T. A. TRAUTMAN. 1982. Lower Tertiary strata on Seymour Island, p. 287–297. *In* C. Craddock (ed.), Antarctic Geoscience. University of Wisconsin Press, Madison.

EMILIANI, C., E. B. KRAUS, AND SHOEMAKER. 1981. Sudden death at the end of the Mesozoic. Earth and Planetary Science Letters, 55:317–334.

FENNER, J. 1977. Cenozoic diatom stratigraphy of the equatorial and southern Atlantic Ocean, p. 491–623. *In* P. R. Supko, K. Perch-Nielsen, et al. (eds.), Initial Reports of the Deep Sea Drilling Project, Supplement to Volumes 38–41. U.S. Government Printing Office, Washington, D.C.

——. 1982. Cretaceous diatoms off New Jersey. Seventh International Symposium on Living and Fossil Diatoms, Philadelphia (abstract), p. 7.

——. 1985. Late Cretaceous and Paleogene planktic diatom stratigraphy, p. 713–762. *In* J. Saunders, H. M. Bolli, and K. Perch-Nielsen (eds.), Plankton Stratigraphy. Cambridge University Press, Cambridge.

FORTI, A. 1909. Studi per una Monographia del genere *Pyxilla* (Diatomee) e dei generi affini. La Nuova Notarisia, 20(12):5–24.

——, AND P. SCHULZ, 1932. Erste Mitteilung über Diatomeen aus dem Hannoverschen Gault. Beihefte zum Botanischen Centralblatt, 50(2):241–246.

FRENCH, F. W., AND P. E. HARGRAVES. 1980. Physiological characteristics of plankton diatom resting spores. Marine Biology Letters, 1:185–195.

FRENGUELLI, J. 1940. Consideraciones sobre los sílicoflagellates fósiles. Revista del Museu de la Plata (Nueva Serie), Seccion Paleontologia, 2(7):37–112.

GARRISON, D. L. 1984. Planktonic diatoms, p. 1–17. *In* K. A. Steidinger and L. M. Walker (eds.), Marine Plankton Life Cycle Strategies. CRC Press, Boca Raton, FL.

GEMEINHARDT, K. 1930. Silicoflagellatae. *In* L. Rabenhorst (ed.), Kryptogamen-Flora von Deutschland, Österreich und der Schweiz. Akademische Verlagsgesellschaft, Leipzig, 10:1–87.

GIVEN, M. M., AND J. H. WALL. 1971. Microfauna from the Upper Cretaceous Bearpaw Formation of south-central Alberta. Bulletin of Canadian Petroleum Geologists, 19(2):502–544.

GLESER, Z. I. 1959. Nekotorye novye dannye o semeystve Vallacertaceae Deflandre (Silicoflagellatae). Informatsionnyi Sbornik Vsesoyuzno Gogeologicheskogo Instituta (VSEGEI), 10:103–113.

——. 1962. Voprosu o filogenese kremnevykh zhgutikovykh vodorosley. Paleontologii Zhurnal, 1:146–156.

——. 1966. Silicoflagellatophyceae. *In* Flora plantarum cryptogamarum USSR. Academiia Nauk S.S.S.R., Institutum Botanicum, 7:1–363.

——, A. P. JOUSÉ, ET AL. (eds.). 1974. Diatoms of the U.S.S.R. fossil and Recent. Nauka, Leningrad, 1:1–403.

GOMBOS, A. M., Jr. 1977. Paleogene and Neogene diatoms from the Falkland Plateau and Malvinas Outer Basin, Deep Sea Drilling Project Leg 36, p. 495–511. *In* P. F. Barker, I.W.D. Dalziel, et al. (eds.), Initial Reports of the Deep Sea Drilling Project, 36. U.S. Government Printing Office, Washington, D.C.

——. 1982. Early and Middle Eocene diatom evolutionary events. Bacillaria, 5:225–242.

——, AND P. F. CIESIELSKI, 1983. Late Eocene to early Miocene diatoms from the southwest Atlantic, p. 583–634. *In* W. I. Ludwig, V. A. Krasheninikov, et al. (eds.), Initial Reports of the Deep Sea Drilling Project, 71. U.S. Government Printing Office, Washington, D.C.

GRESHAM, C. W. 1985. Cretaceous and Paleocene siliceous phytoplankton assemblages from DSDP Sites 216, 214 and 208 in the Pacific and Indian Oceans. Unpublished M.S. thesis, University of Wisconsin, Madison, 233 p.

——. 1986. Cretaceous and Paleocene siliceous phytoplankton assemblages from DSDP Sites 216, 214 and 208 in the Pacific and Indian Oceans. North-Central Section Meeting, Geological Society of America, Kent, Ohio. Abstracts with Programs, 18(4):290.

GREVILLE, R. K. 1861–1866. Descriptions of new and rare diatoms, Ser. I-XX. Transactions of the Microscopical Society of London, N.S.:9–14.

GROVE, E., AND STURT. 1886–1887. On a fossil marine diatomaceous deposit from Oamaru, Otago, New Zealand. Journal of the Quekett Microscopical Club, Series II, 2(16):321–330; 3(17):7–12; 3(18):63–78; 3(19):131–148.

GRUNOW, A. 1882. Beiträge zur Kenntnis der Fossilen Diatomeen Osterreich-Ungarns, p. 136–159. *In* E. von Mojsisovids, and M. Neumayer (eds.). Beiträge zur Paleontologie Osterreich-Ungarns und des Orients, v. 2.

——. 1884. Die Diatomeen von Franz Josef-Land. Denkschriften der Mathematisch-Naturwissenschaftlichen classe der Kaiserlichen Akademie der Wissenschaften, Wien, 48:53–112.

HAJÓS, M. 1976. Upper Eocene and Lower Oligocene diatomaceae, archaeomonadaceae, and silicoflagellatae in southwestern Pacific sediments, Deep Sea Drilling Project Leg 29, p. 817–883. *In* C. Craddock, C. D. Hollister, et al. (eds.), Initial Reports of the Deep Seal Drilling Project, 35: U.S. Government Printing Office, Washington, D.C.

——, AND H. STRADNER. 1975. Late Cretaceous Archaeomonadaceae, Diatomaceae, and Silicoflagellatae from the South Pacific Ocean, Deep Sea Drilling Project Leg 29, Site 275, p. 913–1109. *In* J. P. Kennett, R. E. Houtz, et al. (eds.), Initial Reports of the Deep Sea Drilling Project, 29, U.S. Government Printing Office, Washington, D.C.

HANNA, G. D. 1927a. Cretaceous diatoms from California. Occasional Papers of the California Academy of Sciences, 13:5–49.

——. 1927b. The lowest known Tertiary diatoms in California: Journal of Paleontology, 1:103–126.

——, 1928. Silicoflagellata from the Cretaceous of California: Journal of Paleontology, 1(4):259.

——. 1931. Diatoms and silicoflagellates of the Kreyenhagen Shale. Mining in California, 27(2):197–201.

——. 1934. Additional notes on diatoms from the Cretaceous of California: Journal of Paleontology, 8:352–355.

——, AND A. L. BRIGGER. 1964. Some fossil diatoms from Barbados. Occasional Papers of the California Academy of Science, 45:1–27.

——, AND W. M. GRANT. Expedition to the Revillagigedo Islands, Mexico in 1925. II. Miocene Marine diatoms from Maria Madre Island. Proceedings of the California Academy of Science, 15(2):115–193.

HARGRAVES, P. E., AND F. W. FRENCH. 1983. Diatom resting spores: significance and strategies, p. 49–68. *In* Fryxell, G. A. (ed.), Survival Strategies of the Algae. Cambridge University Press, Cambridge.

HARPER, H. E. 1977. A Lower Cretaceous (Aptian) diatom flora from Queensland, Australia. Proceedings of the Fourth Symposium on Recent and Fossil Diatoms, Nova Hedwigia, Supplement 54:411–412.

HARWOOD, D. M. 1985. Campanian to Eocene Seymour Island siliceous microfossil biostratigraphy. Workshop on Cenozoic Geology of the Southern High Latitudes, Sixth Gondwana Symposium, Institute of Polar Studies, Ohio State University, August 1985 (abstract).

——. 1986. Seymour Cretaceous and Paleogene siliceous microfossil biostratigraphy. Symposium on Polar Research. North-Central Section Meeting, Geological Society of America, Kent, Ohio. Abstracts with Programs, 18(4):292.

HASLE, G. R., AND P. A. SIMS. 1985. The morphology of the diatom resting spores *Syringidium bicorne* and *Syringidium simplex.* British Phycological Journal, 20:219–225.

——. 1986. The Diatom genera *Stellarima* and *Symbolophora* with comments on the genus *Actinoptychus.* British Phycological Journal, 21:97–114.

——, AND E. E. SYVERTSEN. 1980. The diatom genus *Cerataulina:* Morphology and Taxonomy. Bacillaria, 3:79–113.

——. 1985. *Thalassiosiropsis,* a new diatom genus from the fossil records. Micropaleontology, 31:82–91.

HEIBERG, P.A.C. 1863. Conspectus criticus diatomacearum danicarum: Kritisk Oversigt over De Danske Diatomeer, p. 1–135. Wilhelm Priors Forlag, Copenhagen.

HOVASSE, R. 1932. Note préliminaire sur les Ebriacées. Société Zoologique de France Bulletin, 57:118–131.

——. 1932b. Troisieme note sur les Ebriacées. Société Zoologique de France, Bulletin, 57:457–476.

——. 1943. Nouvelles recherches sur les flagellés a squelette siliceux: Ébriidés et silicoflagellés fossiles de la diatomite de Saint-Laurent-La-Vernède (Gard). Bulletin biologique de la France et de la Belgique, 77:271–284.

HOWARTH, M. K. 1958. Upper Jurassic and Cretaceous ammonite faunas from Alexander Land and Graham Land. Falkland Island Dependencies Survey, Scientific Reports, 21:1–16.

——. 1966. Ammonites from the Upper Cretaceous of the James Ross Island group. Bulletin, British Antarctic Survey, 10:55–69.

HUBER, B. T. 1984. Late Cretaceous foraminiferal biostratigraphy, paleoecology, and paleobiogeography of the James Ross Island region, Antarctic Peninsula. Unpublished M. S. thesis, The Ohio State University, 246 p.

——. 1986. Foraminiferal evidence for a terminal Cretaceous oceanic event in Antarctica. Symposium on Polar Research, North-Central Section Meeting, Geological Society of America, Kent, Ohio. Abstracts with Programs, 18(4):309.

——. 1987a. The location of the Cretaceous-Tertiary contact on Seymour Island, Antarctic Peninsula. Antarctic Journal of the United States, 20(5) in press.

——. 1987b. Foraminiferal distribution across the Cretaceous/Tertiary transition on Seymour Island, Antarctic Peninsula, Antarctic Journal of the United States, 21(5) in press.

——, D. M. HARWOOD, AND P.-N. WEBB. 1983. Upper Cretaceous microfossil biostratigraphy of Seymour Island, Antarctic Peninsula. Antarctic Journal of the United States, 18(5):72–74.

HUSTEDT, F. 1930. Die Kieselalgen Deutschlands, Österreichs und der Schweiz mit Berücksichtigung der anderen Länder Europas sowie der angrenzenden Meeresgebiete, p. 1–920. *In* L. Rabenhorst, (ed.), Kryptogamenflora von Deutschland, Österreich und der Schweiz, Band. 7, T. 1. Akademische Verlagsgesellschaft, Leipzig.

JOUSÉ, A. P. 1948. Dotretichnye diatomovye Vodorosli (pre-Tertiary diatom algae). Botanicheskii Zhurnal, 33(3):345–356.

——. 1949a. Novye diatomovye i kremnevye zhgutikovye vodorosli verkhenemelovogo vozrasta iz glinistykl peskov basseyne R. B. Aktay (Vostochnyysklon severnogo Urala). Algae diatomaceae aetatis supernecretaceae ex arenis argillaceis systematis fluminis Bolschoy Aktay in declivitate orientali Ural Borealis. (New diatoms and silicoflagellates of the Upper Cretaceous age from argillaceous sands of the Aktay River Basin [the eastern slope of the North Urals]). Botanicheskie Materialy, Otdela Sporovykh Rastenii, Botanicheskii Institut, Akademiia Nauk S.S.S.R., 6:65–78.

——. 1949b. Diatoms from Mesozoic deposits, p. 109–114. In. Diatom Analisiz. Gosgeolizdat, Leningrad.

——. 1949c. Diatoms from Tertiary deposits, p. 114–153. In, Diatom Analisiz. Gosgeolizdat, Leningrad.

——. 1951a. Diatomeae aetatis Palaeoceani Uralii Septentrionalis [Paleocene diatoms from the north Urals]. Botanicheskie Materialy Otdela Sporovyh Rastenii, Botanicheskii Institut, Akademiia Nauk S.S.S.R., 7:24–42.

——. 1951b. Diatomovye i kremnevye zhgutikovye vodorosli Verkhnemelovogo vozrasta iz severnogo Urala. Diatomomeae et silicoflagellate aetatis cretae superne e montibus uralensibus Septentrionalibus [Late Cretaceous diatoms and silicoflagellatae in the north Urals]. Botanicheskie Materialy Otdela Sporovykh Rastenii, Botanicheskii Institut, Akademiia Nauk S.S.S.R., 7:42–65.

——. 1955. Species novae diatomacearum aetatis palaeogenae [New species of Paleogene diatoms]. Botanicheskie Materialy Otdela Sporovykh Rastenii, Akademiia Nauk S.S.S.R., 10:81–103.

——. 1963. Tip Bacillariophyta. Diatomovye vodorosli [Division Bacillariophyta Diatom algae], p. 55–121. *In* Yu. A. Orlov (ed.), Osnovy Paleontologii, Vodorosli, Mokhoobraznye, Psilofitovye, Plaunovidnye, Chlenistostebel'nye, Paporotniki. Akademiia Nauk S.S.S.R., Moscow, v. 50.

——. 1968. Ancient diatoms and diatomaceous rocks of the Pacific Ocean basin. Lithology and Mineral Resources, 1:11–26. (Translated from Litologiya i Poleznye Iskopaemye).

——. 1978. Diatom biostratigraphy on the generic level. Micropaleontology, 24(3):316–326.

KANAYA, T. 1957, Eocene diatom assemblages from the Kellogg and "Sidney" Shales, Mt. Diablo Area, California. Tohoku University Scientific Reports, Series 2 (geology), 28:27–124.

KILIAN, W., AND P. REBOUL. 1909. Les Céphalopodes néocrétacées de Iles Seymour et Snow Hill. Wissenschaftliche Ergebnisse der Schwedischen Sudpolar-Expedition 1901–1903, Stockholm, 3(6):1–75.

KITCHELL, J. A. 1980. Late Cretaceous and Paleocene diatoms from the central Arctic Ocean. Sixth International Symposium on Living and Fossil Diatoms, Budapest (abstract).

——, AND D. L. CLARK. 1982. Late Cretaceous–Paleogene paleogeography and paleocirculation: evidence of north polar upwelling. Palaeogeography, Palaeoclimatology, Palaeoecology, 40:135–165.

——, AND A. M. GOMBOS, JR. 1986. Biological selectivity of extinction: a link between background and mass extinction. Palaios, 1(5):504–511.

KITTON, T. 1870–1871. Diatomaceous deposits from Jutland. Journal of the Quekett Microscopical Club, 2:99–102; 168–171.

KLEMENT, J. W. 1963. The occurrence of *Dictyocha* (silicoflagellates) in the Upper Cretaceous of Wyoming and Colorado. Journal of Paleontology, 37(1):268–270.

KROTOV, A. I. 1957a. Diatomaceous algae from the Upper Cretaceous and Paleogene deposits on the eastern slope of the Urals and Transurals, p. 298–302. *In* Proceedings of the Interdepartmental Conference on the Development of Stratigraphic Systems for Siberia. Leningrad Gostoptekhizdat.

——. 1957b. The stratigraphy of the Upper Cretaceous and the Paleogene deposits of the eastern slope of the north and middle Urals based on the diatom algae data. Trudy gorno-geologicheskogo Instituta Uralisk. Akademiia Nauk S.S.S.R., 28(4):17–38.

——. 1959. Species novae diatomacearum e sedimentis cretae superioris in montibus Uralensibus [New species of diatom algae from the Upper Cretaceous deposits of the Urals]. Botanicheskie Materialy, Otdela Sporovykh

Rastenii, Botanicheskii Institut, Akademiia Nauk S.S.S.R. 12:106–112.

——, AND K. G. SCHIBKOVA. 1959. Species novae diatomacearum e Paleogeno montium uralensium. Neue diatomeen arten aus dem Paläogen des Urals [New species of diatoms from the Paleogene sediments of the Urals]. Botanicheskie Materialy, Otdela Sporovykh Rastenii, Botanicheskii Institut, Akademiia Nauk U.S.S.R., 12:112–129.

——. 1961. Kompleksi diatomovych i kremnevych zhgutikovych paleogenovych i neogenovych otlozheniiach vostochnogo sklona Urala i Zauralia [Complexes of diatom and silicoflagellate algae in the Upper Cretaceous, Paleogene, and Neogene deposits of the eastern slope of the Urals and Transurals]. Geological and Paleontological Materials, Mineral Resources of the Urals, 9:191–249.

KÜTZING, F. T. 1844. Die kieselschaligen Bacillarien oder Diatomeen. Nordhausen, 1–152, 30 Taf.

LEFÉBURE, P., AND E. CHENEVIÈRE. 1939. Description et iconographia de diatomées rares on Nouvelles. Bulletin de la SociétéFrancaise de Microscopie, 7:8–12; 8:21–26.

LEMMERMANN, E. 1901. Silicoflagellatae. Deutsche Botanische Gesellschaft, Berlin, Berichte, 19:247–271.

LING, H. Y. 1972. Upper Cretaceous and Cenozoic silicoflagellates and ebridians. Bulletins of American Paleontology, 62:135–229.

——. 1981. *Crassicorbisema,* a new silicoflagellate genus, from the Southern Oceans and Paleocene silicoflagellate zonation. Transactions of the Paleontological Society of Japan, New Series, 121:1–13.

——, L. M. MCPHERSON, AND D. L. CLARK. 1973. Late Cretaceous (Maastrichtian?) silicoflagellates from the Alpha Cordillera of the Arctic Ocean. Science, 180:1360–1361.

LOHMAN, K. E. 1960. The ubiquitous diatom—a brief survey of the present state of knowledge. American Journal of Science, 258-A:180–191.

LONG, J. A. D. P. FUGE, AND J. SMITH. 1946. Diatoms of the Moreno Shale. Journal of Paleontology, 20(2):89–118.

MACELLARI, C. E. 1984a. Late Cretaceous stratigraphy, sedimentology, and macropalentology of Seymour Island, Antarctic Peninsula. Unpublished Ph.D. dissertation, The Ohio State University, 599 p.

——, 1984b. Revision of serpulids of the genus *Rotularia* (Annelida) at Seymour Island (Antarctic Peninsula) and their value in stratigraphy. Journal of Paleontology, 58(4):1098–1116.

——. 1986. Late Campanian-Maastrichtian ammonite fauna from Seymour Island (Antarctic Peninsula). Journal of Paleontology, Memoir 18, 60(2):1–55.

——, AND B. T. HUBER. 1982. Cretaceous stratigraphy of Seymour Island (East Antarctic Peninsula). Antarctic Journal of the United States, 17(5):68–70.

——, AND W. J. ZINSMEISTER. 1983. Sedimentology and macropaleontology of the Upper Cretaceous to Paleocene sequence of Seymour Island. Antarctic Journal of the United States, 18(5):69–71.

MANDRA, Y. T. 1960. Fossil silicoflagellates from California, U.S.A., p. 77–89. In T. Sorgenfrel (ed.), International Geological Congress. Report of the 21st session, Part VI, Pre-Quarternary Micropaleontology, Copenhagen.

——. 1968. Silicoflagellates from the Cretaceous, Eocene and Miocene of California, U.S.A. California Academy of Science, Proceedings, 36:231.

MCPHERSON, L. M., AND H. Y. LING, 1973. Surface microstructure of selected silicoflagellates. Micropaleontology 19:475–480.

MILNE, D. H., AND C. P. MCKAY. 1982. Response of marine plankton communities to a global atmospheric darkening, p. 297–303. In L. T. Silver and P. H. Schulz (eds.), Geological Implications of Impacts of Large Asteroids and Comets on the Earth. Geological Society of America Special Paper 190.

MUKHINA, V. V. 1974. Paleocene diatom ooze in the eastern part of the Indian Ocean. Okeanologiva (Oceanography), 14(5):852–858.

MÜLLER, O., 1912. Diatomeenrest aus den Turonschichten der Kreide. Berichte fur Deutsche Botanische Gesellschaft, 29:661–668.

NIKOLAEV, V. A. 1983. On the genus *Symbolophora* (Bacillariophyta). Botanicheskii Zhurnal, 68:1123–1128.

OLIVERO, E. G. 1981. Esquema de Zonoción de ammonites del Cretácio Superior del grupo de islas James Ross, Antártida. VIII Congreso Geológico Argentino, San Luis, 2:897–907.

PALAMARCZUK, S., ET AL. 1984. Las Formaciones López de Bertodano y Sobral en la Isla Vicecomodoro Marambio, Antartida. Noveno Congreso Geologico Argentino, S. C. Barioloche, Actas 1:399–419.

PANTOCSEK, J. 1886. Beiträge zur Kenntnis der Fossilen Bacillarien Ungarns. v. I. Marine Bacillarien, 75 p.

——. 1889. Beiträge zur Kenntnis der Fossilen Bacillarien Ungars. v. II. Brackwasser Bacillarien und Anhang. Analyse der Marinen Depôts von Borny, Bremia Nagy-Kurtös in Ungarn, Ananino und Kusnetkz in Russland, 123 p.

——. 1892. Beiträge zur Kenntnis der Fossilen Bacillarien Ungarns. v. III. Süsswasser Bacillarien. Anhang-analysen 15 neuer Depôts von Platzko, 118 p.

PARAMONOVA, N. V. 1964. The materials of diatoms from Paleogene of northwestern Siberia. Leningrad. Vsesoyuznogo neftyanago nauchnoissledovatel'skogo geologo-razvedochnogo instuta (VNIGRI). Trudy Paleofitol Sbornik, 239:232–247.

PERCH-NIELSEN, K. 1975a. Late Cretaceous to Pleistocene archaemonads, ebridians, endoskeletal dinoflagellates, and other siliceous microfossils from the subantarctic southwest Pacific, Deep Sea Drilling Project, Leg 29, p. 873–907. In J. P. Kennett, R. E. Houtz, et al. (eds.), Initial Reports of the Deep Sea Drilling Project, 29. U.S. Government Printing Office, Washington, D. C.

——. 1975b. Late Cretaceous to Pleistocene silicoflagellates from the southern southwest Pacific, Deep Sea Drilling Project Leg 29, p. 677–721. In J. P. Kennett, R. E. Houtz, et al. (eds.), Initial Reports of the Deep Sea Drilling Project, 19. U.S. Government Printing Office, Washington, D.C.

——. 1976. New silicoflagellates and a silicoflagellate zonation in north European Paleocene and Eocene diatomites. Bulletin of the Geological Society of Denmark, 25:27–40.

——. 1985. Silicoflagellates, p. 713–762. In H. M. Bolli, J. B. Saunders, and K. Perch-Nielsen (eds.), Plankton Stratigraphy. Cambridge University Press, Cambridge.

PRICHARD, A., 1861. A History of Infusoria, Including the Desmidiaceae and Diatomaceae, British and Foreign, 4th ed., rev. and enlarged by J. T. Arlidge, W. Archer, J. Ralfs, W. C. Williamson, and the author. Whittaker and Co., London, 968 p., 40 plates.

PROSCHKINA-LAVRENKO, A. K. 1949. Diatomovyi Analis. Kniga 2. Opredelitel 'Iskopaemykh i Sovremennhykh Diatomomykh Vodorosle, Poryadok Centrales i Mediales, sostavili—A. P. Jousé, p. 14–224, Plates 1–101; I. A. Kisselev, p. 14–209; V. S. Poretzky, p. 14–209; A. I. Proschkina-Lavrenko, p. 210–224; i, V. S. Sheshukova. Botanicheskii Institut im V. L. Komorova Akademii Nauk S. S. S. R., 238 p. Gosudarstvennoe Izdatelystvo Geologicheskoi Literatury, Moskva-Leningrad.

RAMPI, L. 1940. Archaeomonadacae del Cretaceo Americano. Società Italiana di Scienze Naturali, Atti, 79:60–68.

RINALDI, C. A. 1982. The Upper Cretaceous in the James Ross Island Group, p. 281–286. In C. Craddock (ed.), Antarctic Geoscience. University of Wisconsin Press, Madison.

——, A. MASSABIE, J. MORELLI, H. L. ROSENMALL, AND R. DEL VALLE. 1978. Geologia de la Isla Vicecomodoro Marambio: Contribuciones del Instituto Antartico Argentino, 217:1–37.

ROGOZIN, G. S. 1913. Materials on the study of the siliceous clay and infusorial earth of the Simbirsk Province, Karsun District. Simbirsk Oblast Yestestvenno-Istoricheskiy Zapiski, Vypusk 1:1–41.

ROSS, AND P. A. SIMS. 1973. Observations on family and generic limits in the Centrales. Nova Hedwigia, Supplement 45:97–132.

——, AND G. R. HASLE. 1977. Observations on some species of the Hemiauloideae. Proceedings of the Fourth Symposium on Recent and Fossil Diatoms, Nova Hedwigia, Supplement 54:179–214.

RUSSELL D. A. 1979. The enigma of the extinctions of the dinosaurs. Annual Review of Earth and Planetary Science, 7:163–182.

SCHMIDT, A. 1874–1959. Atlas der Diatomaceen-Kunde, begun by A. Schmidt, continued by M. Schmidt, O. Fricke, H. Muller, and F. Hustedt. Reisland, Leipzig, Berlin, 480 plates.

SCHRADER, H.-J. 1969. Die pennaten Diatomeen aus dem Obereozan von Oamaru, Neuseeland. Nova Hedwigia, 28:1–124.

——, AND J. FENNER. 1976. Norwegian Sea Cenozoic diatom biostratigraphy,

p. 921–1099. *In* M. Talwani, G. Udintsev, et al., (eds.), Initial Reports of the Deep Sea Drilling Project, 38. U.S. Government Printing Office, Washington, D.C.

——, AND G. SCHUETTE. 1981. Marine diatoms, p. 1179–1232. *In* C. Emiliani (ed.), The Oceanic Lithosphere. The Sea, v. 7, John Wiley & Sons, New York.

SCHULZ, P. 1928. Beiträge zur Kenntnis fossiler und rezenter Silicoflagellaten. Botanisches Archiv, 21:225–292.

——. 1935. Diatomeen aus senonen Schwammgesteinen der Danziger Bucht. Zugleich ein Beitrag zur Entwicklungsgeschichte der Diatomeem. Botanisches Archiv, 37:383–413.

SHESHUKOVA-PORETZKAKA, V. S., AND Z. I. GLESER. 1964. Diatomeae marinae novae e Paleogeno Ucrainiae. Novitates systematicae plantarum non vascularium, p. 78–92. NAUKA, Moscow.

SCHIBKOVA, K. G. 1961. Nekotovie novie dannie o verchnemelovoi i paleogenovoi flore diatomovych vodoroslei vostochnogo sklona Urala i Zauraliia [Some new data of Upper Cretaceous and Paleogene flora of the diatom algae of the eastern slope of the Urals and the Transurals], p. 239–243. *In* Decisions and Proceedings of the Intraregional Meeting on the "Working Out" of the Stratigraphic Scheme of the West Siberian Depression. Gostoptekhizdat, Leningrad.

SIMONSEN, R. 1972. Ideas for a more natural system of the centric diatoms. Nova Hedwigia, Supplement 39:37–54.

——. 1979. The diatom system: ideas on phylogeny. Bacillaria, 2:9–71.

SPATH, L. 1953. The Upper Cretaceous cephalopod fauna of Graham Land. Falkland Island Dependencies Survey Scientific Report, 3:1–60.

STRELNIKOVA, N. I. 1964. New species of diatom algae from the Upper Cretaceous deposits of the Syny River Basin (West Siberia). Paleophytologii sbornik. "Nedra," Trudy Vsesoyuznogo neftyanogo nauchnoissledovatel's-kogo geologo-razvedochnogo instituta (VNIGRI), 239:229–232.

——. 1965a. Redkie i novye vidy pozneimelovykh diatomovykh vodorosley vostochnogo skona polyarnogo Urala. De diatomeis cretae superioris raris et novis declivis orientalis montium Uralensium Polarium [Rare and new species of Late Cretaceous diatom algae of the eastern slope of the Polar Urals], p. 29–37. Novosti Systematiki Nizshikh Rastenii, Botanicheskii Institut V. A. Komarova, Akademiia Nauk S.S.S.R.

——. 1966a. Revisio specierum generum *Gladius* Schulz and *Pyxilla* Greville (Bacillariophyta) e sedimentis Cretae superioris. Revision of Late Cretaceous representatives of the genera *Gladius* Schulz and *Pyxilla* Greville (Bacillariophyta), p. 23–36. Novosti Systematiki Nizshikh Rastenii; Botanicheskii Institut V. A. Komarova, Akademiia Nauk S.S.S.R.

——. 1966b. New species of the genus *Aulacodiscus* Ehr. from the Late Cretaceous deposits of the pre-polar Urals. p. 29–30. Novisti Systematiki Nizshikh Rastenii, Botanicheskii Institut V. A. Komarova, Akademiia Nauk S.S.S.R.

——. 1968. Pozdnemelovye diatomovye Vodorosli [Late Cretaceous diatom algae]. p. 17–21. Iskopaemye Diatomovye Vodorosli SSSR, Sibirskoe Otdelenie Institut Geologii Geofiziki, Akademiia Nauk S.S.S.R.

——. 1971. Species Novae Bacillariophytorum e sedimentis Cretae posterioris in declivitate orientali partis polaris ac praepolaris montium Uralensium: Novitates Systematicae Plantarum non Vascularium, 8:41–52.

——. 1974. Diatomei pozdnego mela [Late Cretaceous diatoms of western Siberia]. Academiia Nauk U.S.S.R., Roy 8:202.

——. 1975. Diatoms of the Cretaceous Period. Third Symposium on Recent and Fossil Diatoms (Kiel). Nova Hedwigia, Supplement 53:311–321.

TALIAFERRO, N. L. 1933. The relation of volcanism to diatomaceous and associated siliceous sediments. University of California Publication, Department of Geological Sciences Bulletin, 23(1):1–56.

TAPPAN, H. 1962. Foraminifera from the Arctic slope of Alaska, Pt. 3, Cretaceous Foraminifera, p. 91–203. United States Geological Survey Professional Paper 236-C.

——. 1980. The Paleobiology of Plant Protists. W. H. Freeman, San Francisco, 1,028 p.

——, AND A. R. LOEBLICH, JR. 1973. Evolution of the oceanic plankton. Earth

Science Reviews, 9:207–240.

THIERSTEIN, H. R. 1982. Terminal Cretaceous plankton extinctions: a critical assessment, p. 385–399. *In* L. T. Silver and P. H. Schulz (eds.), Geological Implications of Impacts of Large Asteroids and Comets on the Earth. Geological Society of America Special Paper 190.

VAN HEURCK, H. 1880–1885. Synopsis des diatomées de Belgique. Privately published, Antwerp.

——. 1896. A Treatise on the Diatomaceae, trans. by W. E. Baxter. William Wesley & Son, London, 558 p., 35 plates.

VECHINA, V. N. (VEKSCHINA), 1961a. Novyi rod i novye vidy diatomovykh iz Melovyikh i Paleogenovykh Otlozhenii zapadno-sibirskoi nizmennosti [New genus and new species of the Cretaceous and Paleogene diatoms from the West Siberian Depression[, p. 89–93. Trudy Sibirskogo Nauchno-issledovatel'skogo Instituta Geologii, Geofiziki i Mineral'nogo Syr'ja 15.

——. 1961b. The scheme of subdivision of the Cretaceous and Paleogene deposits of the West Siberian Depression based on the analysis data of algae—diatoms, silicoflagellates, Ebriideae, coccolithophorids, p. 223–237. *In* Decisions and Proceedings of the Intraregional Meeting on the "Working Out" of the Stratigraphic Scheme of the West Siberian Depression. Gostoptekhizdat, Leningrad.

VINCENT, J.-S., ET AL. 1983. The late Tertiary–Quaternary stratigraphic record of the Duck Hawk Bluffs, Banks Island, Canadian Arctic Archipelago. Canadian Journal of Earth Sciences, 20(11):1694–1712.

VORONKOV, YU. S. 1959. Melovye otlozheniya vostochnogo Sklona Pripolyarnogo Urala [Cretaceous deposits of the eastern slope of the pre-polar Urals]. Trudy Vsesoyuznogo neftyanogo nauchnoissledovatel'sogo geologo-razvedochnogo instituta [Transaction of the All-Union Petroleum Scientific Research Institute for Geological Prospecting] VNIGRI, 140:120–138.

VOSZHENNIKOVA, T. F. 1960. Paleoalgologic characteristics of the Meso-Cenozoic deposits of the West Siberian Depression. Trudy Institut Geologii Geofiziki, Akademiia Nauk S.S.S.R., 1:7–63.

WALL, J. H. 1975. Diatoms and radiolarians from the Cretaceous system of Alberta, a preliminary report, p. 391–410. *In* W.G.E. Caldwell (ed.), The Cretaceous System in the Western Interior of North America. Geological Association of Canada Special Paper 13.

WEAVER, F. M., AND A. M. GOMBOS, JR. 1981. Southern high-latitude diatom biostratigraphy, p. 445–470. *In* J. E. Warme, R. G. Douglas, and E. L. Winterer (eds.), The Deep Sea Drilling Project: A Decade of Progress. Society of Economic Paleontologists and Mineralogists Special Publication 32.

WEBB, P.-N. 1975. Upper Cretaceous–Paleocene foraminifera from Site 208 (Lord Howe Rise, Tasman Sea), DSDP Leg 21, p. 541–573. *In* R. E. Burns and J. E. Andrews, et al., (eds.), Initial Reports of the Deep Sea Drilling Project, 21. U.S. Government Printing Office, Washington, D.C.

WEIDEMANN, M. 1964. Présence de diatomées dans le Flysch à Helminthoides. Bulletin des Laboratoires de Géologie, Géographie Physique, Minéralogie, Paléontologie, Géophysique and de la Musée Géologique de l'Université de Lausanne, 145:1–5.

WEISNER, H. 1936. Sur la découverte de diatomées et autres microfossiles peu connus dans le Crétacé supérieur de la Bohême. Annales de Protistologie, 5:151–155.

WEISSE, J. F. 1854. Mikroskopische Analyse eines organischen Polirschiefers aus dem Gouvernement Simbirsk. Mélanges biologique tirés du Bulletin de l'Académie des Sciences de St. Petersbourg, 2:237–250.

WIND, F. H. 1979. Maastrichtian-Campanian nannofloral provinces of the southern Atlantic and Indian oceans, p. 123–137. *In* M. Talwani, W. Hay, and W.B.F. Ryan (eds.), Deep Drilling in the Atlantic Ocean: Continental Margins and Paleoenvironment. American Geophysical Union, Washington, D.C.

——. 1983. The genus *Nephrolitus* Gorka, 1957 (Coccolithophoridae). Journal of Paleontology, 57:157–161.

WITT, O. 1885. Ueber den Polierschiefer von Archangelsk-Kurojedowo im Gov. Simbirsk. Verhandlungen, Russisch-Kaiserliche, Mineralogische Gesellschalt zu St. Petersbourg, Series 2, Bd. 22:137–177.

WOLLE, F. 1890. Diatomaceae of North America. Comenius Press, Bethlehem, PA, 47 p., 112 plates.

WORNARDT, W. W., JR. 1972. Stratigraphic distribution of diatom genera in marine sediments in western North America. Palaeogeography, Palaeoclimatology, Palaeoecology, 12:49–72.

ZINSMEISTER, W. J. 1982. Review of the Upper Cretaceous–Lower Tertiary sequence on Seymour Island, Antarctica. Journal of the Geological Society of London, 139(6):779–786.

——— . 1986. Fossil windfall at Antarctica's edge. Natural History, 95(5):60–67.

——— , AND C. E. MACELLARI. 1983. Changes in the macrofossil faunas at the end of the Cretaceous on Seymour Island, Antarctic Peninsula. Antarctic Journal of the United States, 18(5):68–69.

ZITTEL, K. A. 1876. Über einige fossile Radiolarien aus der Norddeutschen Kreide. Zeitschrift fur Deutsche Geológische Gesellschaft, 28(1):75–86.

MANUSCRIPT ACCEPTED BY THE SOCIETY SEPTEMBER 1, 1987
BYRD POLAR RESEARCH CENTER CONTRIBUTION NO. 572

Geological Society of America
Memoir 169
1988

Campanian to Paleocene palynological succession of Seymour and adjacent islands, northeastern Antarctic Peninsula

Rosemary A. Askin
Department of Earth Sciences, University of California, Riverside, California 92521

ABSTRACT

Rich palynomorph assemblages occur throughout the Campanian to Eocene stratigraphic section of Seymour and surrounding islands, northeastern Antarctic Peninsula. The section comprises sediments referred to the López de Bertodano, Sobral, Cross Valley, and La Meseta Formations. Nearshore marine to coastal-deltaic sediments include marine palynomorphs (dinoflagellate cysts, acritarchs, other algae) and diverse land-derived palynomorphs (pollen, spores, fungal spores, fresh-water algae), plus a variety of other organic debris. Palynostratigraphic results are based on a survey of about 530 outcrop samples of Campanian through Paleocene age.

Six palynomorph zones, informally designated 1 through 6 and based on dinocyst species, are recognized in the upper Campanian through Paleocene section on Seymour Island. López de Bertodano sediments include zones 1 to 4 of late Campanian to Maastrichtian age. These Cretaceous zones are characterized by an evolving complex of dinocyst species of *Manumiella* and related genera. Zone 5, of early Paleocene age, occurs in uppermost López de Bertodano Formation sediments and most of the Sobral Formation on Seymour Island. Zone 6, of probable late Paleocene age, occurs in the uppermost Sobral Formation. Paleocene zones 5 and 6 are characterized by dinocysts *Spinidinium* spp., *Deflandrea* and *Ceratiopsis* spp., *Microdinium* sp.; and *Paleoperidinium pyrophorum* in zone 5. Two distinct older assemblages of middle to late Campanian age are recognized from The Naze, situated on northeastern James Ross Island, and Cape Lamb, on Vega Island.

Upper Cretaceous–Paleocene nonmarine palynomorphs reflect a cool, humid, podocarpaceous conifer vegetation with a varied understory of ferns and highly endemic angiosperms.

INTRODUCTION

Fossil palynomorphs are abundant and well-preserved throughout the Upper Cretaceous (Campanian) to lower Tertiary (Eocene) strata of Seymour Island and surrounding islands of the James Ross Island basin (Figs. 1, 2), northeastern Antarctic Peninsula. These predominantly nearshore marine sediments contain both marine (dinoflagellate cysts, acritarchs, other algae) and nonmarine palynomorphs (pollen and spores from land plants, fungal spores, fresh-water algae), plus a plethora of fossils from other floral and faunal groups useful for independent age and comparative paleoenvironmental interpretations.

These results are based on palynomorphs recovered from more than 900 outcrop samples collected during the 1974–1975,

1982, 1983–1984, and 1985 field seasons (Elliot et al., 1975; Askin and Fleming, 1982; Askin, 1984, 1986). The work focuses on the Campanian through Paleocene part of the succession, with only brief comments on Eocene palynology. The Eocene marine palynomorphs are described by Wrenn and Hart (this volume).

This continuing study will provide a palynostratigraphic reference section, for Campanian to Eocene rocks in Antarctica, useful for correlating isolated outcrop and sea-floor samples and for future drilling projects in and around the continent. Other results, with more far-reaching significance, concern interpretation of paleoenvironments, paleoclimates, the record of past Antarctic vegetation, and evolution of southern floras. The James

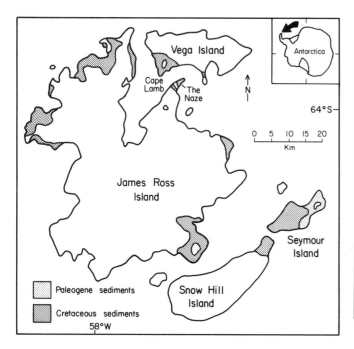

Figure 1. Locality map for the James Ross Island basin, Antarctica.

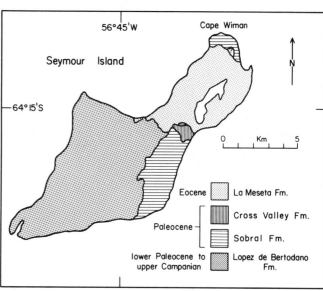

Figure 2. Geological map of Seymour Island.

Ross Island basin is important for reconstruction of Antarctic latest Mesozoic–Cenozoic geologic and climatic history because it has the only exposed sedimentary section known in Antarctica with nearly continuous deposition from the late Cretaceous through much of the Paleogene.

STRATIGRAPHY

Uppermost Cretaceous and Paleogene strata of Seymour and surrounding islands have been referred to the López de Bertodano, Sobral, Cross Valley, and La Meseta Formations (Figs. 2, 3) (Rinaldi et al., 1978; Rinaldi, 1982; Elliot and Trautman, 1982; del Valle et al., 1982; Macellari, 1984b, this volume; Sadler, this volume). Zinsmeister (this volume) presents a detailed account of the earliest expeditions to the region in the 1890s to 1903 that yielded geological results. Subsequent work on Campanian through Paleocene strata has been summarized by the authors listed above.

The Upper Cretaceous and Paleogene sediments of Seymour Island are unconsolidated, apart from occasional calcite-cemented beds. Most of the sequence represents nearshore marine environments. Lithologic character and subdivisions are briefly outlined below, and in Figures 3 and 4.

The López de Bertodano Formation (Rinaldi et al., 1978; Rinaldi, 1982) on Seymour Island comprises 1,190 m of predominantly gray sandy siltstone, with mud-rich units, particularly in the lower part, and some fine-grained, often calcareous, sandstone beds. The uppermost part of the formation records a substantial increase in volcanism, and contains abundant glauconite (Macellari, 1984b). The formation has been divided into 10 lithologic units (Macellari, 1984b, this volume; Figs. 3, 4). Macro- and microfossils are abundant and diverse. In addition to palynomorphs, these include ammonites, bivalves, gastropods, echinoids, corals, serpulid annelids (Macellari, 1984a,b, 1985a, 1986, this volume; Macellari and Zinsmeister, 1983); fish, mosasaur, and plesiosaur bones (Chatterjee and Zinsmeister, 1982; Chatterjee et al., 1984); foraminifera (Huber, 1984, this volume), calcareous nannofossils, silicoflagellates, and diatoms (Huber et al., 1983; Harwood, this volume); and fossil wood. Correlation of the upper Campanian to Danian López de Bertodano Formation units and biostratigraphic zones is shown in Figure 4.

An erosional unconformity separates the López de Bertodano Formation from the overlying Paleocene Sobral Formation. The lower part (~80 m) of the Sobral Formation includes muds and silts, and the rest of the approximately 255-m-thick formation is predominantly sandy. The Sobral Formation (Rinaldi et al., 1978; Rinaldi, 1982) includes five members (Sadler, this volume). Fossils are less diverse and less common than in the underlying López de Bertodano Formation, and are generally absent in the middle and upper parts of the Sobral Formation (Macellari, this volume). Mineralized (calcified) and coalified fossil wood is, however, often abundant.

The ?Paleocene Cross Valley Formation (Elliot and Trautman, 1982) includes five members of mainly coarse-grained sediments infilling a steep-walled channel (Sadler, this volume). Fossils are sparse, except for fossil wood, three specimens of

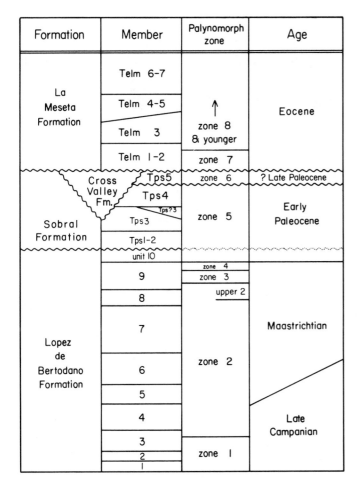

Formation	Member	Polynomorph zone	Age
La Meseta Formation	Telm 6-7	↑ zone 8 & younger	Eocene
	Telm 4-5		
	Telm 3		
	Telm 1-2	zone 7	
Cross Valley Fm.	Tps5	zone 6	? Late Paleocene
Sobral Formation	Tps4	zone 5	Early Paleocene
	Tps?3		
	Tps3		
	Tps1-2		
Lopez de Bertodano Formation	unit 10		
	9	zone 4 / zone 3	Maastrichtian
	8	upper 2	
	7	zone 2	
	6		
	5		
	4		Late Campanian
	3		
	2	zone 1	
	1		

Figure 3. Stratigraphic column for Seymour Island, showing the relationship of formations, members, and palynomorph zones.

which were discussed by Francis (1986). Leaf compressions (Case, this volume) occur in the upper part of the formation.

The Eocene La Meseta Formation (Elliot and Trautman, 1982) fills a wide erosional trough cut into the Sobral, Cross Valley, and López de Bertodano Formations (Sadler, this volume). This unit comprises a complex sequence of sandy and silty sediments that include abundant fossils, with many concentrated in shell beds. Sadler (this volume) has mapped seven members with seven minor variants within the La Meseta Formation.

PREVIOUS PALYNOLOGICAL WORK

Cranwell's (1959, 1969a,b) early studies initiated palynological work on Seymour Island and adjacent islands. Subsequent reports have included Hall (1977), Palamarczuk (1982), Palamarczuk et al. (1984), Wrenn (1982), Wrenn and Hart (this volume), Askin and Elliot (1982), Askin and Fleming (1982),

Fleming and Askin (1982), and Askin (1983, 1984, 1986, 1987, this volume).

Cranwell (1959) discussed palynomorphs found in a single sample of tuffaceous limestone collected by the 1901–1903 Swedish Antarctic Expedition. The sparse assemblage was dominated by conifer (podocarps and araucarians) and southern beech pollen (*Nothofagus* spp., *fusca, brassi,* and *menziesii* types). Other angiosperm pollen were referred to the modern plant families Cruciferae, Myrtaceae, Proteaceae, Loranthaceae, Oenotheraceae, and possibly Winteraceae and Cunoniaceae or Elaeocarpaceae. Sparse pteridophyte spores (Cyatheaceae and Schizaeaceae), fungal teleutospores, "mere traces" of algae and cuticular fragments were also noted. Cranwell reported the Seymour Island deposits as Cretaceous (lower and middle Campanian) and probably lower Miocene, quoting a personal communication of H. J. Harrington based on the work of British geologists. Cranwell indicated, however, that her material had more in common with Southern Hemisphere early Tertiary assemblages, and that it also contained some recycled Cretaceous specimens. Zinsmeister and Camacho (1982) suggested Cranwell's sample probably came from the (?Paleocene) Cross Valley Formation because of abundant volcanic glass, although glass-rich and calcareous rocks are now also known from the Paleocene Sobral Formation (e.g., Macellari, 1984b).

Additional samples from Snow Hill and Seymour Islands were discussed by Cranwell (1969a,b). She concluded that palynomorph assemblages were consistent with a Campanian age, based on ammonites (Howarth, 1966) from coral and *Rotularia*-bearing samples from Snow Hill and Seymour Islands, whereas other younger Seymour Island samples were apparently Maastrichtian to Paleocene. No evidence for Miocene age was found. She suggested shallow inshore or lagoonal paleoenvironments, based on the presence of marine dinocysts, with transported freshwater algae *Botryococcus* and *Pediastrum,* diverse fern spores, and conifer and angiosperm pollen.

Hall (1977) outlined palynological results from 24 samples collected by the expedition of Elliot et al. (1975) to Seymour Island. From dinocyst assemblages, he suggested an age of Paleocene to Late Cretaceous for a section S16 sample from Cape Wiman (?Sobral Formation), Paleocene for S11 samples (Sobral Formation), indeterminate for the upper S13 samples of the Cross Valley Formation, and Late Eocene to Early Oligocene for S3 La Meseta Formation samples (see also Elliot and Zinsmeister, 1978, and Hall's reply).

Palamarczuk (1982) described a dinocyst assemblage dominated (75%) by *Palaeoperidinium pyrophorum* from a single sample of the Sobral Formation, Cape Wiman. She suggested a Danian age and an abnormal-marine, probably estuarine paleoenvironment for the sample. More recently, Palamarczuk et al. (1984) considered several different fossil groups, including dinoflagellate cysts, from sections through the López de Bertodano and Sobral Formations on Seymour Island. They incorrectly concluded that all of the López de Bertodano Formation was Campanian in age, primarily on the basis of ammonite interpreta-

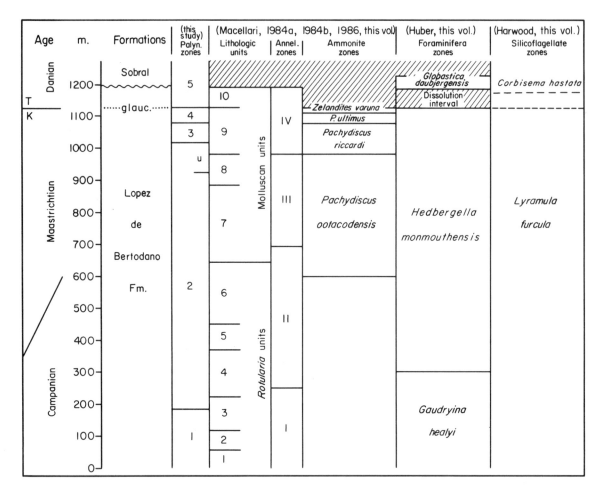

Figure 4. Diagram showing the relationship of López de Bertodano Formation lithologic units and fossil zones on Scymour Island.

tions. Few dinoflagellate cyst species were recovered from these beds, and unfortunately the stratigraphically useful Maastrichtian species *Manumiella druggii* and *M. seelandica* apparently were not recognized. The Sobral Formation samples yielded more diverse dinocyst assemblages, for which a Danian age was postulated. The hiatus recognized between the López de Bertodano and Sobral beds was thus believed to include the entire Maastrichtian Stage and the Cretaceous/Tertiary boundary.

In his dissertation, Wrenn (1982) gave a detailed account of Paleogene dinocysts of Seymour Island. Much of this study is published in Wrenn and Hart (this volume).

Askin and Elliot (1982) discussed geologic implications of Permian and Triassic palynomorphs recycled into Paleogene Seymour Island sediments. The nature of the palynological record across the Cretaceous/Tertiary transition is discussed by Askin (this volume). Other papers on Seymour Island palynomorphs include brief notes by Askin and Fleming (1982), Askin (1983, 1984, 1986, 1987), and Fleming and Askin (1982).

Dettmann and Thomson (1987) have described Cretaceous

palynomorph assemblages from the James Ross Island basin, including eight samples from various localities on James Ross Island, one from Vega Island, two from Dundee Island, and two from Cape Longing.

From elsewhere in Antarctica, other published studies on Late Cretaceous–Paleogene palynomorphs of taxonomic and stratigraphic significance include reports on the Ross Sea region (Cranwell, 1964a; Cranwell et al., 1960; McIntyre and Wilson, 1966; Wilson, 1967, 1968; Kemp, 1975; Kemp and Barrett, 1975; Wrenn, 1981; Wilson and Clowes, 1982), and King George Island, South Shetland Islands (Stuchlik, 1981). Late Cretaceous and Paleogene recycled palynomorphs from various localities around the coast of Antarctica have been described by Kemp (1972a,b), Truswell (1982a, 1983), Truswell and Anderson (1984), and Truswell and Drewry (1984). Some of the angiosperm pollen species illustrated and described by Truswell (1983) from the Ross Sea occur on Seymour Island and are evidently endemic Antarctic species.

Pertinent studies of regional paleoclimatic-paleogeographic

significance include those of Cranwell (1964b, 1969b), Kemp (1978), and Truswell (1982b, 1987).

PALYNOLOGICAL SUCCESSION

General

Palynomorphs are abundant in most samples examined, except where noted, and are generally well preserved. In-place spores are light yellow to yellow in color and indicate that burial or thermal metamorphism on Seymour Island is minimal. The Campanian to Paleocene palynological succession is described here, with paleoenvironmental notes where appropriate. Diagnostic dinocyst species referred to in the text are illustrated in Figures 7 through 11.

The zonation outlined here—although it is preliminary, pending completion of taxonomic studies and further sampling to fill in gaps in the sequence—is based on observations of about 530 samples. The informally numbered biostratigraphic units for Seymour Island (Figs. 3, 5, 6) utilize local range zones (some of which may reflect steps in evolutionary lineages), concurrent range zones, assemblage zones, and acme zones. Some species of restricted stratigraphic range are infrequent to rare and eventually may not prove to be practical zonal markers for the limited sample volume of future drillhole samples. Species selected as primary zonal markers are, when possible, relatively common to abundant components. Most diagnostic species noted are marine dinoflagellate cysts. Unfortunately, many of the more distinctive angiosperm pollen are very rare, and in some cases the highest recorded occurrences of rare species may be reworked specimens. Details of palynomorph assemblage composition and sample data will be presented with taxonomic reports.

Assemblages from The Naze on James Ross Island and Cape Lamb on Vega Island are isolated occurrences on different islands. Their relationships to each other and to those on Seymour Island are not fully understood. As they are separated by unknown lengths of geologic time and possibly additional intervening zones, they are described here as the "Naze assemblage" and "Vega assemblage," rather than as part of the consecutive zonal scheme of Seymour Island.

Ages for the assemblages and zones are derived in part by correlation with other fossil groups, and in part by comparison with palynomorph ranges from elsewhere in the world, particularly from Southern Hemisphere mid- to high latitudes. Stratigraphic ranges for some Seymour Island palynomorph taxa apparently differ from those described from other regions. This, together with the provincial nature of the floras, makes precise age determinations difficult. Heterochroneity of species ranges and endemism are common to many other Seymour Island fossil groups (e.g., Zinsmeister and Feldmann, 1984; Huber, this volume).

Naze assemblage

Description. The Naze assemblage is characterized by the dinoflagellate cysts *Odontochitina operculata, O. spinosa, O. po-*

rifera, Cribroperidinium sp., cf. *Canningiopsis* sp., and a complex of apparently interrelated peridinioid cyst species varying between *Isabelidinium cooksoniae* and cf.*Alterbidinium* sp. (the latter have a distinct pentapartite paracingulum, deltaform to thetaform intercalary archeopyle, and absence of large epicystal shoulders). Infrequent to rarely occurring species include *Operculodinium* sp., *Cyclopsiella* sp., *Palaeocystodinium* spp., *Nelsoniella aceras, Phelodinium* sp., and *Spiniferites* spp.

Occurrence. This assemblage was found in López de Bertodano Formation sediments (two samples only) from The Naze, northeastern James Ross Island.

Age. The age is Campanian, probably middle to late Campanian from associated faunas (e.g., Macellari, 1984b; Huber, this volume), with uncertain correlation to sediments from Cape Lamb, Vega Island. Palynomorphs suggest the two Naze samples are older than those at Cape Lamb (Askin, 1983), although they may represent a facies variant. Interpretation of the geologic structure (del Valle et al., 1982) suggested that The Naze is younger than Cape Lamb sediments. Ammonites (Macellari, 1984b) and foraminifera (Huber, this volume) do not resolve this question, although Huber stated the foraminiferal fauna from The Naze differs somewhat from those at Cape Lamb and Seymour Island. Interestingly, *O. operculata* and the foraminifera *Dorothia elongata* (Huber, this volume), both of which are conspicuous elements of their respective assemblages from The Naze, are absent in Cape Lamb and Seymour samples, yet these forms range throughout the Haumurian (and are Haumurian zonal markers) in New Zealand.

Vega assemblage

Description. This assemblage is marked by the first appearance of *Isabelidinium cretaceum. Cribroperidinium* sp., *Odontochitina porifera,* cf.*Canningiopsis* sp., *Isabelidinium* sp. cf. *I. cretaceum, Isabelidinium cooksoniae*-cf.*Alterbidinium* sp. and *Operculodinium* sp. are characteristic forms. The assemblage includes the first appearance of Dinocyst N. Gen. X. Angiosperm pollen diversity is greater than in The Naze samples, but may reflect a more inshore paleoenvironment for Vega samples (Askin, 1983).

Occurrence. This assemblage was found in the López de Bertodano Formation (10 samples) at Cape Lamb, Vega Island.

Age. The age is Campanian, and probably middle to late Campanian from associated faunas (e.g., Macellari, 1984b; Huber, this volume). The Naze and Vega assemblages correlate, at least in part, with the *Odontochitina porifera* Zone from Piripauan–lower Haumurian rocks (Campanian–?lower Maastrichtian) in New Zealand (Wilson, 1984a). *Odontochitina* species apparently disappeared earlier from the James Ross Island basin than in New Zealand and elsewhere.

Different assemblages of intermediate age between the isolated occurrences of The Naze and Vega Island and those of Seymour Island exist, as illustrated by one rather different assemblage from Vega Island (Dettmann and Thomson, 1987), and

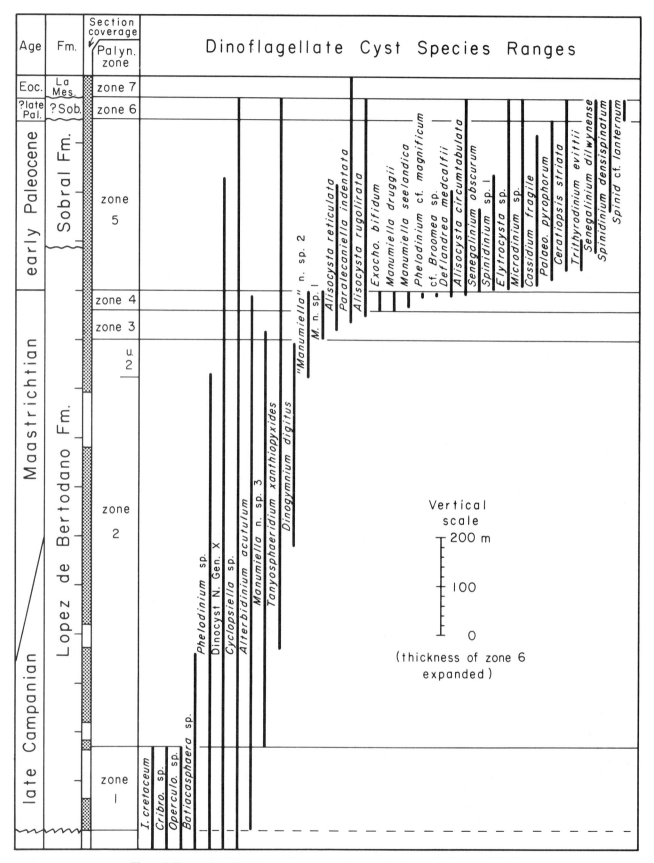

Figure 5. Range chart for selected dinocyst species, zones 1 through 6, Seymour Island.

those at Snow Hill Island (see below). Macellari (1984b) discussed problems with seemingly anomalous co-occurrences of ammonite taxa on Cape Lamb, Vega Island, and suggested either the existence of beds of two different ages at Cape Lamb, or a mixed fauna. In view of the two different palynomorph assemblages, the former suggestion is the more likely.

Snow Hill Island beds

Discussion. Detailed palynological study of these beds has not yet been carried out. Younger beds are evidently equivalent to the oldest Seymour Island units (Macellari, 1984a,b), and two reconnaissance samples from this younger part of the section contain some zone 1 palynomorphs. One sample from the older part of the section contains cf. *Canningiopsis* sp. and members of the *I. cooksoniae*–cf. *Alterbidinium* sp. complex, but no *Odontochitina* spp.

Zone 1

Description. Zone 1 is characterized by *Isabelidinium cretaceum,* with *Operculodinium* sp., and includes the lowest specimens of the dinocyst *Alterbidinium acutulum* and the pollen *Tricolporites lilliei.*

The stratigraphic base of zone 1 is not known, although it may exist on Snow Hill Island. The lowest sample of the Seymour Island section was taken at high tide level on the southwestern coast of the island.

The top of the zone is marked by the first appearance of *Manumiella* n.sp.3, and the last appearances of *Cribroperidinium* sp. and *Operculodinium* sp. Occasional specimens of *I. cretaceum* in lower zone 2 may be reworked.

Occurrence. This zone is in the stratigraphically lowest López de Bertodano Formation in the southwest part of Seymour Island (about 170 m?). Figure 4 shows correlation to López units and *Rotularia* zones of Macellari (1984a,b, 1986), foraminiferal zones of Huber (this volume), and silicoflagellate zones used by Harwood (this volume). This zone is known from seven samples on Seymour Island and requires further sampling to properly define its extent.

Age. Zone 1 is late Campanian from associated fossils. The foraminiferal *Gaudryina healyi* Zone of Huber (this volume) encompasses palynomorph zone 1 on Seymour Island (Fig. 4). The *G. healyi* Zone, which also occurs on The Naze and at Cape Lamb, was assigned a mid(?) to late Campanian age (Huber, this volume), and a late Campanian age at Seymour Island, based in part on evidence from other fossil groups (e.g., Huber et al., 1983). Calcareous nannofossils indicated that the lower López de Bertodano Formation on Seymour Island (0 to 350 or 415 m) is late Campanian (Huber et al., 1983). Macellari (1984b, 1985a, 1986) suggested a late Campanian age for the ammonite *Gunnarites* in unit 1 on Seymour Island. Palynomorph zones 1 through 4 correspond to the Campanian-Maastrichtian silicoflagellate *Lyramula furcula* Zone (Bukry, 1981) used by Harwood (this vol-

ume). Zones 1, 2, and 3 together are broadly comparable to (but of longer duration than) the *"Alterbia acutula"* Zone of late Haumurian age in New Zealand (Wilson, 1984a), which has its base and top marked by the last *Odontochitina* and first *M. druggii,* respectively.

Note on Depositional Environment. Zone 1 and lower zone 2 palynomorphs occur in sediments included in the *"Rotularia* units" of Macellari (1984b, this volume). These sediments contain a high proportion of mud and silt, particularly in massive mudstone units best developed toward the south coast of Seymour Island. The low-diversity invertebrate macrofauna preserved in these sediments is dominated by the serpulid annelid *Rotularia,* which apparently favored these muddy substrates (Macellari, 1984a). Macellari (this volume) discussed the environmental/ecological preferences of the macrofauna and concluded that the *"Rotularia* units" were deposited in very shallow marine, low-energy, delta/estuary-influenced environments, with possible intermittent brackish conditions. Palynomorph assemblages support this interpretation (Askin, 1987). Swarms of acanthomorph acritarchs with long, delicate processes, large brittle pieces of cuticle, and other lines of evidence suggest inshore, low-energy conditions, at least for portions of this interval.

Foraminiferal data agree with Macellari's interpretation for the lower 250 m of section (Huber, this volume); however, foraminiferal assemblage composition and stable isotope data suggest that normal, open marine conditions persisted for the rest of the Cretaceous on Seymour Island (Barrera et al., 1987; Huber, this volume).

Zone 2

Description. The base of zone 2 is marked by the first appearance of *Manumiella* n.sp.3. Other consistently appearing species include *Cyclopsiella* sp., Dinocyst N. Gen. X, *Batiacasphaera* sp., *Spiniferites* spp., *Palaeocystodinium* spp., and *Diconodinium* spp. The first rare specimens of *Tanyosphaeridium xanthiopyxides* appear in the lower part of the zone.

Extremely rare specimens of *Dinogymnium digitus* occur in middle through upper zone 2 samples, the only in-place representatives of the distinctive *Dinogymnium* species that are frequent components of most Maastrichtian assemblages elsewhere in the world.

Diversity is low in the marine component, but the nonmarine component exhibits increasing diversity in angiosperm pollen up through the zone. Recycled, slightly older Cretaceous pollen and spores are common to abundant in many samples; this also holds true for zone 3 samples.

Zone 2 spans a great thickness of López de Bertodano sediments on Seymour Island, with little change in the palynomorph succession, particularly when compared with the turnover in floras through zones 3, 4, and lower 5 (Fig. 5). Huber (this volume) notes that no major changes were taking place in the foraminiferal assemblages through most of this interval and considered this to be indicative of long-term environmental stability, with which the palynomorph data appear to concur.

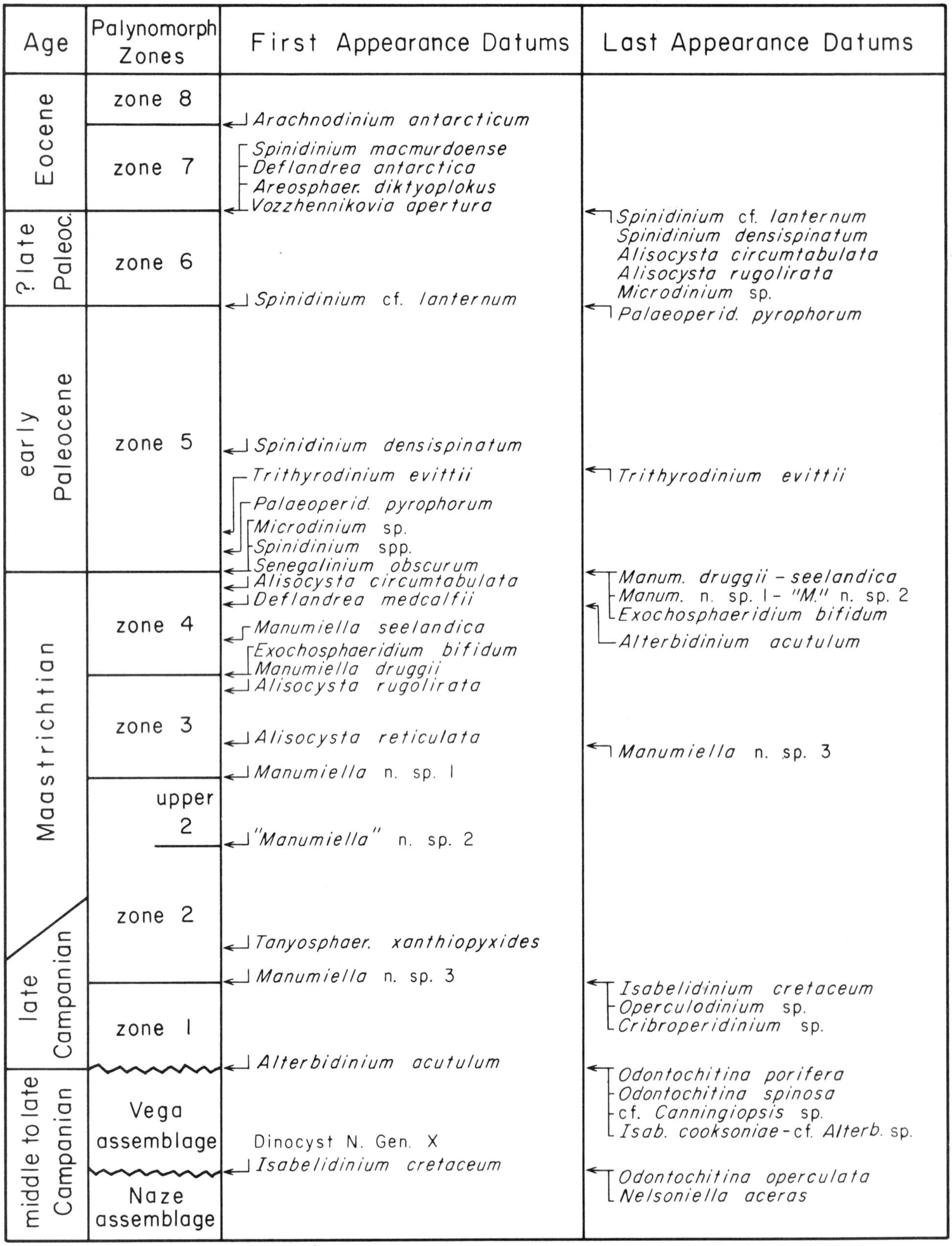

Figure 6. Palynomorph zones for Seymour Island (zones 1 through 8), the Naze assemblage on James Ross Island and Cape Lamb assemblage on Vega Island, showing first-appearance datums and last-appearance datums of selected dinocyst species.

An upper subzone is marked by the first rare specimens of *"Manumiella"* n.sp.2. *M.* n.sp.3 are still predominant, however, and samples from this interval are clearly in zone 2. Appearance of *M.* n.sp.1 and predominance of *"M."* n.sp.2 mark the top of the zone. Occasional specimens of *Manumiella* n.sp.3 from zone 3 may be reworked.

Occurrence. This zone is found in the major part (approximately 830 m) of López de Bertodano Formation on Seymour Island (see Fig. 4 for zone and unit correlations).

Age. Zone 2 is late Campanian and Maastrichtian from associated fossils. Microfossil evidence for the late Campanian age of the lower part of zone 2 is noted above. The upper half of zone 2 corresponds to the ammonite *Pachydiscus ootacodensis* Zone (Fig. 4) of Maastrichtian age (early Maastrichtian for base of *P. ootacodensis* Zone) and lower *Pachydiscus riccardi* Zone of late Maastrichtian age (Macellari, 1984b, 1985a, 1986). Most of zone 2, except its basal 120 m, corresponds to the foraminiferal *Hedbergella monmouthensis* Zone, of late Campanian to late Maastrichtian age (Huber, this volume). Calcareous nannofossils suggested a middle Maastrichtian age for beds approximately 640 m above the base of the Seymour Island section (Huber et al., 1983).

Campanian/Maastrichtian Boundary

Discussion. The Campanian/Maastrichtian boundary occurs between 350 and 600 m above the base of the López de Bertodano beds on Seymour Island (Fig. 4). Macellari (1984b, 1986) considered the *Pachydiscus ootacodensis* Zone Maastrichtian and placed, with some uncertainty, the Campanian/Maastrichtian boundary at the 600-m horizon (at the *P. ootacodensis* Zone base). He stated, however, that environmental conditions may not have been suitable for ammonites and other molluscs during deposition of the *Rotularia* (1–6) units. Harwood (*in* Huber et al., 1983) placed the boundary between 350 m and 415 m, based on calcareous nannoplankton distribution.

Palynomorph data are not diagnostic, but are consistent with a Campanian/Maastrichtian boundary within the lower part of zone 2. The marker species for zone 2, *Manumiella* n.sp.3, is most similar to *Isabelidinium haumuriense* (of more elongate shape with a more stenoform archeopyle) described from the type Haumurian (?upper Campanian–Maastrichtian) of New Zealand by Wilson (1984b). Other zone 2 species range through at least the upper Campanian and Maastrichtian elsewhere. *Alterbidinium acutulum* and *Tricolporites lilliei* are restricted to the upper Campanian–Maastrichtian in New Zealand (Wilson, 1984a; Raine, 1984), as on Seymour Island.

Zone 3

Description. Zone 3 is characterized by the predominance of the *Manumiella* n.sp.1–*"M".* n.sp.2 complex. First appearance of *M.* n.sp.1 and abundance of *"M."* n.sp.2 mark the base of zone 3. The relative abundance of *Palaeocystodinium* spp. increases

through this zone. Lowest occurrences of *Alisocysta reticulata* and *Paralecaniella indentata* are in the middle part of the zone, and *Alisocysta rugolirata* near the top.

Continuous morphologic variation is apparent between the two end members of the *Manumiella* n.sp.1 and *"Manumiella"* n.sp.2 complex, and between the zone 4 *Manumiella druggii–M. seelandica* complex, into which they may have evolved. The intergradational nature of these forms is illustrated in Figure 9.

Occurrence. Zone 3 spans approximately 60 m in the upper part (see Fig. 4) of the López de Bertodano Formation on Seymour Island. It corresponds to part of López unit 9 and much of the *Pachydiscus riccardi* ammonite zone.

Age. Zone 3 is Maastrichtian and probably late Maastrichtian, based on faunas of the associated *Pachydiscus riccardi* Zone (Macellari, 1986) and of the *Hedbergella monmouthensis* Zone (Huber, this volume). Palynomorph assemblages are consistent with this age assignment.

Note on Depositional Environment. Paleoenvironmental interpretations of the different fossil groups indicate shallow (<200 m) shelf marine conditions but are not entirely in agreement on proximity to shoreline for the upper López de Bertodano Formation of Seymour Island. Shelly invertebrate fossils, plus glauconite beds, indicate that "Molluscan" units 7 to 9 represent progressively deeper water, middle to outer shelf facies (Macellari, 1984b, this volume). Macellari interpreted unit 9 as the most offshore facies, probably deposited in an outer shelf environment, although water depths probably did not exceed 150 to 200 m. These paleobathymetric data were derived in part from foraminiferal evidence (Huber, 1984).

Paleoenvironmental evidence from foraminiferal assemblages is somewhat equivocal here, although it is consistent with a transgressive trend in the López de Bertodano section (to within unit 9). Cretaceous foraminifera on Seymour Island seem to represent middle to outer shelf depositional environments, with no evidence of upper shelf neritic conditions for the upper López de Bertodano Cretaceous strata (Huber, 1984, this volume). Plesiosaur remains throughout the Cretaceous sequence suggest shallow-water, nearshore marine environments (S. Chatterjee, personal communication, 1983 *in* Huber, 1984), and diatom assemblages indicate shallow-water shelf conditions (Harwood, this volume).

Unit 9 includes palynomorph zones 3 and 4 and uppermost zone 2 (Fig. 4). Palynomorph assemblage composition changes up through the section, but the general aspect of the assemblages does not change significantly from unit 8 into unit 9. The predominant marine constituents of zones 2 to 4 are the peridinioid dinocysts *Manumiella* and related forms, typical of nearshore/inshore marine environments. The main changes in these low-diversity marine assemblages include an increase in relative abundance of *Palaeocystodinium* up through zones 3 and 4 and the common occurrence (but still secondary to *Manumiella* spp.) of *Exochosphaeridium bifidum* in some zone 4 samples. Chorate dinocysts such as *Spiniferites ramosus* (open marine, often abundant in upper shelf, e.g., Wall et al., 1977), and *Hystricho-*

sphaeridium, Oligosphaeridium, and related forms (open marine, e.g., Downie et al., 1971) are rare (less than 2 percent of marine assemblage) in most samples. Other typically open marine species of chorate dinocysts like *Areoligera* (e.g., Downie et al., 1971) are apparently absent on Seymour Island.

Marine:nonmarine ratios for the palynomorph assemblages on Seymour Island exhibit no significant increasing trend through zones 2 to 4, as might be expected with a depositional trend from middle to outer shelf (units 7, 8, and 9). The palynomorph assemblages found in unit 9 are more compatible with a nearshore depositional environment, possibly restricted or enclosed, rather than offshore, outer shelf, open marine conditions.

Zone 4

Description. The base of zone 4 is defined by the lowest appearances of *Manumiella druggii* and *Exochosphaeridium bifidum*. Interpretation of the first *M. druggii* specimens is somewhat subjective, but appearance of early transitional forms (still with small antapical horns or protrusions) associated with *E. bifidum* serves to define the zone. True *M. seelandica* first appear higher in the zone. Common to abundant specimens of the *M. druggii*-*M. seelandica* complex, plus *Exochosphaeridium bifidum, Palaeocystodinium* spp., *Cordosphaeridium* sp., and *M.* n.sp.1-"*M.*" n.sp.2 complex characterize zone 4. Species with short stratigraphic ranges within zone 4 include *Phelodinium* sp.cf. *P. magnificum* and cf. *Broomea* sp. The first *Deflandrea medcalfii* and *Alisocysta circumtabulata* appear near the top of the zone.

The top of zone 4 is marked by the appearance of *Senegalinium obscurum,* and the last occurrences of the *Manumiella druggii*-*M. seelandica* complex, *Manumiella* n.sp.1-"*M.*" n.sp.2 complex, *E. bifidum* and the pollen *Tricolporites lilliei*; it corresponds approximately with the Cretaceous/Tertiary boundary (Askin, this volume).

Occurrence. Zone 4 spans approximately 40 m of section, near the top of the López de Bertodano Formation on Seymour Island (Fig. 4), corresponding to the *Pachydiscus ultimus* and *Zelandites varuna* ammonite Zones (Macellari, 1984b, 1986; Macellari et al., 1987).

Age. Zone 4 is considered Maastrichtian in age, from palynomorphs and associated macro- and microfossils. The ammonite *P. ultimus* and *Z. varuna* Zones are considered latest Maastrichtian (Macellari, 1986; Macellari et al., 1987). Foraminiferal assemblages of the *H. monmouthensis* Zone (Huber, this volume) and silicoflagellates of the *Lyramula furcula* Zone (Harwood, this volume) are Maastrichtian. The zone 4 dinocyst *E. bifidum* is Maastrichtian (and older), and the *M. druggii*-*M. seelandica* complex is Maastrichtian-Danian (Askin, this volume). Zone 4 correlates with the *Isabelidinium druggii* Zone from Haumurian-Teurian (Cretaceous/Tertiary) transitional rocks in New Zealand (Wilson, 1984a).

Cretaceous/Tertiary transition

Discussion. The Cretaceous/Tertiary (K/T) boundary is provisionally placed at the zone 4/zone 5 boundary, a change

that occurs at a laterally persistent glauconite interval (Askin, this volume). This glauconite interval may also coincide with the stratigraphically highest occurrence of ammonites and numerous other shelly fossils (Macellari, 1984b, 1985b; Macellari et al., 1987; Zinsmeister and Macellari, 1983); and diagnostic Cretaceous planktonic foraminifera and silicoflagellates (Huber, 1984, 1986, 1987; Huber et al., 1985; Harwood, this volume). Most of the observable changes across the K/T transition were taking place in the marine realm, some apparently related to a shallowing trend. Changes in water chemistry, possibly related to the increased volcanism at this level, and possibly as recorded in a "calcite dissolution interval" (Huber, 1987; Huber et al., 1985), may also have affected the dinoflagellate population. The land vegetation, as evaluated from the pollen and spore assemblages, apparently changed little across the K/T transition (Askin, this volume).

Zone 5

Description. The base of zone 5 is marked by first appearance of the small dinocyst *Senegalinium obscurum.* This overlaps by 1 m with the highest occurrence of *M. druggii.* The zone is characterized by *Spinidinium* spp., *Deflandrea* spp. (including *D. medcalfii*), *Ceratiopsis striata, Senegalinium dilwynense, Palaeoperidinium pyrophorum, Alisocysta* spp. (*A. circumtabulata, A. rugolirata, A. reticulata*), *Microdinium* sp., *Elytrocysta* sp., *Cassidium fragile,* and, in the lower part (Fig. 5), *Trithyrodinium evittii.*

Subdivision of zone 5 may be possible following taxonomic study of variable *Spinidinium* and *Deflandrea-Ceratiopsis* populations. *Spinidinium* sp. 1 and *S. pilatum* are common constituents of the lower part (uppermost López de Bertodano, Tps1, Tps2), and abundance of other *Spinidinium* species, including *S. densispinatum,* along with *Microdinium* sp. and *Elytrocysta* sp., characterize the upper shoreline facies (Tps3, Tps4, ?Cross Valley).

Occurrence. Zone 5 spans the uppermost López de Bertodano Formation (unit 10) and almost all of the Sobral Formation (Tps1-4).

Age. Zone 5 is considered early Paleocene (Danian) based on palynomorph, foraminiferal, and silicoflagellate evidence. Dinocyst evidence for the Danian age assignment for zone 5, and in particular the lower part of the zone, was discussed in Askin (this volume). It is possible that the upper part of the zone is as young as early late Paleocene (early Thanetian). The Sobral Formation, which includes the major part of the zone, was also assigned a Danian age by Palamarczuk et al. (1984) based on dinocyst evidence. The foraminiferal *Globastica daubjergensis* Zone, restricted to unit 1 of the Sobral Formation (lower 20 m) plus a sample 1 m below the López de Bertodano/Sobral contact, is Danian in age (Huber, this volume). Silicoflagellates, diatoms, and ebridians occur through much of zone 5, with diagnostic silicoflagellates of the Paleocene (Danian–early Thanetian) *Corbisema hastata* Zone (of Bukry, 1981) first appearing a few meters below the base of the Sobral Formation (Harwood, this volume). Other silicoflagellates that are not considered diagnostic zonal markers, but are nonetheless characteristic of the lower Paleocene, occur in upper López de Bertodano (zone 5) sediments (Harwood, this volume). Harwood considered it unlikely that any of the Sobral Formation (Tps1-4) is as young as late Paleocene.

It is believed, based on a lack of perturbation of dinocyst ranges (Fig. 5), that the erosional unconformity at the base of the Sobral Formation does not represent a major hiatus.

Precise correlation for the Cross Valley Formation and timing of its underlying erosional event have still to be resolved. Marine palynomorphs are rare, and nonmarine palynomorphs are not age diagnostic in the Cross Valley Formation, making it difficult to accurately correlate Cross Valley assemblages to upper zone 5, zone 6, or younger assemblages. Ambiguous outcrop patterns, some of which are debris-covered, complicate interpretation, as exemplified by an assemblage of Eocene (La Meseta Formation) zone 8 dinocysts occurring on the northern flanks of the main Cross Valley outcrop. This possibly explains problems with the stratigraphy referred to in Wrenn and Hart (this volume), where the uppermost part of the formation was considered early late Paleocene and the rest of the formation Eocene. Harwood (this volume) concluded that the Cross Valley Formation was late Paleocene, based on a silicoflagellate assemblage from a sample in the uppermost part of the formation.

Note on Depositional Environment. Fossil macrofaunas and lithology (Macellari, 1984b, this volume), foraminifera (Huber, 1984, this volume), and palynomorphs are all compatible with a regressive phase of deposition up through the uppermost López de Bertodano Formation (unit 10) and the Sobral Formation. Restricted paleoenvironments are indicated at several intervals in lower zone 5 by almost monospecific palynomorph assemblages in the marine component of (from older to younger) either *Senegalinium obscurum, Spinidinium* sp.1 or *Palaeoperidinium pyrophorum* (see also Palamarczuk, 1982). These assemblages may represent dinoflagellate blooms in low-salinity estuarine or enclosed bay conditions. Unusual environmental conditions in the uppermost López de Bertodano beds may be indicated by the "calcite dissolution interval" (if this was not later diagenetic), which may have been caused by changes in water chemistry related to volcanic activity. Stressed, anoxic environmental conditions were suggested (Huber, this volume) for foraminiferal assemblages from the overlying lower member (Tps1) of the Sobral Formation, a unit probably deposited in relatively quiet nearshore marine conditions (Macellari, this volume).

Sobral members Tps3 and Tps4 and the Cross Valley Formation are predominantly sandy, and many samples from this interval contain sparse palynomorph assemblages. Relatively high-energy depositional conditions above wave base, possibly beach facies, have been postulated for at least part of the interval (Macellari, 1984b, this volume), with several intervals of nonmarine deposition. Dinocyst species that exhibit high relative frequencies in these littoral/coastal deltaic facies include *Spinidinium densispinatum* and *Microdinium* sp.

Zone 6

Description. Zone 6 is characterized by a predominance of *Spinidinium* sp. cf. *S. lanternum.* This species, which forms almost monospecific assemblages in some samples, is apparently restricted to zone 6. Many dinocyst species of upper zone 5 also occur in zone 6 (Figs. 5, 6).

Occurrence. Zone 6 assemblages occur in member Tps5 of the Sobral Formation. This member is about 10 m thick and forms the top of three small buttes south of Cross Valley (see Sadler, this volume, geologic map). Zone 6 assemblages also occur in possible Sobral beds near Cape Wiman (mapped as questionable Cross Valley Formation by Sadler, this volume). The erosional unconformity separating these beds from the overlying La Meseta Formation marks the top of the zone at Cape Wiman.

Age. Zone 6 is Paleocene based on palynomorphs. The absence of *Palaeoperidinium pyrophorum* and *Trithyrodinium evittii* suggests an age younger than early late Paleocene, although zone 6 assemblages may reflect a peculiar facies variant. The absence of Eocene species (characterizing zone 7 and higher) support a Paleocene age. If a late late Paleocene age is correct, and if the rest of the Sobral Formation is Danian, it implies a hiatus prior to deposition of uppermost Sobral Formation, member Tps5.

Zones 7, 8, and younger

Discussion. Assemblages from the Eocene La Meseta Formation are characterized by, among other species, *Deflandrea antarctica, Areosphaeridium diktyoplokus, Spinidinium macmurdoensis,* and *Vozzhennikovia apertura.* The basal 200 m (zone 7), representing most of member "Telm 2" of Sadler (this volume), is distinguished by the predominance of chorate dinocysts (particularly *Enigmadinium cylindrifloriferum* of Wrenn and Hart, this volume, with *Areosphaeridium diktyoplokus*) and is possibly an offshore facies variant of zone 8. Common *Vozzhennikovia apertura* and *Deflandrea* spp., with the first appearance of *Arachnodinium antarcticum,* characterize zone 8. Dinocyst assemblages of most of the La Meseta Formation (except basal beds) are described in detail by Wrenn and Hart (this volume) and are not discussed further.

LAND VEGETATION AND PALEOCLIMATES

Five levels of provincialism are apparent for Campanian to Paleocene pollen and spores of Seymour Island: (1) species endemic to the James Ross Island basin, (2) Antarctic endemic species, (3) Weddellian Province, (4) austral, and (5) cosmopolitan species. Most of the James Ross and Antarctic endemic species are angiospermous; these endemic species form a significant proportion (about 50 percent) of the palynoflora. The Antarctic flora includes species present in the James Ross Island area, and reported from the Ross Sea and around the coast of East Antarctica (Truswell, 1983). Many species are restricted to the Weddellian Province of Zinsmeister (1979, 1982), a faunally and florally discrete region encompassing southern South America, Antarctica, New Zealand, and southeastern Australia. Austral pollen and spore taxa occur throughout the southern continents, and some taxa apparently had a more cosmopolitan distribution.

Abundant and diverse podocarpaceous conifer pollen predominate in the Seymour Island Campanian to Paleocene assemblages. Araucarian conifer pollen are also represented. Moss and fern spores are common. All three groups (*brassi, fusca,* and *menziesii* types) of *Nothofagidites,* the pollen of southern beech or *Nothofagus,* are found. Besides these beech pollen, other observed angiosperm pollen may be referred to Proteaceae, Loranthaceae, Myrtaceae, Casuarinaceae, Ericales, and Liliaceae. Unfortunately, the modern affinities of most fossil angiosperm pollen species of Seymour Island remain unknown.

Throughout the Seymour section, the López and Sobral nearshore marine sediments incorporate transported pollen and spores, some probably reworked, from a variety of land habitats. The terrestrial component probably includes representatives from such sources as lowland forests and moist fluvial/lacustrine/swamp areas, drier interfluvial areas, and coastal areas, mixed with palynomorphs transported from upland communities. Thus, reconstructions of past vegetation and inferred paleoclimates are broad generalizations.

If climatic requirements or preferences for modern plants can be presumed for their Campanian-Paleocene palynological counterparts, the podocarp-dominated vegetation of Seymour Island suggests cool temperate, moist conditions. This vegetation has been likened to the present-day lowland rainforest of southern Chile, with its podocarp and araucarian conifers and beeches (e.g., Truswell, 1982b). The distinctive podocarp pollen *Phyllocladidites mawsonii,* common on Seymour Island, are probably allied to the modern conifer *Dacrydium franklinii* (e.g., Cookson, 1953; Playford and Dettmann, 1978), now restricted to the cool temperate, western Tasmanian rainforests. Compressions of fern, *Nothofagus,* and podocarp conifer foliage from the upper part of the Cross Valley Formation support a cool temperate, rainforest

flora (Case, this volume). Four specimens of fossil conifer and angiosperm wood from the Cross Valley and Sobral Formations have uniform growth rings, indicating stable forest environments (Francis, 1986). Narrow growth rings of two of these wood specimens were suggested by Francis as possible reflections of Cretaceous/Tertiary climatic deterioration, in contrast to the warmer temperate climates envisaged for much of the Cretaceous.

From the Campanian through Maastrichtian, terrestrial assemblages exhibit increasing diversity (especially among the angiosperms), with approximately 100 pollen and spore species represented. Paleocene assemblages apparently contain fewer species. This decrease can be explained by several factors (or a combination of factors) including cooler climates, partial denudation of the landscape by the volcanic activity occurring nearby at that time, or meager representation of palynomorphs in relatively coarse-grained coastal Paleocene sediments. There is no obvious abrupt extinction event for terrestrial palynomorphs at the Cretaceous/Tertiary boundary (Askin, this volume), although there is a long-term turnover, possibly related to climate change, in assemblage composition from the Maastrichtian through the Danian.

ACKNOWLEDGMENTS

This work has benefited greatly from discussion and suggestions on various aspects by M. E. Dettmann, D. H. Elliot, D. M. Harwood, B. T. Huber, S. R. Jacobson, C. E. Macellari, P. M. Sadler, M. O. Woodburne, and W. J. Zinsmeister. Helpful reviews of the manuscript were provided by D. Habib, W. S. Drugg, S. R. Jacobson, and M. O. Woodburne. Appreciation is extended to F. C. Barbis, R. H. Fleming, R. T. Kelley, C. H. Robinson, and J. R. Robinson for help with field work, K. G. Pill for laboratory preparations and photography, L. Jankov for photography, and L. Bobbitt and I. Vient for drafting. This research, much of which was carried out at Colorado School of Mines, is supported by National Science Foundation Grant DPP-8314186.

Figure 7. Dinoflagellate cysts from The Naze on James Ross Island and Cape Lamb on Vega Island. Magnification approximately ×500. 1, *Odontochitina operculata* (Wetzel) Deflandre and Cookson, Naze assemblage. 2, *Odontochitina spinosa* Wilson, Naze and Vega assemblages. 3, *Odontochitina porifera* Cookson, Naze and Vega assemblages. 4, *Cribroperidinium* sp., Naze and Vega assemblages, zone 1, = *Cribroperidinium* sp. A of Dettmann and Thomson (1987). 5, cf. *Alterbidinium* sp., Naze and Vega assemblages; 6, cf.*Canningiopsis* sp., Naze and Vega assemblages; apical archeopyle, paracingulum, and parasutural processes; in shape similar to *Pseudoceratium securigerum* (Davey and Verdier) Bint. 7, *Operculodinium* sp., Naze and Vega assemblages, zone 1. 8, *Phelodinium* sp., Naze and Vega assemblages, zones 1 through 2; apical and antapical horns typically broken at bases as in this specimen.

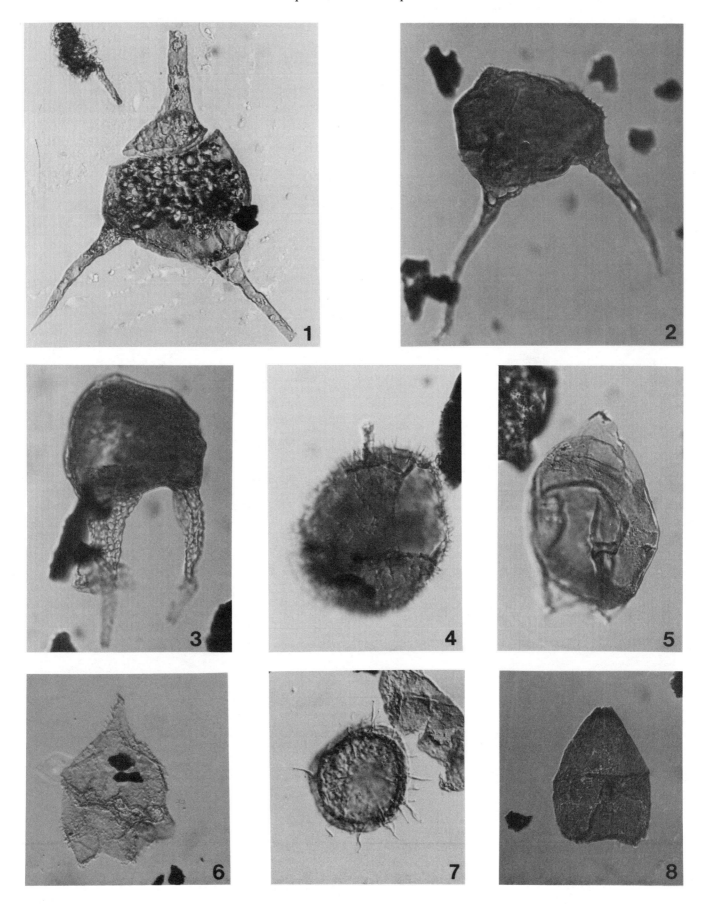

Figure 8. Dinoflagellate cysts from The Naze on James Ross Island (1), Cape Lamb on Vega Island (2, 3), and Seymour Island. Magnification approximately ×500, ×750 for 8. 1, *Nelsoniella aceras* Cookson and Eisenack, Naze assemblage. 2, *Isabelidinium cretaceum* (Cookson) Lentin and Williams, Vega assemblage and zone 1. 3, *Isabelidinium* sp. cf. *I. cretaceum,* Vega assemblage. This more robust form, with substantial thickenings at apex and antapex, and in many examples, almost rectangular endophragm outline, is referred to *Isabelidinium* sp. cf. *I. bakeri* (Deflandre and Cookson) Lentin and Williams by Dettmann and Thomson (1987). 4-6: *Manumiella* n.sp.3, zone 2. 4, Typical example; note very thin, barely discernible endophragm. 5, Specimen with single antapical horn. 6, Specimen with apical and single antapical horn. 7, *Alterbidinium acutulum* (Wilson) Lentin and Williams, zones 1 through 4. 8, *Dinogymnium digitus* (Deflandre) Evitt et al., zone 2. 9, *Cyclopsiella* sp., Naze and Vega assemblages, zones 1 through 6. 10, Dinocyst N.Gen.X, Vega assemblage, zones 1 through 5; apical archeopyle, paracingular, and parasulcal folds.

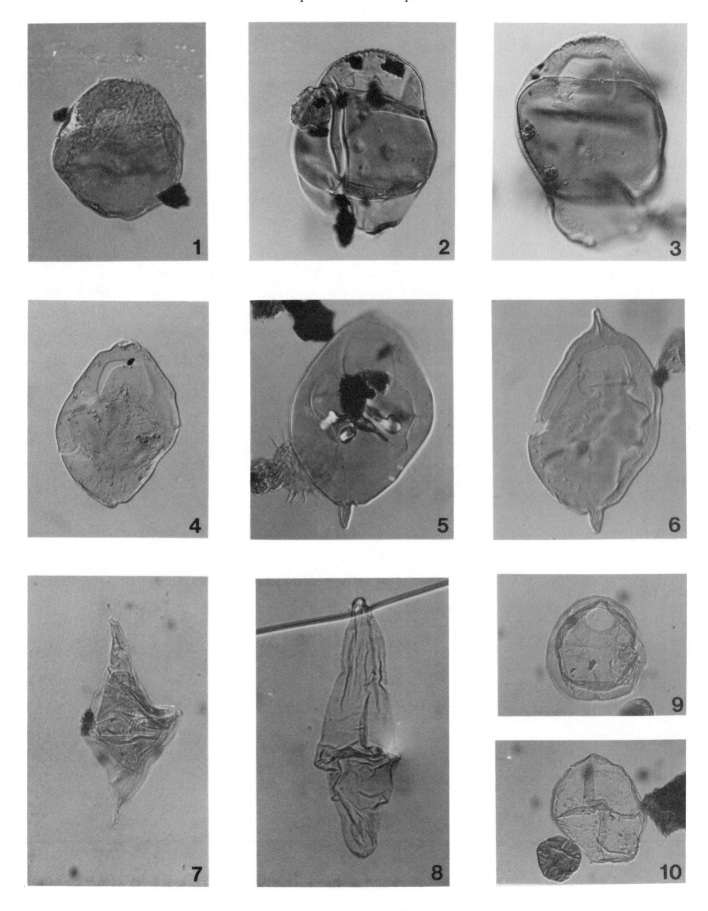

Figure 9. Dinoflagellate cysts from Seymour Island. Magnification approximately ×500. 1, *Manumiella* n.sp.1, zones 3 and 4. Typical specimen exhibiting tendency to wetzelioid shape and slight development of paracingulum. 2, *Manumiella* n.sp.1. Specimen with slight development of paracingulum, slight development of antapical horns, transitional to *M. druggii.* 3, *Manumiella* sp., transitional form, with small apical horn typical of *M. seelandica* and *"M."* n.sp.2. 4, *"Manumiella"* n.sp.2, upper zone 2-zone 4. Typical specimen showing well-developed apical and antapical horns, and the tendency to wetzelioid shape. 5, *"Manumiella"* n.sp.2, specimen with slight development of paracingulum and antapical horns, transitional to *M. seelandica.* 6, ?*Isabelidinium thomasii* (Cookson and Eisenack) Lentin and Williams, zones 3 and 4; specimen with large endophragm and more bicavate form, transitional to *"M."* n.sp.2. 7, *Manumiella druggii* (Stover) Bujak and Davies, zone 4. 8, *Manumiella* sp. cf. *M. druggii,* zone 4; truncate apex. 9, *Manumiella seelandica* (Lange) Bujak and Davies, zone 4.

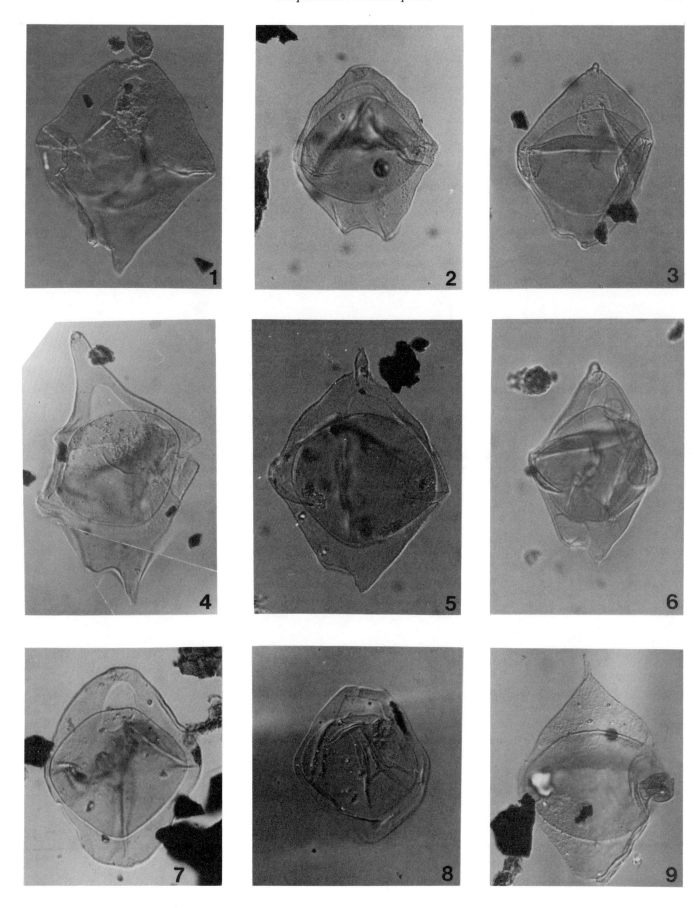

Figure 10. Dinoflagellate cysts from Seymour Island. Magnification approximately ×500, ×750 for 8. 1. *Palaeocystodinium* sp., Naze and Vega assemblages, zones 1 through 8. 2, 3, *Exochosphaeridium bifidum* (Clarke and Verdier) Clarke et al., zone 4. Note differentiated apical process in 2. 4, 5, *Deflandrea medcalfii* Stover, zones 4 through 7. 6, cf. *Broomea* sp., zone 4; intercalary archeopyle, two subequal antapical horns, long apical horn broken off in this specimen. 7, *Senegalinium obscurum* (Drugg) Stover and Evitt, zone 5. 8, *Tanyosphaeridium xanthiopyxides* (Wetzel *em.* Morgenroth) Stover and Evitt, zones 2 through 6. 9, *Paralecaniella indentata* (Deflandre and Cookson) Cookson and Eisenack, zones 3 through 8. 10, *Phelodinium* sp.cf. *P. magnificum* (Stanley) Stover and Evitt, zone 4. 11, 12, *Palaeoperidinium pyrophorum* (Ehrenberg) Sarjeant, zone 5.

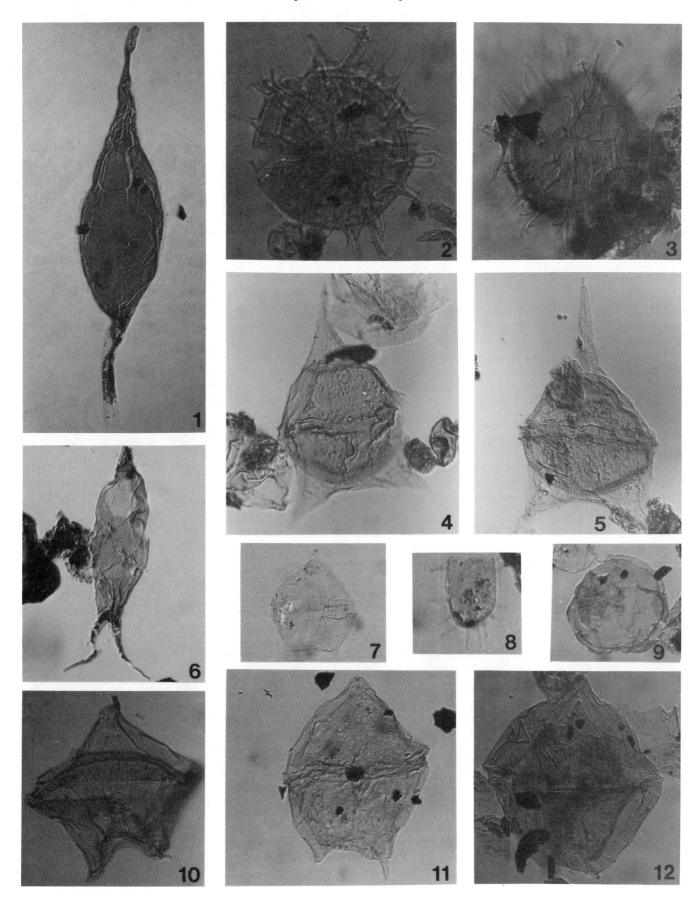

Figure 11. Dinoflagellate cysts from Seymour Island. Magnification approximately ×500 for 1 through 6, 14; ×750 for 7 through 11, 13, 15, 16; ×1,000 for 12. 1, 2, *Alisocysta circumtabulata* (Drugg) Stover and Evitt, zones 4-6. 3, 4, *Alisocysta rugolirata* Damassa, zones 3 through 6, same specimen at different focus. 5, *Alisocysta reticulata* Damassa, zones 3 through 5. 6, *Trithyrodinium evittii* Drugg, zone 5. 7, *Senegalinium dilwynense* (Cookson and Eisenack) Stover and Evitt, zones 5 and 6. 8, *Spinidinium densispinatum* Stanley, zones 5 and 6. 9, *Elytrocysta* sp., zones 5 and 6. Attached operculum, processes capitate, acuminate or blunt, sometimes aligned along paracingulum (= *"Membranosphaera"* sp. of Drugg, 1967). 10, 11, *Microdinium* sp., zones 5 and 6. 12, *Spinidinium* sp. 1, zone 5, capitate processes. 13, *Spinidinium pilatum* (Stanley) Costa and Downie, zone 5. 14, *Ceratiopsis striata* (Drugg) Lentin and Williams, zones 5 through 8. 15, 16, *Spinidinium* sp. cf. *S. lanternum* Cookson and Eisenack, zone 6.

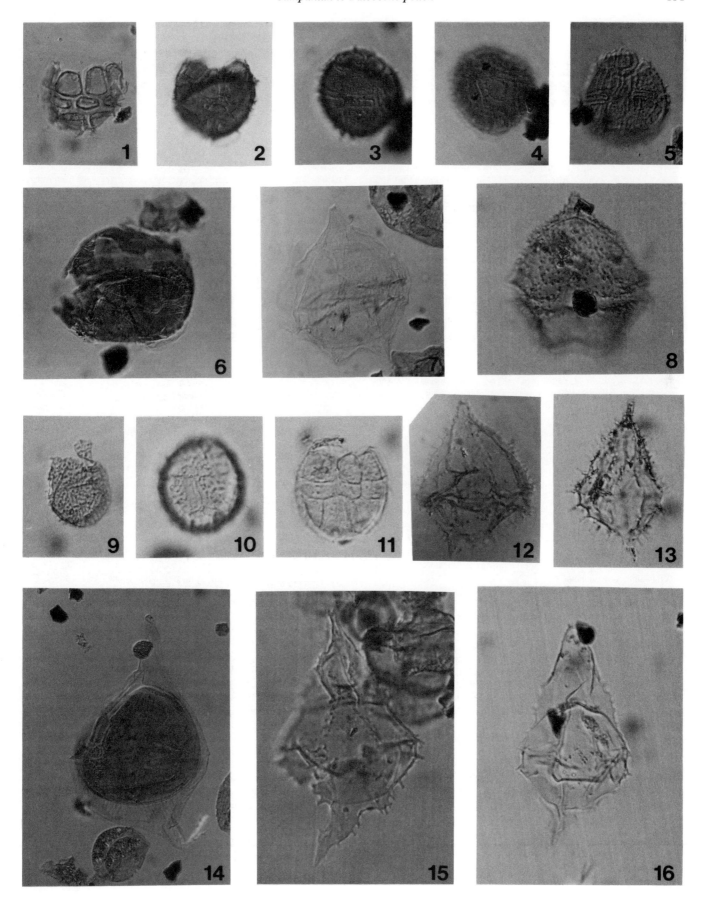

REFERENCES CITED

ASKIN, R. A. 1983. Campanian palynomorphs from James Ross and Vega Island, Antarctic Peninsula. Antarctic Journal of the United States, 18:63–65.

—— . 1984. Palynological investigations of the James Ross Island basin and Robertson Island, Antarctic Peninsula. Antarctic Journal of the United States, 19:6–7.

—— . 1986. Palynological studies in the James Ross Island basin, Antarctic Peninsula: a progress report. Antarctic Journal of the United States, 20:44–45.

—— . 1987. Palynomorphs and depositional environment for upper Campanian sediments on Seymour Island, Antarctica. Antarctic Journal of the United States, 21 (in press).

—— , AND D. H. ELLIOT. 1982. Geologic implications of recycled Permian and Triassic palynomorphs in Tertiary rocks of Seymour Island, Antarctic Peninsula. Geology, 10:547–551.

—— , AND R. F. FLEMING. 1982. Palynological investigations of Campanian to lower Oligocene sediments on Seymour Island, Antarctic Peninsula. Antarctic Journal of the United States, 17:70–71.

BARRERA, E., B. T. HUBER, S. M. SAVIN, AND P. N. WEBB. 1987. Antarctic marine temperatures: late Campanian through early Paleocene. Paleoceanography, 2:21–47.

BUKRY, D. 1981. Synthesis of silicoflagellate stratigraphy for Maastrichtian to Quaternary marine sediment, p. 433–444. In J. E. Warme, et al. (eds.), Deep Sea Drilling Project: A Decade of Progress, Special Publication 32, Society of Economic Paleontologists and Mineralogists.

CHATTERJEE, S., AND W. J. ZINSMEISTER. 1982. Late Cretaceous marine vertebrates from Seymour Island, Antarctic Peninsula. Antarctic Journal of the United States, 17:66.

—— , B. J. SMALL, AND M. W. NICKELL. 1984. Late Cretaceous marine reptiles from Antarctica. Antarctic Journal of the United States, 19:7–8.

COOKSON, I. C. 1953. The identification of the sporomorph Phyllocladidites with Dacrydium and its distribution in southern Tertiary deposits. Australian Journal of Botany, 1:64–70.

CRANWELL, L. M. 1959. Fossil pollen from Seymour Island, Antarctica. Nature, 184:1782–1785.

—— . 1964a. Hystrichospheres as an aid to Antarctic dating with special reference to the recovery of Cordosphaeridium in erratics at McMurdo Sound. Grana palynologica, 5:397–405.

—— . 1964b. Antarctica, Cradle or grave for its Nothofagus, p. 87–93. In L. Cranwell (ed.), Ancient Pacific Floras, the Pollen Story. University of Hawaii Press, Honolulu.

—— . 1969a. Antarctic and Circum-Antarctic palynological contributions. Antarctic Journal of the United States, 4:197–198.

—— . 1969b. Palynological intimations of some pre-Oligocene Antarctic climates, p. 1–19. In van Zinderen Bakker (ed.), Palaeoecology of Africa. S. Balkema, Cape Town.

—— , H. J. HARRINGTON, AND I. G. SPEDEN. 1960. Lower Tertiary microfossils from McMurdo Sound, Antarctica. Nature, 186:700–702.

DEL VALLE, R. A., N. H. FOURCADE, AND F. A. MEDINA. 1982. The stratigraphy of Cape Lamb and The Naze, Vega and James Ross Islands, Antarctica, p.275–300. In C. Craddock (ed.), Antarctic Geoscience. University of Wisconsin Press, Madison.

DETTMANN, M. E., AND M.R.A. THOMSON. 1987. Cretaceous palynomorphs from James Ross Island area, Antarctica – a pilot study. British Antarctic Survey Bulletin (in press).

DOWNIE, C., M. A. HUSSAIN, AND G. L. WILLIAMS. 1971. Dinoflagellate cyst and acritarch associations in the Paleogene of southeast England. Geoscience and Man, 3:29–35.

DRUGG, W. S. 1967. Palynology of the upper Moreno Formation (Late Cretaceous–Paleocene), Escarpado Canyon, California. Palaeontographica, Abt. B, 120:1–71.

ELLIOT, D. H., AND T. A. TRAUTMAN. 1982. Lower Tertiary strata on Seymour Island, Antarctic Peninsula, p. 287–297. In C. Craddock (ed.), Antarctic Geoscience. University of Wisconsin Press, Madison.

—— , AND W. J. ZINSMEISTER. 1978. Biostratigraphy of Seymour Island, Antarctica. Nature, 267:586.

—— , C. RINALDI, W. J. ZINSMEISTER, T. A. TRAUTMAN, W. A. BRYANT, AND R. DEL VALLE. 1975. Geological investigations on Seymour Island, Antarctic Peninsula. Antarctic Journal of the United States, 10:182–186.

FLEMING, R. F., AND R. A. ASKIN. 1982. An early Tertiary coal bed on Seymour Island, Antarctic Peninsula. Antarctic Journal of the United States, 17:67.

FRANCIS, J. E. 1986. Growth rings in Cretaceous and Tertiary wood from Antarctica and their palaeoclimatic implications. Palaeontology, 29:665–684.

HALL, S. A. 1977. Cretaceous and Tertiary dinoflagellates from Seymour Island, Antarctica. Nature, 267:239–241.

—— . 1978. Reply to "Biostratigraphy of Seymour Island." Nature, 271:586.

HOWARTH, M. K. 1966. Ammonites from the Upper Cretaceous of the James Ross Island group. British Antarctic Survey Bulletin, 10:55–69.

HUBER, B. T. 1984. Late Cretaceous foraminiferal biostratigraphy, paleoecology, and paleobiogeography of the James Ross Island region, Antarctic Peninsula. Unpubl. M.S. thesis. The Ohio State University, Columbus, 247 p.

—— . 1986. The location of the Cretaceous-Tertiary contact on Seymour Island, Antarctic Peninsula. Antarctic Journal of the United States, 20:46–48.

—— . 1987. Foraminiferal distribution across the Cretaceous/Tertiary transition on Seymour Island, Antarctic Peninsula. Antarctic Journal of the United States, 21 (in press).

—— , D. M. HARWOOD, AND P. N. WEBB. 1983. Upper Cretaceous microfossil biostratigraphy of Seymour Island, Antarctic Peninsula. Antarctic Journal of the United States, 18:72–74.

—— , ET AL. 1985. Distribution of microfossils and diagenetic features associated with the Cretaceous-Tertiary boundary on Seymour Island, Antarctic Peninsula. GWATT Conference on Rare Events, Switzerland, IGCP 199, Abstracts.

KEMP, E. M. 1972a. Reworked palynomorphs from the West Ice Shelf area, East Antarctica, and their possible geological and paleoclimatological significance. Marine Geology, 13:145–157.

—— . 1972b. Recycled palynomorphs in continental shelf sediments from Antarctica. Antarctic Journal of the United States, 7:190–191.

—— . 1975. Palynology of Leg 28 drillsites, Deep Sea Drilling Project. In D. E. Hays, L. A. Frakes, et al., Initial Reports of the Deep Sea Drilling Project, Leg 28, p. 599–623. U.S. Government Printing Office, Washington, D.C.

—— . 1978. Tertiary climatic evolution and vegetation history in the southeast Indian Ocean region. Palaeogeography, Palaeoclimatology, Palaeoecology, 24:169–208.

—— , AND P. J. BARRETT. 1975. Antarctic glaciation and Early Tertiary vegetation. Nature, 258:507–508.

MACELLARI, C. E. 1984a. Revision of serpulids of the genus Rotularia (Annelida) at Seymour Island (Antarctic Peninsula) and their value in stratigraphy. Journal of Paleontology, 58:1098–1116.

—— . 1984b. Late Cretaceous stratigraphy, sedimentology, and macropaleontology of Seymour Island, Antarctic Peninsula. Unpubl. Ph.D. thesis. The Ohio State University, Columbus, 599 p.

—— . 1985a. Paleobiogeographia y edad de la fauna de Maorites-Gunnarites (Ammonoidea) de la Antartida y Patagonia. Ameghiniana, 21:223–242.

—— . 1985b. El limite Cretacico/Terciario en la Peninsula Antartica y en el Sur de Sudamerica: evidencias macropaleontologicas. Sexto Congreso Latinoamericano de Geologica, Bogota, 1:266–278.

—— . 1986. Late Campanian–Maastrichtian ammonite fauna from Seymour Island (Antarctic Peninsula). Journal of Paleontology Memoir 18, 55 p.

—— , AND W. J. ZINSMEISTER. 1983. Sedimentology and macropaleontology of the Upper Cretaceous to Paleocene sequence of Seymour Island. Antarctic Journal of the United States, 18:69–71.

—— , R. A. Askin, and B. T. Huber. 1987. El limite Cretacico/Terciario en la Peninsula Antartica. X Congreso Geologico Argentino, Tucuman, Actas, 3:167–170.

McIntyre, D. J., and G. J. Wilson. 1966. Preliminary palynology of some Antarctic Tertiary erratics. New Zealand Journal of Botany, 4:315–321.

Palamarczuk, S., 1982, Dinoflagelados de edad Daniana en la Isla Vicecomonoro Marambio, (ex Seymour), Antartida Argentina. Ameghiniana, 19:353–360.

——, et al. 1984. Las Formaciones López de Bertodano y Sobral en la Isla Vicecomodoro Marambio, Antartida, p. 399–419. *In* S. C. Bariloche (ed.), Actas IX Congreso Geologico Argentino.

Playford, G. and M. E. Dettmann. 1978. Pollen of *Dacrydium franklinii* Hook. f. and comparable Early Tertiary microfossils. Pollen et Spores, 10:513–534.

Raine, J. L. 1984. Outline of a palynological zonation of Cretaceous to Paleogene terrestrial sediments in West Coast region, South Island, New Zealand. New Zealand Geological Survey Report, 109, 82 p.

Rinaldi, C. A. 1982. The Upper Cretaceous in the James Ross Island Group, p. 281–286. *In* C. Craddock (ed.), Antarctic Geoscience. University of Wisconsin Press, Madison.

—— , A. Massabie, J. Morelli, H. L. Rosenmall, and R. del Valle. 1978. Geologica de la Isla Vicecomodoro Marambio. Contribuciones del Instituto Antartico Argentino, 217:1–37.

Stuchlik, L. 1981. Tertiary pollen spectra from the Ezcurra Inlet Group of Admiralty Bay, King George Island (South Shetland Islands, Antarctica). Studia Geologica Polonica, 72:109–132.

Truswell, E. M. 1982a. Palynology of seafloor samples collected by the 1911–1914 Australian Antarctic Expedition: implications for the geology of coastal East Antarctica. Journal of the Geological Society of Australia, 29:343–356.

——. 1982b. Antarctica: the vegetation of the past and its climatic implications. Australian Meteorological Magazine, 30:169–173.

——. 1983. Recycled Cretaceous and Tertiary pollen and spores in Antarctic marine sediments: a catalogue. Palaeontographica, Abt. B, 186:121–174.

——. 1987. Antarctica: a history of terrestrial vegetation, *In* Tingey, R. (ed.) (in press).

—— , and J. B. Anderson. 1984. Recycled palynomorphs and the age of sedimentary sequences in the eastern Weddell Sea. Antarctic Journal of the United States, 19:90–92.

—— , and D. J. Drewry. 1984. Distribution and provenance of recycled palynomorphs in surficial sediments of the Ross Sea, Antarctica. Marine Geology, 59:187–214.

Wall, D., B. Dale, G. P. Lohmann, and W. K. Smith. 1977. The environmental and climatic distribution of dinoflagellate cysts in modern marine sediments from regions in the north and south Atlantic oceans and adjacent seas. Marine Micropaleontology, 2:121–200.

Wilson, G. J. 1967. Some new species of Lower Tertiary dinoflagellates from McMurdo Sound, Antarctica. New Zealand Journal of Botany, 5:57–83.

——. 1968. On the occurrence of fossil microspores, pollen grains, and microplankton in bottom sediments of the Ross Sea, Antarctica. New Zealand Journal of Marine and Freshwater Research, 2:381–389.

——. 1984a. New Zealand Late Jurassic to Eocene dinoflagellate biostratigraphy-summary. Newsletters on Stratigraphy, 13:104–117.

——. 1984b. Some new dinoflagellate species from the New Zealand Haumurian and Piripauan Stages (Santonian-Maastrichtian, Late Cretaceous). New Zealand Journal of Botany, 22:549–556.

—— , and C. D. Clowes. 1982. *Arachnodinium,* a new dinoflagellate genus from the Lower Tertiary of Antarctica. Palynology, 6:97–103.

Wrenn, J. H. 1981. Preliminary palynology of the site J-9, Ross Sea. Antarctic Journal of the United States, 16:72–74.

——. 1982. Dinocyst biostratigraphy of Seymour Island, Palmer Peninsula, Antarctica. Unpubl. Ph.D. thesis. Louisiana State University, Baton Rouge, 464 p.

Zinsmeister, W. J. 1979. Biogeographic significance of the late Mesozoic and early Tertiary molluscan faunas of Seymour Island (Antarctic Peninsula) to the final breakup of Gondwanaland, p. 349–355. *In* J. Gray and A. J. Boucot (eds.), Historical Biogeography, Plate Tectonics and the Changing Environment. Oregon State University Press, Corvallis.

——. 1982. Late Cretaceous–Early Tertiary molluscan biogeography of the southern Circum-Pacific. Journal of Paleontology, 56:84–102.

—— , and H. H. Camacho. 1982. Late Eocene (to possibly earliest Oligocene) molluscan fauna of the La Meseta Formation of Seymour Island, Antarctic Peninsula, p. 299–304. *In* C. Craddock (ed.), Antarctic Geosciences. University of Wisconsin Press, Madison.

—— , and R. M. Feldmann. 1984. Cenozoic high latitude heterochroneity of Southern Hemisphere marine faunas. Science, 224:281–283.

—— , and C. E. Macellari. 1983. Changes in the macrofossil faunas at the end of the Cretaceous on Seymour Island, Antarctic Peninsula. Antarctic Journal of the United States, 18:68–69.

Manuscript Accepted by the Society September 1, 1987

Geological Society of America
Memoir 169
1988

The palynological record across the Cretaceous/Tertiary transition on Seymour Island, Antarctica

Rosemary A. Askin
Department of Earth Sciences, University of California, Riverside, California 92521

ABSTRACT

Unconsolidated, fine-grained (sandy silts), shallow marine sediments of the upper López de Bertodano Formation include the Cretaceous/Tertiary boundary on Seymour Island, northeastern Antarctic Peninsula. These strata contain abundant palynomorphs and other micro- and macrofossils.

The Cretaceous/Tertiary boundary is provisionally placed at a dinocyst zonal boundary that occurs within a laterally persistent glauconite-rich interval. This glauconite interval marks the highest, definitely in-place Maastrichtian macro- and microfossils. Strata above this interval are considered Danian, based on palynological evidence, in the absence of other age-diagnostic fossils. Association of Maastrichtian ammonites and microfossils with typically "Danian" dinocysts below the glauconite interval accentuates the transitional nature of the Cretaceous/Tertiary succession on Seymour Island. Changes in Cretaceous/Tertiary dinocyst assemblages may be related to local environmental change, including regression.

Pollen and spore assemblages record little change and no evidence of an abrupt event for the land vegetation across the Cretaceous/Tertiary boundary. Instead, a long-term floral turnover is inferred from nonmarine species. This gradual change is consistent with climatic change, possibly cooling, through much of Maastrichtian and early Paleocene time.

INTRODUCTION

This report outlines the palynological evidence used to locate the Cretaceous/Tertiary boundary on Seymour Island, Antarctica, and the problems associated with this determination. The nature of the palynological transition and some possible causes are discussed.

Seymour Island is the only recorded location in Antarctica with a continuously outcropping sequence of Upper Cretaceous to Paleogene sediments. A discontinuous succession of thin plant-bearing beds representing brief intervals in the Late Cretaceous and Tertiary occurs in the South Shetland Islands (Birkenmajer, 1985). Reworked Cretaceous-Paleogene palynomorphs in sea-floor sediments around the coast of Antarctica (Truswell, 1983) attest to erosion of rocks of that age in areas now covered by the Antarctic ice sheet.

The palynological record at Seymour Island, Antarctica, illustrates gradual disappearances and appearances of marine and nonmarine species across the Cretaceous/Tertiary (K/T) boundary, over a "K/T transitional interval" of at least 12 m. The abundance and generally good preservation of many groups of fossils, combined with the well-exposed, essentially continuous stratigraphic section, make Seymour Island an excellent place to investigate the K/T boundary. Seymour Island is the southernmost outcrop locality with rocks spanning this boundary and should prove to be a useful site for global comparative studies.

STRATIGRAPHY OF THE UPPERMOST CRETACEOUS–BASAL TERTIARY

Sediments that include the K/T boundary are referred to the upper López de Bertodano Formation on Seymour Island, a small (20 km) ice-free island located in the James Ross Island basin adjacent to the northeastern tip of the Antarctic Peninsula (Fig. 1). On Seymour Island the López de Bertodano beds are upper Campanian to lower Danian in age, and are unconforma-

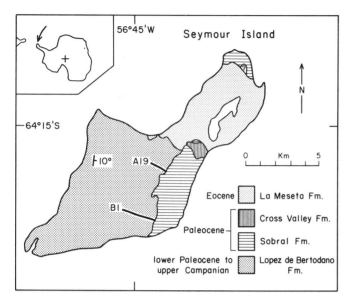

Figure 1. Geologic map of Seymour Island. Stratigraphic sections collected for palynologic study across the K/T boundary are indicated as A19 and B1.

bly overlain by the Paleocene Sobral Formation. The uppermost Cretaceous–basal Tertiary section includes unconsolidated, fine-grained, nearshore marine sediments essentially unaffected by thermal metamorphism. Sandy silts predominate in most of the upper López de Bertodano Formation. Lithology of the López de Bertodano Formation (Rinaldi et al., 1978; Rinaldi, 1982) on Seymour Island was described by Macellari (1984b, this volume).

Macellari (1984b) noted a trend of volcanic material increasing upward through the uppermost López de Bertodano Formation, with a major influx of pumice and glass shards beginning shortly before the top of the Maastrichtian, and a coincident, possibly related, increase of glauconite. Volcanism continued during much of the time of deposition of the Sobral Formation. The uppermost López de Bertodano and Sobral Formations also record a regressive phase of sedimentation. Macellari (this volume) concluded that regressive facies began about 30 m below the K/T boundary, and that no major facies change occurred across that boundary, based on paleontologic and lithologic evidence.

The stratigraphic position of the K/T boundary is discussed here with reference to the glauconite-rich interval that occurs about 90 m (in section B1, Fig. 2) below the erosional unconformity between the López de Bertodano and Sobral Formations. This glauconite-rich interval includes variably resistant beds (some laterally persistent) of glauconitic silty sand and sandy silt, and some more clay-rich glauconite beds. It includes samples B1-103 through B1-112 in section B1 (Fig. 2). At this locality, a distinct green, glauconitic sandy silt bed, <0.1 to 0.4 m thick

(sample B1-103), occurs 2 m below a resistant, ledge-forming (dip slope), silty sandstone bed, 0.4 m thick (overlying B1-107). The section continues above the dip slope with about 2 m of glauconitic, clayey and silty sands (B1-108 to 112). The glauconite-rich interval marks the boundary between units 9 and 10 of Macellari (1984b, this volume), which was placed at the B1-103 glauconite bed at this locality.

K-Ar dates obtained for a much lower glauconite (150 m below K/T), a glauconite bed near the K/T boundary, and a higher (130 m above) glauconite layer are 64.1 ± 0.9, 58.9 ± 1, and 53.2 ± 1.4 m.y., respectively (Macellari, 1985). All three dates are apparently younger than their true age (the accepted age for the K/T boundary is 65 Ma, summarized in Harland et al., 1982).

The upper López de Bertodano–Sobral interval contains abundant palynomorphs (Askin, this volume), including marine microphytoplankton (dinoflagellate cysts, acritarchs, and other algal bodies) and nonmarine palynomorphs (pollen and spores from land plants, fungal remains, and fresh-water algae). The loosely consolidated sediments also contain abundant fossils including wood and other plant debris; ammonites, bivalves, gastropods, brachiopods, corals, echinoids, and serpulid annelids (Macellari, 1984a,b, 1986, this volume); marine reptiles, fish and shark remains (Chatterjee and Zinsmeister, 1982; Chatterjee et al., 1984); and microfossils such as foraminifera, calcareous nannofossils, silicoflagellates, diatoms, and ebridians (Huber et al., 1983; Huber, this volume; Harwood, this volume).

An interval in the upper López de Bertodano Formation (Zinsmeister and Macellari, 1983), marked by a laterally persistent glauconite unit (Macellari, 1984b), was suggested as an approximate position for the K/T boundary because it marked the top of the stratigraphic range of ammonites and numerous other Late Cretaceous shelly fossils. At that time, the highest recorded, definitely in-place ammonite specimens were known from this glauconite interval. Subsequently, in 1984 and 1986–1987, additional ammonite specimens and other typically Maastrichtian molluscs, many in concretions, were found up to 20 m above this glauconite interval at various locations on Seymour Island (Huber, 1986; Sadler, this volume, geological map; Seymour 1986/1987 expedition members' personal communication, 1987). Palynomorphs extracted from six of these specimens collected in 1984 by Zinsmeister and Huber, and from more recently collected concretions, indicate the concretion matrix (both within and encasing the macrofossil) is Maastrichtian (zones 3 to 4 of Askin, this volume). At one location, the sediment adjacent to the concretion contains Danian (zone 5) dinocysts, suggesting that at least this concretion with its macrofossil is reworked or not in place. Evaluation of this situation awaits palynological study of additional material to clarify the relationship of the ammonites to the dinocyst zones.

By the end of the Cretaceous, total species extinction of known Maastrichtian mollusc species, including ammonites, was apparently about 64 percent (Macellari, 1985). Disappearances of these species occurred over many meters of section (Zinsmeister

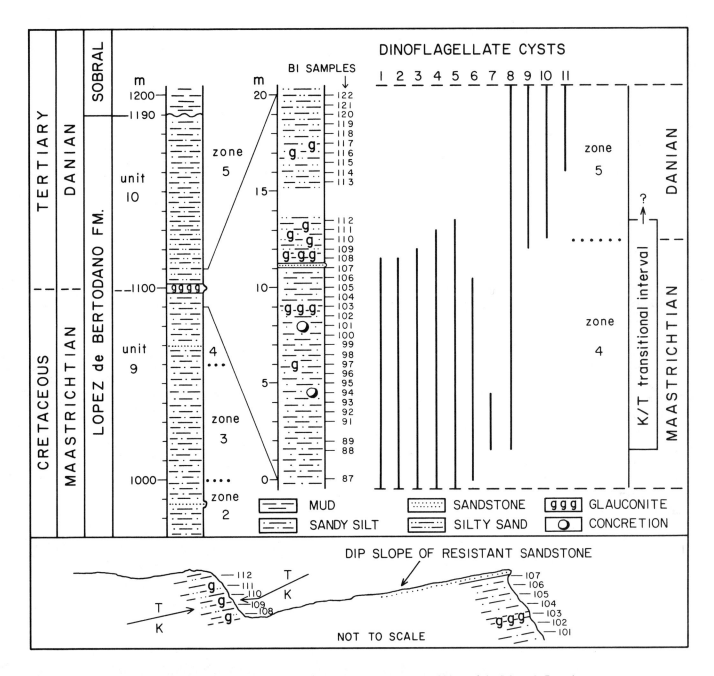

Figure 2. Lithostratigraphic column for section B1 showing upper 220 m of the López de Bertodano Formation. Expanded 20 m section shows lithology, position of samples B1-87 through B1-122, ranges of selected dinocyst species (most other dinocysts occurring in this interval are more long-ranging), zone 4/zone 5 boundary, and suggested location of K/T boundary. The diagrammatic cross section below shows relationship of samples B1-101 through B1-112, dip slope and position of K/T boundary at section B1. Dinoflagellate cysts listed are as follows: 1, *Manumiella* n.sp.1; 2, *M.* n.sp.2; 3, *Exochosphaeridium bifidum*; 4, *M. seelandica*; 5, *M. druggii*; 6, *Phelodinium* sp.cf. *P. magnificum*; 7, cf. *Broomea* sp.; 8, *Deflandrea medcalfii*; 9, *Alisocysta circumtabulata*; 10, *Senegalinium obscurum*; 11, *Spinidinium* sp.1. These species are illustrated in Askin (this volume). (Sample B1-107 is equivalent to 349 of Huber, this volume, and Macellari, this volume; B1-103 to 348; B1-101 to 347; and B1-97 to 346).

and Macellari, 1983; Macellari, this volume, W. J. Zinsmeister, personal communication, 1987). Some bivalve species disappeared at or near the K/T boundary; however, many species evidently disappeared earlier, and others later, during the earliest Tertiary. A shallower water setting was suggested as a possible cause of the decreased macrofaunal diversity in the Paleocene strata (Macellari, this volume).

Maastrichtian microfossils other than palynomorphs occur up to the glauconite interval (Huber, 1986, 1987, this volume; Huber et al., 1983, 1985; Harwood, this volume). No in-place, age-diagnostic, planktonic foraminifera or other calcareous microfossils have been found in the uppermost López de Bertodano Formation between the glauconite interval and 1 m below the upper contact with the Sobral Formation, where Paleocene foraminifera first occur (Huber, this volume). These fossil-poor strata coincide with a "carbonate dissolution event" (Huber, 1986, 1987; Huber et al., 1985). Benthonic foraminifera found in these beds are long-ranging forms.

Diagnostic Paleocene silicoflagellates (including *Corbisema hastata*) first appear about 35 m above the glauconite interval, and a probable Paleocene species, *C. aspera* (although not a widely known or accepted diagnostic form), appears approximately 11 m above the glauconite interval (Harwood, this volume). These occurrences are from the central part of Seymour Island where López de Bertodano unit 10 is only 45 m thick. They cannot be directly correlated to the B1 section.

Two stratigraphic sections spanning the K/T boundary were sampled for palynological study. A 300-m-thick section, B1 (Figs. 1, 2), was measured in 1983–1984 by F. C. Barbis and J. R. Robinson (Askin, 1984), and sampled at 0.5-m intervals across the presumed boundary, and at 1.5-m to 3-m intervals above and below. This section corresponds to "section C," samples 315 to 353, of Macellari (1984b, this volume) and Huber (this volume). The same palynological succession discussed here for section B1 occurs in more widely spaced samples collected in 1982 in a reconnaissance section A19 about 4 km to the north (Askin and Fleming, 1982).

MARINE PALYNOLOGICAL SUCCESSION

The Upper Cretaceous–Paleogene palynological zonation of Seymour Island is based on marine dinoflagellate cyst species (Askin, this volume). In the upper Maastrichtian part of the section (upper zone 2 to zone 4) the marine palynomorph flora includes approximately 50 species, dominated by a biostratigraphically useful, evolving complex of species of *Manumiella* and related forms.

The K/T transition is recorded in zones 4 and 5. Zone 4 (Fig. 2), the top of which may correlate with the K/T boundary, is characterized by the *Manumiella druggii–M. seelandica* complex and *Exochosphaeridium bifidum*. In section B1, zone 4 occupies a stratigraphic thickness of approximately 40 m. The lowest occurrences of dinocyst species elsewhere considered Danian and younger (see below) were found in the upper 11 m of

zone 4 (Fig. 2). They include *Alisocysta circumtabulata* and *Deflandrea medcalfii*, with the former species first occurring at least 5 m below the top of zone 4 in section A19.

The base of zone 5 is marked by an influx of a small dinocyst, *Senegalinium obscurum*, which forms almost monospecific assemblages within the marine component (68 percent in first sample, B1-110, to 97 percent 0.5 m higher, B1-111). The lower part of zone 5 includes the first appearances of species such as *Spinidinium* sp. 1, *Elytrocysta* sp., *Palaeoperidinium pyrophorum*, *Trithyrodinium evittii*, various *Deflandrea* and *Ceratiopsis* species, and *Senegalinium dilwynense* (Askin, this volume). Both *Spinidinium* sp. 1 and *P. pyrophorum* also sometimes occur in essentially monospecific assemblages.

DISTRIBUTION OF STRATIGRAPHICALLY SIGNIFICANT SPECIES

Manumiella druggii, the principal marker species of zone 4 (dinocyst species no. 5 in Fig. 2), occurs in rocks considered upper Maastrichtian and basal Danian elsewhere in the world. At the lower Danian type section at Stevns Klint, Denmark, *M. druggii* first occurs rarely in the uppermost 5 m of the Maastrichtian white chalk below the Danian (Wilson, 1971, 1974), and occurs abundantly in the overlying Danian Fish Clay (Lange, 1969; Hultberg, 1986). A few specimens also were noted in the Fish Clay at Dania at Kjolby Gaard, Denmark (Hultberg, 1986). In the Maastrichtian Marca Shale Member of the upper Moreno Formation, central California, *M. druggii* is common (as *Deflandrea cretacea,* Drugg, 1967), while one specimen (?reworked) was found in the Danian. *M. druggii* occurs in the Latrobe Group of the Gippsland Basin, southeastern Australia (Stover, 1973). The Gippsland Basin *"Deflandrea" druggii* Zone has been correlated with the uppermost Maastrichtian planktonic foraminiferal *Globotruncanella mayaroensis* Zone of van Hinte (1972) by Partridge (1976). The stratigraphic ranges of *M. druggii* and *M. seelandica* define the *"Isabelidinium" druggii* Zone of uppermost Haumurian–basal Teurian age (upper Maastrichtian–lower Danian) in New Zealand (Wilson, 1978, 1984). In New Zealand the lower part of the *"I." druggii* Zone was correlated by Wilson (1984) with the uppermost *Globotruncana circumnodifer* Zone (uppermost Haumurian) of Webb (1971). The upper part of the *"I." druggii* Zone in New Zealand apparently corresponds to a gap in the foraminiferal record below the succeeding *Globigerina pauciloculata* Zone of lower Teurian age. The few other recorded occurrences of *M. seelandica* (Fig. 2, no. 4) include the type Danian Fish Clay at Stevns Klint (Lange, 1969).

Exochosphaeridium bifidum (Fig. 2, no. 3), common in upper zone 4, is widespread in Maastrichtian and older rocks. It occurs in the Maastricht Chalk at several localities in Denmark and the Netherlands-Belgium Maastricht area, including the stratotype section at ENCI Quarry (Wilson, 1974). The specimens in a sample from the Coastal Belt of the Franciscan Complex, northern California, interpreted as Danian from palynomorphs (Damassa, 1979), may represent the only recorded Danian occurrence.

"Typically Paleocene" species of dinoflagellate cysts first appear in upper zone 4 associated with Maastrichtian ammonites and microfossils. They include *Alisocysta circumtabulata* (Fig. 2, no. 9) and *Deflandrea medcalfii* (Fig. 2, no. 8). These two species first appear at the base of the Teurian (base Danian) in New Zealand (Wilson, 1984). *A. circumtabulata* is fairly widespread in Danian (to early Eocene) rocks (e.g., Drugg, 1967). *D. medcalfii* occurs in the Paleocene of southeastern Australia.

Senegalinium obscurum (Fig. 2, no. 10), which is abundant at the base of zone 5, occurs in the Maastrichtian (very rare) and Danian (common) of central California (Drugg, 1967), and Paleocene elsewhere, for example, the lowest Landenian Heersian marls in Belgium (Schumacker-Lambry and Chateauneuf, 1976).

Other species that appear within lower zone 5 (Askin, this volume) include *Palaeoperidinium pyrophorum,* widespread through Upper Cretaceous to lower upper Paleocene rocks; *Ceratiopsis striata* and similar *Ceratiopsis-Deflandrea* species, Paleocene; *Trithyrodinium evittii,* characteristic of Maastrichtian-Danian rocks in southeastern Australia and elsewhere; and *Senegalinium dilwynense* from Paleocene rocks in southeastern Australia, New Zealand, and elsewhere. *Elytrocysta* sp. may be conspecific with *"Membranosphaera"* sp. of Drugg (1967) from the Danian of central California.

AGE OF MARINE PALYNOMORPH ASSEMBLAGES

The late Maastrichtian age for most of zone 4 (lower 30 m) is indisputable. This age can be determined from associated macro- and microfossils (e.g., Macellari, 1986; Macellari et al., 1987; Harwood, this volume; Huber, this volume), as summarized in Askin (this volume).

The upper part of zone 4 contains Maastrichtian ammonites and other molluscs, and Maastrichtian microfossils, up to the glauconite interval at several localities along strike. These beds also contain Maastrichtian dinocysts (e.g., *E. bifidum*), transitional Maastrichtian-Danian dinocysts (*M. druggii–M. seelandica*), and dinocysts that are Danian and younger elsewhere (e.g., *A. circumtabulata* and *D. medcalfii*). The upper part of zone 4 is considered Maastrichtian, and the possibly diachronous nature of first appearances of the latter two species is duly noted.

The age of lower zone 5 and location of the K/T boundary are, however, not as clearly defined.

Not surprisingly, direct palynological correlation of the Seymour Island succession to type Maastrichtian and Danian sections of Europe is not possible. Species composition is entirely different, a result of differences in facies or in paleoenvironments, provinces, etc., and the somewhat isolated, high southern paleolatitude location of the James Ross Island basin. The only similarity is the abundance of *M. druggii* in the Stevns Klint Fish Clay. Interestingly, Hultberg (1986) suggested the Fish Clay was diachronous throughout Denmark. Based on dinocyst evidence, he

assigned the Stevns Klint beds, with their iridium anomaly, to the uppermost Maastrichtian.

The best biostratigraphic correlations for Seymour Island assemblages can probably be made within the faunally and florally discrete Weddellian Province, encompassing southern South America, Antarctica, New Zealand and southeastern Australia (Zinsmeister, 1979, 1982). Palynologic evidence in the province, based on dinocyst species distribution in New Zealand and southeastern Australia, plus the evidence from extraprovincial areas, as outlined above, suggest that the lower part of zone 5 is Danian in age. Well-controlled Maastrichtian-Danian palynologic studies are lacking in historically adjacent southern South America. The Danian age assignment for lower zone 5 is supported to some extent (for at least part of López de Bertodano unit 10) by silicoflagellate evidence (Harwood, this volume), as noted in the Stratigraphy section.

LOCATION OF K/T BOUNDARY

A horizon definitively marking the K/T boundary on Seymour Island is not clearly distinguishable in this transitional sequence. Placement is complicated by many factors including heterochroneity of dinocyst ranges, possible reworking of palynomorphs in this nearshore paleoenvironment, and the lack of other diagnostic microfossils in the lower part of the "carbonate dissolution interval."

Based on palynological evidence, a provisional K/T boundary may be placed at the zone 4/zone 5 boundary. In section B1, this occurs between samples 109 and 110, about 1 m above the resistant sandstone dip slope. It is possible that the glauconite beds represent lower rates or brief cessations of sedimentation, although sedimentation is believed to have been essentially continuous through most of zone 4 and lower zone 5.

How zones 4 and 5 in section B1 correlate to other localities near the southern coast of Seymour Island, where most of the ammonites and ammonite-bearing concretions above the putative K/T boundary glauconite were found, is unclear at present. If ammonites do occur in lower zone 5 sediments, revision of the K/T boundary on Seymour Island may be needed. At least some degree of heterochroneity has already been established for "Danian" dinocysts that appear in upper zone 4 in sections B1 and A19, and first appearances of these (and other) species may be more diachronous than presently supposed.

NATURE OF K/T TRANSITION IN MARINE REALM

Maastrichtian ammonites and dinocysts with Danian affinities (e.g., *A. circumtabulata* and *D. medcalfii*) obviously coexisted for a short period of time in the James Ross Island basin. A "K/T transitional interval," indicated in Figure 2, begins at the first *D. medcalfii* dinocysts associated with ammonites (B1-88). It spans the change to zone 5 assemblages and possibly extends about 20 m above the glauconite if some of the "post-glauconite" ammonites are eventually proven to be in place.

A major change in principal components of the dinocyst populations is evident near the presumed K/T boundary (Fig. 2; Askin, this volume, Fig. 5). This change is even more apparent when relative abundances of dinocyst species are taken into account. An assemblage dominated by *Manumiella* spp. (86 percent of marine component in B1-108) was replaced by a *Senegalinium obscurum*-dominated assemblage within 1 m of stratigraphic section. Stratigraphic ranges for dinocyst species did not all end abruptly at any one horizon, however, and approximately 30 species continued over the boundary. As indicated in Figure 2, changes in principal components took place over 2 m of section in B1, although local reworking may be responsible for extending the ranges of some species upward. The possibility of reworking always exists and may be the case particularly for some sporadic rare occurrences of angiosperm pollen species. Gradual introduction of marine and nonmarine species throughout this part of the stratigraphic section, however, and the gradual extinctions in some macroinvertebrate groups are more consistent with change of a rather gradual nature. This change could be caused primarily by progressive shallowing conditions, and may in part be related to the volcanic activity, with its possible resulting change in water chemistry, which may have affected plankton populations.

It may be difficult to distinguish between the effects of any hypothesized global catastrophic K/T event and the local effects, particularly regarding dinocyst assemblages. That dinocyst populations were little affected by any K/T "extinction event" is well known (e.g., Bujak and Williams, 1979). Hansen (1977) showed for the dinocyst succession in Denmark that the K/T boundary is marked more by the appearance of many new Danian species rather than by species extinction, a trend that is also evident on Seymour Island (Askin, this volume, Fig. 5).

Within the upper part of the glauconite interval in section B1, the influx of small peridinioid dinocysts referred to *Senegalinium obscurum* marks the base of zone 5 and possibly the earliest Danian. Monospecific dinocyst assemblages are generally considered evidence for restricted, inshore-estuarine paleoenvironments (e.g., Morzadec-Kerfourn, 1977; Goodman, 1979), although it could be argued that they represent an invasion by an opportunistic species following devastation of previous marine communities. Independent information about the carbonate dissolution might help clarify some aspects of this problem. Unfortunately, it has not been determined with certainty if loss of calcareous plankton at the glauconite interval ("calcite dissolution event") was a coeval phenomenon or later diagenetic (Huber, 1987; Huber et al., 1985), although Huber (this volume) correlated this dissolution interval followed by anoxic conditions with other similar intervals in the earliest Paleocene in southern mid to high latitudes. The *M. druggii–M. seelandica* complex appears to have survived above this horizon (above B1-103 in section B1), and they may have survived even longer elsewhere in the world. Almost monospecific assemblages of *Spinidinium* sp.1 are first found 15 m above the base of zone 5. *Palaeoperidinium pyrophorum* is common from 22 m above the base of zone 5, sometimes in almost monospecific assemblages. The successive occurrences of mono-

specific dinocyst assemblages in lower zone 5 may record restricted and probably environmentally unstable, shallow coastal/nearshore conditions during the Danian in this basin. Harwood (this volume) emphasized the increase in abundance of diatom resting spores from about 7 percent (of total diatom assemblage) below the presumed K/T boundary to about 35 percent above it, which, together with low diversity of diatom and silicoflagellate assemblages, was explained by stressed environmental conditions in the earliest Paleocene. Huber (this volume) suggested stressed, anoxic bottom-water conditions for benthic foraminiferal assemblages in the overlying lower part of the Sobral Formation.

NONMARINE PALYNOLOGY AND CRETACEOUS/TERTIARY VEGETATION

Based on palynomorph occurrence data, there was apparently little change in land-plant assemblages over the K/T boundary. No more change is seen than that known from any other zonal boundary on Seymour Island. Transported and possibly reworked grains, however, might not be expected to record abrupt change. Survival of seeds, or of plants in refugia, and subsequent reappearances of species (e.g., Tschudy and Tschudy, 1986) might also mask any short-term environmental event. There is, however, a long-term turnover of the land flora of Seymour Island that extends from the early Maastrichtian through the early Paleocene. It is consistent with long-term climatic change not fully understood at present, but possibly including a cooling trend in the Seymour Island area. Apparently lower diversity Paleocene pollen and spore assemblages from higher in zone 5 could be the result of several factors (Askin, this volume), one of which is cooler temperatures. Francis (1986) suggested deteriorating climatic conditions from Cretaceous to Tertiary for two Paleocene wood specimens, also from somewhat higher in the section. Barrera et al. (1987) recorded cooling bottom-water conditions through the middle and late Maastrichtian (from oxygen isotope analyses of benthonic foraminifera on Seymour Island), although two samples suggested possible warmer bottom-water temperatures in the earliest Paleocene.

On Seymour Island, we are fortunate in having a nearshore marine clastic sequence that contains not only marine floras and faunas, but also abundant nonmarine palynomorphs probably representing several different terrestrial plant communities and habitats. About 90 species of pollen and spores have been recognized (taxonomic study in progress) in upper Maastrichtian (zones 3 and 4) sediments on Seymour Island. This assemblage is not particularly diverse by low-latitude standards, but is certainly a good representation of plant life on the adjacent land area. The upper Maastrichtian terrestrial component is dominated by podocarpaceous conifer pollen; the rest of the terrestrial palynoflora includes fern spores and a variety of angiosperm pollen, with relatively low numbers of pollen of southern beech (*Nothofagus*). A humid temperate coniferous forest vegetation can be envisaged for much of the land area during the Maastrichtian and Paleocene, contrasting with the mixed beech-conifer vegetation of the Eocene.

All 15 species of pollen from podocarpaceous conifers, all 6 beech pollen species, and all 15 cryptogam spore species recognized to date in the upper Maastrichtian on Seymour Island continue through into Danian lower zone 5. As far as can be determined from the pollen and spores, this plant association remained essentially unchanged, except for slight differences in relative abundances (which may be facies-related, rather than a reflection of the original vegetation).

The only changes in terrestrial palynomorph species representation are observed in the angiosperm component (excluding *Nothofagus*), in which 51 species are recognized to date. Most species continue over the K/T transition, with only eight species disappearing within zone 4. These extinctions are not simultaneous. Like the dinocysts, these eight angiosperm species disappear at different horizons, though over a longer stratigraphic interval (20 m), assuming highest occurring specimens have not been reworked. Introduction of some new species also occurred through zones 4 and 5. Thus, less than 10 percent of the total preserved terrestrial species disappeared near the K/T boundary, and these species were supplanted by new species throughout this interval.

The apparent lack of disturbance in the Seymour terrestrial vegetation, similar to K/T successions throughout much of the world (e.g., Hickey, 1984; Tschudy, 1984), is consistent with a gradual K/T transition. The Rocky Mountains region of North America records a markedly different sequence of events, including major tectonic, paleogeographic, and paleoclimatic events. In addition to being the most documented possible exception to a gradualism model, it is a source of divergent and conflicting interpretations (e.g., Russell and Singh, 1978; Orth et al., 1981; Nichols et al., 1986; Tschudy and Tschudy, 1986; Wolfe and Upchurch, 1986).

Whatever putative global crisis may have caused the demise of many marine and nonmarine fossil taxa at the end of the Cretaceous, it certainly had little effect on the terrestrial floras of the Seymour Island region. Thus the nonmarine palynomorph succession of Seymour Island seems more compatible with gradual climate change, presumably in response to a general cooling trend through the middle to late Maastrichtian, and possibly through some of the early Paleocene.

ACKNOWLEDGMENTS

This paper has benefited from critical reviews by J. A. Doyle (of an earlier version), S. R. Jacobson, and M. O. Woodburne. Helpful discussions on various aspects of this work have been provided by W. S. Drugg, D. H. Elliot, D. Habib, D. M. Harwood, B. T. Huber, S. R. Jacobson, M. A. Kooser, C. E. Macellari, G. J. Wilson, M. O. Woodburne, and W. J. Zinsmeister. R. F. Fleming helped with field work in 1982, F. C. Barbis and J. R. Robinson did the 1983–1984 field work, K. G. Pill helped with laboratory preparation, and drafting was done by L. Bobbitt. This research, carried out while the author was at Colorado School of Mines, was supported by National Science Foundation Grant DPP-8314186.

REFERENCES

ASKIN, R. A. 1984. Palynological investigations of the James Ross Island basin and Robertson Island, Antarctic Peninsula. Antarctic Journal of the United States, 19:6–7.

——, AND R. F. FLEMING. 1982. Palynological investigations of Campanian to lower Oligocene sediments of Seymour Island, Antarctic Peninsula. Antarctic Journal of the United States, 17:70–71.

BARRERA, E., B. T. HUBER, S. M. SAVIN, AND P. N. WEBB. 1987. Antarctic marine temperatures: late Campanian through early Paleocene. Paleoceanography, 2:21–47.

BIRKENMAJER, K. 1985. Onset of Tertiary continental glaciation in the Antarctic Peninsula sector (West Antarctica). Acta Geologica Polonica, 35:1–31.

BUJAK, J. P., AND G. L. WILLIAMS. 1979. Dinoflagellate diversity through time. Marine Micropaleontology, 4:1–12.

CHATTERJEE, S., AND W. J. ZINSMEISTER. 1982. Late Cretaceous marine vertebrates from Seymour Island, Antarctic Peninsula. Antarctic Journal of the United States, 17:66.

——, J. SMALL, AND M. W. NICKELL. 1984. Late Cretaceous marine reptiles from Antarctica. Antarctic Journal of the United States, 19:7–8.

DAMASSA, S. P. 1979. Danian dinoflagellates from the Franciscan Complex, Mendocino County, California. Palynology, 3:191–207.

DRUGG, W. S. 1967. Palynology of the upper Moreno Formation (Late Cretaceous–Paleocene), Escarpado Canyon, California. Palaeontographica, Abt.B, 120:1–71.

FRANCIS, J. E. 1986. Growth rings in Cretaceous and Tertiary wood from Antarctica and their palaeoclimatic implications. Palaeontology, 29:665–684.

GOODMAN, D. K. 1979. Dinoflagellate "communities" from the Lower Eocene Nanjemoy Formation of Maryland, U.S.A. Palynology, 3:169–190.

HANSEN, J. M. 1977. Dinoflagellate stratigraphy and echinoid distribution in Upper Maastrichtian and Danian deposits from Denmark. Bulletin of the Geological Society of Denmark, 26:1–26.

HARLAND, W. B., ET AL. (eds.). 1982. A Geologic Time-Scale. Cambridge University Press, Cambridge, 131 p.

HICKEY, L. J. 1984. Changes in the angiosperm flora across the Cretaceous–Tertiary boundary, p. 279–313. In W. A. Berggren and J. A. van Couvering (eds.), Catastrophes and Earth History: The New Uniformitarianism. Princeton University Press, Princeton.

HUBER, B. T. 1986. The location of the Cretaceous-Tertiary contact on Seymour Island, Antarctic Peninsula. Antarctic Journal of the United States, 20:46–48.

——. 1987. Foraminiferal distribution across the Cretaceous/Tertiary transition on Seymour Island, Antarctic Peninsula. Antarctic Journal of the United States, 21 (in press).

——, D. M. HARWOOD, AND P. N. WEBB. 1983. Upper Cretaceous microfossil biostratigraphy of Seymour Island, Antarctic Peninsula. Antarctic Journal of the United States, 18:72–74.

——, D. M. HARWOOD, AND P. N. WEBB. 1985. Distribution of microfossils and diagenetic features associated with the Cretaceous-Tertiary boundary on Seymour Island, Antarctic Peninsula. GWATT Conference on Rare Events, Switzerland, IGCP 199, Abstracts.

HULTBERG, S. U. 1986. Danian dinoflagellate zonation, the C-T boundary and the

162 R. A. Askin

stratigraphical position of the fish clay in southern Scandinavia. Journal of Micropalaeontology, 5:37–47.

LANGE, D. 1969. Mikroplankton aus dem Fischton von Stevns-Klint auf Seeland. Beitrage zur Meereskunde, 24–25:110–121.

MACELLARI, C. E. 1984a. Revision of serpulids of the genus *Rotularia* (Annelida) at Seymour Island (Antarctic Peninsula) and their value in stratigraphy. Journal of Paleontology, 58:1098–1116.

———. 1984b. Late Cretaceous stratigraphy, sedimentology, and macropaleontology of Seymour Island, Antarctic Peninsula. Unpubl. Ph.D. thesis. The Ohio State University, Columbus, 599 p.

———. 1985. El limite Cretacico Terciario en la Peninsula Antartica y en el Sur de Sudamerica: evidencias macropaleontologicas. Sexto Congreso Latinoamericano de Geologia, Bogota, 1:266–278.

———. 1986. Late Campanian–Maastrichtian Ammonite Fauna from Seymour Island (Antarctic Peninsula). Journal of Paleontology Memoir 18, 55 p.

———, R. A. ASKIN, AND B. T. HUBER. 1987. El limite Cretacico/Terciario en la Peninsula Antartica. X Congreso Geologico Argentino, Tucuman, Actas, 3:167–170.

MORZADEC-KERFOURN, M.-T. 1977. Les kystes de dinoflagellés dans les sediments Récents le long des côtes Bretonnes. Revue de Micropaleontologie, 20:157–166.

NICHOLS, D. J., D. M. JARZEN, C. J. ORTH, AND P. Q. OLIVER. 1986. Palynological and iridium anomalies at Cretaceous-Tertiary boundary, south-central Saskatchewan. Science, 231:714–717.

ORTH, C. J., J. S. GILMORE, J. D. KNIGHT, C. L. PILMORE, R. H. TSCHUDY, AND J. E. FASSETT. 1981. An iridium abundance anomaly at the palynological Cretaceous-Tertiary boundary in northern New Mexico. Science, 214:1341–1343.

PARTRIDGE, A. D. 1976. The geological expression of eustacy in the Early Tertiary of the Gippsland Basin. APEA Journal, 16:73–79.

RINALDI, C. A. 1982. The Upper Cretaceous in the James Ross Island group, p. 281–286. *In* C. Craddock (ed.), Antarctic Geoscience. University of Wisconsin Press, Madison.

———, A. MASSABIE, J. MORELLI, H. L. ROSENMALL, AND R. DEL VALLE. 1978. Geologica de la Isla Vicecomodoro Marambio. Contribuciones del Instituto Antartico Argentino, 217:1–37.

RUSSELL, D. A., AND C. SINGH. 1978. The Cretaceous-Tertiary boundary in south-central Alberta—a reappraisal based on dinosaurian and microfloral extinctions. Canadian Journal of Earth Sciences, 15:284–292.

SCHUMACKER-LAMBRY, J., AND J.-J. CHATEAUNEUF. 1976. Dinoflagellés et acritarches des marnes Heersiennes de Gelinden (base du Landénien, Paléocène, Belgique). Review of Palaeobotany and Palynology, 21:267–294.

STOVER, L. E. 1973. Paleocene and Eocene species of *Deflandrea* (Dinophyceae)

in Victorian coastal and offshore basins, Australia. Special Publications of the Geological Society of Australia, 4:167–188.

TSCHUDY, R. H. 1984. Palynological evidence for change in continental floras at the Cretaceous–Tertiary boundary, p. 315–337. *In* W. A. Berggren and J. A. van Couvering (eds.), Catastrophes and Earth History: The New Uniformitarianism. Princeton University Press, Princeton.

———, AND B. D. TSCHUDY, 1986. Extinction and survival of plant life following the Cretaceous/Tertiary boundary event, Western Interior, North America. Geology, 14:667–670.

TRUSWELL, E. M. 1983. Recycled Cretaceous and Tertiary pollen and spores in Antarctic marine sediments: a catalogue. Palaeontographica, Abt.B, 186:121–174.

VAN HINTE, J. E. 1972. The Cretaceous time-scale and planktonic foraminiferal zones. Koninklijke Nederlandse Akademie Wetenschappen, Proceedings, Ser. B, 75:1–8.

WEBB, P. N. 1971. New Zealand Late Cretaceous (Haumurian) foraminifera and stratigraphy: a summary. New Zealand Journal of Geology and Geophysics, 14:795–828.

WILSON, G. J. 1971. Observations on European Late Cretaceous dinoflagellate cysts, p. 1259–1275. *In* A. Farinacci (ed.), Proceedings of the II Planktonic Conference, Rome 1970, Edizioni Technoscienza, Rome.

———. 1974. Upper Campanian and Maastrichtian dinoflagellate cysts from the Maastricht region and Denmark. Unpubl. Ph.D. thesis. University of Nottingham, Nottingham, England, 601 p.

———. 1978. The dinoflagellate species *Isabelia druggii* (Stover) and *I. seelandica* (Lange): their association in the Teurian of Woodside Creek, Marlborough, New Zealand. New Zealand Journal of Geology and Geophysics, 21:75–80.

———. 1984. New Zealand Late Jurassic to Eocene dinoflagellate biostratigraphy-summary. Newsletters on Stratigraphy, 13:104–117.

WOLFE, J. A., AND G. R. UPCHURCH. 1986. Vegetation, climatic and floral changes at the Cretaceous-Tertiary boundary. Nature, 324:148–152.

ZINSMEISTER, W. J. 1979. Biogeographic significance of the late Mesozoic and early Tertiary molluscan faunas of Seymour Island (Antarctic Peninsula) to the final breakup of Gondwanaland, p. 349–355. *In* J. Gray and A. J. Boucot (eds.), Historical Biogeography, Plate Tectonics, and the Changing Environment. Oregon State University Press, Corvallis.

———. 1982. Late Cretaceous–Early Tertiary molluscan biogeography of the southern Circum-Pacific. Journal of Paleontology, 56:84–102.

———, AND C. E. MACELLARI. 1983. Changes in the macrofossil faunas at the end of the Cretaceous on Seymour Island, Antarctic Peninsula. Antarctic Journal of the United States, 18:68–69.

MANUSCRIPT ACCEPTED BY THE SOCIETY SEPTEMBER 1, 1987

Geological Society of America
Memoir 169
1988

Upper Campanian-Paleocene foraminifera from the James Ross Island region, Antarctic Peninsula

Brian T. Huber
Byrd Polar Research Center and Department of Geology and Mineralogy, The Ohio State University, Columbus, Ohio 43210

ABSTRACT

Foraminiferal assemblages consisting of 76 genera and 145 species were recovered from Upper Cretaceous and Paleocene sediments in the James Ross Island region, northeastern Antarctic Peninsula. Open taxonomic nomenclature is used for 37 taxa and 10 new species are described, including *Alveolophragmium macellarii*, *Spiroplectammina vegaensis*, *Dorothia paeminosa*, *Buliminella procera*, *Neobulimina digitata*, *Bolivina pustulata*, *Conorbina anderssoni*, *Cibicides nordenskjoldi*, *C. seymouriensis*, and *Anomalinoides larseni*.

Three biostratigraphic range zones are recognized for interregional and regional correlation: (1) the ?mid- to upper Campanian *Gaudryina healyi* Zone, used to correlate the Cape Lamb (Vega Island), The Naze (James Ross Island), Snow Hill, and lowermost Seymour Island beds; (2) the upper Campanian through Maastrichtian *Hedbergella monmouthensis* Zone, occurring only on Seymour Island; and the Danian *Globastica daubjergensis* Zone, also occurring only on Seymour Island. The foraminiferal transition from Upper Cretaceous to Lower Tertiary sediments on Seymour Island is characterized by extinction of all Upper Cretaceous planktonic species and numerous benthic taxa, a lowermost Tertiary dissolution facies that is barren of calcareous microfossils, and the appearance of the Danian marker species *Globastica daubjergensis* with low-diversity, high-dominance calcareous assemblages just above the dissolution facies.

The stratigraphic distribution of foraminifera on Seymour Island attests to long-term environmental stability in the Upper Cretaceous sequence, in contrast to stressed environmental conditions in the lower Paleocene. Foraminiferal distributions indicate that the lowermost 250 m of the Seymour Island sequence was deposited in an inner neritic environment, and the overlying 950 m of Cretaceous and lower Danian sediments in an outer neritic setting. Foraminifera are absent from the younger Paleocene sequence, which was deposited in a relatively high-energy environment.

The foraminiferal assemblages from the James Ross Island region include a mixture of cosmopolitan and provincial species. The Campanian-Maastrichtian fauna is characterized by: (1) the conspicuous absence of specialized benthic and keeled planktonic species, which are used as index fossils in low to middle latitude regions; (2) the predominance of long-ranging, cosmopolitan taxa; and (3) the occurrence of provincial species that were restricted to the southern, extratropical latitudes. These latter taxa are included in the foraminiferal Austral Province. The biogeographic distribution of Austral Province foraminifera supports the postulated trans-Antarctic seaways between East and West Antarctica and/or a marine communication through the Antarctic Peninsula during the Late Cretaceous. The Danian assemblages are characterized by opportunistic, cosmopolitan species and one new buliminellid species, which dominates the calcareous assemblages.

INTRODUCTION

Diverse Upper Cretaceous and Paleocene foraminiferal assemblages are documented from several localities in the James Ross Island region, northeast of the Antarctic Peninsula (Fig. 1). Foraminifera are most abundant on Seymour Island, where the absence of permanent snow and ice cover affords complete stratigraphic exposure and easy lateral correlation of the eastward-dipping, homoclinal marine strata. The 126 benthic and 15 planktonic foraminiferal species, recovered from upper Campanian through Danian sediments, provide the basis for the present taxonomic and biostratigraphic study. This faunal analysis provides important data on the paleoecologic, paleogeographic, and paleoceanographic history of the circum-Antarctic regions. Furthermore, the well-exposed Cretaceous/Tertiary (K/T) boundary sequence on Seymour Island is unique for its essentially continuous stratigraphic record, its high diversity and good preservation of numerous macro- and microfossil groups, and its high-latitude position, representing the southernmost land-based record of terminal Cretaceous extinctions.

Less diverse foraminiferal assemblages were recovered from slightly older and coeval Upper Cretaceous outcrops at Cape Lamb (Vega Island), The Naze (James Ross Island), and Snow Hill Island (Fig. 1). These strata are not as well exposed as the Seymour Island sequence, due to cover by ice, glacial deposits, and upper Tertiary basalts. Diverse foraminiferal assemblages (59 species) were recovered at Cape Lamb, where several diagnostic taxa permit correlation with the oldest upper Campanian beds at Seymour Island. The other localities have yielded mostly agglutinated taxa and are considered ?mid- to late Campanian in age.

METHODS

Field studies conducted during the 1982, 1983/84, and 1985 field seasons included geologic mapping, section measuring, and sediment and fossil collecting, primarily on Seymour Island. Sample stations were precisely located on aerial photographs and later plotted on 1:25,000-scale maps (Macellari, 1984b). Logistical difficulties, harsh weather conditions, and cover by Quaternary alluvium led to uneven sample coverage and variation in sample size. Solifluction of the loose sediments and surficial weathering were factors that had to be considered during sampling. Low-relief outcrops in the island interior have been subjected to these diagenetic processes for extended lengths of time. These factors were negligible only for samples collected along the shore cliff and along the steep walls of interior drainages, where erosion is proceeding rapidly. To minimize the effects of surficial weathering, samples were collected from pits dug into the permafrost layer whenever possible.

Since the 1982 field season, more than 300 samples, each weighing from 150 to 1000 gm, have been processed for microfossils using conventional laboratory methods. The unconsolidated nature of siltstone samples affords easy recovery of the microfauna. Samples were dried, weighed, and washed through a series of sieves (1,000 to 63 μm size), then dried and exhaustively picked for foraminifera and other microfossils, including ostracodes, ichthyoliths, calcispheres, diatoms, radiolarians, and juvenile bivalves and gastropods.

PREVIOUS WORK

The discovery of fossil wood and marine invertebrates on Seymour Island in the late 19th century stimulated considerable interest among European scientists, and prompted subsequent scientific investigations of the James Ross Island region. This began in 1902–03 with the Swedish South Polar Expedition under the leadership of Otto von Nordenskjöld (Nordenskjöld, 1905, 1913; Andersson, 1906). The excellently preserved and abundant fossils collected during that expedition were subsequently described in a series of monographs, including Kilian and Reboul (1909) (ammonites) and Wilckens (1910) (annelids, bivalves, and gastropods). These papers provided a wealth of information about a previously unstudied part of the world, leading to the realization that the Antarctic climate had, in the past, been much warmer.

This region was revisited by geologists from the Falkland Islands Dependencies Survey (now the British Antarctic Survey) between 1945 and 1959. They conducted field studies on Cretaceous outcrops at James Ross, Vega, and Snow Hill Islands, and made two brief visits to Seymour Island. These led to several more reports on this region, including monographs on the cephalopods (Spath, 1953; Howarth, 1958, 1966) and decapods and annelids (Ball, 1960), as well as the first comprehensive stratigraphic summary of the northeast Antarctic Peninsula region (Bibby, 1966).

Activity in this region was renewed when the Instituto Antártico Argentino began a mapping project on Seymour Island (Isla Vicecomodoro Marambio in South American literature) in 1973–74. Several members of the Institute of Polar Studies (Ohio State University) accompanied the Argentine geologists during the 1974–75 field season. Details of these stratigraphic surveys were later published by Rinaldi and others (1978), Rinaldi (1982), Elliot and others (1975), and Elliot and Trautman (1982). Results of more recent reconnaissance visits to James Ross, Vega, and Snow Hill Islands, undertaken by members of the Instituto Antártico Argentino, were reported by Olivero (1981), Rinaldi (1982), and Del Valle and others (1982), among others.

Previous reports of foraminifera from the James Ross Island region are limited to illustration of two agglutinated species collected from an uncertain locality on either Snow Hill or Seymour Island (Holland, 1910), description of several agglutinated taxa collected by Olivero (1975) from Snow Hill Island, and description of 17 poorly preserved taxa from older Cretaceous beds at Stoneley Point, on northwest James Ross Island (Macfadyen, 1966) (Fig. 1). Macfadyen's assemblage was re-examined by Webb (1972a), who found that one species, designated by Macfadyen as a Maastrichtian globotruncanid, actually belonged to

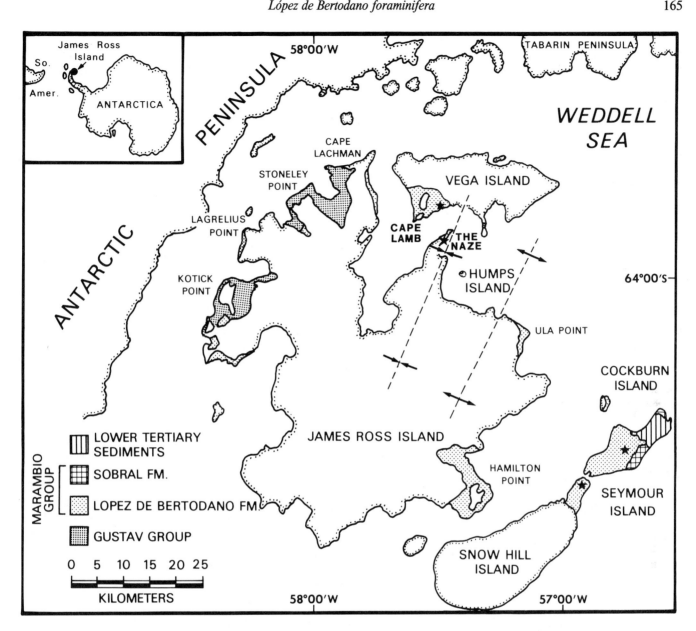

Figure 1. Geologic map of the James Ross Island region, showing the locations of Cretaceous through Lower Tertiary outcrops. Stars indicate localities visited during this study. Fold axes were inferred by Bibby (1966).

the agglutinated family Ataxophragmiidae. Thus, in the absence of diagnostic index fossils, the age of the James Ross Island fauna had not been precisely determined.

GEOLOGIC SETTING

The Cretaceous through Tertiary stratigraphy of the James Ross Island region has undergone several revisions since the initial work of Andersson (1906), as shown in Figure 2. Bibby (1966) provided the first comprehensive stratigraphic framework

for this region. Subsequent studies by Argentine, British, and American geologists have refined the litho- and biostratigraphy to the present level of understanding.

The oldest Cretaceous sediments in the James Ross Island basin crop out on northwestern James Ross Island (Fig. 1). These well-cemented volcaniclastic conglomerates and sandstones make up the Gustav Group, which ranges from ?Aptian through Santonian (Ineson and others, 1986). Overlying the Gustav Group, in ascending order, are the López de Bertodano, Sobral, Cross Valley, and La Meseta Formation (Fig. 2). The latter three units crop

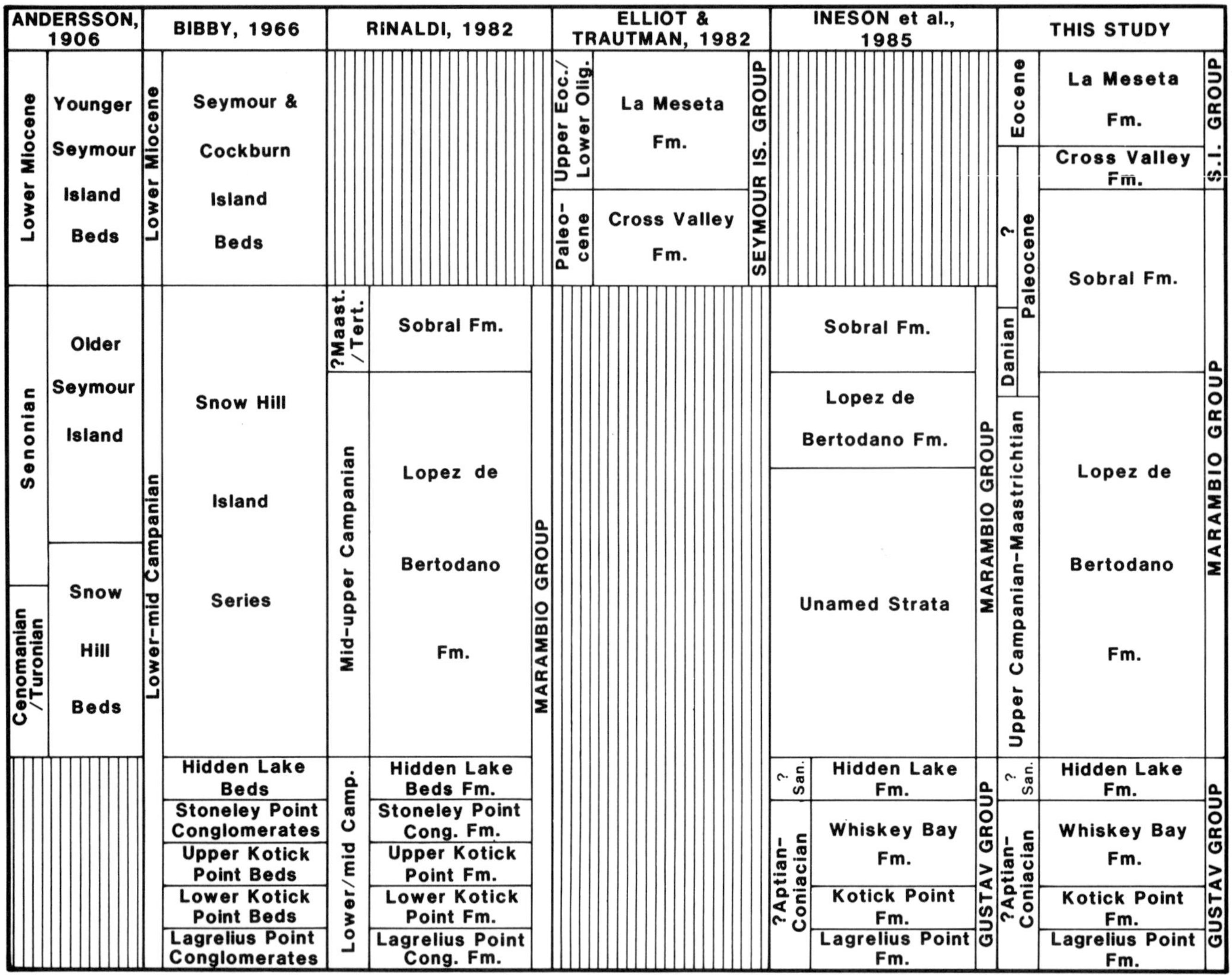

ANDERSSON, 1906	BIBBY, 1966	RINALDI, 1982	ELLIOT & TRAUTMAN, 1982	INESON et al., 1985	THIS STUDY
Lower Miocene — Younger Seymour Island Beds	Lower Miocene — Seymour & Cockburn Island Beds		Upper Eoc./Lower Olig. — La Meseta Fm. / Paleocene — Cross Valley Fm. (SEYMOUR IS. GROUP)		Eocene — La Meseta Fm. / Cross Valley Fm. (S.I. GROUP)
Senonian — Older Seymour Island	Lower–mid Campanian — Snow Hill Island Series	?Maast./Tert. — Sobral Fm. / Mid-upper Campanian — Lopez de Bertodano Fm. (MARAMBIO GROUP)		Sobral Fm. / Lopez de Bertodano Fm. / Unamed Strata (MARAMBIO GROUP)	Paleocene — Danian — Sobral Fm. / Upper Campanian-Maastrichtian — Lopez de Bertodano Fm. (MARAMBIO GROUP)
Cenomanian/Turonian — Snow Hill Beds					
	Hidden Lake Beds / Stoneley Point Conglomerates / Upper Kotick Point Beds / Lower Kotick Point Beds / Lagrelius Point Conglomerates	Lower/mid Camp. — Hidden Lake Beds Fm. / Stoneley Point Cong. Fm. / Upper Kotick Point Fm. / Lower Kotick Point Fm. / Lagrelius Point Cong. Fm.		?San. — Hidden Lake Fm. / ?Aptian-Coniacian — Whiskey Bay Fm. / Kotick Point Fm. / Lagrelius Point Fm. (GUSTAV GROUP)	?San. — Hidden Lake Fm. / ?Aptian-Coniacian — Whiskey Bay Fm. / Kotick Point Fm. / Lagrelius Point Fm. (GUSTAV GROUP)

Figure 2. Summary of the development of the stratigraphic nomenclature and assigned ages for sedimentary units exposed in the James Ross Island region.

out only on Seymour Island (Figs. 1, 3) and are bounded by erosional disconformities. Cretaceous outcrops at Cape Lamb, The Naze, and Snow Hill Island were included in the López de Bertodano Formation by Del Valle and others (1982) and Rinaldi (1982), based on their lithologic and faunal similarity. The Marambio Group is composed of the López de Bertodano and Sobral Formations (Rinaldi, 1982), and ranges from Campanian through Paleocene in age (Huber, 1984; Askin, this volume; Harwood, this volume). The Cross Valley and La Meseta Formations, which occur on northern Seymour Island, are separated from the older Marambio Group by an angular unconformity. The latter unit represents the youngest marine deposit on Seymour Island and is designated as Eocene (Wrenn, 1985) and lower Eocene (Harwood, this volume). Only the López de

Bertodano and lower Sobral Formations are discussed in further detail herein, as no other strata have yielded foraminiferal assemblages during the course of this study.

The sediments of the Marambio Group show little evidence of tectonic disturbance. Homoclinal tilting of the Upper Cretaceous and Paleocene sediments on Seymour and Snow Hill Islands occurred sometime between the late Paleocene and early Eocene. Several diabase dikes of Pliocene age (Rinaldi and others, 1978) cut across the Cretaceous-Paleocene sequence and have affected only the adjacent strata. Evidence of regional stratigraphic repetition by gentle folding of the López de Bertodano Formation sequence was presented by Bibby (1966) (Fig. 1). This was based on the presence of west-dipping beds at Humps Island, horizontal beds present at The Naze, and the recurrence of

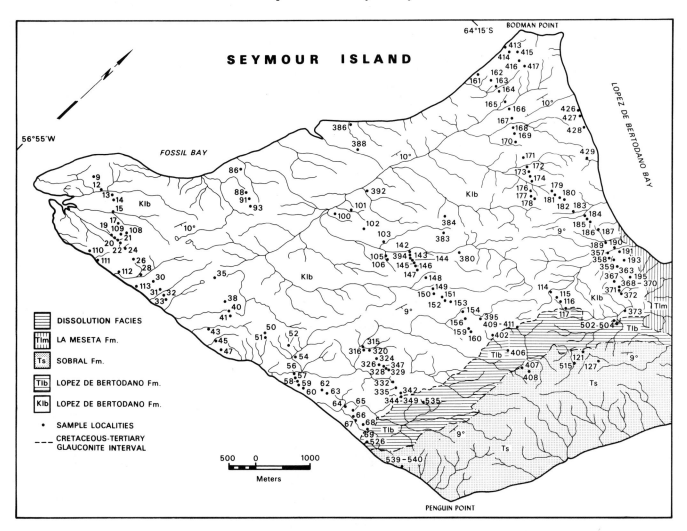

Figure 3. Geologic map and locations of Seymour Island samples that have yielded foraminifera. Note that foraminiferal assemblages are rare in the dissolution facies and absent from most of the Sobral Formation, despite continuous sample coverage.

the Humps Island faunal assemblages at Cape Lamb, Vega Island. Similar faunas and lithologies occur at The Naze, Cape Lamb, Snow Hill Island, and lowermost Seymour Island, attesting to their approximate contemporaneity and analogous depositional environments (Del Valle and others, 1982; Huber, 1984). The unconsolidated nature of these sediments, excellent fossil preservation, and low vitrinite reflectance values (Palamarczuk and others, 1984) indicate a shallow burial depth for the upper Campanian and younger sediments.

López de Bertodano Formation

Sediments of the López de Bertodano Formation consist predominantly of gray, unconsolidated, massive, sandy siltstones that display a high degree of bioturbation (Fig. 4). Grain-size

analyses of Seymour Island samples from these strata indicate that the combined silt and clay percentages range between 40 and 97 percent, showing remarkable lithologic uniformity throughout the sequence on Seymour Island (Macellari, 1984a, b, this volume), as well as in the other beds of this formation. Rounded to irregular calcareous concretions, indurated, fine-grained sandstones, and a few well-bedded horizons occur at sporadic intervals throughout the Cretaceous sequence (Fig. 4). Macrofossils, including ammonites, bivalves, gastropods, decapods, brachiopods, echinoids, serpulids, corals, marine reptiles, and shark vertebrae and teeth, as well as fossil wood, occur in varying abundance and diversity in López de Bertodano Formation deposits on James Ross, Vega, Humps, Snow Hill, and Seymour Island. These are usually preserved in calcareous concretions.

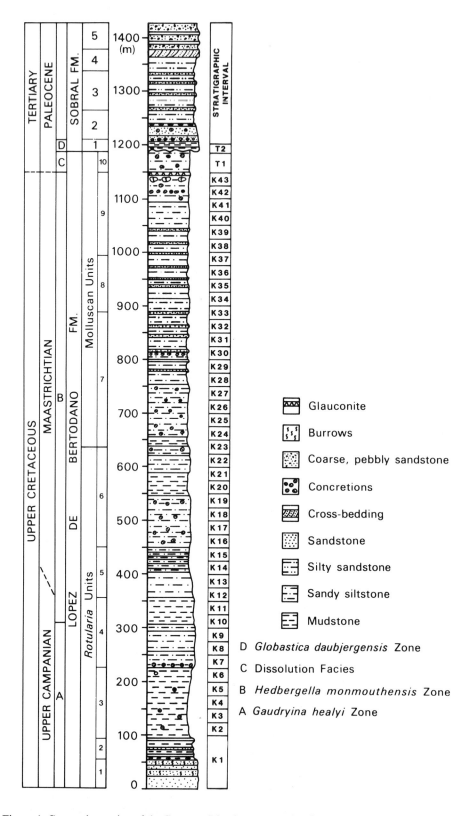

Figure 4. Composite section of the Seymour Island sequence, showing the stratigraphy, foraminiferal zones, and position of stratigraphic intervals defined in this study. Samples included in each stratigraphic interval are listed in Table 1. The informal sedimentary units of the López de Bertodano Formation are from Macellari (1986, this volume) and those of the Sobral Formation are from Sadler (this volume).

**TABLE 1. LIST OF SAMPLE LOCALITIES INCLUDED WITHIN
EACH STRATIGRAPHIC INTERVAL OF FIGURE 4***

Stratigraphic Interval	Sample Localities
T2	121, 351, 407, 408, 540, 515
T1	118-120, 403, 372, 373, 406, 504, 539
K43	114-117, 160, 342, 344-349, 370, 371, 402, 409-411, 502-503, 526, 535
K42	66, 67, 159, 195, 335, 367-369
K41	64, 65, 193, 332, 359, 362, 363, 395
K40	62, 63, 154 156, 191, 328, 329, 357, 358
K39	152, 153, 189, 190, 326, 327, 356
K38	59, 60, 151, 186, 187, 324
K37	58, 150, 185, 320
K36	57, 149a, 149b, 182, 184, 434
K35	56a, 56b, 148, 315, 316, 430, 431
K34	52, 54, 147, 181, 380
K33	51, 146, 180
K32	50, 145, 178, 179
K31	47, 144, 147, 177
K30	142, 143, 176, 394, 429
K29	45, 106, 175, 383
K28	44, 174
K27	43, 105, 173, 384
K26	41, 103
K25	40a, 40b, 171, 172
K24	38, 102, 170, 428
K23	427
K22	
K21	35, 169, 426
K20	33
K19	31, 32, 168
K18	101
K17	100, 167, 392,
K16	30, 166
K15	113
K14	28a-d, 165, 419
K13	26, 112, 164, 418
K12	162, 163, 417
K11	24, 93, 111
K10	22, 415, 416
K9	21
K8	20, 91, 108-110, 161, 413, 414
K7	19, 88, 388
K6	17
K5	15, 86, 386
K4	
K3	13b, 14
K2	12
K1	9

*See Figure 3 for sample localities.

Ineson and others (1986) included only the Cretaceous beds on Seymour Island in the López de Bertodano Formation, and designated Cretaceous outcrops at Cape Lamb, The Naze, and Snow Hill Island as "Unnamed Strata" (Fig. 2). With the exception of the "Older Snow Hill Beds" of Andersson (1906), these exposures are lithologically and sedimentologically indistinguishable from the López de Bertodano Formation on Seymour Island (Huber, personal observation). Microfossils from these localities are also similar (see below; Askin, this volume). Thus, Rinaldi's (1982) definition of the Lopez de Bertodano Formation is retained in this study.

The maximum thickness (1,190 m) of the Lopez de Bertodano Formation on Seymour Island was measured in a section near the southern coast (Macellari and Huber, 1982). The lower 650 m were designated as the *Rotularia* units by Macellari (1984b, 1986), who divided these further into six informal units (Fig. 4). These beds are characterized by the conspicuous presence of the serpulid genus *Rotularia* and the low diversity and poor preservation of molluscs. Molluscan diversity increases in the Molluscan units, which make up units 7 through 10 of the remaining 500 m of the López de Bertodano Formation section (Macellari, 1984b, 1986). This greater diversity in the Molluscan units was attributed by Macellari to paleoenvironmental factors, such as the predominance of restricted, nearshore conditions for the *Rotularia* units and a more offshore, open marine setting for the Molluscan units (Macellari, 1984b, 1986, this volume). However, the transition from the *Rotularia* facies to the Molluscan facies cannot be discerned using foraminiferal stable isotope data and foraminiferal and siliceous microfossil species distributions, which indicate that normal marine conditions persisted at least from above 200 m through to the K/T boundary (Barrera and others, 1987; Harwood, this volume). In the absence of strong lithologic and sedimentologic evidence for marginal marine conditions, factors other than depth or shoreline proximity must be considered to explain the discrepancy in macro- and microfossil distributions for the *Rotularia* units. The paucity of concretions and low-relief physiography in the *Rotularia* units may indicate that loss of some fossil information occurred due to nonpreservation and long-term exposure to weathering.

The López de Bertodano Formation is considered to have been deposited in depths above the shelf/slope break on the basis of sedimentologic and paleontologic data (Huber, 1984; Macellari, this volume; Harwood, this volume). The absence of traction current and slope instability structures, high mud (silt and clay) content, and occurrence of in situ macrofossils attest to the predominance of quiet-water conditions and shelf stability throughout this unit. The sediments within the formation were derived from a combination of volcanic and continental basement sources on the Antarctic Peninsula and deposited in a shelf environment (Macellari, 1984b, this volume).

The "Older Seymour Island Beds" of Andersson (1906), which are included in the López de Bertodano Formation on Seymour Island and southeast Snow Hill Island, were assigned to the Albian through Campanian by Kilian and Reboul (1909).

Howarth (1958, 1966), Rinaldi (1982), Olivero (1981), and Palamarczuk and others (1984) suggested that the entire Cretaceous sequence was Campanian, whereas Spath (1953) interpreted the ammonite-bearing strata on Seymour Island as being upper Campanian to possibly Maastrichtian. After an extensive study and systematic revision of the Seymour Island ammonite fauna, Macellari (1986) concurred with the microfossil data of Huber and others (1983) and Askin (this volume) that the Cretaceous strata of Seymour Island range from upper Campanian through upper Maastrichtian.

The uppermost López de Bertodano Formation is considered Paleocene, based on fossil palynomorphs (Askin, this volume), foraminifera (this chapter), and siliceous microfossils (Harwood, this volume). The location of the Cretaceous/Tertiary (K/T) boundary is not clearly defined, however, because of poor biostratigraphic control and correlation problems, as discussed below. It is placed within a glauconitic interval approximately 9.5 m thick (Askin, this volume), separated from the disconformable base of the Sobral Formation by 50 to 90 m of section (Figs. 3–7).

The K/T glauconite interval coincides with a marked change in fossil content, sediment color, and mineralogy, with no apparent change in grain size or bedding stratification. The underlying beds are massive, gray, sandy silts, with interspersed remains of marine vertebrates and shelly fossils, including echinoid spines, rotularians, ammonites, gastropods, bivalves, and scleractinians. The reddish brown sandy silts above, referred to as the "dissolution facies" (Figs. 3–7), record an increase in volcaniclastic sediments (Macellari, this volume), and are devoid of all but the thick-walled shelly fossils, with dominance of the bivalve *Lahilla*. Calcareous microfossils are completely absent and agglutinated foraminifera are rare and poorly preserved. Siliceous and organic-walled microfossils show a considerable drop in diversity, with a few opportunistic species (mostly cyst forms) occurring in very high abundance (Askin, this volume; Harwood, this volume). Macrofossils show a similar decline (Macellari, 1985a). The first diagnostic Danian (lower Paleocene) foraminiferal assemblage occurs at the top of the López de Bertodano Formation, 1 m below the contact with the Sobral Formation.

Sobral Formation

The Sobral Formation was included in the "Older Seymour Island Beds" of Andersson (1906) and in the "Snow Hill Island Series" of Bibby (1966) (Fig. 2). It was distinguished as a separate lithostratigraphic unit by Rinaldi and others (1978) and subdivided into five informal units by Sadler (this volume). It is characterized by higher relief than the López de Bertodano Formation, preservation of primary sedimentary structures, and sparse macrofauna. The Sobral Formation appears to be conformable with the underlying formation in a cliff exposure on the southeast coast of Seymour Island (see Fig. 3 of Huber, 1987). However, inland exposures show channeling and erosion of as much as 40 m of the uppermost López de Bertodano Formation. Unit 1, which measures 15 to 20 m thick (Fig. 4), is a poorly fossiliferous, finely laminated, silty mudstone with numerous cross-cutting channels. Macellari (this volume) has interpreted this unit as a quiet-water, pro–delta slope facies, probably deposited below effective wave base with a high sediment supply. This is the only unit of the Sobral Formation that yielded foraminifera. Units 2 through 5 are interpreted as a continuation of a prograding delta complex. The coarsening in grain size and occurrence of traction current structures in Unit 3 and above indicate deposition above effective wave base and predominance of high-energy conditions (Macellari, this volume).

The Sobral Formation was judged to be questionably Maastrichtian to Tertiary by Rinaldi and others (1978) and Rinaldi (1982). More recently it was assigned to the lower Paleocene, based on dinoflagellates (Palamarczuk and others, 1984; R. A. Askin, personal communication, 1986). Siliceous microfossils (Harwood, this volume) and foraminifera indicate a definite Danian age for Unit 1 of the Sobral Formation. No calcareous microfossils are known from above Unit 1.

PROBLEMS WITH SOUTHERN HIGH-LATITUDE BIOSTRATIGRAPHY OF THE UPPER CRETACEOUS

Although Cretaceous foraminifera from the López de Bertodano Formation are moderately diverse and well preserved, correlation with lower latitude sites has been difficult. Differences in faunal character between the low- and high- latitude sites are probably related to the effects of environmental instability and physical extremes prevalent in high-latitude regions, such as more seasonal organic productivity and lower marine temperatures. Factors that have inhibited accurate sediment dating and correlation within the Cretaceous circum-Antarctic region include the following:

1. Convergence of morphologic characters among Upper Cretaceous planktonic foraminifera, causing taxonomic uncertainties. This may reflect the relationship between test morphology and the physicochemical constraints imposed by the circum-polar marine environment. The distribution of modern planktonic foraminifera (Bé, 1977) and oxygen isotope data from Upper Cretaceous taxa (Douglas and Savin, 1975) show that "complex" (large, thick-walled, keeled or densely ornamented) species are largely restricted to tropical and warm-temperate waters, generally inhabit greater depths in the water column, and have longer life cycles (Emiliani, 1971). Several authors have noted that these forms are rare to absent in Upper Cretaceous high-latitude deposits (Webb, *in* Hornibrook, 1969; Webb, 1973a; Scheibnerová, 1971, 1973; Sliter, 1976; Krasheninnikov and Basov, 1983, 1986). Conversely, "simple" (small, thin-walled, less ornamented, globular) forms inhabit shallower depths in the water column, have shorter life cycles, and predominate in modern and Late Cretaceous foraminiferal assemblages of extratropical regions. The adaptive advantage of depth stratification may be related to the establishment and maintenance of a favorable ecologic niche (Caron and Homewood, 1983). Morphologi-

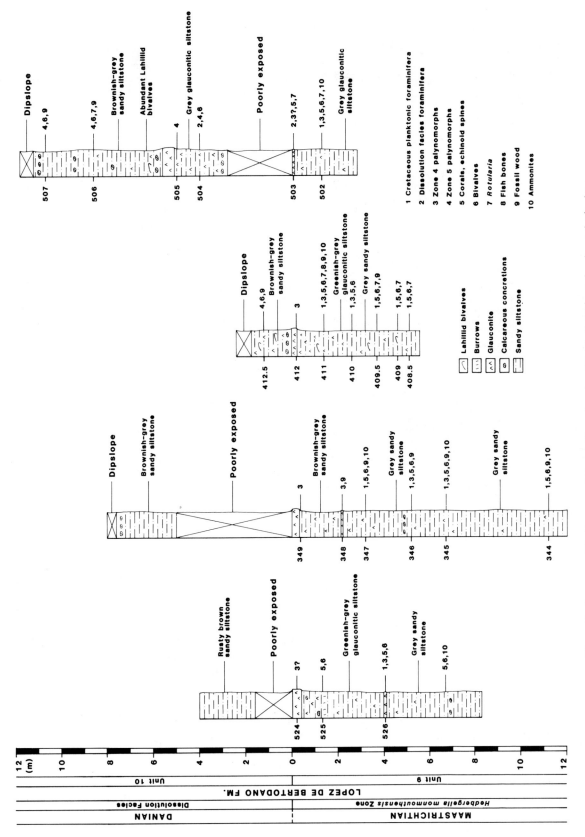

Figure 5. Cretaceous/Tertiary transition sections showing lithostratigrahic and paleontologic data. Sample localities are shown on Figure 3. Note that several glauconite horizons occur above and below the K/T transition interval, and their resistance and thickness change laterally. The palynomorph distribution data is from R. A. Askin (personal communication, 1986). Palynomorph Zone 4 is assigned to the Cretaceous and Zone 4 to the Tertiary (Askin, this volume).

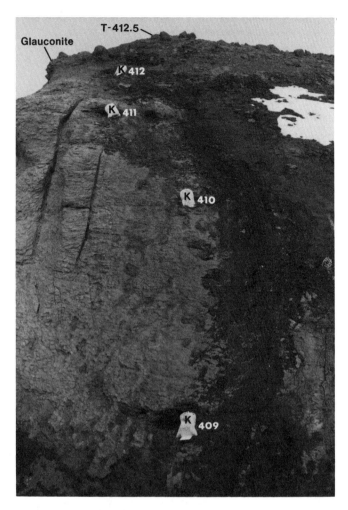

Figure 6. Cretaceous/Tertiary transition section on Seymour Island showing stratigraphic position of samples 409 through 412, all of which yield Cretaceous (Zone 4) palynomorphs (R. A. Askin, personal communication, 1986). Cretaceous planktonic and benthic foraminifera occur in samples 409 through 411. Sample 412.5, which is 1.5 m above the resistant glauconite bed (top of photo), yields Danian (Zone 5) palynomorphs (R. A. Askin, personal communication, 1986), but foraminifera are absent. Thus, the K/T boundary is placed between the resistant glauconite and sample 412.5 (see Fig. 5).

cal adaptations for attainment of increasing depth ranges include addition of ornamentation, such as costellae, keels, and shell thickening.

The poleward loss of complex planktonic morphotypes and test surface ornamentation may be related to a parallel shallowing of the photic zone, a decrease in water density for a given depth, and an increase in seasonal variability of temperature and primary productivity. Although little is known about Cretaceous water-mass structure, analogy with modern oceans would suggest that polar oceans were not as well stratified as in tropical regions,

and thus had a more diffuse density gradient. The Upper Cretaceous complex morphotypes were apparently poorly adapted for habitation of the deeper waters in high-latitude regions because of their presumed long life cycle and requirements for a stable, depth-stratified water mass. On the other hand, simple morphotypes may have dominated in high-latitude regions because of their shorter life cycles and tolerance for seasonally unstable conditions. Symbiont-bearing forms probably needed to maintain a higher position in the water column in polar regions because of the shallower photic zone. The considerable reduction in test density of high-latitude rugoglobigerine morphotypes, which lack well-developed meridional costellae characteristic of lower latitude forms (see discussion in Systematics), may be a good example of an adaptive strategy for water column position maintenance in response to changes in the physicochemical conditions of the upper water mass.

2. Poorly defined Upper Cretaceous stratotypes in the Southern Hemisphere. The most complete Campanian-Maastrichtian land-based marine sequences within the circum-Antarctic, excluding the James Ross Island region, occur in New Zealand and southern South America. Foraminiferal zonations and chronostratigraphic subdivisions proposed for these regions have limited extra-basinal utility for several reasons. First, foraminiferal faunas from southern Chile have been inadequately illustrated and include frequent usage of open taxonomic nomenclature and invalid species names, such that age assignments of the local stages may be inaccurate (see Charrier and Lahsen, 1969; Natland and others, 1974). Second, detailed stratigraphic information is lacking from many of the southern South American Upper Cretaceous outcrops. Third, Upper Cretaceous foraminiferal assemblages of the *Trochammina globigeriniformis* and *Rzehakina epigona* Zones in New Zealand and in the Magallanes (Austral) Basin of southern South America are predominantly facies-controlled (Webb, 1971; Malumian, 1978). Finally, the stratotypes used for definition of the Piripauan and Haumurian Stages (Upper Cretaceous) in New Zealand have been rendered inadequate because of confusion between biostratigraphic and chronostratigraphic concepts and stage boundary definitions (Warren and Speden, 1977). The Marambio Group is better exposed and yields more complete stratigraphic information than any of the other circum-Antarctic land-based sequences. This may warrant its designation as a stratotype for the Upper Cretaceous–Lower Tertiary in the southern high latitudes.

3. Poorly defined Upper Cretaceous stratotypes in Europe. The difficulty of constraining the limits of the Campanian and Maastrichtian stages in the high-latitude regions, where there is a shortage of reliable indicator species, is further compounded by the problems with definition of the European stratotypes. The major hiatus separating the Campanian and Maastrichtian type sections (Jeletzky, 1951; van der Heide, 1954; Berggren, 1962; Surlyk, 1984) and disagreement on which fossil group best defines this boundary level (Birkelund and others, 1984; Harland and others, 1982) render the type boundary definition ambiguous. Furthermore, low-latitude foraminiferal zonations used to

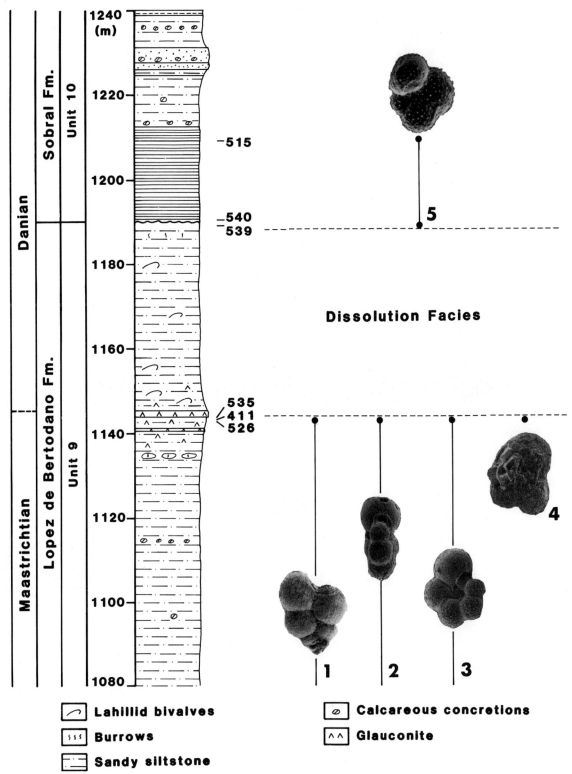

Figure 7. Composite Cretaceous/Tertiary transition section showing K/T glauconite samples yielding Cretaceous planktonic foraminifera (species 1 through 4) and samples bearing Danian planktonics (species 5). 1, *Heterohelix globulosa* (Ehrenberg); 2, *Globigerinelloides multispinatus* (Lalicker); 3, *Hedbergella monmouthensis* (Olsson); 4, *Globotruncanella*? sp.; 5, *Globastica daubjergensis* (Brönnimann).

delineate the Campanian and Maastrichtian stages cannot be applied in the circum-Antarctic regions, because of the lack of correlatable datum events (e.g., see Webb, 1971; Malumian and Masiuk, 1976; Sliter, 1976; Krasheninnikov and Basov, 1983, 1986).

4. High-latitude faunal diachroneity. Temporal changes in latitudinal temperature gradients strongly affect biostratigraphic resolution in high-latitude regions, because of the resultant migration of stenothermal taxa. There is considerable uncertainty as to whether the first and last occurrences of high-latitude species are related to their evolution and extinction, or are paleoecologically controlled. For example, the distribution of the calcareous nannoplankton species *Nephrolithus frequens* Górka, which is traditionally used as a late Maastrichtian index fossil in Tethyan assemblages, has been found to be time-transgressive from the high to low latitudes, with its oldest extra-tropical occurrence reported as late middle Maastrichtian at Deep Sea Drilling Project (DSDP) Site 249 (Wind, 1979). The distribution of *Nephrolithus frequens* and *N. corystus* Wind on Seymour Island and their known stratigraphic range are used as the primary means of dating the Cretaceous portion of the Seymour Island sequence (Huber and others, 1983), but there remains uncertainty as to exactly when these species evolved. Several foraminiferal taxa from the James Ross Island region significantly predate their earliest known occurrences in lower latitude regions: *Cyclammina* spp., *Frondicularia rakauroana, Buliminella* spp., *Ceratolamarckina* cf. *tuberculata, Alabamina westraliensis* (= *A. wilcoxensis*), and *Planispirillina subornata* (see discussion in Systematics). Diachronous high-latitude species with earliest appearance in high latitudes have also been reported for several other Cenozoic macro- and microfossil groups in both hemispheres (Hickey and others, 1983; Zinsmeister and Feldmann, 1984; Backman and others, 1984; Marincovitch and others, 1985).

5. Lower faunal diversity and longer species duration of high latitude taxa. The evolutionary turnover of Upper Cretaceous taxa in extratropical regions was slower than in the tropics, as evidenced by the lower diversity and longer ranges of high-latitude benthic and planktonic foraminifera. The most widely used biostratigraphic schemes for the Upper Cretaceous depend on the presence of stenothermal taxa, which rarely occur in polar and subpolar regions. Thus, the number of correlatable datum events in the polar regions are few, resulting in considerably reduced biostratigraphic resolution.

SEYMOUR ISLAND BIOSTRATIGRAPHY AND PALEOECOLOGY

Composite range chart

Samples collected within and between measured sections on Seymour Island are correlated by aerial photograph interpretation. The K/T glauconite interval in the upper López de Bertodano Formation is used as a reference datum to which the sections are correlated. In order to simplify discussion of their stratigraphic distribution, the samples from all sections are placed within a single composite section that is arbitrarily subdivided into 45 stratigraphic intervals (Fig. 4). The lowermost stratigraphic interval (SI) of the Lopéz de Bertodano Formation (SI-K1) includes about 100 m of section that is barren of foraminifera. Stratigraphic intervals K2 through K43 are 25 m thick. The dissolution facies, SI-T1, includes all strata between the K/T glauconite interval and 1 m below the Sobral Formation. Samples collected from above the dissolution facies to the top of Unit 1 of the Sobral Formation are included in SI-T2. Individual samples included in each stratigraphic interval are listed in Table 1, and their locations are plotted on Figure 3. This stratigraphic scheme is followed by Harwood (this volume) with addition of stratigraphic intervals in the Sobral Formation.

The stratigraphic distribution of foraminiferal species in the Seymour Island López de Bertodano and Sobral Formations is plotted in Table 2, and the species location index is presented in Appendix 1. The data for this composite range chart are an amalgamation of presence/absence information for all samples included in each stratigraphic interval. Presentation of the distribution data in this way provides a means of diminishing bias due to sampling of poorly exposed outcrops, variation in sample size, and large sampling gaps. It also simplifies portrayal of species distribution within the numerous measured sections, which often overlap in stratigraphic coverage. Although information on unit sample presence/absence data is lost by this method, the general trends in stratigraphic species distribution become more readily apparent.

Several generalizations can be made about the species distributions in Table 2. First, most species appear early, in the upper Campanian portion of the Cretaceous sequence, with scattered occurrences up to the Cretaceous/Tertiary transition interval. Therefore, no major biotic changes take place within the Campanian-Maastrichtian portion of the sequence, attesting to long-term environmental stability. Second, there is a dramatic faunal turnover near the Cretaceous/Tertiary boundary (between SI-K43 and SI-T1). Foraminiferal species terminations at this stratigraphic level are 71 percent for agglutinated, 64 percent for calcareous benthic, and 100 percent for planktonic species. Third, several species have their lowest appearance in the Tertiary part of the section. Combining SI-T1 and SI-T2, new appearances are 17 percent for agglutinated, 28 percent for calcareous benthic, and 100 percent for planktonic foraminifera. It is apparent from the considerably reduced number of species present and high species dominance in the Tertiary part of the sequence that stressed environmental conditions prevailed just above the K/T glauconite sequence.

Cretaceous Benthic Species Distributions. Biostratigraphically significant benthic foraminifera are few. Most of the benthic taxa on Seymour Island are cosmopolitan, ranging from the early Senonian through the end of the Cretaceous. Specialized benthic forms that are useful in dating Upper Cretaceous strata in North America and Europe, such as species of *Neoflabellina* and *Bolivinoides,* are conspicuously absent. The most useful benthic species for extra-basinal correlation is *Bolivina incrassata,* which ranges from SI-K24 through SI-K36 (Fig. 4) on Seymour Island. It has been reported from upper Campanian–Maastrichtian deposits in Europe, North America, Mexico, South America, Australia, and New Zealand. The provincial species *Frondicularia rakauroana* and *Gaudryina healyi*, which are used as index fossils for the New Zealand Haumurian Stage (see Webb, 1971, 1972b, 1973b; Huber and Webb, 1986), occur in several Seymour Island samples. Specimens of *G. healyi* were found only in the upper Campanian portion of the Seymour Island sequence (SI-K10), and *F. rakauroana* ranges from SI-K5 through SI-K30 (Table 2, Fig. 4).

The agglutinated taxa *Cyclammina* cf. *complanata* and *Alveolophragmium macellarii* are by far the most common foraminifera of the Cretaceous sequence, both occurring in 91 percent of the stratigraphic intervals (Table 3). These forms, along with *Hyperammina elongata*, may be present in a majority of the studied samples, because (1) they are the most solution resistant and/or (2) they are ecologic generalists, able to tolerate fluctuating environmental conditions (particularly changes in sedimentation rates, salinity, turbidity, and temperature).

Several factors provide evidence that the foraminiferal distribution data presented in Table 2 are largely biased from the effects of diagenesis, and do not reflect primary conditions of the depositional environment. First, the "stratigraphic index" groupings in Table 3 show that the most frequently occurring calcareous benthic species are large, thick walled, and solution resistant. These forms are moderately to poorly preserved in samples that do not bear the thinner walled taxa. Sliter (1975) portrayed a similar ranking among solution-resistant Upper Cretaceous benthic species from the eastern North Pacific. Second, the agglutinated foraminifera show little change in relative abundance among low- and high-diversity samples. Considerable fluctuation in population abundance would be expected if they were responding to changes in the depositional environment.

Finally, comparison of the "miscellaneous microfossils" presence/absence data with the foraminiferal distributions (Table 2) shows a very strong correlation among high foraminiferal diversity intervals and occurrence of calcareous-walled miscellaneous microfossils. The latter group is rare or absent in stratigraphic intervals that yield wholly agglutinated faunas and low-diversity, poorly preserved benthic faunas (including a few thick-walled, calcareous benthic forms). Particular examples of this occur in stratigraphic intervals SI-K1 through SI-K3, SI-K5 through SI-K7, SI-K9, SI-K11, SI-K15 through SI-K16, SI-K19, and SI-28 (Table 2). No samples were collected from SI-K4 and SI-K22, so these are not considered. The complete absence of the miscellaneous group from these intervals—due to competitive exclusion or paleoenvironmental stress—seems unlikely, since they include a variety of trophic levels, which would not all respond simultaneously to the same external constraints. Their absence is more likely related to dissolution on the sea floor, burial diagenesis, increased sedimentation rates, or loss due to subaerial exposure. Unfortunately, it is not possible to determine which of these is responsible for the resulting thanatocoenosis.

It is noteworthy that several agglutinated specimens, including *Spiroplectammina laevis, Gaudrying heayli, Dorothia paeminosa,* and *Marssonella oxycona,* occur only in samples that also bear a diverse calcareous benthic assemblage. This again indicates that the presence/absence data for these species may be the result of diagenesis, as these forms are less solution-resistant than are the more robust agglutinated taxa.

Cretaceous Planktonic Species Distributions. The stratigraphic distribution of planktonic foraminifera is plotted in Table 2, and the relative stratigraphic abundance indices are presented in Table 4. Planktonic taxa are either absent or occur in small numbers in the studied Cretaceous samples. The most common planktonic species in the Seymour Island sequence include *Hedbergella monmouthensis, Globigerinelloides multispinatus, Heterohelix globulosa, Rugoglobigerina?* sp. 1, and *Rugoglobigerina rugosa.* Less common are *Globotruncanella minuta, Hedbergella* sp. 1, *Heterohelix glabrans,* and *Rugoglobigerina rotundata.* The remaining taxa shown in Table 4 are very rare, occurring only in one or two samples. The Seymour Island planktonic assemblage strongly resembles upper Campanian–Maastrichtian planktonic faunas reported from the Falkland Plateau by Sliter (1976) and Krasheninnikov and Basov (1983, 1986). In both localities, species diversity is quite low, and there is a conspicuous absence of thermophilic keeled taxa used in correlation of lower latitude zones.

Species with cosmopolitan ranges of upper Campanian through Maastrichtian include *Guembelitria cretacea, Heterohelix glabrans,* and *Globotruncanella havanensis. Rugoglobigerina rotundata* occurs in upper Campanian through Maastrichtian strata at the Falkland Plateau (Sliter, 1976; Krasheininnikov and Basov, 1983) but is restricted to the mid- to upper Maastrichtian in lower latitude regions (Robaszynski and others, 1984). Robaszynski and others (1984, p. 168) report that *Globotruncanella minuta* may be the phylogenetic link between *Hedbergella holmdelensis-monmouthensis* and *Globotruncanella* spp., although its known range in the low latitudes is reported as mid- to upper Maastrichtian. The earlier occurrence on Seymour Island supports their contention. *Hedbergella monmouthensis* (=*H. holmdelensis* of Sliter, 1976) was suggested to have evolved from *H. holmdelensis* Olsson in the North Atlantic region during the early to mid-Maastrichtian (Olsson, 1964), but several other authors have reported this species in older Campanian sediments (Sliter, 1968; Hanzlikova, 1972; Robaszinski and others, 1984). This species has often been wrongly identified (R. Olsson, personal communication, 1986), so that its earliest appearance is equivocal at present. No specimens of *Hedbergella holmdelensis* have been recognized in the López de Bertodano Formation.

Campanian/Maastrichtian boundary. Placement of the Campanian/Maastrichtian boundary between SI-K11 and SI-K15 on the composite section (Fig. 4) is based primarily on the overlapping ranges of the calcareous nannoplankton species *Nephrolithus corystus* Wind and *Nephrolithus frequens* Górka on Seymour Island, and correlation of these species with their ranges at Falkland Plateau DSDP Sites 327A and 511 (Wind, 1979,

B. T. Huber

TABLE 2. DISTRIBUTION CHART OF SEYMOUR ISLAND FORAMINIFERA*

Column headers (left to right): K1 K2 K3 K4 K5 K6 K7 K8 K9 K10 K11 K12 K13 K14 K15 K16 K17 K18 K19 K20 K21 K22 K23 K24 K25 K26 K27 K28 K29 K30 K31 K32 K33 K34 K35 K36 K37 K38 K39 K40 K41 K42 K43 T1 T2

BENTHONIC FORAMINIFERA

1 *Cyclammina* cf. *complanata*
2 *Alabamina westraliensis*
3 *Allomorphina cretacea*
4 *Anomalinoides larseni*
5 *Anomalinoides piripaua*
6 *Anomalinoides rubiginosus*
7 *Cibicides nordenskjoldi*
8 *Cibicides seymouriensis*
9 *Cibicides* cf. *beaumontianus*
10 *Dentalina basiplanata*
11 *Gyroidinoides nitidus*
12 *Gyroidina* cf. *globosa*
13 *Haplophragmoides eggeri*
14 *Alveolophragmium macellarii*
15 *Lagena sphaerica*
16 *Marssonella oxycona*
17 *Marssonella? sp.*
18 *Nonionella* cf. *robusta*
19 *Psammosphaera sp.*
20 *Rhizammina algaeformis*
21 *Ammobac. fragmentarius*
22 *Bathysiphon californicus*
23 *Haplophragmoides platus*
24 *Hoeglundina supracretacea*
25 *Reophax subfusiformis*
26 *Reophax texanus*
27 *Rzehakina epigona*
28 *Spiroplectammina laevis*
29 *Troch. globigeriniformis*
30 *Trochammina ribstonensis*
31 *Spiroplectammina spectabilis*
32 *Dentalina gracilis*
33 *Dentalina marcki*
34 *Frondicularia rakauroana*
35 *Hyperammina elongata*
36 *Lenticulina ovalis*
37 *Pseudonodosaria parallela*
38 *Rhabdammina sp.*
39 *Saracenaria triangularis*
40 *Cyclammina sp.*
41 *Saccammina sphaerica*
42 *Karreriella aegra*
43 *Marginulina bullata*
44 *Praebulimina midwayensis*
45 *Stilostomella impensia*
46 *Alabamina creta*
47 *Ammodiscus pennyi*
48 *Bolivina sp.*
49 *Charltonina sp.*
50 *Hiltermannella kochi*
51 *Lagena apiculata*
52 *Lenticulina spissocostata*
53 *Neobulimina aspera*
54 *Neobulimina digitata*
55 *Osangularia cordieriana*
56 *Praebulimina cushmani*
57 *Praebulimina* cf. *cushmani*
58 *Praebulimina* cf. *reussi*
59 *Trochammina triformis*
60 *Budashevaella multicamerata*
61 *Buliminella creta*
62 *Buliminella* cf. *fusiforma*
63 *Eouvigerina hispida*
64 *Epistominella glabrata*
65 *Fissurina orbignyana*
66 *Gaudryina healyi*
67 *Gavelinella sandidgei*
68 *Globulina lacrima*
69 *Guttulina trigonula*
70 *Lagena acuticosta*
71 *Lagena semiinterrupta*
72 *Lagena semilineata*
73 *Nodosaria aspera*
74 *Nodosaria navarroana*
75 *Nodosaria obscura*
76 *Pullenia cretacea*
77 *Pyrulina cylindroides*
78 *Quad. allomorphinoides*
79 *Ramulina pseudoaculeata*
80 *Serovaina orbicella*
81 *Stilostomella pseudoscripta*
82 *Dentalina lorneiana*

*See Appendix 1 for cross-reference.

TABLE 2. (CONTINUED)

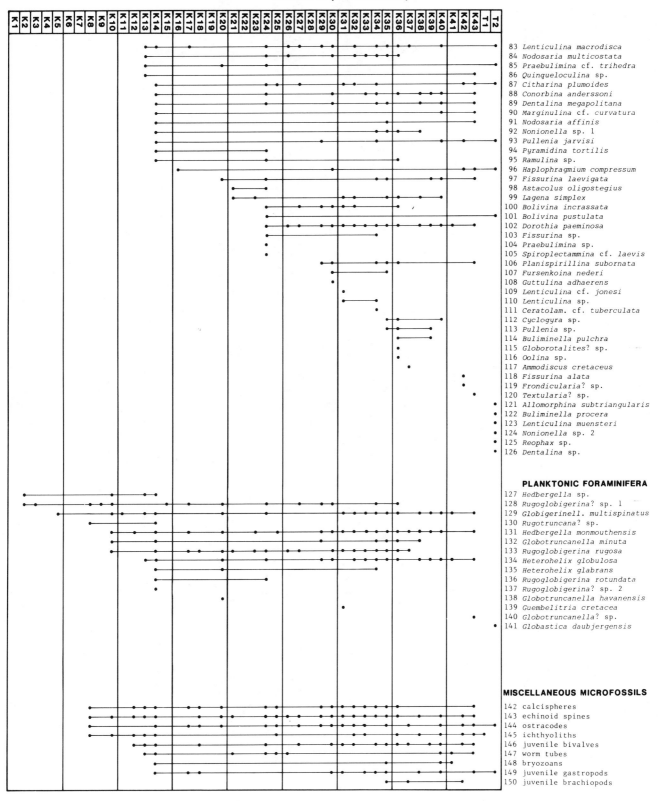

83 *Lenticulina macrodisca*
84 *Nodosaria multicostata*
85 *Praebulimina* cf. *trihedra*
86 *Quinqueloculina* sp.
87 *Citharina plumoides*
88 *Conorbina anderssoni*
89 *Dentalina megapolitana*
90 *Marginulina* cf. *curvatura*
91 *Nodosaria affinis*
92 *Nonionella* sp. 1
93 *Pullenia jarvisi*
94 *Pyramidina tortilis*
95 *Ramulina* sp.
96 *Haplophragmium compressum*
97 *Fissurina laevigata*
98 *Astacolus oligostegius*
99 *Lagena simplex*
100 *Bolivina incrassata*
101 *Bolivina pustulata*
102 *Dorothia paeminosa*
103 *Fissurina* sp.
104 *Praebulimina* sp.
105 *Spiroplectammina* cf. *laevis*
106 *Planispirillina subornata*
107 *Fursenkoina nederi*
108 *Guttulina adhaerens*
109 *Lenticulina* cf. *jonesi*
110 *Lenticulina* sp.
111 *Ceratolam.* cf. *tuberculata*
112 *Cyclogyra* sp.
113 *Pullenia* sp.
114 *Buliminella pulchra*
115 *Globorotalites?* sp.
116 *Oolina* sp.
117 *Ammodiscus cretaceus*
118 *Fissurina alata*
119 *Frondicularia?* sp.
120 *Textularia?* sp.
121 *Allomorphina subtriangularis*
122 *Buliminella procera*
123 *Lenticulina muensteri*
124 *Nonionella* sp. 2
125 *Reophax* sp.
126 *Dentalina* sp.

PLANKTONIC FORAMINIFERA

127 *Hedbergella* sp.
128 *Rugoglobigerina?* sp. 1
129 *Globigerinell. multispinatus*
130 *Rugotruncana?* sp.
131 *Hedbergella monmouthensis*
132 *Globotruncanella minuta*
133 *Rugoglobigerina rugosa*
134 *Heterohelix globulosa*
135 *Heterohelix glabrans*
136 *Rugoglobigerina rotundata*
137 *Rugoglobigerina?* sp. 2
138 *Globotruncanella havanensis*
139 *Guembelitria cretacea*
140 *Globotruncanella?* sp.
141 *Globastica daubjergensis*

MISCELLANEOUS MICROFOSSILS

142 calcispheres
143 echinoid spines
144 ostracodes
145 ichthyoliths
146 juvenile bivalves
147 worm tubes
148 bryozoans
149 juvenile gastropods
150 juvenile brachiopods

TABLE 3. STRATIGRAPHIC INDEX VALUES FOR BENTHIC FORAMINIFERAL SPECIES ON SEYMOUR ISLAND RANGING FROM SI-K1 THROUGH SI-K43*

Stratigraphic Index Values = No. of Stratigraphic Occurrences
43 Stratigraphic Intervals

VERY COMMON (70 to 100%) / COMMON (45 to 69%)	MODERATELY COMMON (30 to 44%)	RARE (15 to 29%)	VERY RARE (1 to 14%)	VERY RARE (1 to 14%)
VERY COMMON (70 to 100%)	Allomorphina cretacea	Buliminella creta	Ammobaculites fragmentarius	N. multicostata
Cyclammina cf. complanata	Neobulimina aspera	Pyrulina cylindroides	Lenticulina ovalis	Pullenia sp.
Alveolophragmium macellarii	Rhizammina algaeformis	Quinqueloculina sp.	Planispirillina subornata	Ramulina sp.
Hyperammina elongata	Cibicides nordenskjoldi	Serovaina orbicella	Praebulimina cf. reussi	Rzehakina epigona
	Dentalina gracilis	Conorbina anderssoni	Ammodiscus pennyi	Buliminella sp.
COMMON (45 to 69%)	Epistominella glabrata	Globulina lacrima	Anomalinoides rubiginosus	Fissurina sp.
Cibicides seymouriensis	Dorothia paeminosa	Nodosaria obscura	Spiroplectammina spectabilis	Fursenkoina nederi
Cibicides cf. beaumontianus	Lenticulina spissocostata	Saracenaria triangularis	Budashevaella multicamerata	Gyroidina cf.globosa
Reophax texanus	Quadrimorphina allomorphinoides	Dentalina megapolitana	Guttulina trigonula	Lagena sulcata
Anomalinoides piripaua	Spiroplectammina laevis	Haplophragmoides platus	Nodosaria navarroana	Nodosaria aspera
Alabamina westraliensis	Lagena apiculata	Lagena acuticosta	Nonionella sp. 1	Osangularia cordieriana
Hoeglundina supracretacea	Psammosphaera sp.	Praebulimina midwayensis	Pullenia cretacea	Praebulimina cf. cushmani
Alabamina creta	Haplophragmoides eggeri	Citharina plumoides	P. jarvisi	Pyramidina tortilis
Cyclammina sp.	Dentalina lorneiana	Dentalina marcki	Charltonina? sp.	Ramulina pseudoaculeata
Dentalina basiplanata	Praebulimina cf. trihedra	Lagena seminterrupta	Fissurina orbignyana	Saccammina sphaerica
Anomalinoides larseni	Stilostomella impensia	L. semilineata	Haplophragmium compressum	Spiroplectammina cf. laevis
Gyroidinoides nitidus	Gavelinella sandidgei	L. simplex	Reophax subfusiformis	Ammodiscus cretaceus
Rhabdammina sp.	Karreriella aegra	Bolivina incrassata	Trochammina ribstonensis	Buliminella cf. fusiforma
Hiltermannella kochi	Lagena sphaerica	Bolivina sp.	T. triformis	Frondicularia? sp.
Lenticulina macrodisca	Marssonella oxycona	Nonionella cf. robusta	Bolivina pustulata	Guttulina adhaerens
Praebulimina cushmani	Stilostomella pseudoscripta		Cyclogyra sp.	Lenticulina cf. jonesi
	Trochammina globigeriniformis		Eouvigerina hispida	Lenticulina sp.
			Fissurina laevigata	Oolina sp.
			Frondicularia rakauroana	Praebulimina sp.
			Marginulina cf. curvatura	Textularia? sp.
			Nodosaria affinis	

*Foraminifera are listed from bottom to top of each column in order of least to most common.

TABLE 4. STRATIGRAPHIC INDEX VALUES FOR PLANKTONIC FORAMINIFERAL SPECIES ON SEYMOUR ISLAND RANGING FROM SI-K1 THROUGH SI-K43

COMMON (45 to 69%) *Hedbergella monmouthensis*	**VERY RARE** (1 to 14%) *Heterohelix glabrans* *Hedbergella* sp.
MODERATELY COMMON (30 to 44%) *Heterohelix globulosa* *Globigerinelloides multispinatus* *Rugoglobigerina rugosa*	*Rugoglobigerina rotundata* *Rugotruncana?* sp. *Globotruncanella havanensis* *Globotruncanella?* sp. *Guembelitria cretacea*
RARE (15 to 29%) *Globotruncanella minuta*	*Rugoglobigerina?* sp.

*Foraminifera are listed from bottom to top of each column in order of least to most common.

1983), as discussed in Huber and others (1983). Poor definition of the stratotype boundary, lack of age diagnostic foraminiferal taxa, and sparse recovery of calcareous nannoplankton preclude a more accurate definition of this boundary on Seymour Island. These age determinations are consistent with biostratigraphic distribution data for ammonites (Macellari, 1984a, b, 1986), siliceous microfossils (Harwood, this volume), and palynomorphs (Askin, this volume).

Tertiary foraminifera and the Cretaceous/Tertiary boundary. The K/T boundary is placed within 9.5 m above a resistant glauconite bed in localities where in situ paleontologic data are available, about 50 m below the Sobral Formation (Figs. 3–7). Diagnostic Cretaceous fossils, including ammonites and several gastropod and bivalve taxa (Zinsmeister and Macellari, 1983, this volume; Macellari, 1985a, 1986), silicoflagellates (Harwood, this volume), and foraminifera range within several meters below a consolidated glauconite horizon (Fig. 5). Askin (this volume) reports that the transition from Cretaceous Zone 4 to Tertiary Zone 5 palynomorph assemblages occurs between the top of a resistant glauconite bed and within 9.5 m of the immediately overlying strata. At this stratigraphic level, the palynomorph and siliceous microfossil assemblages show a marked reduction in species diversity and increase in abundance of a few opportunistic species, which are mostly cyst forms (Askin, in review; Harwood, this volume). Foraminifera are very poorly represented in the dissolution facies (between the K/T glauconite bed and 1 m below the Sobral Formation), and age-diagnostic planktonic species are absent. This lower Paleocene dissolution facies is widespread in the southern mid- to high latitudes, as it has been reported from several New Zealand sections (Webb, 1971; Strong, 1981, 1984), the Falkland Plateau (Krashennikov and Basov, 1986), and the Walvis Ridge (Hsü, 1984).

Although the K/T faunal turnover consistently occurs at the level of a resistant glauconite unit in the detailed measured sections (Fig. 5), there have been problems with its correlation and

recognition in several areas of Seymour Island. Precise mapping of this glauconite horizon has been difficult because of lateral changes in its lithification and thickness, dissection from stream drainages and a diabase dike, and the presence of more than one resistant glauconite bed within several meters above and below the faunal transition (Fig. 5) (Sadler, this volume). One example of the confusion this has caused is the reported occurrence of ammonites on a dipslope above a resistant glauconite bed (Huber, 1987). The concretions in which the ammonites occur bear Cretaceous palynomorphs, indicating that they have not been reworked (R. A. Askin, personal communication, 1985). Several meters of section above the ammonite-bearing dipslope are covered by Quaternary alluvium, preventing accurate recognition of the K/T boundary.

The highest stratigraphic occurrence of Cretaceous planktonic foraminifera is in sample nos. 411, 526, and 535 of SI-K43, all located from within 1 to 4 m below a glauconite bed (Figs. 3, 5–7). These samples yield the Upper Cretaceous planktonic species *Heterohelix globulosa, Globigerinelloides multispinatus, Hedbergella monmouthensis,* and *Globotruncanella?* sp., in addition to diverse Cretaceous benthic assemblages. The overlying stratigraphic interval (SI-T1) contains few poorly preserved agglutinated foraminifera (Huber, 1986). Other calcareous microfossils (e.g., ostracodes, juvenile molluscs, echinoid spines, calcispheres, calcareous nannoplankton) are either absent or poorly preserved within this sequence (Table 2).

Foraminifera diagnostic of the Danian were recovered from sea cliff sample no. 539, collected 1 m below the López de Bertodano/Sobral Formation contact, and samples nos. 515 and 540, from Unit 1 of the basal Sobral Formation (Fig. 7). The planktonic species *Globastica daubjergensis,* found in all three samples, is widely recognized as a Danian index fossil (e.g., see Troelsen, 1957; Berggren, 1962; Hornibrook, 1972; Hansen, 1970; Blow, 1979). No other age diagnostic foraminifera have been found above the K/T glauconite bed.

Other specimens making their first stratigraphic appearance

TABLE 5. DANIAN FORAMINIFERA RECOVERED FROM SAMPLES 539 AND 540 OF THE UPPERMOST LÓPEZ DE BERTODANO AND SOBRAL FORMATIONS ON SEYMOUR ISLAND*

Sample 539 (n = 81, 1,200 gm of sediment)	Sample 540 (n = 463, 1,150 gm of sediment)
Buliminella procera n. sp. (32%)+	*Buliminella procera* n. sp. (82%)
Allomorphina subtriangularis (Kline) (22%)†	*Praebulimina midwayensis* (Cushman and Parker) (3%)
Nonionella sp. 2 (10%)†	*Nonionella* sp. 2 (3%)
Globastica daubjergensis (Brönnimann) (6%)†	*Lenticulina macrodisca* (Reuss) (3%)
Reophax sp. (5%)†	*Pullenia jarvisi* Cushman (2%)
Hyperammina elongata Brady (4%)	*Quadrimorphina allomorphinoides* (Reuss) (2%)
Dentalina basiplanata Cushman (4%)	*Pyrulina cylindridoides* (Roemer) (1%)
Stilostomella impensia (Cushman) (4%)	*Alabamina creta* (Finlay) (1%)
Lenticulina muensteri (Roemer) (2%)†	*Anomalinoides piripaua* (Finlay) (1%)
Bolivina pustulata n. sp. (2%)	*Globoconusa daubjergensis* (Brönnimann) (1%)
Trochammina globigeriniformis (Parker and Jones) (1%)†	*Dentalina basiplanata* Cushman (<1%)
Dentalina lorneiana d'Orbigny (1%)	*Praebulimina trihedra* (Cushman) (<1%)
Lagena apiculata (Reuss) (1%)	*Bolivina pustulata* n. sp. (<1%)
Saracenaria triangularis (d'Orbigny) (1%)	*Lagena apiculata* (Reuss) (<1%)
Cibicides seymouriensis n. sp. (<1%)	
Stilostomella pseudoscripta (Cushman) (<1%)	
Citharina plumoides (Plummer) (<1%)	
Dentalina sp. (<1%)†	

*Percentage abundance in each sample are given.
†Absent from Cretaceous strata.

in SI-T2 (Table 5) include cosmopolitan species that range from the Upper Cretaceous through Lower Tertiary and several new or indeterminant species. The calcareous benthic *Buliminella procera* n. sp. is conspicuously abundant in the Danian samples. Agglutinated taxa are very rare above the K/T glauconite bed, contrary to their consistent occurrence in the Cretaceous assemblages.

Paleoecology

Introduction. Ecology studies of Recent foraminifera have shown that the geographic and bathymetric distribution of benthic assemblages are controlled by a complex interaction of numerous physicochemical parameters, which may be indiscernible in the fossil record (Douglas, 1979; Douglas and Heitman, 1979; Buzas and Culver, 1980). There are few living species known to be restricted to absolute depth and temperature ranges in sublittoral environments (Culver and Buzas, 1983). Among fossil assemblages, the number of taxa used as reliable paleoecologic sensors is further reduced. The taxonomic uniformitarian approach to paleoenvironmental analysis cannot be applied with confidence above the species level, since adaptations to particular environments may have changed through time. Interpretations of ancient environments are further complicated by taphonomic and diagenetic processes, which often result in considerable alteration of the biocoenosis.

Many of the Seymour Island samples have been strongly

affected by diagenesis, as shown by the higher stratigraphic index values of solution-resistant species (Table 3). Because of nonuniform sample size and variation in the degree of outcrop weathering throughout the measured sections, detailed paleoenvironmental study based on the present data is not possible. Nevertheless, some generalizations can be made on a broader scale, based on the proportions of agglutinated, calcareous benthic, and planktonic species and their relative abundance.

Species diversity in individual samples. As discussed earlier, most samples yielded wholly agglutinated assemblages or agglutinated assemblages with few poorly preserved calcareous benthic taxa (e.g., no. 539 in Fig. 8), and no other calcareous microfossils. The relative abundance of the represented foraminiferal species in these samples is the same as in the high-diversity samples. Sliter (1975) designated these types of faunas as "residue assemblages." Solution of the thinner walled taxa could have occurred at the sediment/water interface, after burial, or during recent exposure to surficial weathering. Cretaceous samples yielding high-diversity assemblages, such as no. 415 (Fig. 8), more accurately represent the biocoenosis. They are typically dominated by more than 30 calcareous benthic species, fewer than 10 agglutinated species occur in low to moderate abundance, and planktonic species diversity is low (fewer than five species), with very low abundance.

The preservation potential of the calcareous microfossils may have depended on the bulk carbonate availability in the surrounding sediments, which probably varied with changes in sedimentation rates. The high-diversity foraminiferal faunas generally occur in carbonate-rich sediments that bear numerous shelly fossils, whereas shelly material is sparse or absent in beds yielding foraminiferal residue assemblages. The highest concentration of shelly fossils occurs along hard ground horizons, which are inferred to represent periods of slow sedimentation (Sadler, this volume). On the other hand, intervals of low shelly fossil abundance may correspond with periods of high sedimentation rates. Pore waters flowing through these poorly fossiliferous beds were probably undersaturated in calcium carbonate, resulting in dissolution of the calcareous microfossil constituents.

Samples showing significant deviation from the typical high-diversity and residue assemblages are rare in the Cretaceous sections on Seymour Island. Several samples (nos. 14, 386, 414, 349) have an unusually high abundance and high diversity of coarse-grained, agglutinated specimens, which co-occur with few or no calcareous taxa. Sample nos. 14, 386, and 414 occur in *Rotularia* units 1 through 4, which are interpreted as nearshore, estuarine deposits (Macellari, 1986; this volume). Sample no. 349 occurs 40 cm below the K/T boundary. These assemblages are predominantly composed of *Hyperammina, Ammodiscus, Reophax, Bolivinopsis, Rzehakina, Haplophragmoides, Alveolophragmium, Cyclammina, Trochammina, Budashevaella,* and *Karreriella,* and may represent lag accumulates from sediment winnowing, or opportunistic blooms occurring during periods of high turbidity and sedimentation. There are, however, no apparent differences in grain size or bedding structure at these sample localities, as compared to sample localities bearing the high-diversity and residue assemblages.

Paleocene samples collected from Unit 1 of the Sobral Formation differ from the Cretaceous samples in the near-absence of agglutinated specimens and the low species equitability. Specimens of *Buliminella procera* dominate the Sobral samples, composing as much as 82 percent of the assemblage in sample no. 540 (Table 6; no. 540 in Fig. 8). Several factors indicate that Unit 1 was deposited in a low-oxygen, chemically reducing environment: (1) preservation of fine-scale primary sedimentary structures (thin, horizontal laminae); (2) absence of a bioturbating infauna; (3) occurrence of pyrite framboids within foraminiferal tests and in the sediments; and (4) occurrence of low-diversity, high-dominance foraminiferal assemblages with a large standing stock of small, thin-walled, and high-spired buliminellids.

In a study of foraminiferal morphologies in anoxic environments, Bernhard (1986) demonstrated that small, thin-walled tests with a high surface area-to-volume ratio often dominate in anoxic environments because of an enhanced ability for oxygen acquisition. The foraminifera occur in large standing stocks with low-diversity assemblages. The Sobral Formation foraminifera share strong similarities with Bernhard's characteristic assemblages from low-oxygen environments.

Composite diversity plot. A diversity plot for agglutinated, calcareous benthic, and planktonic foraminifera, compiled from the composite range chart (Table 2), is depicted in Figure 9. The

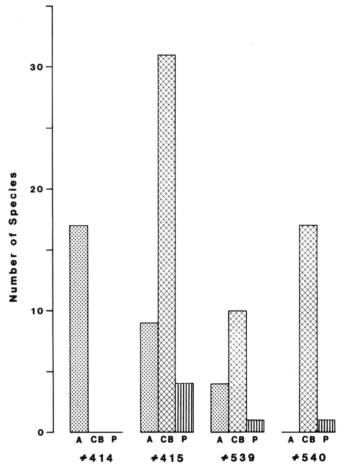

Figure 8. Foraminiferal species diversity graph for individual Seymour Island samples, showing examples of the variation in faunal composition. Sample 414 bears a wholly agglutinated fauna, 415 yields a high diversity assemblage, 539 bears a moderately diverse fauna, and 540 yielded no agglutinated taxa. A, agglutinated; CB, calcareous benthic; P, planktonic.

histograms show three orders of magnitude, including high diversity (>40 calcareous benthic species), moderate diversity (15–40 benthic species), and low diversity or wholly agglutinated assemblages (<15 calcareous benthic species). The diversity fluctuations are not necessarily reflective of primary paleoenvironmental conditions, as bias due to sample collection from weathered outcrops, small sample size, and sampling gaps must be considered (e.g., compare the number of samples per stratigraphic interval in Table 1 with Fig. 9). Nevertheless, several stratigraphic trends can be recognized. First, stratigraphic intervals K1 through K7 yield very low diversity planktonic assemblages and low- to moderate-diversity benthic faunas. This portion of the sequence corresponds with Macellari's (1986) *Ro-*

TABLE 6. DISTRIBUTION AND AGE RANGES OF BENTHIC FORAMINIFERA RESTRICTED TO SEVERAL LOCALITIES WITHIN THE AUSTRAL PROVINCE

Species Names	Age Ranges	References
Cyclammina cf. *complanata* Chapman	Late Campanian to Maastrichtian Middle(?) Campanian to Paleocene Paleocene to early(?) Eocene	New Zealand (Webb, 1971) James Ross Island region (Huber and Webb, 1986; this study) Southern Australia (Ludbrook, 1977)
Gaudryina healyi Finlay	Late Campanian to Maastrichtian Early(?) to late Maastrichtian Late Campanian Early Maastrichtian	New Zealand (Webb, 1971) Lord Howe Rise (Webb, 1973b) James Ross Island region (Huber and Webb, 1986; this study) Southern Argentina (Malumian and Masiuk, 1976)
Dorothia elongata Finlay	Late Campanian to late Maastrichtian Early(?) to late Maastrichtian Middle(?) to late Campanian	New Zealand (Webb, 1971) Lord Howe Rise (Webb, 1973b) James Ross Island region (Huber and Webb, 1986; this study)
Karreriella aegra Finlay	Late Campanian to late Maastrichtian	New Zealand (Webb, 1971), James Ross Island region (Huber and Webb, 1986; this study)
Frondicularia rakauroana (Finlay)	Late Maastrichtian Early(?) to late Maastrichtian Late Campanian to early Maastrichtian	New Zealand (Webb, 1971) Lord Howe Rise (Webb, 1973b) James Ross Island region (Huber and Webb, 1986; this study)
Buliminella creta (Finlay)	Late Campanian to late Maastrichtian Paleocene	James Ross Island region (Huber and Webb, 1986; this study) New Zealand (Webb, 1971), Lord Howe Rise (Webb, 1973b)
Bolivinoides draco (Marsson) *doreeni* Finlay	Late Maastrichtian	New Zealand (Webb, 1971), Lord Howe Rise (Webb, 1973b), Southern Chile (Charrier and Lahsen, 1969)
Alabamina creta (Finlay)	Late Campanian to late Maastrichtian Early(?) to late Maastrichtian	New Zealand (Webb, 1971), James Ross Island region (Huber and Webb, 1986; this study) Lord Howe Rise (Webb, 1973b)

tularia units 1 through 4, which have very low molluscan diversity. The high relative abundance of acanthomorph acritarchs and nonmarine palynomorphs (R. A. Askin, this volume) and predominance of large, coarse-grained agglutinated foraminifera are consistent with Macellari's (this volume) interpretation of nearshore conditions for these stratigraphic intervals. The report of high diatom diversity within SI-K3 and SI-K7 (Harwood, this volume) indicates normal, open marine conditions for these units.

Stratigraphic intervals K8-K14, which correspond with Macellari's *Rotularia* units 4 through 6, yield moderate- to high-diversity foraminiferal assemblages (Fig. 9). From SI-K15 through SI-K28, foraminiferal diversity is predominantly in the low to moderate range, with the exception of SI-K24, which shows a high-diversity peak. The low-diversity assemblages of these intervals probably reflect poor sample coverage and collec-

tion from weathered outcrops in low-relief, poorly exposed areas. The remaining stratigraphic intervals, from SI-K29 through SI-K43, all yielded moderate- to high-diversity faunas, with maximum diversity occurring just below the Cretaceous/Tertiary transition. The foraminiferal and siliceous microfossil distributions indicate a mid- to outer shelf depositional setting for SI-K8 through SI-K43.

Foraminifera are quite rare in the dissolution facies, SI-T1 (Fig. 9). Of 41 samples processed for microfossils, 20 were barren of foraminifera and 21 yielded few poorly preserved agglutinated specimens. Samples from Unit 1 of the Sobral Formation (SI-T2) contain higher diversity faunas than in SI-T1, but the diversity levels reached in the underlying Cretaceous intervals are never attained, and agglutinated taxa are rare. No foraminifera have been recovered from above SI-T2.

FORAMINIFERAL DIVERSITY

NUMBER OF SPECIES

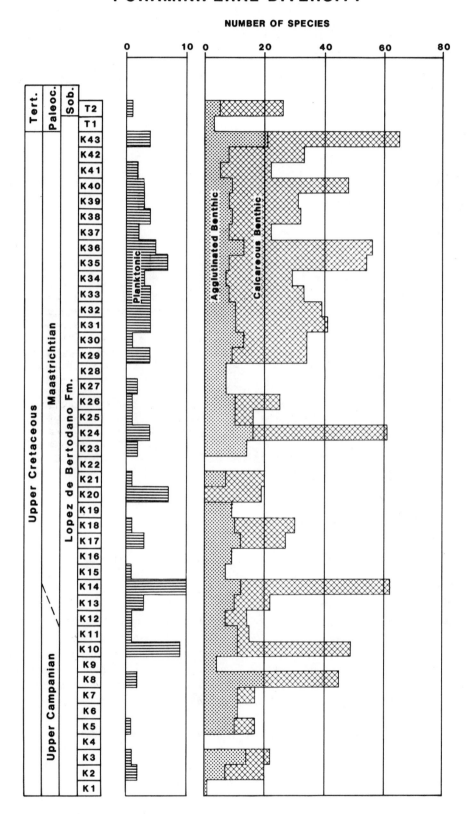

Figure 9. Composite foraminiferal species diversity graph for agglutinated, calcareous benthic, and planktonic species in the Seymour Island Marambio Group sequence.

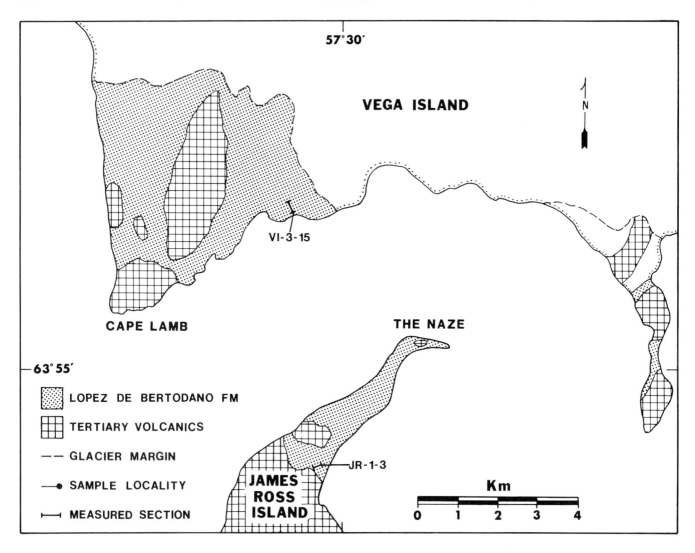

Figure 10. Geologic map and sample station localities for The Naze and Cape Lamb sections.

FORAMINIFERA FROM CAPE LAMB, VEGA ISLAND

The López de Bertodano Formation outcrop at Cape Lamb, Vega Island (Fig. 10) has yielded well-preserved and diverse foraminiferal assemblages very similar to the Cretaceous fauna from Seymour Island. Two species, *Lenticulina williamsoni* and *Spiroplectammina vegaensis,* do not occur on Seymour Island, whereas the remaining 15 agglutinated, 38 calcareous benthic, and 3 planktonic taxa are common to both areas (Appendix 2). The faunal character of the Cape Lamb foraminifera primarily differs from the Seymour Island assemblages by the high relative abundance of *Spiroplectammina vegaensis, Gaudryina healyi,* and the Nodosariidae, which together make up as much as 60 percent of the total fauna.

The Cape Lamb foraminifera share the highest degree of similarity to assemblages recovered from the lower part of the López de Bertodano Formation on Seymour Island. This is primarily based on correlation of agglutinated assemblages with *Gaudryina healyi, Spiroplectammina spectabilis,* and *Rzehakina epigona.* Planktonic species of *Heterohelix globulosa, Globigerinelloides multispinatus,* and *Rugoglobigerina*? sp. 1, which range through most of the Cretaceous Seymour Island sequence, also occur in the Cape Lamb beds. Evidence that the Cape Lamb beds are no older than middle Campanian is afforded by the range of *Gaudryina healyi* in New Zealand and the Lord Howe Rise, where this species is used as an index fossil for the Haumurian Stage (mid-upper Campanian through Maastrichtian) (Webb, 1971, 1973b). According to Askin (1983), palynomorphs recovered from the Cape Lamb sequence also indicate a late Campa-

Figure 11. López de Bertodano Formation outcrop at The Naze showing fossiliferous sediments (lower part of photo) and Pliocene basalt cliff.

nian to possibly earliest Maastrichtian age. These floral assemblages compare closely with the lowermost Seymour Island palynofloras (R. A. Askin, personal communication, 1986).

In the absence of more precise chronostratigraphic data, there remains the possibility that the Cape Lamb strata are slightly older than the lowest Seymour Island beds. A detailed study of the Snow Hill Island sequence, which includes continuous exposure of sediments older than those preserved on Seymour Island, may reveal the relative stratigraphic position of the Cape Lamb beds.

FORAMINIFERA FROM THE NAZE, JAMES ROSS ISLAND

Foraminifera recovered from a López de Bertodano Formation exposure at The Naze, James Ross Island (Figs. 10, 11) are represented by 19 agglutinated species and 2 calcareous benthic species preserved as broken internal molds (Appendix 3). The fauna differs from the other localities in that it is dominated by numerous specimens of *Cyclammina* cf. *C. complanata, Reophax texanus,* and *Karreriella aegra,* with specimens of *Alveolophragmium macellarii, Trochammina triformis,* and *Dorothia elongata* occurring in moderate abundance. Foraminifera that do not occur elsewhere in the James Ross Island region include *Thurammina papillata* and *Dorothia elongata.* The biocoenosis of The Naze assemblage probably included a much higher percentage of hyaline taxa, but these have been lost to postdepositional solution, as indicated by the poor preservation of the several calcareous specimens represented.

The foraminiferal fauna from The Naze shows some similarity to those from Cape Lamb and the basal section on Seymour Island, with occurrences of *Rzehakina epigona, Spiroplectammina spectabilis,* and *Gaudryina healyi* at all three localities and the presence of *Spiroplectammina vegaensis* at The Naze and Cape Lamb. Although *Dorothia elongata* is used as an index fossil for the New Zealand Haumurian Stage (Webb, 1971, 1973b), its restricted occurrence in The Naze beds may indicate that it is either facies-controlled or it has a diachronous range

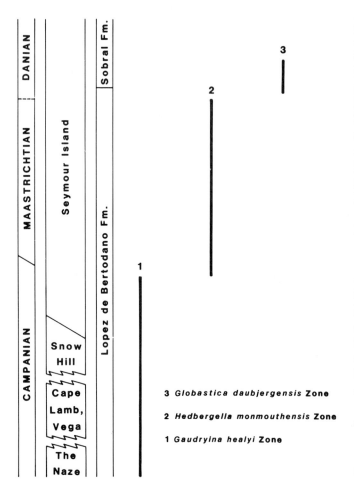

3 *Globastica daubjergensis* Zone

2 *Hedbergella monmouthensis* Zone

1 *Gaudryina healyi* Zone

Figure 12. Foraminiferal zonation for López de Bertodano Formation sequences at The Naze (James Ross Island), Cape Lamb (Vega Island), Snow Hill Island, and López de Bertodano and Sobral Formations on Seymour Island.

relative to its distribution in New Zealand. The Naze beds are approximated as ?mid- to late Campanian in age, based on foraminiferal correlation with the basal Seymour Island beds and palynomorph data (Askin, 1983). The stratigraphic position of The Naze beds relative to the Cape Lamb strata cannot be ascertained on the basis of foraminifera.

FORAMINIFERA FROM SNOW HILL ISLAND

Samples taken from Snow Hill Island were collected from weathered surface outcrops, and as a result, the recovered foraminiferal faunas are nondiagnostic residue assemblages, consisting only of agglutinated taxa (Appendix 4). None of the Snow Hill Island foraminifera provides a basis for correlation with any particular stratigraphic interval on Seymour Island, as they are all

long-ranging species. It is apparent from correlation of the Snow Hill and Seymour Island ammonite faunas (Macellari, 1985b) and physiographic correlation of the Cretaceous sediments (Fig. 1) that both islands share a thick sequence of the same stratigraphic intervals (approximately SI-K1 through SI-K8). Nevertheless, several hundred meters of the lowermost strata on Snow Hill Island are not represented on Seymour Island. A more careful sampling of this interval, leading to better microfossil recovery, may extend the stratigraphic ranges of a number of foraminiferal taxa; moreover, this may lead to a better understanding of the stratigraphic relationships of The Naze and Cape Lamb beds.

FORAMINIFERAL ZONATION

The following biostratigraphic zones are defined for the James Ross Island region based on the ranges of foraminifera in the upper Campanian through Paleocene sequence on Seymour Island. The zonal boundaries are defined by the first and last occurrences of selected taxa, and do not necessarily represent the total stratigraphic range of any particular species. None of the zonal boundaries correspond with a recognizable stratigraphic break in the sequence (Fig. 12). The ranges and relative positions of all species are shown on the range chart (Table 2).

Gaudryina healyi Assemblage-zone

Base. The lowest occurrence of *Cyclammina* cf. *C. complanata* Chapman.

Top. The lowest occurrence of *Hedbergella monmouthensis* Olsson.

Association. It is defined by the association of *Gaudryina healyi* Finlay, *Rzehakina epigona* (Rzehak), *Spiroplectammina spectabilis* Grzybowski, and large, coarsely agglutinated specimens of *Reophax texanus* Cushman & Waters, *Alveolophragmium macellarii* n. sp., and *Cyclammina* cf. *C. complanata* Chapman. The calcareous benthic component of this zone includes mostly large nodosariids and specimens of *Cibicides seymouriensis* n. sp. *Anomalinoides pirpuaua* Finlay, and *Gyroidinoides nitidus* (Reuss). Planktonic foraminifera include rare specimens of *Hedbergella* sp., *Rugoglobigerina*? sp. 1, *Rugotruncana*? sp. and *Globigerinelloides multispinatus* (Lalicker).

Age. ?Mid to late Campanian.

Occurrence. This zone includes all López de Bertodano Formation strata at The Naze (James Ross Island), Cape Lamb (Vega Island), and SI-K1 through SI-K10 on Seymour Island.

Discussion. The *Gaudryina healyi* Zone of Malumian and Masiuk (1976) was defined by the range of the nominal taxon and its association with two indeterminate species of *Ramulina*. It is modified in this study because of the better preservation and higher diversity of the associated fauna in the López de Bertodano Formation. Malumian and Masiuk (1976) tentatively assigned their *Gaudryina healyi* Zone to the Campanian–lower Maastrichtian. It is correlated with the *Rzehakina epigona* Zone of Webb

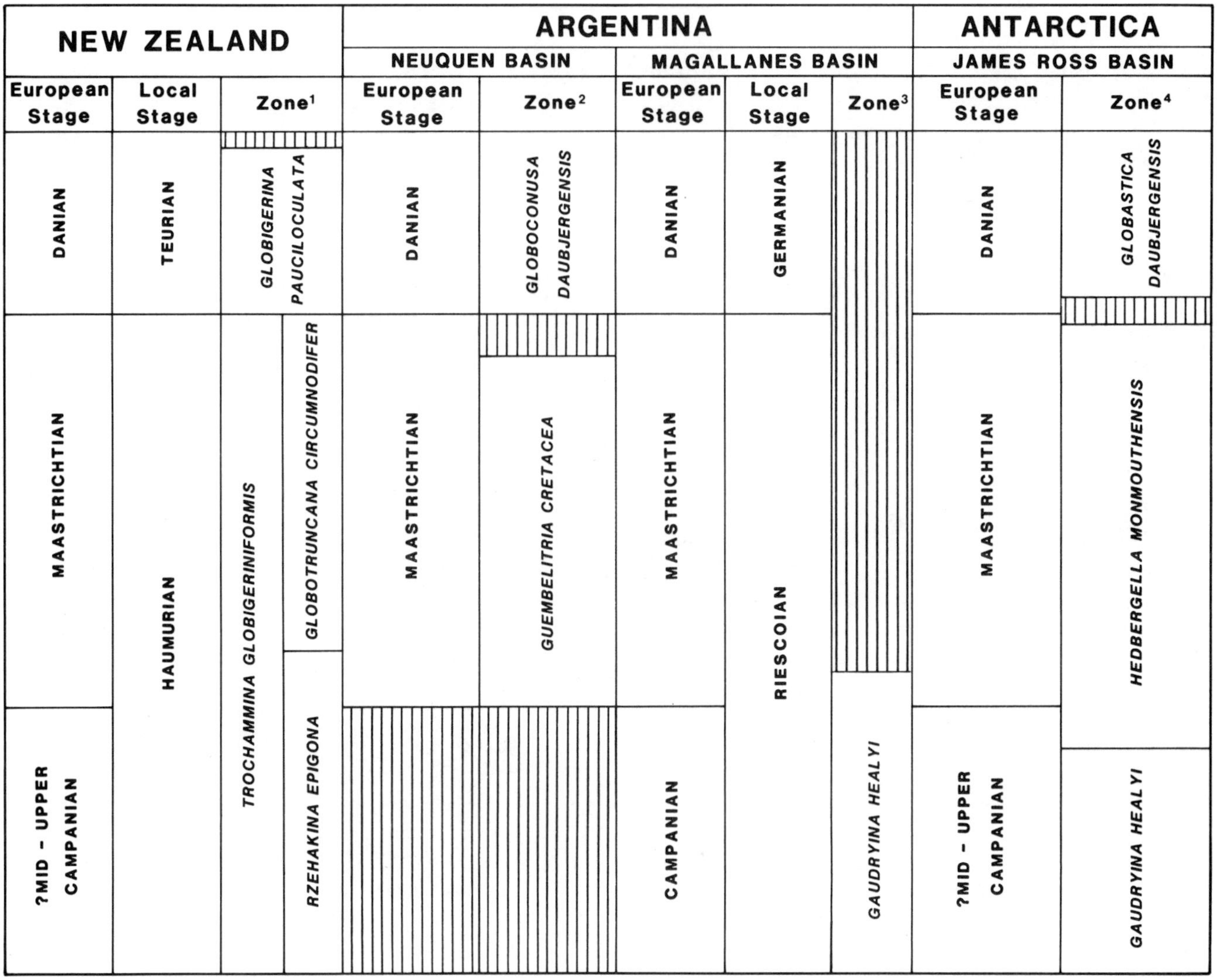

NEW ZEALAND			ARGENTINA					ANTARCTICA	
			NEUQUEN BASIN		MAGALLANES BASIN			JAMES ROSS BASIN	
European Stage	Local Stage	Zone[1]	European Stage	Zone[2]	European Stage	Local Stage	Zone[3]	European Stage	Zone[4]
DANIAN	TEURIAN	GLOBIGERINA PAUCILOCULATA	DANIAN	GLOBOCONUSA DAUBJERGENSIS	DANIAN	GERMANIAN		DANIAN	GLOBASTICA DAUBJERGENSIS
MAASTRICHTIAN	HAUMURIAN	TROCHAMMINA GLOBIGERINIFORMIS / GLOBOTRUNCANA CIRCUMNODIFER	MAASTRICHTIAN	GUEMBELITRIA CRETACEA	MAASTRICHTIAN	RIESCOIAN	GAUDRYINA HEALYI	MAASTRICHTIAN	HEDBERGELLA MONMOUTHENSIS
?MID – UPPER CAMPANIAN		RZEHAKINA EPIGONA			CAMPANIAN			?MID – UPPER CAMPANIAN	GAUDRYINA HEALYI

Figure 13. Comparison of Campanian through Danian foraminiferal zonations developed for several southern high latitude sites. 1, Webb, 1971; Jenkins, 1966; 2, Bertels, 1970a; 3, Malumian and Masiuk, 1976; 4, this study.

(1971) (Fig. 13). The nominal species ranges throughout the Haumurian Stage (upper Campanian through upper Maastrichtian) in New Zealand and occurs in Maastrichtian sediments at the Lord Howe Rise (Webb, 1971, 1973b). Its absence from Maastrichtian sediments on Seymour Island indicates that its distribution was probably environmentally controlled.

Hedbergella monmouthensis Range-zone

Base. The lowest occurrence of *Hedbergella monmouthensis.*

Top. The highest occurrence of *Hedbergella monmouthensis.*

Age. Late Campanian through late Maastrichtian.

Occurrence. This zone includes all strata on Seymour Island from SI-K10 through SI-K43.

Discussion. This zone is characterized by the restricted occurrence of *Bolivina incrassata* Reuss, *Dorothia paeminosa* n. sp., *Heterohelix glabrans* Cushman, *Guembelitria cretacea* Cushman, *Rugoglobigerina rugosa* (Plummer), and *Globotruncanella minuta* Caron & Gonzalez Donoso. Diverse assemblages of long-ranging agglutinated and calcareous taxa occur throughout this zone on Seymour Island. It is correlated with the *Globotruncana circumnodifer* Zone of Webb (1971) and the *Guembelitria cretacea* Zone of Bertels (1970a) (Fig. 13). Olsson (1964, personal communication, 1986) has reported that *Hedbergella*

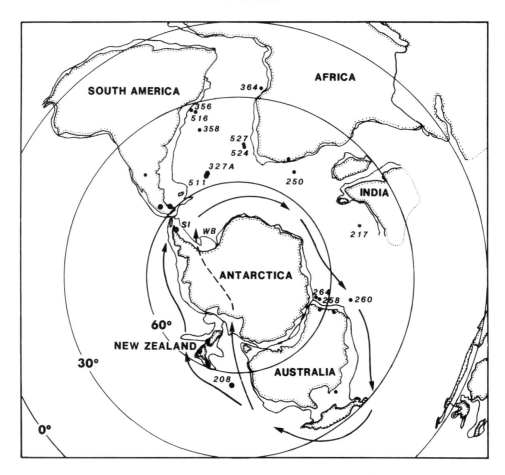

Figure 14. Paleogeographic distribution of southern latitude land-based and Deep Sea Drilling Project sites that have yielded Campanian-Maastrichtian Austral Province foraminiferal faunas (large dots) or Transitional Province assemblages (small dots). Arrows indicate oceanic surface circulation routes as inferred from foraminiferal biogeography. Continental reconstruction map for 80 Ma from Smith and others (1981). SI, Seymour Island; WB, Weddell Basin.

monmouthensis is confined to the Maastrichtian in the North Atlantic region.

Globastica daubjergensis Range-zone

Base. Lowest occurrence of *Globastica daubjergensis* Brönnimann.

Top. Highest occurrence of *Globastica daubjergensis* Brönnimann.

Age. Danian (early Paleocene).

Occurrence. This zone is restricted to SI-K2 of the uppermost López de Bertodano Formation and Unit 1 of the Sobral Formation on Seymour Island.

Discussion. The *Globastica daubjergensis* Zone is correlated with the *Globigerina daubjergensis* Zone of Berggren (1962), the *Globoconusa daubjergensis* Zone of Bertels (1970a), the *Globigerina pauciloculata* Zone of Jenkins (1966), and the

P1 and P2 zones of Blow (1979). On Seymour Island, the nominal taxon occurs with abundant specimens of *Buliminella procera* n. sp.

FORAMINIFERAL BIOGEOGRAPHY

Southern Hemisphere paleogeography

Southern Hemisphere continental reconstructions for the Late Cretaceous indicate that the James Ross Island region occupied a latitudinal position similar to that of today, i.e., between 60 and 64°S (Norton, 1982; Lawver and others, 1985). The paleogeographic reconstruction of the Gondwana continents for 80 Ma by Smith and others (1981) show that all of Antarctica, southern Australia, and New Zealand were positioned near or within the Antarctic Circle (Fig. 14). Changes in the Cretaceous paleogeography of the Southern Hemisphere are associated with the breakup of the Gondwana continents. This began with the separation of South America from Africa, and India from Antarctica-Australia during the Early Cretaceous (Larson and Ladd, 1973; Markl, 1974; Larson, 1977). This was followed by the migration of New Zealand from 80°S, at 95 Ma (Oliver and others, 1979) to 62°S by 75 Ma (Grindley and others, 1977). Australia began separating from Antarctica about 80 Ma (Cande and Mutter, 1982) and moved northward as New Zealand continued its northward drift.

The Cretaceous paleogeography of the Antarctic continent is largely conjectural because of the lack of information from the continental interior. Drewry (1983) showed that a considerable portion of the Antarctic interior would still lie below sea level after removal of all ice and subsequent isostatic uplift. Webb and others (1984) suggested that marine sedimentation occurred in these basins during parts of the late Mesozoic and Cenozoic. Late Cretaceous biogeographic distribution patterns of southern high latitude marine molluscs (Zinsmeister, 1982; Macellari, 1985b) and foraminifera (Sliter, 1976; Huber and Webb, 1986) also indicate the presence of marine communication routes between East and West Antarctica (Fig. 14). Occurrence of reworked Upper Cretaceous foraminifera in glaciogene sediments from above 75°S in the McMurdo region (Webb and Neall, 1972; Webb and others, 1984), and at DSDP Site 270 in the Ross Sea (Leckie and Webb, 1986) strengthens the argument for the former existence of trans-Antarctic seaways.

The Antarctic Peninsula is believed to have been an emergent arc terrane during the Late Cretaceous (Farquharson, 1982; Elliot, 1983; Macellari, this volume). It is uncertain whether this existed as a continuous landmass, with substantial subaerial relief, or a series of islands separated by shallow marine basins. Woodburne and Zinsmeister (1984) favored a relatively continuous land connection, from southern South America to Australia, during the Late Cretaceous to explain the origin of Seymour Island and Australian marsupials. On the other hand, Sliter (1976) and Huber and Webb (1986) suggested the presence of marine connections across the Antarctic Peninsula as possible communication routes for Campanian-Maastrichtian, southern high-latitude foraminifera. In the absence of direct geologic evidence, both reconstructions are viable hypotheses.

Campanian-Maastrichtian Austral Province

The biogeographic distribution of modern planktonic foraminiferal provinces roughly parallels the latitudinal arrangement of climatic belts (Bradshaw, 1959; Bé and Tolderlund, 1971; Bé, 1977). In modern oceans, taxonomic diversity is highest at the equator (15 dominant species) and lowest at the poles (1 dominant species) (Bé, 1977). Morphology varies similarly with complex (flattened, strongly ornamented, thick-walled) forms in warmer waters and primitive (globular, weakly ornamented, thin-walled) types in the higher latitudes.

Late Cretaceous planktonic foraminifera also show a poleward decrease in taxonomic diversity (Stehli and others, 1969;

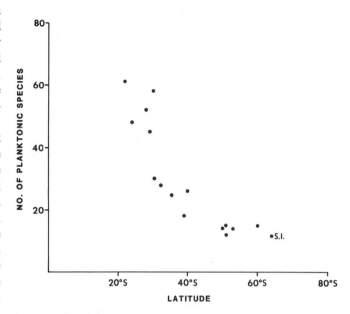

Figure 15. Planktonic foraminiferal species diversity gradient for the late Campanian-Maastrichtian in the Southern Hemisphere. The dramatic drop in diversity at about 30°S may delineate the boundary between the Tethyan and Transitional Provinces. Austral Province faunal diversity is between 14 and 20 species. S.I., Seymour Island.

Douglas, 1972). An example for some upper Campanian-Maastrichtian sites in the Southern Hemisphere is shown in Fig. 15. Highest diversity assemblages occur in low paleolatitudes and are dominated by species of *Globotruncana, Abathomphalus, Archaeoglobigerina, Rugoglobigerina, Pseudoguembelina, Pseudotextularia, Raceguembelina, Plummerita,* and *Trinitella* (e.g., see Premoli-Silva and Boersma, 1977). Scheibnerová (1971, 1973) and Sliter (1976) included these in the tropical Tethyan Province. These same authors included lower diversity planktonic faunas of similar composition in the warm-temperate Transitional Province. Planktonic species diversity shows a rapid drop at about 30°S (Fig. 15). Douglas (1972) recognized a similar rapid decrease between 40 and 60°N and suggested that the diversity break may reflect the limit of the Late Cretaceous subtropics. This approximates the boundary between the Tethyan and Transitional Provinces, as defined by Scheibnerová (1973) and Sliter (1976).

Low-diversity primitive assemblages, dominated by species of *Heterohelix, Globigerinelloides, Hedbergella,* and *Rugoglobigerina,* ranged into the circum-Antarctic regions during the Late Cretaceous (Webb, *in* Hornibrook, 1969; Webb, 1973a, b). Scheibnerová (1971, 1973), Sliter (1976), and Krasheninnikov and Basov (1983, 1986) included these in the cool-temperate Austral Province. These authors defined the Austral Province on the absence of Tethyan planktonic indicator species, rather than the presence of provincially restricted species. Localities considered to fall within the Campanian-Maastrichtian Austral Province include the Falkland Plateau (South Atlantic), southern South America, New Zealand, and Antarctica (Sliter, 1976). Webb (1973a) noted the faunal similarity of DSDP Site 208 (Tasman Sea) foraminifera to other southern high latitude assemblages. It is included within the Austral Province in this study, although Sliter (1976) previously placed it in the Transitional Province. The boundaries between all three Southern Hemisphere bioprovinces are considered to parallel the Late Cretaceous climatic belts, but their geographic limits are gradational, and are presently not well defined.

Similar morphovariants of *Rugoglobigerina, Globotruncanella,* and *Hedbergella,* occur at several Austral Province sites, but these forms need further study to evaluate their taxonomic status before their biogeographic distribution can be determined. The absence of globotruncanid species from the James Ross Island region is probably due to the subpolar location as well as the prevalent nearshore environmental conditions, which would have inhibited their reproductive cycle (see Caron and Homewood, 1984).

Benthic foraminifera restricted to the Austral Province are shown on Table 6, with their reported age ranges and geographic distribution. In addition to the James Ross Island region, several of the benthic species occur in New Zealand (Webb, 1971), the Lord Howe Rise (Webb, 1973a, b), and southern South America (Malumian and Masiuk, 1976; Charrier and Lahsen, 1969). Endemic species that do not occur outside a single Austral locality are not listed. The benthic foraminiferal species closely parallel the limits of the macrofossil Weddellian Province, which also composed the Late Cretaceous–Early Tertiary southern circum-Pacific margin (Zinsmeister, 1982). Isolation due to the distribution of Gondwana land masses, prevailing ocean circulation, and cool-temperate conditions of the southern high latitudes probably contributed to development of the Austral Province in the Late Cretaceous.

SYSTEMATIC PALEONTOLOGY

The following species descriptions are systematically arranged according to the classification of Loeblich and Tappan (1964), with some modifications from Loeblich and Tappan (1984) and other subsequent taxonomic studies. The reader is referred to Loeblich and Tappan (1964) for generic authorship, descriptions, and synonymies, except where more recent classifications are cited. In the interest of saving space, synonymies and descriptions of well-known species are kept to a minimum.

Museum numbers, dimensions, and sample localities are presented for holotypes of new species and each figured holotype. Sample localities with the prefix of JR- and VI- refer to James Ross and Vega Island, respectively; those without a prefix occur on Seymour Island. Relative stratigraphic index values from Tables 3 and 4 are given for each species, as are the range and distribution in the James Ross Island region. Relative abundances within individual samples are also stated.

The study material is deposited in the Orton Museum of Geology at The Ohio State University. The holotypes and two paratypes of each new species will be deposited in the U.S. National Nuseum, Washington, D.C. Identification of the foraminiferal species described herein has been conducted by means of comparative studies with previously published literature on Upper Cretaceous and Early Tertiary foraminifera, examination of primary and secondary type material at the U.S. National Museum, and comparison with contemporaneous circum-Antarctic faunas. These include material from Deep Sea Drilling Project Sites 208 (Lord Howe Rise), 327A, 511 (both from the Falkland Plateau), the Magellanes Basin (southern Chile), and New Zealand. Abbreviations used are as follows: OSU, Orton Museum, The Ohio State University; USNM, U.S. National Museum, Washington, D.C.; DSDP, Deep Sea Drilling Project; and SI, stratigraphic interval.

Rhabdammina sp.
Figure 16.1
Figured Hypotype. OSU 39702: length, 1.59 mm; width, 0.91 mm; López de Bertodano Formation, locality 19, Seymour Island.
Remarks. Form similar to *Hyperammina elongata* Brady except that it is larger and coarser grained.
Occurrence. Moderately common. Occurs in very low abundance from SI-K5 through SI-K43 on Seymour Island, as well as at Cape Lamb, The Naze, and Snow Hill Island.

Rhizammina algaeformis Brady, 1879
Figure 16.2
Rhizammina algaeformis BRADY 1879, p. 20, Plate 4, Figures 16, 17.
Figured Hypotype. OSU 39703: length, 0.91 mm; width, 0.19 mm; López de Bertodano Formation, locality 17, Seymour Island.
Occurrence. Common. Occurs in low abundance in Seymour Island samples, ranging from SI-K2 through SI-K43. Specimens also present in the Cape Lamb and The Naze samples.

Bathysiphon californicus Martin, 1964
Figure 16.3
Bathysiphon californicus MARTIN 1964, p. 43, Plate 1, Figure 2.
Figured Hypotype. OSU 39704: length, 1.19 mm; width, 0.31 mm; López de Bertodano Formation, locality 14, Seymour Island.
Occurrence. Rare. Occurs in very low abundance from SI-K3 through SI-K43 on Seymour Island, as well as on Snow Hill Island.

Hyperammina elongata Brady, 1878
Figure 16.4
Hyperammina elongata BRADY 1878, p. 433, Plate 20, Figure 2a-b.
Figured Hypotype. OSU 39705: length, 1.38 mm; width, 0.48 mm; López de Bertodano Formation, locality 101, Seymour Island.
Occurrence. Very common. Ranges from SI-K5 through SI-T2 on Seymour Island; also occurs in samples from Cape Lamb and The Naze.

Psammosphaera sp.
Figure 16.5

Figured Hypotype. OSU 39706: diameter, 0.41 mm; López de Bertodano Formation, locality 14, Seymour Island.

Remarks. Composed of large, irregular mafic grains; has an indefinite aperture.

Occurrence. Common. Occurs in low abundance from SI-K2 through SI-K43 on Seymour Island.

Saccammina sphaerica Sars, 1872
Figure 16.9

Saccammina sphaerica SARS 1872, p. 250.

Figured Hypotype. OSU 39707: diameter, 0.91 mm; López de Bertodano Formation, locality 17, Seymour Island.

Occurrence. Very rare. Several specimens occur in SI-K6 and SI-K8 on Seymour Island.

Thurammina papillata Brady, 1879
Figure 16.10

Thurammina papillata BRADY 1879, p. 45, Plate 5, Figures 4-8.

Figures Hypotype. OSU 39708: diameter, 1.21 mm; López de Bertodano Formation, locality JR-2, The Naze, James Ross Island.

Occurrence. Very rare. Single specimen recovered from The Naze. Holotype of this species originally described from Recent sediments; similar form reported by Webb (1971) from Haumurian (upper Campanian–Maastrichtian) Whangai Formation of New Zealand.

Ammodiscus cretaceus (Reuss, 1845)
Figure 16.8

Operculina cretaceous REUSS 1845, p. 35, plate 13, Figures 64-65.

Figured Hypotype. OSU 39709: diameter, 0.62 mm; breadth, 0.13 mm; López de Bertodano Formation, locality 185, Seymour Island.

Occurrence. Very rare. Two specimens recovered from SI-K37, sample 185, on Seymour Island; one specimen found at The Naze.

Ammodiscus pennyi Cushman & Jarvis, 1928
Figure 16.7

Ammodiscus pennyi CUSHMAN & JARVIS 1928, p. 37, Plate 12, Figures 4-5.

Figured Hypotype. OSU 39710: diameter, 4.12 mm; breadth, 0.62 mm; López de Bertodano Formation, locality 24, Seymour Island.

Occurrence. Very rare. Occurs in low abundance in samples ranging from SI-K8 through SI-K43 on Seymour Island, and samples from Cape Lamb and The Naze.

Reophax subfusiformis Earland, emend. Höglund, 1947
Figure 16.6

Reophax subfusiformis EARLAND, emend. Höglund, 1947, p. 77, 82.

Figured Hypotype. OSU 39711: length, 1.02 mm; width, 0.35 mm; López de Bertodano Formation, locality 12, Seymour Island.

Occurrence. Very rare. Occurs in very low abundance from SI-K3 through SI-K18 on Seymour Island.

Reophax texanus Cushman & Waters, 1927
Figure 16.14-15

Reophax texanus CUSHMAN & WATERS 1927, p. 82, Plate 10, Figure 2.

Figured Hypotypes. OSU 39712 (Fig. 16.14): length, 2.23 mm; width, 1.44 mm; López de Bertodano Formation, Loc. 17, Seymour Island; OSU 39713 (Fig. 15): length, 3.71 mm, width, 1.59 mm, López de Bertodano Formation, locality 19, Seymour Island.

Remarks. Specimens included in this species are quite variable in size, ranging from 0.87 to 3.80 mm in length. Degree of chamber overlap also varies considerably, with some specimens showing spherical to elongate chambers and indistinct sutures, and others having strongly overlapping chambers and depressed sutures.

Occurrence. Moderately common. Present in low to moderate abundance in Seymour Island samples ranging from SI-K3 through SI-K43; moderately abundant at Cape Lamb and abundant in The Naze samples.

Reophax sp.
Figure 32.1

Figured Hypotype. OSU 39714: length, 1.04 mm; width, 0.68 mm; López de Bertodano Formation, locality 539, Seymour Island.

Remarks. These forms are elongate, thin, irregular, and coarsely agglutinated with strongly constricted sutures; range in size from 0.68 to 1.17 mm in length and 0.44 to 0.54 mm in width.

Occurrence. Very rare. Occur in SI-T2 in low abundance, within samples 539, 407, 408, and 120, which are all Danian in age.

Rzehakina epigona (Rzehak, 1895)
Figure 17.8

Silicina epigona RZEHAK 1895, p. 214, Plate 6, Figure 1.

Figured Hypotype. OSU 39715: length, 0.45 mm; width, 0.31 mm; breadth, 0.05 mm; López de Bertodano Formation, locality 14, Seymour Island.

Occurrence. Very rare. Occurs in low to moderate abundance within samples from lowermost Seymour Island strata, SI-K3 through SI-K10, Cape Lamb, and The Naze. Considered diagnostic of the *Gaudryina healyi* Zone in the James Ross Island region.

Haplophragmoides eggeri Cushman, 1926
Figure 16.11

Haplophragmoides eggeri CUSHMAN 1926b, p. 583, Plate 15, Figure 1.

Figured Hypotype. OSU 39716: diameter, 0.79 mm; breadth, 0.48 mm; López de Bertodano Formation, locality 24, Seymour Island.

Occurrence. Common. Occurs in low abundance from SI-K2 through SI-K38.

Haplophragmoides platus Loeblich, 1946
Figure 16.12-13

Haplophragmoides platus LOEBLICH 1946, p. 134–135, Plate 22, Figure 5a-b.

Figured Hypotype. OSU 39721: diameter, 0.71 mm; breadth, 0.16 mm; López de Bertodano Formation, locality 28b, Seymour Island.

Occurrence. Rare. Occurs in very low abundance in several samples from SI-K3 through SI-K43 on Seymour Island.

Alveolophragmium macellarii n. sp.
Figures 16-16-19, 17.1-5

Figured Holotype. USNM 415748: (Fig. 17.2-4): diameter, 3.04 mm; breadth, 1.49 mm; López de Bertodano Formation, locality 19, Seymour Island.

Figured Hypotypes. OSU 39717 (Fig. 16.16-17): diameter, 1.13 mm; breadth, 0.41 mm; López de Bertodano Formation, locality 12, Seymour Island; OSU 39718 (Fig. 16.18-19): diameter, 1.38 mm; breadth, 0.49 mm; López de Bertodano Formation, locality 414, Seymour Island; OSU 39719 (Fig. 17.1): diameter, 0.94 mm, breadth, 0.62 mm; López de Bertodano Formation, locality 26, Seymour Island; OSU 39720 (Fig. 17.5): diameter, 1.72 mm; López de Bertodano Formation, locality 86, Seymour Island.

Etymology. Dedicated to Carlos Macellari for his careful field studies and paleontologic research on Seymour Island and in southern South America.

Diagnosis. Adult test very large, subject to considerable distortion, planispirally or sometimes streptospirally enrolled, 4 to 10 chambers composing the final whorl, interiomarginal aperture with bordering lip, multiple areal apertures sometimes discernable from exterior, wall coarsely arenaceous, hypodermis perforated by unbranching alveoles that do not penetrate thin epidermis.

Description. The adult test is very large, subject to considerable distortion, often resulting in axial or equatorial compression, umbilici strongly depressed in well-preserved specimens, planispirally or sometimes streptospirally coiled, biconvex, and involute. The microspheric forms have 3 whorls and 5 to 10 chambers in the final whorl; megalospheric forms have 2½ whorls and 4 to 6 chambers in the final whorl, and the sutures are poorly to well defined, straight, and radial.

The apertural face is broad, high, and rounded in axially compressed specimens, high and subrounded in equatorially compressed specimens, with small, supplementary areal apertures sometimes visible from the exterior. A narrow, interiomarginal aperture at the base of the apertural face is bordered by a narrow lip, and extends from the equatorial periphery part way toward the coiling axis.

The wall is composed mainly of scattered medium grains of quartz, with a coarsely finished exterior. The hypodermis is perforated by unbranching, coarse alveoles that do not penetrate the thin epidermis and do not interconnect. The septal walls are about one-third the interseptal width of the chamber lumina and are occasionally perforated by the areal apertures, which may bear short tubes directed toward the previous chamber lumina. The distal ends of the alveoli are sometimes visible through the thin epidermis of well-preserved specimens.

Remarks. This species differs from cyclamminids in having fewer chambers in the final whorl and a more coarsely finished wall exterior. The presence of only primary alveoles (*sensu* Brönnimann, 1951) in the outer walls of adult specimens and poorly developed areal apertures (Fig. 17.4) suggests it may be the phylogenetic link between the simply structured, nonalveolar *Haplophragmoides* and *Cyclammina*. Detailed ontogenetic study of the antarctic material, which is hampered by the rarity of undistorted specimens, may provide further information on the classification of this species.

Occurrence. Very common. Specimens restricted to the López de Bertodano Formation and occur in low to moderate abundance from SI-K2 through SI-K43 (upper Campanian through Maastrichtian) on Seymour Island; also moderately abundant in samples from The Naze, Cape Lamb, and Snow Hill Island. Occurrence of *Alveolophragmium macellarii* and *Cyclammina* cf. *C. complanata* in Campanian sediments marks the oldest reported occurrence of reticulate-walled lituolids.

Cyclammina cf. *C. complanata* Chapman, 1904
Figure 17.6

cf. *Cyclammina complanata* CHAPMAN 1904, p. 228, Plate 12, Figure 12. LUDBROOK 1977, p. 187–188, Plate 2, Figures 28-30, Plate 4, Figures 41–44.
Figured Hypotype. OSU 39722: diameter, 2.10 mm; breadth, 1.11 mm; López de Bertodano Formation, locality 19, Seymour Island.
Remarks. The antarctic specimens resemble *Cyclammina complanata* in their large size, with a tendency of becoming evolute in the final whorl (which has from 10 to 16 chambers), an alveolar hyperdermis, and a smoothly finished epidermis which is not penetrated by the alveolae. It differs in having a lower apertural face that lacks distinct supplementary areal apertures. The interiomarginal and areal apertures were recognized only in thin sectioned specimens. The test size of the antarctic material ranges from 0.85 to 3.10 mm in diameter.
Occurrence. Very common. Longest ranging and most common of the James Ross Island region foraminifera. Abundant in The Naze and Cape Lamb samples; occurs in low to moderate abundance in Snow Hill and

Seymour Island samples. On Seymour Island, it ranges from SI-K2 through SI-T2. Identical specimens, referred to as *Cyclammina* cf. *elegans* Cushman and Jarvis by Webb (1971), occur in the New Zealand upper Haumurian (Maastrichtian). Taylor (1965) and Ludbrook (1977) reported that *Cyclammina complanata* occurs only in the Paleocene to ?early Eocene in the Australian Otway Basin. *Cyclammina* occurrence in ?mid- to upper Campanian sediments of the James Ross Island region represents the oldest known record of this genus.

Cyclammina sp.
Figure 17.7

Figured Hypotype. OSU 39723: diameter, 0.49 mm; breadth, 0.21 mm; López de Bertodano Formation, locality 14, Seymour Island.
Remarks. Specimens are much smaller (between 0.40 and 0.85 mm) and more finely agglutinated than *Cyclammina* cf. *C. complanata,* with fewer chambers (7 to 10) in the final whorl. The more compressed forms resemble *Haplophragmoides platus,* but have more pronounced umbilici and are not as thin. This group may represent a juvenile form of *C.* cf. *C. complanata.*
Occurrence. Moderately common. Occurs in low to moderate abundance from SI-K6 through SI-K43 on Seymour Island and in moderate abundance in Cape Lamb and The Naze samples.

Ammobaculites fragmentarius Cushman, 1927
Figure 17.12

Ammobaculites fragmentarius CUSHMAN 1927, p. 130, Plate 1, Figure 8.
Figured Hypotype. OSU 39724: length, 0.69 mm; width, 0.38 mm; López de Bertodano Formation, locality 28b, Seymour Island.
Occurrence. Very rare. Ranges throughout much of the Seymour Island sequence, from SI-K3 through SI-K43, occurring in very low abundance.

Haplophragmium compressum Beissel, 1886
Figure 17.9-10

Haplophragmium compressum BEISSEL 1886, p. 138.
Figured Hypotype. OSU 39725: length, 2.78 mm; width, 1.98 mm, breadth, 0.54 mm; López de Bertodano Formation, locality 349, Seymour Island.
Description. The tests are large, compressed to globular, with a subrounded periphery and a streptospiral coil that may compose most of the test, later followed by one to three uniserial, equitant chambers. The chambers are indistinct to slightly inflated, and the sutures are flush to slightly depressed. The wall is composed of large, irregularly arranged, subangular quartz and mafic grains, and the surface is rough. The aperture is terminal and usually indistinct.
Remarks. The size range of the studied specimens is between 1.2 and 3 mm.
Occurrence. Very rare. Single specimens found in SI-K16 and SI-K30; occurs in low to moderate abundance in samples ranging from SI-K42 through SI-T2.

Spiroplectammina laevis (Roemer, 1841)
Figure 17.13, 18

Textularia laevis ROEMER 1841, p. 97, Plate 15, Figure 17.
Figured Hypotype. OSU 39727: length, 0.72 mm; width, 0.43 mm; breadth, 0.28 mm; López de Bertodano Formation, locality 110, Seymour Island.
Occurrence. Common. Occurs in low to moderate abundance throughout the Seymour Island sequence, from SI-K3 through SI-K43, as well as at Cape Lamb and The Naze.

Spiroplectammina cf. *S. laevis* (Roemer, 1841)
Figure 17.16-17

cf. *Textularia laevis* ROEMER 1841, p. 97, Plate 15, Figure 17.

Figured Hypotype. OSU 39728: length, 0.43 mm; width, 0.22 mm; breadth, 0.17 mm; López de Bertodano Formation, locality 38, Seymour Island.

Remarks. This form differs from *S. laevis* in that the sutures are more depressed, the test is less tapering, and the aperture is a more distinctive, low, broad arch.

Occurrence. Very rare. Single specimen found in sample 38 of SI-K24, which is Maastrichtian in age.

Spiroplectammina spectabilis Grzybowski, 1897
Figure 17.11

Spiroplectammina spectabilis GRZYBOWSKI 1897, p. 293, Plate 12, Figure 12.

Figured Hypotype. OSU 39726: length, 0.59 mm; width, 0.19 mm; López de Bertodano Formation, locality 14, Seymour Island.

Occurrence. Very rare. Restricted to lower López de Bertodano Formation, occurring in low to moderate abundance from SI-K5 through SI-K17 on Seymour Island, as well as at Cape Lamb and The Naze.

Spiroplectammina vegaensis n. sp
Figures 17.14-15, 33.1

Figured Holotype. USNM 415749 (Fig. 17.14-15): length, 0.90 mm; width, 0.54 mm; breadth, 0.38 mm; López de Bertodano Formation, locality VI-8, Cape Lamb, Vega Island.

Figured Hypotype. OSU 39893 (Fig. 33.1): length, 0.93 mm; width, 0.68 mm; López de Bertodano Formation, locality IV-8, Cape Lamb, Vega Island.

Etymology. Species named for Vega Island, the type locality.

Diagnosis. Test large, slowly to moderately tapering in side view, strongly tapered in edge view, planispiral section compressed, biserial section broad with tapered periphery, sutures indistinct, wall surface roughly finished.

Description. The tests are large, slowly tapering in edge view, slowly to moderately tapering in side view, the greatest width and breadth are at the apertural end, and the peripheral margin is compressed. The earliest chambers are indistinct, planispiral, and usually make up less than one-fourth of the test length. This is followed by two to five pairs of indistinct to distinct biserial chambers. The chambers increase gradually in size and inflation. The sutures are initially flush and indistinct, later slightly to moderately depressed, and strongly arcuate throughout (Fig. 33.1). The wall is composed of medium-size sand grains, the surface is slightly roughened, except for the apertural face, which is smooth, and the wall interior is simple. The aperture is an interiomarginal, small arch that may be slightly offset from the center of the final chamber face.

Remarks. This form is distinguished from *S. laevis* (Roemer) by its larger, less tapering, more coarsely agglutinated test. In addition, its final chambers are more inflated and the sutures are less distinct and form a less oblique angle to the long axis. The test size ranges from 0.39 to 1.35 mm in length, 0.33 to 0.67 mm in width, and 0.22 to 0.47 mm in breadth.

Occurrence. Occurs in moderate to high abundance in samples from Cape Lamb, Vega Island and The Naze, James Ross Island, within the *Gaudryina healyi* Zone, which is assigned to the upper Campanian. No specimens found on Seymour Island.

Textularia? sp.
Figure 18.1-2

Figured Hypotype. OSU 39729: length, 0.64 mm; width, 0.49 mm; breadth, 0.29 mm; López de Bertodano Formation, locality 115, Seymour Island.

Remarks. The test of this form appears to be biserial throughout, is moderately tapered and has a subangular periphery. It appears to be most similar to *Spiroplectammina vegaensis* n. sp., but the initial part of the test of *Textularia?* sp. is much more reduced in size. It is more coarsely agglutinated, lacks a planispiral coil, and is more elongate than *Spiroplectammina laevis* (Roemer).

Occurrence. Very rare. Restricted to upper portion of the López de Bertodano Formation on Seymour Island, occurring in low abundance within SI-K43.

Trochammina globigeriniformis (Parker & Jones, 1865)
Figure 18.3-4

Lituola nautiloidea Lamarck var. *globigeriniformis* PARKER & JONES 1865, p. 407, Plate 15, Figures 46-47, 96-98.

Figured Hypotype. OSU 39730: diameter, 0.79 mm; breadth, 0.57 mm; López de Bertodano Formation, locality 19, Seymour Island.

Occurrence. Common. Occurs in low abundance throughout the Seymour Island sequence, ranging from SI-K3 through SI-T2.

Trochammina ribstonensis Wickenden, 1932
Figure 18.5, 9

Trochammina ribstonensis WICKENDEN 1932, p. 90, Plate 1, Figure 12a-c.

Figured Hypotype. OSU 39731: diameter, 0.21 mm; breadth, 0.06 mm; López de Bertodano Formation, locality 14, Seymour Island.

Occurrence. Very rare. Occurs in very low abundance at The Naze and ranges from SI-K3 through SI-K27 on Seymour Island.

Trochammina triformis Sliter, 1968
Figure 18.6

Trochammina triformis SLITER 1968, p. 47, Plate 3, Figure 2.

Figured Hypotype. OSU 39732: length, 1.88 mm; width, 1.42 mm; López de Bertodano Formation, locality 414, Seymour Island.

Occurrence. Very rare. Species reported to range from Campanian through Maastrichtian on the southwest coast of North America. Sparsely distributed on Seymour Island from SI-K8 through SI-K43, occurring in low to moderate abundance. Also found in The Naze and Cape Lamb samples.

Budashevaella multicamerata (Voloshinova & Budasheva, 1961)
Figure 18.7-8

Circus multicameratus VOLOSHINOVA & BUDASHEVA 1961, p. 201, Plate 7, Figure 6, Plate 8, Figures 1a-b, 7.

Figured Hypotype. OSU 39733: diameter, 0.91 mm; breadth, 0.32 mm; López de Bertodano Formation, locality 24, Seymour Island.

Occurrence. Very rare. Restricted to lower López de Bertodano Formation on Seymour Island, occurring in low abundance, and ranging from SI-K10 through SI-K17. Also occurs in low abundance at Cape Lamb.

Gaudryina healyi Finlay, 1939
Figure 18.10-11

Gaudryina healyi FINLAY 1939, p. 311, Plate 25, Figures 34-35. MALUMIAN & MASIUK 1976, p. 186, Plate 2, Figure 2.

Figures Hypotype. OSU 39734: length, 0.77 mm; width, 0.38 mm; breadth, 0.36 mm; López de Bertodano Formation, locality VI-8, Cape Lamb, Vega Island.

Remarks. This species is characterized by its initial triangular to sub-rounded triserial portion of the test, which is followed by two to four increasingly inflated biserial chambers that are somewhat compressed in the direction of the long axis. The sutures are initially indistinct, later depressed, at about 60° from the long axis. The aperture is a low, elongate slit at the base of the final chamber, and the wall is composed of fine silt- to sand-size grains with a slightly roughened surface texture. The antarctic forms are identical to the New Zealand specimens.

Occurrence. Very rare on Seymour Island, moderate to high abundance at Cape Lamb and The Naze. Species is diagnostic of the Haumurian Stage (upper Campanian-Maastriachtian) of New Zealand and the Lord Howe Rise (Webb, 1971, 1973b). Also reported from the Campanian-lower Maastrichtian of Tierra del Fuego (Malumian and Masiuk, 1976). Occurs on Seymour Island, in low abundance within several samples of SI-K10 and is the nominal taxon for *Gaudryina healyi* Zone.

Dorothia elongata Finlay, 1940
Figure 18.17

Dorothia elongata FINLAY 1940, p. 450, Plate 62, Figures 16-17.

Figured Hypotype. OSU 39735: length, 1.13 mm; width, 0.35 mm; breadth, 0.44 mm; López de Bertodano Formation, locality JR-2, The Naze, James Ross Island.

Remarks. The antarctic specimens have between one and six pairs of biserial chambers following the initial trochospiral portion of the test; they show varying degrees of test compression; and they are finely agglutinated with a smooth test surface. Their size ranges from 0.33 to 1.13 mm in length, 0.25 to 0.35 mm in width, and 0.19 to 0.44 mm in breadth. The aperture is a low, rounded opening at the base of the final chamber. The antarctic forms are identical to the New Zealand topotypes.

Occurrence. Abundant in The Naze samples but not found elsewhere in James Ross Island region. Has been recovered from Haumurian (upper Campanian–Maastrichtian) sediments in New Zealand (Finlay, 1940; Webb, 1971) and the Lord Howe Rise, DSDP Leg 21 (Webb, 1973b).

Dorothia paeminosa n. sp.
Figures 18.12-14, 33.2-4

Figured Holotype. USNM 415750 (Fig. 18.12-13): length, 0.74 mm; width, 0.46 mm; breadth, 0.39 mm; López de Bertodano Formation, locality 148, Seymour Island.

Figured Hypotypes. OSU 39736 (Fig. 18.14): length, 0.44 mm; width, 0.37 mm; breadth, 0.24 mm; López de Bertodano Formation, locality 38, Seymour Island; OSU 39737 (Fig. 33.2): length, 0.48 mm; width, 0.34 mm; López de Bertodano Formation, locality 315, Seymour Island: OSU 39738 (Fig. 33.4): length, 0.29 mm; width, 0.23 mm; López de Bertodano Formation, locality 315, Seymour Island.

Etymology. Specific name from the Latin *paeminosus,* uneven, rough, referring to the test surface texture.

Diagnosis. Test small to medium, conical, rounded to oval in end view, biserial portion slightly compressed, sutures indistinct except on last chambers, wall of medium to coarse grains, surface roughly finished, wall interior canaliculate, aperture a low, arched slit.

Description. The test is small to medium in size, and varies in shape depending on the growth stage. Juveniles are small, globular, trochospiral throughout, with three to four chambers per whorl, or with one pair of biserial chambers; the chambers increase moderately in size, and the sutures are initially indistinct, later slightly depressed. The aperture is a small, low arch at the base of the final chamber. The adult specimens are moderately to strongly tapering, broadening from a somewhat blunt apex, initially trochospiral with three to four chambers per whorl, later with two to four pairs of biserial chambers, the chambers gradually increase in size. The sutures are initially indistinct and flush, later becoming slightly depressed, or they remain indistinct throughout, arcuate in the biserial portion. The wall is composed of medium to coarse, angular, clastic grains, and the exterior surface is roughly finished. The wall interior is more smoothly finished with randomly arranged, fine perforations that are not visible from the outside of the test (Fig. 33.3). The aperture is a low, arched opening at the base of the final chamber.

Remarks. D paeminosa resembles *D. conicula* Belford in outline, but differs in that it is much smaller and coarser grained, the chambers are

less distinct, and it is more broadly ovate in cross-sectional view. Some elongate specimens possess biserial chambers that are somewhat compressed in the direction of the long axis so that they resemble *Gandryina healyi* Finlay. The latter form differs by its less tapering test, more strongly depressed sutures; the biserial chambers are more compressed in the axial direction, and the early portion of the test is subtriangular in cross section. Loeblich and Tappan (1985) found that the wall interior of the type species for this genus, *Dorothia bulletta* (Carsey), is also finely canaliculate. The test size of *D. paeminosa* ranges from 0.13 to 0.78 mm in length and 0.20 to 0.50 mm in breadth.

Occurrence. Common. Species restricted to Maastrichtian portion of Seymour Island sequence, occurring in low to high abundance, and ranges from SI-K24 through SI-K43.

Marssonella oxycona (Reuss, 1860)
Figure 18.15

Gaudryina oxycona REUSS 1860, p. 329, plate 12, Figure 3.

Figured Hypotype. OSU 39739: length, 0.44 mm; width, 0.35 mm; López de Bertodano Formation, locality 41, Seymour Island.

Occurrence. Common. Occurs in low abundance in samples ranging from SI-K2 through SI-K43 on Seymour Island and has been recorded from Coniacian through Maastrichtian sediments elsewhere.

Marssonnella? sp.
Figure 18.16

Figured Hypotype. OSU 39740: length, 0.43 mm; width, 0.38 mm; López de Bertodano Formation, locality 41, Seymour Island.

Remarks. The strongly tapering, finely agglutinated test of this form appears to be trochospiral throughout, though its sutures are indistinct. The flattened final chambers are suggestive of an attached mode of life.

Occurrence. Very rare. Only three specimens of this distinctive form have been recognized, occurring in SI-K2, SI-K26, and SI-K43.

Karreriella aegra Finlay, 1940
Figure 18.18

Karreriella (Karrerulina) aegra FINLAY 1940, p. 451, Plate 62, Figures 21-22, 25-26.

Figured Hypotype. OSU 39741: length, 1.47 mm; width, 0.36 mm; López de Bertodano Formation, locality 145, Seymour Island.

Remarks. The antarctic specimens display wide morphologic variability, as do the New Zealand topotypes. Younger growth forms have the early chambers arranged in several trochspiral whorls followed by one or two pairs of biserial chambers. Larger adult forms have as many as 10 pairs of rectilinear to slightly twisted biserial chambers, which become progressively more staggered, and may be followed by one or several uniserial chambers. The aperture is a nearly terminal round opening on the biserially arranged chambers and a terminal, round opening with a slight lip on tests with uniserial final chambers. The wall is constructed of coarse silt and sand grains, and the surface is roughly finished.

Occurrence. Common. Taxon ranges from the Late Cretaceous through Eocene in New Zealand. Present in moderate to high abundance within samples from The Naze and Cape Lamb and occurs in low abundance from SI-K7 through SI-T2 on Seymour Island.

Planispirillina subornata (Brotzen, 1940)
Figure 24.8, 11-13

Spirillina subornata BROTZEN 1940, p. 26, Figure 6a-c. HILTERMANN & KOCH 1962, p. 337, Plate 46, Figure 15.

Figured Hypotypes. OSU 39900 (Fig. 24.11-13): diameter, 0.38 mm; breadth, 0.12 mm; López de Bertodano Formation, locality 115, Seymour Island; OSU 39851 (Fig. 24.8): diameter, 0.42 mm; breadth, 0.13 mm; López de Bertodano Formation, locality 56a, Seymour Island.

Description. The test is discoidal with a subrectangular peripheral margin, an indistinct to distinct proloculus, and is planispirally coiled into two and one-half whorls, with a flat, evolute, coarsely perforate dorsal side. The ventral side is flat to slightly concave and is variably ornamented by a secondary accumulation of nodes and pustules such that the spiral suture is obscured, except for the last whorl. The aperture is a moderately high arch at the end of the coiled tube. The wall is composed of single-layered, microcrystalline needles arranged perpendicular to the test surface.

Remarks. The antarctic material shows considerable variability in the coarseness and density of the test perforations on the dorsal side, as well as in the arrangement of nodes and pustules on the ventral side. Rare specimens have wall perforations and irregularly arranged nodes on both sides of the test. Brotzen (1940) did not illustrate or mention variability in test surface ornament. His described species is the most common form among the antarctic specimens, and variations from the original description are here considered to be intraspecific.

Because of the strong similarity of wall microstructure and overall test morphology, Brotzen's species is placed within *Planispirillina*, which was separated by Bermudez (1952) based on the presence of a lamella that covers the spiral suture. Piller (1983) assigned *Planispirillina* to the Involutinidae based on investigations of test wall ultrastructure.

Occurrence. Rare. Specimens range from SI-K29 through SI-K43 on Seymour Island, occurring in low abundance. Also occurs at Cape Lamb. Brotzen's holotype was recovered from lower Danian sediments in Sweden. Hiltermann and Koch (1962) extended the range of this species into upper Maastrichtian in northwest Germany. Antarctic forms represent earliest known occurrence of this species.

Cyclogyra sp.
Figure 18.19-20

Figured Hypotype. OSU 39742: diameter, 0.50 mm; breadth, 0.09 mm; López de Bertodano Formation, locality 184, Seymour Island.

Remarks. The tests are discoidal with a small prolocular chamber followed by a partially evolute, planispirally coiled, tubular chamber. The wall is calcareous, imperforate, and porcellaneous with a smooth surface. The aperture is a broad opening at the end of the chamber.

Occurrence. Very rare. Occurs in low abundance in samples ranging from SI-K35 through SI-K40 on Seymour Island; also occurs at Cape Lamb.

Quinqueloculina sp.
Figure 19.1-4

Figured Hypotypes. OSU 39743 (Figs. 19.1-2): length, 0.36 mm; width, 0.25 mm; breadth, 0.19 mm; López de Bertodano Formation, locality 184, Seymour Island; OSU 39744 (Fig. 19.3-4): length, 0.50 mm; width, 0.41 mm; breadth, 0.28 mm; López de Bertodano Formation, locality 179, Seymour Island.

Remarks. The antarctic forms of this genus are somewhat variable in the degree of chamber inflation and test outline. Its terminal, rounded aperture bears a simple tooth that extends from the base of the final chamber.

Occurrence. Rare. One or two specimens occur in samples ranging from SI-K13 through SI-K43 on Seymour Island.

Nodosaria affinis Reuss, 1845
Figure 19.5-7

Nodosaria affinis REUSS 1845, p. 26, Plate 13, Figure 16.

Figured Hypotypes. OSU 39745 (Fig. 19.5): length, 0.88 mm; width, 0.23 mm; López de Bertodano Formation, locality 67, Seymour Island; OSU 39746 (Fig. 19.6): length, 1.38 mm; width, 0.22 mm; López de Bertodano Formation, locality VI-8, Cape Lamb, Vega Island; OSU 39903 (Fig. 19.7): length, 0.30 mm; width, 0.90 mm; López de Bertodano Formation, locality 38, Seymour Island.

Occurrence. Very rare. Occurs in low abundance within SI-K14, SI-K35, and SI-K43 on Seymour Island, as well as at Cape Lamb.

Nodosaria aspera Reuss, 1845
Figure 19.14

Nodosaria aspera REUSS 1845, p. 26, Plate 13, Figures 14-15.

Figured Hypotype. OSU 39747: length, 0.92 mm; width, 0.31 mm; López de Bertodano Formation, locality 415, Seymour Island.

Occurrence. Very rare. Single specimens occur in samples 415, 346, and 411, ranging from SI-K10 through SI-K43 on Seymour Island.

Nodosaria multicostata (d'Orbigny, 1840)
Figure 19.8

Dentalina multicostata D'ORBIGNY 1840, p. 15, Plate 1, Figures 14-15.

Figured Hypotype. OSU 39748: length, 0.96 mm; width, 0.56 mm; López de Bertodano Formation, locality 38, Seymour Island.

Occurrence. Very rare. Found only on Seymour Island, ranging from SI-K13 through SI-K36.

Nodosaria navarroana Cushman, 1937
Figure 19.9-10

Nodosaria navarroana CUSHMAN 1937, p. 103, Plate 15, Figure 11.

Figured Hypotypes. OSU 39749 (Fig. 19.9): length, 0.28 mm; width, 0.09 mm; López de Bertodano Formation, locality 38, Seymour Island; OSU 39750 (Fig. 19.10): length, 0.21 mm; width, 0.07 mm; López de Bertodano Formation, locality 38, Seymour Island.

Occurrence. Very rare. Occurs in low abundance, ranging from SI-K10 through SI-K36 on Seymour Island; also found at Cape Lamb.

Nodosaria obscura Reuss
Figure 19.11-13

Nodosaria obscura REUSS 1845, p. 26, Plate 13, Figures 7-9.

Figured Hypotypes. OSU 39751 (Fig. 19.11): length, 0.23 mm; width, 0.09 mm; López de Bertodano Formation, locality 45, Seymour Island; OSU 39752 (Fig. 19.12): length, 0.20 mm; width, 0.09 mm; López de Bertodano Formation, locality VI-8, Cape Lamb, Vega Island; OSU 39753 (Fig. 19.13): length, 0.43 mm; width, 0.11 mm; López de Bertodano Formation, locality 148, Seymour Island.

Occurrence. Rare. Occurs in low abundance in numerous samples from SI-K10 through SI-K43 on Seymour Island; also occurs at Cape Lamb.

Astacolus oligostegius (Reuss, 1860)
Figure 21.7

Cristellaria oligostegia REUSS 1860, p. 213, Plate 8, Figure 8.

Figured Hypotype. OSU 39754: diameter, 0.51 mm; breadth, 0.30 mm; López de Bertodano Formation, locality 38, Seymour Island.

Occurrence. Very rare. Several specimens occur in samples 426 and 38, ranging from SI-K21 through SI-K24 on Seymour Island.

Citharina plumoides (Plummer, 1927)
Figure 19.21-24

Vaginulina plumoides PLUMMER 1927, p. 113, Plate 6, Figure 6.
Planularia whangaia FINLAY 1939, p. 317, Plate 26, Figures 63-65.

Figured Hypotypes. OSU 39755 (Fig. 19.21): length, 0.61 mm; width, 0.71 mm; breadth, 0.12, López de Bertodano Formation, locality 38, Seymour Island; OSU 39756 (Fig. 19.22-23): length, 1.38 mm; width, 0.33 mm; breadth, 0.12 mm; López de Bertodano Formation, locality VI-8, Cape Lamb, Vega Island; OSU 39757 (Fig. 19.24): length, 0.29 mm, width, 0.10 mm, breadth, 0.08 mm, López de Bertodano Formation, locality 38, Seymour Island.

Remarks. The New Zealand and antarctic forms of this species exhibit wide morphological variation in test outline, from being long and slender to subtriangular. All tests possess a striate ornamentation that varies from faint and discontinuous to thin and sharply raised. Webb (1966) determined that variants of *Citharina whangaia* (Finlay) were identical to

Plummer's Gulf Coast topotypes. Consequently, Finlay's species was placed in the synonymy of *C. plumoides*. The antarctic material compares very well with the New Zealand and Gulf Coast topotypes.
Occurrence. Rare. Plummer's species originally described from the upper Midway of Texas, which is Paleocene in age. New Zealand material occurs in the Haumurian (upper Campanian–Maastrichtian) Laidmore and Whangai Formations. Antarctic forms occur in low abundance at Cape Lamb, Vega Island, and from SI-K14 through SI-T2 on Seymour Island.

Dentalina basiplanata Cushman, 1938a
Figure 19.15

Dentalina basiplanata CUSHMAN 1938a, p. 38, Plate 6, Figures 6-8.
Figured Hypotype. OSU 39758: length, 0.98 mm, width, 0.22 mm, López de Bertodano Formation, locality 38, Seymour Island.
Occurrence. Moderately common. Ranges from SI-K2 through SI-T2 on Seymour Island.

Dentalina gracilis (d'Orbigny, 1840)
Figure 19.16-17

Nodosaria (Dentalina) gracilis D'ORBIGNY 1840, p. 14, Plate 1, Figure 5.
Figured Hypotypes. OSU 39759 (Fig. 19.16): length, 1.09 mm, width, 0.23 mm, López de Bertodano Formation, locality 167, Seymour Island; OSU 39760 (Fig. 19.17): length, 0.59 mm, width, 0.10 mm, López de Bertodano Formation, locality 28a, Seymour Island.
Occurrence. Common. Occurs throughout Seymour Island sequence, ranging from SI-K5 through SI-K43; also found in Cape Lamb samples.

Dentalina lorneiana d'Orbigny, 1840
Figure 19.18

Dentalina lorneiana D'ORBIGNY 1840, p. 14, Plate 1, Figures 8-9.
Figured Hypotype. OSU 39761: length, 1.22 mm, width, 0.21 mm, López de Bertodano Fm., locality 164, Seymour Island.
Occurrence. Common. Occurs in a number of samples, from SI-K11 through SI-T2 on Seymour Island.

Dentalina marcki Reuss, 1860
Figure 19.20

Dentalina marcki REUSS 1860, p. 188, Plate 2, Figure 7.
Figured Hypotype. OSU 39762: length, 0.59 mm, width, 0.30 mm, López de Bertodano Formation, locality 110, Seymour Island.
Occurrence. Rare. Occurs in low abundance from SI-K5 through SI-K42 on Seymour Island.

Dentalina megapolitana Reuss, 1855
Figure 19.19

Dentalina megapolitana REUSS 1855, p. 267, Plate 8, Figure 10.
Figured Hypotype. OSU 39763: length, 0.77 mm, width, 0.24 mm, López de Bertodano Formation, locality 145, Seymour Island.
Occurrence. Rare. Occurs in samples ranging from SI-K14 through SI-K43 on Seymour Island; also occurs in samples from Cape Lamb.

Dentalina sp.
Figure 32.2

Figured Hypotype. OSU 39764: length, 0.38 mm, width, 0.08 mm, López de Bertodano Formation, locality 539, Seymour Island.
Remarks. This form looks similar to *Ellipsonodosaria exilis* Cushman (Cushman, 1936, p. 51, Plate 9, Figures 1-2), although no apertural tooth is discernable.
Occurrence. Single specimens recovered from samples 539 and 540 of SI-T2, in the uppermost López de Bertodano and lowermost Sobral Formations.

Frondicularia rakauroana (Finlay, 1939)
Figure 20.1-2

Palumula rakauroana FINLAY 1939, p. 314, Plate 26, Figures 51-52; WEBB 1972b, p. 94–100, Plate 1, Figures 1-27, Plate 2, Figures 1-5; HUBER & WEBB 1986, p. 135–140, Plate 1, Figures 1-9.
Figured Hypotype. OSU 38551: length, 1.15 mm, width, 0.71 mm, breadth, 0.30 mm, López de Bertodano Formation, locality 101, Seymour Island.
Remarks. The distribution of this distinctive taxon on Seymour Island and its restriction to the southern high latitudes has been discussed by Huber and Webb (1986). The New Zealand specimens express much variability in test outline, but they all display costellae that are initially longitudinal, then bifurcating and transverse. The three antarctic specimens show little variability, but compare well with the New Zealand forms.
Occurrence. Very rare. Species restricted to upper Haumurian (Maastrichtian) of New Zealand; has been found in samples 386, 101, 394, and 429, ranging from SI-K5 through SI-K30 on Seymour Island.

Frondicularia? sp.
Figure 20.3

Figured Hypotype. OSU 39765: length, 0.84 mm, width, 0.36 mm, breadth, 0.11 mm, López de Bertodano Formation, locality 367, Seymour Island.
Remarks. This elongate, palmate form is biserially symmetric about the long axis, but the absence of the prolocular chamber makes its identification tentative. It is ornamented with fine, longitudinal costae.
Occurrence. Very rare. A single specimen recovered from sample 367 of SI-K42 on Seymour Island.

Lagena acuticosta Reuss, 1862
Figure 20.4

Lagena acuticosta REUSS 1862, p. 305, Plate 1, Figure 4.
Figured Hypotype. OSU 39766: length, 0.24 mm, width, 0.16 mm, López de Bertodano Formation, locality 165, Seymour Island.
Occurrence. Rare. Has been recovered in low abundance in a number of samples, ranging from SI-K10 through SI-K43 on Seymour Island; also found at Cape Lamb.

Lagena apiculata (Reuss, 1851)
Figure 20.6

Oolina apiculata REUSS 1851, p. 22, Plate 2, Figure 1.
Figured Hypotype. OSU 39767: length, 0.52 mm, width, 0.26 mm, López de Bertodano Formation, locality 28a, Seymour Island.
Occurrence. Common. Occurs in low abundance in Seymour Island samples ranging from SI-K8 through SI-T2.

Lagena semiinterrupta Berry, 1929
Figure 20.8

Lagena sulcata (Walker and Jacob) var. *semiinterrupta* BERRY, *in* Berry and Kelley, 1929, p. 5, Plate 3, Figure 19.
Figured Hypotype. OSU 39768: length, 0.24 mm, width, .13 mm, López de Bertodano Formation, locality 179, Seymour Island.
Remarks. The antarctic forms all bear between seven and nine longitudinal costae, which coalesce at or near a basal ring and form a thickened area just below the slender necked aperture.
Occurrence. Rare. Occurs in low abundance on Seymour Island, from SI-K10 through SI-K43. Several specimens also found at Cape Lamb. This species originally described from the Maastrichtian of Tennessee.

***Lagena semilineata* Wright, 1885**
Figure 20.7

Lagena semilineata WRIGHT 1885, p. 320.
Figured Hypotype. OSU 39769: length, 0.60 mm, width, 0.33 mm, López de Bertodano Formation, locality 154, Seymour Island.
Occurrence. Rare. Occurs in very low abundance in samples ranging from SI-K10 through SI-T2 on Seymour Island.

***Lagena simplex* (Reuss, 1851)**
Figure 20.9

Oolina simplex REUSS 1851, p. 22, Plate 2, Fig. 2a-b.
Figured Hypotype. OSU 39770: length, 0.76 mm, width, 0.57 mm, López de Bertodano Formation, locality 177, Seymour Island.
Occurrence. Rare. Several Seymour Island samples, ranging from SI-K21 through SI-K40, have yielded this species; also occurs at Cape Lamb.

***Lagena sphaerica* Marie, 1941**
Figure 20.5

Lagena sphaerica MARIE 1941, p. 81, Plate 9, Figure 100.
Figured Hypotype. OSU 39771: length, 0.38 mm, width, 0.29 mm, López de Bertodano Formation, locality 110, Seymour Island.
Occurrence. Common. Occurs in low abundance on Seymour Island, ranging from SI-K2 through SI-T2; also occurs at Cape Lamb.

***Lenticulina* cf. *L. jonesi* Sandidge, 1932**
Figure 20.10-11

cf. *Lenticulina jonesi* SANDIDGE 1932, p. 273, Plate 43, Figures 1, 2.
Figured Hypotype. OSU 39772: diameter, 1.38 mm, breadth, 0.46 mm, López de Bertodano Formation, locality 177, Seymour Island.
Remarks. It is similar to this species because of the rounded final chamber face, having six chambers in the final whorl and its slightly keeled periphery. It differs from the Gulf Coast forms in having costae that are subparallel to the periphery from the third to fifth chamber in the final whorl.
Occurrence. Very rare. Single specimen found in sample 177 of SI-K31. The holotype described from the mid-Paleocene Ripley Formation of Alabama.

***Lenticulina macrodisca* (Reuss, 1863)**
Figure 20.12-13

Cristellaria macrodisca REUSS 1863, p. 78, Plate 9, Fig. 5a-b.
Figured Hypotype. OSU 39773: diameter, 0.46 mm, breadth, 0.23 mm, López de Bertodano Formation, locality 177, Seymour Island.
Occurrence. Moderately common. Occurs throughout Seymour Island sequence in low to moderate abundance, ranging from SI-K13 through SI-T2; abundant in several of the Cape Lamb samples.

***Lenticulina muensteri* (Roemer, 1839)**
Figure 32.3

Robulina münster ROEMER 1839, p. 48, Plate 22, figure 29.
Figured Hypotype. OSU 39774: diameter, 1.42 mm, breadth, 0.46 mm, López de Bertodano Formation, locality 539, Seymour Island.
Occurrence. Very rare. Single specimen recovered from sample 539 of SI-T2 on Seymour Island.

***Lenticulina ovalis* (Reuss, 1845)**
Figure 21.3

Cristellaria ovalis REUSS 1845, p. 34, Plate 8, Figure 49, Plate 12, Figure 19; Plate 13, Figures 60-63.
Figured Hypotype. OSU 39775: length, 0.30 mm, width, 0.19 mm, López de Bertodano Formation, locality 177, Seymour Island.
Remarks. The antarctic specimens have a large proloculus that is fol-

lowed by two to three chambers. As Sliter (1968, p. 67) stated, it is uncertain whether this form represents a distinct species or a juvenile form of another species.
Occurrence. Very rare. Occurs in very low abundance on Seymour Island, ranging from SI-K5 through SI-K43.

***Lenticulina spissocostata* (Cushman, 1938a)**
Figure 20.14-15

Robulus spisso-costatus CUSHMAN 1938a, p. 32, Plate 5, Figure 2.
Figured Hypotypes. OSU 39776 (Fig. 20.14): diameter, 2.93 mm, breadth, 1.04 mm, López de Bertodano Formation, locality 369, Seymour Island; OSU 39777 (Fig. 20.15): diameter, 2.78 mm, breadth, 1.24 mm, López de Bertodano Formation, locality 430, Seymour Island.
Remarks. Some antarctic specimens are quite large, reaching as much as 3 mm in diameter. The sutures are limbate, showing variation in elevation and thickness toward the umbo among different specimens. A peripheral keel occurs on some of the antarctic forms.
Occurrence. Common. Taxon ranges from SI-K8 through SI-K43 on Seymour Island; several specimens also occur at Cape Lamb.

***Lenticulina williamsoni* Reuss, 1867**
Figure 20.16-17

Cristellaria williamsoni REUSS 1862, p. 327, Plate 6, Figure 4.
Figured Hypotype. OSU 39778: diameter, 1.11 mm, breadth, 0.57 mm, López de Bertodano Formation, locality VI-8, Cape Lamb, Vega Island.
Occurrence. Single specimen recovered from Cape Lamb.

***Lenticulina* sp.**
Figure 21.1-2

Figured Hypotype. OSU 39779: diameter, 0.99 mm, breadth, 0.37 mm, López de Bertodano Formation, locality 177, Seymour Island.
Remarks. This form resembles *Saracenaria triangularis* (d'Orbigny) in that it is partially evolute and has a broad, flat final chamber face. It is distinguished by its raised, limbate sutures and distinct striae, which are transverse to the sutures, extending from the early portion of the final whorl to the penultimate chamber.
Occurrence. Very rare. Single specimens recovered from samples 147 and 177, ranging from SI-K31 through SI-K34 on Seymour Island.

***Marginulina bullata* Reuss, 1845**
Figure 21.5, 8

Marginulina (Marginulina) bullata REUSS 1845, p. 29, Plate 13, Figures 34-38.
Figured Hypotypes. OSU 37980 (Fig. 21.5): length, 0.96 mm, breadth, 0.40 mm, López de Bertodano Formation, locality 148, Seymour Island; OSU 39781 (Fig. 21.8): length, 0.20 mm, breadth, 0.10 mm, López de Bertodano Formation, locality 38, Seymour Island.
Occurrence. Common. Occurs in low abundance on Seymour Island, ranging from SI-K7 through SI-K43; also occurs at Cape Lamb.

***Marginulina* cf. *M. curvatura* Cushman, 1938a**
Figure 21.9-10

cf. *Marginulina curvatura* CUSHMAN 1938a, Plate 5, Figures 13, 14.
Marginulina cf. *M. curvatura* Cushman, SLITER 1968, p. 70, Plate 8, Figures 8-9.
Figured Hypotype. OSU 39782: length, 0.49 mm, breadth, 0.36 mm, López de Bertodano Formation, locality 110, Seymour Island.
Remarks. The antarctic forms are identical to Sliter's illustrated hypotype, which was designated as *Marginulina* sp. cf. *M. curvatura.* It is mostly involute. The chambers increase very rapidly in size, and it lacks an apertural neck.
Occurrence. Very rare. Several specimens occur in SI-K14, SI-K40, and SI-K43 on Seymour Island.

Pseudonodosaria parallela (Marsson, 1878)
Figure 21.11
Glandulina parallela MARSSON 1878, p. 124, Plate 1, Figure 4a-b.
Figured Hypotype. OSU 39783: length, 0.93 mm, width, 0.50 mm, López de Bertodano Formation, locality 177.
Occurrence. Very rare. Occurs in very low abundance in López de Bertodano Formation on Seymour Island, ranging from SI-K5 through SI-K43.

Saracenaria triangularis (d'Orbigny, 1840)
Figure 21.4
Cristellaria triangularis D'ORBIGNY 1840, p. 27, Plate 2, Figures 21-22.
Figured Hypotype. OSU 39784: length, 1.83 mm, width, 1.17 mm, breadth, 0.38 mm, López de Bertodano Formation, locality VI-11, Cape Lamb, Vega Island.
Occurrence. Rare. Found in Seymour Island samples ranging from SI-K5 through SI-T2; also recovered from Cape Lamb sequence.

Globulina lacrima (Reuss, 1845)
Figure 21.12-13
Polymorphina (Globulina) lacrima REUSS 1845, p. 40, Plate 12, Figure 6; Plate 13, Figure 83.
Figured Hypotypes. OSU 39785 (Fig. 21.12): length, 0.32 mm, width, 0.24 mm, López de Bertodano Formation, locality 146, Seymour Island; OSU 39904 (Fig. 21.13): length, 0.76 mm, breadth, 0.41 mm, López de Bertodano Formation. Locality VI-8, Cape Lamb, Vega Island.
Occurrence. Rare. Ranges from SI-K10 through SI-K43 on Seymour Island, occurring in low abundance; several specimens also found in Cape Lamb samples.

Guttulina adhaerens (Olszewski, 1875)
Figure 21.14
Polymorphina adhaerens OLSZEWSKI 1875, p. 119, Plate 1, Figure 11.
Figured Hypotype. OSU 39786: length, 0.40 mm, width, 0.28 mm, López de Bertodano Formation, locality 164, Seymour Island.
Occurrence. Very rare. Occurs in very low abundance in SI-K30 on Seymour Island; also occurs at Cape Lamb.

Guttulina trigonula (Reuss, 1845)
Figure 21.15
Polymorphina trigonula REUSS 1845, p. 40, Plate 13, Figure 84.
Figured Hypotype. OSU 39787: length, 0.39 mm, width, 0.31 mm, López de Bertodano Formation, locality 164, Seymour Island.
Occurrence. Very rare. Occurs in low abundance on Seymour Island, ranging from SI-K10 through SI-K43.

Pyrulina cylindroides (Roemer, 1838)
Figure 21.16
Polymorphina cylindroides ROEMER 1838, p. 385, Plate 3, Figure 26.
Figured Hypotype. OSU 39788: length, 0.46 mm, width, 0.29 mm, López de Bertodano Formation, locality 148, Seymour Island.
Occurrence. Rare. On Seymour Island, it ranges from SI-K10 through SI-T2; also occurs in the Cape Lamb sequence.

Ramulina pseudoaculeata (Olsson, 1960)
Figure 21.17
Dentalina pseudoaculeata OLSSON 1960, p. 14, Plate 3, Figures 1-2.
Figured Hypotype. OSU 39789: length, 0.56 mm, width, 0.34 mm, López de Bertodano Formation, locality VI-11, Cape Lamb, Vega Island.
Occurrence. Very rare. Several specimens recovered from Cape Lamb beds, as well as SI-K10 and SI-K24 on Seymour Island.

Ramulina sp.
Figure 23.18
Figured Hypotype. OSU 39790: length, 0.54 mm, width, 0.14 mm, López de Bertodano Formation, locality 38, Seymour Island.
Remarks. This form is distinguished by its small, slender test which is broken at both ends. A nodular ornamentation covers much of the wall surface.
Occurrence. Very rare. Has been found in SI-K14 and SI-K36 on Seymour Island, occurring in low abundance.

Oolina sp.
Figure 21.19
Figured Hypotype. OSU 39791: length, 0.33 mm, width, 0.16 mm, López de Bertodano Formation, locality 184, Seymour Island.
Remarks. This form is distinguished by its asymmetric, single-chambered test, which is ornamented by numerous longitudinal costae. The aperture is terminal and radiate.
Occurrence. Very rare. Single specimen found in sample 184 of SI-K36 on Seymour Island.

Fissurina alata Reuss, 1851
Figure 22.5-6
Fissurina alata REUSS 1851, p. 58, Plate 3, Figure 1.
Figured Hypotype. OSU 39792: length, 0.27 mm, width, 0.15 mm, breadth, 0.14 mm, López de Bertodano Formation, locality 415, Seymour Island.
Occurrence. Very rare. Several specimens recovered from samples 367 and 368 of SI-K42 on Seymour Island.

Fissurina laevigata Reuss, 1850
Figure 22.1
Fissurina laevigata REUSS 1850, p. 336, Plate 46, Figure 1.
Figured Hypotype. OSU 39793: length, 0.15 mm, width, 0.11 mm, breadth, 0.08 mm, López de Bertodano Formation, locality 38, Seymour Island.
Occurrence. Very rare. Occurs in low abundance from SI-K20 through SI-K43 on Seymour Island; also occurs in Cape Lamb strata.

Fissurina orbignyana Seguenza, 1862
Figure 22.2-3
Fissurina orbignyana SEGUENZA 1862, p. 66, Plate 2, Figures 25-26.
Figured Hypotype. OSU 39794: length, 0.22 mm, width, 0.20 mm, breadth, 0.12 mm, López de Bertodano Formation, locality 182, Seymour Island.
Occurrence. Very rare. Occurs in low abundance from SI-K10 through SI-K43 on Seymour Island.

Fissurina sp.
Figure 22.4
Figured Hypotype. OSU 39795: length, 0.08 mm, width, 0.06 mm, breadth, 0.04 mm, López de Bertodano Formation, locality 110, Seymour Island.
Remarks. This form is distinguished by the presence of two marginal keels and one that extends up the central portion of the test exterior.
Occurrence. Very rare. Single specimens found in SI-K24 and SI-K34 on Seymour Island.

Ceratolamarckina cf. *C. tuberculata* (Brotzen, 1948)
Figure 23.13

cf. *Ceratobulimina tuberculata* BROTZEN 1948, p. 124–125, Plate 19, Figures 2, 3.

Figured Hypotype. OSU 39849: diameter, 0.37 mm, breadth, 0.23 mm, López de Bertodano Formation, locality 380, Seymour Island.

Remarks. The antarctic specimens are very similar to Brotzen's (1948) type description. They have a broadly rounded periphery and no more than six chambers in the final whorl, with only five visible on the umbilical side. The most distinguishing characteristics include a dentate apertural margin on the final chamber face and the presence of distinct tubercles, which project outward from the final chamber in the umbilical region. The antarctic forms differ from the type specimens by having less depressed sutures, spiral sutures that are more oblique, a lower spire, and a more sharply angled peripheral margin.

Occurrence. Very rare. Several specimens recovered from SI-K34, sample 380, on Seymour Island. Brotzen's (1948) type specimens of *C. tuberculata* recovered from lower Paleocene sediments in Denmark.

Hoeglundina supracretacea (ten Dam, 1948)
Figure 25.15-16

Epistomina supracretacea TEN DAM 1948, p. 163, Plate 1, Fig. 8.

Figured Hypotype. OSU 39850: diameter, 0.61 mm, breadth, 0.40 mm, López de Bertodano Formation, locality 148, Seymour Island.

Occurrence. Moderately common. Found in low to high abundance in many López de Bertodano Formation samples, ranging from SI-K3 through SI-T2 on Seymour Island; also occurs in Cape Lamb samples. This species recognized in many Campanian-Maastrichtian localities worldwide.

Buliminella creta (Finlay, 1939)
Figure 23.1

Elonglobula creta FINLAY 1939, p. 322, Plate 27, Figures 88-91.

Figured Hypotype. OSU 39796: length, 0.29 mm, width, 0.12 mm, López de Bertodano Formation, locality 28d, Seymour Island.

Remarks. Finlay (1939) erected the genus *Elonglobula* to include elongate buliminids, which are intermediate in character between *Buliminella* and *Buliminoides.* According to Finlay, forms included in *Elonglobula* are more loosely coiled and have lower chambers than *Buliminella,* whereas they differ from *Buliminoides* only in the absence of ornamentation. Although the New Zealand paratypes and antarctic specimens of *B. creta* show irregular chamber size and frequently have a distorted appearance, unlike typical species of *Buliminella,* the author follows Loeblich and Tappan (1964, p. 544) in placing Finlay's species in *Buliminella.* The antarctic specimens are identical in size, as well as morphologic variability, to the New Zealand topotypes.

Occurrence. Rare. Abundant in several samples from Cape Lamb; occurs in low to moderate abundance in Seymour Island samples, ranging from SI-K10 through SI-K39. In New Zealand, occurs in Haumurian through Teurian (upper Campanian through Paleocene) sediments in a number of localities.

Buliminella cf. *B. fusiforma* Jennings, 1936
Figure 22.8

cf. *Buliminella fusiforma* JENNINGS 1936, p. 30, Plate 3, Figure 18.

Figured Hypotype. OSU 39797: length, 0.14 mm, width, 0.12 mm, López de Bertodano Formation, locality 8, Cape Lamb, Vega Island.

Remarks. The Seymour Island specimens most closely resemble this species in test form, number of chambers per whorl (four), and number of whorls (three). They differ in that the apertures of the antarctic forms are subparallel to the interiomarginal suture, with a surrounding lip, and their sutures are flush rather than depressed.

Occurrence. Very rare. Several specimens recovered from upper Campanian beds at Cape Lamb and SI-K10 on Seymour Island.

Buliminella procera n. sp.
Figures 32.5-6, 33.11

Holotype. USNM 415751: length, 0.30 mm, breadth, 0.10 mm, Sobral Formation, locality 540, Seymour Island.

Figured Hypotypes. OSU 39798 (Fig. 32-5-6): length, 0.33 mm, breadth, 0.12 mm, Sobral Formation, locality 540, Seymour Island; OSU 39894 (Fig. 33.11): length, 0.30 mm, breadth, 0.15 mm, Sobral Formation, locality 540, Seymour Island.

Etymology. Specific name from the Latin *procerus,* tall, slender, referring to the test outline.

Diagnosis. Test small, elongate, fusiform, high trochospiral, composed of two to three chamber whorls, the last forming more than three-fourths of the test length, aperture bordered by a narrow, raised lip, parallel to the spiral suture.

Description. The test is small, elongate, fusiform, sometimes irregularly rectilinear, and is circular in cross section. It has two to three chamber whorls, the last forming more than three-fourths of the test length. The chambers are distinct, increasing slightly as added, five or six occur in the final whorl. The sutures are distinct, slightly curved, and flush, and the spiral suture is slightly depressed. The wall is very finely perforate, and the surface is smooth, except for the final chamber face, which bears fine pustules (Fig. 32.6). The aperture is a distinct loop-shaped opening, parallel to the spiral suture, bordered by a narrow, raised lip, with a simple internal toothplate connecting successive apertures. The apertural face is flat, narrow, and extends well down the side of the test.

Remarks. This form is similar to *B. creta* Finlay in having an elongate, sometimes irregular test, but differs by having very elongate chambers, such that the final whorl makes up over three-fourths of the test length in adult specimens. It differs from *B. pseudoelegantissima* Bertels in having fewer chambers in the final whorl and an aperture that parallels the spiral suture. The size ranges from 0.11 to 0.40 mm in length, and 0.07 to 0.14 mm in breadth.

Occurrence. Very rare. Very abundant in samples in which it occurs, composing as much as 82 percent of the total assemblage. Restricted to the *Globastica daubjergensis* Zone, occurring in samples 539, 540, 514, and 515 in SI-T2 on Seymour Island.

Buliminella pulchra (Terquem, 1882)
Figure 22.7

Bulimina pulchra TERQUEM 1882, p. 114, Plate 12, Figures 8-12.

Figured Hypotype. OSU 39799: length, 0.15 mm, width, 0.10 mm, López de Bertodano Formation, locality 57, Seymour Island.

Remarks. This form differs from *B.* cf. *B. fusiforma* Jennings in that there are seven rather than four chambers in the final whorl, one and one-half rather than three whorls, all sutures are flush with the chamber surface, and the final chamber extends all the way to the apex rather than half way.

Occurrence. Very rare. Occurs in very low abundance within samples ranging from SI-K36 through SI-K39 on Seymour Island.

Neobulimina aspera (Cushman & Parker, 1940)
Figure 22.12-14

Bulimina aspera CUSHMAN & PARKER 1940, p. 44, Plate 8, Figures 18-19.

Neobulimina aspera (Cushman & Parker), BERTELS 1972, p. 335–336, Plate 1, Figures 7-9.

Figured Hypotypes. OSU 39800 (Fig. 22.12): length, 0.89 mm, width, 0.33 mm, López de Bertodano Formation, locality 148, Seymour Island; OSU 39801 (Fig. 22.13-14): length, 0.52 mm, width, 0.21 mm, López de Bertodano Formation, locality 115, Seymour Island.

Remarks. This taxon is characterized by its elongate, slightly tapering test with medium perforations that are linearly arranged and a slightly roughened wall surface. The antarctic forms are similar to *B. kickapooensis* Cole, except the latter has a smooth test surface with randomly

arranged perforations. It is placed in *Neobulimina* because of its tendency toward a biserial chamber arrangement. No specimens bearing apical spines were encountered in the antarctic samples.
Occurrence. Common. Occurs in a number of samples from SI-K8 through SI-K43 on Seymour Island. Originally described from Taylor and Navarro strata of the American Gulf; has also been found in Maastrichtian sediments of the South American Neuquen Basin (Bertels, 1972, 1979).

Neobulimina digitata n. sp.
Figures 22.15-16, 33.12
Holotype. USNM 415742: length, 0.21 mm, width, 0.07 mm, locality 28d, López de Bertodano Formation, Seymour Island.
Figured Hypotypes. OSU 39801 (Fig. 22.15-16): length, 0.24 mm, width, 0.10 mm, López de Bertodano Formation, localion 148, Seymour Island; OSU 39895 (Fig. 33.12): length, 0.21 mm, width, 0.10 mm, López de Bertodano Formation, locality 380, Seymour Island.
Etymology. Species name from the Latin *digitatus,* having fingers, referring to the retral processes that overlap previous chambers.
Diagnosis. Test small, elongate, with one to four pairs of biserial chambers following triserial portion, wall ornamented with pustules and retral processes that overlap the previous chamber.
Description. The test is small, elongate, straight to slightly arcuate, initially with two to five triserial whorls, later with one to four pairs of biserial chambers, with a tendency toward uniserial chamber arrangement in more elongate specimens. The sutures are slightly depressed in the triserial portion, more distinct and depressed in the biserial portion, and form a strongly oblique angle to the long axis (Fig. 33.12). The chambers gradually increase in inflation and size as added. The wall is finely perforate and sparsely covered with medium-size pustules, which are irregularly arranged on the test surface. Distinct retral processes extend from the peripheral sides of the biserial chambers and overlap the previous chamber. The aperture is nearly terminal, loop-shaped, with a simple internal tooth plate and narrow bordering lip.
Remarks. This form is characterized by its small, slender test, pustulose ornamentation. and the distinct chamber extensions that overlap the previous biserial chambers. It is distinguished from *N. canadensis* Cushman & Wickenden and *N. irregularis* Cushman & Parker by its smaller size, more slender test, surface ornamentation, and nearly terminal position of the aperture. It is not placed within *Loxostomoides* because of the initial triserial chamber arrangement. It ranges from 0.17 to 0.25 mm in length, 0.07 to 0.09 mm in width, and 0.05 to 0.07 mm in breadth.
Occurrence. Common. Occurs in low to moderate abundance in Seymour Island samples, ranging from SI-K8 through SI-K43 (upper Campanian through Maastrichtian).

Praebulimina cushmani (Sandidge, 1932)
Figure 22.18
Buliminella cushmani SANDIDGE 1932, p. 280, Plate 42, Figures 18-19.
Figured Hypotype. OSU 39803: length, 0.20 mm, width, 0.16 mm, López de Bertodano Formation, locality 28b, Seymour Island.
Occurrence. Moderately common. Occurs in low to moderate abundance on Seymour Island from SI-K8 through SI-K43.

Praebulimina cf. *P. cushmani* (Sandidge, 1932)
Figure 22.17
cf. *Buliminella cushmani* SANDIDGE 1932, p. 280, Plate 42, Figures 18-19.
Figured Hypotype. OSU 39804: length, 0.16 mm, width, 0.07 mm, López de Bertodano Formation, locality 38, Seymour Island.
Description. The test is small, one and one-half times long as broad, subglobular, trochospiral with three chambers in the final whorl,

chambers gradually increasing in size and inflation. The sutures are distinct and depressed throughout. The wall is finely perforate and smooth except for the final chamber face, which is densely covered with small- to medium-size pustules. The aperture is loop-shaped, extending halfway up the final chamber face with a simple tooth plate.
Remarks. This form resembles *P. cushmani* except the test is more elongate and the final chamber face is covered with pustules. It differs from *P. intermedia* Cushman & Parker in that its chambers are less inflated, the test is less tapering, and the aperture is lower in position on the final chamber face.
Occurrence. Very rare. Occurs in low abundance within samples 110 and 38 of SI-K8 and SI-K24 on Seymour Island.

Praebulimina midwayensis (Cushman & Parker, 1936)
Figure 23.2
Bulimina arkedelphiliana Cushman & Parker var. *midwayensis* CUSHMAN & PARKER 1936, p. 42, Plate 7, Figures 9-10.
Bulimina rakauroana FINLAY 1940, p. 454, Plate 64, Figures 75-76.
Figured Hypotype. OSU 39805: length, 0.28 mm, width, 0.19 mm, López de Bertodano Formation, locality 110, Seymour Island.
Occurrence. Rare. Occurs in low to moderate abundance from SI-K7 through SI-T2 on Seymour Island.

Praebulimina cf. *P. reussi* (Morrow, 1934)
Figure 22.19
cf. *Bulimina reussi* MORROW 1934, p. 195, Plate 29, Figure 12.
Figured Hypotype. OSU 39806: length, 0.30 mm, width, 0.19 mm, López de Bertodano Formation, locality 110, Seymour Island.
Remarks. The antarctic specimens are similar to the original description of this species in that the test is ovate and globular, the chambers are triserial throughout, and the aperture is nearly terminal. It differs in that the test is less tapering and the chambers are more inflated.
Occurrence. Very rare. On Seymour Island, ranges from SI-K8 through SI-K40.

Praebulimina cf. *P. trihedra* (Cushman, 1926b)
Figure 22.20
cf. *Bulimina trihedra* CUSHMAN 1926b, p. 591, Plate 17, Figure 6a,b.
Figured Hypotype. OSU 39807: length, 0.20 mm, width, 0.09 mm, López de Bertodano Formation, locality 28b, Seymour Island.
Occurrence. Common. Ranges from SI-K13 through SI-T2 on Seymour Island.

Praebulimina sp.
Figure 23.3
Figured Hypotype. OSU 39808: length, 0.21 mm, width, 0.09 mm, López de Bertodano Formation, locality 110, Seymour Island.
Remarks. This form is characterized by its elongate test that bears a very small initial chamber and inflated later chambers that gradually increase in size. The aperture is nearly terminal with a surrounding lip. It may represent the earliest, triserial portion of *Hiltermannella kochi* (Bertels), with the biserial portion missing.
Occurrence. Very rare. Several specimens recovered from sample 38 of SI-K24.

Pyramidina tortilis (Reuss, 1862)
Figure 23.11
Bulimina tortilis REUSS 1862, p. 338, Plate 8, Figures 3a-b; CUSHMAN & PARKER 1947, p. 85, Plate 20, Figure 14a-b.
Figured Hypotype. OSU 39809: length, 0.18 mm, width, 0.08 mm, López de Bertodano Formation, locality 38, Seymour Island.
Description. The test is small, moderately tapering, sometimes slightly

twisted, triangular in cross section, with acute angles and concave sides. The chambers are distinct, triserially arranged throughout, gradually increasing in size with as many as six whorls. The sutures are slightly depressed and curved toward the apex. The wall is calcareous and finely perforate, and the surface is smooth. The aperture is nearly terminal, extending up from the base of the final chamber, with a simple internal tooth plate.

Remarks. These antarctic forms strongly resemble the original description and subsequent descriptions of *P. tortilis,* except they are smaller, ranging between 0.16 and 0.23 mm in length and 0.07 and 0.08 mm in breadth. New Zealand specimens referred to as *Pyramidina* sp. by Webb (1971) are included in this species. Specimens referred to as *Bulimina referata* Jennings (Jennings, 1936, p. 189, Plate 3, Figures 21a-b) and Olsson (1960, p. 32, Plate 5, Figures 3-4) are similar to this species, except they are more robust with less concavity to the test sides and have more rounded peripheral angles.

Occurrence. Very rare. Several specimens were found in samples 165, 28d, and 38, ranging from SI-K14 through SI-K24 on Seymour Island. Similar specimens, referred to as *Pyramidina* sp. by Webb (1971), are restricted to the New Zealand *Globotruncana circumnodifer* Zone, which is considered as upper Haumurian (Maastrichtian). The holotype was described from the Upper Cretaceous of Europe. Cushman and Parker reported this species from the "Senonian Greensands" of New Jersey.

Eouvigerina hispida Cushman, 1931
Figure 23.12

Eouvigerina hispida CUSHMAN 1931, p. 45, Plate 7, Figures 12-13.
Figured Hypotype. OSU 39810: length, 0.25 mm, width, 0.13 mm, López Bertodano Formation, locality 28b, Seymour Island.
Occurrence. Very rare. It occurs in very low abundance within samples ranging from SI-K10 through SI-K17. This species is well known in Gulf Coast, southern California, and Atlantic Coastal Plain sediments of Campanian-Maastrichtian age.

Stilostomella impensia (Cushman, 1938a)
Figure 23.14

Ellipsonodosaria alexanderi Cushman var. *impensia* CUSHMAN 1938a, p. 48, Plate 8, Figures 4-5.
Figured Hypotype. OSU 39811: length, 1.41 mm, width, 0.25 mm, López de Bertodano Formation, locality 28b, Seymour Island.
Occurrence. Common. Occurs in samples ranging from SI-K7 through SI-T2 on Seymour Island, as well as at Cape Lamb. Originally described from Gulf Coast Navarro strata of North America; also occurs in Campanian-Maastrichtian sediments of California.

Stilostomella pseudoscripta (Cushman, 1937)
Figure 23.15

Ellipsonodosaria pseudoscripta CUSHMAN 1937, p. 103, Plate 15, Figure 14.
Figured Hypotype. OSU 39812: length, 0.59 mm, width, 0.11 mm, López de Bertodano Formation, locality 164, Seymour Island.
Occurrence. Common. Occurs in low abundance in samples ranging from SI-K10 through SI-T2. Originally described from upper Taylor Marl and has a cosmopolitan distribution. The known range of this taxon is Santonian to Maastrichtian.

Bolivina incrassata Reuss, 1851
Figure 23.4-5

Bolivina incrassata REUSS 1851, p. 29, Plate 5, Figure 13.
Figured Hypotype. OSU 39813: length, 0.59 mm, width, 0.20 mm, breadth, 0.11 mm, López de Bertodano Formation, locality 38, Seymour Island.

Remarks. The size range of the antarctic specimens is from 0.38 to 0.59 mm in length and 0.18 to 0.20 mm width. The large forms strongly resemble *B. incrassata gigantea* Wicker, which, according to Sliter (1968, p. 88), is an ecophenotypic variant occurring in an upper shelf depositional environment. The smaller specimens more closely resemble *B. incrassata* s.s. There are too few specimens to determine whether the antarctic material can be separated at the subspecific level.
Occurrence. Rare. Single or few specimens occur in samples ranging from SI-K24 to SI-K36 on Seymour Island. This species occurs in upper Campanian-Maastrichtian localities worldwide.

Bolivina pustulata n. sp.
Figures 23.10, 32.7-9

Figured Holotype. USNM 415756 (Fig. 23.10): length, 0.18 mm, width, 0.10 mm, breadth, 0.06 mm, López de Bertodano Formation, locality 38, Seymour Island.
Figured Hypotype. OSU 39907 (Fig. 32.7-9): length, 0.17 mm, width, 0.10 mm, breadth, 0.07 mm, López de Bertodano Formation, locality 539, Seymour Island.
Etymology. Species name derived from the Latin *pustulosus,* meaning full of blisters, referring to the test surface ornamentation.
Diagnosis. Test small, tapered toward apex in side and edge views, chambers subglobular, wall surface moderately to densely covered with fine pustules, except for upper surface of last pair of chambers, aperture loop-shaped, extending up from base of final chamber, bordered by narrow lip.
Description. The test is small, planar to slightly twisted, moderately to strongly tapered toward the apex in side view, slightly tapered toward the apex in edge view, biserial with a rounded periphery. The chambers are subglobular, distinct, increasing moderately in size as well as inflation, and the sutures are distinct and depressed throughout. The wall is finely perforate, and ornamented by few to numerous fine pustules on all bu the upper portion of the last pair of chambers. The aperture is nearly terminal, extends up the final chamber face as an elongate loop, and is bordered by a narrow lip.
Remarks. The antarctic specimens of *B. pustulata* are distinguished from other bolivinids by their subglobular chambers and distinctive, pustulose ornament. The pustules on the Cretaceous forms are more pronounced and abundant than those of the Danian specimens.
Occurrence. Very rare. Several specimens occur in very low abundance in samples 38, 539, and 540, ranging from SI-K24 through SI-T2 (upper Campanian through Danian).

Bolivina sp.
Figure 23.6-9

Bolivina sp. BERTELS 1972, p. 340, Plate 2, Fig. 6a-b.
Figured Hypotypes. OSU 39815 (Fig. 23.6): length, 0.16 mm, width, 0.11 mm, breadth, 0.06 mm, López de Bertodano Formation, locality 110, Seymour Island; OSU 39816 (Fig. 23.7-8): length, 0.22 mm, width, 0.09 mm, breadth, 0.06 mm, López de Bertodano Formation, location 148, Seymour Island; OSU 39905 (Fig. 23.9): length, 0.22 mm, width, 0.12 mm, breadth, 0.07 mm, López de Bertodano Formation, locality 165, Seymour Island.
Description. The tests are elongate, biserial, sometimes slightly twisted with a rounded to subrounded periphery. The chambers are initially compressed, later inflated and reniform, gradually increasing in size. The sutures are distinct, initially flush, later depressed, and sigmoidal throughout. The wall is finely perforate with a smooth surface, and the aperture is an elongate loop extending from the base of the final chamber.
Remarks. The antarctic specimens are distinguished from *B. incrassata* Reuss by their smaller size, rounded periphery, and reniform chambers. They are very similar to specimens referred to as *Bolivina* sp. by Bertels

(1972), recovered from the Maastrichtian Jagüel Formation, Rio Negro Province, Argentina.

Occurrence. Rare. Few specimens recovered from samples ranging from SI-K8 to SI-K40 on Seymour Island.

Fursenkoina nederi Sliter, 1968
Figure 24.16-17

Fursenkoina nederi SLITER 1968, p. 112, Plate 20, Figure 1.

Figured Hypotype. OSU 39828: length, 0.84 mm, width, 0.33 mm, breadth, 0.24 mm, López de Bertodano Formation, locality 145, Seymour Island.

Remarks. The antarctic specimens are slightly larger, but very similar to the southern California forms described by Sliter (1968). The tests are finely perforate, biserial, and twisted about the long axis.

Occurrence. Very rare. Several specimens recovered from Seymour Island samples 394 and 148, ranging from SI-K30 through SI-K35. The California specimens have reported range of middle to late Campanian.

Hiltermannella kochi (Bertels, 1970b)
Figure 22.9-11

Hiltermannia kochi BERTELS 1970b, p. 170, Plate 1, Figures 1-6.

Figured Hypotypes. OSU 39817 (Fig. 22.9): length, 0.20 mm, width, 0.12 mm, breadth, 0.08 mm, López de Bertodano Formation, locality 28b, Seymour Island; OSU 39818 (Fig. 22.10-11): length, 0.29 mm, width, 0.13 mm, breadth, 0.09 mm, López de Bertodano Formation, locality 182, Seymour Island.

Remarks. This species is characterized by its elongate, strongly twisting, biserial test, which becomes nearly uniserial in adult specimens. The aperture is an elongate loop extending from the base of the final chamber, becoming areal and nearly terminal in adult specimens, with a bordering lip and simple tooth plate. The size range of the antarctic specimens is between 0.13 and 0.33 mm in length and 0.10 and 0.17 mm in width. They are identical to the Argentine topotypes. Specimens referred to as *Pseudouvigerina cretacea* Cushman by Webb (1971) may belong to this species. However, these New Zealand forms were unavailable for comparison.

Occurrence. Moderately common. Quite common in the López de Bertodano Formation, occurring in low to high abundance from SI-K8 through SI-K43 on Seymour Island; also occurs in moderate abundance within the Cape Lamb samples. Originally described from Maastrichtian sediments in upper Jagüel Formation, Rio Negro Province, Argentina (Bertels, 1970b). New Zealand forms resembling this species, referred to as *Pseudouvigerina cretacea* Cushman by Webb (1971), occurs in the Laidmore and Whangai Formations, which are Haumurian (upper Campanian-Maastrichtian) in age.

Conorbina anderssoni n. sp.
Figures 23.16-17, 33.7

Holotype. USNM 415758: diameter, 013 mm, breadth, 0.06 mm, López de Bertodano Formation, locality 28d, Seymour Island.

Figured Hypotypes. OSU 39892 (Figs. 23.16-17): diameter, 0.16 mm, breadth, 0.05 mm, López de Bertodano Formation, locality 57, Seymour Island; OSU 39897 (Fig. 33.7): diameter, 0.12 mm, López de Bertodano Formation, locality 380, Seymour Island.

Etymology. Species name honors John G. Andersson, geologist and participant in the Swedish South Polar Expedition to the James Ross Island region (1901-1903).

Diagnosis. Test small, plano-convex, moderately high-spired, chambers four to five in final whorl, aperture a narrow reentrant at base of final chamber with narrow bordering lip.

Description. The test is small, plano-convex, moderately high-spired, with a subangular periphery and distinct, narrow umbilicus; two to three whorls are visible on the spiral side, and only the last chambers of the final whorl are visible on the umbilical side. The chambers are distinct, lunate as seen from spiral side and internal umbilical view (Fig. 33.7), increasing gradually in breadth as added, having greater breadth than height, four to five occur in the final whorl, final chamber occupying as much as one-third of the umbilical side. The spiral side sutures are flush and strongly oblique; the umbilical sutures are slightly depressed, radial, and slightly curved. The wall has a few coarse perforations on the umbilical side, is finely perforate on the opposite side, and the surface is smooth. The aperture is a low slit at the base of the final chamber, in small re-entrant, midway between the umbilicus and periphery, with a narrow bordering lip.

Remarks. This form differs from *Conorbina* sp., figured by Tappan (1962, Plate 53, Fig. 1), in having coarse perforations on the umbilical side, more oblique spiral side sutures, and an extra-umbilical aperture. The antarctic specimens range in size from 0.12 to 0.16 mm in diameter and 0.05 to 0.10 mm in breadth.

Occurrence. Rare. Occurs in low abundance within Seymour Island samples, ranging from SI-K14 through SI-K43; low to moderate abundance within several Cape Lamb samples.

Epistominella glabrata (Cushman, 1938b)
Figure 23.18-20

Pulvinulinella glabrata CUSHMAN 1938b, p. 66, Plate 11, Figure 4.

Figured Hypotype. OSU 39821: diameter, 0.20 mm, breadth, 0.10 mm, López de Bertodano Formation, locality 184, Seymour Island.

Remarks. The antarctic specimens are somewhat smaller than Cushman's Gulf Coast specimens, ranging from 0.11 to 0.16 mm in diameter and 0.06 to 0.08 mm in breadth. Otherwise, they compare very well with the figured holotype.

Occurrence. Common. Occurs in low to moderate abundance in Seymour Island samples, ranging from SI-K10 through SI-K43. Its distribution in Gulf Coast region reported as upper Taylorian through Navarroan.

Serovaina orbicella Bandy, 1951
Figure 26.7-8

Gyroidina globosa (Hagenow) var. *orbicella* BANDY 1951, p. 505, Plate 74, Figure 2.

Figured Hypotype. OSU 39819: diameter, 0.20 mm, breadth, 0.11 mm, López de Bertodano Formation, locality 101, Seymour Island.

Remarks. The antarctic forms of this species have chambers that increase gradually in size, two whorls, and six to eight chambers in the final whorl. They are distinguished from species of *Gyroidina* by the more rounded final chamber face and a more gradual chamber size increase. The size ranges between 0.11 and 0.20 mm in diameter and 0.06 and 0.11 mm in breadth.

Occurrence. Rare. Few to abundant specimens occur in Seymour Island samples ranging from SI-K10 through SI-K43. A few specimens also found in the Cape Lamb samples.

Cibicides cf. *C. beaumontianus* (d'Orbigny, 1840)
Figures 23.21-22, 33.5-6

cf. *Truncatulina beaumontiana* D'ORBIGNY 1840, p. 35, Plate 3, Figures 17-19.

Figured Hypotypes. OSU 39822 (Fig. 23.21-22): diameter, 0.29 mm, breadth, 0.16 mm, López de Bertodano Formation, locality 148, Seymour Island; OSU 39896 (Fig. 33.5): diameter, 0.24 mm, López de Bertodano Formation, location 315, Seymour Island.

Remarks. The antarctic specimens have a plano-convex test that is subrounded in outline; a noncarinate, subrounded periphery; and six to seven chambers in the final whorl. The spiral side sutures are flush, strongly curved, and limbate, whereas those of the umbilical side are radial, slightly curved, initially flush, and later slightly depressed. The wall is microgranular, bilamellar (Fig. 33.5-6), finely perforate on both sides with a smooth surface. A low aperture with a narrow bordering lip extends from near the periphery on the umbilical side, part way onto the spiral side. The size ranges between 0.19 and 0.42 mm in diameter and 0.12 and 0.24 mm in breadth.

The antarctic material resembles *C. beaumontianus,* illustrated by Cushman (1946, Plate 65, Fig. 12). They also bear strong resemblance to the New Zealand forms referred to be Webb (1971) as *C.* aff. *beaumontianus,* except the latter have a carinate periphery and are larger in size (as much as 0.95 mm in diameter). Webb (1966) found that the European topotypes are not identical to Cushman's specimens nor to the New Zealand material. Sliter (1968, p. 109) placed *C. beaumontianus* within *Falsocibicides* because of the noncarinate periphery. His illustrated specimens show little resemblance to the antarctic and New Zealand forms. A further study of the wall ultrastructure and morphologic variability, as well as comparison to other forms referred to as this species, will be necessary to ascertain the correct taxonomic placement of the antarctic forms.

Occurrence. Moderately common. Occurs in low to moderate abundance within Seymour Island samples, ranging from SI-K2 through SI-T2. Several specimens also found in Cape Lamb samples.

Cibicides nordenskjoldi n. sp.
Figures 24.5-7, 33.9-10

Holotype. USNM 415754: diameter, 0.12 mm, breadth, 0.07 mm, López de Bertodano Formation, locality 28d, Seymour Island.
Figured Hypotypes. OSU 39822 (Fig. 24.5-7): diameter, 0.10 mm, breadth, 0.05 mm, López de Bertodano Formation, locality 38, Seymour Island, OSU 39823 (Fig. 33.9): diameter, 0.12 mm, López de Bertodano Formation, locality 315, Seymour Island; OSU 39824 (Fig. 33.10): diameter, 0.19 mm, López de Bertodano Formation, locality 380, Seymour Island.
Etymology. Species name honors Otto von Nordekskjöld, leader of the Swedish South Polar Expedition to the James Ross Island region (1901-1903).
Diagnosis. Test small, irregular oval in outline, plano-convex, megalospheric, evolute on both sides, periphery subrounded, noncarinate, three to four chambers follow proloculus, wall coarsely perforate on last two chambers of planar side, finely perforate near periphery of convex side, aperture a low interiomarginal slit extending part way along spiral suture on flattened side, with a narrow bordering lip.
Description. The test is small, megalospheric, evolute on both sides, single-coiled planispiral, plano-convex, irregularly oval-shaped in outline, with a noncarinate, subrounded periphery. The chambers are few, with an inflated prolocular chamber followed by three to four chambers that gradually increase in size (Fig. 33.9-10); the final chamber has a tendency to uncoil on the planar side. The sutures are flush and curved on the planar side, slightly depressed and curved on the convex side. The wall is microgranular, monolamellar, finely perforate near the periphery on the convex side, coarsely perforate on last two chambers of the planar side, with a smooth surface. The aperture is a low, asymmetric, interiomarginal opening extending from the periphery, part way along the spiral suture of the flattened side, bordered by a faint, narrow lip.
Remarks. The plano-convex test, apertural characteristics, and coarse perforations warrant placement in *Cibicides,* although a peripheral keel is absent. It resembles the Seymour Island specimens of *C.* cf. *C. beaumontianus,* but the latter does not have coarse perforations on the spiral side and has a bilamellar, rather than monolamellar wall microstructure. No other microspheric forms among the antarctic material show close resemblance to the megalospheric *C. nordenskjoldi.*
Occurrence. Common. Occurs in low to moderate abundance within samples ranging from SI-I2 to SI-K40 (upper Campanian through Maastrichtian) on Seymour Island.

Cibicides seymouriensis n. sp.
Figures 24.1-4, 33.13-15

Holotype. USNM 415755: diameter, 0.48 mm, breath, 0.25 mm, López de Bertodano Formation, locality 28d, Seymour Island.
Figured Hypotypes. OSU 39825 (Fig. 24.1-3): diameter, 0.41 mm, breadth, 0.18 mm, López de Bertodano Formation, locality 38, Seymour Island; OSU 39826 (Fig. 24.4): diameter, 0.16 mm, breadth, 0.10 mm, López de Bertodano Formation, locality 38, Seymour Island; OSU 39827 (Fig. 33.13): diameter, 0.35 mm, López de Bertodano Formation, locality 315, Seymour Island.
Etymology. Species name refers to type locality, Seymour Island.
Diagnosis. Test plano-convex, subrounded in outline, carinate, deep umbilicus surrounded by distinctive nodes protruding from umbilical side chambers, wall coarsely perforate on umbilical side, imperforate on spiral side, aperture extending from near periphery on spiral side, part way along the spiral suture, with a narrow bordering lip.
Description. The test is plano-convex, subrounded in outline, flattened to excavated on the spiral side, moderately to deeply umbilicate on opposite side, with a carinate periphery. The chambers increase gradually in size, six to seven occur in the final whorl, and there are two to two and one-half whorls. Distinct nodes are produced on the umbilical side of all or just the final two chambers of the last whorl, near the umbilical margin. The spiral side sutures are limbate, flush, and curved, and the sutures of the umbilical side are initially slightly depressed, later strongly depressed, radial, and slightly curved. The wall is microgranular with an irregular blocky, bilamellar microstructure (Fig. 33.13-15), it is moderately to coarsely perforate on the umbilical side, and imperforate on the peripheral keel, umbilical nodes, as well as all the spiral side. The aperture is a low, narrow opening extending from near the periphery on the umbilical side, across the periphery, part way along the spiral suture on the spiral side, with a narrow bordering lip.
Remarks. The flattened to excavated spiral side, peripheral keel, and apertural characteristics are typical of other cibicidids. However, it differs from other *Cibicides* species in having a deep umbilicus and coarsely perforate, nodular umbilical side. There is some variability in the perforation coarseness and degree of node development. The size variation is from 0.16 to 0.48 mm in diameter and 0.09 to 0.23 mm in breadth. All specimens bear a moderately- to well-developed keel.
Occurrence. Moderately common. The most common calcareous benthic of the López de Bertodano Formation. Occurs in low to high abundance, ranging from SI-K2 through SI-T2 on Seymour Island, with assigned age range of late Campanian through Danian; also occurs in moderate abundance in Cape Lamb samples.

Allomorphina cretacea Reuss, 1851
Figure 25.6-7

Allomorphina cretacea REUSS 1851, p. 42, Plate 5, Figure 6.
Figured Hypotype. OSU 39829: diameter, 0.48 mm, breadth, 0.39 mm, López de Bertodano Formation, locality 148, Seymour Island.
Occurrence. Moderately common. Occurs throughout Seymour Island sequence from SI-K2 through SI-K41; also occurs at Cape Lamb. Found in Upper Cretaceous sediments worldwide.

Allomorphina subtriangularis (Kline, 1943)
Figure 32.4

Chilostomella subtriangularis KLINE 1943, p. 56, Plate 6, Figure 3.
Figured Hypotype. OSU 39830: diameter, 0.90 mm, breadth, 0.63 mm, López de Bertodano Formation, locality 539, Seymour Island.
Remarks. This form differs from *A. cretacea* Reuss in being more elongate, with the final chamber occupying more than two-thirds the length of the test. The size ranges up to 0.97 mm in diameter.
Occurrence. Very rare. Quite abundant in SI-T2 sample 539 of the uppermost López de Bertodano Formation, which is Danian in age. This species was originally described from the Paleocene Midway Group in

Mississippi; has also been reported from upper Campanian–Maastrichtian sediments in southern California (Sliter, 1968).

Quadrimorphina allomorphinoides (Reuss, 1860)
Figure 25.4-5

Valvulinaria allomorphinoides REUSS 1860, p. 223, Plate 11, Figure 6.
Figured Hypotype. OSU 39831: diameter, 0.15 mm, breadth, 0.10 mm, López de Bertodano Formation, locality 184, Seymour Island.
Occurrence. Moderately common. On Seymour Island, ranges from SI-K10 through SI-T2. Found in Santonian-Maastrichtian strata in numerous Northern and Southern Hemisphere localities.

Nonionella cf. *N. robusta* Plummer, 1931
Figure 24.9-10

cf. *Nonionella robusta* PLUMMER 1931, p. 175, Plate 14, Figure 12.
Figured Hypotype. OSU 39832: diameter, 0.10 mm, breadth, 0.08 mm, López de Bertodano Formation, locality 184, Seymour Island.
Remarks. The antarctic forms are similar to Plummer's species in that the test is broader in peripheral view and there are fewer chambers than other species of *Nonionella.* They differ in that they are smaller (0.14 mm maximum), there are four to five rather than eight chambers in the final whorl, the chambers increase more rapidly in size, and the periphery is more broadly rounded. Further comparison of the antarctic material with other noninellids is necessary to determine if they represent a new species.
Occurrence. Rare. Very few specimens occur in Seymour Island samples ranging from SI-K2 through SI-K36.

Nonionella sp. 1
Figure 24.14-15

Figured Hypotype. OSU 39833: diameter, 0.18 mm, breadth, 0.10 mm, López de Bertodano Formation, locality 182, Seymour Island.
Description. The test is small, nearly planispiral, involute and somewhat compressed, with a rounded periphery. The chambers increase moderately in size and inflation, four to five occur in the final whorl, and the final chamber face is in the form of a slightly asymmetric broadly rounded, equilateral triangle. The sutures are slightly depressed, straight, and radial. The wall is finely perforate with a smooth surface, and the umbilical region is slightly depressed, free of secondarily deposited calcite. The aperture is a narrow, interiomarginal, equatorial opening, bordered by a thin raised lip and a row of multiple, areal apertures.
Remarks. The antarctic forms have fewer chambers in the final whorl and a lower final chamber face than other species of *Nonionella.* It differs from *Nonionella* cf. *robusta* Plummer in that the test is nearly planispiral and less broadly rounded in axial cross section.
Occurrence. Very rare. Occurs as single or few specimens in several samples, ranging from SI-K14 through SI-K38 on Seymour Island.

Nonionella sp. 2
Figure 32.12-13

Figured Hypotypes. OSU 39834 (Fig. 32.12-13): diameter, 0.15 mm, breadth, 0.08 mm, Sobral Formation, locality 540, Seymour Island.
Remarks. These forms are very small, low trochospiral with six to eight chambers in the final whorl, increasing moderately in size. They differ from *Nonionella* cf. *robusta* Plummer in that the chamber size increase is more gradual, the final chamber face is more rounded, and there are more chambers in the final whorl.
Occurrence. Very rare. Occurs in moderate abundance within samples 539 and 540 in SI-T2 of the uppermost López de Bertodano Formation and lowermost Sobral Formation.

Pullenia cretacea Cushman, 1936
Figure 25.2-3

Pullenia cretacea CUSHMAN 1936, p. 75, Plate 13, Figure 8.
Figured Hypotype. OSU 39835: diameter, 0.40 mm, breadth, 0.28 mm, López de Bertodano Formation, locality 184, Seymour Island.
Occurrence. Very rare. Occurs in very low abundance in Seymour Island samples, ranging from SI-K10 through SI-K43. A common form in Upper Cretaceous deposits worldwide.

Pullenia jarvisi Cushman, 1936
Figure 24.18-19

Pullenia jarvisi CUSHMAN 1936, p. 77, Plate 13, Figure 6.
Figured Hypotype. OSU 39836: diameter, 0.30 mm, breadth, 0.14 mm, López de Bertodano Formation, locality 28a, Seymour Island.
Remarks. This form differs from *P. cretacea* Cushman in having a more compressed test, a higher, less broadly rounded chamber face, and radial sutures that are slightly curved and depressed. The antarctic specimens differ from the lower latitude forms in being more compressed and lacking deep umbilici.
Occurrence. Very rare. On Seymour Island, species ranges from SI-K14 through SI-T2, occurring in low abundance.

Pullenia sp.
Figure 25.1

Figured Hypotype. OSU 39837: diameter, 0.17 mm, breadth, 0.13 mm, López de Bertodano Formation, locality 182, Seymour Island.
Description. The test is small, planispiral, and involute, with a broadly rounded periphery. The chambers increase rapidly in size; five are in the final whorl; the final chamber face is flat, low, and broad; and the sutures are initially flush, later slightly depressed, radial, and straight. The wall is calcareous, with a smooth surface, and the umbilici are very shallow. The aperture is a low arch restricted to the equatorial periphery at the base of the final chamber face.
Remarks. The small test size, rapidly increasing chamber size, and small, distinctive aperture distinguish this form from other species of *Pullenia.*
Occurrence. Very rare. Single specimens occur in samples 148, 182, and 356, ranging from SI-K35 through SI-K39 on Seymour Island.

Alabamina creta (Finlay, 1940)
Figure 25.8-9

Pulvinulinella creta FINLAY 1940, p. 463, Plate 66, Figures 187-192.
Figured Hypotype. OSU 39838: diameter, 0.27 mm, breadth, 0.16 mm, López de Bertodano Formation, locality 110, Seymour Island.
Remarks. The antarctic forms are identical to the New Zealand topotypes. The size range is considerable, from 0.12 to 0.37 mm in diameter and 0.09 to 0.25 mm in breadth. In a comparative study of New Zealand and European topotypes, Webb (1966) determined that *A. creta* and *A. dorsoplana* Brotzen are synonymous.
Occurrence. Moderately common. On Seymour Island, it ranges from SI-K8 through SI-T2, occurring in low to high abundance; also occurs at Cape Lamb. In New Zealand, ranges from Haumurian through Teurian stages (upper Campanian–Paleocene).

Alabamina westraliensis (Parr, 1938)
Figure 25.13-14

Pulvinulinella obtusa (Burrows and Holland), var. *westraliensis* PARR 1938, p. 84, Plate 3, Figure 1.
Figured Hypotype. OSU 39839: diameter, 0.48 mm, breadth, 0.27 mm, López de Bertodano Formation, locality 38, Seymour Island.
Remarks. This species is distinguished from *A. creta* (Finlay) by its bluntly acute periphery and greater diameter/height ratio. An identical form, *A. wilcoxensis* Toulmin, described from several North American lower Tertiary localities, was placed in the synonymy of *A. westraliensis*

by McGowran (1965). Webb (1987) discussed the occurrence of *A. westraliensis* in Upper Cretaceous deposits in southeast Otago, New Zealand. His topotypes are identical to the antarctic material. The antarctic specimens show considerable size variation, from 0.21 to 0.47 mm in diameter and 0.13 to 0.31 mm in breadth, as well as some variation in the degree in spiral side convexity, from planar to moderately convex.

Occurrence. Moderately common. *A. westraliensis* originally described from the Paleocene King's Park shale, in Perth, western Australia. *A. wilcoxensis* (=*A. westraliensis*) first described from the lower Eocene Salt Mountain Limestone (lower Wilcox Group) of Alabama (Toulmin, 1941). Only reported Late Cretaceous distribution is in Haumurian (upper Campanian–Maastrichtian) Brighton Limestone and Barrons Hill Shelly Greensand member in New Zealand (Webb, 1988), and the present study. On Seymour Island, ranges from SI-K2 through SI-T2, occurring in low to high abundance; also found in several samples collected at Cape Lamb. Earlier occurrence in New Zealand and Antarctic Peninsula is a significant example of the origination of some foraminiferal taxa in southern high latitudes.

Gyroidina cf. *G. globosa* (Hagenow, 1842)
Figure 26.3-4

Figured Hypotype. OSU 39844: diameter, 0.44 mm, breadth, 0.34 mm, López de Bertodano Formation, locality 13b, Seymour Island.

Remarks. The Seymour Island specimens are small, with a sharply rounded peripheral margin, six chambers in the final whorl, a slightly depressed spiral suture, and flush chamber sutures. It differs from the USNM type material by its smaller size and fewer chambers in the final whorl.

Occurrence. Single specimens were recovered from samples 12 and 13b of SI-K2 and SI-K3, respectively, on Seymour Island.

Gyroidinoides nitidus (Reuss, 1845)
Figure 26.5-6

Rotalina nitida REUSS 1845, p. 35, Plate 8, Figure 52, Plate 12, Figures 8, 20.

Figured Hypotype. OSU 39840: diameter, 0.32 mm, breadth, 0.21 mm, López de Bertodano Formation, locality 115, Seymour Island.

Remarks. The antarctic specimens show considerable variability in test size, ranging from 0.21 to 0.47 mm in diameter and 0.13 to 0.31 mm in breadth, and test convexity. In addition, there is a wide range of variation in flatness of the final chamber face, development of an apertural flap, and sutural depression. In the antarctic material, megalospheric specimens have a single whorl of six to seven chambers following the proloculus, and the proloculus of microspheric specimens is followed by two and one-half whorls, with eight chambers in the final whorl.

Occurrence. Moderately common. Ranges from SI-K2 through SI-K43 on Seymour Island, occurring in low to high abundance; also occurs in moderate abundance in several Cape Lamb samples.

Osangularia cordieriana (d'Orbigny, 1840)
Figure 25.17-18

Rotalina cordieriana D'ORBIGNY 1840, p. 33, Plate 3, Figures 9-11.

Figured Hypotype. OSU 39841: diameter, 0.47 mm, breadth, 0.23 mm, López de Bertodano Formation, locality 177, Seymour Island.

Occurrence. Very rare. Single specimens recovered from SI-K8 and SI-K31 on Seymour Island.

Charltonina? sp.
Figure 25.10-12

Figured Hypotype. OSU 39842: diameter, 0.28 mm, breadth, 0.15 mm, López de Bertodano Formation, locality 110, Seymour Island.

Remarks. The figured specimen is small, inequally biconvex, with an acute peripheral margin, five chambers in the final whorl, strongly,

oblique spiral side sutures, and radial, slightly curved umbilical sutures. It resembles *Charltonina acutimarginata* Finlay, except it is smaller, its aperture does not extend parallel to the periphery, and it has a less developed peripheral keel. Until better preserved specimens are found, identification of this form will remain tentative.

Occurrence. Very rare. Several poorly preserved specimens occur on Seymour Island, ranging from SI-K8 through SI-K34.

Globorotalites? sp.
Figure 26.1-2

Figured Hypotype. OSU 39843: diameter, 0.23 mm, breadth, 0.12 mm, López de Bertodano Formation, locality 184, Seymour Island.

Remarks. The figured specimen cannot be positively identified because of its poor preservation. The apertural characteristics are not discernable. It is placed in *Globorotalites* because of the strongly oblique dorsal side sutures and deep umbilicus on the ventral side.

Occurrence. Very rare. Single specimen recovered from sample 184 of SI-K36 on Seymour Island.

Anomalinoides larseni n. sp.
Figures 26.12-13, 33.8

Holotype. USNM 415757: diameter, 0.20 mm, breadth, 0.11 mm, López de Bertodano Formation, locality 28d, Seymour Island.

Figured Hypotypes. OSU 39845 (Fig. 26.12-13): diameter, 0.22 mm, breadth, 0.13 mm, López de Bertodano Formation, locality 28b, Seymour Island; OSU 39908 (Fig. 33.8): diameter, 0.28 mm, López de Bertodano Formation, locality 380, Seymour Island.

Etymology. Species name honors Captain C. A. Larsen, who was among the first to recognize fossils from Seymour Island.

Diagnosis. Test subrounded in outline, periphery rounded, seven to nine chambers in final whorl, wall with loosely arranged, coarse perforations, aperture a low interiomarginal arch extending from near spiral side periphery, part way toward umbilicus, with narrow bordering lip.

Description. The test is small, subrounded in outline, nearly biconvex to plano-convex, axial periphery rounded, very low trochospiral, evolute to partially involute on the spiral side, involute on the umbilical side. The chambers increase moderately in size, becoming slightly inflated, seven to nine compose the final whorl, with two to two and one-half whorls total. The sutures on both sides are limbate but not elevated, slightly curved, initially flush, later somewhat depressed. The wall is microgranular, with moderate to coarse perforations and a smooth surface. The aperture is a low interiomarginal arch that extends from near the periphery on the spiral side, part way toward the umbilicus, with a narrow bordering lip.

Remarks. This species is most similar to *A. piripaua* Finlay, but is distinguished by its smaller size, more rapid increase of chamber size, lower test breadth/diameter ratio, less densely perforated wall, and more involute spiral side. It ranges in size from 0.11 to 0.34 mm in diameter and 0.07 to 0.21 mm in breadth.

Occurrence. Moderately common. Occurs in low to high abundance, ranging from SI-K2 through SI-K43 on Seymour Island. Specimens also recovered from the Cape Lamb samples.

Anomalinoides piripaua (Finlay, 1939)
Figure 26.14-16

Anomalina piripaua FINLAY 1939, p. 325–326, Plate 128, Figures 141-143.

Figured Hypotype. OSU 39846: diameter, 0.47 mm, breadth, 0.24 mm, López de Bertodano Formation, locality 148, Seymour Island.

Remarks. The antarctic specimens are identical to the New Zealand topotypes. The size range of the antarctic forms is from 0.20 to 0.65 mm in diameter and 0.11 to 0.31 mm in breadth.

Occurrence. Moderately common. Occurs throughout López de Bertodano Formation; also found in basal Sobral Formation. Occurs in low to high abundance from SI-K2 through SI-T2 on Seymour Island, and is abundant in the Vega Island samples. Has reported range of Haumurian-Teurian (upper Campanian–Paleocene) in New Zealand (Webb, 1971).

Anomalinoides rubiginosus (Cushman, 1926b)
Figure 26.9-11

Anomalina rubiginosa CUSHMAN 1926b, p. 607, Plate 21, Figure 6a-c.
Figured Hypotype. OSU 39847; diameter, 0.41 mm, breadth, 02.5 mm, López de Bertodano Formation, locality 13b, Seymour Island.
Occurrence. Very rare. Ranges from SI-K2 through SI-K43 on Seymour Island, occurring in very low abundance; common in Upper Cretaceous sediments of Gulf Coastal region.

Gavelinella sandidgei (Brotzen, 1936)
Figure 26.17-18

Cibicides sandidgei BROTZEN 1936, p. 191, Plate 14, Figures 2-4.
Figured Hypotype. OSU 39848: diameter, 0.12 mm, breadth, 0.06 mm, López de Bertodano Formation, locality 184, Seymour Island.
Remarks. The antarctic specimens are much smaller than the North American and European forms of this species. Otherwise, they compare quite well.
Occurrence. Common. Occurs in very low abundance from SI-K10 through SI-K43 on Seymour Island. Originally described from the Campanian-Maastrichtian of Sweden.

Guembelitria cretacea Cushman, 1933
Figure 27.1

Gümbelitria cretacea CUSHMAN 1933, p. 37, Plate 4, Figure 12.
Figured Hypotype. OSU 39852: length, 0.11 mm, width, 0.10 mm, López de Bertodano Formation, locality 177, Seymour Island.
Occurrence. Very rare. Only one specimen found, occurring in sample 177 of SI-K31. This species commonly occurs in mid- to outer shelf marine facies of late Campanian to Maastrichtian age.

Heterohelix glabrans (Cushman, 1938b)
Figure 27.2-3

Gümbelina glabrans CUSHMAN 1938b, p. 15, Plate 3, Figures 1-2.
Figured Hypotype. OSU 39853: length, 0.27 mm, width, 0.17 mm, breadth, 0.06 mm, López de Bertodano Formation, locality 33, Seymour Island.
Occurrence. Very rare. Occurs as single specimens in several Seymour Island samples, ranging from SI-K14 through SI-K34; range elsewhere has been recorded as late Campanian–Maastrichtian.

Heterohelix globulosa (Ehrenberg, 1839)
Figure 27.4-5

Textularia globulosa EHRENBERG 1839 [1840], p. 135, Plate 4, Figure 4b.
Figured Hypotypes. OSU 39854 (Fig. 27.4): length, 0.20 mm, width, 0.12 mm, breadth, 0.07 mm, López de Bertodano Formation, locality 148, Seymour Island; OSU 39906 (Fig. 27.5): length, 0.21 mm, width, 0.18 mm, breadth, 0.11 mm.
Occurrence. Common. Ranges from SI-K13 through SI-K43 on Seymour Island, occurring in low abundance. Several specimens have also been recovered from the Vega Island samples.

Globigerinelloides multispinatus (Lalicker, 1948)
Figure 27.6-7, 10-13

Biglobigerinella multispina LALICKER 1948, p. 624, Plate 92, Figures 1-3.
Figured Hypotypes. OSU 39855 (Fig. 27.6-7): diameter, 0.14 mm, breadth, 0.05 mm, López de Bertodano Formation, locality 165, Sey-

mour Island; OSU 39856 (Fig. 27.10-11): diameter, 0.16 mm, breadth, 0.08 mm, López de Bertodano Formation, location 28a, Seymour Island; OSU 39857 (Fig. 27.12): diameter, 0.15 mm, breadth, 0.08 mm, López de Bertodano Formation, locality 28b, Seymour Island; OSU 39858 (Fig. 27.13): diameter, 0.23 mm, breadth, 0.17 mm, López de Bertodano Formation, locality 165, Seymour Island.
Remarks. These forms are small, with finely hispid to moderately pustulose surface ornament and highly arched apertures that are bordered by a narrow lip. Biapertural forms are rare.
Occurrence. Common. Occurs in low to moderate abundance on Vega Island and in samples from SI-K5 through SI-K43 on Seymour Island. A common species in Santonian-Maastrichtian assemblages worldwide.

Hedbergella monmouthensis (Olsson)
Figure 27.14-17

Globorotalia monmouthensis OLSSON 1960, p. 47, Plate 9, Figures 22-24.
Hedbergella holmdelensis Olsson, SLITER 1976, p. 542, Plate 2, Figures 1-3.
Hedbergella monmouthensis (Olsson), WEBB 1973b, p. 552, Plate 4, Figures 1-4.
Figured Hypotypes. OSU 39860 (Figs. 27.14-16): diameter, 0.19 mm, breadth, 0.11 mm, López de Bertodano Formation, locality 28b, Seymour Island; OSU 39861 (Fig. 27.17): diameter, 0.30 mm, breadth, 0.16 mm, López de Bertodano Formation, locality 28b, Seymour Island.
Remarks. This species differs from *Hedbergella holmdelensis* Olsson primarily in having more globular chambers and a more rapid increase in chamber size. The wall surface of the antarctic specimens varies from being nearly smooth to moderately pustulose.
Occurrence. Moderately common. Occurs in low abundance in Seymour Island samples ranging from SI-K10 through SI—K43 (upper Campanian through Maastrichtian), and is the nominal species of the *Hedbergella monmouthensis* Zone. This species confined to the Maastrichtian in the North Atlantic region (R. K. Olsson, personal communication, 1986) and has been reported from upper Campanian through Maastrichtian deposits elsewhere.

Hedbergella sp.
Figure 27.8-9

Figured Hypotypes. OSU 39859: diameter, 0.23 mm, breadth, 0.12 mm, López de Bertodano Formation, locality 26, Seymour Island.
Remarks. This form differs from *Hedbergella monmouthensis* (Olsson) by its lower trochospiral coiling, broader umbilicus, and nearly equatorial position and lower arch of the aperture. It is distinguished from *H. holmdelensis* Olsson by its more globular chambers and broader umbilical region.
Occurrence. Rare. Few specimens recovered from samples 12, 26, 415, 165, and 28a-d, ranging from SI-K2 through SI-K14 on Seymour Island.

Rugoglobigerina rotundata Brönnimann, 1952
Figure 28.12-14

Rugoglobigerina rugosa rotundata BRÖNNIMANN 1952, p. 34–36, Plate 4, Figures 7-9, text-figures, 15a-e, 16a-c. SLITER 1976, p. 542, Plate 11, Figures 1-3. KRASHENINNIKOV AND BASOV 1983, p. 807, Plate 11, Figures 7-11.
Figured Hypotype. OSU 39868: diameter, 0.25 mm, breadth, 0.23 mm, López de Bertodano Formation, locality 28b, Seymour Island.
Remarks. This species is distinguished by its high trochospiral test, which bears coarse, randomly situated pustules. The Seymour Island specimens have a reduced final chamber and poorly preserved tegilla.
Occurrence. Very rare. Sliter (1976) and Krasheninnikov and Basov (1983) reported species range as late Campanian–Maastrichtian in the Falkland Plateau region, whereas Robaszynski and others (1984) sug-

gested mid- to late Maastrichtian range. On Seymour Island, single specimens of *R. rotundata* occur in samples 415, 28d, and 38, ranging from SI-K14 through SI-K24.

Rugoglobigerina rugosa (Plummer, 1927)
Figure 28.1-11

Globigerina rugosa PLUMMER 1927, p. 38, Plate 2, Figure 10.
Figured Hypotypes. OSU 39869 (Figure 28.1-3): diameter, 0.29 mm, breadth, 0.16 mm, López de Bertodano Formation, locality 28a, Seymour Island; OSU 39870 (Fig. 28.4-5): diameter, 0.39 mm, breadth, 0.20 mm, López de Bertodano Formation, locality 148, Seymour Island; OSU 39871 (Fig. 28.6-7, 11): diameter, 0.49 mm, breadth, 0.31 mm, López de Bertodano Formation, locality 182, Seymour Island; OSU 39872 (Figs. 28.8-10): diameter, 0.34 mm, breadth, 0.19 mm, López de Bertodano Formation, location 415, Seymour Island.
Remarks. Most of the antarctic specimens show a faint development of meridionally aligned costellae on at least one chamber (e.g., Fig. 28.1-3, 8-10). Those that do not display this biocharacter have instead irregularly situated, fine to medium pustules (e.g., Fig. 28.4-7, 11). The number of chambers in the final whorl ranges from 4.5 to 5, and the rate of chamber size increase varies between moderate and rapid. The umbilical primary aperture of well-preserved specimens is often covered by thin to broad tegilla, which are perforated by infra- and intralaminal accessory apertures. The Seymour Island specimens are large (between 0.25 and 0.40 mm in diameter), and sometimes have an axially compressed final chamber. Lower latitude forms of this species have a coarser surface ornament and more axially compressed chambers in the final whorl.
Occurrence. Common. Robaszynski and others (1984) stated this taxon ranges from the Campanian through the Maastrichtian. On Seymour Island, occurs in very low abundance from SI-K10 through SI-K37.

Rugoglobigerina? sp. 1
Figures 29.1-14, 30.5-10

Hedbergella monmouthensis (Olsson), KRASHENINNIKOV & BASOV 1983, p. 804–805, Plate 6, Figures 5-8.
Rugoglobigerina pilula Belford, SLITER 1976, p. 542, Plate 10, Figures 7-9. KRASHENINNIKOV & BASOV 1983, p. 807, Plate 11, Figures 3-6.
Rugoglobigerina pustulata Brönnimann, KRASHENINNIKOV & BASOV 1983, p. 806, Plate 10, Figures 10-13.
Figured Hypotypes. OSU 39862 (Fig. 29.1-3): diameter, 0.25 mm, breadth, 0.14 mm, López de Bertodano Formation, locality 13b, Seymour Island; OSU 39863 (Fig. 29.4): diameter, 0.23 mm, breadth, 0.14 mm, López de Bertodano Formation, locality 164, Seymour Island; OSU 39864 (Fig. 29.5-7): diameter, 0.19 mm, breadth, 0.12 mm, López de Bertodano Formation, locality 28c, Seymour Island; OSU 39865 (Fig. 29.8-9): diameter, 0.17 mm, breadth, 0.09 mm, López de Bertodano Formation, locality 8, Cape Lamb, Vega Island; OSU 39866 (Fig. 29.10-11): diameter, 0.17 mm, breadth, 0.07 mm, López de Bertodano Formation, locality 28c, Seymour Island; OSU 39867 (Figs. 29.12-14): diameter, 0.23 mm, breadth, 0.13 mm, López de Bertodano Formation, locality 164, Seymour Island; OSU 39873 (Fig. 30.5-7): diameter, 0.34 mm, breadth, 0.19 mm, López de Bertodano Formation, locality 415, Seymour Island; OSU 39874 (Fig. 30.8): diameter, 0.35 mm, breadth, 0.21 mm, López de Bertodano Formation, locality 415, Seymour Island; OSU 39875 (Fig. 30.9-10): diameter, 0.21 mm, breadth, 0.11 mm, López de Bertodano Formation, locality 415, Seymour Island.
Remarks. The forms included in *Rugoglobigerina?* sp. 1 show considerable variability in: (1) rate of chamber size increase in the final whorl, (2) number of chambers in the final whorl, (3) test surface ornamentation, (4) apertural size and position, and (5) size of the ultimate chamber. Identical upper Campanian and lower Maastrichtian specimens occur in much higher abundance at Falkland Plateau DSDP Sites 327A and 511 and have been studied for comparison. Several methods were used to

determine that the observed variability is caused by ontogenetic, and not genetic, differences (B. T. Huber, in preparation). These include test dissection combined with study of the ontogeny using the scanning electron microscope and x-ray image analysis. This study has found that chamber size increase is more rapid, the apertural position is more extraumbilical, and the test surface is less densely ornamented in the dissected juvenile chambers of larger specimens. Thus, morphotypes with between 4.5 and 5.25 chambers in the final whorl that gradually increase in size (see Fig. 29.12-14, this study; Sliter, 1976, Plate 10, Figs. 7-9; Krasheninnikov and Basov, 1983, Plate 11, Figs. 3-6, Plate 10, Figs. 10-13), a moderately to coarsely pustulose surface with rare occurrence of meridionally arranged costellae, and an umbilical aperture are considered as "adult morphovariants." Smaller specimens with between 3.75 and 4.75 chambers in the final whorl which increase moderately in size (see Figs. 29.1-11, 30.5-12, this study; Krasheninnikov and Basov, 1983, Plate 6, Figs. 5-8), a finely to moderately pustulose surface lacking meridionally arranged costae, and an umbilical-extraumbilical aperture are considered as "juvenile morphovariants." Specimens bearing kummerform chambers are identical to nonkummerform specimens once the ultimate chambers are removed. Portical structures are rarely preserved in the Seymour Island material, but they occur more frequently among the Falkland Plateau specimens. No tegilla have been observed on specimens from either locality.

The absence of tegilla and meridionally arranged costae makes the generic assignment to *Rugoglobigerina* dubious based on the original diagnosis of this genus. However, Blow's (1979, p. 1365–1366) emended definition permits considerably greater variability. He noted that ". . . the degree of serial organisation of the muricae is both variable in ontogeny of the individual specimen of a taxon and variable ecophenotypically within a population in response to habitat. Thus, *Rugoglobigerina* spp. from tropical/equatorial regions show a massive development of so-called 'meridionally" arranged chamber costae but in extra-tropical regions these costae, in the same taxon, become more discontinuous, less coherently fused as linear structures and with the test muricae frequently more randomly situated over the surfaces of the primary chambers." He further stated; "Thus, even in specimens from fully tropical regions, the costae are poorly organised in juveniles and the muricae more randomly situated over the early primary chamber walls of the tests as compared to the adults whose costal organization may be very considerable." The morphologic similarity of specimens included in *Rugoglobigerina rugosa* and *Rugoglobigerina?* sp. 1 indicates their close relationship. A detailed study of specimens of *Rugoglobigerina* spp. from the low to high latitudes is presently underway (B. T. Huber, in preparation) to further evaluate the generic and specific status of the high latitude taxa.
Occurrence. Moderately common. Occurs in low abundance in samples ranging from SI-K2 through SI-K36 (upper Campanian through Maastrichtian) on Seymour Island and in several Vega Island samples. Falkland Island specimens included in *Rugoglobigerina?* sp. 1 range from upper Campanian through lower Maastrichtian.

Rugoglobigerina? sp. 2
Figure 31.12, 16

Figured Hypotypes. OSU 39886 (Pl. 16, Figs. 12, 16): diameter, 0.32 mm, breadth, 0.23 mm, López de Bertodano Formation, locality 165, Seymour Island.
Remarks. Assignment of the figured specimen to *Rugoglobigerina* is tentative since it shows little resemblance to typical forms of this genus. It is biconvex, globular, with a quadrate axial periphery, and has four and one-half axially compressed chambers in the final whorl, with randomly distributed pustules on the chamber surface. Its aperture is indistinguishable since it is covered by a thick umbilical bulla that bears infralaminal accessory apertures.
Occurrence. Very rare. Figured hypotype the only representative recognized; occurs in SI-K14 (sample 165).

Rugotruncana? sp.
Figures 30.1-4; 31.5-6, 9-11

Figured Hypotypes. OSU 39876 (Figs. 30.1-5): diameter, 0.31 mm, breadth, 0.15 mm, López de Bertodano Formation, locality 110, Seymour Island; OSU 39877 (Fig. 31.5): diameter, 0.23 mm, breadth, 0.14 mm, López de Bertodano Formation, locality 415, Seymour Island; OSU 39878 (Fig. 31.6): diameter, 0.47 mm, breadth, 0.22 mm, López de Bertodano Formation, locality 415, Seymour Island; OSU 39879 (Fig. 31.9-11): diameter, 0.25 mm, breadth, 0.15 mm, López de Bertodano Formation, locality 415, Seymour Island; OSU 39887 (Figs. 31-13-15): diameter, 0.15 mm, breadth, 0.09 mm, López de Bertodano Formation, locality 165, Seymour Island.

Remarks. Included in this group are forms that are low trochospiral, inequally biconvex to plano-convex with a strongly lobate equatorial periphery and a rounded to subrounded axial periphery. Several specimens (e.g., Fig. 30.1-4) show a faint meridional alignment of pustules and spines, some of which coalesce into rugae. An equatorial keel is weakly developed on the penultimate and earlier chambers along an imperforate peripheral band (see Fig. 30.4), and a second keel is faintly discernable on the umbilical side of some specimens (see Figs. 30.1; 31.5, 9-10). Some forms (e.g., Fig. 31.5, 6, 9) show an axially thickened imperforate band on the umbilical sides of the second and third chambers of the final whorl, extending from the peripheral margin toward the earlier chamber and hooking back near the suture. Portici occur on some specimens, whereas others show thin tegilla with infralaminal accessory apertures. The figured antarctic forms are similar to specimens of *Rugotruncana circumnodifer* (Finlay) (see Finlay, 1940, Plate 65, Figs. 150–157; Webb, 1973b, Plate 4, Figs. 1-4), but the latter show more prominent peripheral keels and rugosities. Olsson's (1964, Plate 6, Figs. 7-8) figured hypotypes of *Globotruncana subrugosa* Gandolfi show some resemblance to *Rugotruncana*? sp.

Occurrence. Very rare. Occurs in very low abundance in samples ranging from SI-K8 through SI-K14 on Seymour Island.

Globotruncanella havanensis (Voorwijk, 1937)
Figure 31.3-3

Globotruncana havanensis VOORWIJK 1937, p. 195, Plate 1, Figures 25, 26, 29.

Figured Hypotype. OSU 39880: diameter, 0.31 mm, breadth, 0.15 mm, López de Bertodano Formation, locality 33, Seymour Island.

Remarks. This form is distinguished by its inequally biconvex, lobate test that bears a subacute equatorial periphery and a faint single keel. The figured specimen has an apertural cover in the form of a tegilum with infralaminal accessory apertures. The chamber surface is covered with pustules that are randomly situated except along the periphery, where they are aligned in the form of a keel. The antarctic specimen differs from lower latitude forms of this species in that its test is smaller, less compressed, less petaloid, and the peripheral keel is poorly developed.

Occurrence. Very rare. Single specimen was recovered from Seymour Island within SI-K20; the cosmopolitan range of this species is upper Campanian through Maastrichtian.

Globotruncanella minuta Caron & Gonzalez Donoso, 1984
Figure 30.11-17

Hedbergella monmouthensis (Olsson), PESSAGNO 1967, Plate 61, Figures 1-3. SLITER 1976, p. 542, Plate 3, Figures 1-3.

Globotruncanella minuta CARON & GONZALEZ DONOSO, *in* Robaszynski and others, 1984, p. 266, Plate 43, Figures 5-8.

Figured Hypotypes. OSU 39882 (Fig. 30.11-12): diameter, 0.20 mm, breadth, 0.11 mm, López de Bertodano Formation, locality 28c, Seymour Island; OSU 39883 (Fig. 30.13-14): diameter, 0.11 mm, breadth, 0.08 mm, López de Bertodano Formation, locality 149b, Seymour Island; OSU 39884 (Fig. 30.15): diameter, 0.17 mm, breadth, 0.09 mm, López de Bertodano Formation, locality 415, Seymour Island; OSU

39885 (Fig. 30.16-17): diameter, 0.27 mm, breadth, 0.15 mm, López de Bertodano Formation, locality 148, Seymour Island.

Remarks. This species was erected to include small, globular forms with chambers that increase rapidly in size, showing a strong resemblance to *Hedbergella monmouthensis* (Olsson) but bearing well-developed, merging portici. The surface ornamentation varies between being finely spinose to densely pustulose, with a more finely ornamented test surface on the final chamber than on the previous chambers of the last whorl. The figured specimen shown in Figure 30.13-14 is identical to Sliter's (1976, Plate 3, Figs. 1-3) figured hypotype of *Hedbergella monmouthensis*. Both of these forms appear to have broken portici in their umbilical views. The antarctic specimen shown in Figure 30.15 shows well-developed portici that merge along the chamber margins in the umbilical region. The umbilicus of the specimen shown in Figure 30.16-17 is completely covered by a thin tegilum that bears infralaminal accessory apertures.

Occurrence. Rare. Antarctic specimens assigned to this species occur in very low abundance, ranging from SI—K10 through SI-K38. Stratigraphic distribution of *G. minuta* given as middle to upper Maastrichtian (Robaszynski and others, 1984). It was noted by these authors, however, that its range is insufficiently documented. They suggested that it may provide the "missing link" between the *Hedbergella monmouthensis-holmdelensis* group and typical globotruncanellids. Therefore this taxon's earlier occurrence in Antarctica may lend further credence to their phylogenetic interpretation.

Globotruncanella? sp.
Figure 31.4, 7-8

Figured Hypotype. OSU 39881: diameter, 0.18 mm, breadth, 0.10 mm, López de Bertodano Formation, locality 411, Seymour Island.

Remarks. The figured specimen is similar to *G. havanensis* in having an inequally biconvex, strongly lobate test with trapezoidal, slightly compressed chambers that are covered with randomly situated pustules. The antarctic specimens are more globular than typical forms of *G. havanensis*, however, and lack a peripheral keel. Because of the broken final chamber on the figured hypotype, the position and size of the aperture cannot be determined. The umbilical region appears to be covered by a tegilum, as an infralaminal accessory aperture is apparent, bordering the second chamber of the final whorl.

Occurrence. Very rare, This form is represented by several specimens recovered from samples 411, 526, and 535 in SI-K43 on Seymour Island.

Globastica daubjergensis (Brönnimann, 1953)
Figure 32.10-11, 14-17

Globigerina daubjergensis BRÖNNIMANN 1953, p. 340, Plate 10, Figures 1-3; Plate 22, Figures 1-6; Plate 23, Figures 1-2.

Globastica daubjergensis (Brönnimann), BLOW 1979, p. 1235– 1240, Plate 74, Figures 7-9, Plate 256, Figures 1-9, Plate 257, Figures 3-4.

Figured Hypotype. OSU 39901 (Fig. 32.10): diameter, 0.11 mm, breadth, 0.08 mm, López de Bertodano Formation, locality 539, Seymour Island; OSU 39902 (Fig. 32.11): diameter, 0.07 mm, breadth, 0.05 mm, López de Bertodano Formation, locality 539, Seymour Island; OSU 39888 (Fig. 32.14-16): diameter, 0.13 mm, breadth, 0.08 mm, Sobral Formation, locality 540, Seymour Island; OSU 39889 (Fig. 32.17): diameter, 0.08 mm, breadth, 0.05 mm, Sobral Formation, locality 540, Seymour Island.

Remarks. The antarctic specimens are in good agreement with specimens described from the Danian beds of Denmark. They are very small, high spired, have a spinose wall, have three to three and one half chambers in the final whorl, and have supplementary apertures along the spiral sutures. The earliest occurring Seymour Island forms are distinctly less pustulose than the younger morphotypes (compare Figure 32.10-11 with 32.14-17).

Occurrence. Very rare. Several specimens recovered from samples 539, 540, and 515 of SI-T2, ranging from the top of the López de Bertodano Formation (sample 539) to the top of Unit 1 in the (sample 515) Sobral Formation on Seymour Island. This species used as an index fossil for the Danian worldwide. It is the nominal taxon for the *G. daubjergensis* Zone on Seymour Island.

ACKNOWLEDGMENTS

This study would not have been possible without the support and encouragement from P. N. Webb, D. H. Elliot, and W. J. Zinsmeister. I have greatly profited from my association with C. E. Macellari, who assisted with collection of the samples. I thank W. V. Sliter, N. Malumian, R. A. Askin, P. Sadler, D. M. Harwood, and S. E. Ishman for helpful discussions. I gratefully acknowledge P. N. Webb, D. H. Elliot, and C. E. Macellari for reviewing early drafts of this manuscript, and D. M. Harwood, R. M. Leckie, and W. V. Sliter for reviewing the final draft. I appreciate the assistance of K. Beatley, M. Grant, M. Marks, and M. Karrer for help in some of the sample processing, and J. Fortner for assistance with typing. M. Buzas and S. Richardson are thanked for their assistance with the USNM collections.

Funding for this research was provided by National Science Foundation Division of Polar Programs Grant DPP-8214174-A01 to P. N. Webb and Grant DPP-8213985-A01 to W. J. Zinsmeister; and by grants from Sigma Xi, the Geological Society of America, the American Association of Petroleum Geologists, AMOCO Production Company, and the Friends of Orton Hall (OSU) (to B.T.H.).

APPENDIX 1. SPECIES LOCATION INDEX FOR THE RANGE CHART OF SEYMOUR ISLAND FORAMINIFERA SHOWN IN TABLE 2

Index Number	Species	Index Number	Species
46	*Alabamina creta* (Finlay)	42	*Karreriella aegra* Finlay
2	*Alabamina westraliensis* (Parr)	70	*Lagena acuticosta* Reuss
3	*Allomorphina cretacea* Reuss	51	*Lagena apiculata* (Reuss)
121	*Allomorphina subtriangularis* (Kline)	71	*Lagena semiinterrupta* Berry
14	*Alveolophragmium macellarii* n. sp.	72	*Lagena semilineata* Wright
117	*Ammodiscus cretaceus* (Reuss)	15	*Lagena sphaerica* Marie
47	*Ammodiscus pennyi* Cushman and Jarvis	99	*Lagena simplex* (Reuss)
4	*Anomalinoides larseni* n. sp.	83	*Lenticulina macrodisca* (Reuss)
5	*Anomalinoides piripaua* (Finlay)	123	*Lenticulina muensteri* (Roemer)
6	*Anomalinoides rubiginosus* (Cushman)	36	*Lenticulina ovalis* (Reuss)
98	*Astacolus oligostegius* (Reuss)	52	*Lenticulina spissocostata* (Cushman)
22	*Bathysiphon californicus* Sliter	109	*Lenticulina* cf. *L. jonesi* Sandidge
100	*Bolivina incrassata* Reuss	110	*Lenticulina* sp.
101	*Bolivina pustulata* n. sp.	43	*Marginulina bullata* Reuss
48	*Bolivina* sp.	90	*Marginulina* cf. *M. curvatura* Cushman
60	*Budashevaella multicamerata* (Voloshinova and Budasheva)	16	*Marssonella oxycona* (Reuss)
61	*Buliminella creta* (Finlay)	17	*Marssonella?* sp.
122	*Buliminella procera* n. sp.	53	*Neobulimina aspera* (Cushman and Parker)
62	*Buliminella* cf. *B. fusiforma* Jennings	54	*Neobulimina digitata* n. sp.
114	*Buliminella pulchra* (Terquem)	91	*Nodosaria affinis* Reuss
111	*Ceratolamarckina* cf. *C. tuberculata* (Brotzen)	73	*Nodosaria aspera* Reuss
49	*Charltonina?* sp.	84	*Nodosaria multicostata* (d'Orbigny)
7	*Cibicides nordenskjoldi* n. sp.	74	*Nodosaria navarroana* Cushman
8	*Cibicides seymouriensis* n. sp.	75	*Nodosaria obscura* Reuss
9	*Cibicides* cf. *C. beaumontianus* (d'Orbigny)	18	*Nonionella* cf. *N. robusta* Plummer
87	*Citharina plumoides* (Plummer)	92	*Nonionella* sp. 1
88	*Conorbina anderssoni* n. sp.	124	*Nonionella* sp. 2
1	*Cyclammina* cf. *C. complanata* Chapman	116	*Oolina* sp.
40	*Cyclammina* sp.	55	*Osangularia cordieriana* (d'Orbigny)
112	*Cyclogyra* sp.	106	*Planispirillina subornata* (Brotzen)
10	*Dentalina basiplanata* Cushman	56	*Praebulimina cushmani* (Sandidge)
32	*Dentalina gracilis* (d'Orbigny)	57	*Praebulimina* cf. *P. cushmani* (Sandidge)
82	*Dentalina lorneiana* d'Orbigny	44	*Praebulimina midwayensis* (Cushman and Parker)
33	*Dentalina marcki* Reuss	85	*Praebulimina* cf. *P. trihedra* (Cushman)
89	*Dentalina megapolitana* Reuss	58	*Praebulimina* cf. *P. reussi* (Morrow)
126	*Dentalina* sp.	104	*Praebulimina* sp.
102	*Dorothia paeminosa* n. sp.	19	*Psammosphaera* sp.
63	*Eouvigerina hispida* Cushman	37	*Pseudonodosaria parallela* (Marsson)
64	*Epistominella glabrata* (Cushman)	76	*Pullenia cretacea* Cushman
118	*Fissurina alata* Reuss	93	*Pullenia jarvisi* Cushman
97	*Fissurina laevigata* Reuss	113	*Pullenia* sp.
65	*Fissurina orbignyana* Seguenza	94	*Pyramidina tortilis* (Reuss)
103	*Fissurina* sp.	77	*Pyrulina cylindroides* (Roemer)
34	*Frondicularia rakauroana* (Finlay)	78	*Quadrimorphina allomorphinoides* (Reuss)
119	*Frondicularia?* sp.	86	*Quinqueloculina* sp.
107	*Fursenkoina nederi* Sliter	79	*Ramulina psuedoaculeata* (Olsson)
66	*Gaudryina healyi* (Finlay)	95	*Ramulina* sp.
67	*Gavelinella sandidgei* (Brotzen)	25	*Reophax subfusiformis* Earland, emend. Hoeglund
129	*Globigerinelloides multispinatus* (Lalicker)	26	*Reophax texanus* Cushman and Jarvis
141	*Globastica daubjergensis* (Brönnimann)	125	*Reophax* sp.
115	*Globorotalites?* sp.	38	*Rhabdammina* sp.
138	*Globotruncanella havanensis* (Voorwijk)	20	*Rhizammina algaeformis* Brady
132	*Globotruncanella minuta* Caron and Gonzalez Donoso	136	*Rugoglobigerina rotundata* Brönnimann
140	*Globotruncanella?* sp.	133	*Rugoglobigerina rugosa* (Plummer)
68	*Globulina lacrima* (Reuss)	128	*Rugoglobigerina?* sp. 1
139	*Guembelitria cretacea* Cushman	137	*Rugoglobigerina?* sp. 2
108	*Guttulina adhaerens* (Olszewski)	130	*Rugotruncana circumnodifer* (Finlay)
69	*Guttulina trigonula* (Reuss)	27	*Rzehakina epigona* (Rzehak)
12	*Gyroidina* cf. *G. globosa* (Hagenow)	41	*Saccamina sphaerica* Sars
11	*Gyroidinoides nitidus* (Reuss)	39	*Saracenaria triangularis* (d'Orbigny)
96	*Haplophragmium compressum* Beissel	80	*Serovaina orbicella* Bandy
13	*Haplophragmoides eggeri* Cushman	28	*Spiroplectammina laevis* (Roemer)
23	*Haplophragmoides platus* Loeblich	105	*Spiroplectammina* cf. *S. laevis* (Roemer)
131	*Hedbergella monmouthensis* (Olsson)	31	*Spiroplectammina spectabilis* Grzybowski
127	*Hedbergella* sp.	45	*Stilostomella impensia* (Cushman)
135	*Heterohelix glabrans* (cushman)	81	*Stilostomella pseudoscripta* (Cushman)
135	*Heterohelix globulosa* (Ehrenberg)	120	*Textularia?* sp.
50	*Hiltermannella kochi* (Bertels)	29	*Trochammina globigeriniformis* (Parker and Jones)
24	*Hoeglundina supracretacea* (ten Dam)	30	*Trochammina ribstonensis* Wickenden
35	*Hyperammina elongata* Brady	59	*Trochammina triformis* Sliter

APPENDIX 2. SYSTEMATICALLY ARRANGED LIST OF FORAMINIFERA RECOVERED FROM THE LÓPEZ DE BERTODANO FORMATION AT CAPE LAMB, VEGA ISLAND

Rhabdammina sp.
Hyperammina elongata Brady
Reophax texanus Cushman and Waters
Ammodiscus pennyi Cushman and Jarvis
Rzehakina epigona (Rzehak)
Alveolophragmium macellarii n. sp.
Haplophragmoides rugosa Cushman and Waters
Haplophragmoides platus Loeblich
Cyclammina cf. *C. complanata* Chapman
Cyclammina sp.
Budashevaella multicamerata Voloshinova and Budasheva
Spiroplectammina laevis (Roemer)
Spiroplectammina vegaensis n. sp.*
Trochammina globigeriniformis (Parker and Jones)
Gaudryina healyi Finlay
Karerriella aegra Finlay
Cyclogyra sp.
Nodosaria affinis Reuss
Nodosaria navarroana Cushman
Nodosaria obscura Reuss
Citharina plumoides (Plummer)
Dentalina gracilis (d'Orbigny)
Dentalina marcki Reuss
Dentalina megapolitana Reuss
Lagena acuticosta Reuss
Lagena simplex (Reuss)
Lagena sphaerica Marie
Lagena semiinterrupta Berry
Lenticulina spissocostata (Cushman)

Lenticulina williamsoni Reuss*
Marginulina bullata Reuss
Marginulina cf. *M. curvatura* Cushman
Saracenaria triangularis (d'Orbigny)
Globulina lacrima (Reuss)
Guttulina adhaerens (Olszewski)
Pyrulina cylindroides (Roemer)
Ramulina pseudoaculeata (Olsson)
Buliminella creta (Finlay)
Buliminella cf. *B. fusiforma* Jennings
Stilostomella impensia (Cushman)
Hiltermannella kochi (Bertels)
Conorbina anderssoni n. sp.
Epistominella glabrata (Cushman)
Cibicides cf. *C. beaumontianus* (d'Orbigny)
Cibicides seymouriensis n. sp.
Allomorphina cretacea Reuss
Quadrimorphina allomorphinoides (Reuss)
Pullenia cretacea Cushman
Alabamina westraliensis Toulmin
Gyroidinoides nitidus (Reuss)
Anomalinoides piripaua (Finlay)
Anomalinoides larseni n. sp.
Gavelinella sandidgei (Brotzen)
Hoeglundina supracretacea (ten Dam)
Planispirillina subornata (Brotzen)
Heterohelix globulosa (Ehrenberg)
Globigerinelloides multispinatus (Lalicker)
Rugoglobigerina? sp. 1

*Absent from Seymour Island strata.

APPENDIX 3. FORAMINIFERA RECOVERED FROM THE LÓPEZ DE BERTODANO FORMATION AT THE NAZE, JAMES ROSS ISLAND

Rhabdammina sp.
Rhizammina algaeformis Brady
Thurammina papillata Brady*
Ammodiscus cretaceus (Reuss)
Ammodiscus pennyi Cushman and Jarvis
Reophax texanus Cushman and Waters
Rzehakina epigona (Rzehak)
Alveolophragmium macellarii n. sp.
Cyclammina cf. *C. complanata* Chapman
Spiroplectammina laevis (Roemer)
Spiroplectammina vegaensis n. sp.
Spiroplectammina spectabilis (Grzybowski)
Trochammina globigeriniformis (Parker and Jones)
Trochammina ribstonensis Wickenden
Trochammina triformis Sliter
Gaudryina healyi Finlay
Dorothis elongata Finlay*
Dorothia paeminosa n. sp.
Karreriella aegra Finlay
Lagena apiculata (Reuss)
Dentalina basiplanata Reuss

*Absent from the Seymour Island strata.

APPENDIX 4. FORAMINIFERA RECOVERED FROM THE LÓPEZ DE BERTODANO FORMATION ON SNOW HILL ISLAND

Rhabdammina sp.
Bathysiphon californicus Sliter
Haplophragmoides eggeri Cushman
Alveolophragmium macellarii n. sp.
Cyclammina cf. *C. complanata* Chapman
Budashevaella multicamerata (Voloshinova and Budasheva)

Figure 16. Agglutinated foraminifera. See introduction to the systematics for explanation of abbreviations. 1, *Rhabdammina* sp., OSU 39702, locality 19, x26; 2, *Rhizammina algaeformis* Brady, OSU 39703, locality 17, ×46; 3, *Bathysiphon californicus* Martin, OSU 39704, locality 14, ×28; 4, *Hyperammina elongata* Brady, OSU 39705, locality 101, ×30; 5, *Psammosphaera* sp., OSU 39706, locality 14, ×102; 6, *Reophax subfusiformis* Earland, emend. Hoglund, OSU 39711, locality 12, ×47; 7, *Ammodiscus pennyi* Cushman & Jarvis, OSU 39710, locality 24, ×10; 8, *Ammodiscus cretaceus* Reuss, OSU 39709, locality 185, ×66; 9, *Saccammina sphaerica* Sars, OSU 39707, locality 17, ×45; 10, *Thurammina papillata* Brady, OSU 39708, locality JR-2, ×31; 11, *Haplophragmoides eggeri* Cushman, OSU 39716, locality 24, ×52; 12-13, *Haplophragmoides platus* Loeblich, side and edge views, OSU 39721, locality 28b, ×59; 14, *Reophax texanus* Cushman & Jarvis, OSU 39712, locality 17, ×19; 15, *Reophax texanus* Cushman & Jarvis, OSU 39713, locality 19, ×13; 16-17, *Alveolophragmium macellarii* n. sp., side and edge veiws, OSU 39717, locality 12, ×31; 18-19, *Alveolophragmium macellarii* n. sp., side and edge views, OSU 39718, locality 414, ×30.

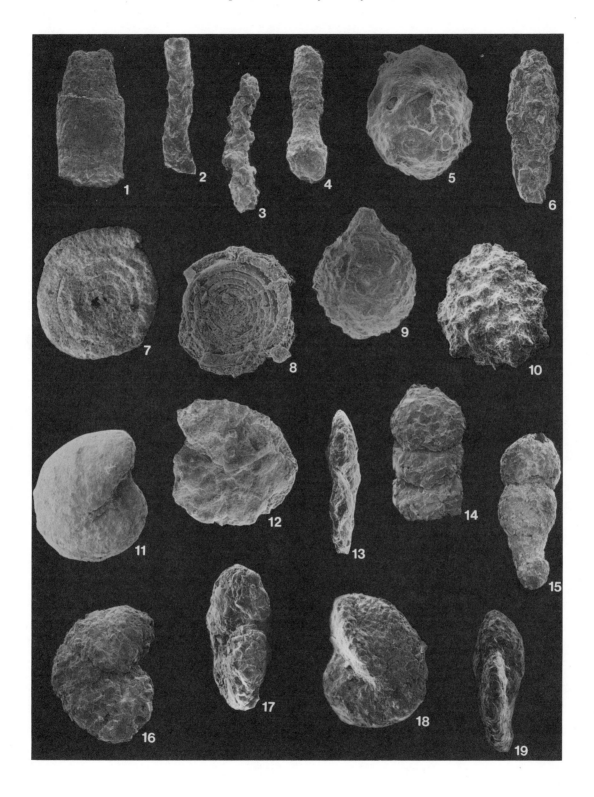

Figure 17. Agglutinated foraminifera. *Alveolophragmium macellarii* n. sp. 1, Edge view, OSU 39719, locality 26, ×46; 2-4, *Alveolophragmium macellari* n. sp., edge, side, and enlarged views, USNM 415748, locality 19, ×14; 4, same specimen; magnified view of interiomarginal aperture with bordering lip and multiple areal apertures, ×101; 5, *Alveolophragmium macellarii* n. sp., internal view showing well-developed alveolar wall structure, OSU 39720, locality 86, ×25; 6, *Cyclammina* cf. *C. complanata* Chapman, OSU 38722, locality 19, ×20; 7, *Cyclammina* sp., OSU 39723, locality 14, ×88; 8, *Rzehakina epigona* (Rzehak), OSU 39715, locality 14, ×93, 9-10, *Haplophragmium compressum* Beissel, OSU 39725, locality 349, ×16; 11 *Spiroplectammina spectabilis* Grzybowski, OSU 39726, locality 14, ×70; 12, *Ammobaculites fragmentarius* Cushman, OSU 39724, locality 28b, ×70; 13, 18, *Spiroplectammina laevis* (Roemer), side and edge views, OSU 39727, locality 110, ×78; 14-15, *Spiroplectammina vegaensis* n. sp., side and edge views, USNM 415749, locality VI-8, ×78; 16-17, *Spiroplectammina* cf. *S. laevis* (Roemer), side and edge views, OSU 39728, locality 38, ×102.

Figure 18. Agglutinated and calcareous benthic foraminifera. 1-2, *Textularia*? sp., side and edge views, OSU 39729, locality 115, ×67; 3-4, *Trochammina globigeriniformis* (Parker & Jones), umbilical and edge views, OSU 39730, locality 19, ×56; 5, 9, *Trochammina ribstonensis* Wickenden, umbilical and edge views, OSU 39731, locality 14, ×210; 6, *Trochammina triformis* Sliter, OSU 39732, locality 414, ×23; 7-8, *Budashevaella multicamerata* (Voloshinova & Budasheva), umbilical and edge views, OSU 39733, locality 24, ×91; 10-11, *Gaudryina healyi* (Finlay), side and edge views, ISU 39734, locality VI—8, ×58; 12-13, *Dorothia paeminosa* n. sp., side and edge views, USNM 415750, locality 148, ×58; 14, *Dorothia paeminosa* n. sp., OSU 39736, locality 38, ×98; 15, *Marssonella oxycona* (Reuss), OSU 39739, locality 41, ×80; 16, *Marssonella*?, OSU 39740, locality 41, ×102; 17, *Dorothia elongata* Finlay, OSU 39735, locality 39735, locality JR-2, ×36; 18, *Karreriella aegra* Finlay, OSU 39741, locality 145, ×30; 19-20, *Cyclogyra* sp., side and edge views, OSU 39742, locality 184, ×86.

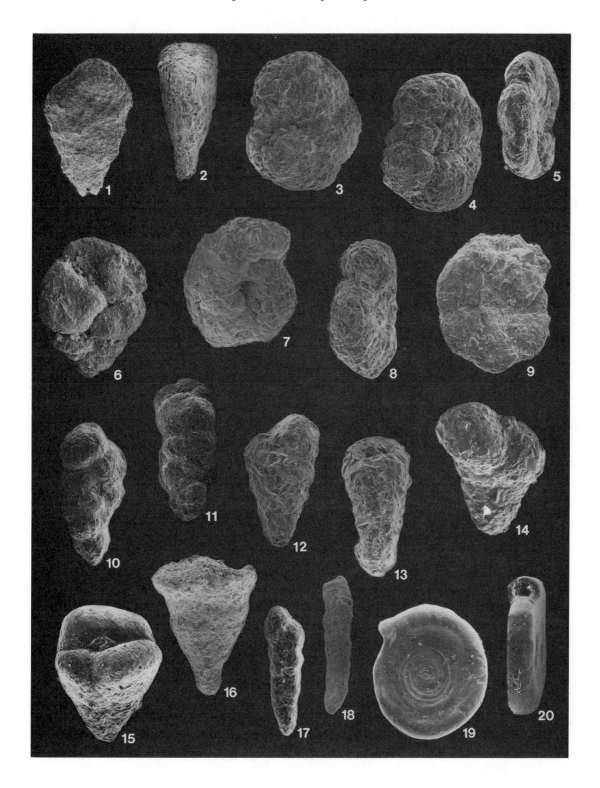

Figure 19. Calcareous benthic foraminifera. 1-2, *Quinqueloculina* sp., side and apertural views, OSU 39743, locality 184, 1 = ×125, 2 = ×140; 3-4, *Quinqueloculina* sp., side and apertural views, OSU 39744, locality 179, ×86; 5, *Nodosaria affinis* Reuss, OSU 39745, locality 67, ×50; 6, *Nodosaria affinis* Reuss, OSU 39746, locality VI-8, ×31; 7, *Nodosaria affinis* Reuss, OSU 39903, locality 38, ×48; 8, *Nodosaria multicostata* (d'Orbigny), OSU 39748, locality 38, ×154; 9, *Nodosaria navarroana* Cushman, OSU 39749, locality 38, ×157; 10, *Nodosaria navarroana* Cushman, OSU 39750, locality 38, ×210; 11, *Nodosaria obscura* Reuss, OSU 39751, locality ×187; 12, *Nodosaria obscura* Reuss, OSU 39752, locality VI-8, 225; 13, *Nodosaria obscura,* OSU 39753, locality 148, ×105; 14, *Nodosaria aspera* Reuss, OSU 39747, locality 415, ×50; 15, *Dentalina basiplanata* Cushman, OSU 39758, locality 38, ×45; 16, *Dentalina gracilis* (d'Orbigny), OSU 39759, locality 167, ×183; 17, *Dentalina gracilis,* (d'Orbigny), OSU 39760, locality 28a, ×76; 18, *Dentalina lorneiana* d'Orbigny, OSU 39761, locality 164, ×38; 19, *Dentalina megapolitana* Reuss, OSU 39763, locality 145, ×60; 20, *Dentalina marcki* Reuss, OSU 39762, locality 110, ×78; 21, *Citharina plumoides* (Plummer), OSU 39755, location 38, ×69; 22-23, *Citharina plumoides* (Plummer), side and edge views, OSU 39756, location VI-8, ×33; 24, *Citharina plumoides* (Plummer), juvenile specimen, OSU 39757, locality 38, ×152.

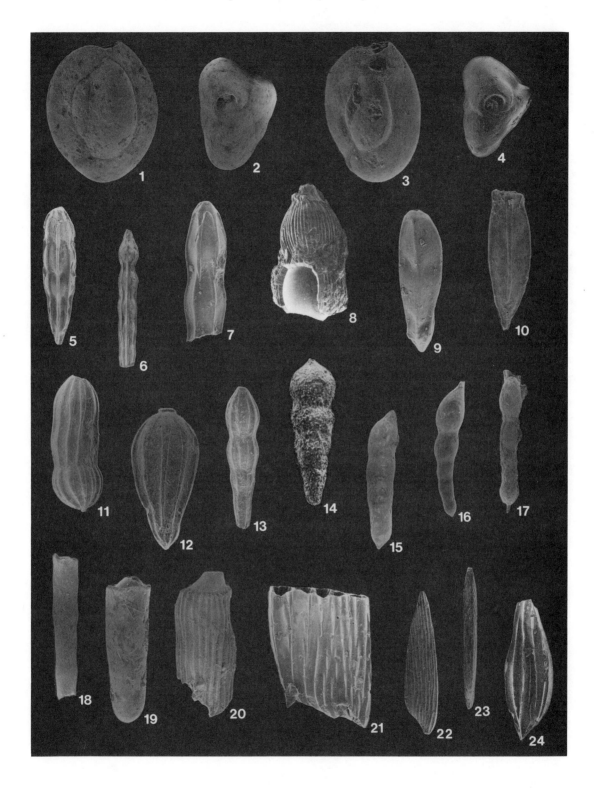

Figure 20. Calcareous benthic foraminifera. 1-2, *Frondicularia rakauroana* (Finlay), side and edge views, OSU 38551, locality 101, ×190; 3 *Frondicularia*? sp., OSU 39765, locality 367, ×54; 4, *Lagena acuticosta* Reuss, OSU 39766, locality 165, ×179; 5, *Lagena sphaerica* Marie, OSU 39771, locality 110, ×116; 6, *Lagena apiculata* (Reuss), OSU 39767, locality 28a, ×89; 7, *Lagena semilineata* Wright, OSU 39769, locality 154, ×72; 8, *Lagena semiinterrupta* Berry, OSU 39768, locality 179, ×180; 9, *Lagena simplex* (Reuss), OSU 39770, locality 177, ×60; 10-11, *Lenticulina* cf. *L. jonesi* Sandidge, side and edge views, OSU 39772, locality 177, ×31; 12-13, *Lenticulina macrodisca* (Reuss), side and edge views, OSU 39773, locality 177, ×93; 14, *Lenticulina spissocostata* (Cushman), OSU 39776, locality 369, ×15; 15, *Lenticulina spissocostata* (Cushman), OSU 39777, locality 430, ×16; 16-17, *Lenticulina williamsoni* Reuss, side and edge views, OSU 39778, locality VI-8, ×37.

Figure 21. Calcareous benthic foraminifera. 1-2, *Lenticulina* sp., side and edge views, OSU 39779, locality 177, ×44; 3, *Lenticulina ovalis* (Reuss), OSU 39775, locality 177, ×147; 4, *Saracenaria triangularis* (d'Orbigny), OSU 39784, locality VI-11, ×25; 5, *Marginulina bullata* Reuss, OSU 39780, locality 148, ×44; 6-7, *Astocolus oligostegius* (Reuss), OSU 39754, locality 38, × 84; 8, *Marginulina bullata* Reuss, OSU 39781, locality 38, ×220; 9-10, *Marginulina* cf. *M. curvatura* Cushman, edge and side views, OSU 39782, locality 110, ×88; 11, *Pseudonodosaria parallela* (Marsson), OSU 39783, locality 177, ×47; 12, *Globulina lacrima* (Reuss), OSU 39785, locality 146, ×138; 13, *Globulina lacrima* (Reuss), OSU 39904, locality VI-8, ×58; 14, *Guttulina adhaerens* (Olszewski), OSU 39786, locality 164, ×108; 15, *Guttulina trigonula* (Reuss), OSU 39787, locality 164, ×113; 16, *Pyrulina cylindroides* (Roemer), OSU 39788, locality 148, ×96; 17, *Ramulina pseudoaculeata* (Olsson), OSU 39789, locality VI-11, ×79; 18, *Ramulina* sp., OSU 39790, locality 38, ×83; 19, *Oolina* sp., OSU 39791, locality 184, ×145.

Figure 22. Calcareous benthic foraminifera. 1, *Fissurina laevigata* Reuss, OSU 39793, locality 38, ×293; 2-3, *Fissurina orbignyana* Seguenza, side and edge views, OSU 39794, locality 182, ×209; 4, *Fissurina* sp., OSU 39795, locality 110, ×538; 5-6, *Fissurina alata* Reuss, edge and side views, OSU 39792, locality 415, ×163; 7, *Buliminella pulchra* Terquem, OSU 39799, locality 57, ×287; 8, *Buliminella* cf. *B. fusiforma* Jennings, OSU 39797, VI-8, ×321; 9, *Hiltermannella kochi* Bertels), OSU 39817, locality 28b, ×225; 10-11, *Hiltermannella kochi* (Bertels), side and edge views, OSU 39818, locality 182, ×155; 12, *Neobulimina aspera* (Cushman & Parker), OSU 39800, locality 148, ×50; 13-14, *Neobulimina aspera* (Cushman & Parker), side views, OSU 39801, locality 115, ×83; 15-16, *Neobulimina digitata* n. sp., apertural and opposite views, OSU 39801, locality 148, ×188; 17, *Praebulimina* cf. *P. cushmani* (Sandidge), OSU 39804, locality 38, ×306; 18, *Praebulimina cushmani* (Sandidge), OSU 39803, locality 28b, ×220; 19, *Praebulimina* cf. *P. reussi* (Morrow), OSU 39806, locality 110, ×150; 20, *Praebulimina* cf. *P. trihedra* (Cushman), OSU 39807, locality 28b, ×210.

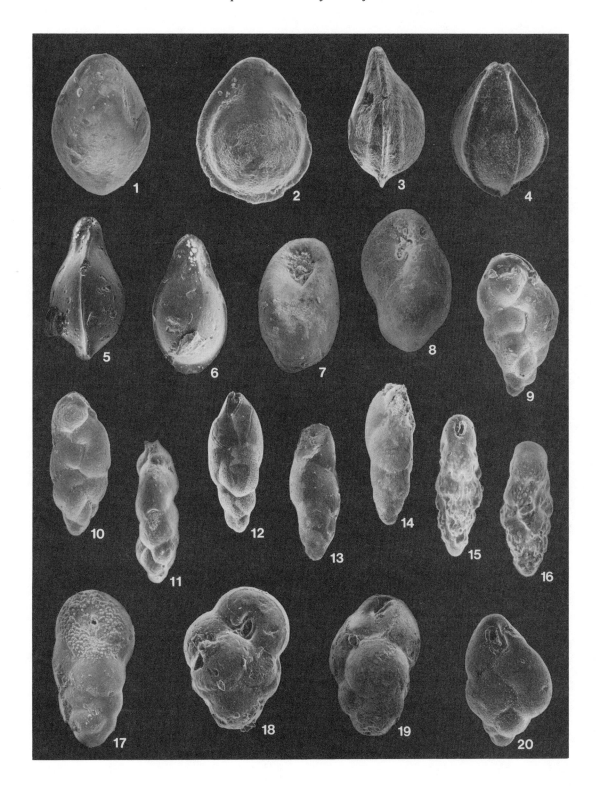

Figure 23. Calcareous benthic foraminifera. 1, *Buliminella creta* (Finlay), OSU 39796, locality 28d, ×159; 2, *Praebulimina midwayensis* (Cushman & Parker), OSU 39805, locality 110, ×146; 3, *Praebulimina* sp., OSU 39808, locality 110, ×200; 4-5, *Bolivina incrassata* Reuss, OSU 39813, locality 38, ×75; 6, *Bolivina* sp., OSU 39815, locality 110, ×269; 7-8, *Bolivina* sp., side and edge views, OSU 39816, locality 148, ×200; 9, *Bolivina* sp., OSU 39905, locality 165, ×191; 10, *Bolivina pustulata* n. sp., USNM 415756, locality 38, ×259; 11, *Pyramidina tortilis* (Reuss), OSU 39809, locality 38, ×239; 12, *Eouvigerina hispida* Cushman, OSU 39810, locality 28b, ×184; 13 *Ceratolamarckina* cf. *C. tuberculata* (Brotzen), OSU 39849, locality 380, ×111; 14, *Stilostomella impensia* (Cushman), OSU 39811, locality 28b, ×31; 15, *Stilostomella pseudoscripta* (Cushman), OSU 38912, locality 164, ×73; 16-17, *Conorbina anderssoni* n. sp., umbilical and edge views, OSU 39892, locality 380, ×263; 18-20, *Epistominella glabrata* (Cushman), umbilical, edge, and spiral views, OSU 39821, locality 184, ×215; 21-22, *Cibicides* cf. *C. beaumontianus* (d'Orbigny), umbilical and edge views, OSU 39822, locality 148, ×145.

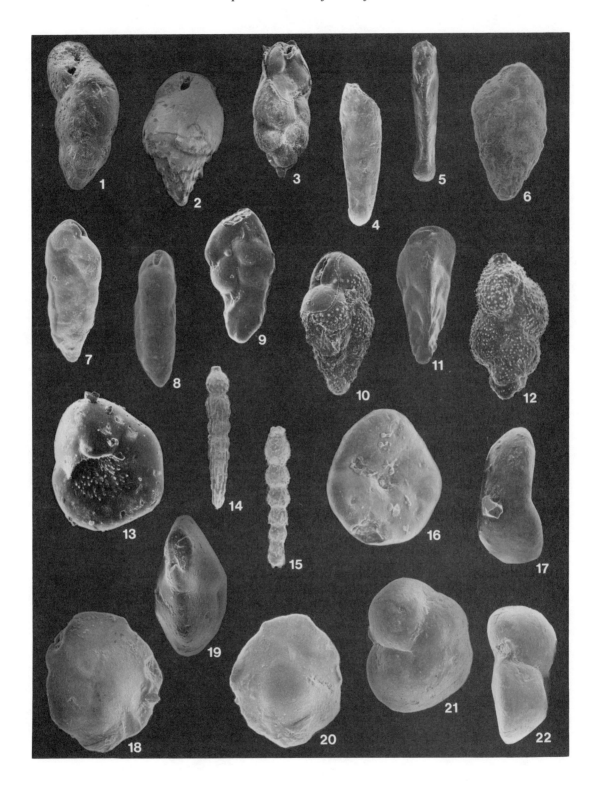

Figure 24. Calcareous benthic foraminifera. 1-3, *Cibicides seymouriensis* n. sp., umbilical, edge, and spiral views, OSU 39825, locality 38, ×107; 4, *Cibicides seymouriensis* n. sp., OSU 39826, locality 38, ×294; 5-7, *Cibicides nordenskjoldi* n. sp., convex, edge, and planar views, OSU 39822, locality 38, ×420; 8, *Planispirillina subornata* (Brotzen), OSU 39900, locality 115, ×116; 9-10, *Nonionella* cf. *N. robusta* Plummer, edge and umbilical views, OSU 39832, locality 184, ×430; *Planispirillina subornata* (Brotzen), 11-13, side, edge and opposite views, OSU 49851, locality 56a, ×98; 14-15, *Nonionella* sp. 1, edge and side views, OSU 39833, locality 182, ×239; 16-17, *Fursenkoina nederi* Sliter, apertural and side views, OSU 39828, locality 145, ×51; 18-19, *Pullenia jarvisi* Cushman, edge and side views OSU 39836, locality 28a, ×150.

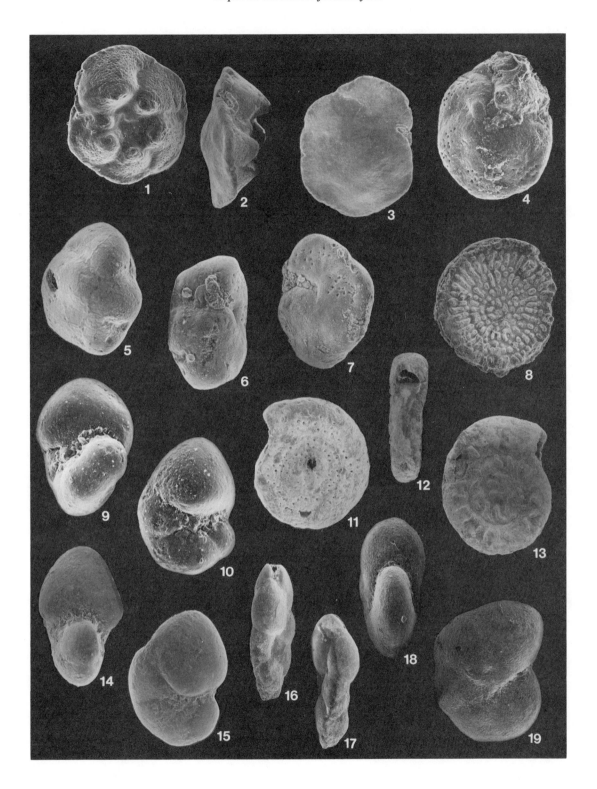

Figure 25. Calcareous benthic foraminifera. 1, *Pullenia* sp., OSU 39837, locality 182, ×250; 2-3, *Pullenia cretacea* Cushman, side and edge views, OSU 39835, locality 184, ×100; 4-5, *Quadrimorphina allomorphinoides* (Reuss), edge and umbilical views, OSU 39831, locality 184, ×287; 6-7, *Allomorphina cretacea* Reuss, umbilical and edge views, OSU 39829, locality 148, ×88; 8-9, *Alabamina creta* (Finlay), OSU 39838, locality 110, ×163; 10-12, *Charltonina*? sp., umbilical, edge, and spiral views, OSU 39842, locality 110, ×150; 13-14, *Alabamina westraliensis* (Parr), umbilical and edge views, OSU 39839, locality 38, ×90; 15-16, *Hoeglundina supracretacea* (ten Dam), umbilical and edge views, OSU 39850, locality 28d, ×69; 17-18, *Osangularia cordieriana* (d'Orbigny), umbilical and edge views, OSU 39841, locality 177, ×89.

Figure 26. Calcareous benthic foraminifera. 1-2, *Globorotalites*? sp., umbilical and edge views, OSU 39843, locality 184, ×178; 3-4, *Gyroidina* cf. *G. globosa* (Hagenow), umbilical and edge views, OSU 39844, locality 13b, ×98; 5-6, *Gyroidinoides nitidus* (Reuss), umbilical and edge views, OSU 39840, locality 115, ×131; 7-8, *Serovaina orbicella* (Bandy), umbilical and edge views, OSU 39819, locality 101, ×210; 9-11, *Anomalinoides rubiginosus* (Cushman), umbilical, edge, and spiral views, OSU 39847, locality 13b, ×105; 12-13, *Anomalinoides larseni* n. sp., edge and umbilical views, OSU 39845, locality 28b, ×186; 14-16 *Anomalinoides piripaua* (Finlay), umbilical, edge, and spiral views, OSU 39846, locality 148, ×87; 17-18, *Gavelinella sandidgei* (Brotzen), umbilical and edge views, OSU 39848, locality 184, ×350.

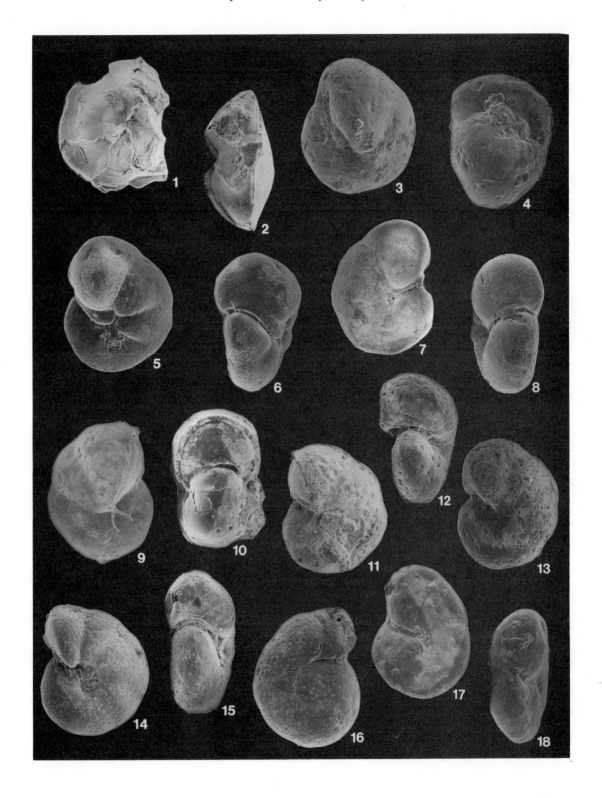

Figure 27. Planktonic foraminifera. 1, *Guembelitria cretacea* Cushman, OSU 39852, locality 177, ×390; 2-3, *Heterohelix glabrans* (Cushman), OSU 39853, locality 33, ×167; 4, *Heterohelix globulosa* (Ehrenberg), OSU 39854, locality 148, ×215; 5, *Heterohelix globulosa* (Ehrenberg), OSU 39906, locality 148, ×219; 6-7 *Globigerinelloides multispinatus* (Lalicker), edge and umbilical views of juvenile form with a reduced ultimate chamber and an indistinguishable aperture, OSU 39855, locality 165, ×358; 8-9, *Hedbergella* sp., OSU 39859, locality 26, ×190; 10-11, *Globigerinelloides multispinatus* (Lalicker), side and edge views, OSU 39856, locality 28a, ×288; 12, *Globigerinelloides multispinatus* (Lalicker), OSU 39857, locality 28b, ×300; 13, *Globigerinelloides multispinatus* (Lalicker), edge view of specimen bearing double apertures in the ultimate chamber, OSU 39858, locality 165, ×196; 14-16, *Hedbergella monmouthensis* (Olsson), umbilical edge, and spiral views, OSU 39860, locality 28b, ×242; 17, *Hedbergella monmouthensis* (Olsson), oblique umbilical view, OSU 39861, locality 28b, ×147.

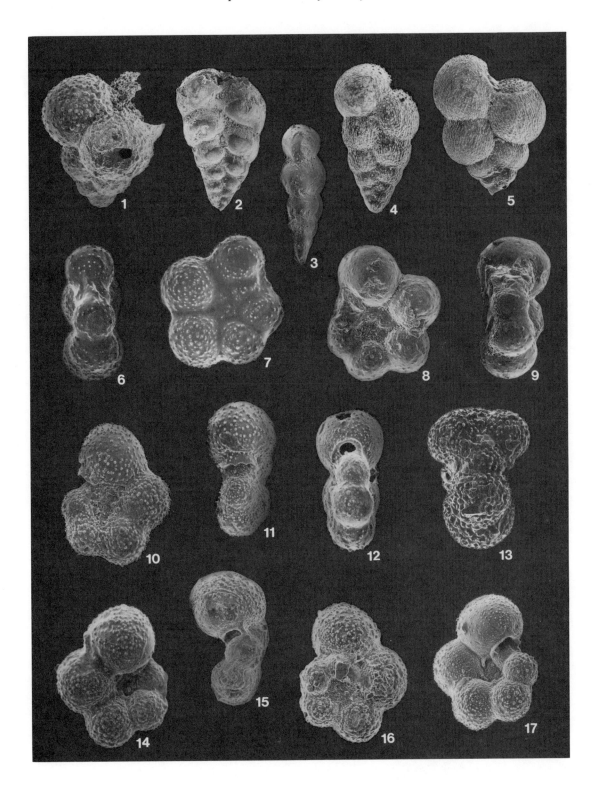

Figure 28. Planktonic foraminifera. 1-3, *Rugoglobigerina rugosa* (Plummer), umbilical, edge, and spiral views; note the faint development of meridionally aligned costellae on umbilical side, but their absence on the spiral side, OSU 39869, locality 28a, ×150; 4-5, *Rugoglobigerina rugosa* (Plummer), umbilical and spiral views; note the slight axial compression of ultimate chamber and absence of meridionally aligned costellae, OSU 39870, locality 148, ×115; 6-7, 11, *Rugoglobigerina rugosa* (Plummer), umbilical, spiral, and edge views; note the absence of meridional costellae, OSU 39871, locality 182, ×92; 8-10, *Rugoglobigerina rugosa* (Plummer), umbilical, edge, and spiral views; note the meridional arrangement of costellae on the fourth from last chamber on spiral side, OSU 39872, locality 415, ×126; 12-14, *Rugoglobigerina rotundata* Brönnimann, umbilical, edge, and spiral view, OSU 39868, locality 38, ×176.

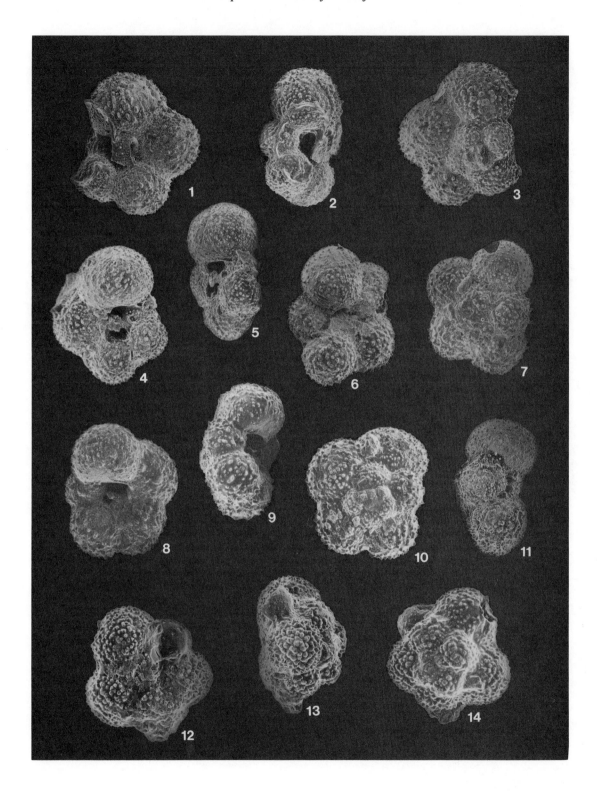

Figure 29. Planktonic foraminifera. 1-14, *Rugoglobigerina*? sp. 1, 1-3, umbilical, edge, and spiral views of high-spired form with indistinguishable aperture, OSU 39862, locality 13b, ×184; 4, umbilical view of pustulose form with reduced ultimate chamber, OSU 39863, locality 164, ×204; 5-7, umbilical, edge, and spiral views, OSU 39864, locality 28c, ×242; 8-9, umbilical and edge views, OSU 39865, locality VI-8, ×; 10-11, umbilical and edge views of specimen bearing an umbilical cover plate, OSU 39866, locality 28c, ×240; 12-14, umbilical, edge, and spiral views, OSU 39867, locality 164, ×187.

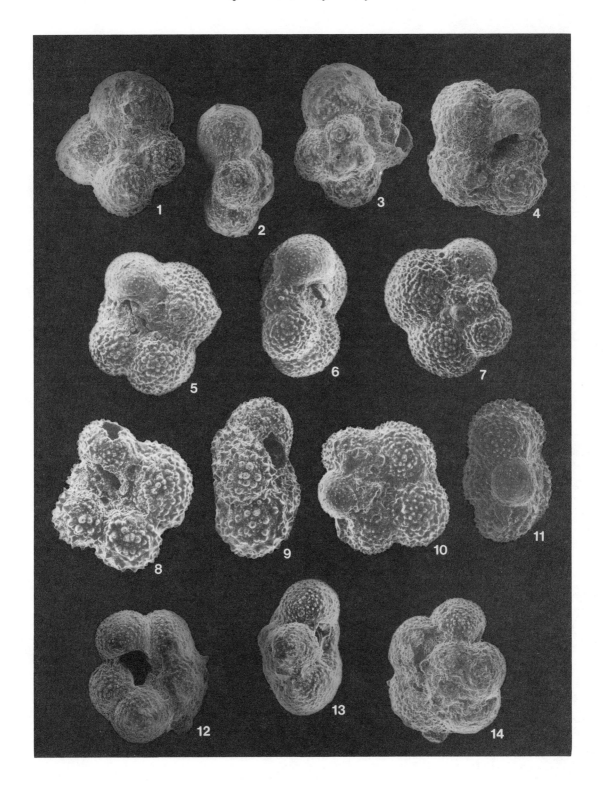

Figure 30. Planktonic foraminifera. 1-3, *Rugotruncana*? sp., Umbilical, edge, spiral, and magnified edge views; note the faintly developed peripheral keel and fused, meridionally arranged pustules on umbilical side, OSU 39876, locality 110, ×145; 4, same specimen showing enlarged view of imperforate peripheral band in the equatorial periphery, ×425; 5-7, *Rugoglobigerina*? sp. 1, umbilical, edge, and spiral views; note the complete absence of fused surface pustules, moderate chamber size increase, and similarity to *Rugoglobigerina rugosa* (Plummer), OSU 39873, locality 415, ×129; 8, *Rugoglobigerina*? sp. 1, OSU 39874, locality 415, ×195; 9-10, *Rugoglobigerina*? sp. 1, umbilical and edge views, OSU 39875, locality 415, ×195; 11-12, *Globotruncanella minuta* Caron & Gonzalez Donoso, umbilical and peripheral views; note the rapid chamber size increase and portical flap that covers the entire umbilical region, OSU 39882, locality 28c, ×215; 13-14, *Globotruncanella minuta* Caron & Gonzalez Donoso, edge and umbilical views, OSU 39883, locality 149b, <391; 15, *Globotruncanella minuta* Caron & Gonzalez Donoso, umbilical view showing well-developed portical flap, OSU 39884, locality 415, ×253; 16-17, *Globotruncanella minuta* Caron & Gonzalez Donoso, umbilical and peripheral views showing a portical flap (tegillum?) that completely covers the umbilicus, OSU 39885, locality 148, ×167.

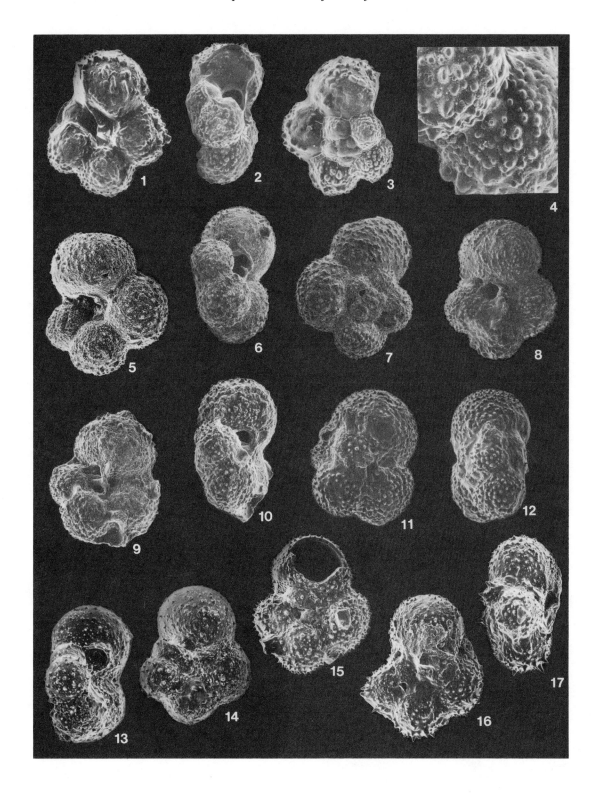

Figure 31. Planktonic foraminifera. 1-3, *Globotruncanella havanensis* (Voorwijk), umbilical, edge, and spiral views, OSU 39880, locality 33, ×148; 4, 7, 8, *Globotruncanella*? sp., umbilical, spiral, and edge views; note the dissimilarity to typical rugoglobigerine forms (e.g., higher spire, fewer chambers in the last whorl, and low, narrow aperture with a bordering lip), OSU 39881, locality 411, ×260; 5, *Rugotruncana*? sp., umbilical view showing peripheral keel, OSU 39877, locality 415, ×200; 6, *Rugotruncana*? sp., edge view showing rare double-keeled ornament, OSU 39878, locality 415, ×87, 9-11, *Rugotruncana*? sp. 1, umbilical, edge, and spiral views; note the single peripheral keel, OSU 39879, locality 415, ×180; 12-15, *Rugoglobigerina*? sp. 2, umbilical, spiral, and edge views; note the thick bulla covering the umbilical region and the axially compressed final chamber, OSU 39886, locality 165, ×128; 13-14, *Rugotruncana*? sp., umbilical and edge views; note the small umbilical aperture, OSU 39887, locality 165, ×287.

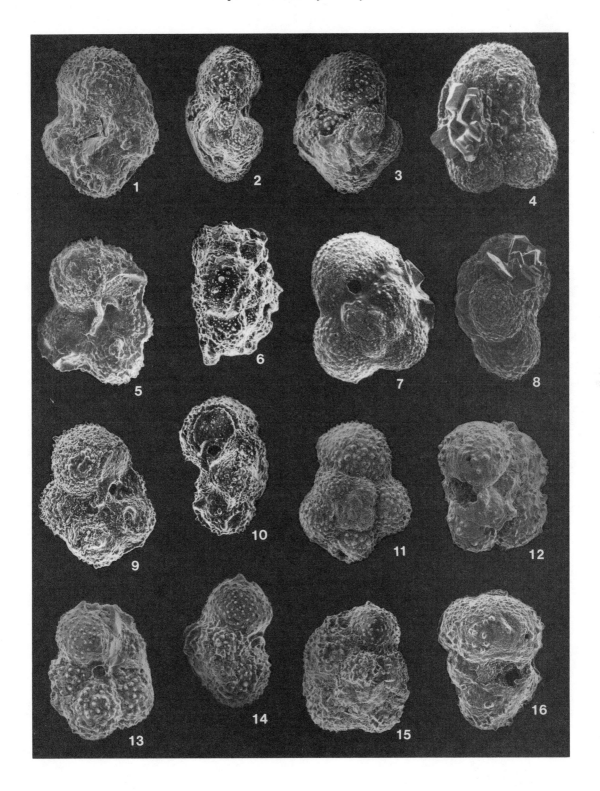

Figure 32. Danian foraminifera from Seymour Island. 1, *Reophax* sp., OSU 39714, locality 539, ×47; 2, *Dentalina* sp., OSU 39764, locality 539, ×111; 3, *Lenticulina muensteri* (Roemer), OSU 39774, locality 539, ×30; 4, *Allomorphina subtriangularis* (Kline), OSU 39830, locality 539, ×52; 5, *Buliminella procera* n. sp., OSU 39798, locality 540, ×133; 6, same specimen, showing enlarged view of aperture and narrow bordering lip, ×317; 7-8, *Bolivina pustulata* n. sp.; 7-8, side and edge views, OSU 39907, locality 539, ×259; 9, same specimen, showing enlarged loop-shaped aperture, ×629; 10, *Globastica daubjergensis* (Brönnimann), umbilical view showing pustulose ornament restricted to the sutural regions, OSU 39901, locality 539, ×400; 11, *Globastica daubjergensis* (Brönnimann), umbilical view, OSU 39902, locality 539, ×586; 12-13, *Nonionella* sp. 2, OSU 39834, locality 540, ×280; 14-16, *Globastica daubjergensis* (Brönnimann), umbilical, edge, and peripheral views; note the multiple sutural apertures on the spiral side, OSU 39888, locality 540, ×362; 17, *Globastica daubjergensis* (Brönnimann), OSU 39889, locality 540, ×538.

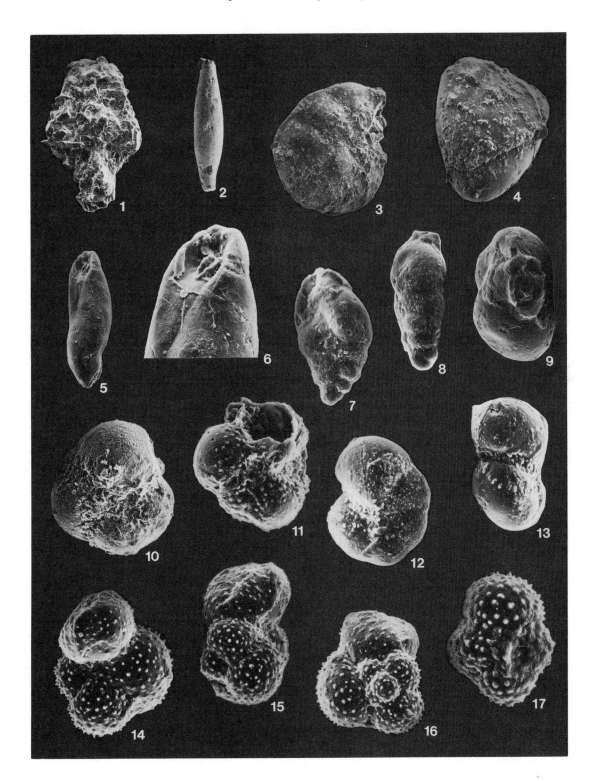

Figure 33. Internal morphology and wall structure of described new species and *Cibicides* cf. *C. beaumontianus* (d'Orbigny). 1, *Spiroplectammina vegaensis* n. sp., half section, side view, OSU 39893, locality VI-8, ×49; 2, *Dorothia paeminosa* n. sp., half section; note the canaliculate wall structure, OSU 39737, locality 315, ×155; 3, *Dorothia paeminosa* n. sp., enlarged broken specimen, showing internal view of alveolar wall structure; note the smooth appearance of chamber internal surface, ×760, 4, *Dorothia paeminosa* n. sp., half section, OSU 39738, locality 315, ×155; 5, *Cibicides* cf. *C. beaumontianus* (d'Orbigny), half section, OSU 39896, locality 315, ×183; 6, *Cibicides* cf. *C. beaumontianus* (d'Orbigny), thin section; note the bilamellar wall structure, ×170; 7, *Conorbina anderssoni* n. sp., half section, OSU 39897, locality 380, ×350; 8, *Anomalinoides larseni* n. sp., half section, OSU 39908, locality 380, ×161; 9, *Cibicides nordenskjoldi* n. sp., half section, OSU 39822, locality 315, ×392; 10, *Cibicides nordenskjoldi* n. sp., enlarged view showing wall microstructure; note internal growth of pyrite framboids, OSU 39824, locality 380, ×980; 11, *Buliminella procera* n. sp., half section, OSU 39894, locality 540, ×153; 12, *Neobulimina digitata* n. sp., half section, OSU 39895, locality 380, ×219; 13, *Cibicides seymouriensis* n. sp., half section, OSU 39827, locality 315, ×129; 14, *Cibicides seymouriensis* n. sp., thin section; note the irregular, blocky wall microstructure, locality 148, ×100; 15, detail of blocky wall microstructure, ×1,651.

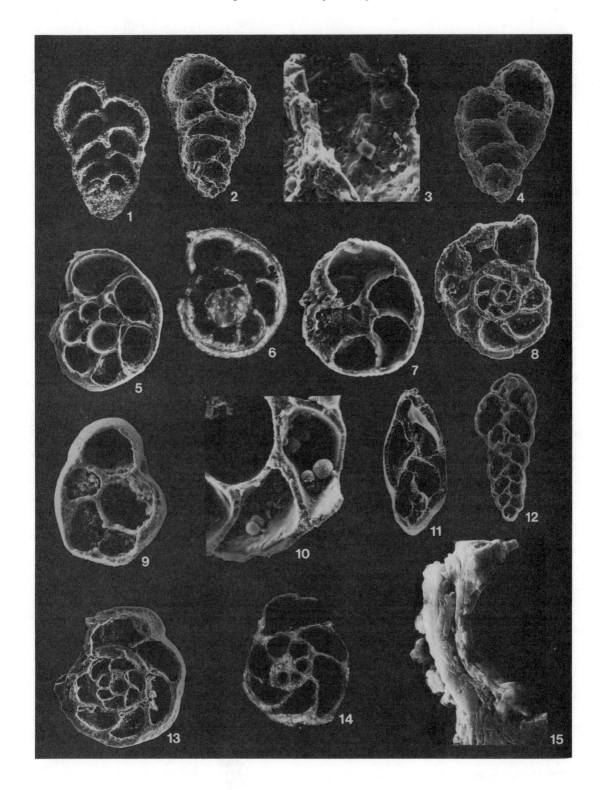

REFERENCES

ANDERSSON, J. G., 1906. On the geology of Graham Land. Geological Institute, University of Uppsala, Bulletin, 7:19-71.

ASKIN, R. A., 1983. Campanian palynomorphs from James Ross and Vega Islands, Antarctic Peninsula. Antarctic Journal of the United States, 18(5):63-64.

BACKMAN, J., M. J. WESTBERG-SMITH, AND J. G. BALDAUF, ET AL. 1984. Biostratigraphy of Leg 81 sediments: a high latitude record, p. 855–860. *In* D. G. Roberts, D. Schnitker, et al. (eds.), Initial Reports of the Deep Sea Drilling Project, Vol. 81. U.S. Government Printing Office, Washington, D.C.

BALL, H. W., 1960. Upper Cretaceous Decapoda and Serpulidae from James Ross Island, Graham Land. Falkland Dependencies Survey, Scientific Reports, 24:1-30.

BANDY, O. 1951. Upper Cretaceous Foraminifera from the Carlsbad area, San Diego County, California. Journal of Paleontology, 25:488-513.

BARRERA, E., B. T. HUBER, S. M. SAVIN, and P. N. WEBB. 1987. Antarctic marine temperatures: late Campanian through early Paleocene. Paleoceanography, 2(1):21-47.

BE, A.W.H. 1977. An ecological, zoogeographic, and taxonomic review of Recent planktonic foraminifera, p. 1–100. *In* A.T.S. Ramsay (ed.), Oceanic Micropaleontology. Academic Press, London.

——, AND D. S. TOLDERLUND. 1971, Distribution and ecology of living planktonic foraminifera in surface waters of the Atlantic and Indian Oceans, p. 105–149. *In* B. M. Funnell and W. R. Riedel (eds.), The Micropaleontology of the Oceans. Cambridge University Press, Cambridge.

BEISSEL, I. 1886. Der Aachener Sattel und die aus demselben vorbrechchenden Thermalquellen. Aachen: Naturwissenschaftliche gesellschaft, pt. 1, 138.

BERGGREN, W. A., 1962. Some planktonic Foraminifera from the Maestrichtian and type Danian stages of southern Scandanavia. Stockholm Contributions to Geology, 9(1):1-106.

BERMUDEZ, P. J. 1952. Estudio sistemático de los Foraminíferos rotaliformes. Ministerio de Minas e Hidrocarburos, Boletin de Geologia, 2(4):1-230.

BERNHARD, J. M. 1986. Characteristic assemblages and morphologies of benthic foraminifera from anoxic, organic-rich deposits: Jurassic through Holocene. Journal of Foraminiferal Research, 16(3):207-215.

BERRY, E. W., AND L. KELLEY. 1929. The Foraminifera of the Ripley Formation of Coon Creek, Tennessee. United States National Museum, 76(19):1-20.

BERTELS, A. 1970a. Los foraminíferos planctónicos de la cuenca Cretácico-Tertiaria en Patagonia Septentrional (Argentina), con consideraciones sobre la estratigrafía de Fortín General Roca (Provincia de Rio Negro). Ameghiniana, 7(1):1-56.

——. 1970b. *Hiltermannia* n. gen. (Foraminiferida) del Cretácico Superior (Maastrichtiano) de Argentina. Ameghiniana, 7(2):167-172.

——. 1972. Buliminacea y Cassidulinacea (Foraminiferida) guias del Cretácico Superior (Maastrichtiano Medio) y Terciario Inferior (Danian Inferior) de la Republica Argentina. Revista Española de Micropaleontologia, 4(3): 327-353.

——. 1979. Paleobiogeografía de los foraminíferos del Cretácico Superior y Cenozoico de America del Sur. Ameghiniana, 16(3):273-356.

BIBBY, J. S. 1966. The stratigraphy of part of north-east Graham Land and the James Ross Island Group. British Antarctic Survey, Scientific Report, 53:1-37.

BIRKELUND, T., ET AL. 1984. Cretaceous stage boundaries: proposals. Bulletin of the Geological Society of Denmark, 33:3-20.

BLOW, W. H. 1979. The Cainozoic Globigerinida, Vols. 1-3. E. J. Brill, Leiden.

BRADSHAW, J. S. 1959, Ecology of living planktonic foraminifera in the north and equatorial Pacific. Cushman Foundation for Foraminiferal Research, Contributions, 10:25-64.

BRADY, H. B. 1878. On the reticularian and radiolarian Rhizopoda (Foraminifera and Polycystina) of the North Polar Expedition of 1875–76. Annals and Magazine of Natural History, series 5, 1:425-440.

——. 1879. Notes on some of the reticularian Rhizopoda of the *Challenger* Expedition. Quarterly Journal of Microscopical Science, 19:51.

BRÖNNIMANN, P. 1951. Internal structure of *Cyclammina cancellata.* Journal of Paleontology, 25(6):756-761.

——. 1952. Globigerinidae from the Upper Cretaceous (Cenomanian-Maestrichtian) of Trinidad, B.W.I. Bulletin of American Paleontology, 34(140):1-70.

——. 1953. Note on planktonic foraminifera from Danian localities of Jutland, Denmark. Ecologae Geological Helvetiae, 45(2):339-341.

BROTZEN, F. 1936. Foraminifera aus dem schwedischen untersten senon von Eriksdal in Schonen. Sveriges Geologiska Undersokning, ser. C, no. 396, 30(3):1-206.

——. 1940. Flintrannans och trindelrannans Geologi (Oresund). Sveriges Geologiska Undersokning, ser. C, no. 435, 34(5):1-33.

——. 1948. The Swedish Paleocene and its foraminiferal fauna. Sveriges Geologiska Undersokning, ser. C. no. 493, 42(2):1-140.

BUZAS, M. A., AND S. J. CULVER. 1980. Foraminifera: distribution of provinces in the western North Atlantic. Science, 209:687-689.

CANDE, S. C., AND J. C. MUTTER. 1982. A revised identification of the oldest seafloor spreading anomalies between Australia and Antarctica. Earth and Planetary Science Letters, 58:151-160.

CARON, M., AND P. HOMEWOOD. 1983. Evolution of early planktic foraminifers. Marine Micropaleontology, 7:453-462.

CHAPMAN, F. 1904. On some Cainozoic Foraminifera from Brown's Creek, Otway Coast. Records of the Geological Survey of Victoria, 1:227-230.

CHARRIER, R., AND A. LAHSEN. 1969. Stratigraphy of Late Cretaceous–Early Eocene, Seno Skyring–Strait of Magellan Area, Magallanes Province, Chile. American Association of Petroleum Geologists, Bulletin, 53(3):568-590.

CULVER, S. J., AND M. A. BUZAS. 1983. Benthic foraminifera at the shelfbreak: North American Atlantic and Gulf margins. Society of Economic Paleontologists and Mineralogists, Special Publication no. 33, 359–371.

CUSHMAN, J. A. 1926a. Some Foraminifera from the Mendez shale of eastern Mexico. Cushman Laboratory for Foraminiferal Research, Contributions, 2:16-26.

——. 1926b. The Foraminifera of the Velasco shale of the Tampico embayment. American Association of Petroleum Geologists, Bulletin, 10:581-612.

——. 1927. Some foraminifera from the Cretaceous of Canada. Royal Society of Canada, Transactions, ser. 3, 21(2):127-132.

——. 1931. A preliminary report on the Foraminifera of Tennessee. Tennessee Division of Geology, Bulletin, 41:5-112.

——. 1933. Some new foraminiferal genera. Cushman Laboratory for Foraminiferal Research, Contributions, 9:32-38..

——. 1936. Cretaceous Foraminifera of the family Chilostomellidae. Cushman Laboratory for Foraminiferal Research, Contributions, 12:71-78.

——. 1937. A few new species of American Cretaceous Foraminifera. Cushman Laboratory for Foraminiferal Research, Contributions, 13:100-105.

——. 1938a. Additional new species of American Cretaceous Foraminifera. Cushman Laboratory for Foraminiferal Research, Contributions, 14:31-52.

——. 1938b. Cretaceous species of *Gümbelina* and related genera. Cushman Laboratory for Foraminiferal Research, Contributions, 14:2-28.

——. 1946. Upper Cretaceous Foraminifera of the Gulf Coastal region of the United States and adjacent areas. U.S. Geological Survey Professional Paper 206, 241 p.

——, AND P. W. JARVIS. 1928. Cretaceous Foraminifera from Trinidad. Cushman Laboratory for Foraminiferal Research, Contributions, 4:85-103.

——. 1932. Upper Cretaceous foraminifera from Trinidad. U.S. National Museum, Proceedings, 80(14):1-60.

——, AND F. L. PARKER. 1936. Some American Eocene buliminas. Cushman Laboratory for Foraminiferal Research, Contributions, 12:39-45.

——. 1940. The species of the genus *Bulimina* having Recent types: Cushman Laboratory of Foraminiferal Research, Contributions, 16:7-23.

——. 1947. Bulimina and related foraminiferal genera. U.S. Geological Survey Professional Paper 210-D, 55-176.

——, AND J. A. WATERS. 1927. Some arenaceous Foraminifera from the Upper

Cretaceous of Texas. Cushman Laboratory for Foraminiferal Research, Contributions, 2:81-85.

DAM, A., TEN. 1948. Les espèces du genre *Epistomina* Terquem, 1883. Institut Francais Pétrole, Revue, 3:161-170.

DEL VALLE, R. A., N. H. FOURCADE, AND F. A. MEDINA. 1982. The stratigraphy of Cape Lamb and The Naze, Vega, and James Ross Islands, Antarctica, p. 275-280. *In* C. Craddock (ed.), Antarctic Geoscience. University of Wisconsin Press, Madison.

DOUGLAS, R. G. 1969. Upper Cretaceous planktonic foraminifera in northern California. Micropaleontology, 15(2):151-209.

——. 1972. Paleozoogeography of Late Cretaceous planktonic foraminifera in Nortrh America. Journal of Foraminiferal Research, 2:14-34.

——. 1979. Benthic foraminiferal ecology and paleoecology: a review of concepts and methods. *In* Society of Economic Paleontologists and Mineralogists, Shortcourse no. 6, 21-53.

——, AND H. L. HEITMAN. 1979. Slope and basin benthic foraminifera of the California borderland. Society of Economic Paleontologists and Mineralogists, Special Publication no. 27, 231-246.

——, AND S. M. SAVIN. 1975. Oxygen isotopic evidence for the depth stratification of Tertiary and Cretaceous planktic foraminifera. Palaeogeography, Palaeoclimatology, Palaeoecology, 3:175-196.

DREWRY, D. J. 1983. Antarctica: Glaciological and Geophysical Folio. University of Cambridge, Cambridge.

EHRENBERG, C. G. 1840. Uber die Bildung der Kriedefelsen und des Kreidemergels durch unsichtbare Organismen. K. Akademie der Wissenschaften, Berlin. Physikalische Abhandlungen (1830), p. 59-147.

ELLIOT, D. H. 1983. The mid-Mesozoic to mid-Cenozoic active plate margin of the Antarctic Peninsula, p. 347-351. *In* R. L. Oliver, P. R. James, and J. B. Jago (eds.), Antarctic Earth Science. Cambridge University Press, Cambridge.

——, ET AL. 1975. Geological investigations on Seymour Island, Antarctica. Antarctic Journal of the United States, 10:182-186.

——, AND T. A. TRAUTMAN. 1982. Lower Tertiary strata on Seymour Island, p. 287-297. *In* C. Craddock (ed.), Antarctic Geoscience. University of Wisconsin Press, Madison.

EMILIANI, C. 1971. Depth habitats of growth stages of pelagic foraminifera. Science, 173:1122-1124.

FARQUHARSON, G. W. 1982, Late Mesozoic sedimentation in the northern Antarctic Peninsula and its relationship to the southern Andes. Journal of the Geological Society of London, 139:721-727.

FINLAY, H. J. 1939. New Zealand Foraminifera: key species in stratigraphy. Transactions of the Royal Society of New Zealand, 69(3):309-329.

——. 1940. New Zealand Foraminifera: key species in stratigraphy. Transactions of the Royal Society of New Zealand, 69(4):448-472.

GRINDLEY, G. W., C.J.D. ADAMS, J. T. LUMB, AND W. A. WATERS. 1977. Paleomagnetism, K-Ar dating and tectonic interpretation of Upper Cretaceous and Cenozoic volcanic rocks of the Chatham Islands, New Zealand. New Zealand Journal of Geology and Geophysics, 20:425-467.

GRZYBOWSKI, J. 1897. Otwornice pokladow naftonosynch okolicy Krosna (Foraminifera of oil-bearing strata in the neighborhood of Krosno). Akademija umiejetnosci, Rozprawy. Wydzial matematyczno-pozyrodniczy, v. 33, p. 257-305.

HAGENOW, F., VON. 1842. Monographie der Rügen schen Kreide-Versteinerungen, Abt. III-Mollusken: Nues jarbuch für mineralogie, geologie, und paleontologie, p. 528-575.

HAGN, H. 1953. Die Foraminiferen der Penswanger Schichten (unteres Obercampan); ein Beitrag zur Mikropalaeontologie der helvetischen Oberkreide Sudbayerns. Palaeontographica, v. 104, Abt. A, p. 1-119.

HANSEN, H. J. 1970. Biometric studies on the stratigraphic evolution of *Globoconusa daubjergensis* (Brönnimann) from the Danian of Denmark. Meddelelser fra Dansk Geologisk Forening, 19:341-360.

HANZLIKOVÁ, E. 1972. Carpathian Upper Cretaceous foraminifera of Moravia (Turonian-Maestrichtian). Rozpravy Ustredniho ústavu geologisckeho, 39:5-160.

HARLAND, W. B., ET AL. 1982. A Geologic Time Scale. Cambridge University Press, Cambridge, 131 p.

HEIDE, S., VAN DER. 1954. The original meaning of the term Maastrichtian (Dumont 1849). Geologie en Mijnbouw, 16:509-511.

HICKEY, L. J., ET AL. 1983. Arctic terrestrial biota: paleomagnetic evidence of age disparity with mid-north latitudes during the Late Cretaceous and early Tertiary. Science, 221:1153-1156.

HILTERMANN, H., AND W. KOCH. 1962. Oberkreide des nordlichen Mitteleuropa, p. 337. *In* Leitfossilen der Mikropalaontologie. Berlin, Nikolassee, Borntraeger.

HÖGLUND, H. 1947. Foraminifera in the Gullmar Fjord and the Skagerak. Zoologiska Bidrag Uppsala, 26:1-328.

HOLLAND, R. 1910. The fossil foraminifera: Wissenschaftliche Ergebnisse der Schwedischen Südpolar-Expedition 1901-1903, Stockholm, 3(9):1-11.

HORNIBROOK, N., DE B. 1969. News reports, New Zealand. Micropaleontology, 15:128-129.

——. 1972. *Globoconusa daubjergensis* (Foraminifera) at the base of the stratotype of the Teurian Stage, New Zealand (Note). New Zealand Journal of Geology and Geophysics, 15:178-181.

HOWARTH, M. K. 1958. Upper Jurassic and Cretaceous ammonite faunas from Alexander Land and Graham Land. Falkland Islands Dependencies Survey, Scientific Reports, 21:1-16.

——. 1966. Ammonites from the Upper Cretaceous of the James Ross Island group. Bulletin, British Antarctic Survey, 10:55-69.

HUBER, B. T. 1984. Late Cretaceous foraminiferal biostratigraphy, paleoecology, and paleobiogeography of the James Ross Island region, Antarctic Peninsula. Unpublished M.Sc. thesis, The Ohio State University, 246 p.

——. 1986. Foraminiferal evidence for a terminal Cretaceous oceanic event. Geological Society of America, Abstracts with Programs, 18(4):309-310.

——. 1987. The location of the Cretaceous-Tertiary contact on Seymour Island, Antarctic Peninsula. Antarctic Journal of the United States, 20(5) (in press).

——, D. M. HARWOOD, AND P. N. WEBB. 1983. Upper Cretaceous microfossil biostratigraphy of Seymour Island, Antarctic Peninsula. Antarctic Journal of the United States, 18(5):72-74.

——, AND P. N. WEBB. 1986. Distribution of *Frondicularia rakauroana* (Finlay) in the southern high latitudes. Journal of Foraminiferal Research, 16(2):135-140.

HSÜ, K. J. 1984. A scenario for the terminal Cretaceous event, p. 755-763. *In* K. J. Hsü, J. L. LaBrecque, et al., Initial Reports of the Deep Sea Drilling Project, Vol. 13. U.S. Government Printing Office, Washington, D.C..

INESON, J. R., J. A. CRAME, AND M.R.A. THOMSON. 1986. Lithostratigraphy of the Cretaceous strata of west James Ross Island. Cretaceous Research, 7(2):141-159.

JELETZKY, J. A. 1951. Die Stratigraphie und Belemniten-fauna des Obercampan und Maastricht Westgalens, Nordwest-deutschlands und Danemarks sowie einige allgemeine Gliederungs-Probleme der jungeren borealen Oberkreide Eurasiens. Geologische Jahrbuch Beiheft, 1:1-144.

JENKINS, D. G., 1966. Planktonic foraminiferal zones and new taxa from the Danian to lower Miocene of New Zealand. New Zealand Journal of Geology and Geophysics, 9(6):1088-1126.

JENNINGS, P. H. 1936. A microfauna from the Monmouth and basal Rancocas groups of New Jersey. Bulletins of American Paleontology, 23(78):161-232.

KILIAN, W., AND P. REBOUL. 1909. Les Céphalopodes néocrétacées des Îles Seymour et Snow Hill. Wissenschaftliche Ergebnisse der Schwedischen Südpolar-Expedition 1901-1903, Stockholm, 3(6):1-75.

KLINE, V. H. 1943. Midway foraminifera and Ostracoda. Mississippi Geologic Survey Bulletin, 53:5-98.

KRASHENINNIKOV, V. A., AND I. A. BASOV. 1983. Stratigraphy of Cretaceous sediments of the Falkland Plateau based on planktonic foraminifera, Deep Sea Drilling Project, Leg 71, p. 789-820. *In* W. J. Ludwig and V. A. Krasheninnikov et al., Initial Reports of the Deep Sea Drilling Project, Vol. 7. U.S. Government Printing Office, Washington, D.C.

——. 1986. Late Mesozoic and Cenozoic stratigraphy and geological history of the South Atlantic high latitudes: Palaeogeography, Palaeoclimatology, Palaeoecology, 55:145-188.

LALICKER, C. G. 1948. A new genus of Foraminifera from the Upper Cretaceous. Journal of Paleontology, 22(5):624.

LARSON, R. L. 1977. Early Cretaceous break up of Gondwanaland off western Australia. Geology, 5:57-60.

——, AND J. W. LADD. 1973. Evidence for opening of the South Atlantic in the Early Cretaceous and Late Jurassic. Nature, 246:209-212.

LAWVER, L. A., J. G. SCLATER, AND L. MEINKE. 1985. Mesozoic and Cenozoic reconstructions of the South Atlantic. Tectonophysics, 114(1-4):233-254.

LECKIE, R. M., AND P. N. WEBB. 1985. Late Paleogene and early Neogene foraminifers of Deep Sea Drilling Project Site 270, Ross Sea, Antarctica, p. 1093-1142. *In* J. P. Kennett, C. C. von der Borch, et al., Initial Reports of the Deep Sea Drilling Project, Vol. 90. U.S. Government Printing Office, Washington, D.C.

LOEBLICH, A. R., JR. 1946. Foraminifera from the type Pepper shale of Texas. Journal of Paleontology, 20(2):130-139.

——, AND H. TAPPAN. 1964. Sarcodina, chiefly "Thecamoebians" and Foraminiferida, p. C1–C900. *In* R. C. Moore (ed.), Treatise on Invertebrate Paleontology, Protista 2, pt. C, v. 1, University of Kansas Press, Lawrence.

——. 1984. Suprageneric classification of the Foraminiferida (Protozoa). Micropaleontology, 30(1):1-70.

——. 1985. Some new and redefined genera and families of agglutinated foraminifera II. Journal of Foraminiferal Research, 15(3):175-217.

LUDBROOK, N. H. 1977. Early Tertiary *Cyclammina* and *Haplophragmoides* (Foraminiferida: Lituolidae) in southern Australia: Transactions of the Royal Society of South Australia, 101(7):165-198.

MACELLARI, C. E. 1984a. Revision of serpulids of the genus *Rotularia* (Annelida) at Seymour Island (Antarctic Peninsula) and their value in stratigraphy. Journal of Paleontology, 58(4):1098-1116.

——. 1984b. Late Cretaceous stratigraphy, sedimentology, and macropaleontology of Seymour Island, Antarctic Peninsula. Unpub. Ph.D. dissertation, The Ohio State University, 599 p.

——. 1985a. El límite Cretácico Terciario en la Península Antártica y en el sur de Sudamerica: evidencias macropaleontológicas. Memorias Sexto Congreso Latinoamericano de Geología, Bogotá, Colombia, 1:267-278.

——. 1985b. Paleobiogeografía y edad de la fauna de *Maorites-Gunnarites* (Amonoidea) de la Antártida y Patagonia. Ameghiniana, 21(2):131-150.

——. 1986. Late Campanian–Maastrichtian ammonite fauna from Seymour Island (Antarctic Peninsula). Journal of Paleontology, Memoir 18, 60(2):1-55.

——, AND B. T. HUBER. 1982. Cretaceous stratigraphy of Seymour Island (east Antarctic Peninsula). Antarctic Journal of the United States, 17(5):68-70.

MACFADYEN, W. A. 1966. Foraminifera from the Upper Cretaceous of James Ross Island. Bulletin, British Antarctic Survey, 8:75-87.

MALUMIAN, N. 1978. Aspectos paleoecológicos de los foraminíferos del cretácico de la cuenca Austral. Ameghiniana, 15(1-2):149-160.

——, AND V. MASIUK. 1976. Foraminíferos de la Formacion Cabeza de León (Cretacico Superior), Tierra del Fuego, Rep. Argentina. Asociacion Geológica Argentina, Revista, 31(3):180-221.

MARIE, P. 1941. Les foraminifères de la craie a Belemnitella mucronata du bassin de Paris. Museum National d'Histoire Naturelle, Mémoires, 12:1-296.

MARINCOVITCH, L., JR., E. M. BROUWERS, and L. D. CARTER. 1985. Early Tertiary fossils from northern Alaska: implications for Arctic Ocean paleogeography and faunal evolution. Geology, 13(11):770-773.

MARKL, R. G. 1974. Evidence for the breakup of eastern Gondwanaland by the Early Cretaceous. Nature, 251:196-200.

MARSSON, T. 1878. Die Foraminiferen der weissen Schreibkreide der Inseln Rügen. Mittheilungen aus dem natur-wissenschaftlichen verein von Neu-Vorpommern und Rügen, 10:115-196.

MARTIN, L. 1964. Upper Cretaceous and Lower Tertiary Foraminifera from Fresno County, California. Geologische Bundesanstalt, Wien, Jahrbuch, Sonderband 9:1-128.

MCGOWRAN, B. 1965. Two Paleocene foraminiferal faunas from the Wangerrip Group Pebble Point coastal section, Western Victoria. Proceedings of the Royal Society of Victoria, 79(1):9-74.

MORROW, A. L. 1934. Foraminifera and Ostracoda from the Upper Cretaceous of Kansas. Journal of Paleontology, 8:186-205.

NATLAND, M. L., P. E. GONZALEZ, A. CAÑON, AND M. ERNST. 1974. A system of stages for correlation of Magallanese Basin sediments. Geological Society of America Memoir, 139:1-26.

NORDENSKJÖLD, O. 1905. Petrographische Unterschungen aus dem westantarktischen Gebiete. Geological Institute, University of Uppsala, Bulletin, 6(2):234-246.

——. 1913. Antarktis, p. 1–28. *In* Handbuch der Regionalen Geologie, Vol. 8. Heidelberg, Carl Winter's Universitats-Buchhandlung.

NORTON, I. O. 1982, Paleomotion between Africa, South America, and Antarctica, and implications for the Antarctic Peninsula, p. 99–106. *In* C. Craddock (ed.), Antarctic Geoscience. University of Wisconsin Press, Madison.

OLIVER, P. J., T. C. MUMME, G. W. GRINDLEY, AND P. VELLA. 1979. Upper Cretaceous Mt. Somers volcanics, Canterbury, New Zealand. New Zealand Journal of Geology and Geophysics, 22:199-212.

OLIVERO, E. B. 1975. Perfil geológico, descripción de la fauna de ammonites y geomorfología del extremo N.E. de las Isla Cerro Nevado, Grupo de Islas James Ross, sector Antártico Argentino. Unpub. thesis, Facultad Ciencias Exactas y Naturales, Universidad de Buenos Aires, 48 p.

——. 1981. Esquemo de zonación de ammonites del Cretácico Superior del grupo de islas James Ross, Antártida: VIII Congreso Geológico Argentino, San Luis, 2:897-907.

OLSSON, R. K. 1960. Foraminifera of latest Cretaceous and earliest Tertiary age in the New Jersey Coastal Plain. Journal of Paleontology, 34:1-58.

——. 1964. Late Cretaceous planktonic Foraminifera from New Jersey and Delaware. Micropaleontology, 10:157-188.

OLSZEWSKI, S. 1875. Zapinski paleontologiczne. Akademija umiejetnosci, Krakowie, Komisja fizyograficzna, Sprawozdania, 9:95-149.

ORBIGNY, A. D., D'. 1840. Mémoire sur les Foraminifères de la craie blanche du bassin de Paris. Société géologique de France, Mémoires, 4(1):1-51.

PALAMARCZUK, S., ET AL. 1984. Las Formaciones Lopez de Bertodano y Sobral en la Isla Vicecomodoro Marambio, Antártida: Actas IX Congreso Geológico Argentino, 1:399-419.

PARKER, W. K., AND T. R. JONES. 1865. On some Foraminifera from the north Atlantic and Arctic Oceans, including Davis Straits and Baffin's Bay. Royal Society of London, Philosophical Transactions, 155:325-441.

PARR, W. J. 1938. Upper Eocene Foraminifera from deep borings in King's Park, Perth, western Australia. Journal of the Royal Society of Western Australia, 24:67-101.

PESSAGNO, E. A., JR. 1967. Upper Cretaceous planktonic foraminifera from the western Gulf Coastal Plain. Palaeontographica Americana, 5(37):245-445.

PILLER, W. E. 1983. Remarks on the suborder Involutinina Hohenegger and Piller, 1977. Journal of Foraminiferal Research, 13(3):191-201.

PLUMMER, H. J. 1927. Foraminifera of the Midway Formation in Texas. University of Texas Bulletin, 2644:1-206.

——. 1931. Some Cretaceous Foraminifera in Texas: University of Texas Bulletin, 3101:109-236.

PREMOLI-SILVA, I., AND A. BOERSMA. 1977. Cretaceous planktonic foraminifers-DSDP Leg 39 (South Atlantic), p. 615–641. *In* P. R. Supko et al., Initial Reports of the Deep Sea Drilling Project, Vol. 39, U.S. Government Printing Office, Washington, D.C.

REUSS, A. E. 1845. Die Versteinerungen der bohmischen Kreideformation: Stuttgart, E. Schweizerbart, Abt. 1, 58 p.

——. 1850. Nene Foraminiferen ous den Schichten des Osterreichischen Tertiarbeckens: K. Akademie der Wissenschaften, Mathematisch-naturwissenschaft-liche klasse, Denkschriften, 1:1-366.

——. 1851. Die Foraminiferen and Entomostraceen des Kreidemergels von Lemberg, p. 17–52. *In* Haidinger, W. H., Naturwissenschaftliche ahandlungen, gesammelt und durch subscription, Vol. 1. Braumüller und Seidel, Wien.

——. 1855. Ein Beitrag zur genaueren Kenntniss der Kreide-schichten Meckenburgs: Zeitschrift der Deutschen geologischen gesellschaft, 7: 261-292.

——. 1860. Die Foraminiferen der Westphaelischen, Kreideformation: K. Akademie der Wissenschaften, Wien, Mathematische-naturwissenschaftliche klasse, Sitzungsberichte, 40: 147-238.

——. 1862. Palaeontologische Beitrage: K. Akademie der Wissenschaften, Wien, Mathematische-naturwissenschaftliche klasse, Sitzungsberichte, 44(1):301-342.

——. 1863. Die Foraminiferen des norddeutschen Hils und Gault: K. Akademie der Wissenschaften, Wien, Mathematische-naturwissenschaftliche klasse, Sitzungsberichte, 46:5-100.

RINALDI, C. A. 1982. The Upper Cretaceous in the James Ross Island Group, p. 281–286. *In* C. Craddock (ed.), Antarctic Geoscience. University of Wisconsin Press, Madison.

——, ET AL. 1978. Geología de la Isla Vicecomodoro Marambio: Contribuciones del Instituto Antártico Argentino, 217:1-37.

ROBASZYNSKI, F., ET AL. 1984. Atlas of Late Cretaceous globotruncanids. Revue de Micropaléontologie, 26(3-4):145-305.

ROEMER, F. A. 1838. Die Cephalopoden des Nord-Deutschen tertiären Meersandes, p. 381–394. Neues jahrbuch für mineralogie, geognosie, geologie, und petrefaktenkunde. Kunde, Stuttgart.

——. 1839. Die Versteinerungen des norddeutschen Oolithen-Gebirges, p. 1–59. Ein Nachtrag. Hahnschen Hofbuchhandlung, Hannover, 1-59.

——. 1841. Die Versteinerungen des norddeuschen Kreide-gebirges: Hannover, Im verlage der Hahn'schen hofbuchhandlung, 145 p.

RZEHAK, A. 1891. Die Foraminiferenfauna der alttertiaeren Ablagerungen von Bruderndorf in Nieder-Oesterreich, mit Beruchsichtigung des angeblichen Kreidevorkommens von Leitzersdorf. Annalen der Naturhistoriches Hofmuseum, Wien, 6:1-12.

——. 1895. Ueber einige merkwürdige Foraminiferen aus oesterreichischen Tertiaer. Annalen der Naturhistoriches Hofmuseum, Wien, 10:213-230.

SANDIDGE, J. R. 1932. Foraminifera from the Ripley Formation of western Alabama. Journal of Paleontology, 6:265-287.

SARS, G. O. 1872. Undersogelser over Hardangerfjordens Fauna. Videnskabs-Selskabet i Christiania Forhandlinger, 1871:246-255.

SCHEIBNEROVÁ, V. 1971. Foraminifera and their Mesozoic biogeoprovinces. Records of the Geological Survey of New South Wales, 13(3):135-174.

——. 1973. Non-tropical Cretaceous foraminifera in Atlantic deep-sea cores and their implications for continental drift and paleooceanography of the South Atlantic Ocean. Records of the Geological Survey of New South Wales, 15(1):19-46.

SEGUENZA, G. 1862. Die terreni terziarii del distretto di Messina, Parte II. Descrizione dei foraminiferi monotalamici delle marne mioceniche del distretto di Messina: T. Capra, Messina, 84 p.

SLITER, W. V. 1968. Upper Cretaceous foraminifera from Southern California and Northwest Baja California, Mexico. The University of Kansas Paleontological Contributions—Protozoa, art. 8, 141 p.

——. 1973, Upper Cretaceous foraminifers from the Vancouver Island area, British Columbia, Canada. Journal of Foraminiferal Research, 3(4):167-186.

——. 1975. Foraminiferal life and residue assemblages from Cretaceous slope deposits. Geological Society of America Bulletin, 86:897-906.

——. 1976. Cretaceous foraminifera from the southwest Atlantic Ocean, Leg 36, Deep Sea Drilling Project, p 519–573. *In* P. F. Barker and I.W.D. Dalziel et al., Initial Reports of the Deep Sea Drilling Project, Vol. 36. U.S. Government Printing Office, Washington, D.C.

SMITH, A. G., A. M. HURLEY, AND J. C. BRIDEN. 1981. Phanerozoic paleocontinental world maps. Cambridge University Press, Cambridge, 102 p.

SPATH, L. 1953. The Upper Cretaceous Cephalopod Fauna of Graham Land. Falkland Island Dependencies Survey, Scientific Report, 3:1-60.

STEHLI, F. G., R. G. DOUGLAS, AND N. D. NEWELL. 1969. Generation and maintenance of gradients in taxonomic diversity. Science, 164:947-949.

STRONG, C. P. 1981. Cretaceous-Tertiary boundary at Woodside Creek revisited. Geological Society of New Zealand Newsletter, 51:4-5.

——. 1984. Cretaceous-Tertiary boundary, Mid–Waipara River section, North Canterbury, New Zealand. New Zealand Journal of Geology and Geophysics, 27:231-234.

SURLYK, F. 1984. The Maastrichtian stage of NW Europe, and its brachiopod zonation. Bulletin of the Geological Society of Denmark, 33:217-223.

TAPPAN, H. 1962. Foraminifera from the Arctic Slope of Alaska. U.S. Geological Survey Professional Paper 236-C, 209 p.

TAYLOR, D. J. 1965. Preservation, composition and significance of Victorian Lower Tertiary "*Cyclammina* faunas." Proceedings of the Royal Society of Victoria, 78(2):143-160.

TERQUEM, O. 1882. Les foraminifères de l'Eocène des environs de Paris. Société géologique de France, Mémoires, Paris, France, sér. 3, 2(3):1-193.

TOUMLIN, L. D. 1941. Eocene smaller Foraminifera from the Salt Mountain Limestone of Alabama. Journal of Paleontology, 15:567-611.

TROELSEN, J. C. 1957. Some planktonic foraminifera of the type Danian and their stratigraphic importance. U.S. National Museum Bulletin, 215:125-132.

VOLOSHINOVA, N. A., AND A. I. BUDASHEVA. 1961. Lituolidy i trochamminidy iz tretichonykh othlozheniy ostrova Sakhalina i poluostrova Kamchatki: Mikrofauna SSSR, Sbornik 12, VNIGRI, Trudy, no. 170, p. 169–233.

VOORWIJK, G. H. 1937. Foraminifera from the Upper Cretaceous of Habana, Cuba. K. Nederlandsche akademie van wetenschappen, Proceedings of the Section of Sciences, 40:190-198.

WARREN, G., AND I. SPEDEN. 1977. The Piripauan and Haumurian stratotypes (Mata Series, Upper Cretaceous) and correlative sequences in the Haumuri Bluff District, South Marlborough. New Zealand Geological Survey, Bulletin, 92:1-60.

WEBB, P. N. 1966. New Zealand Late Cretaceous foraminifera and stratigraphy. Unpub. Ph.D. dissertation, Utrecht University, 320 p.

——. 1971. New Zealand Late Cretaceous (Haumurian) foraminifera and stratigraphy: a summary. New Zealand Journal of Geology and Geophysics, 14(4):795-828.

——. 1972a. Comments on the reported occurrence of ?*Globotruncana contusa* (Cushman) from the Upper Cretaceous of James Ross Island, Grahamland, Antarctica. New Zealand Journal of Geology and Geophysics, 15(1):183.

——. 1972b. A redescription of *Frondicularia rakauroana* (Finlay) from the Late Cretaceous (Maastrichtian) of New Zealand. Micropaleontology, 18(1):94-100.

——. 1973a. Preliminary comments on Maastrichtian-Paleocene foraminifera from Lord Howe Rise, Tasman Sea. Proceedings of Oceanography of the South Pacific (Wellington, February, 1972). UNESCO, p. 144–146.

——. 1973b. Upper Cretaceous–Paleocene foraminifera from Site 208 (Lord Howe Rise, Tasman Sea), DSDP, Leg 21, p. 541–573. *In* R. E. Burns, J. Andrews, et al., Initial Reports of the Deep Sea Drilling Project, Vol. 21. U.S. Government Printing Office, Washington, D.C.

——, ET AL. 1984, Cenozoic marine sedimentation and ice-volume variation on the East Antarctic craton. Geology, 12(5):287-291.

——, AND V. E. NEALL. 1972. Cretaceous foraminifera in Quaternary deposits from Taylor Valley, Victoria Land, p. 653–657. *In* R. J. Adie (ed.), Antarctic Geology and Geophysics. Universitetsforlaget, Oslo.

——. 1988. Upper Cretaceous–Eocene stratigraphy and micropaleontology of the Dunedin area (S163), southeast Otago, New Zealand. New Zealand Geological Survey Paleontological Bulletin (in press).

WICKENDEN, R.T.D. 1932. New species of Foraminifera from the Upper Cretaceous of the Prairie provinces. Royal Society of Canada Transactions, 3rd ser., 26(4):85-91.

WILCKENS, O. 1910. Die Anneliden, Bivalven, und Gastropoden der Antarktischen Kreideformation: Wissenschtliche Ergebnisse der Schwedischen Südpolar-Expedition 1901–1903, Stockholm, 3(12):1-132.

WIND, F. H. 1979. Maastrichtian-Campanian nannofloral provinces of the southern Atlantic and Indian Oceans, p. 123–137. *In* M. Talwani, W. Hay, and W.B.F. Ryan (eds.), Deep Sea Drilling in the Atlantic Ocean: Continental Margins and Paleoenvironment. American Geophysical Union, Washington, D.C.

——. 1983. The genus *Nephrolithus* Górka, 1957 (Coccolithophoridae). Journal of Paleontology, 57(1):157-161.

WOODBURNE, M. O., AND W. J. ZINSMEISTER. 1984. The first land mammal from Antarctica and its biogeographic implications. Journal of Paleontology, 58(4):913-948.

WRENN, J. H. 1985. Paleogene dinoflagellate cyst biostratigraphy, Seymour Island, Antarctica, p. 36. Workshop on Cenozoic Geology of the Southern High Latitudes (August 16-17, 1985), Sixth Gondwana Symposium, The Ohio State University, Columbus.

WRIGHT, J. 1885. Foraminifera of the Belfast Naturalists' Field Club's cruise off Belfast. Proceedings, p. 315–325.

ZINSMEISTER, W. J. 1982. Late Cretaceous–Early Tertiary molluscan biogeography of the southern circum-Pacific. Journal of Paleontology, 56(1):84-102.

ZINSMEISTER, W. J., AND R. M. FELDMANN. 1984. Cenozoic high latitude heterochroneity of Southern Hemisphere marine faunas: Science, 224:281-283.

MANUSCRIPT ACCEPTED BY THE SOCIETY SEPTEMBER 1, 1987
BYRD POLAR RESEARCH CENTER CONTRIBUTION NO. 592

Geological Society of America
Memoir 169
1988

Bivalvia (Mollusca) from Seymour Island, Antarctic Peninsula

William J. Zinsmeister
Department of Earth and Atmospheric Sciences, Purdue University, West Lafayette, Indiana 47907
Carlos E. Macellari
Earth Sciences and Resources Institute, University of South Carolina, Columbia, South Carolina 29208

ABSTRACT

The Upper Cretaceous–lowermost Tertiary López de Bertodano and Sobral Formations Seymour Island, Antarctic Peninsula, contain one of the most important marine faunas known for this interval of the Earth's history. Faunal data from this sequence are providing important new understanding of the origin and biogeographic history of the marine biota of the Southern Hemisphere and insight into the faunal transition at the end of Cretaceous time.

The bivalves described herein where collected during four expeditions (1975, 1982, 1983–1984, 1985) to Seymour Island. In addition, the collections made by the Swedish South Polar Expedition, 1901–1903, housed in the Naturhistoriska Riksmuseet, Stockholm, were also examined during the course of this study. This chapter describes 13 new species and 2 genera of bivalves: *Nucula (Leionucula) hunickeni* n. sp., *Australoneilo casei* n. sp., *Austrocucullaea* n. gen. *oliveroi* n. sp., *Cucullaea ellioti* n. sp., *Pinna freneixae* n. sp., *Phelopteria feldmanni* n. sp., *Entolium seymourensis* n. sp., *E. sadleri* n. sp., *Acesta shackletoni* n. sp., *A. webbi* n. sp., *Seymourtula* n. gen. *antarctica* (Wilckens), *Lahillia huberi* n. sp., *Marwickia woodburnei* n. sp., *Cyclorisma chaneyi* n. sp., Surobula n. gen. *nucleus* (Wilckens), *Thracia askinae* n. sp. Twenty-one previously described species are redescribed and figured, and their taxonomy revised.

INTRODUCTION

Seymour Island, located on the northeast tip of the Antarctic Peninsula (Fig. 1), contains one the most important Upper Cretaceous sequences known. Although this fauna has been known since the late 19th century, very little work has been done since the publication of Wilckens' (1910) report on the fossils collected by Otto Nordenskjöld's Swedish South Polar Expedition (1901–1903). Except for several brief visits by parties from the British Antarctic Survey, there has been no systematic study of the Seymour Island sequence until the early part of the 1970s, when Argentina established a permanent station (Marambio) near the north tip of the island. Seymour Island is referred to in Argentine literature as Isla Vice Comodoro Marambio.

Field work on Seymour Island in the mid-1970s and early 1980s revealed that the faunas from both the Cretaceous and Tertiary are considerably more diverse than previously thought. Taxa described herein are based on new material from more than 600 localities collected during the course of four field seasons (1975, 1982, 1983–1984, and 1985) and reexamination of Nordenskjöld's collections described by Wilckens (1910). We have

restricted this chapter to the Bivalvia because of a number of taxonomic problems that remain to be addressed with the Gastropoda. In light of the detailed stratigraphic studies conducted during the field programs, we are able to provide, for the first time, the stratigraphic distribution of all the taxa discussed in this paper (Fig. 2).

GEOLOGY

The geology of Seymour Island has been discussed by Andersson (1906), Bibby (1966), Rinaldi et al. (1978), Elliot and Trautman (1982), Zinsmeister (1982), Huber (1984), Macellari (1984, this volume), and Palamarczuk et al. (1984). The bivalve fauna discussed herein was collected from the Late Cretaceous- to Paleocene-age beds of the López de Bertodano and Sobral Formations exposed on the southern two-thirds of the island.

The López de Bertodano Formation is composed of 1,190 m of friable, clayey sandy siltstones with intercalations of more indurated concretion-bearing sandstone horizons. The sediments

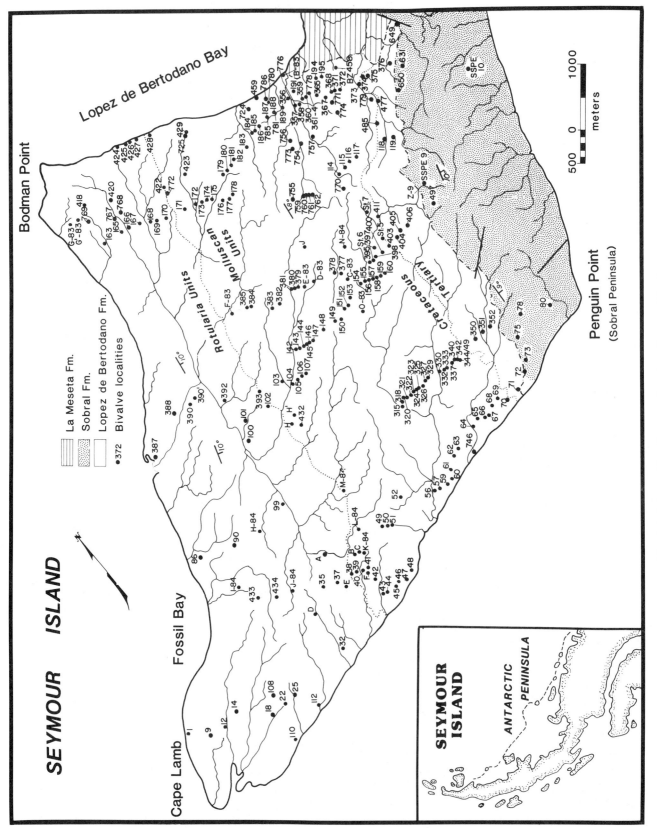

Figure 1. Index and locality map of southern Seymour Island.

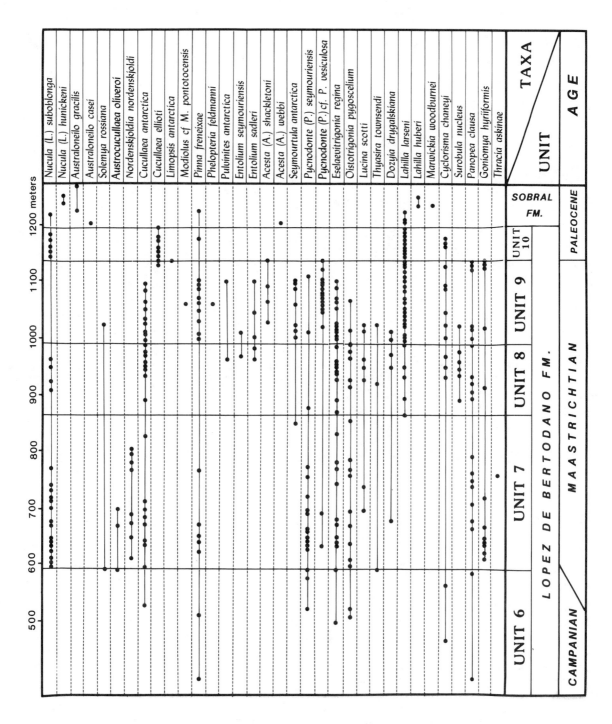

Figure 2. Stratigraphic ranges of the bivalve taxa from the López de Bertodano and Sobral Formations.

have a massive appearance due to extensive bioturbation. Surface accumulations of large numbers of concretions would seem to suggest that concretions are abundant, but fresh exposures along the seacliffs indicate that concretions are, in fact, rare in the formation. Although most of the fauna discussed in this paper was obtained from concretions, occasional fossilferous, silty sand horizons contained abundant specimens of the smaller taxa. The fauna is believed to represent a shelf environment that gradually shallowed during the early Paleocene (Huber et al., 1983; Askin, 1984; Macellari, 1986, this volume).

The Sobral Formation reaches a maximum thickness of 255 m, and is composed of well-bedded, dark, clayey sandy siltstones that gradually change to cross-bedded glauconitic sands near the top of the formation. The Sobral Formation has been interpreted as a prograding facies of a deltaic system, including both delta top and delta front facies (Macellari, this volume). The age of the Sobral Formation has been established as Paleocene on the basis of micropaleontological studies (Hall, 1977; Palamarczuk et al., 1984; Askin, 1984) and on glauconite dating (Macellari, 1984).

SYSTEMATIC PALEONTOLOGY

The material used in this study is housed in the United States National Museum (USNM), Washington, D.C.; Museo de La Plata (MLP), La Plata; Institute of Geological Sciences (IGS), London; Orton Geological Museum (OGM), The Ohio State University, Columbus; and the Naturhistoriska Riksmuseet (Mo), Stockholm. Additional material from Seymour Island is housed in the British Museum of Natural History, London; British Antarctic Survey, Cambridge; Centro de Investigaciones en Recursos Geologicos, Buenos Aires; and Purdue University, West Lafayette, Indiana. All the localities listed in the systematic section are plotted on the topographic map of Seymour, which has been drawn by Henry Brecker, Institute of Polar Studies, The Ohio State University.

Phylum MOLLUSCA Linnaeus, 1758
Class BIVALVIA Linnaeus, 1758
Subclass PALAEOTAXODONTA Korobkov, 1954
Order NUCULOIDA Dall, 1889
Superfamily NUCULACEA Gray, 1824
Family NUCULIDAE Gray, 1824

Genus *Nucula* Lamarck, 1799
Type species.—*Arca nucleus* Linnaeus, 1758 (by monotypy).

Subgenus *Leionucula* Quenstedt, 1930
Type species.—*Nucula albensis* d'Orbigny, 1844 (by original description)

Nucula (Leionucula) suboblonga (Wilckens, 1907)
Figure 3.1–5

Nucula suboblonga Wilckens, 1907, p. 53; 1910, 22–24, Plate 2, Figs. 1a,b,2.

Nuculoma (Palaeonucula) poyaensis Freneix, 1956, pp. 157–158, Plate 1, Fig. 1a,b.
Leionucula poyaensis (Freneix), Freneix, 1980, pp. 75–77, Plate 1, Figs. 1–4.
Supplementary description. Shell medium-sized, thick (5 mm), inflated; umbones small, extremely posteriorly located; dorsal anterior margin long, nearly stratight to slightly convex; anterior margin bluntly rounded; anterior half of ventral margin broadly rounded, posterior quarter straight to slightly concave; posterior margin short, subtruncate; posterior dorsal margin straight to slightly convex, steeply sloping; escutcheon broad, open, bounded by posterior umbonal ridges; rostral sinus poorly developed; lunule elongated, defined by shallow furrow, elongated, surface smooth except for very faint closely spaced radial ribs; resilifer oblique; nine posterior hinge teeth strongly developed, subchevronshaped, elevated (5 mm); inner margin smooth; adductor scars deeply sunken; medial muscle scars small, punctiform, located ventral to anterior half of hinge.
Dimensions. USNM 404809; length, 44 mm; height, 30 mm; width of paired valves, 23 mm.
Types. Lectotype, MO 1424a (herein designated); hypotypes, USNM 404809, USNM 404810, USNM 404811, USNM 404812, USNM 404813.
Type locality. SSPE locality 9, Wilckens (1910).
Stratigraphic range. Top of unit 6 of the López de Bertodano Formation to middle of unit 1 of the Sobral Formation.
Localities. K-37, K-39, K-41, K-42, K-43, K-71, K-72, K-105; K-143, K-385, K-423, K-426, Z-9, Z-463, Z-477, Z-497, Z-459, Z-514, Z-725, Z-754, Z-757, Z-762, Z-776, Z-777, B, E, F.
Material. 48 specimens (most specimens broken).
Discussion. This species was originally described from the Upper Cretaceous sequence at Cerro Cazador in southern Patagonia (Wilckens, 1907). In the same paper, Wilckens described a second species of *Leionucula (N. (L.) oblonga)* and separated the two by the more centrally located umbones of *N. (L.) suboblonga*. Unfortunately he did not figure *N. (N.) suboblonga* because he did not have a "perfect" specimen. Wilckens (1910) subsequently reported the occurrence of this species on Seymour Island and included the first figure of the species. Freneix (1958) described a similar species *(N. (L.) poyaensis)* from New Caledonia. Because the measurements of *N. (L.) poyaensis* and *N. (L.) suboblonga* are nearly identical, we believe that the two are conspecific. An early Miocene species *(N. (L.) grandis* Mulumian, Camacho, and Gorrono, 1979) from Tierra del Fuego is very close to the Seymour Island species, differing only in the absence of radial ornamentation.

Nucula (Leionucula) hunickeni n. sp.
Figure 3.10,11
Description. Shell small, moderately inflated; umbones small, moderately elevated; anterior dorsal margin gently sloping, slightly convex; anterior margin bluntly rounded; ventral margin broadly rounded; posterior mar-

Figure 3. 1-5, *Nucula (Leionucula) suboblonga* (Wilckens), hypotypes, 1, USNM 404809, ×1. 2, Hyptotype, USNM 404810, 1×. 3, Hypotype, USNM 404811, 1×. 4, Hypotype, USNM 404812, 1×. 5. Paratype, USNM 404813, 1×. 6-7, *Australoneilo gracilis* (Wilckens), hypotype, USNM 404817, 1×. 6, Hypotype, USNM 404816, 1×. 8-9, *Australoneilo casei* n. sp., holotype, USNM 404818, 1×. 10-11 *Nucula (Leionucula) hunickeni* n. sp., holotype, USNM 404814, 1×. 12-15, *Solemya rossiana* Wilckens, 12, Hypotype, USNM 405827, 1×. 15, Hypotype, USNM 404821, 1×. 13-14, Hypotype, USNM 404820, 1×. 16, *Pinna freneixae* n. sp., holotype, USNM 404846, .75×.

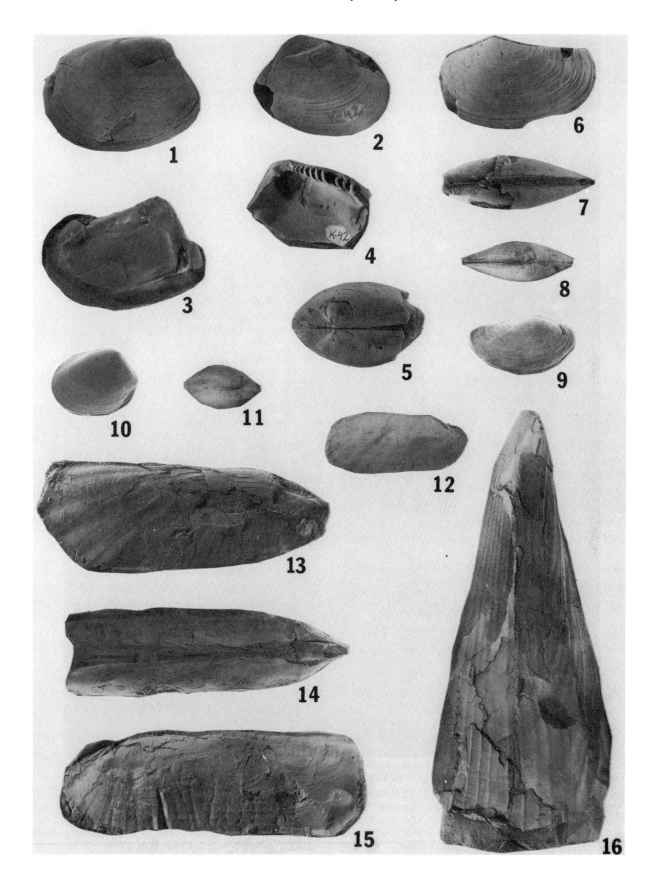

gin short, rounded; dorsal posterior margin slightly concave, steeply sloping; posterior umbonal ridge low, rounded, defining shallow escutcheon; lunule faint and shallowly depressed; surface smooth except for fine, closely spaced growth increments and fine radial striae that are restricted to anterior quarter of shell; inner margin smooth; adductor muscle scars slightly sunken.

Dimensions. Holotype, USNM 404814; length, 23 mm; height, 17 mm; width of paired valves, 12 mm.

Types. Holotype, USNM 404814; paratype, USNM 404815.

Type locality. Z-9.

Stratigraphic range. Middle part of unit 1 of the Sobral Formation.

Localities. Z-9, Z-497.

Material. 7 specimens.

Discussion. This species of *Leionucula* may be distinguished from *N. (L.) suboblonga* by its smaller size and straighter anterior, which slopes at a steeper angle. *N. (L.) hunickeni* n. sp. has only been encountered in a dark brown silty sandstone of unit 1 of the Sobral Formation.

Etymology. This species named for Dr. Mario Hunicken, Universidad de Cordoba, to recognize his years of study of the Patagonian Cretaceous.

Superfamily NUCULANACEA H. and A. Adams, 1858
Family MALLETIIDAE H. and A. Adams, 1858

Genus *Australoneilo* Zinsmeister, 1984
***Type species. Australoneilo rossi* Zinsmeister, 1984 (by original designation).**

***Australoneilo gracilis* (Wilckens, 1907)**
Figure 3.6,7

Malletia gracilis Wilckens, 1907, p. 35, Plate 5, Fig. 10; Wilckens, 1910, p. 25, Plate 2, Fig. 4.

Supplementary description. Shell medium-sized, elongated, anterior half of valve slightly inflated; posterior half slightly compressed; beaks small, located one-third of the length of the value from anterior margin; posterior margin long, slightly concave; rostrum poorly developed; ventral margin broadly rounded; anterior margin bluntly rounded; anterior dorsal margin straight to slightly sloping; umbonal ridges poorly developed; surface smooth except for closely spaced growth increments; hinge two-thirds the length of valve, with approximately 30 chevron-shaped teeth on either side of beak; pallial sinus deep, moderately narrow with a gentle ventral sloping orientation; anterior margin of sinus rounded; inner margin smooth; posterior adductor muscle scar elongated, slightly impressed.

Dimensions. Hypotype, USNM 404817; length, 42 mm; height, 25 mm; width of paired valves, 16 mm.

Types. Lectotype MLP 9127 (herein designated); hypotypes, USNM 404816, 404817.

Type locality. Southern Patagonia (Wilckens, 1907).

Localities. K-80, Z-9.

Stratigraphic range. Units 1–3 of the Sobral Formation.

Material. 7 specimens.

Discussion. Australoneilo gracilis from Seymour Island appears to be identical to the species Wilckens (1907) described from Cerro Cazador and Sierra de los Baguales in southern Patagonia. Although the exact location of Wilckens' locality is uncertain, the sequence at Cerro Cazador and Sierra Baguales ranges in age from Late Cretaceous to early Tertiary and is approximately coeval with the López de Bertodano Formation. The morphologic similarities and age of the two deposits strongly suggest that the Seymour Island specimens are the same as *A. gracilis* from southern Patagonia. The specimen figured by Wilckens cannot be located in the collections of the Museo de La Plata and is considered to be lost; we are designating MLP 9127 as the lectotype.

The genus *Australoneilo* represents one of the paleoaustral elements that were restricted to the southern circum-Pacific during the Late Cretaceous and Cenozoic, with *A. gracilis* representing the earliest known occurrence of the genus. *A. rossi* Zinsmeister, which is very common in the Upper Eocene deposits on Seymour Island, is remarkably similar to *A. gracilis*. The only differences between the two species are the more elongated profile and a stronger umbonal ridge of *A. rossi*.

***Australoneilo casei* n. sp.**
Figure 3.8,9

Description. Shell medium-sized, elongated, moderately inflated; umbones low rounded, located approximately 40 percent of the value length from anterior margin; anterior dorsal margin of medium length, straight; anterior margin short, bluntly rounded, merging with broadly rounded ventral margin; posterior narrow, rounded, slightly rostrate; posterior margin long, concave; anterior hinge with 14 to 16 chevron-shaped teeth becoming very small near beak; approximately 25 slightly chevron-shaped teeth on posterior half of hinge.

Dimensions. Holotype, USNM 404818; length, 32 mm; height, 17 mm; width of paired valves, 11 mm.

Types. Holotype, USNM 404818; paratype, USNM 404819.

Type locality. Z-746.

Stratigraphic range. Known only from one locality in unit 10 of the López de Bertodano Formation.

Locality. Z-746.

Material. 2 specimens.

Discussion. Although the overall shape of *Australoneilo casei* n. sp. is typical of the genus *Nuculana,* the absence of a resilifer clearly places this species in the family Malletiidae. *A. casei* n. sp. can easily be separated from *A. gracilis* by its broad convex ventral margin and narrow posterior margin. The ventral margin of *A. gracilis* is nearly straight and the rostrum is more strongly developed. Wilckens (1910) figured a malletiid-like bivalve *(M. pencanoides)* from Snow Hill Island. The posterior margin of *M. pencanoides* is considerably more truncate and the valves are not as inflated as *A. casei* n. sp.

Etymology. This species named after Judd Case, University of California at Riverside, to recognize his work on Early Tertiary mammal faunas from Antarctica.

Subclass CRYPTODONTA Neumayr, 1884
Order SOLEMYOIDA Dall, 1899
Family SOLEMYIDAE Adams and Adams, 1857 (1840)

Genus *Solemya* Lamarck, 1818
***Type species. Tellina togata* Poli, 1795 (subsequent designation, Children, 1823**

***Solemya rossiana* Wilckens**
Figure 3.12–15

Solenomya rossiana Wilckens, 1910, pp. 65–67, Plate 3, Fig. 9.

Supplementary Description. Shell medium-sized, elongated, thin, compressed; beaks small with extreme anterior location; anterior dorsal margin short, straight; anterior bluntly rounded; ventral margin long, straight; posterior margin bluntly rounded merging with the long straight dorsal ventral margin; surface ornamented with two sets of radiating ribs—16 to 20 low rounded ribs on anterior portion of valve, having wavy intersections with anterior ribs, interspace width subequal, rib width on posterior portion of valve wide, flat, approximately twice the width of interspaces.

Dimensions. USNM 404820; length, 76 mm; height, 26 mm.

Type. Holotype, Mo 1602; hypotypes, USNM 404820, 404821.
Type locality. Seymour Island (Wilckens, 1910).
Stratigraphic range. Upper part of unit 6 to lower part of unit 9 of the López de Bertodano Formation.
Localities. N84, Z-459, E.
Material. 33 specimens.
Discussion. The specimens from locality E are considerably larger than Wilckens' figured specimen. Several specimens from locality Z-459 conform to Wilckens' dimensions of *S. rossiana,* but unfortunately, preservation of material from locality Z-459 is too poor for proper identification. Solemyas are fairly rare in the López de Bertodano Formation, but occur in large numbers at the few localities where they have been recorded.

Subclass PTERIOMORPHA Beurlen, 1944
Order ARCOIDA Stoliczka, 1871
Superfamily ARCACEA Lamarck, 1809
Family CUCULLAEIDAE Stewart, 1930

The phylogenetic relationships of the Parallelodontidae and the Cucullaeidae are poorly understood; the two families tend to grade into one another. Since no precise limits have been defined to separate the two families, placement of some genera in these families is somewhat arbitrary. The phylogenetic uncertainties reflect, in part, the convergence of morphologic features within the many genera that make up both the Parallelodontidae and the Cucullaeidae.

Members of the Parallelodontidae tend to be elongated, inflated, and inequilateral, with a nondenticulate shell margin. The hinge varies considerably, but tends to be long, with few elongated pseudolateral teeth and numerous medial teeth. The Cucullaeidae is characterized by subtrigonal to subquadrate shell with nearly medial located umbones with valve margins frequently denticulate. The dentition of the Cucullaeidae is similar to the Parallelodontidae; it consist of elongated psuedolaterals and short medial teeth that are frequently contorted. The Parallelodontidae has been divided into the subfamilies Parallelodontinae Dall and Grammatodontinae Branson. The Grammatodontinae has been further divided into the Grammatodon, Cucullaria, and Catella groups.

Newell (1969) placed *Nordenskjoldia* within the Grammatodon group, which is characterized by having a trapezoidal outline and short nonhorizontal teeth. With the exception of *Nordenskjoldia,* the anterior pseudolaterals of the Grammatodon group converge ventrally below the hinge toward the midline of the valve. Although the degree of convergence may vary, all possess the ventral converging orientation. In contrast, the anterior pseudolaterals of *Nordenskjoldia* are short and converge dorsally above the hinge line. The Cucullaeidae is the only group within the Arcacea that displays diverging laterals. Newell's placement of *Nordenskjoldia* within the Grammatodontinae apparently was based on the short length of the anterior teeth and the relatively long posterior teeth. The familial placement of *Nordenskjoldia* depends on the weight given to the length and significance of the orientation of the lateral teeth. We believe that the extreme divergence of the anterior pseudolateral teeth of *Nordenskjoldia* was in response to the extreme anterior location of the beak. The anterior location of the beak probably resulted in a shortening and marked steeping of the divergence of the anterior pseudolateral teeth. If this is the case, *Nordenskjoldia* would be better placed within the Cucullaeidae rather than in the Parallelodontidae. This placement of *Nordenskjoldia* within the Cucullaeidae is strengthened by the denticulate margin, which is not characteristic of the Parallelodontidae.

Genus *Nordenskjoldia* Wilckens, 1910
Diagnosis. Shell trapezoidal; beaks near anterior margin; anterior pseudolateral teeth short, steeply diverging.
Type species. *Arca disparilis* d'Orbigny, 1846 (by original designation)

***Nordenskjoldia nordenskjoldi* Wilckens, 1910**
Figure 4.8–12
Nordenskjoldia nordenskjoldi Wilckens, 1910, pp. 26–30, Plate 2, Figs. 8a–c, 9a,b, 11.
Supplementary description. Shell medium-sized, subquadrate, elongated, thick walled; anterior dorsal margin short, sloping sharply; anterior margin bluntly rounded; ventral margin straight to slightly concave; posterior margin produced, bluntly rounded; posterior dorsal margin long, gently sloping, concave; umbones prominent, located near anterior third of hinge line length; umbonal ridge, high, rounded, strongly developed, with dorsal side forming a flat ridge; beak small; surface nearly smooth, marked by irregularly developed concentric growth increments; subsurface with strong flat-topped ribs, subequal in width on central region of valve, broader posterior to umbonal ridge; hinge approximated two-thirds the length of valve; central teeth small, irregular, contorted; anterior three to four pseudolateral teeth short, sloping steeply anteroventrally, some teeth crenulated and distorted; two to three long, gently posterioventrally sloping posterior pseudolaterals, mildly deformed with crenulations; ligamental area moderately broad, slightly concave, chevrons slightly offset anteriorly; posterior adductor 50 percent larger than anterior scar; numerous raised radial striae along inner margin of pallial sinus; inner margin crenulated.
Dimensions. USNM 404830; length, 67 mm; height, 47 mm; width of single valve, 22.5 mm.
Types. Lectotype, Mo 1552a (Wilckens, 1910, Plate 2, Fig. 8a), (herein designated); hypotypes, USNM 404825, 404826, 404827, 404828, 404829, 404830.
Type locality. Seymour Island (Wilckens, 1910).
Stratigraphic range. Unit 3 through the middle part of unit 7 of the López de Bertodano Formation.
Localities. K-12, K-47, K-49, K-143, K-145, K-216, K-384, K-394, K-423, Z-37, Z-725, F.
Material. 34 specimens.
Discussion. The genus *Nordenskjoldia* has been reported from the Upper Cretaceous of Madagascar (Collignon, 1951), and New Caledonia (Freneix, 1980). The distribution of *Nordenskjoldia* is believed to reflect the existence of a broad, southern faunal province that dominated the high southern latitudes during the Late Cretaceous and early Tertiary.

***Austrocucullaea* Zinsmeister n. gen.**

Diagnosis. Shell subquadrate, posterior and anterior pseudolaterals subequal; anterior pseudolateral moderately long converging dorsally above hinge line; margin of valve, denticulate.
Type species. Austrocucullaea oliveroi n. sp.
Discussion. Although the exterior ornamentation of *Austrocucullaea* superficially resembles *Indogrammatodon,* the relatively long dorsally converging anterior pseudolateral teeth place this genus within the Cucullaeidae. The anterior pseudolateral teeth of *Indogrammatodon* are short and converge ventrally below the hinge line. The subequal development of the pseudolateral and dorsally converging nature of the anterior teeth of *Austrocucullaea* is similar to a number of other cucullaeids from the Southern Hemisphere (*Cucullaea raea* Zinsmeister and *C. donaldi* Sharman and Newton), but may be separated by the presence of narrow, rounded radial ribs with relatively wide interspaces with fine secondary riblets. The development of denticulate ventral margin of *Austrocucullaea* also separates it from *Indogrammatodon.*

Austrocucullaea oliveroi n. sp.
Figures 4.1–7

Description. Shell medium-sized, subquadrate, equivalve, inequilateral; slightly inflated; umbones located at one-third the length of shell, moderately elevated; umbonal ridge moderately developed, intersecting edge of shell at the posterior ventral margin; anterior dorsal margin of medium length, slightly sloping; anterior margin subrounded; ventral margin long, slightly convex, merging with posterior margin at angulation formed by the intersection of umbonal ridge with posterior ventral margin of valve; posterior margin nearly straight, sloping anteriorly; posterior dorsal margin straight, nearly twice the length of anterior dorsal margin; surface of central portion of shell ornamented with approximately 16 narrow radial ribs with very wide flat interspaces having secondary riblets, ribs near posterior and anterior margin becoming very closely spaced, ribs of left valve more widely spaced and lower than on left valve; intersection of regularly spaced growth increments and radial ribs marked by small nodes; ligamental area approximately 70 percent of length of valve, ligamental grooves closely spaced, apices of chevrons inclined slightly posteriorly; vertical central hinge teeth numerous, irregular; four posterior horizontal teeth sloping slightly ventrally; three thick anterior horizontal teeth slightly inclined ventrally; posterior adductor scar larger than anterior scar; inner margin crenulated.

Dimensions. USNM 404823; length, 60 mm; height, 43 mm.

Types. Holotype, USNM 404822; paratypes, USNM 404823, 404824.

Type locality. K-43.

Stratigraphic range. Top of unit 8 to middle part of unit 9 of the López de Bertodano Formation.

Localities. K-43, Location F.

Material. 3 specimens.

Discussion. The Seymour Island specimens are similar in shape, size, and ornamentation to the New Caledonian species (*Indogrammatodon lormandi* Freneix). The characteristic ornamentation of *Austrocucullaea* n. gen. easily separates this species from other cucullaeas from the López de Bertodano Formation. Wilckens (1910) figured a small *Cucullaea*-like bivalve from Snow Hill Island, which he referred to as *Nordenskjoldia nordenskjoldi.* This specimen is superficially similar to *A. oliveroi,* but has a more quadrate outline, nearly centrally located umbones and the narrow radial ribs are absent. It is also much more quadrate than other known species of *Nordenskjoldia.*

Genus *Cucullaea* Lamarck, 1801
Type species. Cucullaea auriculifera Lamarck, 1801 (by subsequent designation, Children, 1823).

Cucullaea ellioti n. sp.
Figure 5.1–10

Description. Shell medium-sized, moderately elongated, subquadrate to suboval, moderately inflated; dorsal margin straight; anterior margin nearly at a right angle with the dorsal margin; ventral margin straight to slightly rounded; posterior margin bluntly rounded; height approximately 75 percent of the length; umbones moderately elevated, located midway between the anterior margin and center of hinge plate; hinge plate narrow; number of ligamental grooves variable; apices of chevrons posteriorly offset; pseudolateral numbers three and four, sloping slightly ventrally; medially located hinge denticles twisted and variable in number; surface nearly smooth except for finely spaced concentric growth increments, which are crossed by flat-topped radial ribs, interspaces narrow, approximately one-quarter the width of rib; inner margin crenulated; adductor myophoric flange poorly developed.

Dimensions. USNM 404838; length, 51 mm; height, 35 mm; width of single valve, 16 mm (see Fig. 6 for additional measurements).

Types. Holotype, USNM 404838; paratypes, USNM 404839, 404840, 404841, 404842.

Type locality. Z-9.

Stratigraphic range. Upper part of unit 9 through unit 1 of the Sobral Formation.

Localities. Z-9, Z-477, Z-485, Z-746, K-70, K-118, K-119.

Material. 32 specimens.

Discussion. Cucullaea ellioti n. sp. is easily separated from *C. antarctica* by its smaller size, greater elongation, and narrower ligamental region; also, it is not as inflated (Fig. 4). The absence of a myophoric flange also serves to distinguish *C. ellioti* from *C. antarctica.* E. Kauffman (personal communication, 1979) stated that, in smaller individuals of cucullaeas, the myophoric ridge is frequently lost because of abrasion. The absence of a myophoric flange on *C. ellioti* n. sp. may reflect abrasion during transport prior to burial.

The lowest occurrence of *C. ellioti* n. sp. is in uppermost Maastrichtian beds (top of unit 9), but it becomes abundant only above the Cretaceous/Tertiary boundary (unit 10 of the López de Bertodano Formation and basal part of the Sobral Formation).

Etymology. This species of *Cucullaea* is named in honor of Dr. David Elliot, director of the Institute of Polar Studies, The Ohio State University, for his lifetime of research in the Antarctic.

Cucullaea antarctica Wilckens
Figures 5.11,12; 7.1–7

Cucullaea antarctica Wilckens, 1907, p. 36–37, Plate 6, Figs. 5a–b, 6.

Supplementary description. Shell moderately large, subquadrate inflated; umbones mesogyrous, medially located, relatively elevated; anterior and ventral margins evenly rounded; posterior margin moderately truncated; surface smooth except for irregularly spaced growth increments; prominent radial ribs with narrow interspaces developed just below shell surface, exposed on abraded surfaces; ligamental area wide with 5 to 10 ligamental grooves, apices of ligamental grooves slightly offset to the anterior; hinge composed of four anterior and five posterior subhorizontal pseudolateral teeth with irregular medial denticles; internal margin crenulated; posterior muscle scar with prominent myophoric ridge.

Dimensions. USNM 404831; length, 76 mm; height, 65 mm; width of single valve, 35 mm (see Fig. 6 for additional measurements).

Types. Lectotype, MLP 9116, Wilckens, 1907, Plate 6, Figure 6 (herein designated); hypotypes, USNM 404831, 404832, 404833, 404834, 404835, 404836, 404837.

Type locality. Sierra de Los Baguales, Argentina.

Stratigraphic range. Top of unit 6 to middle part of unit 9 of the López de Bertodano Formation.

Localities. K-42, K-63, K-103, K-118, K-119, K-131, K-132, K-151, K-153, K-179, K-186, K-356, K-380, K-423, Z-37, Z-459, Z-725, Z-754, Z-756, Z-757, Z-758, Z-762, Z-772, K-776, Z-780, Locations B, E, F, H.

Material. 60 specimens.

Discussion. The better of the two specimens figured by Wilckens (Fig. 5a,b) could not be located in the Museo de La Plata and is considered lost. As a consequence, we have designated Wilckens' specimen (Fig. 6) as the lectotype. *C. antarctica* is superficially similar to the illustrations of *C. argentina* Feruglio from Patagonia, but examination of the type specimen currently housed in the invertebrate collections at the

Figure 4. 1-7, *Austrocucullaea* n. gen. *oliveroi* n. sp. 1, 3, Paratype, USNM 404823, 1×. 2, 4, 6, Holotype, USNM 404822, 1×. 5, 7, Paratype, USNM 404824, 1×. 8-12, *Nordenskjoldia nordenskjoldi* Wilckens. 8, Hypotype, 404826, 1×. 9, Hypotype, USNM 404828, 1×. 10, Hypotype, USNM 404827, 1×. 11, Hypotype, USNM 404830, 1×. 12, Hypotype, USNM 404825, 1×.

University of Bologna, Italy, revealed that Fergulio's types of *C. argentina* consisted of poorly preserved internal casts and should not have been used to define a valid species. Wilckens (1910) described *C. grahamensis* from a single fragmentary specimen in a concretion. Although no specimens referrable to *C. grahamensis* have been found on Seymour Island, several well-preserved specimens referrable to this species from the James Ross Island region are in the collections of the British Museum of Natural History. The outline of these individuals of *C. grahamensis* is more quadrate, and the posterior carina is more developed than in *C. antarctica.*

Family LIMOPSIDAE Dall, 1895

Genus LIMOPSIS Sassi, 1827
Type species. Arca aurita Brocchi, 1814 (by original designation).

Subgenus *Limopsis*
Type species. Arca aurita Brocchi, 1814 (by original designation).

Diagnosis. Limopsids having the valve surface without radial ornamentation and the inner margins smooth.

Limopsis antarctica Wilckens
Figure 8.1
Limopsis antarctica Wilckens, 1910, p. 31–32, Plate 2, Figs. 14, 15.
Supplementary description. Shell medium-sized, subtrigonal, oblique, orthogyrate; valve height exceeds width; anterior margin narrow; posterior margin long slight convex, ventral margin broadly rounded, anterior margin nearly straight; height slightly exceeding width; beak small, pointed; surface ornamented with evenly spaced growth increments; adductor muscle scars positioned with posterior scar more ventrally located; posterior hinge teeth more strongly developed than anterior teeth.
Dimension. USNM 405774; length, 13.2 mm; height, 14.5 mm; width of single valve, 3 mm.
Types. Holotype, Mo 1462; hypotype, USNM 405774.
Type locality. SSPE Locality 8.
Stratigraphic range. Known only from single specimen from unit 9 of the López de Bertodano Formation.
Localities. St.-5.
Material. One specimen.
Discussion. Although the hinge of this specimen could not be observed, its shape is nearly identical to Wilckens' figure (1910, Plate 2, Fig. 14). The only apparent difference between our specimen and that of Wilckens is the absence of radial lines present on our new material. Wilckens' specimen shows a very faint trace of radial sculpture, which would place it within the subgenus *Pectunculina.* The margin of the Seymour Island species is smooth in contrast to the crenulate margin of *Pectunculina.* Because of the absence of marginal crenulation and the extremely poor development of the radial ornamentation, we place the antarctic specimens within the subgenus *Limopsis.*

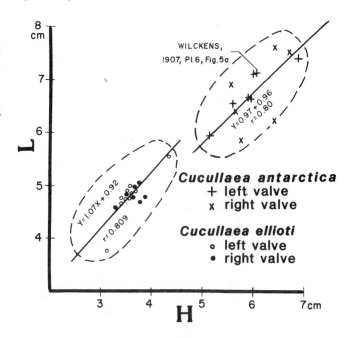

Figure 6. Comparison of length/height ratios between *Cucullaea antarctica* Wilckens and *Cucullaea elliot* n. sp.

L. antarctica is extremely rare in the López de Bertodano Formation and is represented by only one specimen. Normally, *Limopsis* is present in large numbers where it occurs, and its rarity on Seymour Island is unusual. Heinberg (1979) showed that the life habitat of *Limopsis* is closely related to the orientation and position of the adductor muscle scars. Based on Heinberg's study, *L. antarctica* lived attached to a hard substrate.

Order MYTILOIDA Ferussac, 1822
Superfamily MYTILACEA Rafinesque, 1815
Subfamily MODIOLININAE Keen, 1958

Genus *Modiolus* Lamarck, 1799
Type species. Mytilus modiolus Linnaeus (by subsequent designation, Gray, 1847).

Modiolus cf. *M. pontotocensis* del Valle and Medina, 1980
Figure 8.14,15

Modiolus pontotocensis del Valle and Medina, 1980, p. 53–54, Plate 2, Figs. 3, 4, 5.
Supplementary Description. Shell medium-sized, elongated modioliform; beak subterminal; posterior dorsal margin straight for approximately 70 percent of length of valve; posterior asymmetrically rounded; ventral margin straight, slightly concave; anterior margin short, blunt; umbonal ridge rounded, extending to posterioventral margin of shell; region ventral to umbonal ridge slightly depressed; surface ornamented with fine, closely spaced radial ribs; concentric growth increments irregularly spaced.
Dimensions. USNM 404843; length, 42 mm; height of widest portion of valve, 17 mm.
Type. Holotype, CIRGEO 3731; hypotype, USNM 404843, 404844.

Figure 5. 1-10, *Cucullaea ellioti* n. sp. 1, Paratype, USNM 404841, 1×. 2, 3, Paratype, USNM 405828, 1×. 4, 10, Paratype, USNM 404842, 1×. 5, Paratype, USNM 404839, 1×. 6, 9, Holotype, USNM 404838, 1×. 7, 8, Paratype, USNM 404840, 1×. 11-12, *Cucullaea antarctica* Wilckens, hypotype, USNM 404837, 1×.

Locality. Z-754.
Stratigraphic range. Known from only one locality in unit 9 of the López de Bertodano Formation on Seymour Island.
Material. 2 opened, paired valved specimens.
Discussion. These poorly preserved specimens of *Modiolus* are found at a single locality in unit 9. We tentatively refer our material from Seymour Island to *M. pontotocensis,* described by del Valle and Medina (1980) from Cape Lamb on James Ross Island, because of the similarity in size, fine radial ornamentation, and shell outline.

Superfamily PINNACEA Leach, 1819
Family PINNIDAE Leach, 1819

Genus *Pinna* Linnaeus, 1758
Type species. *Pinna rudis* Linnaeus (by subsequent designation, Childrens, 1823).

Pinna freneixae n. sp.
Figure 3.16

Description. Shell large, triangular, elongated, thin-walled; anterior end forming an apical angle ranging from 23 to 26°; umbones terminal, medially carinated; surface ornamentation of ventral half of valve initially consisting of five to six low, rounded, straight, longitudinal ribs decreasing to four or five on posterior three-fourths of valve, interspaces broadening posteriorly; three longitudinal ribs dorsal to medial carina, intersected by moderately spaced, inclined, curved ribs imparting a wavy appearance; ornamentation becoming obsolete near posterior margin on exceptionally large specimens.
Dimensions. USNM 404846; length, 299 mm; height of posterior margin, 90 mm.
Types. Holotype, 404846; paratype, USNM 404845, 404847.
Type locality. K-384.
Stratigraphic range. Units 5 through 9 of the López de Bertodano Formation.
Localities. K-116, K-143, K-157, K-333, K-334, K-340, K-360, K-361, K-420, K-427, K-428, Z-763, Z-764, Z-769, Z-772, Z-776, G-83, Locations B, F.
Discussion. *Pinna freneixae* n. sp. is easily separated from *P. anderssoni* Wilckens from Snow Hill Island by its fewer longitudinal ribs on both the dorsal and ventral portions of the valve. The dorsal half of *P. anderssoni* has approximately nine longitudinal ribs, while *P. freneixae* n. sp. has no more than six. *P. arata* Forbes from the Trichinopoly Group of India has a similar ornamentation, shape, and number of longitudinal ribs, but the similarity in the number of ribs of *P. arata* (Stoliczka, 1871, Plate 25, Fig. 1) results from the development of secondary ribs. *P. noesmoeni* Freneix is similar to *P. freneixae* n. sp., differing only by a larger apical angle (35°) and the presence of secondary longitudinal ribs.
Etymology. This species is named after Dr. Suzanne Freneix, Museum National d'Histoire Naturelle, France, for her lifelong research on Upper Cretaceous bivalves.

Figure 7. 1-7, *Cucullaea antarctica* Wilckens. 1, Hypotype, USNM 404836, 1×. 2, Hypotype, USNM 404832, 1×. 3, 5, Hypotype, USNM 404831, 1×. 4, Hypotype, USNM 404835, 1×. 6, Hypotype, 404833, 1×. 7, Hypotype, USNM 404834.

Order PTERIOIDA Newell, 1965
Suborder PTERIINA Newell, 1965
Superfamily PTERIACEA Gray, 1847
Family PBAKEVELLIIDAE King, 1850

Genus *Phelopteria* Stephenson, 1952
Type species. *Pteria? dalli* Stephenson (by original designation).

Phelopteria feldmanni n. sp.
Figure 8.6–8

Descrption. Shell small, suboval, compressed, with posterior wing-like projection; anterior margin straight, with small anterior auricle; posterior wing well developed, extending slightly past the posterior margin of valve; shell smooth; hinge width at least two ligamental pits; adductor scar narly circular, centrally located.
Dimensions. USNM 404848; length, 24 mm; height, 19 mm.
Type. Holotype, USNM 404848; paratype, USNM 405763, 405764, 405773.
Type locality. K-398.
Stratigraphic range. Known from only a single horizon in Unit 9 of the López de Bertodano Formation.
Material. 19 specimens in a single concretion.
Discussion. The presence of ligamental pits separates this species from the genus *Pteria.* All the specimens occur in a single concretion at locality K-398. The specimens are concentrated around a large specimen of *Amberleya,* to which they were probably attached.
Etymology. This species is named in honor of Dr. Rodney M. Feldmann, Kent State University, in recognition of his work on the fossil arthropod faunas of Seymour Island.

Family PULVINITIDAE Stephenson, 1941
Genus *Pulvinites* Defrance, 1824
Type species. *Pulvinites adansoni* Defrance (by monotypy).

Pulvinites antarctica Zinsmeister
Figure 8.2–5

Pulvinites antarctica Zinsmeister, 1978, p. 565–569, Plate 1, Figs. 1–4.
Dimensions. Holotype, OGM 32878; height, 49 mm; width, 50 mm.
Type. Holotype, OGM 32878.
Type locality. Z-37.
Stratigraphic range. Middle part of Unit 9 of the López de Bertodano Formation.
Locality. Z-37, K-322, K-340.
Material. 3 specimens.
Discussion. Palmer (1984) presented a fairly detailed revision of the Mesozoic members of the family Pulvinitidae and included a discussion of the Recent Australian species, but unfortunately omitted any discussion of the Tertiary members of the family except for inclusion of the measurements of *P. pacifica* Zinsmeister and *P. californica* Zinsmeister in his Table 1. He questioned the validity of *P. antarctica* and synonymized it with *P. anansoni,* indicating that the morphologic differences discussed by Zinsmeister (1978) were insufficient for the erection of a separate species. A number of consistent differences, however, exist between the two species, such as the number of ligamental pits (10 to 12 for *P. antarctica* versus 7 to 8 for *P. adansoni*) and the fact that *P. antarctica* has a longer ligamental region. These features are in themselves sufficient to warrant specific separation. Furthermore, *P. antarctica* has a prominent ridge extending along and above the posterior portion of the ligamental area. Such a ridge has not been reported to occur on any other species of *Pulvinites.* The shell outline of the two species is also consistent-

ly different. *P. antarctica* is nearly always suborbicular in outline, but on occasion the posterior margin may flatten a bit and become almost straight. In no instance does the margin become concave and produced as in *P. adansoni* (Palmer, 1984, Plate 72, Fig. 1).

Originally the age of *P. antarctica* was given as ranging from the Campanian into the Maastrichtian, but subsequent work has shown that it is restricted to the Maastrichtian part of the López de Bertodano Formation.

Superfamily PECTINACEA Rafinesque, 1815
Family ENTOLIIDAE Teppner, 1922

Genus ENTOLIUM Meek, 1865

Type species. Pecten demissus Phillips, 1858 (by original designation).

Diagnosis. Subcircular, convex byssal notch lacking auricles projecting above hinge margin in left valve.

Entolium seymourensis n. sp.
Figure 8.11

Description. Shell medium-sized, thin, subcircular, width, 81 percent of height; umbonal angle approximately 100°; auricles of left valve even with hinge margin; anterior auricles straight, margin of auricles slightly inclined; auricles of right valve extending dorsally beyond hinge margin with posterior margin of anterior auricle slightly rounded and with very shallow byssal notch; surface ornamented with narrowly and evenly spaced raised growth increments (three per millimeter); growth lines on auricles parallel to anterior and posterior edges.

Dimensions. USNM 405765; height, 16 mm; width, 13 mm.

Type. Holotype, USNM 405765; paratype, USNM 405766.

Type locality. Z-459.

Stratigraphic range. Middle of unit 9 of the López de Betodano Formation.

Localities. Z-459, Z-754, Z-757.

Material. 15 specimens.

Discussion. Entolium seymourensis n. sp. may be separated from *Entolium sadleri* n. sp. by its smaller size and narrower width to height ratio, whereas *Entolium sadleri* n. sp. is nearly circular in outline. Wilckens (1910) figured a poorly preserved entolium-like pectinid from Cockburn Island and referred it to *Syncyclonema* aff. *S. membranaceus* (Nills). The elevated nature and equal size of the auricles above the hinge margin is typical of *Entolium*. The poor preservation of Wilckens' figured specimens makes it nearly impossible to adequately compare *S.* aff.*S. membranaceus* to either of the two new species of *Entolium* from Seymour Island.

Figure 8. 1, *Limopsis antarctica* Wilckens, hypotype, 405774, 1×. 2-5, *Pulvinites antarctica* Zinsmeister. 2, 3, Hypotype 405820, 1×. 4, 5, Holotype, OGM 32879, 1×. 6-8, *Phelopteria feldmanni* n. sp., 6, holotype, USNM 404848, 1×. 7, Paratype, USNM 405764, 1×. 8, Paratype, USNM 404773, 1×. 9-10, *Entolium sadleri* n. sp. 9, Paratype, USNM 405768, 1×. 10, Holotype, USNM 405767, 1×. 11, *Entolium seymourensis,* holotype, USNM 405765, 1×. 12-13, *Seymourtula* n. gen. *antarctica* (Wilckens). 12, Hypotype, USNM 405790, 1×. 13, Hypotype, USNM 405788, 1×. 14, 15, *Modiolus* cf. *M. pontotocensis,* hypotype, USNM 404844.

Entolium sadleri n. sp.
Figure 8.9,10

Description. Shell large, thin, nearly circular; umbonal angle approximately 120°; auricles relatively small for size of valve; auricles of right valve extending dorsally beyond above hinge margin, byssal notch absent; margins of auricles inclined; surface smooth except for very fine, regular, broadly spaced, incised growth increments (one per 5 mm).

Dimensions. USNM 405767; height, 52 mm; length, 48 mm.

Types. Holotype, USNM 405767; paratype, USNM 405768.

Type locality. K-377.

Stratigraphic range. Known only from unit 9 of the López de Bertodano Formation.

Localities. K-377, Z-754.

Material. 11 specimens.

Discussion. The nearly circular outline, smooth surface, and the proportionally smaller auricles easily separate *Entolium sadleri* from *E. seymourensis.* Wilckens (1907) described a similar pectinid (*Pecten molestus* Wilckens) from the Late Cretaceous near Cerro Cazador in southern Patagonia. Although the two species are similar, the auricles of *E. sadleri* are larger, and the outline of its valve is more nearly circular.

Etymology. Named for Dr. Peter Sadler, University of California at Riverside, in recognition of stratigraphic studies made on Seymour Island.

Superfamily LIMACEA Rafinesque, 1815
Family LIMIDAE Rafinesque, 1815

Genus *Acesta* Adams and Adams, 1858
Type species. Ostrea excavata Fabricius, 1779 (by monotypy).

Acesta shackletoni n. sp.
Figure 9.4

Description. Shell large, ovate, equivalve, inequilateral, oblique, moderately inflated; length, 93 percent of height; moderate byssal gape; auricles unequal, anterior nearly absent, posterior moderately broad; anterior margin depressed; surface ornamented with low, closely spaced (two to three per millimeter), wavy, radial ribs that are more strongly developed near anterior and posterior margin; central portion of valve nearly smooth; cardinal area wide; chondrophore broad, sloping slightly to the posterior.

Dimension. USNM 405771; length, 83 mm; height, 92 mm; width of paired valves, 55 mm.

Type. Holotype, USNM 405771; paratype, USNM 405772.

Type locality. K-427.

Stratigraphic range. Known from top of Unit 9 of López de Bertodano Formation.

Localities. K-427.

Material. 5 specimens.

Discussion Acesta shackletoni n. sp. is similar to *A. snowhillensis* Wilckens, but may be distinguished by its greater length to height ratio (93 percent for *A. shackletoni* versus 73 percent for *A. snowhillensis*). The anterior dorsal margin slopes sharply and varies from nearly straight to slightly concave. The beak of *A. snowhillensis* is not as sharp as that of *A. shackletoni* n. sp.

A. shackletoni n. sp. is fairly common in the upper part of unit nine and frequently occurs at or just below the Cretaceous/Tertiary boundary.

Etymology. This species named in honor of Sir Ernest Shackleton, one of Antarctica's greatest explorers.

Acesta webbi n. sp.
Figure 9.1–3

Description. Shell medium-sized, ovate, equivalve, inequilateral, moderately inflated, length 71 percent of height; anterior dorsal margin steeply

Figure 9. 1-3, *Acesta webbi* n. sp., 1, 2, holotype, USNM 405779, 1×. 3, paratype, USNM 405770, 1×. 4, *Acesta shackletoni* n. sp., holotype, USNM 405771, 1×. 5, 6, *Lucina scotti* (Wilckens), hypotype, USNM 404837, 1×. 7, 8, *Thyasira townsendi* (White), hypotype, USNM 405773, 1×.

sloping, straight to slightly concave; beak small, elevated; anterior auricle obsolete, posterior broad; surface smooth except for irregularly spaced concentric growth increments; very fine radial striae present near margins of shell; cardinal area wide and open; chondrophore broad and shallow.

Dimensions. USNM 405769; length, 56 mm; height, 71 mm; width of paired valves, 30 mm.

Types. Holotype, USNM 405769; paratype, USNM 405770.

Type locality. K-72.

Stratigraphic range. Unit 1 of the Sobral Formation. Single, poorly preserved specimen questionably referred to this species was found in Unit 9 of the López de Bertodano Formation.

Localities. K-72, N-84?

Material. 3 specimens.

Discussion. Acesta webbi n. sp. is easily separated from *A. shackeltoni* n. sp. by its smaller size, more elongated shell outline, and near-absence of radial sculpture. The almost total absence of radial ornamentation is uncommon among members of the family Limidae. *"Lima"* cf. *L. latens* Feruglio (1936, Plate 14, Figs. 10,11) from the Maastrichtian of Lago Argentino in Patagonia is very close to *A. webbi* n. sp., but the posterior auricle is not as broad and the shell not as inflated. *Acesta marlburiensis* (Woods) (Woods, 1917, Plate 3, Fig. 3) from New Zealand has a similar shape, but the posterior auricle is more rounded and the radial ornamentation is well developed. The ornamentation of *A. webbi* is very similar to that of *Lima (Plagiostoma?) derbyi* White from the Cretaceous of Brazil (White, 1924), but it has a more developed anterior auricle.

Etymology. This species is named in honor of Dr. Peter N. Webb, The Ohio State University, in recognition of his years of work in Antarctica.

Genus *Seymourtula* Zinsmeister n. gen.

Diagnosis. Posterior submargin smooth and considerably broader than anterior submargin, resulting in a distinctly inequilateral profile.

Type species. Seymourtula antarctica Wilckens, 1910.

Discussion. The exceptionally wide posterior submargin gives *Seymourtula antarctica* a distinctly modioliform outline that easily separates it from other limatulas. Although the development and placement of the radial ribs are typical of *Limatula,* the two may be separated by the broad posterior margin of *Seymourtula.*

Seymourtula antarctica (Wilckens, 1910)
Figure 8.12,13

Lima (Limatula) antarctica Wilckens, 1910, p. 16–17, Plate 1, Fig. 8; Fleming, 1978, p. 52, Fig. 26 (not *S. antarctica* Wilckens).

Supplementary description. Shell small- to medium-sized, opisthocline, inflated; posterior margin broad, convex; anterior margin narrow; ornamentation consists of five to eight centrally located radial ribs; margins of valve smooth.

Dimensions. USNM 405787; length, 32 mm; height, 18 mm; width of paired valves, 16 mm.

Types. Holotype, Mo 1636; hypotypes, USNM 405787, 405788, 405789, 405790.

Type locality. Seymour Island (Wilckens, 1910).

Stratigraphic range. Middle part of unit 7 to middle part of unit 9 of the López de Bertodano Formation.

Localities. K-180, K-342, Z-458, Z-757, Z-763, Z-776, N-84.

Material. 33 specimens.

Discussion. Wilckens' type specimen consists of a partially preserved internal cast and does not show the development of the broad posterior margin of the valve. Fleming (1978) described a steinkern from the Upper Cretaceous mid-Waipara Group of the South Island of New Zealand in the species *Limatula antarctica.* The New Zealand specimen shows the typical broad development of the posterior margin of *Seymourtula.* Whether it is conspecific with the Seymour Island species is difficult to determine because it is known from a single, inadequately

preserved specimen. The New Zealand specimen has weaker radial ribbing. Fleming remarked upon the similarity of the Waipara specimen to *S. antarctica* and suggested they constitute another representative of the paleoaustral element in the Late Cretaceous fauna of the southern Pacific margin. *L. (L.) parisi* described by Freneix (1980) from the Upper Cretaceous of New Caledonia appears to be referrable to *Seymourtula,* but it may be distinguished from *S. antarctica* by its very small size. Freneix (1980) described a second species of limatula *(L. (L.) australis)* based on several deformed specimens. Two of the figured specimens of *L. (L.) australis* (Plate 3, Figs. 5, 6) have a distinctly circular profile rather than the elongated typical form of *Limatula.* The circular outline of these two specimens may reflect only the crushed nature of the material.

Suborder OSTREINA Ferussac, 1822
Superfamily OSTREACEA Rafinesque, 1815
Family GYPHAEIDAE Vyalov, 1936
Subfamily PYCNODONTEINAE Stenzel, 1959

Genus *Pycnodonte* Fisher von Waldheim, 1835
Type species. Pycnodonte radiata Fisher von Waldheim (by original designation).

Subgenus *Phygraea* Vyalov, 1936
Type species. Phygraea frauscheri Vyalov, 1936 (by original designation).

Pycnodonte (Phygraea) cf. *P. (P.) vesiculosa* (Sowerby, 1816)
Figure 10.1–10

Gryphaea cf. *vesicularis* Lamarck, Wilckens, 1910, p. 21–22, Plate 1, Figs. 14a,b, 15.

Supplementary description. Shell medium-sized, gryphaeoid, subtrigonal, strongly convex, thick, with prominent incurved beak; left valve inflated, beak pointed, projecting with slight twist; surface smooth except for irregularly spaced growth lamellae; faint, delicate radial striae occasionally present; chomata limited to dorsal fourth of valve; adductor scar nearly circular, located near anterior margin; Quenstedt scar small; right valve concave, moderately thick, smaller than left valve; rounded adductor scar located near anterior margin.

Dimensions. USNM 405791; length, 52 mm; height, 52 mm.

Types. Hypotypes, USNM, 405791, 405792, 405793, 405794, 405831, 405835.

Type locality. Seymour Island (Wilckens, 1910).

Stratigraphic range. Unit 5 through unit 9 of the López de Bertodano Formation.

Localities. K-37, K-66, K-68, K-166, K-142, K-160, K-319, K-340, K-343, K-368, K-399, K-427, Z-37, Z-468, Z-683, Z-724, Z-762, Z-763, Z-770, Z-774, Z-775, Z-778, Location B.

Material. 98 specimens.

Discussion. The material, described by Wilckens (1910) as *Gryphaea* cf. *G. vesicularis,* clearly belongs to the *Pycnodonte vesicularis* group, but is related to the gryphaeoid forms that are included in *P. vesiculosa.* The Seymour Island specimens lack the typical furrow that separates the posterior part of the shell from the remainder of the valve (Woods, 1913). Juvenile specimens of *P. seymourensis* and *P.* cf. *P. vesiculosa* are difficult to differentiate. Future collections of these two species, coupled with a biometric study, should clarify the relationships between these two species.

Pycnodonte (Phygraea) seymourensis (Wilckens, 1910)
Figures 10.11,12; 11.1–6

Ostrea seymourensis Wilckens, 1910, p. 19–12, Plate 1, Figs. 11, 12a,b. *Ostrea* ex. aff. *lesueuri* d'Orbigny, Wilckens, 1910, p. 18–19, Plate 11, Fig. 10a,b.

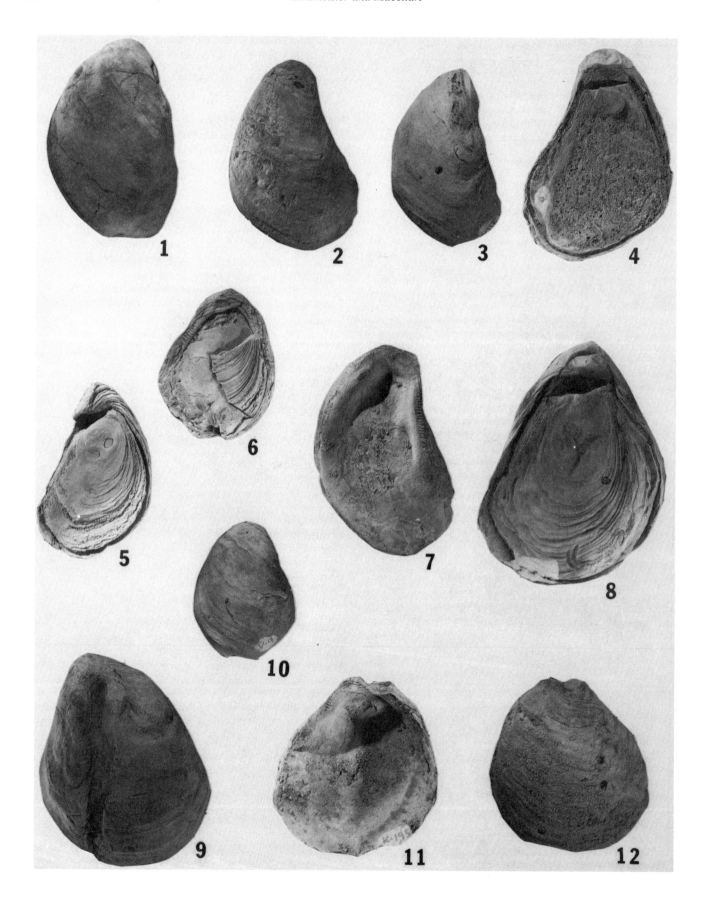

Supplementary description. Shell medium- to large-sized, suborbicular, compressed; left valve larger than right, convex; umbone raised, slightly curved; posterior flange weakly developed on large individuals; surface smooth except for irregularly spaced undulating ridges; vermicular chomata developed only near dorsal margin; adductor scar relatively small, subcentral toward dorsal margin, Quenstedt scar just below hinge; auricles greatly reduced to absent; right valve subtrigonal, smaller than left flat; adductor scar small, oval, slightly pointed posterodorsally.

Dimensions. USNM 405795; length, 63 mm; height, 69 mm.

Types. Holotype, Mo 1644; hypotypes, USNM 405795, 405796, 405795, 405832.

Type locality. No specific locality designated by Wilckens (1910) except southern part of Seymour Island.

Stratigraphic range. Unit 7 through unit 9 of the López de Bertodano Formation.

Localities. K-41, K-47, K-84, K-104, K-106, K-116, K-173, K-195, K-342, K-384, K-394, K-428, Z-38, Z-724, Z-757, Z-764, Z-722, Z-774, Z-775, Z-776, Locations B, C, F.

Material. 96 specimens.

Discussion. Although *Pycnodonte (Phygraea) seymourensis* varies considerably in shape and size, it is easily separated from *P. (P.)* cf. *P. (P.) vesiculosa* (Sowerby) by its larger size and suborbicular outline. The shell of *P. (P.)* cf. *P. (P.) vesiculosa* is proportionally thicker and shows no tendency to develop a posterior flange.

Pycnodonte (Phygraea) seymourensis is possibly identical to *P. vesicularis* (Lamarck), but it is preferable to maintain the Seymour Island specimens in a different species until more detailed quantitative studies can be conducted. *P. vesicularis* (Lamarck) displays a well-known polymorphism (Woods, 1913; Muller, 1970; Freneix, 1972; Pugaczewska, 1977) that ranges from gryphaeoid to flatter forms. Different criteria have been used to differentiate this species and the closely related species *P. vesiculosa* (Sowerby) and *P. subvesiculosa* (Renngarten). Freneix (1972) included all of these forms in *P. vesicularis,* ranking them as subspecies. The typical subspecies *P. vesicularis vesicularis* has a less pointed umbo with a larger attachment surface and has a larger size than *P. vesicularis vesiculosa.* Additionally, the area in the former subspecies is lower, and usually the height of the shell is relatively small in proportion to the length (Woods, 1913, p. 375). However, *P. seymourensis* lacks the typical radial ornamentation on the right valve, clearly displayed in the specimens of Woods (1913) and Pugaczewska (1977).

Order TRIGONIOIDA Dall, 1889
Superfamily TRIGONIACEA Lamarck, 1819
Subfamily NOTOTRIGONIINAE Skwarko, 1963

Genus *Eselaevitrigonia* Kobayashi and Mori, 1954
Type species. *Trigonia meridiana* Woods, 1917 (by original designation).

Figure 10. 1-10, *Pycnodonte (Phygraea)* cf. *P. (P.) vesiculosa* (Sowerby). 1, Hypotype, USNM 405791, 1×. 2, 4, Hypotype, USNM 405793, 1×. 3, 5, Hypotype, 405794, 1×. 6, 10, Hypotype, 405831, 1×. 7, Hypotype, USNM 405792, 1×. 8, 9, Hypotype, USNM 405835, 1×. 11-12, *Pycnodonte (Phygraea) seymourensis* (Wilckens), hypotype, USNM 405832.

Eselaevitrigonia regina (Wilckens, 1910)
Figure 12.1–7

Trigonia cf. *T. ecplecta* Wilckens, 1907, pp. 39–40, Plate 7, Fig. 5a,b.
Trigonia regina Wilckens, 1910, p. 41, Plate 2, Figs. 22–26; Feruglio, 1936, pp. 103–104, Plate 12, Fig. 9; Medina, 1980, pp. 109–110, Plate 2, Figs. 3–4, Plate 3, Figs. 1–3.

Supplementary description. Shell medium-sized, thick, trigonal to trigonally ovate, inequilateral, moderately inflated; beark small; umbones located on anterior quarter of valve; dorsal anterior margin nearly straight, steeply sloping; ventral margin broadly rounded; dorsal posterior margin long, straight, moderately sloping, passing evenly into narrow, slightly produced posterior margin; escutcheon nearly obsolete, marked only by flattened region; carina faint, present in some individuals; surface smooth except for low, regularly placed concentric ribs which become slightly offset forming elongated pustules near the anterior edge of valve; hinge—LV with broad, triangular, shallowly grooved medial tooth (2), base of 2 moderately concave, 4b of medium length, 4a lower than 2 and of medium length; RV with 3a and 3b diverging at approximately 70°, teeth of moderate thickness, crest of 3b narrower than 3a; thin, poorly developed nymph; anterior adductor muscle scar sunken in well-developed myophore that extends anteroventrally from 3a; posterior adductor scar impressed, shallow subcentral radial groove across scar; anterior pedal retractor small, deeply sunken, at base of 3a; posterior pedal retractor larger than anterior pedal retractor and located on distal ventral edge of 3b, well-developed pedal protractor scar just ventral of anterior adductor scar, single pedal elevator scar located underneath hinge directly below beak.

Dimensions. USNM 405786; length, 53 mm; height, 42 mm.

Types. Lectotype, MLP N. 9005 (herein designated); hypotypes, USNM 405783, 405784, 405785, 405786.

Type locality. Sierra de los Baguales, southern Chile.

Stratigraphic range. Base of unit 7 to K/T boundary.

Localities. K-37, K-42, K-116, K-150, K-o151, K-153, K-157, K-160, K-173, K-178, K-179, K-181, K-182, K-186, K-189, K-191, K-221, K-340, K-357, K-380, K-385, Z-458, Z-459, Z-463, Z-469, Z-725, Z-754, Z-757, Z-758, Z-763, Z-773, Z-776, Locations B, E, F, J.

Material. 172 specimens.

Discussion. Wilckens (1907) described a large *Trigonia (T. ecplecta)* from southern Patagonia. He also included external and internal figures of a second specimen of trigonia, which he referred to as *T.* cf. *ecplecta.* Wilckens distinguished this specimen from *T. ecplecta* because of a kink in the anterior ornamentation, smaller size, and weaker concentric ribbing. Later, when describing *T. regina* from Antarctica, Wilckens (1910) concluded that the specimen he referred to as *T.* cf. *ecplecta* from Patagonia was identical to the Antarctic species. The measurements of the patagonian specimen fall well within the observed range of variation of *T. regina* from Seymour Island. The large collection recently obtained from Seymour Island enables us to obtain a better understanding of the morphologic range of *T. regina* (Fig. 13). The kink in the anterior concentric ribbing varies considerably, from being nearly absent to moderately strong; in no case is it very prominent. The strength of the ribbing seems to be independent of the prominence of the kink. *E. ecplecta* may be separated from *E. regina* by its considerably larger size and its strongly developed concentric ornamentation.

The generic status of this species has been a matter of debate. This species has been alternatively included in the genera *Rutitrigonia* (Cox, 1952), *Pacitrigonia* (Cox, 1952; Nakano, 1961; Perez and Reyes, 1978), *Nototrigonia* (Reyes and Perez, 1979), and *Laevitrigonia (Eselaevitrigonia)* (Medina, 1980). Nakano (1961) believed that *"Trigonia" regina* could be an immature form of *"Trigonia" ecplecta* and included this species in the genus *Nototrigonia.* Medina (1980) pointed out that this species lacks the antecarinal depression, typical of both *Nototrigonia* and *Pacitrigonia* (which also have an oblique ornamentation on the flank), rather than the concentric ornamentation of *Eselaevitrigonia regina.* The

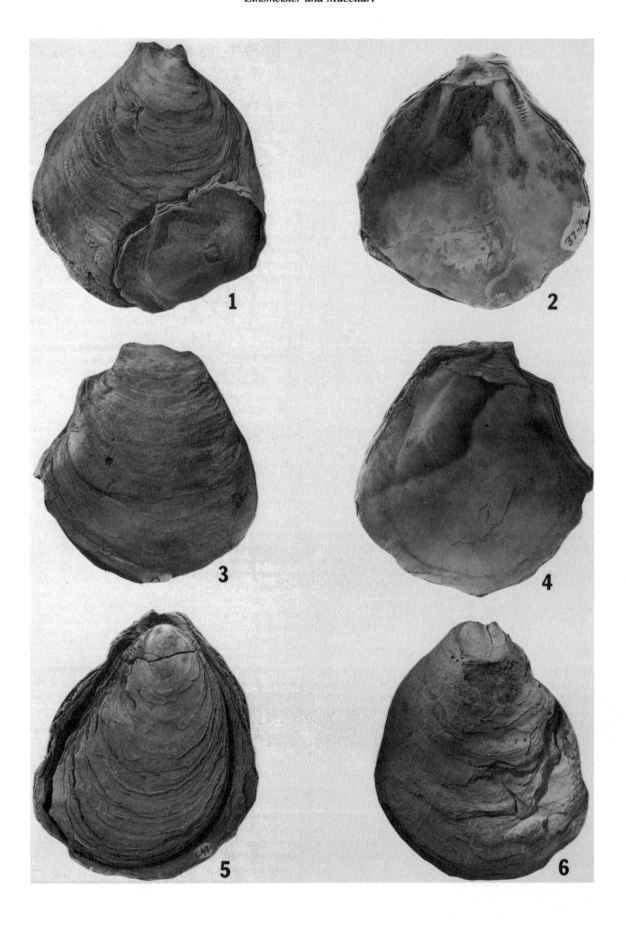

area of this species is, however, more rounded and lacks the antecarinal depression.

Subfamily MYOPHORELLINAE Kobayashi and Tamura, 1955

Genus *Oistotrigonia* Cox, 1952
Type species. Trigonia spinosa Parkinson, 1811 (by original designation).

Oistotrigonia pygoscelium (Wilckens, 1910)
Figure 12.10–17

Trigonia pygoscelium Wilckens, 1910, pp. 39–41, Plate 2, Fig. 21a–b.
Linotrigonia (Oistotrigonia) pygoscelium (Wilckens), Freneix, 1980, p. 100; Medina, 1980, p. 114.
Supplementary description. Shell medium-sized, subtriangular, trigonally ovate, inequilateral, moderately inflated; beak small, located near anterior margin; umbonal ridge moderately developed, extending almost to posterior ventral margin; anterior dorsal margin short, merging into steep, slightly rounded anterior margin; ventral margin long, slightly rounded; posterior margin bluntly rounded, slightly produced; posterior dorsal margin straight to slightly concave; escutcheon marked by flattening of shell; ornamentation consisting of two prominent sets of radial ribs intersecting at an angle on posterior quarter of shell and forming chevrons, anterior two-thirds of valve with 21 to 25 ribs, posterior set of ribs short, varying in number from 16 to 20; ribs tending to have regularly developed closely spaced nodes; hinge—LV with broad, massive, triangular tooth, broad, shallow medial groove, ventral margin sharply concave, 4b medium length, 4a lower than 2, medium length; RV with 3a and 3b widely diverging, moderately thick; anterior muscle scar sunken in well-develped myophore extending anteroventrally from 3a; posterior scar moderately developed, depressed; well-developed crenulation along interior margin of shell.
Dimensions. USNM 405780; length, 42.5 mm, height, 37 mm.
Types. Holotype, Mo 1500 a,b; hypotypes, USNM 405775, 405776, 405777, 405778, 405779, 405780, 405781, 405781, 405782.
Type locality. Only a general locality south of Cross Valley given by Wilckens (1910).
Stratigraphic range. Unit 5 through unit 9 of the López de Bertodano Formation.
Localities. K-37, K-39, K-41, K-43, K-46, K-99, K-144, K-147, K-150, K-166, K-380, K-382, Z-468, Z-623, Z-724, Z-756, Z-757, Z-768, Z-776.
Material. 63 specimens.
Discussion. Wilckens (1910) described two species of *Oistotrigonia: O. antarctica* from Snow Hill Island and *O. pygoscelium* from Seymour Island. Although adult specimens of *O. pygoscelium* are quite distinctive, juvenile specimens vary from subtrigonal to nearly circular in outline and are very similar to *O. antarctica.* The number of ribs and the development of the nodes along the ribs vary considerably. However, *O. antarctica* has a more rounded outline because its posterior and ventral margins meet transitionally and not at a sharp angle as in *O. pygoscelium.* It appears that the two species represent a single lineage of *Oistotrigonia* in which Campanian species *(O. antarctica)* from Snow Hill Island gave rise to the Maastrictian *O. pygoscelium* of Seymour Island.

Figure 11. 1-6, *Pycnodonta (Phygraea) seymourensis* (Wilckens). 1, 2, Hypotype, USNM 405795, 1×. 3, 4, Hypotype, USNM 405796, 1×. 5, 6, Hypotype, USNM 405797, 1×.

Subclass HETERODONTA Neumayr, 1884
Order VENEROIDA H. and A. Adams, 1856
Family LUCINIDAE Fleming, 1818

Genus *Lucina* Bruguiere, 1797
Type species. Venus jamaicensis Spengler, 1784 (by subsequent designation, Gray, 1847).

Lucina scotti (Wilckens, 1910)
Figure 9.5,6

Phacoides scotti Wilckens, 1910, p. 57, Plate 3, Figs. 2a,b.
Supplementary description. Shell medium-sized, nearly circular, slightly inflated; beaks small, centrally located; anterior dorsal margin concave; anterior, ventral, and posterior margins broadly rounded; lunule well developed, asymmetrical; escutcheon elongated, narrow, deep; surface ornamented with well-developed concentric growth increments; adductor scar, medium length.
Dimensions. USNM 404837; length, 27 mm; height, 25 mm; width of paired valves, 12 mm.
Types. Lectotype, Mo 1563 (herein designated); hypotype, USNM 404837.
Type locality. Only a general locality description south of Cross Valley given by Wilckens (1910).
Stratigraphic range. Upper part of unit 8 to upper part of unit 9 of the López de Bertodano Formation.
Localities. K-99, Z-459, Z-758, Z-762, Z-763, Z-776, N-84.
Material. 59 specimens.
Discussion. All the specimens of *L. scotti* consist of paired valves; consequently, no hinges were available for study. *L. scotti* is locally abundant in units throughout most of the upper part of the López de Bertodano Formation.

Family THYASIRIDAE Dall, 1901

Genus *Thyasira* Leach in Lamarck, 1818
Type species. Amphidesma flexuosa Lamarck, 1818 (by original designation).

Thyasira townsendi (White, 1990)
Figure 9.7,8

Lucina? townsendi White, 1890, p. 14, Plate 3, Figs. 1,2; Weller, 1903, p. 415, Plate 1, Figs. 2,3.
Thyasira townsendi (White), Wilckens, 1910, p. 53, Plate 2, Fig. 31a,b,c; Plate 3, Fig. 1.
Supplementary description. Shell medium to large, obliquely subquadrangular; thin-walled; beak small, acute; umbones prominent, rounded; posterior dorsal margin slightly convex, steeply sloping; ventral margin short, bluntly rounded; anterior margin nearly straight to slightly concave; prosocline forming a slightly produced, moderately sharp anterodorsal angle; anterior dorsal margin medium length, nearly straight; moderately broad, very deep sulcus extending along posteroventral margin; fold narrow, blade-like, extending well above sulcus; lunule ill defined with flatten region; shallow anterior sulcus parallel to lunule; small raised area formed at intersection of anterior sulcus with anterior margin near anterodorsal angle; medial area flattened, poorly developed on posterior quarter of valve; ornamentation consisting of closely spaced concentric growth increments that tend to form low, irregularly spaced undulations; muscle scars ill defined.
Dimensions. USNM 405773; length, 53 mm; height, 47 mm; width of paired valves, 26 mm.
Types. Location of holotype unknown; hypotypes, Mo 1560, USNM 405773.

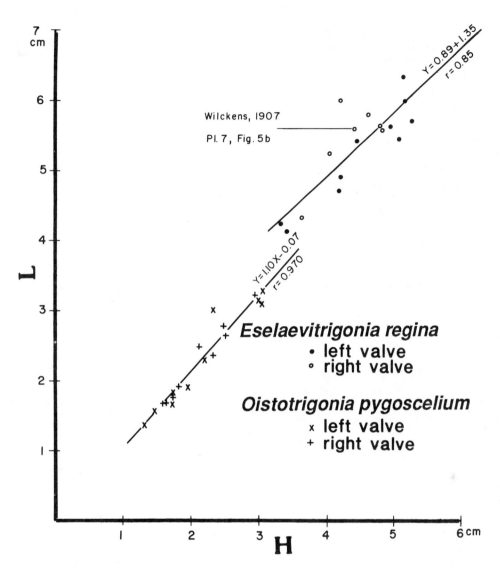

Figure 13. Comparison of length ratios of *Eselaevitrigonia regina* (Wilckens) and *Oistotrigonia pygoscelium* (Wilckens).

Figure 12. 1-7, *Eselaevitrigonia regina* (Wilckens). 1, 2, Hypotype, USNM 405783, 1×. 3, Hypotype, USNM 405785, 1×. 4, 6, Hypotype, USNM 405786, 1×. 5, 7, USNM 405784, 1×. 8-17, *Oistotrigonia pygoscelium* (Wilckens). 8, 9, Hypotype, USNM 405776, 1×. 10, Hypotype, USNM 405777, 1×. 11, Hypotype, USNM 405778, 1.5×. 12, 13, Hypotype, 405775, 1.5×. 14, Hypotype, 405779, 1×. 15, 16, Hypotype, 405881, 1×. 17, Hypotype, 405780, 1×.

Stratigraphic range. Upper part of unit 8.
Localities. Location E, Z-459, N-84.
Material. 12 specimens.
Discussion. Although the obliquely subquadrangular outline of the shell is similar to the subgenus *Conchocele* proposed by Gabb (1869) for a large species *(T. (Conchocele) disjuncta)* from the late Pliocene of California, the posterior sulcus of *T. townsendi* is considerably deeper and has a prominently raised fold. It would be surprising if the two species are related, as there is no known species from the Tertiary with a similar morphology. Freneix (1980) described *T. collingnoni* from the Late Cretaceous of New Caledonia, but unfortunately her material consists only of internal molds, which makes comparisons with *T. townsendi* difficult. Although *T. collignoni* is not as oblique, the posterior sulcus appears to be deeply impressed similar to that of *T. townsendi.*

Superfamily CRASSATELLACEA Ferussac, 1822
Family ASTARTIDAE d'Orbigny, 1844
Subfamily ERIPHYLINAE Chavan, 1952

Genus *Dozyia* Bosquet *in* Dewalque, 1868
Type species. Lucina lenticularis Goldfuss, 1840 (by monotypy).

Dozyia drygalskiana (Wilckens, 1910)
Figures 14.1–4

Astarte (Eriphyla) drygalskiana Wilckens, 1910, pp. 51–53, Plate 3, Fig. 3a,b.
Astarte (Eriphyla) meridiana Woods, 1917, p. 28–29, Plate 15, Figs. 2–7; del Valle and Medina, 1980, p. 57, Plate 3, Figs. 3, 4; Speden and Keyes, Plate 23, Fig. 2.
Supplementary description. Shell medium-sized, lenticular, slightly inflated; prominent prosogyral umbones; margins uniformly rounded, except for concave dorsal anterior margin; lunule sunken; moderately deep, narrow escutcheon; surface ornamented with closely spaced, raised, concentric growth increments; hinge with two prominent blunt cardinals and one weakly developed posterior in right valve; single, blunt central tooth in left valve; edge of lunule in left valve extended into anterior cardinal tooth, which fits into narrow, shallow socket; posterior portion of left hinge broad, flattened; posterior half of right valve hinge not as broad; well-developed adductor scars; small sunken anterior pedal scar located beneath hinge directly below anterior edge of lunule; three to five accessory scars located across base of hinge plate; pallial sinus varies from a broad, shallow to a moderately angular, upward pointing indentation; inner margin smooth.
Dimensions. USNM 405798; length, 34 mm; height, 37 mm; width of single valve, 12 mm.
Types. Holotype, Mo 1501; paratypes, USNM 405798, 405799.
Type locality. SSPE Location 4.
Stratigraphic range. Lower part of unit 9 to 25 m below K/T boundary.
Localities. K-42, K-99, K-360, K-377, K-378, K-382, Z-757, Z-762, Z-773.
Material. 40 specimens.
Discussion. Wilckens' figured specimen (Mo 1510) is slightly deformed, which results in prominent angularity of the ventral margin of the valve. Several small cracks—along which the shell was deformed—were not illustrated on Wilckens' figures. We have collected 40 specimens of *D. drygalskiana* from Seymour Island, and except for several individuals that are slightly deformed, the outline of the shell is nearly circular. Wilckens (1910) originally referred this species to the subgenus *Eriphyla,* but comparison of the hinges of *D. drygalskiana* and *Eriphyla* reveals significant differences. The antarctic species has two strong cardinal teeth, whereas *Eriphyla* is characterized by only a single cardinal tooth. The nearly circular shape of the shell and the dentition of Late Cretaceous genus *Dozyia* are very similar to Wilckens' antarctic species. As a consequence, we are placing *drygalskiana* within the genus *Dozyia.*

Woods (1917) described a similar species *(Eriphyla meridiana)* from the Upper Cretaceous of New Zealand. Although no figure of the hinge was included in his description, Woods stated that only one cardinal tooth was present in each valve of *E. meridiana.* He mentioned the presence of a lateral tooth next to the lunule in the right valve. Whether this lateral tooth corresponds to the tooth to which we are referring a an anterior cardinal is not known. Because of the very strong development of the anterior tooth in the antarctic species and its location next to the posterior cardinal, we believe that the anterior tooth should be referred to as a cardinal tooth rather than a lateral. We believe that, because of the close similarity of *D. drygalskiana* and the New Zealand species, the two are conspecific, with *D. meridiana* being a junior synonym. Woods (1917) figured a second species, which he referred to as *Eriphyla lenticularis* (Goldfuss). This species has the characteristic circular shell profile, but differs from *D. drygalskiana* by having a moderately deep, well-developed pallial sinus. It should be noted that the type species of *Dozyia* is *lenticularis,* which apparently has an exceptionally broad geographic distribution, extending from Europe to India and into high southern latitudes of Africa and New Zealand. Whether all species referred to as *D. lenticularis* actually belong to *lenticularis* is not known and is well beyond the scope of this paper. It should be noted that a number of Late Cretaceous bivalves have a nearly circular outline with elevated umbones; without information about the hinge, these bivalves are virtually impossible to separate. Woods (1908) figured another *Dozyia*-like bivalve from Africa, which he referred to *lenticularis.* It appears to be identical to the species he referred to *lenticularis* from New Zealand. Unfortunately, he did not present any information concerning the internal morphology of the african specimen. The only apparent difference between the african *lenticularis* and *D. drygalskiana* is a slightly more circular outline and smaller beak of the former.

Superfamily CARDIACEA Lamarck, 1809
Family LAHILLIDAE Finlay and Marwick, 1937

Genus *Lahillia* Cossmann, 1899
Type species. Amathusia angulata Philippi, 1887 (by subsequent designation, Finlay and Marwick, 1937).

Lahillia larseni (Sharman and Newton, 1897)
Figure 15.1–5

Cyprina larseni Sharman and Newton, 1898, p. 59–60, pl. 1.
Lahillia luisa Wilckens, 1910, p. 58–63, Plate 3, Figs. 4, 5, 6, 7a–c, 11.
Supplementary description. Shell large, thick-walled, inflated, subtrigonal to suboval, prominent subcentral prosogyral umbones; anterior dorsal margin concave; anterior margin bluntly rounded; ventral margin broadly rounded, gradually merging with posterior margin; posterior dorsal margin short, nearly straight; lunule faint, with flattened region marked by slight angulation; escutcheon marked by narrow groove on either side of massive, blunt blade nymph that pjrojects above margin of valve; ornamentation nearly smooth except for regularly spaced concentric growth increments; hinge with blunt peg-like cardinal tooth with corresponding deep socket in each valve; right valve with second diverging cardinal; posterior cardinal of right valve small, rounded, projecting above hinge and nearly touching beak, fused to nymph; apex of posterior deflected anteriorly and nearly forming an inverted V with anterior cardinal; blunt, asymmetrical, elongated posterior lateral in each valve; basal margin of hinge with prominent concavity beneath nymph; nymph, massive, flat, posterior edge abruptly truncated just anterior to posterior lateral; adductor scars subequal; anterior pedal scar, just beneath anterior edge of hinge plate, sunken; pallial sinus shallow; inner margin smooth.
Dimensions. USNM 405800; length, 105 mm; height, 94 mm; width of paired valves, 72 mm.
Types. Holotype, IGS 4053; hypotypes, USNM 405800, 405801, 405802, 405803.

Figure 14. 1-4, *Dozyia drygalskiana* (Wilckens). 1, 2, Hypotype, USNM 405798, 1×. 3, 4, Hypotype, USNM 405799, 1×. 5-9, *Cyclorisma chaneyi* n. sp. 5, Paratype, RV, USNM 405806, 1×. 6, 7, Holotype, USNM 405825, 1×. 8, Paratype, USNM 405807, 1×. 9, Paratype, USNM 405808, 1×. 10-12, *Surobula* n. gen. *nucleus* (Wilckens). 10-11, Hypotype, USNM 405811, 3×. 12, Hypotype, USNM 405810, 3×. 13-16, *Marwickia woodburnei* n. sp. 13, 14, Paratype, USNM 405834, 1×. 15, Holotype, USNM 405804, 1×. 16, Paratype, USNM 405805, 1×.

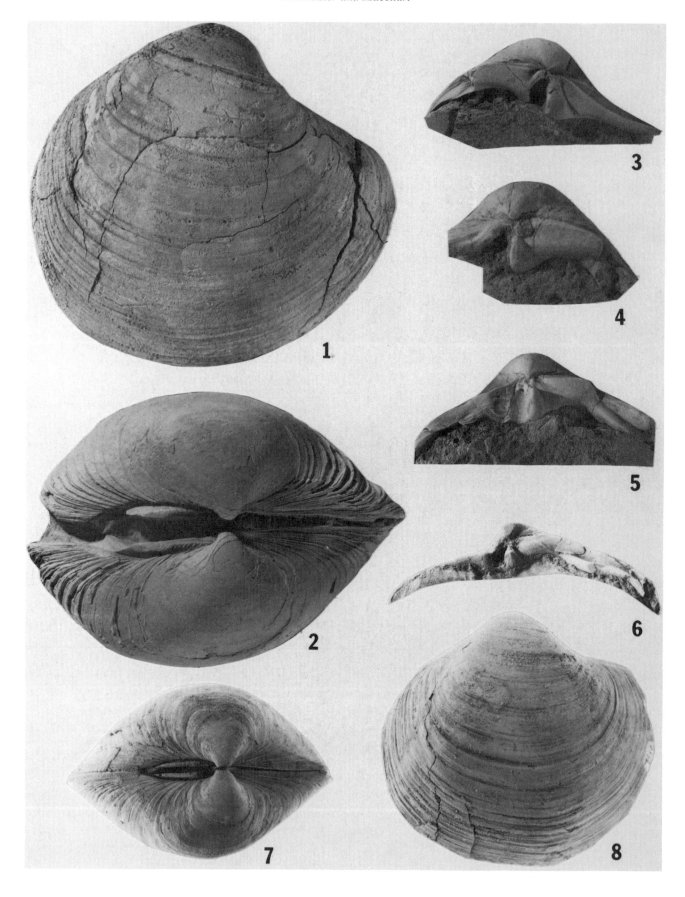

Type locality. No locality given by Sharman and Newton (1897) other than Seymour Island.

Stratigraphic range. Top of unit 7 to middle part of unit 10 of the López de Bertodano Formation.

Localities. K-69, K-71, K-181, K-189, K-360, Z-485, Z-496, Z-477, Z-631, Z-757, Z-776, Z-777, Z-778.

Material. 72 specimens.

Discussion. Lahillia larseni was described by Sharman and Newton (1987) from a medium-sized paired-valve specimen of *Lahillia* collected by Captain A. Larsen during his first visit to Seymour Island during the austral summer of 1892–1893. Since the material Sharman and Newton had previously described from Seymour Island was Tertiary, they assumed that the undescribed species of *Lahillia* was also of Tertiary age. They were not aware of the presence of Cretaceous deposits on Seymour Island. The existence of Cretaceous sediments was not recognized until the Swedish South Polar Expedition several years later. As a consequence, they referred *L. larseni* to the Tertiary, which Wilckens (1911) followed. Wilckens referred the Cretaceous species from Seymour Island to the South American species *L. luisa.* Examination of Larsen's specimen and the collection of a large number of additional specimens from the López de Bertodano Formation (Zinsmeister, 1984) indicated that *L. larseni* was described from Upper Cretaceous, not Tertiary beds. *L. larseni* may be distinguished from *L. luisa* by its prominent elevated prosogyral umbones that impart a distinct concave profile to the anterior dorsal margin, in contrast to the centrally located umbones of *L. luisa.* The hinge plate and the nymph of *L. larseni* are considerably more massive than those of *L. luisa.*

L. larseni first appeared in the lower part of the Maastrichtian part of the López de Bertodano Formation and continued into the lower part of the Sobral Formation. Its abundance increases upward, and is most common in the uppermost part of the López de Bertodano Formation, above the K/T boundary. This species does not display any morphologic variation or changes in abundance across the K/T boundary. A smaller form of *Lahillia* occurs in unit 1 of the Sobral Formation. The average size is about two-thirds that of those that occur in the López de Bertodano Formation, the umbones are slightly more centrally located, and the anterior dorsal margin is not a markedly concave. Although the hinge development of the two forms differs, particularly in the strength of the nymph and thickness of the hinge, such variation of the hinge within one species is common for the genus. Examination of large numbers of well-preserved specimens of the late Eocene *L. wilckensi* from the La Meseta Formation revealed similar hinge variations. In some individuals, the thickness of the hinge was nearly twice that of other individuals. The larger size, thicker shell, and pronounced concave anterodorsal margin separate *L. larseni* from *L. huberi* n. sp. from the Sobral Formation.

Lahillia huberi n. sp.
Figure 15.6–8

Description. Shell medium- to large-sized, inflated, subtrigonal to suboval; beaks prominent subcentral, orthogyral, nearly touching; anterior dorsal margin moderately long, concave; moderately well-developed angulation on anterodorsal edge of valve; anterior margin bluntly rounded; ventral margin broadly convex; posterior margin rounded, merging with

short posterior dorsal margin; lunule region flattened, not inscribed; escutcheon narrow; surface smooth except for broad, low, irregularly spaced undulations formed by bunching of growth increments; hinge plate moderately narrow, posterior arch well developed on ventral margin of hinge plate just posterior to cardinal tooth; central cardinal tooth peg-like; posterior lateral tooth moderately elongated, bluntly rounded; nymph of medium thickness.

Dimensions. USNM 405817; length, 71 mm; height, 64 mm; width of paired valves, 44 mm.

Types. Holotype, USNM 405817; paratypes, USNM 405818, 405819.

Type locality. Z-746.

Stratigraphic range. Unit 1 of the Sobral Formation.

Localities. Z-9, Z-496, Z-631, Z-746.

Material. 18 specimens.

Discussion. Lahillia huberi n. sp. may be distinguished from *L. larseni* by its smaller size, subtrigonal shell outline, orthogyrate beaks, and an only moderately concave anterior margin. The nymph of *L. larseni* is considerably more massive, longer, and extends above the margin of the valve. *L. huberi* is a common element in unit 1 of the Sobral Formation. Only one specimen, which we tentatively refer to *L. huberi,* was found approximately 20 m below the Cretaceous/Tertiary boundary.

Etymology. This species of *Lahillia* is named after Brian Huber, in recognition of his work on the foraminifera fauna of the López de Bertodano Formation.

Superfamily VENERACEA Rafinesque, 1815
Family VENERIDAE Rafinesque, 1815
Subfamily PITARINAE Stewart, 1930

Genus *Marwickia* Finlay, 1930
Type species. Finlaya parthiana Marwick, 1930 (by original designation).

Marwickia woodburnei n. sp.
Figure 14.13–16

Astarte cf. *venatorum* Wilckens, 1910, pp. 49–50, Plate 2, Fig. 28a,b; not Fig. 29a,b.

Description. Shell medium-sized, ovate, moderately inflated; umbones moderately elevated; lunule broad, weakly incised; anterior dorsal margin short, concave; anterior dorsal margin short, concave; anterior margin broadly rounded, merging with broad, rounded ventral margin; posterior dorsal margin long, slightly convex, moderately sloping; surface smooth except for regularly spaced growth increments; hinge of right valve with three cardinals, 3a blade-like, sloping anteriorly, 1 blunt, peg-like, 3b thick triangular with shallow medial groove, socket for AII elongated; pallial sinus bluntly V-shaped, pointing slightly toward hinge; inner margin smooth.

Dimensions. USNM 405804; length, 30 mm; height, 29 mm; width of paired valves, 16 mm.

Type. Holotype, USNM 405804, paratypes, 405805, 405834.

Type locality. SSPE Location 9.

Stratigraphic range. Middle part of unit 1 of Sobral Formation.

Localities. Z-9.

Material. 11 specimens.

Discussion. Venerid bivalves are rare in the López de Bertodano and Sobral Formations; nearly all have a similar slightly inflated ovate shell. They are separated by their hinge structure. Wilckens (1910) tentatively referred specimens from SSPE Location 9 to *Astarte* cf. *A. venatorum.* He also figured two incomplete bivalve hinges from SSPE Location 8 on Snow Hill Island. We believe that Wilckens' decision to refer the specimens from SSPE Location 9 to *A. venatorum* was influenced by two astarte-like hinge fragments from Snow Hill Island. The presence of an anterior lateral tooth on specimens collected at locality Z-9, which is the same as SSPE Location 9, clearly shows that it belongs in the venerid

Figure 15. 1-5, *Lahillia larseni* (Sharman and Newton). 1, 2, Hypotype, USNM 405803, 1×. 3, LV hinge, hypotype, USNM 405801, 1×. 4, RV hinge, hypotype, USNM 405800, 1×. 5, Hypotype, RV hinge, USNM 405802, 1×. 6-8, *Lahillia huberi* n. sp. 6, RV hinge, paratype, USNM 405719, 1.25×. 7, 8, Holotype, USNM 405817, 1×.

subfamily Pitarinae and not in the family Astartidae. *M. woodburnei* is easily separated from *Cyclorisma chaneyi* n. sp. by the presence of an elongate anterior lateral and by more centrally located umbones.

Etymology. This species is named in honor of Dr. Michael O. Woodburne, discoverer of the first mammal fossils from the continent of Antarctica.

Subfamily TAPETINAE Adams and Adams, 1857

Genus *Cyclorisma* Dall, 1902
Type species. *Cyclothyris carolinensis,* Conrad, 1875
(by monotypy).

Cyclorisma chaneyi n. sp.
Figure 14.5–9

Description. Shell medium-sized, ovate, inflated; umbones low; beaks prosogyral; umbonal ridge broad; anterior dorsal margin short, concave; anterior margin evenly rounded, slightly produced; ventral margin broadly rounded; posterior dorsal margin broadly convex; escutcheon narrow; surface smooth except for closely spaced growth increments; hinge of right valve with three cardinals, 3a moderately thick blade-like; 1 asymmetrically triangular posterior half of tooth sloping, 3b moderately thick, shallowly bifid, elongated along margin of valve; hinge plate immediately anterior to 3a flat; pallial sinus bluntly V-shaped; inner margin smooth.

Dimensions. USNM 405807; length, 27 mm; height, 23 mm; width of single valve, 10 mm.

Types. Holotype, USNM 405806; paratypes, USNM 405807, 405808, 405826.

Type locality. Z-762.

Stratigraphic range. Units 8 through 10 of the López de Bertodano Formation.

Localities. K-66, K-119, K-60, K-167, K-356, Z-458, Z-459, Z-477, Z-757, Z-760, Z-762, Z-768, Z-776, Z-777.

Material. 35 specimens.

Discussion. Although the general external morphology of *Cyclorisma chaneyi* n. sp. is similar to *M. woodburnei* n. sp., the two may be easily separated by the absence of an anterior lateral in the former. The more elevated prosogyrate umbones of *Marwickia woodburnei* n. sp. also serve to separate the two.

Etymology. This species is named for Dan Chaney, to recognize his work on the Early Tertiary mammal fauna of Antarctica.

Order MYOIDA Stoliczka, 1870
Suborder MYINA Stoliczka, 1870
Superfamily MYACEA Lamarck, 1809
Family CORBULIDAE Lamarck, 1818

Surobula Zinsmeister n. gen.

Diagnosis. Shell small, subovate; left valve only slightly smaller than right valve; ornamentation consists of faint concentric undulations; pallial line interrupted, forming a series of oval scars; pallial sinus absent.

Type species. Sphaerium? nucleus Wilckens, 1910.

Discussion. Members of the Corbulidae are characterized by unequal valve size, with the right valve smaller. In some genera the differences may be extreme, with the posterodorsal edge of the right valve fitting into a narrow, socket-like groove along the posterior dorsal margin of the left valve. Most members of the family also have a tendency to have a rostrate dorsal margin. The hinge structure also varies within the group from strong to very peg-like cardinals with or without a chondrophore. The *Surobula* outline is nearly subtrigonal without any rostral development; the left valve is only slightly smaller than the right, but does fit snugly into an elongated groove along the posterior dorsal margin of the

right valve. Although the smooth surface of the shell is somewhat atypical of corbulas, several subgenera (e.g., *Bicorbula* and *Paramicobula*) are characterized by a smooth shell. The left valve (only slightly smaller) and the absence of a pallial sinus lead us to believe that the antarctic species represents a distinct genus within the Corbulidae.

Surobula nucleus (Wilckens, 1910)
Figure 14.10–12

Sphaerium? nucleus Wilckens, 1910, pp. 69–68, Plate 3, Fig. 19.

Supplementary Description. Shell small, subtrigonal, inflated umbones moderately elevated; left valve slightly smaller than right; anterior dorsal margin moderately sloping, gradually merging with round ventral margin; posterior dorsal margin slightly convex and longer than anterior dorsal margin, sloping at approximately the same angle as the anterior dorsal margin; surface nearly smooth except for occasional concentric undulations; RV hinge with anteriorly sloping peg-like cardinal; broad; open socket posterior to cardinal, elongated socket-like groove along posterior dorsal margin of valve; LV hinge with deep, slightly curved socket anterior to blunt, posteriorly sloping cardinal tooth; adductor scars nearly equal in size; pallial line complete, irregularly broken into a series of pits; inner margin smooth.

Dimensions. USNM 405809; length, 9 mm; height, 7 mm; width of paired valves, 5 mm.

Type. Holotype, Mo 1583a; hypotypes, USNM 405809, 405810, 405811.

Type locality. Seymour Island (Wilckens, 1910).

Stratigraphic range. Middle of unit 8 to middle of unit 9 of the López de Bertodano Formation.

Localities. K-356, Z-459, Z-724, Z-754, Z-757, Z-759, Z-760, Z-776, Z-777.

Material. Several hundred specimens.

Discussion. *S. nucleus* is fairly common in the lower part of unit 9 and frequently occurs in vast numbers in small pockets. It appears to have been a nestling species in small depressions on the surface of the sea floor. Such a habitat is supported by the fact that nearly all of the specimens occur as paired valves, which suggests that these small pockets of *S. nucleus* were frequently quickly smothered by shifting sediments.

Superfamily HIATELLACEA Gray, 1824
Family HIATELLIDAE Gray, 1824

Panopea Menard, 1807
Type species. *Mya glycimeris* Born, 1778
(by subsequent designation, Fleming, 1818)

Panopea clausa Wilckens, 1910
Figure 16.1–3

Panopea (Pleuromya?) clausa Wilckens, 1910, p. 68–69, Plate 3, Fig. 10a,b.

Panopea clausa Wilckens, Woods, 1917, p. 33, Plate 18, Figs. 6a,b, 7; Freneix, 1958, p. 343, Plate 3, Fig. 6; Warren and Speden, 1977, p. 40, Fig. 26–13.

Supplementary description. Shell medium-sized, thin-shelled, subrectangular, inflated; umbones moderately elevated, approximately of anterior 25 percent of valve; beaks small mesogyrate; umbonal ridges at anterior 20 percent of valve, becoming obsolete near anteroventral margin; anterior dorsal margin short, concave; anterior margin bluntly rounded; ventral margin flat; posterior bluntly truncated with prominent open gape; posterior dorsal margin straight; valves open along posterior ventral toward gape; sides of valves tapered evenly posteriorly, resulting in posterior half of the shell being moderately compressed; surface ornamentation of irregularly spaced, low, rounded, undulating concentric ribs that occasionally form irregular bifurcations; each valve with well-developed

Figure 16. 1-3, *Panopea clausa* Wilckens. 1-2, Hypotype, USNM 405813, 1×. 3, Hypotype, USNM 405812, 1×. 4, *Thracia askinae* n. sp., holotype, 405764, 1×. 5-7, *Goniomya hyriiformis* (Wilckens). 5, Hypotype, USNM 405815, 1×. 6, Hypotype, USNM 405814, 1×. 7, Hypotype, USNM 405816, 1×.

perpendicularly oriented tooth that fits into corresponding narrow socket, a second poorly developed sloping cardinal on posterior side of socket; ligament resting on blunt protuberance; pallial line complete; pallial sinus deep, bluntly rounded.

Dimensions. USNM 405813; length, 48 mm; height, 30 mm; width of paired valves, 25 mm.

Type. Lectotype, Mo 1608 (herein designated); hypotype, USNM 405812, 405813.

Type locality. SSPE Location 8.

Stratigraphic range. Unit 7 through lower part of unit 9 of the López de Bertodano Formation.

Localities. K-42, K-46, K-56, K-105, K-175, K-177, K-182, K-185, K-409, Z-458, Z-459, Z-757, Z-769, Z-776, Locations B, F, N-84.

Material. 38 specimens.

Discussion. *Panopea*-like bivalves have not received a great deal of attention. The thin shell, weak hinge, and frequent occurrence as paired valves of most fossil specimens has resulted in the lumping of all subrectangular inflated bivalves with a prominent gape into the genus *Panopea.* The subrectangular shape with broad gape is characteristic of many groups of deep-burrowing bivalves. Because of their deep-burrowing habit, most specimens occur as paired valves, and the hinge is not visible. Without access to the hinge, generic identification is very difficult, if not impossible.

The hinge of Cenozoic panopeas varies from nearly edentulous to possessing moderately developed teeth. Commonly, a fairly well-developed, peg-like tooth is present in the right valve and fits into a broad, shallow socket in the left valve. The central tooth of the left valve is not as prominent as the right valve tooth and slopes posteriorly. In many species, the central tooth of the left valve is nearly obsolete. *P. clausa* differs in having a well-developed central blade-like cardinal in the left valve. Whether this type of dentition is common in other panopeas is not known, but it is strikingly different than the typical *Panopea.* At this time we defer proposing a new generic name for this species from Seymour Island.

Subclass ANOMALODESMATA Dall, 1889
Order PHOLADOMYOIDA Newell, 1965
Superfamily PHOLADOMYACEA Gray, 1847
Family PHOLADOMYIDAE Gray, 1847

Genus *Goniomya* Agassiz, 1842
Type species. Mya angulifera J. Sowerby, 1819 (by subsequent designation, Herrmannsen, 1847)

Goniomya hyriiformis (Wilckens, 1910)
Figure 16.5–7

Trigonia hyriiformis Wilckens, 1910, pp. 47–49; Plate 2, Fig. 27.
Iotrigonia hyriiformis (Wilckens), Perez and Reyes, 1978, p. 11, Plate 2, Fig. 2.

Supplementary description. Shell large, thin-shelled, subrectangular, elongated, inflated; umbones slightly prosogyrate, moderately elevated, at anterior quarter of shell; irregular umbonal ridge developed along intersection of inverted V's of concentric ribs; ligament external; anterior dorsal margin moderately short, concave, merging with bluntly rounded anterior margin; ventral margin long, nearly flat; posterior margin bluntly rounded; posterior dorsal margin long, flat, sloping upward near umbones; well-developed posterior gape; surface ornamented with two sets of obliquely sloping ribs which intersect to form an inverted V pattern along umbonal ridge, posterior set numbering approximately 20, anterior set numbering approximately 15, tending to droop and become wavy near intersection with posterior set; hinge edentulous.

Dimensions. USNM 405815; length, 114 mm; height, 52 mm; width of paired valves, 46 mm.

Type. Lectotype Mo 1487 (herein designated); hypotype, USNM 405814, 405815, 405816.

Type locality. Snow Hill Island (Wilckens, 1910).

Stratigraphic distribution. Lower part of unit 7 to upper part of unit 9 of the López de Bertodano Formation.

Localities. K-41, K-42, K-423, Z-459, Z-772, Locations B, H.

Material. 10 specimens.

Discussion. Wilckens' (1910) original description was based on one incomplete specimen that he referred to *Trigonia.* During the course of three field seasons, a number of moderately well-preserved specimens have been collected. All the specimens consist of paired valves and show some degree of deformation. The deformed nature of the material reflects the fact that the shells are very thin, and thus are easily deformed by compaction. Sectioning along the hinge of one of the specimens clearly showed that the hinge is edentulous and not of a trigoniid type.

Superfamily PANDORACEA Rafinesque, 1815
Family THRACIIDAE Stoliczka, 1870

Genus *Thracia* Sowerby, 1823
Thracia pubescens Lamarck, 1779 (by subsequent designation, Anton, 1839).

Thracia askinae n. sp.
Figure 16.4

Thracia (Thracia) sp. del Valle and Medina, 1980, p. 54–55, Plate 3, Figs. 1, 2.

Description. Shell medium-sized, compressed, inequivalve, right valve slightly larger than left; umbones centrally located, low, slightly opisthogyrate; anterior dorsal margin of moderate length, straight to slightly convex, merging with bluntly rounded anterior margin, ventral margin moderately convex, posterior margin rostrate, nearly flat, slightly inclined anteriorly, posterior dorsal margin broadly concave; umbonal ridge well developed, extending to posteroventral margin of valve, forming a prominent angulation; escutcheon narrow, moderately deep; ornamentation consisting of irregularly spaced concentric growth increments.

Dimensions. USNM 405764; length, 39.5 mm; height, 31 mm.

Type. Holotype, USNM 405764.

Type locality. K-46.

Stratigraphic range. Known from upper part of unit 5 to lower part of unit 8 of the López de Bertodano Formation.

Localities. K-30, K-46.

Material. 2 specimens.

Discussion. Although the type of *T. askinae* n. sp. consists of an external cast, the preservation is sufficient to warrant a specific designation. We consider the Seymour Island species to be identical to an undescribed species of *Thracia* figured by del Valle and Medina (1980) from Cape Lamb on Vega Island. *T. askinae* n. sp. is also similar to *T. haasti* Woods (1917) from the Late Cretaceous of New Zealand, but may be distinguished by its more compressed valves and greater elongation. *T. lenticularis* from southern Patagonia (Wilckens, 1907) may be easily separated from *T. askinae* n. sp. by the rounded posterior margin.

Etymology. This species is named in honor of Dr. Rosemary Askin for her years of study of the fossil floras of Antarctica.

ACKNOWLEDGMENTS

We particularly express our appreciation to Dr. David H. Elliot, director of the Institute of Polar Studies, The Ohio State University, for his recognition of the scientific potential of the Seymour Island region and for initiating the American field program on Seymour Island in 1975. Without his foresight, this chapter, along with the others in this volume, would not have been developed. Special thanks are also extended to Sir Charles Fleming for his comments and information concerning the Trigonidae; to LouElla Saul, Los Angeles County Museum of Natural History, for her help, comments, and casts of *Eriphyla* and *Dozyia;* to Drs. Valdar Jaanusson and Henri Mutei, Naturhistoriska Riksmuseet, for providing access to the Swedish South Polar collections; to Dr. Thomas R. Waller, Smithsonian Institution, for his comments concerning *Entolium;* to Dr. Adriano Ferrari, Institute di Geologia e Paleontologica, Universidad di Bologna, for loaning several of Fergulio's Patagonian types for comparative studies; and to the members of the four expeditions to Seymour Island for finding many of the fossils described herein. This work would not have been possible without the long-term support of the Division of Polar Programs, National Science Foundation Grants DPP-7421509, 7721585, 7920215, 8020096, and 8213985.

REFERENCES

ANDERSSON, G. 1906. On the geology of Graham Land. Geological Institute, University of Uppsala Bulletin, 7:19–71.

ASKIN, A. R. 1984. Palynological studies in the James Ross Island basin and Roberston Island, Antarctic Peninsula. Antarctic Journal of the United States, 19(5):6–7.

BIBBY, J. S. 1966. The stratigraphy of part of north-east Graham Land and the James Ross Island Group. British Antarctic Survey Scientific Report, 53:1–37.

COLLIGNON, M. 1951. Faune Mäestrichtienne de la Cote d'Ambatry (Province de Betioky) Madagascar. Madagascar Service des Mines: Annales Géologiques, 19:45–69.

COX, L. R. 1937. The Jurassic lamellibranch fauna of Cutch (Kackk): no. 3 Families Pectinidae, Amussiidae, Plicatulidae, Limidae, Ostreidae and Trigonidae (Supplement). Indian Geological Survey Member, Palaeontologia, Indica, Series 9, 3(4):1–128.

——— . 1952. Notes on the Trigoniidae, with outlines of a classification of the family. Malacological Society of London, 29:45–70.

DEL VALLE, R., AND F. A. MEDINA. 1980. Nuevos invertebrados fósiles de Cabo Lamb (Isla Vega) y Cabo Morro (Isla James Ross). Contribuciones del Instituto Antártico Argentino, 228:51–67.

ELLIOT, D. E., AND T. A. TRAUTMAN. 1982. Lower Tertiary strata on Seymour Island, p. 287–297. *In* C. Craddock (ed.), Antarctic Geoscience. University of Wisconsin Press, Madison.

FERUGLIO, E. 1936. Paleontographia Patagonica. Memoir of the Instituto di Geologie, University Padova, 9:1–384.

FLEMING, C. A. 1978. The bivalve mollusc genus *Limatula:* a list of described species and a review of living and fossil species in the southwest Pacific. Journal of the Royal Society of New Zealand, 8:17–91.

FRENEIX, S. 1958, Contribution à l'étude des Lamellibranches du Crétacé de Nouvelle-Calédonie. Science Terre, 4(3-4):153–207.

——— . 1972. Les mollusque bivalves Crétacés du Bassin côier de Tarfaya (Marc méridional). Notes et Memoires, Service Geologique du Maroc, 228:49–225.

——— . 1980. Bivalves Néocrétacés de Nouvelle-Calédonie, signification biogéographique, biostratigraphique, paléoécologique. Annales de Paléontologie (Invertébrés), 66:67–134.

GABB, W. M. 1869. Cretaceous Tertiary fossils. Geological Survey of California, 2:1–299.

HALL, S. A. 1977. Cretaceous and Tertiary dinoflagellates from Seymour Island Antarctica. Nature, 267:239–241.

HEINBERG, C. 1979. Evolutionary ecology of nine sympatric species of the pelecypod *Limopsis* in Cretaceous Chalk. Lethaia, 12(4):325–340.

HUBER, B. T. 1984. Late Cretaceous foraminiferal biostratigraphy, paleoecology, and paleobiogeography of the James Ross Island regions, Antarctic Peninsula. Unpublished M.Sc. thesis, The Ohio State University, 246 p.

——— , D. M. HARWOOD, AND P. N. WEBB. 1983. Upper Cretaceous microfossil biostratigraphy of Seymour Island, Antarctic Peninsula. Antarctic Journal of the United States, Annual Review, 18(5):72–74.

MACELLARI, C. E. 1984. Late Cretaceous stratigraphy, sedimentology, and macropaleontology of Seymour Island, Antarctic Peninsula. Unpublished Ph.D. dissertation. The Ohio State University, 599 p.

——— . 1985. Paleobiogeografía y edad de la fauna de *Maorites-Gunnarites* (Ammonoidea) de la Antártida y Patagonia. Amehiniana, 21(2-4):223–242.

——— . 1986. Late Campanian-Maastrichtian ammonite fauna from Seymour Island (Antarctic Peninsula). Journal of Paleontology, 60(2 of 2):1–55.

MALUMIAN, N., H. H. CAMACHO, AND R. GORRONO. 1978. Moluscos del Terciario Inferior ("Magallanense") de la Isla Grande de Tierra del Fuego (Republica Argentina). Ameghiniana, 15(3-4):265–284.

MEDINA, F. A. 1980. Revisión y origen de las trigonias del Grupo de Islas James Ross. Contribuciones Instituto Antártico Argentino, 247:107–126.

MULLER, A. J. 1970. Zur funktionellen morphologie, taxiologie und okologie von Pycnodonte (Ostreina, Lamellibranchiata). Momatsberichte Deutsche Akademie der Wissenschasten Berlin, 112, 1:902–903.

NAKANO, M.. 1961. On the Trigoniidae. Journal of Science, Hiroshima University, Series C, 4(1):71–94.

NEWELL, N. 1969. Family Parallelodontidae Dall, 1898. *In* R. C. Moore (ed.), Treatise on Invertebrate Paleontology. Mollusca 6, N(1):N256–N259.

PALMER, T. J. 1984. Revision of the bivalve Family Pulvinitidae Stephenson, 1941. Paleontology, 27(4):815–824.

PEREZ, E. d'A., AND R. B. REYES. 1978. Las Trigonias del Cretácico superior de Chile y su valor cronoestratigráfico. Boletín, Instituto de Investgaciones Geológicas, Chile, 34, 67 p.

PALAMARCZUK, S., ET AL. 1984. Las Formaciones Lopez de Bertodano y Sobral en la Isla Vicecomodoro Marambio, Antártida. Ameghiniana, 17(4):323–333.

PUGACZEWSKA, H. 1977. The Upper Cretaceous Ostreidae from the Middle Vistula region (Poland). Acta Palaontologica Polonica, 22(2):187–204.

REYES, R., AND E. PEREZ D'A. 1979. Estado actual del conocimento de la familia Trigonidae (Mollusca: Bivalia) en Chile. Revista Geologica de Chile, 8:13–64.

RINALDI, C. A., ET AL. 1978. Geologí de la Isla Vicecomodoro Marambio. Contribuciones Instituto Antártico Argentino, 217:1–37.

SHARMAN, G., AND E. T. NEWTON. 1894. Notes on some fossils from Seymour Island in the Antarctic region obtained Dr. Donald. Transactions of the Royal Society of Edinburgh, 37, Part 3(30):707–709.

——— . 1897. Notes on some additional fossils collected at Seymour Island, Graham's Land, by Dr. Donald and Captain Larsen. Proceedings of the Royal Society of Edinburgh, 22(1):58–61.

STOLICZKA, F. 1871. Cretaceous fauna of southern India. Geological Survey of India Memoir, Palaeontological Indica, Series 6, 3:223–537.

WILCKENS, O. 1907. Die lamellibranchiaten gastropoden u.s.w. der oberen Kreide Südpatagoniens. Berichte Naturforsh Gesellshaft Freiburg, 15:97–166.

——— . 1910. Die anneliden, bivalven, und gastropoden der Antarktischen Kreideformation. Wissenschaftliche Ergebnisse der Schwedischen Südpolarexpedition 1901-1903, 3(12):1–132.

——— . 1911. Die mollusken der antarktischen Tertiarformation. Wissenschaftliche

Ergebnisse der Schwedischen Südpolar-expedition, 1901–1903, 3(13):1–62.

WOODS, H. 1908. The Cretaceous fauna of Pondoland. Annals of the South African Museum, 4:275–350.

——— . 1913. A monograph of the Cretaceous lamellibranchia of England. Paleontolographia, 2:1–473.

——— . 1917. The Cretaceous faunas of the north-eastern part of the South Island of New Zealand. New Zealand Geological Survey, Palaeontological Society of London Bulletin, 4, 41 p.

ZINSMEISTER, W. J. 1978. Three new species of *Pulvinites* (Mollusca: Bivalvia) from Seymour Island (Antarctic Peninsula) and southern California. Journal of Paleontology, 52:565–569.

——— . 1982. Review of the Upper Cretaceous–Lower Tertiary sequence on Seymour Island, Antarctica. Journal of the Geological Society of London, 139(6):779–796.

——— . 1984. Late Eocene bivalves (Mollusca) from the La Meseta Formation, collected during the 1974-1975 joint Argentine-American expedition to Seymour Island, Antarctic Peninsula. Journal of Paleontology, 58(6):1497–1527.

MANUSCRIPT ACCEPTED BY THE SOCIETY SEPTEMBER 1, 1987

Geological Society of America
Memoir 169
1988

The new dimitobelid belemnite from the Upper Cretaceous of Seymour Island, Antarctic Peninsula

Peter Doyle
Department of Palaeontology, British Museum (Natural History), Cromwell Road, London SW7 5BD, United Kingdom
William J. Zinsmeister
Department of Earth and Atmospheric Sciences, Purdue University, West Lafayette, Indiana 47907

ABSTRACT

Dimitobelus (Dimitocamax) seymouriensis n. sp., a new subgenus and species of Dimitobelidae, a belemnite family restricted to the Cretaceous Southern Hemisphere, is described from the Campanian-Maastrichtian López de Bertodano Formation of Seymour Island, Antarctic Peninsula. The subgenus *D. (Dimitocamax)* n. subgen. is erected for those species of *Dimitobelus* possessing a regular hastate form and *Actinocamax*-like alveolar regions. *D. (Dimitocamax)* n. subgen. had a restricted stratigraphic and geographic distribution in contrast to the wider distribution of *D. (Dimitobelus)* Whitehouse.

INTRODUCTION

This chapter describes a new subgenus and species of dimitobelid belemnite from strata of late Campanian or early Maastrichtian age on Seymour Island, Antarctic Peninsula. The specimens were collected by one of us (W.J.Z.) and by B. J. Huber during the Antarctic summer of 1985. The majority of these specimens are housed in the National Museum of Natural History, Washington (USNM), although a smaller number, collected by Huber (Institute of Polar Sciences, Ohio State University), are in the collections of the British Museum (Natural History) (BMNH). Comparative Albian *Dimitobelus* from the collections of the British Antarctic Survey, Cambridge (BAS), and the Hunterian Museum, Glasgow (HM), were also studied.

The belemnites were all collected from a single locality in Seymour Island (locality 768, see Fig. 1), at a level approximately 620 m below the Cretaceous-Tertiary boundary (see Macellari, 1984, 1985) within the López de Bertodano Formation. The sediments of this formation are the oldest rocks exposed on Seymour Island; they consist largely of bioturbated silty sandstones (maximum thickness, 1,190 m) that have yielded a rich variety of fauna, including mollusks, echinoderms, and serpulids (Rinaldi et al., 1978; Macellari and Huber, 1982; Zinsmeister, 1982a; Macellari and Zinsmeister, 1983; Macellari, 1984, 1985, 1986). The belemnites occur in a poorly consolidated, medium-brown fossiliferous sandy siltstone exposed in a low cliff (3 m high) that

extends along the strike of the bed for approximately 0.5 km at locality 768 (Fig. 1). Although the belemnites are common, it is surprising that this is the only horizon where they have been found on Seymour Island. The associated fauna is dominated by serpulid worms (*Rotularia*) and large, spatulate echinoid spines (*Cyathocidaris*). In addition, rare crinoid ossicles and corals also occur with the belemnites. The age of the López de Bertodano Formation has been variously ascribed to mainly Campanian (Howarth, 1966; Olivero, 1981) or Campanian-Maastrichtian (Spath, 1953; Huber et al., 183; Macellari, 1984, 1985, 1986). B. J. Huber (personal communication, 1985) has suggested that the belemnite horizon is of basal Maastrichtian age, based on microfaunal evidence (in unit K15 of Huber et al., 1983).

The belemnites discussed below are the first to be described from Seymour Island. Cretaceous belemnites are locally common in the earlier Cretaceous sequences exposed on James Ross Island (64°05′S, 58°25′W) and Alexander Island (71°30′S, 68°15′W) (Willey, 1972, 1973; Doyle, 1985, 1987b), and the Dimitobelidae are well represented in the Aptian-Albian sediments of these areas (Doyle, 1987b). Records of belemnites from late Cretaceous sediments of the Antarctic Peninsula and the Southern Hemisphere in general are sparse (Doyle, 1987a), making this new discovery particularly valuable. The following systematic discussion is the responsibility of the senior author

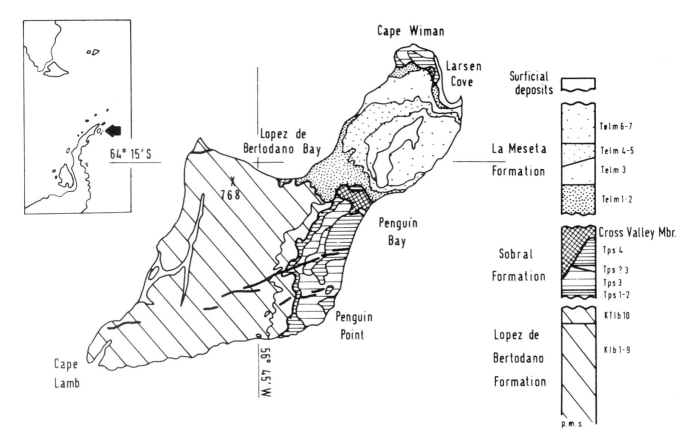

Figure 1. Geologic sketch map of Seymour Island, Antarctica, with the belemnite locality (locality 768) indicated.

(P. D.). The terminology used below is discussed in Stevens (1965) and Doyle (1987b). Abbreviations used herein and in Table 1 are as follows: L indicates the total preserved length of the rostrum; X, the length from apex to position of maximum inflation; Dvmax, the maximum dorso-ventral diameter; and Dlmax, the maximum lateral diameter.

SYSTEMATIC DESCRIPTIONS

Order BELEMNITIDA Zittel, 1895
Suborder BELEMNOPSEINA Jeletzky, 1965
Family DIMITOBELIDAE Whitehouse, 1924

Genus *Dimitobelus* Whitehouse, 1924
Type species. *Belemnites canhami* Tate, 1880 (junior subjective synonym of *Belemnitella diptycha* M'Coy, 1867), by original designation.
Diagnosis. See Whitehouse (1924), Stevens (1965), and Doyle (1987b).
Remarks. Species of this genus are common in the Cretaceous sediments of the Southern Hemisphere; the majority are attributable to the subgenus *D. (Dimitobelus)* Whitehouse, 1924. Typical species of *D. (Dimitobelus)* include *D. (Dimitobelus) diptychus* (M'Coy) (Figs. 2.16,17), *D.*

(Dimitobelus) stimulus Whitehouse, *D. (Dimitobelus) praelindsayi* Doyle and *D. (Dimitobelus) lindsayi* (Hector), and the type species of *Cheirobelus* Whitehouse, 1924 (see Stevens, 1965; Doyle, 1987b). The subgeneric distinction of *D. (Dimitobelus)* from *D. (Dimitocamax)* n. subgen. is discussed below.

Subgenus *Dimitocamax* Doyle, n. subgen.
Diagnosis. Small- to medium-sized *Dimitobelus,* outline and profile symmetrical, hastate; transverse sections circular, elliptical or subquadrate, generally only weakly depressed; lateral lines (*Doppellinien*) incised, central to flanks; ventro-lateral alveolar grooves, short, slightly deflected dorsally; alveolus lost due to imperfect calcification alveolar region attenuated; apical line central and ortholineate.
Type species. *Dimitobelus hectori* Stevens, 1965 (Figs. 2.1–4).
Included species. *D. (Dimitocamax) hectori* Stevens, *D. (Dimitocamax) seymouriensis* Doyle, n. sp., and possibly the forms described by Doyle (1987b) as *Dimitobelus* sp. aff. *diptychus* (M'Coy).
Range. ?Campanian to Maastrichtian of New Zealand and the Antarctic Peninsula.
Discussion. The species described from the Maastrichtian of New Zealand by Stevens (1965) as *Dimitobelus hectori* n. sp. had long puzzled earlier workers (including F. W. Whitehouse), who assigned typical specimens of this species to the hastate early Cretaceous genera *Neohibo-*

lites Stolley and *Hibolithes* Montfort (Stevens, 1965, p. 126), possibly as they also possess decayed, imperfectly calcified alveolar regions (e.g., Swinnerton, 1936–1955, p. xiv, xxxix). Stevens himself noted superficial resemblances of this species to those of the late Cretaceous boreal genus *Actinocamax* Miller (Stevens, 1965, p. 127), but determined its true affinity by the discovery of a unique specimen of *D. hectori* with ventro-lateral alveolar grooves preserved intact. Although possessing alveolar grooves typical of *Dimitobelus, D. hectori* differs from most other species of this genus in possessing a remarkably regular hastate form with an undepressed transverse section, a central, ortholineate apical line, and an attenuated alveolar region. The subsequent discovery of a second species from Antarctica bearing similar features justifies their separation into a new subgenus, *D. (Dimitocamax)* n. subgen. In addition, it is possible that the slightly more depressed, but nevertheless similar, *Dimitobelus* sp. aff. *diptychus* (M'Coy) described by Doyle (1987b) from James Ross Island, Antarctica, could be assigned to *Dimitobelus (Dimitocamax)*, but further data are required to confirm this, as the specimens described here were incomplete.

D. (Dimitobelus) differs from *D. (Dimitocamax)* in that its species possesses asymmetrical profiles, strongly depressed transverse sections, and weakly cyrtolineate apical lines. Also, species of *D. (Dimitobelus)* generally have less attenuated alveolar regions (with pseudalveoli) than are found in species of *D. (Dimitobelus)*; they are never developed to the extent found in species of *D. (Dimitocamax)*. Species of *Actinocamax* are much more slender, and where preserved, possess only a single alveolar groove or slit. The resemblance of some members of the Dimitobelidae to those of the Belemnitellidae (which includes *Actinocmax*) could be due to their possible derivation from a common ancestor (Stolley, 1927; Doyle, 1987a).

Derivation of name. A hybrid of the generic names *Dimitobelus* and *Actinocamax,* indicating generic affinity to the former and morphologic similarity to the latter.

Dimitobelus (Dimitocamax) seymourensis Doyle, n. sp.
Figure 2.5–15

Diagnosis. Medium sized; outline and profile similar, hastate to almost cylindrical with an acute apex; transverse sections weakly depressed, subquadrate to elliptical.

Description. Medium sized *D. (Dimitocamax),* with a total length of approximately six times the maximum dorso-ventral diameter (Dvmax) (Table 1); outline symmetrical and hastate, with Dlmax in the apical third of the rostrum; alveolar region very attenuated (in both outline and profile); alveolus lost, due to imperfect calcification; apex acute in outline, commonly mucronate and often exfoliated; profile generally similar to the outline, being symmetrical and hastate, venter and dorsum inflated to the same degree, giving an acute apex; transverse sections of the rostrum slightly depressed and subquadrate, especially in the anterior of the rostrum, becoming rounded and elliptical adapically in the stem region; *Doppellinien* (double lateral lines), generally well developed, extend for the whole length of the rostrum become notably less sharp over the stem region; where the *Doppellinien* are incised, the subquadrate shape of the transverse section is enhanced; no traces of the ventro-lateral alveolar grooves or the phragmocone are found, due to decay and loss of the alveolar region; apical line centrally placed, ortholineate; apical canal rarely developed.

Types. Holotype. USNM. 404809; *Paratypes.* USNM. 404810, 404811.
Type locality. Locality 768 (locality 166 of Macellari, 1984), Seymour Island, Antarctic Peninsula (Fig. 1).
Material. Sixteen complete and eight fragmentary rostra (Zinsmeister Collection, USNM 404809 through 404832); one almost complete and six fragmentary rostra (Huber Collection, BMNH C. 59279 through 59285), all from the ?Maastrichtian of the López de Bertodano Formation, locality 768, Seymour Island, Antarctic Peninsula.
Measurements. See Table 1.

TABLE 1. DIMENSIONS OF *DIMITOBELUS (DIMITOCAMAX) SEYMOURIENSIS* n. sp. FROM SEYMOUR ISLAND

Specimen	L (mm)	X (mm)	Dvmax (mm)	Dlmax (mm)
USNM 404809	46.15	19.05	7.90	8.60
USNM 404810	46.65	15.00	7.80	8.85
USNM 404811	41.65	15.80	7.40	8.15
USNM 404812*	44.15	15.25	8.75	8.90
USNM 404813*	65.05	22.55	6.85	7.05
USNM 404814	34.94	15.05	6.95	7.80
USNM 404815	40.60	20.85	6.85	8.00
USNM 404816	36.40	16.35	5.90	6.75
USNM 404817	43.90	15.25	4.50	5.80
USNM 404818	42.75	16.80	6.35	7.30
USNM 404819	34.95	15.05	5.15	5.50
USNM 404820	28.10	14.95	6.20	6.40
USNM 404821	41.00	12.85	6.10	—
USNM 404822	30.50	12.15	6.45	6.85
USNM 404824	30.02	13.85	5.25	5.95
USNM 404825*	30.55	16.00	6.80	7.60
BMNH C.59279	26.00	12.60	6.30	7.55

*Pathologically deformed specimens.

Derivation of name. Specimens all collected from Seymour Island, Antarctic Peninsula.
Remarks. Three of the rostra described are pathologically deformed (USNM. 404812, 404813, 404825). The first (USNM. 404813) (Fig. 2.12,13) is unusually well preserved, having a very elongate alveolar region, and possesses a distorted apex that is dorsally incurved just posteriorad, a pathologic constriction at the position of maximum inflation of the rostrum. The second (USNM. 404825) is similar, with the apex asymmetrical in outline, although in this case with a curvature toward the left flank. The final specimen (USNM. 404812) Fig. 2.14,15) possesses a constriction similar to that in the first specimen, occurring just after the position of maximum inflation. In this case the apex remains abnormally bulbous. Such pathologic conditions are fairly common in belemnites, and similar examples have been described by Naef (1922), Pugaczewksa (1961), and Stevens (1965). It is likely that the deformities described were caused by damage to the shell secreting the epithelium, either by infection or by predation.

D. (Dimitocamax) seymouriensis n. sp., although lacking the paired ventro-lateral alveolar grooves diagnostic of the Dimitobelidae (see Whitehouse, 1924; Glaessner, 1957; Stevens, 1965; Doyle, 1985, 1987a,b) is assigned to *Dimitobelus* for two reasons: (1) Although clearly a distinct species (see discussion below), it closely resembles *Dimitobelus hectori* Stevens. This New Zealand species is usually found without its alveolar region preserved (e.g., Fig. 2.1–4), the holotype being the only specimen known with its alveolar grooves preserved intact. Consequently, this species has been mistaken in the past for other Belemnopseina (see Stevens, 1965, p. 127). (2) At this time (Campanian-Maastrichtian), most Belemnopseina—other than the Dimitobelidae—had died out in the Southern Hemisphere (Stevens, 1973; Doyle, 1987a). No Belemnitellidae (common in the boreal regions at this time) have been recorded from the Southern Hemisphere.

D. (Dimitocamax) seymouriensis can be readily distinguished from *D. (Dimitocamax) hectori* Stevens, as this New Zealand species is much shorter, and has an extremely bulbous apex and circular cross section. In

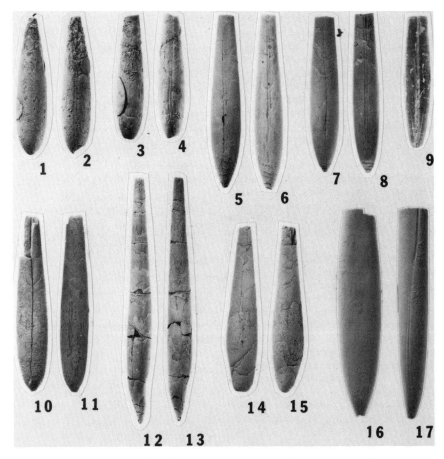

Figure 2. Representative *Dimitobelus*. Each specimen is illustrated by a ventral outline and left profile (venter to the left), unless otherwise stated. 1-4, *Dimitobelus (Dimitocamax) hectori* Stevens, 1965, Haumurian (Maastrichtian), Brighton, south of Dunedin, New Zealand. 1-2, BMNH C.22665, ×1. 3-4, BMNH C.57071, ×1. 3-15, *Dimitobelux (Dimitocamax) seymouriensis* n. sp., ?Maastrichtian, López de Bertodano Formation, Seymour Island, Antarctica. 5-6, Holotype, USNM. 404809, ×1. 7-8, Paratype, USNM. 404811, ×1. 9, Long section showing the central (ortholineate) apical line, BMNH C.59282, ×1. 10-11, Paratype, USNM. 404810, ×1. 12-13, Specimen with well-preserved attenuated alveolar region and pathologically deformed apex, USNM. 404813, ×1. 14-15, Pathologically deformed specimen, USNM. 404812, ×1. 16-17, *Dimitobelus (Dimitobelus) diptychus* (M'Coy, 1867), Albian, Lake Eyie, south Australia; Ventral outline and right profile, HM S.5581, ×1.

addition, the Antarctic species is generally more depressed than *d. (Dimitocamax) hectori,* especially in the alveolar region, although it is not as depressed as species of *D. (Dimitobelus).* However, *D. (Dimitocamax) seymouriensis* does resemble at least one species of *D. (Dimitobelus); D. (Dimitobelus) stimulus* Whitehouse is relatively undepressed, long, and slender, but *D. (Dimitocamax) seymouriensis* is distinguished by its more hastate shape (Dlmax is closer to the apex) and very attenuated alveolar region.

DISTRIBUTION OF LATE CRETACEOUS DIMITOBELIDAE

The Dimitobelidae are of interest to biogeographers due to their apparent restriction to the Southern Hemisphere, south of the 30°S Cretaceous paleolatitude. Pre-Aptian belemnite faunas in this region are notable for their Tethyan affinities, with the presence of such belemnopseid genera as *Belemnopsis* Bayle and *Hibolithes* Montfort. The recognition of a contrasting endemic post-Aptian austral belemnite fauna was attributed by Stevens (1971, 1973) to the southward movement of Gondwana, away from a warm Tethyan influence and toward a cooler polar one. Thus the Dimitobelidae are first recognized in the Aptian, probably having evolved from endemic stocks of an originally Tethyan genus such as *Hibolithes* (Doyle, 1987a). This family was the southern counterpart of the boreal Belemnitellidae, a family confined to the Northern Hemisphere. The dimitobelids reached their maximum development in the Aptian-Albian, occurring in ?Patagonia, Antarctica, Australasia, southern India, and southern Africa (Stevens, 1973; Doyle, 1987a,c). After the Albian, with the demise of the genera *Peratobelus* Whitehouse and *Tetrabelus* Whitehouse, the Dimitobelidae—solely represented by *Dimitobelus*—was restricted to the regions contained within the Weddellian Province of Zinsmeister (1979, 1982b).

In the Albian, the only subgenus of *Dimitobelus* known was *D. (Dimitobelus)* Whitehouse, occurring in the Antarctic Peninsula and Australasia (New Zealand, Australia, and New Guinea). There are few records of Southern Hemisphere belemnites in the Cenomanian, but a distribution similar to that of the Albian is still apparent. In the Turonian to Santonian interval only a single species of *D. (Dimitobelus)* is known, *D. (Dimitobelus) superstes* (Hector) endemic to New Zealand. Finally, *D. (Dimitobelus)* continued into the Companian in New Zealand, with the single species *D. (Dimitobelus) lindsayi* (Hector); there are no records of this subgenus from Maastrichtian sediments.

The subgenus *D. (Dimitocamax)* probably first appeared in the Campanian, possibly with the somewhat depressed form *D.* sp. aff. *diptychus,* described by Doyle (1987b) from James Ross Island, Antarctic Peninsula. *D. (Dimitocamax) seymouriensis* appeared in the early Maastrichtian, and the last *D. (Dimotocamax)* was *D. (Dimitocamax) hectori* Stevens, known exclusively from the Maastrichtian of New Zealand. There are no records of this subgenus from the late Cretaceous sediments of other regions (such as Australia). Thus, all known species of the Campanian-Maastrichtian subgenus *D. (Dimitocamax)* are endemic to their specific and separate geographic regions (James Ross Island, Seymour Island, New Zealand), perhaps matching the increasing endemism in the benthic mollusks on the Pacific coast of the fragmenting Gondwana at this time (Zinsmeister, 1979, 1982b). This is in contrast with the Albian, when it was common for species of *D. (Dimitobelus)* to be widely distributed; for example, *D. (Dimitobelus) diptychus* (M'Coy) is known from such widely separated regions as James Ross and Alexander Islands in the Antarctic Peninsula, and the Great Artesian Basin of Australia.

ACKNOWLEDGMENTS

We thank G. R. Stevens and B. J. Huber for their assistance, M. K. Howarth and H. G. Owen for critical reading, and J. A. Jeletzky and R. A. Hewitt for their review comments. J. A. Crame, M.R.A. Thomson (BAS), and W.D.I. Rolfe (HM) kindly allowed access to their collections. One of us (P.D.) carried out this work while in receipt of a NERC research fellowship at the British Museum (Natural History). Field work was supported by National Science Foundation Grant DDP-8213985 (to W.J.Z.).

REFERENCES

DOYLE, P. 1985. "Indian" belemnites from the Albian (Lower Cretaceous) of James Ross Island, Antarctica. British Antarctic Survey Bulletin, 69:23–35.

——. 1987a. The belemnite family Dimitobelidae in the Cretaceous of Gondwana. International Union of Geological Sciences, Series A (in press).

——. 1987b. The Cretaceous Dimitobelidae (Belemnitida) of the Antarctic Peninsula region. Palaeontology, 30 (in press).

——. 1987c. Early Cretaceous belemnites from southern Mozambique. Palaeontology, 30 (in press).

GLAESSNER, M. F. 1957. Cretaceous belemnites from Australia, New Zealand, and New Guinea. Australian Journal of Science, 20:88–89.

HOWARTH, M. K. 1966. Ammonites from the Upper Cretaceous of the James Ross Island Group. British Antarctic Survey Bulletin, 10:55–69.

HUBER, B. J., D. M. HARWOOD, AND P. N. WEBB. 1983. Upper Cretaceous microfossil biostratigraphy of Seymour Island, Antarctic Peninsula. Antarctic Journal of the United States, Annual Review, 18(5):72–74.

MACELLARI, C. E. 1984. Revision of serpulids of the genus *Rotularia* (Annelida) at Seymour Island (Antarctic Peninsula) and their value in stratigraphy. Journal of Paleontology, 58:1098–1116.

——. 1985. Paleobiogeografia y edad de la fauna de *Maorites-Gunnarites* (Ammonoidea) del Cretacico Superior de la Antartida y Patagonia. Ameghiniana, 21:223–242.

——. 1986. Late Campanian–Maastrichtian ammonite fauna from Seymour Island (Antarctic Peninsula). Journal of Paleontology, 60(2), Supplement, Paleontological Society Memoir 18:55 p.

——, AND B. J. HUBER. 1982. Cretaceous stratigraphy of Seymour Island, East Antarctic Peninsula. Antarctic Journal of the United States, Annual Review, 17(5):68–70.

——, AND W. J. ZINSMEISTER. 1983. Sedimentology and macropaleontology of the Upper Cretaceous to Paleocene sequence of Seymour Island. Antarctic Journal of the United States, Annual Review, 18(5):69–71.

NAEF, A. 1922. Die Fossilien Tintenfische: eine palaozologisches monographie. Gustav Fischer, Jena, 322 p.

OLIVERO, E. B. 1981. Esquenna de zonación de ammonites del Cretácico Superior del grupo de islas James Ross, Antártida. VIII Congreso Geológico Arentino, San Luis, 2:897–907.

PUGACZEWSKA, H. 1961. Belomnoids from the Jurassic of Poland. Acta Palaeontologica Polonica, 6:105–236.

RINALDI, C. A., A. MASSABIE, J. MORELLI, H. L. RROSENMAN, AND R. DEL VALLE. 1978. Geologia de la Isla Vicecomodoro Marambio. Contribuciones del Instituto Antártico Argentino, 217:5–43.

SPATH, L. F. 1953. The Upper Cretaceous cephalopod fauna of Graham Land. Falkland Islands Dependencies Survey Scientific Reports, 3:60 p.

STEVENS, G. R. 1965. The Jurassic and Cretaceous belemnites of New Zealand and a review of the Jurassic and Cretaceous belemnites of the Indo-Pacific region. New Zealand Geological Survey Paleontological Bulletin, 36:283 p.

——. 1971. Relationship of isotopic temperatures and faunal realms to Jurassic-Cretaceous palaeogeography, particularly of the S.W. Pacific. Journal of the Royal Society of New Zealand, 1:145–148.

——. 1973. Cretaceous belemnites, p. 387–401. *In* A. Hallam (ed.), Atlas of palaeobiogeography. Elsevier, Amsterdam.

STOLLEY, E. 1927. Zur Systematik und Stratigraphie median gefurchter Belemniten. Jahresbericht Niedersachsischen geologischen Vereins 20:111–136.

SWINNERTON, H. H. 1936–1955. A monograph of the British Lower Cretaceous belemnites. Palaeontographical Society of London Monographs, 86 p. 18 pl.

WHITEHOUSE, F. W. 1924. Dimitobelidae: a new family of Cretaceous belemnites. Geological Magazine, 61:410–416.

WILLEY, L. E. 1972. Belemnites from southeastern Alexander Island: I. The occurrence of the family Dimitobelidae in the Lower Cretaceous. British Antarctic Survey Bulletin, 28:29–42.

——. 1973. Belemnites from southeastern Alexander Island: II. The occurrence of the family Belemnopsidae in the Upper Jurassic and Lower Cretaceous. British Antarctic Survey Bulletin, 36:33–59.

ZINSMEISTER, W. J. 1979. Biogeographic significance of the late Mesozoic and early Tertiary molluscan faunas of Seymour Island (Antarctic Peninsula) to the final break-up of Gondwanaland, p. 349–355. *In* J. Gray and A. Boucot (eds.), Historical Biogeography, Plate Tectonics, and the Changing Environment. Oregon State University Press, Corvallis.

—— . 1982a. Review of the Upper Cretaceous–Lower Tertiary sequence on Seymour Island, Antarctica. Journal of the Geological Society of London, 139:779–786.

—— . 1982b. Late Cretaceous–early Tertiary molluscan biogeography of the southern circum-Pacific. Journal of Paleontology, 56:86–102.

MANUSCRIPT ACCEPTED BY THE SOCIETY SEPTEMBER 1, 1987

Geological Society of America
Memoir 169
1988

Macruran decapods, and their epibionts, from the López de Bertodano Formation (Upper Cretaceous), Seymour Island, Antarctica

Dale M. Tshudy and Rodney M. Feldmann
Department of Geology, Kent State University, Kent, Ohio 44242

ABSTRACT

Two species of macruran decapods are reported from Upper Cretaceous rocks of the López de Bertodano Formation on Seymour Island, Antarctica. Eighty-eight specimens of decapods were collected from the unit; all are from concretions. *Hoploparia stokesi* has been collected previously on Seymour Island, Cockburn Island, and James Ross Island, in Antarctica. A new species of palinurid lobster is described, *Linuparus macellarii* n. sp. The genus has not been reported previously from Antarctica, and its only published reports in the Southern Hemisphere have been from two localities in the Upper Cretaceous of Africa and Madagascar. Seven specimens of *H. stokesi* are encrusted with epibionts, including the oyster *Pycnodonte* cf. *P. vericosa* (Forbes) and the tube-forming worm *Rotularia* sp.

INTRODUCTION

The upper Campanian–Paleocene López de Bertodano Formation contains abundant macruran decapods. In the 1983–1984 austral summer, one of us (R.M.F.) collected 88 concretions bearing macruran decapods from two sites in the López de Bertodano. These specimens, and their epibionts, constitute the bulk of the decapod material available for study from this unit, but the collection also includes five specimens collected on other expeditions.

To date, only one macruran, *Hoploparia stokesi* (Weller) has been reported from the Upper Cretaceous of Seymour Island, this by Weller (1903) and Del Valle and Rinaldi (1975). All but a one of the 93 specimens in the study collection have been identified as *H. stokesi.* The remainder of the collection contains a single decapod specimen not previously reported from Seymour Island. A new species of palinurid lobster is designated, *Linuparus*

macellarii n. sp. This is the first report of *Linuparus* from Antarctica, and only its third report in the Southern Hemisphere. Its occurrence represents a significant extension of geographic range.

This chapter describes a new species of lobster and reports on occurrences of *Hoploparia stokesi.* Also included is a description of the epibionts that encrusted eight of the specimens, and a discussion of whether their attachment occurred before or after the death of the lobsters. The bulk of the specimens were collected from sites A and B (Fig. 1) in the late Campanian to Paleocene (Macellari, this volume) López de Bertodano Formation. Sites A and B occur in Units 7 and 9b, respectively. Macellari (this volume) describes Units 7 through 10 as consisting of "a thick sedimentary sequence of generally monotonous, sandy siltstone," and notes these "Molluscan Units" as containing abundant and diverse bivalves, gastropods, and ammonites.

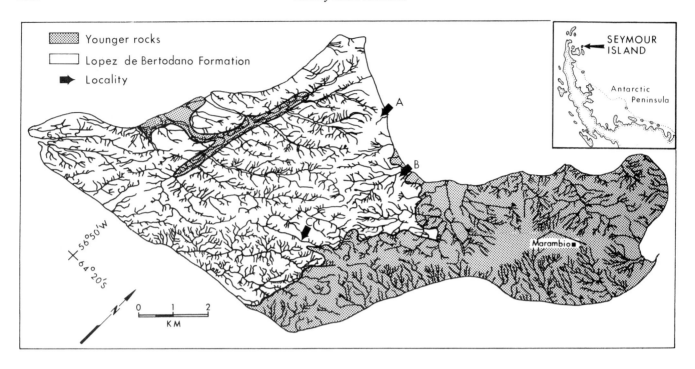

Figure 1. Location map showing collecting Stations A, B, and C in the López de Bertodano Formation.

SYSTEMATIC PALEONTOLOGY

Order DECAPODA Latreille, 1803
Suborder PLEOCYEMATA Burkenroad, 1963
Infraorder ASTACIDEA Latreille, 1803
Family NEPHROPIDAE Dana, 1852
Subfamily HOMARINAE Huxley, 1879

Genus *Hoploparia* McCoy, 1849
Type species. *Astacus longimana* Sowerby, 1826

***Hoploparia stokesi* Weller**
Figures 2.1–3; 3.1–6

Glyphea stokesi Weller, 1903, p. 418–419; Plate 1, Fig. 1.
Hoploparia stokesi Weller. Ball, 1960, p. 6–12, Plate 1, Figs. 1–5,
Plate 3, Figs. 1–2. Del Valle and Rinaldi, 1975, p. 4, Figs. 4–9.
Diagnosis. "Rostrum curved and strongly denticulate. Carapace grooves
strongly marked, anterior of cephalothorax spinose. Abdomen strongly
spinose. Heterochelous; carpus long, about equal in length to the merus"
(Ball, 1960).
Type Material. "Holotype; right side of cephalothorax and incompletely
preserved abdomen and appendages, part of the Stokes Collection of
Antarctic Fossils in the Walker Museum, Chicago; Palaeontological
Collection No. 9705" (Ball, 1960).
Studied Material. Plaster casts of holotype and additional specimens
from the British Museum (Natural History), including In. 51772,
51777A,B, 51778–81, and 60125; five specimens from Ohio State Uni-
versity, including OSU 37-3, K-150, K-179, and two unnumbered

specimens; and 86 specimens from the Feldmann collection, USNM
410841 through 410887 and 410889 through 410928.
Distribution. Snow Hill Island, Lower to Middle Campanian, Holotype,
Stokes Collection, Field Museum of Natural History (Weller, 1903).

James Ross Island, Lower to Middle Campanian, Croft Collection,
British Museum (Natural History) (Ball, 1960).

Southwestern part of Seymour Island, Cockburn Island, Snow Hill
Island Series, Swedish Museum of Natural History (Weller, 1903).

Bodman Point, Seymour Island, Snow Hill Island Series, Del Valle
and Rinaldi Collection, repository unknown (Del Valle and Renaldi,
1975).

Seymour Island, Stations, A, B, and C (Fig. 1) in northeastern
exposure of López de Bertodano Formation, uppermost Cretaceous
Units 7 and 9, Feldmann collection, U.S. National Museum of Natural
History.
Remarks. Pereiopods 2 and 3, shown conjecturally in Ball's (1960)
reconstruction of *H. stokesi,* are indeed chelate, as they are on all astacids
(Fig. 3.3). The relative proportions of the pereiopod segments are accurate as well. The right chela of an extraordinarily large individual, tentatively identified as *H. stokesi,* is 10.0 cm in length. If the identification of
this individual is correct, the maximum size of the species will be significantly emended. The only apparent inaccuracy in Ball's description is in
the reconstruction of the eye, which is drawn as being supported on a
long stalk. A remarkably well-preserved specimen, in which even the
facets on the corneal surface are distinct (Fig. 3.4), demonstrates that the
eye stalk was rather short and the eye well protected by the orbit.

The large collection of *Hoplopoaria stokesi* material reveals greater

Figure 2. *Hoploparia stokesi* (Weller, 1903). USNM 410928. 1, Dorsal view of latex cast of right manus and carpus, both incomplete. 2, Dorsal view of left manus and incomplete pollex. 3, Ventral view of left manus and incomplete pollex. Scale bar = 1.0 cm.

intraspecific variation in chela morphology, particularly with regard to ornamentation, than was previously observed. Most chelae in the study collection are coarsely granulose, as Ball (1960) described. However, a few are far more coarsely ornamented (Figs. 2 and 5). In contrast to Ball's specimens, on which propodus ornamentation coarsens posteriorly and marginally (Ball, p. 11), the more coarsely ornamented chelae bear nodes which become slightly coarser distally, and are coarsest on the broad medial ridge, where they are as large as 4.0 mm in diameter. Longitudinal rows of similar-sized nodes are situated on the upper surface of the chelae with coarse ornamentation. Ornamentation is slightly less ordered on the lower surface. Dentition on the inner margin is nearly as coarse as on the medial ridges. Nodes, and particularly dentition, are directed distally. A large boss is located on the inner margin at the propodus-dactylus boundary. Although the specimens that Ball described appear to have become noticeably heterochelous with age, several of the larger individuals in the study collection are isochelous.

The abdominal pleurae are dimorphic, as first noted by Ball (1960); they are reflected principally in the outline of pleurae 2 through 5. In one type, "Type A," the pleural margin of the second segment has a rounded anterolateral angle, and terminates in a posteriorly directed tooth, which is situated posterior of the posterior margin of the tergite (Fig. 3.5). The pleural margins of somites 3 through 5 terminate in a similar fashion, although the terminal teeth are situated approximately in line with the posterior margin of the tergum. In "Type B," the second pleurae are generally quadrate, the anterior corner is sharply rounded, and the posterior corner is squared but indented toward the anterior (Fig. 3.6). Pleurae 3 through 5 are bilaterally symmetrical, ovate, and terminate approxi-

mately midway between the anterior and posterior tergal margins. Additionally, Type B individuals display a prominent lateral spine above the termination of the pleuron; this spine is absent in Type A forms.

The modern lobster *Homarus gammarus* (Linnaeus) displays a similar dimorphism, though less well-pronounced, which is clearly sexual; Type A and B individuals are female and male, respectively (Ball, 1960). Ball (1960) tentatively extended the relationship between sex and pleuron shape seen in *H. gammarus* to *H. stokesi*, although—as first observed by Taylor (1979, p. 20)—he inadvertently transposed the sexual symbols in his Figure 2. Glaessner (1969, Fig. 265, 4b–c, p. R459) showed essentially the same reconstruction but with the correct sexual symbols. A similar dimorphism has been observed and tentatively proposed as sexual in *Schleuteria menabensis* (Secretan, 1964, Fig. 79) and the Mecochiridae (Taylor, 1979, p. 18).

If the sex of the fossil lobsters could be determined from some feature independent of epimere shape, then verifying or rejecting a correlation between sex and pleuron type would be very straightforward. Unfortunately, primary sex characteristics, such as those seen in the modern lobster, *Homarus americanus* H. Milne-Edwards, which include an accessory copulatory appendage on the male, a longer second and fourth swimmeret, and smaller sternal spines on the female (Templeman, 1944) are rarely, if ever, preserved. The cause of this pleuron dimorphism in fossil lobsters has been proposed only by analogy with *H. gammarus*.

The number of well-preserved individuals in the study collection facilitated qualitative and simple statistical tests of the hypothesis that the dimorphism in *H. stokesi* is sexual. The abdomens of 15 individuals,

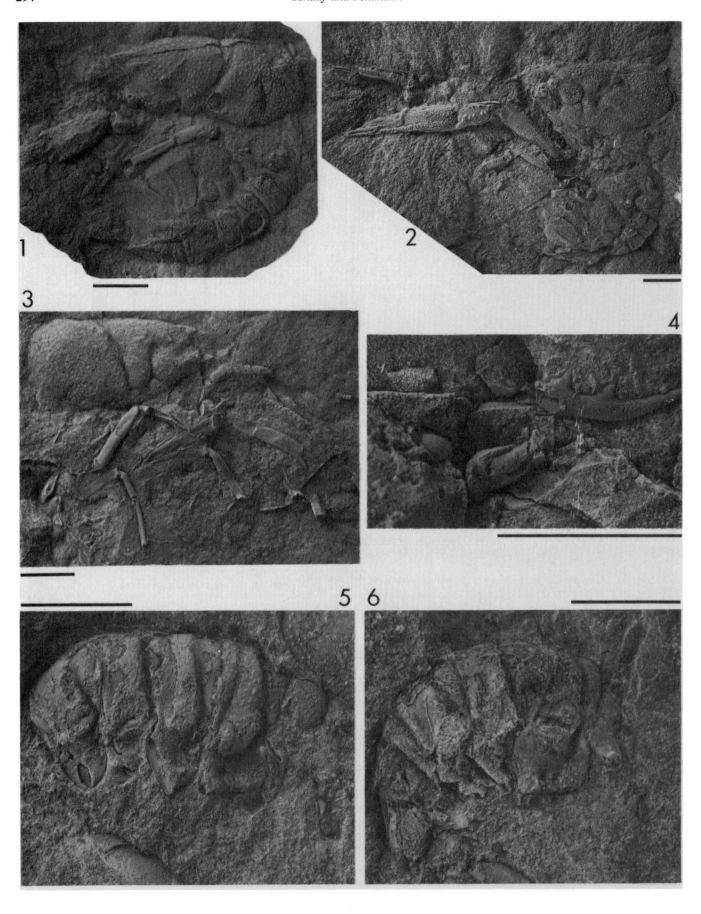

including British Museum (Natural History) In. 60125, have been preserved well enough to be designated as Type A or B, totalling 6 and 9 members, respectively. Because two discrete pleuron morphologies with no intermediate forms were observed, random genetic variation was rejected. The possibility of ontogenetic variation as the cause of the dimorphism was examined by comparing abdomen type to abdomen size, more precisely the length of the second somite. Ontogenetic variation was rejected on the basis of the complete overlap of the size distributions for Types A and B. Sexual dimorphism would explain both the lack of intermediate pleuron morphologies and the fact that both pleuron types are observed in individuals of varying size; therefore, this was considered the most likely interpretation.

Infraorder PALINURA Latreille, 1803
Superfamily PALINUROIDEA Latreille, 1803
Family PALINURIDAE Latreille, 1802

Genus *Linuparus* White, 1847
Type species. *Palinurus trigonis* von Siebold, 1824

Linuparus macellarii n. sp.
Figure 4.1–5

Description. Specimen lacks anterior portion of carapace. Carapace large for genus, almost twice as long (6.1 cm) as wide (3.1 cm) at approximate position of cervical groove. Lateral margins generally parallel, converge slightly near anterior corner, and where intersected by cervical groove, quadrate transversely.

Three longitudinal keels, higher median dorsal keel and two marginal keels, extend length of carapace. Median dorsal keel about 6 mm higher than lateral keels. Cervical groove crosses midline at midpoint on carapace fragment, recurved at midpoint, extends laterally and anteriorly, incising lateral margin in about 25 percent distance from anterior to posterior. Median dorsal keel divided by cervical groove into broader, flatter, cephalic section, and higher, narrower, thoracic section. Gastric carinae lined with about 20 circular to ovate pustules, pustules ranging from less than 0.5 to nearly 3.0 mm maximum diameter, arranged in two curved rows bilaterally symmetrical about lyrate depression with axial striation. Maximum distance between curving rows of pustules about 15 percent maximum width of carapace. Thoracic section of median dorsal keel densely pustulose, apex of keel possibly spinose but poorly preserved, pustules less than 0.5 to slightly more than 1.5 mm in diameter. Marginal keels rather uniformly shaped throughout lengths, highest around midpoint, densely pustulose, pustules coarser than on thoracic median keel, approach 3.0 mm diameter. Steep outer sides of marginal keels constitute lateral margins of carapace. Sides densely pustulose and generally coarser than elsewhere on carapace. Pustules directed slightly anteriorly and on distinctly coarser and denser posteriorad cervical groove.

Sternum subtriangular, 2.8 cm long, and 2.1 cm at widest point near posterior margin, four ventral axial nodes, subpyramidal, 0.3 to 0.5 cm along elongate longitudinal axes. Anteriormost node protrudes sharply from sternum. Nodes on sternal elements 1 through 3 elongate,

increasing in size posteriorly; node on element 4 smaller than former, nearly equidimensional; sternal element 5 without axial node. Lateral margins narrow, raised as high as medial nodes, sutured and fluted in four places, at approximate level of basal articulation. Genital openings not observed. Lateral margins appear as slightly concave-outward segments. Basal segments of pereiopods 1 through 4 preserved, attached slightly anteriorad of fluted sutures. Ischia wider anteriorly. Ischium nearly 1½ times wider than merus on right fourth pereiopod. Remainder of thoracic appendages unknown.

Abdomen cylindrical, vaulted, some dorsoventral flattening; lacking first somite, second somite incomplete dorsally. Abdomen tapers gradually to sixth somite. Somite width: 3, 2.7 cm; 4, 2.3 cm; 5, 2.1 cm; and 6, 2.6 cm. Somite lengths measured along dorsal midline: 3, 1.2 cm; 4, 1.0 cm; 5, 1.0 cm; and 6, 1.40 cm. Abdomen bears medial dorsal pustulose keel. Terga and pleurae, except for pleural spines, sparsely decorated with pustules 1 to 3 cm in diameter. Sixth somite bears coarse longitudinal ridges and furrows. Telson incomplete. Tergal-pleural junction on somites 2 through 5 arcuate, curving dorsally anteriorally. Pleurae with proximal, variably shaped polygonal regions, and distal dentate regions. Pleural border smooth-surfaced, dentate, inset relative to pleural field, which is swollen, irregularly surfaced. Pleurae bear two spines each; posterior pleural spines reduced on somites 2 through 5, others indeterminate. Longest anterior spine, on fifth somite, approaches 1.0 cm long. Posterior spine on same pleuron about 0.3 cm.

Etymology. Name honors Dr. Carlos E. Macellari, Earth Sciences and Resources Institute, University of South Carolina, Columbia, South Carolina, discoverer of collecting site, in 1983.

Age. Specimen collected from Unit 9b (Macellari, 1986) of late Campanian to Paleocene age López de Bertodano Formation. Top of Unit 9b shows marked decrease in macrofaunal diversity and total disappearance of the ammonites, therefore is considered to represent Cretaceous/Tertiary boundary (Macellari, this volume). Holotype thus suggested to be of latest Cretaceous age.

Repository. Holotype and only specimen, USNM 410888, deposited in U.S. National Museum of Natural History, Washington, D.C.

Type locality. Holotype collected by Feldmann in 1983 from Seymour Island Station B (Fig. 1), in northeastern sector of exposed López de Bertodano Formation.

Remarks. This species is referable to the genus *Linuparus* on the basis of its flattened carapace, traversed longitudinally by three keels (Woods, 1925-1931, p. 26). The genus has been subdivided by Mertin (1941, p. 215), and later by Remy (1954, *in* Secretan, 1964) into four subgenera, based on the texture of the carapace carinae, the development of the orbital spines, and the morphology of the abdominal pleurae. *Linuparus macellarii* is best referred to the subgenus *Podocratus* Schlüter because of the coarsely ornamented carapace keels and the dentate abdominal pleurae. However, the definition of *Podocratus* states that the carapace keels and carinae are spinose and that the abdominal pleurae are trifid. On *L. macellarii* the carapace keels are not spinose, but are pustulose, and the third, or posteriorad, pleural spine is absent or greatly reduced. Thus, placement in the subgenus *Podocratus* cannot be made with certainty.

Linuparis grimmeri and *L. watkinsi* are both from the lower Turonian of Texas (Stenzel, 1944, p. 408); *L. dentatus* is from the upper Cenomanian of France (van Straelen, 1936, p. 4). They are most similar to *L. macellarii* in carapace morphology in that these species bear three coarsely ornamented, longitudinal, carapace keels, and are divided transversely by a cervical groove that is recurved about the median keel. The cephalic and thoracic regions are nearly equal in length on *L. dentatus,* but the thoracic region is longer than the cephalic region on *L. grimmeri* and *L. watkinsi* (Stenzel, 1945, Plate 34; van Straelen, 1936, Plate 1). Further, the lateral margins on both species curve inward where incised by the cervical groove and at the anterior corner (Stenzel, 1945, Plate 34). The thorax of *L. japonicus* from the Senonian of Japan (Nagao, 1931, Plate 14) is similar to the new species with regard to the shape, but

---

Figure 3. *Hoploparia stokesi* (Weller, 1903). 1, USNM 410850, latex cast showing carapace spinosity, pleurae, and uropods. 2, USNM 410883, latex cast showing carapace sculpture, chelipeds 3, USNM 410843, showing subchelate pereiopods. 4, USNM 410851, showing position of short-stalked eye relative to carapace and rostrum. 5, USNM 410842, showing Type A pleurae expressed on abdominal somites 1-5. 6, USNM 410848, showing Type B pleurae expressed on abdominal somites 1-6. Scale bar = 1.0 cm.

Figure 4. *Linuparus macellarii* n. sp., USNM 410888 4, Dorsal view of cephalothorax. 3, Left lateral view of cephalothorax, emphasizing coarseness of pustules. 5, Sternum and six pereiopods. Note singular axial nodes. 1, Dorsal view of abdomen, showing segments 2 (incomplete), 3, 4, 5, and telson. 2, Right lateral view of abdomen, showing two-spined abdominal pleura. Scale bar = 1.0 cm.

not ornamentation, of the lateral and median keels, yet it is distinctly different anteriorly. On *L. japonicus* and on *L. vancouverensis,* from the Upper Cretaceous of Vancouver (Whiteaves, 1903, Plate 40), the recurving of the cervical groove is less obvious, and the groove is far broader than that of *L. macellarii.* Because the carapace of the type specimen of *L. macellarii* is incomplete anteriorally and the apex of the median dorsal keel is poorly preserved, a comparison of the medial and orbital spines of the new species to other linuparids is not possible.

One distinct difference between these species lies in the texture of the carapace integument. The carapaces of *L. watkinsi* and especially *L. grimmeri* (Stenzel, 1945, Plate 34), *L. dentatus* (van Straelen, 1936, Plate 1), and *L. japonicus* (Nagao, 1931, Plate 14), are much smoother than that of *L. macellarii.* The coarsely ornamented integument of *L. tuberculosa* from the Aptian of England (Reed, 1911, Plate 7) is by far the most similar texturally to *L. macellarii,* although the "tubercles" on *L. tuberculosa* are, "sharply pointed and conical" (Reed, 1911, p. 119), whereas the "pustules" on *L. macellarii* are rounded. Except for the similarity in integument texture, the carapace morphologies of the two species are distinctly different. The ornamentation on the gastric carinae of *L. macellarii* is similar to that of *L. grimmeri* (Stenzel, 1945, Plate 34), but pustules on the gastric carinae are greater in number than those on *L. dentatus* (van Straelen, 1936, Plate 1), and *L. watkinsi* (Stenzel, 1945, Plate 34) and *L. japonicus* (Nagao, 1931, Plate 14).

A comparison of the sternal and abdominal elements of *L. macellarii* reveals features quite distinct from those of any other linuparid. The two spines on the abdominal pleurae on the new species are very different from the trifid pleurae of *L. watkinsi* (Stenzel, 1945, Plate 34), *L. grimmeri,* and *L. dentatus* (Secretan, 1964, p. 121), or the nonspinose pleurae of *L. japonicus* (Nagao, 1931, p. 213) and *L. tuberculatus* (Reed, 1911, p. 120). In the latter two species the pleurae are minutely serrated posteriorly.

The sternum of *L. macellarii* is equally distinctive in that it bears a prominent axial node on each of the sternal elements. The sternum is smooth on *L. japonicus* (Nagao, 1931, p. 213), and bears axial nodes only on the anterior tip and on the fifth sternal element of *L. grimmeri* and *L. watkinsi* (Stenzel, 1945, p. 407, 409).

The most diagnostic features of *L. macellarii* are the pustulose nature of the carapace, the recurved cervical groove, the single axial nodes on the sternal elements that are absent, or double, on other linuparids, and the nature of the marginal spines on the abdominal pleurae. Pleurae with two spines are unique to the new species; other linuparids have either rounded or trifid pleural margins.

Distribution. Linuparus appeared in the Neocomian (lower Cretaceous) in the Northern Hemisphere and flourished there throughout the remainder of the Cretaceous, with scattered occurrences continuing into the Eocene (Woodward, 1900; Whiteaves, 1903; Woods, 1925-1931; Nagao, 1931; Rathbun, 1935; Mertin, 1941; Stenzel, 1945; Secretan, 1964; Feldmann, 1981; Bishop and Williams, 1986). The genus is not known from the Southern Hemisphere in rocks older than the Turonian, where it appears in Niger, and is subsequently reported from only two Southern Hemisphere localities: the Campanian of Madagascar (Secretan, 1964), and this occurrence in the Late Cretaceous of Seymour Island. An additional, as yet unpublished, observation of the genus has been made in the Eocene of New Zealand.

EPIBIONTS ON LOBSTERS

Rotulariids

Five specimens of *Hoploparia stokesi,* USNM 410903, 410912, 410917, 410918, and one specimen from the Ohio State University collection, have served as the substratum upon which what appear to be juvenile rotulariid worms have attached (Fig. 5.1–6). All but the Ohio State specimen were collected from Station B. The fossil lobster most densely encrusted, USNM 410903, has broken away from a concretion, leaving approximately 30 attached worm tubes preserved as steinkerns embedded in the mold of the exterior of the lobster. The structures are distributed on the mold singularly and in groups of up to five. The individuals within these groups are often nearly contiguous. The most closely spaced are 0.2 mm apart, or about 15 percent of their diameter.

The attachment surface of the tube is generally coiled dextrally in one plane, though the plane of coiling may change to conform to irregular attachment surfaces. The outline of the coil is typically circular, or nearly so. The tightness of coiling varies between specimens and within individual specimens. In those specimens not tightly coiled, coiling is loosest adapically. In at least one specimen the outer whorl is separated from the previous whorl by approximately the thickness of the previous whorl. In other specimens, outer whorls are contiguous with adjacent inner whorls. The largest individuals are composed of three tubular whorls gradually increasing in diameter. Because the tube increases in diameter adapically, the same whorls that form a plane on the "attachment side" appear to spiral up and away from the plane of coiling on the umbilical side. Whorls seldom overlap. Maximum tube width and height are approximately 0.4 mm. The largest coils are commonly 0.5 mm high and 1.3 mm in diameter. These figures seem to represent natural maxima.

The apex of the coil is a curving, pointed termination of the tube. The aperture of the tube is broken on all individuals. The periphery varies from nearly flat to slightly convex, and bears transverse growth lines of variable coarseness. A nodose carina is evident on the growth side of some individuals, particularly nearest the apex. The sides of the tube appear finely pitted. Tube walls appear structureless in thin section at x200.

These specimens are morphologically similar to the serpulid taxon that Ball (1960, p. 25) described from James Ross Island, *Rotularia dorsolaevis* Ball, in which the tube is discoidal, spirally coiled, and commonly composed of three whorls. However, even the most minute specimen of *R. dorsolaevis,* the smallest serpulid described by Ball, is more than twice the diameter of the largest studied specimen. *Rotularia dorsolaevis* ranges in diameter from 3 to 13 mm, and varies in height from 1 to 4 mm. This disparity in size would suggest that these specimens do not belong to *R. dorsolaevis,* presuming the specimens from the López de Bertodano were mature forms.

Although the mode of life of *Rotularia* is uncertain because the genus has no recent counterparts for comparison (Macellari, 1984), the observations by previous workers on serpulids suggest that the attached coils are not mature forms; rather, they represent early stages in the ontogeny of *R. dorsolaevis* or another species. Both Wilckens (1922, *in* Macellari, 1984) and Ball (1960) suggested that individual rotulariids were initially attached but were subsequently free-living benthos. According to Ball (p. 27), writing on *R. dorsolaevis,* "In no instance is the initial apical part of

Figure 5. *Rotularia* sp. 6, *Rotularia* sp. (Ball, 1960) encrusting carpus and chela of *Hoploparia stokesi*, USNM 410903. Scale bar = 1.0 cm. 1, Scanning electron microscope (SEM) photograph of planar attachment surface of *Rotularia* sp. 2, SEM photograph of *Rotularia* sp. attachment surface conforming to a convex spot on *Hoploparia stokesi* integument. 4, SEM photograph of umbilical side of *Rotularia* sp. 3, SEM photograph of oblique view of umbilical side of *Rotularia* sp. All SEM photos x28. 5, Longitudinal thin section showing apical termination.

the spiral preserved and it is probable that, as with *R. callosa* (Stoliczka), it consisted of a fragile tube cemented to some suitable object above the muds of the sea-floor. With increased growth, the weight of the tube became too great for the apical portion to support and the spiral portion became detached, assuming a random position on the sea-bed." Examples of the broken apices and circular, apical voids found in rotulariids were illustrated by Ball (1960, Plate 3, Figs. 4, 7) and Macellari (1984, Fig. 8). All apertures on the tiny coils and all apices on the mature rotulariids are broken. The tiny juvenile coils were attached to lobster integument, whereas the mature rotularids probably "assumed a free-living existence on the sea-floor" (Ball, 1960, p. 23). These observations suggest that the attached coils may represent the initial growth lacking on mature specimens. The two types of coils may represent stages in the ontogeny of a single taxon.

If the tiny, attached coils are indeed ontogenetically related to a free-living form, it is certain that any subsequent coils did not grow in the same plane as the initial coils, even though the initial coils are similar in size and shape to the apical voids in *R. dorsolaevis*. There is not enough room between those coils that are found in gregarious masses on the lobster host to have allowed the growth of subsequent whorls. Therefore, some other growth orientation for the subsequent forms must have developed if these two forms are, indeed, ontogenetic stages of a single taxon.

Ball (1960, p. 19) observed that three juvenile specimens of *R. callossa*—a species much larger than *R. dorsolaevis*—showed remnants of an apical growth, "a short, straight, thin-walled tube continuing tangentially and slightly upward from the first whorl." Macellari (1984, p. 1103) stated that some of his specimens had "a large apical tube," but his illustrations indicate that he meant "apertural" instead of "apical."

From Ball's description of an apical tube, and the preceding discussion, it seems possible that the apical tube served as a stalk connecting the nuclear whorls on the lobster chela and the free-

Figure 6. *Pycnodonte vesiculosa* (Forbes) 1, *P.* cf. *P. vesiculosa* encrusting carapace and chela of *Hoploploparis stokesi*, USNM 410855. 2, Enlargement of manus. 3, *Pycnodonte* cf. *P. vesiculosa* encrusting chela of *Hoploploparis stokesi*, OSU 82-K-179. Scale bars = 1.0 cm.

living adult forms. The outer whorl of the attached coil probably grew upward, away from the host, in a stalk-like fashion. At some distance from the attachment surface, this stalk began to coil, giving rise to a larger adult coil. When the adult coil became too heavy or hydrodynamically too large for the connecting stalk to support, it separated from the last whorl of the attached coil, and then assumed a random position on the sea floor, as described by Ball. With subsequent movement on the sea floor, the fragile connecting stalk usually also broke off of the adult coil. This hypothetical ontogeny could explain the broken apertures on the tiny attached coils and the broken apices on the larger mature coils. This must remain speculative until the attachment coil, connecting stalk, and adult coil can be found together. *Rotularia dorsolaevis,* the species most size-compatible with the atttached coils in question, was reported, not from Seymour Island, but from neighboring James Ross Island. The smallest rotulariid reported from Seymour Island, *R. (Austrorotularia) fallax* (Wilck-

ens), has an average diameter of 18.45 mm (Macellari, 1984, p. 1108), and the diameter of the tube at the broken apex seems far too large to have been connected to the tiny tube in question.

The identification of these tiny forms to the species level is not possible until more complete evidence is found. Examinations of thin sections by Ball and by the authors indicate that tube wall structure, a useful criterion in taxonomic classification (Ball, 1960), may change through the ontogenetic sequence, becoming more complex with growth. For this reason, the comparison of tube wall structure and coil morphology between attached and adult coils is insufficient proof for species-level classification of these small tests.

Oysters

Juvenile oysters, probably referrable to *Pycnodonte* cf. *P. vesiculosa* (Forbes) (Macellari, 1986), are found encrusting three specimens of *Hoploparia stokesi* (Fig. 6.1 through 6.3). Two specimens, USNM 410855 and 410870, were collected at Station A. The other, which also served as the substratum for rotulariids, is an unnumbered specimen in the collections at Ohio State University. All of the oysters were preserved intact. The maximum diameter of the largest individual is 6.0 mm. Each individual is nearly circular in outline, the beaks protrude slightly from the outline, and the surface of the oysters mimics the attachment surface.

Timing of epibiont attachment

Both the oysters and the rotulariids are found, among other places, on the chelae, including the dactyli. On one lobster an oyster is fixed to the second abdominal somite, overlapping the articulating ridge. The positions of these epibionts on the lobsters raise the question of whether they attached to live or dead lobsters or to exuviae. Taylor (1979, p. 7), observed serpulid coils on the chelae of *Glyphea* and *Paleastacus* and wrote, "It is difficult to determine whether the serpulids (which are particularly common epizoans on live Macrura and Brachyura) encrusted the fossil decapods when the hosts were alive or dead." There is evidence for both conditions.

It has been shown that modern lobsters, while normally well-groomed animals (A.B. Williams, 1986, personal communication), are occasionally host to epibionts, such as tube-forming annelids (Herrick, 1911; Taylor, 1979) and mussels (Herrick, 1911; Stewart, *in* Cobb and Phillips, 1980). Infestations typically occur when the lobster is compelled for any reason to lead a sluggish life, particularly when confined in enclosures in captivity. Herrick added that the lobster rids itself of these parasites with each successive molt, and that older lobsters, in which the molting periods have become infrequent, are the worst sufferers from epibionts. The seven encrusted individuals studied may serve to reinforce Herrick's observation. All are adults; the three oyster-bearing individuals are large and presumably old. In light of these observations on modern lobsters, the attachment of the epibionts to live lobsters seems plausible. The sites of attachment, however, suggest postmortem attachment by both the oysters and the worms.

Whereas oysters encrusting the carapace would probably cause the host little problem, oysters on the fingers of the chelae and the abdominal somites would logically inhibit its mobility. It is doubtful that the lobsters would have tolerated this.

The sites of rotulariid attachment are also suggestive of postmortem attachment. The coils on the infested chela are minute, but as with the oysters, it is doubtful that the typically well-groomed lobster would have allowed such development. Furthermore, the broken apertures indicate that the worms developed upward apertural projections following completion of the attached outer coil as discussed above. It is very doubtful that this fragile stalk could have developed on the chela of a live lobster. This argument must be taken with caution, however. It is possible that the worm successfully developed the observed planispiral coil on a live lobster, but was unsuccessful in developing the stalk.

Thus the positions of the oysters and rotulariids suggest postmortem attachment. There are, however, two remaining lines of evidence suggesting that the epibionts may have attached during the life of the lobster. First, two of the oyster-bearing lobsters are remarkably well-preserved, indicating rapid burial. Preservation is so extraordinary that it seems doubtful that sufficient time would have elapsed between death and burial, or during the burial of exuviae, to have permitted development of the 6-mm individuals. Second, only 7 of the 93 specimens studied bear epibionts. This suggests that attachment occurred during the life of the lobster, due to some peculiarity in its behavior; with postmortem attachment, we might have expected that encrustation of carcasses lying on the sea floor would have been more universal.

ACKNOWLEDGMENTS

Michael O. Woodburne, Department of Geology, University of California at Riverside, and Barry B. Miller, Department of Geology, Kent State University, read portions of the manuscript and offered numerous helpful suggestions. Gale Bishop and Murray Copeland provided valuable reviews. Several specimens of *H. stokesi* were loaned by William J. Zinsmeister of Purdue University, by Carlos E. Macellari, and by Tonianne Pezzetti of Ohio State University, who collected specimens on previous expeditions to Seymour Island. Field work (of R.M.F.) was supported through a National Science Foundation grant to Zinsmeister. The laboratory work was supported by National Science Foundation Grant DPP-8411842 (to R.M.F.). Contribution 338, Department of Geology, Kent State University.

REFERENCES

BALL, H. W. 1960. Upper Cretaceous Decapoda and Serpulidae from James Ross Island, Graham Land. Falkland Islands Dependencies Survey Scientific Reports, 24: 1–30.

BISHOP, G. A., AND A. B. WILLIAMS. 1986. The fossil lobster *Linuparus canadensis,* Carlile Shale (Cretaceous), Black Hills. National Geographic Research, 2(3):372–387.

COBB, J. S., AND PHILLIPS, B. F. (eds.), 1980. The Biology and Management of Lobsters: Physiology and Behavior. Academic Press, New York, 463 p.

DEL VALLE, R. A., AND RINALDI, C. A. 1975. Sobre la presencia de *Hoploparia stokesi* (Weller) en las "Snow Hill Island Series," de la Isla Vicecomodora Marambio, Antartida. Contribucion Del Instituto Antartico Argentino, 190:1–19.

FELDMANN, R. M. 1981. Paleobiogeography of North American lobsters and shrimps (Crustacea, Decapoda). Geobios, 14:449–468.

HERRICK, F. H. 1911. Natural history of the American lobster. Bulletin of the Bureau of Fisheries, 29:1–408.

MACELLARI, C. E. 1984. Revision of serpulids of the genus *Rotularia* (Annelida) at Seymour Island (Antarctic Peninsula) and their value in stratigraphy. Journal of Paleontology, 58(4):1099–1116.

——. 1986. Late Campanian–Maastrichtian ammonite fauna from Seymour Island (Antarctic Peninsula). Paleontological Society Memoir, 18:1–55.

MERTIN, H. 1941. Decapode krebse aus dem subhercynen und Braunschweiger Emscher und Untersenon. Nova Acta Leopold, 10:150–264.

NAGAO, T. 1931. Two new decapod species from the Upper Cretaceous Deposits of Hokkaido, Japan. Journal of Faculty of Science, Hokkaido Imperial University Series 4, 1(2):1–214.

RATHBUN, M. J. 1935. Fossil Crustacea of the Atlantic and Gulf Coastal Plain, p. 1–160. Geological Society of America Special Paper 2, Boulder, CO.

REED, F.R.C. 1911. New Crustacea from the Lower Greensand of the Isle of Wight. Geological Magazine, 8:115–120.

SECRETAN, S. 1964. Les crustacés décapodes du Jurassique Supérieur et du Crétacé de Madagascar. Mémoires du Muséum National d'Histoire Naturelle, 14:1–223.

STENZEL, H. B. 1945. Decapod crustaceans from the Cretaceous of Texas. University of Texas Contributions to Geology, 1944, 4401:401–477.

TAYLOR, B. J. 1979. Macrurous Decapoda from the Lower Cretaceous of southeastern Alexander Island. British Antarctic Survey Scientific Reports, 81:1–39.

TEMPLEMAN, W. 1944. Sexual dimorphism in the lobster (*Homarus americanus*). Journal of the Fisheries Board of Canada, 6:228–232.

VAN STRAELEN, V. 1936. Crustacés décapodes nouveaux ou peu connus de l'époque Cretacique. Bulletin du Museé royal d'Histoire naturelle de Belgique, 12(45):1–49.

WATERMAN, T. H. (ed.). 1960. The Physiology of Crustacea: Metabolism and Growth. Academic Press, New York, 670 p.

WELLER, S. 1903. The Stokes collection of antarctic fossils. Journal of Geology, 11:413–419.

WHITEAVES, J. F. 1903. Mesozoic Fossils, Volume 1, Part 5: On Some Additional Fossils from the Vancouver Cretaceous, with a Revised List of the Species Therefrom. Geological Society of Canada, Ottawa.

WOODS, H. 1925-1931. A Monograph of the Fossil Macrurous Crustacea of England. Paleontographical Society, London, 122 p.

WOODWARD, H. 1900. Further notes on podophthalmous crustaceans from the Upper Cretaceous Formation of British Columbia, etc. Geological Magazine, 7:392–393.

MANUSCRIPT ACCEPTED BY THE SOCIETY SEPTEMBER 1, 1987

Geological Society of America
Memoir 169
1988

Geometry and stratification of uppermost Cretaceous and Paleogene units on Seymour Island, northern Antarctic Peninsula

Peter M. Sadler
Department of Earth Sciences, University of California at Riverside, Riverside, California 92521

ABSTRACT

The Cretaceous-Paleogene sequence on Seymour Island has been mapped in terms of four previously named units: the López de Bertodano Formation (Cretaceous-Paleocene), the Sobral Formation (Paleocene), the Cross Valley Formation (late Paleocene), and the La Meseta Formation (Eocene). The basal surfaces of the Sobral and Cross Valley Formations include portions with clear evidence of erosional unconformity. The base of the La Meseta Formation laps onto a steep buttress unconformity at both of its exposed margins.

The scale of lenticularity of the units increases upward throughout the sequence. The López de Bertodano Formation and most of the Sobral Formation are built of essentially tabular units, and have been effectively characterized by isolated measured sections. The Cross Valley Formation is a channel-filling lens. The La Meseta Formation has a complex, asymmetric, trough-like form. It is built of large-scale lenses characterized by different shelly facies that do not all extend across the full width of the trough. Consequently, mapping between measured sections is essential to establish the lateral relationships. Even then, the geometric relationships do not identify time lines; establishing the real age relationships will require very carefully located faunas. The observed relationships between the length and height of lenticular units can be used to estimate the expected vertical error when faunas are projected into a composite section.

The whole sequence is essentially a gently dipping homocline. The youngest units have slightly gentler dips, indicating progressive or repeated tilting. But a larger component of the variance in bedding attitude is contributed by steeper dips in the sedimentary filling of channels and slump scars. The regional tilt, the development of growth faults, facies changes, and the orientation of erosional surfaces have a symmetrical arrangement from which a common cause is inferred. The best estimates of the timing of the major unconformities and erosional episodes are only partially compatible with current models of global sea-level change.

INTRODUCTION

Seymour Island is a small, relatively ice-free island in the James Ross group, located east of the northern end of the Antarctic Peninsula (Fig. 1). The very fossiliferous sedimentary sequence exposed there offers a unique opportunity for studying late Cretaceous and Paleogene faunal changes at high latitude. The collection of its rich, well-preserved faunas and the imposition of stratigraphic nomenclature began in rather haphazard fashion at the turn of the century.

Some of the most recent, and more systematic, faunal analyses examine the Cretaceous/Paleogene extinction patterns in a wide variety of taxa. Others are beginning to document high-latitude heterochroneity in the Paleogene faunas; that is, the Seymour Island faunas have been shown to contain an admixture of taxa that are present in significantly younger faunas at lower latitudes. Such studies clearly demand the best possible lithostratigraphic framework and some assessment of the completeness of the sedimentary sequences. In the absence of a detailed topographic base map, however, the lithostratigraphic nomenclature for Seymour Island developed around a few measured sections. The suitability of isolated measured sections as a basis for lithostratig-

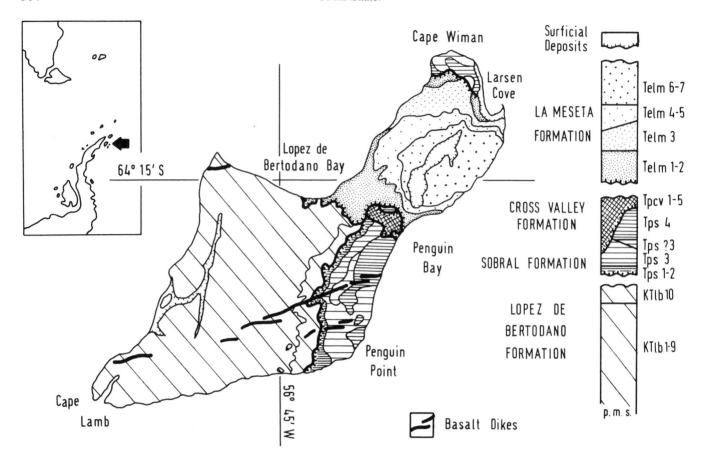

Figure 1. Location map of major stratigraphic units on Seymour Island. Heavy lines indicate basalt dikes.

raphy and for the compositing of faunal studies depends on the degree to which units are lenticular.

The use of four lithostratigraphic units, named López de Bertodano (Rinaldi et al., 1978), Sobral (Rinaldi et al., 1978), Cross Valley (Trautman, 1976), and La Meseta (Elliot and Trautman, 1977), stabilized in the late 1970s. The names appeared regularly in literature of the early 1980s, but the variety of accompanying maps and sections showed that the boundaries of the units remained somewhat controversial. More extensive field mapping was necessary to determine the unit geometries and stabilize the stratigraphic subdivision.

The Sobral Formation was known to overlie the López de Bertodano Formation. However, all other exposed boundaries were typically shown as faults and were, therefore, not used as evidence of the original stratigraphic sequence. Nevertheless, the relative ages of the four named units had been clarified in the late 1970s, using biostratigraphic evidence. The López de Bertodano sections are characterized by Upper Cretaceous molluscs, but Zinsmeister (1982) recognized that faunal ranges, published by Argentinian geologists between 1975 and 1978, anticipated the more recent interpretation that the Cretaceous-Tertiary boundary

lies near the top of that unit. The Cross Valley type section included sites that had produced Paleocene microflora (Cranwell, 1959; Hall, 1977), whereas the microfloral and macrofaunal evidence (summarized by Zinsmeister, 1982) had indicated a late Eocene age for beds later assigned the name La Meseta Formation.

Detailed mapping and refinement of the lithostratigraphic units became feasible in the austral summer of 1983–1984 when the first large-scale topographic map had been completed. Map units defined on the basis of their sedimentary character and their shelly macrofauna allowed detailed subdivision within the framework of the existing nomenclature. The outcrop pattern and boundary relationships of these map units show an up-section trend from dominantly tabular to dominantly lenticular geometry. Thus, the existing stratigraphic type sections are least representative in the youngest parts of the sequence.

The account of the lithostratigraphy that follows will concentrate on the nature, scale, and continuity of stratification; it identifies hiatuses, condensed parts of the section, and mixed intervals that represent limiting factors in the resolution of biostratigraphic studies. The short, unpredictable field seasons in Antarctica demanded an unusual concentration on the task of completing the map. Units were characterized to an extent that allowed them to be consistently recognized and traced in detail around the island. Time and priorities did not permit the exhaustive description of sedimentological details upon which a respectable interpretation of depositional environments should be based.

LÓPEZ DE BERTODANO FORMATION

The oldest part of the sedimentary sequence on Seymour Island is a homoclinal stack of 10, apparently tabular units (Klb1 to K/Tlb10) that constitute the López de Bertodano Formation (Macellari, 1984). The upper limit of ammonites and the micro-fossil biostratigraphy indicate that the base of the Paleogene may lie within this formation, above the base of unit 10. That unit and the upper part of unit 9, the *Pycnodonte* beds of Macellari (1984), were mapped in detail and confirmed to be essentially tabular. The mappability of the units depends on subtle benches in the surface topography, color differences enhanced by weathering, and lag concentrates of the macrofauna. These secondary enhancements need to be carefully reconciled with the rather different appearance of the formation in fresh vertical exposures.

Very poorly indurated, gray, silty sandstones with intense bioturbation characterize most of the top two units of the López de Bertodano Formation. Several generations of cross-cutting burrows have effectively homogenized much of the sediment. In favorable coastal exposures, however, less completely disrupted patches can be identified that preserve some original alternation of mud and fine sand layers. Relict parallel lamination and flaser bedding represent the depositional structures. Calcareous nodules and the shells of molluscs are typically very sparsely dispersed in these sediments. More rarely, there are diffuse bands, marked by a slight concentration of shell debris, or layers defined by nodules. The shelly layers run parallel to the relict bedding, but oblique to some nodule layers. On the gentle inland slopes, both shells and nodules are misleadingly concentrated by deflation into conspicuous bands and pockets of surficial lag gravel. The composition of these concentrates shows characteristic up-section changes that are persistent across the outcrop. Away from fresh, steep sections the intensity of bioturbation is best indicated by the trace fossils preserved in the concretions.

Hard ground surfaces and thin, glauconitic horizons record periods of slower or interrupted sediment accumulation, and provide more reliable evidence of the large-scale tabular stratification that is apparent inland. The hard grounds are very sharply defined surfaces at the top of thin zones, in which the abundance of infaunal shells increases geometrically. They resemble Kidwell's (1986) "Type I" shell beds, and can be interpreted in terms of a gradual shutdown of sediment supply. For a few centimeters above the hard-ground surface, the sediment may be unusually calcareous, but shelly fossils are very sparse. This better cemented zone above the hard ground and the mechanical contrast at the hard ground surface often produce a subtle bench on the gentle inland slopes.

The glauconite-rich horizons are typically thin silty sands with sharp bases and a weak internal grading. They can be mapped by tracing the accompanying wider zone of variegated yellow and brown limonitic staining of the surficial sediment. The concentration of glauconite suggests some condensation of the section in these intervals.

The boundary between units 9 (Klb9) and 10 (KTlb10) is marked by a glauconite horizon that coincides with a particularly distinctive change in the shelly facies. The silty gray sands beneath are characterized by very rich lag concentrates that contain numerous spines of cidarid echinoids and rotularian worm tubes. Ammonites are abundant and bones of marine reptiles are not uncommon. The overlying silty sands and sandstones have a rubbly surface lag of concretions that is almost totally devoid of echinoderm spines, worm tubes, and ammonites. These overlying sediments have a brownish hue, which is often distinct from those in unit 9, and the shells in the surface lag are dominated by the pelecypod *Lahilla*. At the southeast end of the outcrop, comparable glauconite beds are not seen within unit 10. The northwest end of the outcrop is not quite so simple, since unit 10 there includes limonitic horizons. Furthermore, Huber (this volume, Fig. 5) suggests that some fossils, which are elsewhere indicative of unit 10, were recovered *below* the distinctive uppermost glauconite at the northern end of its outcrop.

Unit 10 retains primarily the large and thick-shelled elements of the macrofauna of unit 9, as might be expected as a result of dissolution. Huber (this volume) attributes the loss of planktonic foraminifera to dissolution, but the loss of the abundant cidarid echinoid spines—which have a very stable, single-crystal, calcite microstructure and quite robust shapes—is unexpected. Macellari's (1984, this volume) sandstone analyses record an increase in pumice and glass shards at, or just below, this level.

A strong case has been made that the macrofaunal and microfaunal changes at the base of unit 10 also mark the base of the Paleogene (Macellari, 1984; Askin, this volume; Huber, this volume; see Huber, 1986, for a summary of other possibilities). Plate 1 marks several sites where ammonites were found above this boundary glauconite. One lies near the north end of the outcrop; the others are toward the south end of the island. The ammonites extend at least 10 m above the base of unit 10. They are quite sparse relative to their abundance in unit 9, but include several genera, a variety of sizes, and a range of quality of preservation. Two large individuals were found in place, in vertical faces; one of these projected from the boundary glauconite into the beds above, the other was wholly above the glauconite. Most individuals were collected at the surface; the matrix adhering to these ammonites did not obviously suggest that they were out of place. Two specimens were hammered out of concretions that gave no outward evidence of their fossil content. These circumstances make the case against the possibility that the ammonites were dropped up-section by careless collectors.

It should also be remembered that the majority of the macro-fossils from Seymour Island have been similarly collected from surface lags. Along the outcrop of unit 10 the ground generally slopes up toward the base of the Sobral; the only slopes on which lag materials might be washed, or roll, up-section are dip slopes that are nearly parallel to bedding and consequently traverse very little stratigraphic distance. Even on such slopes, the relatively small rotularians and echinoid spines retain their pattern of much greater abundance below the glauconite layer. The overwhelming effect of the concentration of macrofossils during the winnowing and washing out of the finer sediment matrix will be to shift faunas down-section.

Where microfossils have been recovered from these ammonite fillings, or their host concretions, in unit 10, they are Cretaceous (R. A. Askin and B. T. Huber, written communications). I am not aware of any other floras or faunas from unit 10 that have been interpreted as Cretaceous, although for some groups the existence of a nondiagnostic interval has been admitted. The ammonites and their sediment fillings seem to be out of place, but sedimentary reworking and bioturbation hardly seem adequate to account for them. They are by far the largest sedimentary clasts and include the additional weight of internal sediment and cement. The ammonites in unit 10 remain an enigma. It seems prudent to allow the possibility that some of the apparent biostratigraphic mismatch is real. Seymour Island is remote from definitive Cretaceous-Tertiary sections, so the biostratigraphic zonations applied there may be extrapolations of diachronous events. Indeed, high-latitude heterochroneity has been demonstrated in its Paleogene faunas. Seymour Island may present an unusually complete sequence across the Cretaceous-Tertiary boundary, allowing the vertical separation of events that are condensed to a single level in most sections.

The distinctive glauconite bed and macrofaunal changes at the boundary of units 9 and 10 can be traced across the island (Plate 1, in pocket) between and beyond the four measured sections established by Macellari (1984) to characterize this portion of the formation. Lateral continuity is maintained except for stretches of a few meters where obscured by surficial cover. Thus, the correlations proposed between the measured sections have been corroborated by establishing the physical continuity of a marker bed. To the north and south, beyond the control of measured sections, the boundary glauconite was found to crop out farther east ("up-section") than suggested by photoreconnaissance (Macellari, 1984). The precise stratigraphic position of any isolated faunas sampled earlier in these two regions needs careful review.

Discussion

Detailed measured sections provide an excellent lithostratigraphic framework for this part of the sequence, since the large-scale units are tabular. Projecting biostratigraphic data into the nearest section and the generation of composite sections can probably be safely accomplished using marker beds or strike lines. But precision will ultimately be limited by two factors. The hard ground surfaces and glauconitic horizons represent condensed portions of the sequence. In the intervening bioturbated intervals,

the precise vertical relationships have been destroyed. The length of the vertical burrow sections indicates that the mixing depth is at least 10 cm. Estimates of the accumulation rate of the López de Bertodano Formation (see Summary and Discussion section below) suggest that this translates to a limiting precision due to bioturbation of more than 1,000 yr.

The boundary between units 9 and 10 seems to record a dramatic biotic change primarily in terms of fossil abundance. This change may not coincide with either the extinction of ammonites or the close of the Mesozoic era. The paleontologic change coincides with a reduced sedimentation rate, as evidenced by the glauconite layer. There are also color and provenance differences, but the boundary between units 9 and 10 is not accompanied by any changes in the style of stratification as marked as those at the base of the overlying Sobral Formation.

SOBRAL FORMATION

The formal use and scope of the term Sobral Formation follows Rinaldi (1982), but the definition of the base is taken from the detailed section descriptions made by Macellari (1984, this volume). The base of the Sobral Formation is marked by a pronounced steepening of the topography and the survival of fine-scale lamination. This diagnostic change in stratification results from a dramatic reduction in bioturbation. The relict patches of primary stratification that have survived the widespread obliteration by bioturbation in units 9 and 10 of the underlying López de Bertodano Formation are not very different from the lowermost Sobral.

The base of the Sobral Formation is marked by subtle but unmistakable evidence of unconformity. At a scale of a few tens of meters, the lowermost Sobral includes lenticular packets with erosional bases that locally cut into the underlying formation. Map patterns at the southern end of the island show that the basal contact cuts gradually but progressively downward. The basal surface exposed in that area carries a thin lag conglomerate with clasts of the López de Bertodano Formation, basaltic clasts, glauconite-coated cobbles, and a concentration of phosphatic clasts, especially shark's teeth.

The bulk of the Sobral Formation was mapped in terms of five units (Tps1 through Tps5; Fig. 2) that are essentially tabular at the scale of the whole outcrop, but include smaller scale lenticular stratification. One larger lenticular unit is tentatively identified within this sequence. The Cross Valley Formation is a radically lenticular unit that may be equivalent to a horizon near the top of the exposed Sobral Formation.

Sobral 1 (Tps1)

Three nodule-free, poorly fossiliferous facies with well-preserved fine-scale stratification were combined with the basal conglomerate lenses into a single mapping unit 20 to 30 m thick. The three facies are a dark brown silt ("brown chocolate layer" of Macellari, 1984), a flaser-bedded alternation of pale gray and

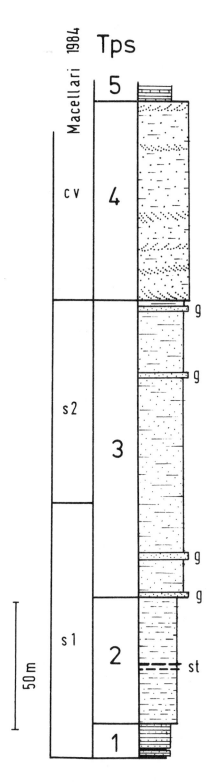

Figure 2. Mapped subdivisions of the Sobral Formation compared with usage of Macellari (1984) in his measured section E; g indicates prominent glauconitic sandstones; st, sanidine tuffs. For lithologic descriptions, see text.

buff, fine sandstones and silty mudstones, and pale gray laminated mudstones and sandstones. The dark brown silt occurs as thin lenses that persist for hundreds of meters. One or two of these may be encountered in a given vertical section. The laminated and flaser-bedded sands are arrayed on a scale of tens of meters, into lenses with cross-cutting, erosional bases.

The macrofauna of Tps1 is sparse, except at the conglomeratic base of some lenses. Large *Pinna* shells and shark's teeth are locally abundant and distinctive. Bryozoa and corals are less common. Otherwise, the macrofaunas include gastropods (especially *Perissoptera*) and pelecypods (especially *Lahilla* and *Cucullaea*) that are not distinguishable from the underlying López de Bertodano faunas, and abundant cidarid spines and fossil wood comparable with Klb9. Trace fossils are sparsely represented throughout Tps1. Thus, although the base of the Sobral probably represents a significant hiatus and a major change in stratification, the faunal change is primarily a matter of abundances.

Sobral 2 (Tps2)

The base of Tps2 is marked by the presence of nodular concretions in laminated sandstones and mudstones similar to those of Tps1. The dark brown silty mudstones are absent, and Tps2 has a less sparse and more uniformly distributed fauna. Crinoid ossicles are conspicuously more abundant than in the underlying units. One to three thin, persistent white tuff beds occur in the lower third of this 40- to 60-m-thick unit.

Sobral 3 (Tps3)

The base of the third mapped subdivision of the Sobral is drawn below the first of a series of thick, coarse, indurated, glauconitic sandstones. These very conspicuous beds form parallel crag lines near the top of the steep slopes formed by the lower Sobral. The interbeds are softer, gray, glauconitic, silty sands and yellow limonitic mudstones. The macrofauna is very scarce, but fossil wood is abundant. Trace fossils are common, but nowhere sufficiently abundant to obliterate the primary stratification.

Macellari (1984) recognized that the lower Sobral is rich in volcaniclastic materials, whereas the upper Sobral is relatively quartz-rich. This petrologic boundary falls within Tps3 (Fig. 2), but is difficult to determine consistently using field criteria, especially on dip slopes. The 100- to 120-m-thick Tps3 is differentiated on the map (Plate 1) by showing the outcrop of the thicker, persistent sandstones.

Sobral ?3 (Tps?3)

In the center of the Sobral outcrop, near the basalt dike set (Fig. 1), a coarse and relatively massive sandstone dominates the talus and crops out between units 3 and 4. The diagnostic cross-bedding of Tps4 is not apparent, but such sandstone is not so abundant elsewhere in Tps3. Poor weather conditions prevented proper mapping of this unit, but a major lenticular body may interrupt the normal sequence in unit 3.

Sobral 4 (Tps4)

The base of unit 4 is placed at the first occurrence of gray cliff-forming sandstones with pervasive tabular cross-bedding. This definition corresponds to the base of the Cross Valley Formation as drawn by Macellari (1984, but not this volume) in measured sections. Retaining these beds as a unit of the Sobral Formation preserves the original scope of that formation (Rinaldi et al., 1978) and allows separate use of the term Cross Valley for beds more strictly related to its type section (see below).

This unit is 75 to 100 m thick, but has an extensive dip-slope outcrop descending to Penguin Bay (Plate 1). Some of the more resistant packets are pebbly, cross-bedded sandstones with a channel-like lenticular form. The interbeds are more friable sandstones, which locally have bright green and purple colors that highlight ripple-drift stratification and trace fossils. Fossil wood fragments are common, but shelly macrofossils are very rare. Yellow, cross-bedded sands that crop out below the La Meseta Formation, between Cape Wiman and Larson Cove, may belong to this unit, but appear to have a discordant lower contact with other Sobral strata.

Sobral 5 (Tps5)

The top of the Sobral Formation was not recognized on Seymour Island. The youngest unit crops out on the top of three small buttes on the dip slope near the northern end of the formation. It is characterized by indurated, gray, calcareous mudstones that contain nodular sandstones in lenses, sparse trace fossils, and crinoidal debris. Fewer than 10 m of this unit is visible.

Discussion

The base of the Sobral Formation coincides with the onset of well-preserved stratification, cross-lamination, and lenticular units. The top is more coarsely stratified, and is lenticular on a much larger scale. The erosional event that preceded Sobral deposition probably falls into the early Danian (Askin, this volume).

Existing measured sections (Macellari, 1984; and this volume) provide an excellent stratigraphic summary for the tabular units of the bulk of the Sobral Formation (Tps1–Tps4). The glauconite-rich sandstones in the Sobral Formation indicate levels at which accumulation may have been slow and the sections condensed.

CROSS VALLEY FORMATION

A lenticular unit of large amplitude certainly interrupts the northern end of the main Sobral outcrop. Its marginal unconformities are so distinct that they can be mistaken for faults. The unit encompasses the measured section that Elliot and Trautman (1982) named the Cross Valley Formation. Much of this type section consists of facies not encountered in the Sobral; but there are coarse, varicolored, trough cross-bedded sandstones that are rich in glauconite and limonite and resemble beds in Tps3 and Tps4. This similarity is sufficiently striking that Macellari (1984, but not in this volume) extended the Cross Valley Formation to include the upper third of the Sobral. The differences, and the erosional contacts, are better reflected, however, if the Cross Valley Formation is restricted to beds that are embraced within, or conformably continuous with, the original type section.

Five lithologically distinct facies were distinguished within the Cross Valley Formation. They are numbered in the sequence of first occurrence in an axial section. The axis of the lens provides the thickest section and some repetition of the facies. Marginal sections are thinner and less complete. The rock bodies within the lens typically thin toward its edges and may pinch out.

The erosional margin of the lens generates a markedly angular unconformity. Beds in the center of the lens dip gently southeast and would be conformable with bedding in the upper Sobral (see Structure section). Toward the edges of the lens, and particularly near its base, the beds dip more steeply toward the axis. The lower surface of the lens was certainly cut down into Tps2, and probably originally eroded as deep as the López de Bertodano Formation. In the lower part of the lens the basal surface is a narrow, altered, and better cemented (?weathered) zone in the Sobral deposits, succeeded by conglomerate as much as 25 cm thick (Tpcv1). The pale gray conglomerate contains small pebbles of glauconitic and nodular sandstones.

The basal surface of the Cross Valley Formation loses its distinctive conglomerate and discordance when traced upward and/or toward the edge of the lens. In the south-southeast portion of its outcrop the basal surface is essentially conformable with the Sobral, near the top of Tps4. The Cross Valley Formation is tentatively interpreted to be the fill of an erosional channel that developed contemporaneously with the deposition of limonitic sands between Tps4 and Tps5.

Varicolored, yellow and orange, coarse, limonitic, crossbedded sands and sandstones (Tpcv2) dominate the lower part of the Cross Valley Formation and are repeated in smaller bodies up section. They are the most striking aspect of the Cross Valley Formation at the type section. The sands are weakly consolidated, poorly sorted and locally pebbly. Wood fragments are the only obvious macrofossils. Similar beds occur between Cape Wiman and Larsen Cove (Fig. 1), but can also be compared with unit 4 of the Sobral Formation (see above).

Friable buff sandstones and siltstones (Tpcv3) occupy much of the middle of the lens, but are typically covered by colluvium from succeeding bluff-forming units. Bedding in this facies is marked by thin lime-cemented beds near the base, or by brown silts higher in the sequence. Macrofossils are not evident.

The most resistant, bluff-forming component of the Cross Valley Formation is a very coarse, gray sandstone and conglomerate (Tpcv4). The clasts are dominated by angular volcanic fragments, mudstone intraclasts, and glauconitic sandstones. Detrital mica is abundant. Fine parallel lamination and low-angle cross-bedding are evident in a layering produced by grain-size differences. Mechanical partings along bedding planes are quite subordinate to vertical joints.

The uppermost part of the Cross Valley Formation is exposed around a small mesa and was described mainly in terms of thinner layers of the facies seen lower in the type section. A distinctive, silvery gray paper shale with fossil leaves is treated as a distinct fifth facies (Tpcv5). The 1986–1987 field party reports that this uppermost part of the Cross Valley lens includes sediments and faunas that more closely resemble the younger La Meseta Formation. W. J. Zinsmeister (personal communication) could project beds from the La Meseta Formation onto the Cross Valley Formation without any indication of displacement. Further investigation may well complete the case for including some areas mapped as Tpcv or Tps5 in the La Meseta Formation.

Discussion

The Cross Valley Formation is lenticular at a larger scale than any part of the Sobral or López de Bertodano Formations. It is interpreted to be the fill of a channel. The lateral variation and the full range of facies are incompletely described by the type section (Elliot and Trautman, 1982), but that section does lie close to the channel axis. The erosional event that produced the Cross Valley channel is probably of early Paleocene age (Askin, this volume). The channel may have been cut across an emergent shelf, but the microflora retains marine elements (R. A. Askin, personal communication).

LA MESETA FORMATION

The La Meseta Formation crops out around the flanks of the meseta at the north end of Seymour Island. It was formally named by Elliot and Trautman (1982) on the basis of one long and four very short measured sections, none of which extended to the base of the formation. The local base of the formation was later identified at an angular unconformity, 250 m below the long measured section in beds initially referred to the Cross Valley Formation (Zinsmeister and DeVries, 1982). This unconformable base has now been mapped all along the south side of Cross Valley, dipping northeast toward the meseta (Plate 1). The same basal unconformity surface emerges with a southwest dip to crop out north of the meseta, between Cape Wiman and Larson Cove. Thus the La Meseta Formation has an overall lenticular form that can be regarded as a small trough or large channel.

The La Meseta Formation was mapped in terms of seven numbered units (Telm1 through Telm7) and seven minor variants. All include small-scale lenticularity, and most can be shown to be large-scale lenses. No one vertical section can incorporate all of these units. This departure from the simpler scheme proposed by Elliot and Trautman (1982) for their measured section (Fig. 3) results partly from the greater differentiation needed for detailed mapping. But it also reflects the greater vertical extent and lateral variability now known to characterize the La Meseta Formation.

La Meseta 1 (Telm1)

The basal unconformity at Cape Wiman and west of Cross Valley is succeeded by very localized units characterized by the presence of large *Ostrea* and *Pecten* shells and shell fragments. These taxa are not as abundant anywhere else in the La Meseta Formation. In other respects the two occurrences of Telm1 are rather different.

At Cape Wiman the *Ostrea* shells are very large (some exceed 20 cm), abundant, and irregular. They are associated with abundant *Cucullaea* shells in a red-brown silt and fine sand matrix. The contact with the steep erosional surface of the Sobral is quite irregular in detail; some of the shells appear to have had an encrusting habit on that surface. At this locality, Unit 1 is closely associated with a megabreccia facies (Telmm) in which large blocks of Sobral and Cross Valley are abundant in a red-brown matrix. Both clasts and matrix are rather friable, so these are not immediately obvious as breccias. They can be traced away from the unconformity surface and seen to grade laterally into shelly, granule sandstones.

In 1985, at the western limit of the La Meseta Formation above López de Bertodano Bay, W. J. Zinsmeister discovered a second, very small patch of basal facies with large *Ostrea* shells and fragments of *Pecten*. The diagnostic shells occur in a sandy pebble-conglomerate that rests unconformably on red-stained Cretaceous rocks and is succeeded by fissile sandstones. The conglomeratic interval is less than 2 m thick. It contains basement clasts and soft, limonitic clasts that resemble the Cross Valley sandstones. The sandy matrix is rich in shelly hash. Better preserved faunal elements include bryozoa, a variety of brachiopods, serpulid worm tubes, rare crinoids, and the pelecypod *Cucullaea*.

At both the northern and southern outcrops of the basal surface, this *Ostrea-Pecten* facies typifies only the lowest exposures. As the contact is traced east, it rises and is overlain by successively higher levels of the La Meseta sequence. Most commonly, the bottom of the La Meseta Formation is in a sparsely fossiliferous, laminated mud and sand sequence, differentiated as Telm2, that oversteps Telm1.

La Meseta 2 (Telm2)

A high mud content, a paucity of shell beds, well-preserved fine stratification, and accompanying low relief characterize Telm2, which is well exposed in low cliffs at the apex of López de Bertodano Bay. Thin, graded sandstones and flaser-bedded and convolute-laminated sands alternate with dark gray, friable mudstones in a sequence that coarsens upward. Burrows are common but do not obscure the fine stratification that is the most obvious difference between this unit and the Cretaceous-Paleocene muds on which it rests. Rare, calcareous siltstones locally contain abundant molluscs, echinoderms, and leaves. Small wood fragments are often concentrated in the muds or at thin ferruginous, concretionary seams.

On inland surfaces the sands and muds weather to buff, yellow, and rarer orange colors. Packets of beds can typically be traced only tens or hundreds of meters. Each packet has a curved strike and terminates unconformably under the succeeding packets. Thus, Telm2 is built of a stack of concave-upward lenses. The

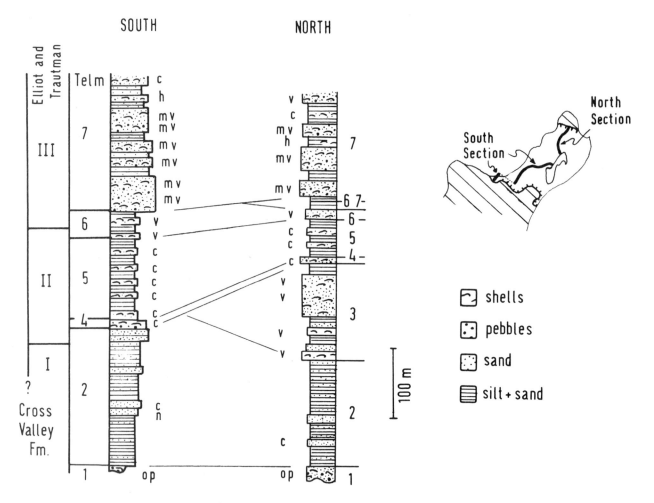

Figure 3. Mapped subdivisions of the La Meseta Formation compared with usage of Elliot and Trautman (1982). Small letters refer to characteristic pelecypods in shell beds: p, *Pecten*; o, *Ostrea*; c, *Cucullaea*; n, *Nucula*; v, veneroids; m, *Modiolus*; h, *Hiatella*. The two sections are based on determinations of the thicknesses of whole mapped units; they are not simplifications of bed-by-bed measured sections.

basal surfaces of the lenses typically lack any coarse lag, and the structures have been interpreted as a series of rotational slump scars (L. Krissek and T. F. Pezzetti, personal communication, 1985).

Two local variations have been mapped separately on the basis of higher sand and/or shell contents. In the sandy aspect (Telm2s), massive or graded, clean gray sands account for more than half the thickness. More rarely, sands are combined with basement cobbles and coarse shells into a channel-fill aspect. The pelecypod *Cucullaea* typically dominates the coarse shelly fauna. *Nucula* shells are the most abundant element in a finer grained shelly variation mapped in Cross Valley.

The upper limit of the coarsening upward sequence of Telm2 is drawn at the appearance of persistently shelly beds. In the north, veneroid pelecypods dominate the overlying unit

(Telm3); in the south, *Cucullaea* shell beds (Telm4 and 5) succeed Telm2. Thus, Telm2 is a coarsening upward, muddy sequence that oversteps locally developed basal facies, buttresses the steep, erosional margins and is succeeded by coarser, shallower water shelly facies. It has all the hallmarks of a prograding fill in the erosional relief created by an earlier interval of emergence. In the axis of the earlier Cross Valley channel fill, Telm2 has been overstepped, and a shelly conglomerate of uncertain, younger affinities rests directly upon the Cross Valley units (Plate 1).

La Meseta 3 (Telm3)

Buff-weathering, cross-bedded sands and silts, abundant shell beds and lenses, and a dominance of veneroid pelecypods are diagnostic for Telm3. The upper limit of the unit is marked by the first *Cucullaea*-bearing, purple and green sands. The background to this sedimentologically complex unit consists of large-scale, low-amplitude lenses of silty sands with a shell-rich layering. Superimposed upon this are thick, unfossiliferous packets of ripple-drift sands, shell-rich channel fillings, and laterally persistent veneroid shell beds.

The channels are lenses 1 to 50 m thick of imbricated, cross-stratified sands that make a composite fill above an erosional surface. Burrowed horizons and internal lag deposits indicate that the fill was accomplished in several stages. The sands become cleaner and parallel laminated toward the top of the lenses. Shells are concentrated as lags along the channel bottoms and sides, and as imbricated fragments along the cross-strata. Cobbles, intraclasts, fossil wood, and gastropod shells are abundant in the basal lags. The shells are typically disarticulated, whereas beyond the channels, articulation and living position are often preserved.

The local shelly lag gravels, associated with cut-and-fill cycles, contrast with the persistent shell layers, which reach thicknesses as much as 1 m and extend for more than 1 km. The beds are dominated by veneroids with single-valve preservation, and have fewer pebbles and fragmentary shells than the channel lags. The basal surfaces show limited erosional scouring. These shell beds appear to become thinner and less numerous up-section and to the north.

Telm3 crops out across a zone of monoclinal flexure that appears to be draped over growth faults (Fig. 4). When traced northeastward across this zone, the veneroid shell beds are replaced by shelly, weakly cemented, conglomeratic sands (Telm3s). The sands contain sparse nodules and well-rounded basement cobbles. The shells are less concentrated than in the equivalent beds to the southwest. A more significant, but still small, admixture of *Cucullaea* shells appears up-section, but the overlying beds are not seen. The transition from the underlying Telm2 to this sandy aspect is a thin unit of purple and green silts with cannon-ball concretions.

Telm3 is a very localized unit of breccias characterized by large intraformational blocks in a coarse conglomerate matrix with veneroid shells. The breccias and conglomerates occupy small lenses, and crop out near the northern limit of Telm3.

La Meseta 4 (Telm4)

A unit dominated by *Cucullaea* shell beds (Telm5) crops out on the lower flanks of most of the meseta. On the west side the base of this unit is marked by a particularly thick, distinctive *Cucullaea* bed that succeeds veneroid shell beds (Telm3) in the north and buff silty sands (Telm2) in the south. This bed was mapped as a separate unit (Telm4) and can be distinguished from overlying *Cucullaea* shell beds by its thickness, coarseness, and relatively high content of phosphatic teeth and bones. Shark and skate teeth, in particular, are sufficiently abundant to be used as a mapping criterion. Telm4 typically caps low cuestas where the shells are further concentrated by surficial processes to produce a coarse white lag on the dip slopes.

The unit is as thick as 3 m and has a composite construction, with shells distributed unevenly through a gray to buff, conglomeratic coarse sand. The base exhibits irregular scour and fill structures with as much as 80 cm of vertical relief and is succeeded by a sequence characterized by crude cross-bedding defined by shelly layers. The upper part of the unit has a subtle parallel layering. Large whole valves dominate poorly stratified layers separated by layers in which finer grained shell fragments define a finer stratification. Shells are most commonly disarticulated and oriented with the maximum projection area parallel to bedding. Many valves lie concave-down, sheltering a calcite-filled vug. The shells may have accumulated over several construction phases, but the deposit has the overall form of Kidwell's (1986) type IV shell beds. As such, it probably results from slow sediment accumulation following a period of erosion.

Shells of the pelecypod *Cucullaea* and darwinellid gastropods dominate the macrofauna. Both are large, robust forms. Fish teeth, natissids, perissodontids, and brachiopods are common. Veneroid pelecypods, which dominate the underlying unit, are common only in the basal scours. This suggests reworking and winnowing of the underlying deposits during an interval in which critical ecologic factors changed.

La Meseta 5 (Telm5)

This unit occupies the lower slopes of the west side of the meseta. It is characterized by rhythmic alternations of purple and gray-green silts and sands, punctuated by beds and lenses of *Cucullaea* shells.

The purple, silty sand layers, 1 to 2 cm thick, have a faint parallel lamination and often a basal concentration of plant fragments. Thin, coaly seams and large, fossilized logs are common. The interbedded gray-green, medium to fine sands are 2 to 10 cm thick and show ripple cross-lamination. The regular alternation is disrupted by sparse burrow mottling and by convolute liquefaction structures. Rarer, larger scale slump folds involve packets of these beds as much as 10 m thick. Calcareous nodules in the form of large tablets or small spheres are concentrated at some horizons. In contrast with the shell beds, the faunas are sparse in these rhythmic sequences. Veneroid pelecypods, brachiopods, and echinoderms are most common. The shells are typically isolated, articulated, and in life position, but they are extensively altered to limonite and consequently very fragile. In nodule layers, often found beneath the shell beds, small pelecypods, fossil wood, and crabs of the genus *Lyreidus* are common.

The *Cucullaea* shells are concentrated in two configurations: persistent beds that are traceable for distances of kilometers, and lenses that pinch out in a few meters. The lenses contain abundant intraclasts and are localized on the floors and margins of erosional channels or at the bases of individual cross-bedded packets within composite channel fills. The shell beds include a wide selection of rounded cobbles but few intraclasts. The shells are better preserved, and often articulated, but in a grain-supported texture with no semblance of life positions. Some portions of the shell

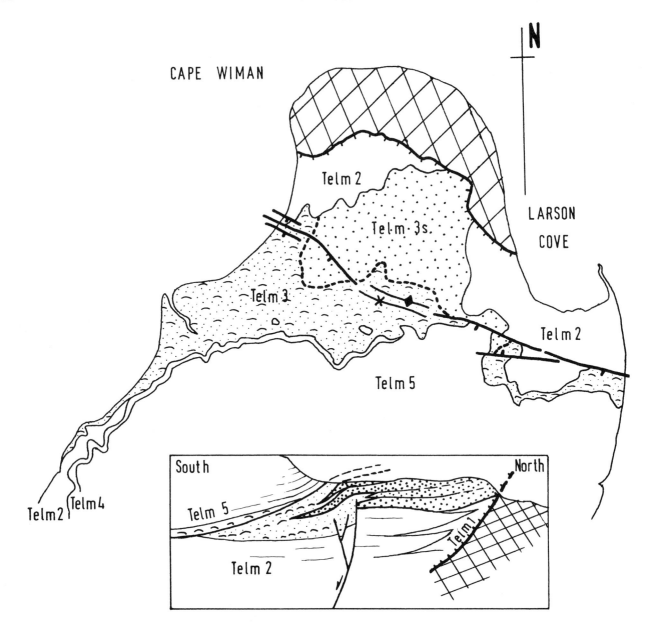

Figure 4. Spatial relationship of growth fault zone and facies changes in Telm3 at the north end of Seymour Island. Coarse stipple indicates sandstone facies, with dispersed veneroid shells, Telm3s; fine stipple, silty sand facies with well-defined shell beds. Fold axes mark a monoclinal fold in the fault zone. Schematic cross section has two-fold vertical exaggeration.

beds have a clearly imbricate fabric. The shell beds frequently thin out laterally to an orange-brown, limonitic seam with very poor fossil preservation, and then thicken again along strike. Both the persistent beds and the more restricted lenses are typically associated with surfaces of erosion. Although some shell concentrates were observed to pass up into a surface of omission or erosion (types I and II of Kidwell, 1986), most lie on such surfaces.

Massive and cross-laminated cleaner sands locally form lenses as much as 5 m thick and are mapped separately as Telm5s. The lenses have sharp bases on top of intensely burrowed intervals.

The overall form of this unit, as defined by its shell types and color, is a large lens (Fig. 5). The time significance of the boundary with Telm3 needs careful evaluation. It may be a diachronous, lateral facies change. The scoured base of the lowest *Cucullaea* shell bed (Telm4), and its concentration of cobbles and phosphatic fossils, however, indicate that the basal surface of

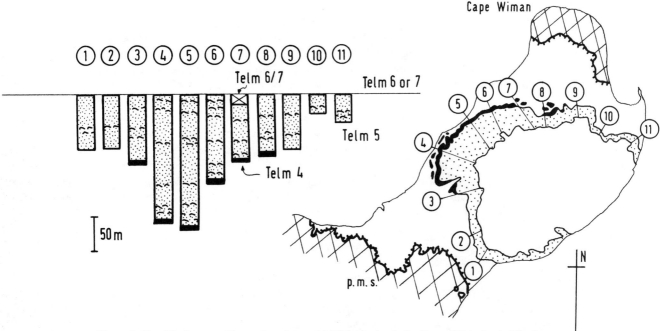

Figure 5. Simplified map and located sections to highlight the lenticular form of Telm5 unit. The datum for the 11 sections is the base of the succeeding unit, Telm6 or Telm7. Unit Telm 6/7, atop section 7, has a pronounced erosional base, and is plotted below the datum to compensate for the loss of the top of Telm5. Notice that the Telm5 lens is arranged symmetrically between the marginal unconformities of the La Meseta Formation. Telm4 can clearly be seen to be the major shell bed associated with the base of the lens.

Telm4 and 5 is more likely to be a significant hiatus. Telm4 occupies the lowest part of the bottom surface of Telm5, and mimics the occurrence of smaller shell lenses in small channels.

La Meseta 6 (Telm6)

This unit closely resembles Telm3, but is located above Telm5. It is capped to the north by a smaller lens of barren, brown, silty sands (Telm6/7).

La Meseta 7 (Telm7)

Relatively resistant, buff, pebbly sandstones with a shelly macrofauna dominated by veneroids and *Modiolus* form the upper part of the meseta and the top of the La Meseta Formation. Several shell concentrations found near the top of the unit are dominated by *Hiatella* or *Cucullaea* and could form the basis of definition of a higher unit (Fig. 3), but are too poorly exposed to be usefully mapped.

The veneroid and *Modiolus* shells are very abundant, but not as clearly concentrated into shell beds as in the lower units. Individual shell beds could not be mapped readily, but the unit is differentiated by mapping bands of more resistant sandstones that form steeper slopes (Telm7s).

The more and less resistant aspects (Telm7 and Telm7s) appear to form a rhythmic couplet. The couplets commence with the more friable, purplish sands and silts above a slightly discordant base that can often be seen to cut across the more resistant sands at the top of the couplet below. The basal cutout is typically accomplished on a scale of hundreds of meters, but the basal surface has only very local shelly lag gravels. The resistant sands often have a somewhat nodular cementation, and typically include better cemented, stick-like burrows that weather out to dominate the surficial lags along with the shells.

La Meseta Megabreccia Unit (Telmm)

On the coast southwest of Cape Wiman, a spectacular channel breccia crops out. Very coarse shelly conglomerates and intraclast boulders fill a lens with an erosional base. The south side of the lens includes three slide blocks of the wall material that have been buried by channel fill after slipping a short way from the steep, eroded margins. Shelly gravels fill in the curved pull-away fissures and cap the slide blocks.

Discussion

The base of the La Meseta Formation is an angular unconformity at Cross Valley and Cape Wiman, but the angular relationships are unusual (Plate 1). The beds just above the unconformity often dip more steeply than those below (especially in Cross Valley), and the unconformity is typically steeper than both. The angularity of the unconformity results because it is a steep erosional surface buttressed by the subsequent fill. Close to the erosion surface, the fill tends to dip away from the surface, with steeper dips and anomalous strikes relative to the bulk of the formation. The angular relationships are a variety of what Vail and others (1980) termed "onlap." The ages of units in contact along the unconformity place the erosional event(s) in the early Eocene.

The steep, erosional margins of the La Meseta Formation presumably had appreciably more than the 70 m of vertical relief currently exposed and are 5 km apart. A topographic low of these dimensions is best understood as the result of subaerial erosion of an emergent shelf. The red coloration of the wall rocks may be the result of soil-forming processes, but the burrows and the attached fauna are evidence of a significant interval during which the valley was drowned but not yet filled. The marginal facies include shallow, subtidal, and possible intertidal faunal elements, but the base of the valley fill is not exposed. Subaerial sediments might be expected at the base of the fill, below the meseta. Although the exposed terrigenous sedimentary fill is mostly fine grained, and contains a thoroughly marine macrobiota, very well-rounded basement cobbles occur in the shell beds throughout the La Meseta Formation. These cobbles indicate that mainland beaches were never very far away. Their influence is strongest in Telm7.

The occurrence of possible rotational slide masses is essentially confined to Telm2; convolute slump deposits occur in Telm3 and Telm5. The influence of growth faults on facies is seen clearly only in Telm3. Much of the valley relief had probably been subdued by the sedimentary fill prior to the deposition of the first really extensive shell bed (Telm4), although the base of this bed still has about 60 m of relief. The sedimentary fill includes erosional channels, which were probably cut by tidal currents and have only a few meters of vertical relief. The larger scale lenticular character of the mapped units might be attributed to deposition as a complex of lagoons and barriers, but there is little evidence of aeolian deposition.

The architecture of the whole formation is a stack of lenticular units that fill the earlier erosional valley. The stacking of the lenses is asymmetrical about the axis. Thus, at the very least, two type sections are needed, one beginning at each of the exposed margins (Fig. 3). The small, filled channels seen in cliff exposures of Telm3 mimic the geometry and stratification of the whole formation. The shelly lag conglomerates and shell beds clearly represent condensed parts of the sequence (Kidwell, 1982, 1984, 1986), and probably indicate hiatuses. Telm4 has the highest concentrations of phosphatic clasts and may succeed the most extreme hiatus.

Lateral projection of faunas into composite sections are more risky here than in any other part of the Seymour Island stratigraphy. Ideally, each of the lenticular units needs its own axial section, and faunas need to be collected along the base of lenses to check for overlap. Certainly the base of the La Meseta Formation is diachronous. Field observations show successively higher levels in contact with the steep marginal unconformities, proceeding from west to east. Telm4 best illustrates the need for careful dating of shell beds. This unit may be a diachronous transition between two shelly facies (Telm3 and Telm5) that are partly lateral equivalents. Sedimentologic features suggest that it is more likely to be a lag concentrate that marks the base of a fill cycle following a hiatus. In the latter case the shell bed may be isochronous, but include faunal elements that lived during the span of the hiatus, or were scoured from underlying units.

STRUCTURE

The structure of the Paleogene beds of Seymour Island is simple. There are growth faults with associated folding and facies changes, angular unconformities, and a regional tilt. The growth faults are best seen near Cape Wiman in Telm3. This unit is flexed into a narrow monocline, trending west-northwest, which passes downward into normal faults of small separation (Plate 1). The trace of the fault zone coincides with the sandstone facies change discussed above (Fig. 4). The fault zone appears to parallel the buttress-like edge of the La Meseta Formation, which may owe its position to earlier faults.

Surfaces of distinct, angular unconformity mark the base of the Cross Valley and La Meseta Formations. But these surfaces are steep, had appreciable primary relief, and are to some extent buttressed by the younger fill. They do not fix tectonic events in any simple way; rather, they record episodes of emergence and channeling, or sea-floor scouring. Furthermore, the bedding attitudes close to the unconformities are not representative of the bulk of the overlying units. The unconformity at the base of the Sobral is relatively subtle.

Figure 6 summarizes bedding attitudes on Seymour Island. The Cretaceous and early Paleogene strata have a simple, gentle, homoclinal dip to the southeast. Attitudes in the axis of the Cross Valley Formation are not distinguishable from the older homocline that it cuts, but poles to bedding throughout the lens define part of a wide-angle, small-circle girdle. Attitudes from the La Meseta Formation define a much tighter small-circle girdle. The modal orientation for bedding in the La Meseta Formation dips in the same general direction as the older units but is 3 to 5° less steep. The small circle girdles are generated by conical, synform attitudes in the fill of erosional lows, by draping over growth faults, and by slump-type deformation. It is evident in the field, and from the stereographic plots, that local bedding attitudes in the La Meseta Formation are much more variable than the average attitudes that influence the outcrop pattern.

The modal dip lines, the small-circle axes, and the modal strike of the growth faults all trend west-northwest and lie within

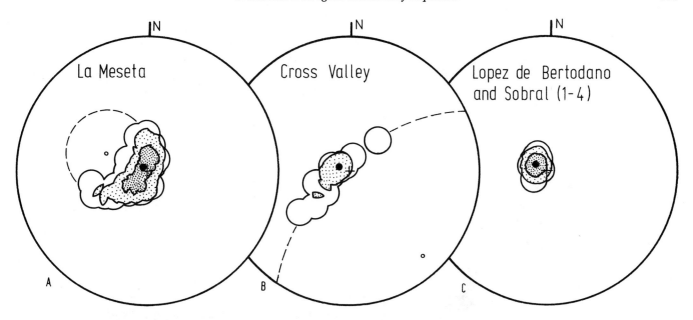

Figure 6. Bedding attitudes on Seymour Island. Poles to bedding contoured by the Mellis (1942) method at 1, 3, and 10 points per 1 percent area. Cross indicates center of stereogram; solid circle, modal attitude; dashed arcs, best fit, small-circle girdles; open circle, girdle axis. Stereogram C (López de Bertodano and Sobral Formations) shows the well-defined maximum to be expected from a homoclinal sequence. Stereograms A (La Meseta Formation) and B (Cross Valley Formation) show variable bedding attitudes that define small-circle arcs indicative of portions of conical surfaces. In these units the bedding attitudes drape up against the marginal unconformity surfaces and are locally influenced by lenticular cut-and-fill structures, rotational slides, and slump folds. The stereonets provide the more reliable indication of the regional tilt.

25° of one another (Fig. 7). The regional tilts, growth faults, channel orientations, lens forms, and facies changes appear to be part of a unified sedimentary and structural response. Slight tilting of the Cretaceous-Paleocene beds appears to have been accomplished prior to deposition of the Eocene sequence, but the same sense of tilt continues later. The geometrically related lenticular stratification and channeling commences in the Paleocene. The geometry of the erosional surfaces beneath the Cross Valley and La Meseta units are reflected in the conical bedding attitudes above them. The cone axes are essentially parallel, suggesting comparable causes for both erosional events. Furthermore, the Cross Valley channel is very nearly coincident with the margin of the La Meseta Formation.

SUMMARY AND DISCUSSION

The Paleogene sequence on Seymour Island includes a transition from dominantly tabular to dominantly lenticular stratigraphic units (Fig. 8). Thus, isolated measured sections become progressively less reliable as the sole basis for stratigraphic studies as one works up-section.

The uppermost López de Bertodano Formation is simply tabular (Fig. 8a), and suitable for the projection of biostratigraph-

ic data into a few isolated sections. Limits to biostratigraphic precision have been set by the mixing action of the infauna, and the reduced accumulation rates at hard-ground surfaces and glauconite-rich layers.

The base of the Sobral Formation appears to be a significant hiatus. Improved preservation of stratification is a hallmark of the Sobral units, but the tabular units here are interbedded with lenticular units from scales of tens to thousands of meters. Successive units in the Sobral show coarser stratification, and the Cross Valley Formation marks a radical increase in the height of the lenticular units (Fig. 8b–c). The troublesome nature of lenticular units is particularly evident whenever an attempt is made to project scattered faunas into one composite section and preserve their proper vertical sequence. Such a composite section would be valuable for the relatively sparsely fossiliferous Sobral Formation. The problems associated with projecting stratigraphic position across lenticular units are examined for the La Meseta Formation.

The La Meseta Formation is dominated by large-scale lenticular units with smaller scale internal lenticularity (Fig. 8d). The small-scale features relate to channeling and gravity sliding. The large-scale lenses are bounded by facies changes, but appear to be separate units in the composite fill of a large erosional low. Thus,

P. M. Sadler

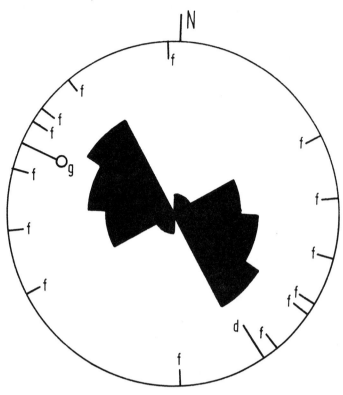

Figure 7. A rose diagram of the orientation of small-scale channel axes in the La Meseta Formation (Telm3 through 6), compared with the strike directions of growth faults (f), the modal tilt direction of the formation (d), and the azimuth of the axis of the small-circle girdle that best fits the variation of local bedding attitudes (g, see small open circle in Fig. 6A for derivation).

the lower surfaces of the lenticular units may be erosional at all scales. Certainly faunas cannot be projected laterally into composite sections across such boundaries without the risk of considerable loss of vertical precision. The loss of precision is a function of the vertical erosional relief on the boundary. Figure 8d shows that more than 100 m of vertical imprecision might result from inadvertently projecting an isolated fauna across the lower boundary of just one major lenticular mapping unit. We must also reckon with lenticular units with widths of only a few meters, that have vertical amplitude of several meters, and can therefore introduce vertical errors of at least this magnitude whenever faunal positions are projected across their boundaries.

Of course, the danger of crossing the erosional lower surfaces of lenticular units increases with the distance of projection between faunas and sections. So it is reasonable to ask how vertical precision might deteriorate as a function of projection distance. Since the La Meseta Formation is pervasively lenticular, a rather simple answer to this question may be seen in Figure 8. The reasoning is detailed in the appendix to this chapter. The solution is that the distance between the projected faunas and sections is half the "critical lens length." The regression of observed lens amplitude on length provided by Figure 8 may be used to translate this critical horizontal distance into the corresponding vertical lens amplitude. The average vertical error will be half this amplitude. Where this procedure suggests an average error of the same scale as the vertical separation between faunas in the composite section, that section is unreliable. To express the error as a time span, the duration of the erosional hiatuses would have to be known, but it is unlikely that the relationship between stratigraphic distance and time is at all simple.

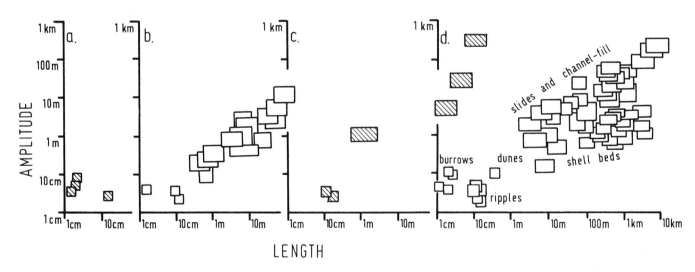

Figure 8. The horizontal and vertical dimensions of nontabular bedding units from the scale of burrows to whole formations in: a, the López de Bertodano Formation; b, the Sobral Formation; c, the Cross Valley Formation; and d, the La Meseta Formation. The boxes reflect the range of dimensions for families of small-scale features, and uncertainty of dimensions of large-scale features. Shaded boxes are used to distinguish data belonging to plots a and c, in spite of the overlap of axes. Graph d is annotated to show the typical origin of lenticular bedding units at different scales. These lens dimensions are not necessarily determined perpendicular to a trough axis, since this was not always determinable.

Major shell beds provide the best lithostratigraphic markers in the La Meseta Formation, but are also one possible shallow, subtidal expression of the transgressions that follow low stands in sea level (Kidwell, 1984). Thus, the richest fossil sites are potentially systematically associated with hiatuses and the levels of most extreme stratigraphic condensation. Each shell bed should be regarded as potentially diachronous, condensed, mixed, and adjacent to a hiatus. The only thoroughly reliable evidence of relative age of the La Meseta faunas will be relative position in a single vertical section. If the biostratigraphic framework is built upon a more adventurous interpretation of the time significance of lithostratigraphic units, then the interplay of lateral facies changes and hiatuses may never be resolvable.

Geometric relationships (Figs. 6, 7) suggest that the erosional events and the progressive, or repeated, regional tilting are closely related. The regional tilt is basinward. It is possible that the Paleogene phase of basin development simply exaggerated the paleoslope, causing progressive shallowing and increased channeling and lenticularity up-section. Alternatively, the main hiatuses may be related to discrete events, but there is no indication that these were local tectonic events. The main erosional events, which suggest shallowing or emergence, occur in the early Paleocene, late Paleocene, and early Eocene. The pre–La Meseta event would have been large. The timing and relative magnitude of these erosional events permits comparison with major, global sea-level lows postulated by Vail and others (1980) to account for interregional unconformities in northern latitudes.

The sea-level history hypothesized by Vail and others (1980) showes little change in the Late Cretaceous and a more complex pattern of major and minor cycles in the Paleogene. This mirrors the lenticularity trend in the Seymour Island sequence, but the simplicity of the late Cretaceous sea-level curve may be an artifact of insufficient data. Certainly the large, late Ypresian (49.5 m.y.) low stand of sea level could account for the erosion prior to deposition of the La Meseta Formation. The ages of Telm1 and Telm2 are crucial. The base of the Cross Valley Formation might correspond to the latest Danian (60 m.y.) low stand or to smaller sea-level fluctuations proposed in the Thanetian.

The more subtle erosional event at the base of the Sobral Formation is apparently Danian (Askin, this volume; Huber, this volume) and does not correspond to any low stand hypothesized by Vail and others (1980). Their terminal Cretaceous low stand might have to be referred to the stratigraphic condensation represented by the glauconite layer at the base of KTlb10 (see Huber, 1986, which discusses an alternative). The glauconite could be taken to mark the earliest phase of a relatively rapid rise that starved the outer shelf of terrigenous sediment, but the argument has been made above that the end of the Cretaceous lies farther up-section. We should not necessarily contrive to fit Antarctic erosional episodes to a model of low stands in sea level set up to explain North Atlantic seismic stratigraphy.

Evidently, continued refinement of the Paleogene biostratigraphy on Seymour Island can test the global aspirations of the

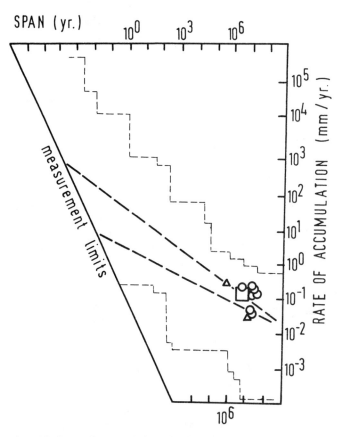

Figure 9. Rate of accumulation of the Upper Cretaceous–Paleogene sequence on Seymour Island compared with other clastic shelf sections. Light dashed lines indicate envelope of data from other sections, after Sadler (1981, and unpublished data); heavy dashed lines, range of modal values from other sections; circles, López de Bertodano Formation; triangles, Sobral Formation; boxes, La Meseta Formation. Symbol sizes reflect uncertainty of unit duration.

model (Vail and others, 1980) linking inter-regional unconformities to sea-level fluctuations. Should the correspondence of hiatuses and hypothesized sea-level low stands be confirmed for the Paleogene, then the hard grounds and glauconitic condensed intervals in the López de Bertodano Formation might provide a valuable data set for refining the late Cretaceous sea-level curve.

If the Seymour Island sequence is to test hypotheses of evolution and extinction, it needs to be a relatively complete record of late Cretaceous and Paleogene events. If it is to test models of sea-level change, its hiatuses and condensed sections need precise dating. Proper attention to either test will, therefore, improve the rigor of the other; the two notions should be pursued together. The rates of accumulation of stratigraphic sections, at

318 P. M. Sadler

different time scales, may be used as a guide to expected completeness (Sadler, 1981). In a very general way, it is also reasonable to expect thicker sections to be more complete recorders of a given time span in a given environment.

Figure 9 shows accumulation rates calculated from radiometric glauconite dates (Macellari, 1984) and biostratigraphy (Askin, this volume), and compares them with rates compiled for other siliciclastic shelf accumulations. The glauconite ages supply the shorter term rates. As absolute ages they appear to be too young; the differences between these ages, as used to determine rates, are subject to less error, but still not entirely trustworthy. The Seymour Island rates are generally at or above average for comparable environments, so the sequence is unlikely to be less complete than average.

ACKNOWLEDGMENTS

Assistance from several individuals was vital to the completion of this mapping project. Henry Brecker crafted the topographic map from a very trying set of aerial photographs. The field mapping was supported by National Science Foundation Grant DPP-8215493 to M. O. Woodburne, who encouraged the project at all stages. W. J. Zinsmeister kindly led me to many sites on the island where he had previously recognized crucial stratigraphic relationships. B. T. Huber, S. R. May, R. M. Feldmann, C. E. Macellari, M. A. Kooser, J. A. Case, D. Chaney, L. A. Krissek, and T. F. Pezzetti shared many important insights during discussions in the field and afterward. N. A. Wells and D. F. Palmer reviewed the manuscript.

APPENDIX: PROBABLE VERTICAL ERROR IN COMPOSITE SECTIONS

A basic element in biostratigraphy is the vertical section that records the faunal succession. Since crucial faunas are not always found in strict superposition in a single vertical section, it is not uncommon that faunas from different sections are combined into a hypothetical composite section. Individual faunas must be projected into the composite section in such a way that their real temporal sequence is preserved. Typically, independent evidence such as magnetostratigraphy or radiometric dates is not available, and the faunas are projected parallel to marker beds or the inferred bedding surfaces. There is an implicit and usually undefended assumption that such markers are not diachronous at the scale of the projection distance. Where lithostratigraphic units are dominantly tabular, this method might generate a valid faunal succession in the composite section. But where units are lenticular, and especially where they have erosional bases, the projected position of a fauna in the composite section is subject to a vertical error. Fastovsky and Dott (1986) have shown how the most enlightened attempts to compensate for channels during lateral correlation can be frustrated by lack of field evidence of channel geometry.

Erosional surfaces with appreciable relief permit overlying beds to be physically lower than older beds cut by that surface (Fig. 10). Thus, beds within a simple erosional channel fill should be projected into the interchannel stratigraphy at the highest level at which the channel margin unconformity can be recognized. Where the channel fill and marginal levees build simultaneously (Fig. 10b), it can become more difficult to combine the two adjacent sections into a valid composite.

It is clear, however, that the magnitude of the potential vertical error in such correlation is partly dependent on the vertical relief of any unrecognized channel margins that are crossed. It is also reasonable to suppose that the error will generally increase with the horizontal distance over which projection is attempted. This follows because a longer projection line is more likely to cross larger channel margins. If the projection intersects multiple channels (Fig. 10c), the error can be increased or reduced. It is therefore useful to estimate the average vertical error associated with crossing a single channel margin.

The average or expected error can be estimated where the variation of lens amplitude as a function lens length (e.g., Fig. 8) is known for a lenticular unit. It was impossible in the construction of Figure 8 to ensure that all measurements were made in planes perpendicular to channel axes. But the lines of projection of faunas into composite sections may cross channel axes at any angle, so the unrefined data of Figure 8 are exactly what is needed. Any given fauna may be regarded as a point within a characteristic lens. That lens must be imagined to be within a nested set of lenses of all scales recognized for the given stratigaphic unit. When that fauna is projected laterally, errors arise whenever it is projected across a lens margin. Clearly the projection distance, and the distance to each lens margin is critical. Fortunately, the average distance to the margin is given by a fauna in the center of the lens, and thus equidistant from both margins. The average, or expected, error is to be derived from a model in which each scale of lens is nested at the center of the next larger lens. If the lens geometry is simplified to a rectangle, then the average distance to the lens margin is simply half the lens width, in the direction of projection. By a similar argument, the average vertical error is given by half the lens height.

The geometric model of the lenses might be improved in some instances by using semicircles, trapezoids, or triangles. But the lens proportions (e.g., Fig. 8) should ideally be collected only in the direction of projection. Unless the latter condition is met, there is probably little point to fine adjustments of the geometric model. Thus, the critical lens length can be taken as twice the projection distance. The expected vertical error is then half the height of the critical lens. Inspection of Figure 8 shows that, for a given critical length, one could use half the average height, or, for a more critical test, half the maximum lens height. If the critical length is greater than any known lenses in the unit, then the expected error is given by the largest known lens.

In a unit such as the La Meseta Formation, where lenticularity is pervasive, we may allow the simplifying assumption that all points lie within a lens. If lenses are more sparsely distributed, then we must recognize that many points will lie between lenses. For such points, the preceding analysis is too simple, and we would need to allow for average lens separations.

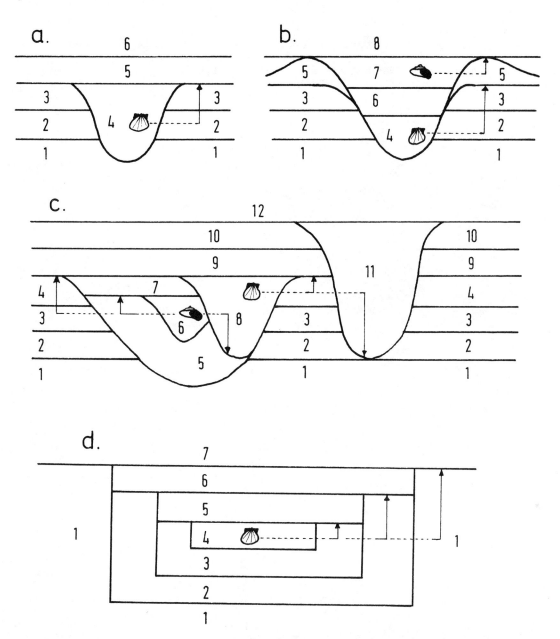

Figure 10. Sources of error in lateral correlation across channel margins. Numbers indicate the depositional sequence. Horizontal dashed lines represent unenlightened projection parallel to strike. Vertical arrows show the error compared with projection parallel to isochronous surfaces. a, Projection across the margin of a simple erosional channel; b, the added complications of a channel with levees; c, the added complication of multiple channels; d, the simplified geometric model of an average case for a formation with pervasive lenticularity, which allows Figure 8 to be used to estimate expected vertical error.

REFERENCES

CRANWELL, L. M. 1959. Fossil pollen from Seymour Island, Antarctica. Nature, 184:1782–1785.

ELLIOT, D. H., AND T. A. TRAUTMAN. 1977. Lower Tertiary strata on Seymour Island, Antarctic Peninsula [Abstract]. 3rd Symposium on Antarctic Geology and Geophysics. Madison, Wisconsin, p. 168.

——. 1982. Lower Tertiary strata on Seymour Island, Antarctic Peninsula, p. 287–297. *In* C. Craddock (ed.), Antarctic Geoscience. University of Wisconsin Press, Madison.

FASTOVSKY, D. E., AND R. H. DOTT. 1986, Sedimentology, stratigraphy, and extinctions during the Cretaceous-Paleogene transition at Bug Creek, Montana. Geology, 14: 279–282.

HALL, S. A. 1977. Cretaceous and Tertiary dinoflagellates from Seymour Island, Antarctica. Nature, 267:239–241.

HUBER, B. T. 1986. The location of the Cretaceous-Tertiary contact on Seymour Island, Antarctic Peninsula. Antarctic Journal of the United States, 20:46–48.

KIDWELL, S. M. 1982. Time scales of fossil accumulation: patterns from Miocene benthic assemblages. Proceedings of the Third North American Paleontology Convention, 1:295–300.

——. 1984. Outcrop features and origin of the basin margin unconformities in the Lower Chesapeake Group (Miocene), Atlantic coastal plain, 37–38. *In* J. S. Schlee (ed.), Interregional Unconformities and Hydrocarbon Accumulation. American Association of Petroleum Geologists Memoir 36.

——. 1986. Models for fossil concentrations: paleobiologic implications. Paleobiology, 12:6–24.

MACELLARI, C. E. 1984. Late Cretaceous stratigraphy, sedimentology, and macropaleontology of Seymour Island, Antarctic Peninsula. Unpublished Ph.D. thesis, The Ohio State University, 599 p.

MELLIS, O. 1942. Gefügediagramme in stereographischer Projecktien. Zeitschrift für mineralogische und petrographische Mitteilungen, 53:330–353.

RINALDI, C. A. 1982. The Upper Cretaceous in the James Ross Island Group, p. 281–286. *In* C. Craddock (ed.), Antarctic Geoscience. University of Wisconsin Press, Madison.

——, ET AL. 1978. Geologia de la Isla Vicecomodoro Marambio. Contribuciones Instituto Antarctico Argentino, 217:1–37.

SADLER, P. M. 1981. Sediment accumulation rates and the completeness of stratigraphic sections. Journal of Geology, 89:569–584.

TRAUTMAN, T. A. 1976. Stratigraphy and petrology of Tertiary clastic sediments, Seymour Island. Unpublished M.S. thesis, The Ohio State University, 170 p.

VAIL, P. R., R. M. MITCHUM, T. H. SHIPLEY, AND R. T. BUFFLER. 1980. Unconformities of the North Atlantic. Philosophical Transactions of the Royal Society of London, A294:137–155.

ZINSMEISTER, W. J. 1982. Review of the Upper Cretaceous–Lower Tertiary sequence on Seymour Island, Antarctica. Journal of the Geological Society of London, 139:776–786.

——, AND T. DEVRIES. 1982. Observations on the stratigraphy of the Lower Tertiary Seymour Island Group, Seymour Island, Antarctic Peninsula. Antarctic Journal of the United States, 17:71–72.

MANUSCRIPT ACCEPTED BY THE SOCIETY SEPTEMBER 1, 1987

Geological Society of America
Memoir 169
1988

Paleogene dinoflagellate cyst biostratigraphy
of Seymour Island, Antarctica

John H. Wrenn
Amoco Production Company, P.O. Box 3385, Tulsa, Oklahoma 74102
George F. Hart
Department of Geology, Louisiana State University, Baton Rouge, Louisiana 70803

ABSTRACT

Diverse and abundant organic walled microplankton have been recovered from the only known outcrops of lower Tertiary marine sediments in the Antarctic area. The 70 samples collected from the Paleogene Seymour Island deltaic complex yielded 43 genera and 74 species of dinoflagellate cysts; 2 genera and 9 species are new, and a number of new combinations and taxonomic emendations are proposed. Specimens attributable to the Acritarcha, Pterospermatales, and Chlorococcales also were noted.

Sediments of the Cross Valley Formation in the Cape Wiman area are early late Paleocene in age. The exact source of samples reputedly collected from the Cross Valley Formation at the type locality in Cross Valley is uncertain. This is because faults and surface slumps in the area were only partially mapped at the time of sampling. The lower 100 m of Section 12-13 are Eocene in age, whereas the overlying 20 m are no younger than early late Paleocene. The deposits of the stratigraphically higher La Meseta Formation can be subdivided into beds of late early and middle to late Eocene age.

Deposition occurred within the subenvironments of the Seymour Island Paleocene-Eocene deltaic complex. Specific paleoenvironments identified include distributary channels, interdistributary bays, lagoons, and perhaps, prodelta areas.

The Paleocene dinoflagellate cyst floras are composed of cosmopolitan taxa. Eocene floras include cosmopolitan and provincial taxa. The cosmopolitan taxa have been reported from regions as widely separated as Europe, North America, and Australia. The provincial taxa belong to the hypothesized transantarctic flora that was well developed in and around Antarctica during the Eocene. The distribution of the transantarctic flora suggests that a marine seaway across West Antarctic connected the southwest Atlantic and southwest Pacific oceans during Eocene time.

Rare, extra-Antarctic occurrences of some species included in the transantarctic flora suggest that such species should not be considered part of this flora, that the distribution of these taxa was bipolar, or that the complete distribution of the transantarctic flora is yet to be documented.

INTRODUCTION

Seymour Island lies in the Weddell Sea, near the northern end of the Antarctic Peninsula. Two-thirds of the island consists of Cretaceous deposits. The lower Tertiary marine sediments that crop out on the remaining third of Seymour Island are the only such exposures known from the Antarctic area. Previous palynological studies recovered rich dinoflagellate cyst (dinocyst) and spore-pollen assemblages (Cranwell, 1959, 1969a, b; Hall, 1977). These unique palyniferous marine deposits contain the only known in situ early Tertiary dinocyst floras from Antarctica and are of fundamental importance for understanding early Tertiary biostratigraphy of that ice-bound continent. This chapter (1) documents the types and distribution of the dinocysts in the lower

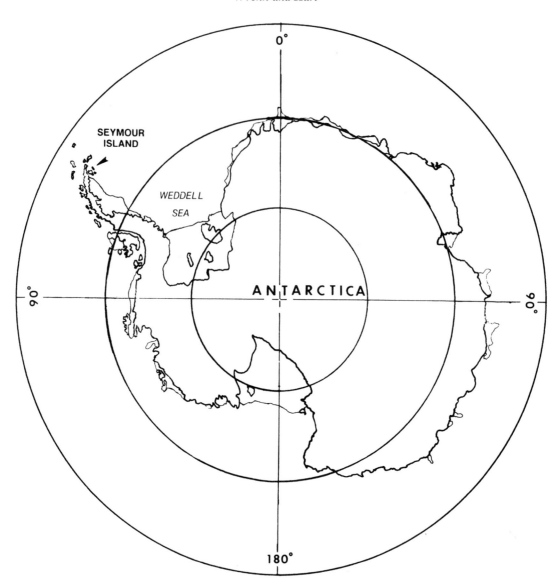

Figure 1. Map of Antarctica showing the general location of Seymour Island.

Tertiary deposits, (2) establishes a biostratigraphic framework based on dinocrysts, and (3) compares the Seymour Island lower Tertiary dinocysts with correlative assemblages.

SEYMOUR ISLAND

Introduction

Seymour Island lies approximately 100 km southeast of the Antarctic Peninsula (Fig. 1) at 64°17′S latitude, 56°45′W longitude, some 25 km east of James Ross Island (Fig. 2). The island is 20.5 km long and 9.6 km wide; it rises approximately 200 m above sea level at its highest elevation (Elliot et al., 1975), al-

though most of it is much less. The geology of the island is well exposed because it is one of the few islands in the area devoid of permanent ice, and the frigid climate precludes the growth of most terrestrial plants.

Seymour Island is separated into two unequal physiographic areas by a northwest-southeast-trending valley, Cross Valley (Fig. 3). The low-lying terrain southwest of Cross Valley is hummocky and has been highly dissected by running water. The smaller, northeastern portion of the island is dominated by a flat-topped, undissected meseta 180 to 200 m in elevation (Andersson, 1906; Elliot et al., 1975). Northeast of the meseta is a smaller, highly dissected area physiographically similar to the southwest portion of the island.

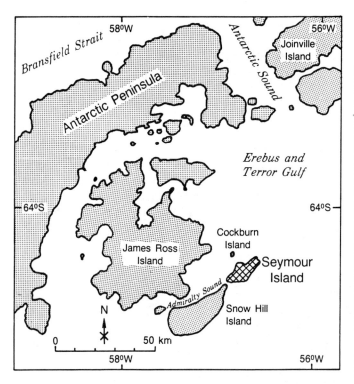

Figure 2. Map of the northern Antarctic Peninsula and the James Ross Island group showing the location of Seymour Island. (After Trautman, 1976).

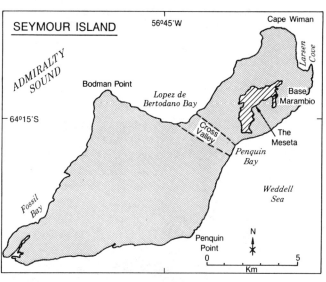

Figure 3. Map of Seymour Island showing the location of geographic features.

Steep sea cliffs border the island except where meltwater streams debouch into the sea (Andersson, 1906; Zinsmeister, 1976b). The beach and intertidal morphology is controlled by sea currents and the movement of sea ice (Zinsmeister, 1976b).

Sir James Clark Ross discovered Seymour Island, part of the James Ross Island group (Fig. 2), while exploring the northwestern portion of the Weddell Sea in 1843. He landed on neighboring Cockburn Island on January 6 and claimed the entire island group for the British Crown (Ross, 1847). Originally, Seymour Island was thought to be a peninsula of Snow Hill Island; accordingly, it was named "Cape Seymour" after Rear Admiral Sir George Francis Seymour. Subsequent exploration during the early 1890s established its insular nature and it was renamed Seymour Island (Donald, 1893).

There is no record of any landing being made on Seymour Island until December 4, 1892, when Carl A. Larsen, captain of the Norwegian whaling ship *Jason,* explored a portion of the island, collected geologic samples, and claimed the whole of it for Norway. Larsen returned to the island during the following whaling season to collect additional fossils. The Larsen collection of fossilized plants and Tertiary invertebrates is the earliest collection of Antarctic fossils still in existence (Bertrand, 1971). The Swedish South Polar Expedition (1901–1903) recovered a greater variety of fossils than did Larsen, including Tertiary

plants, invertebrates, penguin and whale bones, and Cretaceous invertebrates and plants.

Seymour Island was explored by expeditions of the Falkland Island Dependency Survey during the 1940s and 1950s. W. N. Croft (1947) mapped the island and collected fossils from Cretaceous and Tertiary deposits during the 1946 field season. Among the fossils recovered were well-preserved penguin bones (Marples, 1953). Seymour Island was remapped in 1953, and the unconformable relationship between the Cretaceous and Tertiary systems was clearly established (Adie, 1958).

Argentina established a year-round research station, named Base Marambio (Fig. 3), on Seymour Island during the austral summer of 1969. Geologists of the Instituto Antarctico Argentino began extensive mapping and sampling of the Cretaceous beds in the southern two-thirds of the island during 1973. At the invitation of Argentinian scientists, geologists from the Institute of Polar Studies at The Ohio State University (OSU) and from Northern Illinois University (NIU) participated in the 1974–1975 field season (Elliot et al., 1975). The group from the United States began a detailed study of the stratigraphy, sedimentary petrology, and paleontology of the Tertiary beds, while the Argentine geologists continued their studies of the Cretaceous deposits. This chapter presents the palynological results of an investigation of the samples collected for NIU in 1975.

Geologic Setting

Tectonic uplift of the northern Antarctic Peninsula during the Andean Orogeny created a foreland basin in the James Ross Island area by at least the middle Cretaceous. Coarse conglomeratic beds of Turonian age or older crop out in this region and pass upward into Upper Cretaceous sandstones and shales (Elliot, 1982).

The Cretaceous-Tertiary angular unconformity on Seymour Island suggests that a phase of the Andean Orogeny led to uplift and erosion, following the deposition of the Cretaceous beds (Trautman, 1976). Subsequently, a depositional basin again was established in the Seymour Island area, and sedimentation commenced anew during the early Tertiary. Sediments were carried from northwestern highlands into the foreland basin by southeastward-flowing rivers. Petrologic studies indicate the source terrain was an emerging Cordilleran belt, the remnants of which crop out on the Antarctic Peninsula. Rapid sedimentation kept up with subsidence, and all deposits accumulated under strandline or deltaic conditions (Trautman, 1976).

The sedimentary sequence on Seymour Island was divided into the "Older Seymour Island beds" of the Cretaceous System and the "Younger Seymour Island beds" of the Tertiary System by Andersson (1906). Although Andersson (1906) referred these beds to the "Cretaceous Series" and the Tertiary "Seymour Island Series," he did not formally name or describe the units.

The Seymour Island deposits are currently divided into the Cretaceous López de Bertodano Formation and the Paleogene Sobral, Cross Valley, and La Meseta Formations (Elliot and Trautman, 1982; Rinaldi, 1977, 1982; Sadler, this volume). Thin gravel beds cap the lower Tertiary deposits of the meseta in the northeastern part of the island. These beds are no older than ?late Tertiary or Quaternary in age. They have received little attention beyond the recognition of their existence by Trautman (1976). Detailed discussions of the stratigraphy and lithology can be found in Trautman (1976), Elliot and Trautman (1982), Rinaldi (1982), Macellari (1984), and Sadler (this volume).

The López de Bertodano Formation occupies the majority of the southwestern two-thirds of the island. These Cretaceous deposits consist of approximately 1,100 m of loosely consolidated tan to medium gray intercalated siltstone and concretionary sandstone beds. No samples of the López de Bertadano Formation were studied during this investigation.

The Cretaceous deposits are unconformably overlain by the Sobral Formation (Rinaldi et al., 1978; Sadler, this volume; Askin, this volume). The Sobral Formation type section crops out inland from and along the southeastern coast, between Penguin Point and Cross Valley. This formation consists of tabular silts, mudstones, and sandstones intercalated with lenticular units (Sadler, this volume). No samples of the Sobral Formation were studied.

The Cross Valley Formation is strongly lenticular, and its sediments range from mudstones to conglomerates. The type section of the Cross Valley Formation (Fig. 4) consists of 106 m of poorly sorted, pebbly coarse sands and sandstones that pass upward into carbonaceous silts and silty sandstone beds (Trautman, 1976). Large carbonized logs occur in the terrestrial sandstone beds that compose approximately the lower 80 m of the section. Leaf impressions and carbonized plant debris are present in the overlying 20 m or so of tuffaceous silts and silty sandstones. The uppermost beds are thin, resistant sandstones containing poorly preserved marine molluscs (Zinsmeister, 1982). Lithologically similar deposits occur in the Cape Wiman area.

The La Meseta Formation (Fig. 4) crops out in the northeastern third of the island and was established by Elliot and Trautman (1982). The formation includes at least 620 m of intercalated conglomerates, sandstones, shell banks and unconsolidated silts, and sands and clays (Trautman, 1976; Sadler, this volume).

Previous Paleontologic Investigations Pertaining to the Age of the Palaeogene Deposits

The early Tertiary fossils of Seymour Island have attracted the interest of geologists since the early 1890s. Although most of the fossils collected by the 1893 Larsen whaling expedition were considered to be "mostly Jurassic forms" (Donald, 1893), the molluscs were subsequently determined to be early Tertiary in age (Sharman and Newton, 1894, 1898; Fig. 5).

The first extensive paleontological investigations of the Seymour Island Paleogene deposits were conducted on the large fossil collection recovered by the 1901–1903 Swedish South Polar Expedition (Wiman, 1905; Buckman, 1908; Dusen, 1908; Smith-Woodward, 1908; Wilckens, 1911). Fossil leaf and mollusc assemblages indicated a late Oligocene–early Miocene age for these deposits (Dusen, 1980; Wilckens, 1911). The sparse fish remains indicated a Tertiary age for the samples studied by Smith-Woodward (1908).

Palynological study of a rock sample collected by the Swedish Expedition shed little light on the age of the Seymour Island deposits. Cranwell (1959) noted the presence of reworked Cretaceous miospores and stated, "For the younger component an upper Tertiary limit cannot as yet be hazarded: it contains types represented in early Tertiary deposits elsewhere in the southern hemisphere and seems to lack others, such as grasses, sedges, and composites, which are usually present by the Miocene." This noncommital statement was subsequently interpreted to indicate an early Tertiary age (Hall, 1977) and, more precisely, a Paleocene age (Elliot et al., 1975; Zinsmeister, 1976a, 1977, 1982). Both interpretations seem to be in error because, Cranwell (1959, p. 3) reported that Harrington had informed her that the Cretaceous System was separated from probable early Miocene deposits by an angular unconformity. She then stated, "This clears up the longstanding confusion and makes it clear that no early Tertiary deposits have been discovered."

Cranwell (1969a) subsequently suggested an age equivalent to the early Tertiary (mainly Eocene), as well as to the Maastrichtian to Paleocene (Cranwell, 1969b) "for the younger Seymour

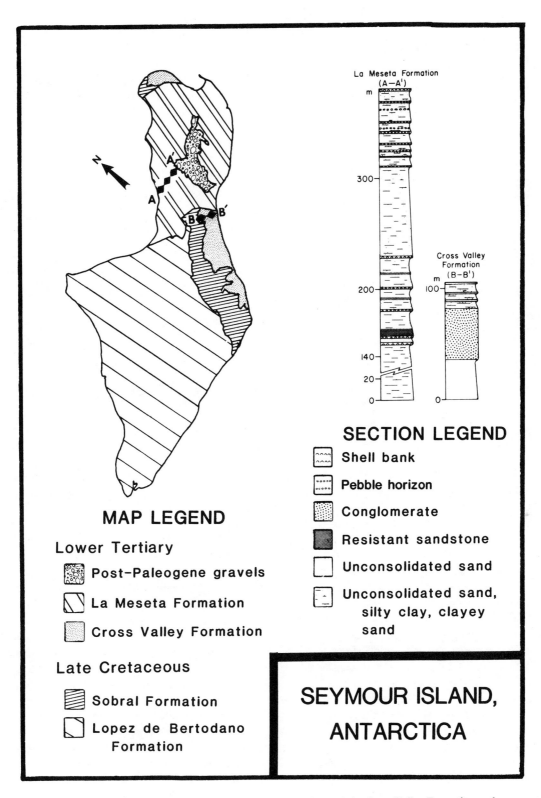

Figure 4. Stratigraphic columns of the La Meseta Formation and the Cross Valley Formation and a geologic location map. (Map after C. Macellari, personal communication, 1984; sections after Trautman, 1976, and Welton and Zinsmeister, 1980.)

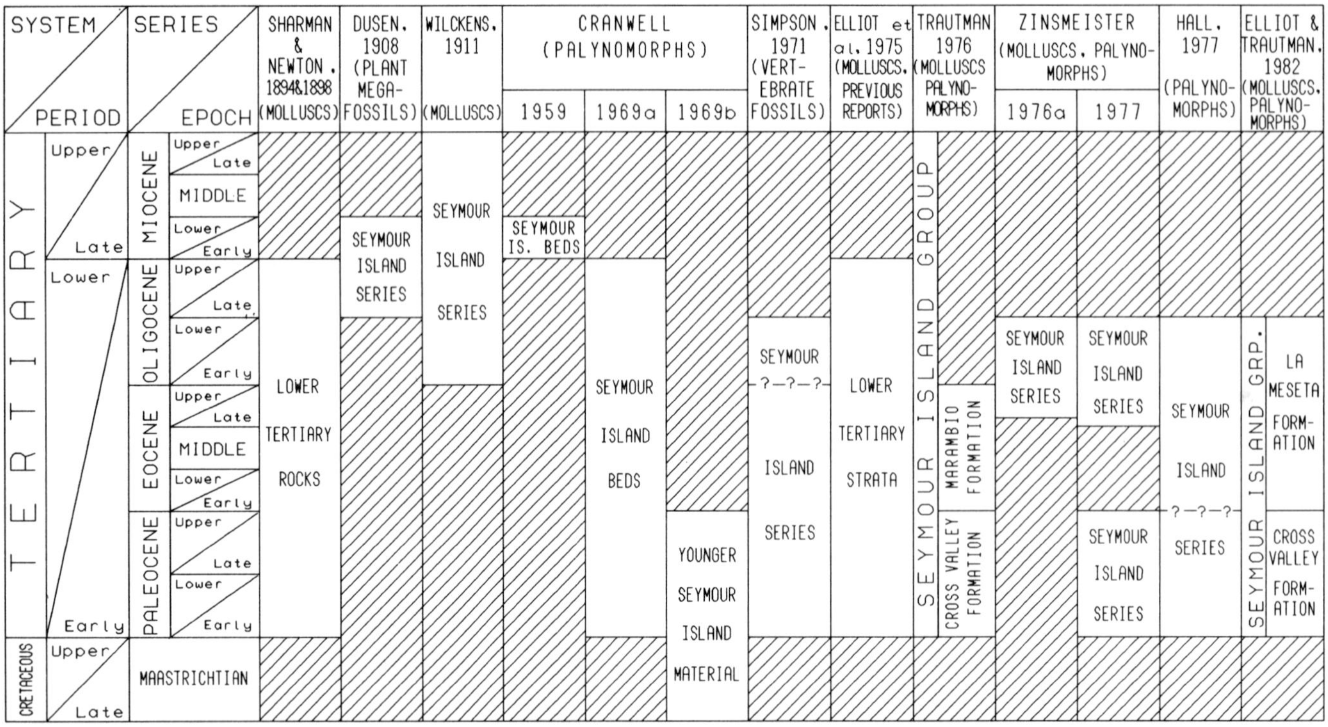

SYSTEM / SERIES — PERIOD / EPOCH	SHARMAN & NEWTON, 1894&1898 (MOLLUSCS)	DUSEN, 1908 (PLANT MEGA-FOSSILS)	WILCKENS, 1911 (MOLLUSCS)	CRANWELL (PALYNOMORPHS) 1959	1969a	1969b	SIMPSON, 1971 (VERTEBRATE FOSSILS)	ELLIOT et al. 1975 (MOLLUSCS, PREVIOUS REPORTS)	TRAUTMAN 1976 (MOLLUSCS, PALYNOMORPHS)	ZINSMEISTER (MOLLUSCS, PALYNOMORPHS) 1976a	1977	HALL, 1977 (PALYNOMORPHS)	ELLIOT & TRAUTMAN 1982 (MOLLUSCS, PALYNOMORPHS)
TERTIARY — MIOCENE Upper/Late, MIDDLE, Lower/Early		SEYMOUR ISLAND SERIES	SEYMOUR ISLAND SERIES	SEYMOUR IS. BEDS	SEYMOUR ISLAND BEDS				SEYMOUR ISLAND GROUP	SEYMOUR ISLAND SERIES	SEYMOUR ISLAND SERIES	SEYMOUR ISLAND SERIES	SEYMOUR ISLAND GRP.
OLIGOCENE Upper/Late, Lower/Early	LOWER TERTIARY ROCKS					YOUNGER SEYMOUR ISLAND MATERIAL	SEYMOUR -?-?-?- ISLAND SERIES	LOWER TERTIARY STRATA	MARAMBIO FORMATION		SEYMOUR ISLAND SERIES		LA MESETA FORMATION
EOCENE Upper/Late, MIDDLE, Lower/Early									CROSS VALLEY FORMATION		SEYMOUR ISLAND SERIES		CROSS VALLEY FORMATION
PALEOCENE Upper/Late, Lower/Early													
CRETACEOUS Upper/Late — MAASTRICHTIAN	MAASTRICHTIAN												

Figure 5. Comparison of the stratigraphic unit names and ages assigned to the lower Tertiary deposits on Seymour Island. (The fossil group(s) on which the age determinations were made are shown beneath each reference.)

Island material." The latter age determination also was reported to be simply Paleocene (Zinsmeister, 1976; Zinsmeister and Camacho, 1982).

Simpson (1971) assigned a middle Eocene to early Oligocene age to the deposits from which he had extracted an assemblage of large fossil penguins. However, he recognized that the penguin fossils were derived from a very limited part of the section and he accepted Cranwell's (1969b) determination that some of the deposits were Paleocene in age.

Elliot et al. (1975) made the first stratigraphically oriented fossil collections from measured sections on Seymour Island. An age as young as Eocene was considered probable for at least part of the early Tertiary deposits. Trautman (1976) considered the Cross Valley Formation and the La Meseta Formation (the Marambio Formation of Trautman, 1976) to be Paleocene and Eocene in age, respectively. The mollusc assemblages indicated a late Eocene–early Oligocene age (Zinsmeister, 1976a).

A palynological study of some of the samples collected for NIU resulted in the recovery of abundant palynomorphs and preliminary age determinations for the lower Tertiary beds (Hall, 1977). Palynomorphs indicated a Paleocene age for the Cross Valley Formation and a late Eocene–early Oligocene age for the La Meseta Formation (Hall, 1977). Zinsmeister (1977) agreed with Hall's findings.

The confusion and the divergence of opinion regarding the age of the Seymour Island lower Tertiary deposits is due in part to the nature of the sample material studied. All studies conducted prior to 1975 were based on grab samples, the precise stratigraphic and geographic location of which were unknown. Various lithologies were collected from various parts of the island, often on scree slopes. As a result, all of the geologic ages reported in the literature could well be represented by deposits somewhere on Seymour Island.

In addition, most previous studies were based on macrofossils; these are not widely nor uniformly distributed through the Tertiary deposits. Often they occur in shell banks, localized concentrations, or as scattered individuals; consequently, their stratigraphic control is discontinuous.

Calcareous microfossils are rare or absent in the Paleogene beds. Some siliceous floras have been recovered, but the entire sequence has yet to be studied (Harwood, 1985).

Only the palynomorphs are known to be abundant, widely distributed, and well preserved. It is upon them that detailed stratigraphic control must be constructed in the complex lower Tertiary deposits of Seymour Island.

Paleoenvironmental Setting

Trautman (1976) postulated that the Paleogene deposits were laid down in a tide-dominated deltaic complex. The nonmarine beds of the Cross Valley Formation at the type section were deposited in high-energy distributary channels and low-energy marshes or swamps.

The informal lithologic units established for the La Meseta Formation also reflect the depositional environments (Elliot and Trautman, 1977). The unfossiliferous, finely laminated silt and silty sandstone beds designated as Unit I by Elliot and Trautman (1977) accumulated in a low-energy environment landward of tidal influence. The alternating fine sand and silty clay beds, sedimentary and biogenic structures, and macrofossils accumulations of their Unit II were considered to be tidally bedded delta front platform deposits (Trautman, 1976). Unit III, on the other hand, was envisioned to have been deposited in a shallow, low-energy marine environment.

The terrestrial to estuarine deposits of the Cross Valley Formation were considered to be stratigraphically lower than the delta plain (Unit I), delta front platform (Unit II), and shallow marine (Unit III) deposits of the La Meseta Formation (Trautman, 1976). Transitional paleoenvironments are probably obscured by the discontinuous contact between the two formations. The overall sequence of paleoenvironments, according to Trautman (1976), indicates a local marine transgression. These paleoenvironmental interpretations are not completely supported by the results of our investigation.

SAMPLE MATERIAL AND PROCEDURES

The 70 samples studied were collected by Elliot et al. (1975) from eight outcrop sections located in the northeastern one-third of Seymour Island. The location of each section and the distribution of the samples within the sections are shown in Figures 6 and 7, respectively. Samples from Sections 12-13, 15, and 16 were collected from the Cross Valley Formation, whereas samples from Sections 3, 17, 18, and 19 are from the La Meseta Formation. Detailed descriptions and stratigraphic sections for the Seymour Island Paleogene deposits have been presented by Trautman (1976), Mascellari (1984), and Sadler (this volume). The lithologic characteristics of each sample studied in this investigation have been reported elsewhere (Wrenn, 1982).

Preparation of the palynologic samples was conducted according to the procedures described by Barss and Williams (1973) and Barss and Crilley (1976). The first 50 microplankton were counted on each of six slides per sample, yielding a total of 300 specimens. However, microplankton assemblages in some sample preparations were so sparse that the total count fell short of 300. Dinocysts, exclusive of those counted, were studied on additional slides.

Slide numbers with an "E" prefix were prepared by Exxon Production Research Company, those prefaced by a "P" were prepared by Amoco Production Company, and those prefaced by a "W" or "M" were prepared for this study by J.H.W. The slide number of each illustrated specimen is followed by the microscope coordinates and the alphanumeric Microlocator square reference (e.g., 8451E/1 [124.5 × 4.2, D30]). The coordinates are

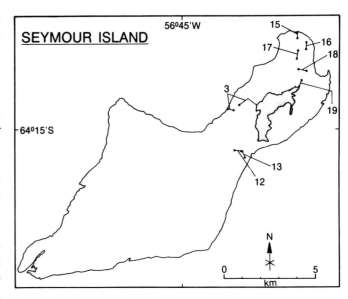

Figure 6. Map depicting the locations of the stratigraphic sections on Seymour Island.

from Zeiss universal microscope (serial no. 038295) belonging to Amoco Production Company, Tulsa, Oklahoma. The Microlocator slide is stored with the Seymour Island sample slides in the Botanical Micropaleontology Laboratory, Department of Geology, Louisiana State University, Baton Rouge.

RESULTS OF THE PALYNOLOGICAL INVESTIGATION

General Results

Of the 70 samples studied, 67 yielded microplankton; of these, concentrations were good to excellent in 49 samples. Microplankton were poorly represented in the remaining 18 samples.

Forty-three dinocyst genera containing 74 species and 4 subspecies were recovered, of which 2 genera and 9 species are new. Six genera and 4 species of acritarchs, 2 genera of Pterospermatales, and 2 genera and 3 species belonging to the Chlorococcales were recovered. Rare specimens of diatoms, silicoflagellates, and scolecodonts were observed in some samples. Miospores and terrestrial plant and/or fungal debris occurred in most samples, often in great abundance (Wrenn, 1982). Nineteen samples were processed for nannofossil examination; all were barren (R. W. Pierce, personal communication, 1981).

Table 1 shows the microplankton species recovered from each section. Figures 8 through 11 record the stratigraphic distri-

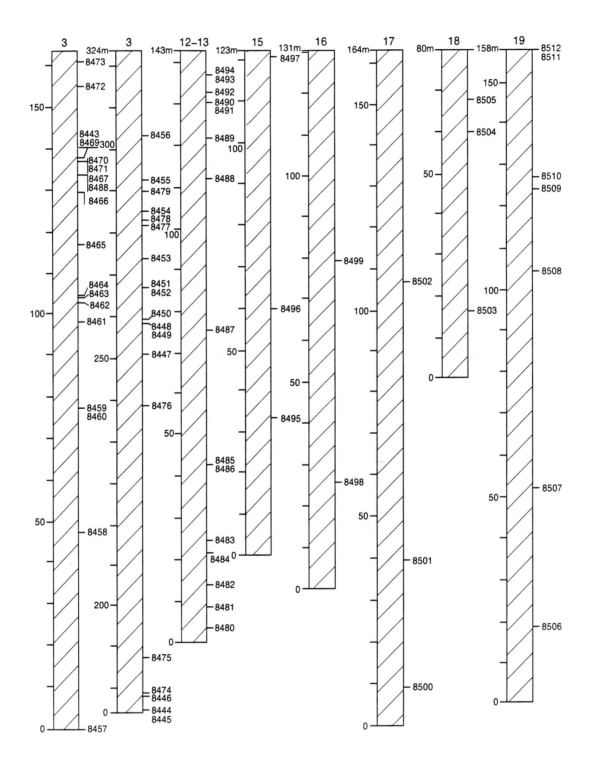

Figure 7. Location of the samples within the sections.

EPOCH	Section	Sample Number	Comasphaeridium cometes	Impletosphaeridium lorum sp. nov.	Spinidinium essol	Spiniferites ramosus subsp. granomembranaceus	Eyrea nebulosa	Spinidinium densispinatum	Paralecaniella indentata	Phthanoperidinium echinatum	Impletosphaeridium clavus sp. nov.	Palaeoperidinium pyrophorum	Deflandrea sp. indet.	Alisocysta circumtabulata	Spiniferites ramosus subsp. reticulatus	Palaeocystodinium golzowense	Cometodinium cf. ?whitei	Cyclopsiella trematophora	Palambages morulosa	Vozzhennikovia apertura	Areosphaeridium diktyoplokus	Palambages Forma B	Spinidinium colemanii sp. nov.	Impletosphaeridium ligospinosum	Spinidinium sp. indet.	Indeterminate	Total Microplankton	POLLENITES Striatiti
EARLY LATE PALEOCENE		8494	X	1	4	X	1	7	13	X																11	37	
		8493	1				1	16	38	X	3	1	1													25	88	
		8492		2	2	X	6	10	153	X	4					1	2	1								25	206	2
		8490				8		25	6	X	1	1					1	1	X	2	5					19	89	
EOCENE (?)	12 – 13	8491																										
		8489																										
		8488						2	X											1	1						4	
		8487																										
		8485					1																1				2	
		8486						4						1						2						9	16	
		8483						1												1			1				3	
		8484																								1	1	
		8482						1	X																	1	2	
		8481																		2			1			1	4	1
		8480																							1	1	2	

Figure 8. Distribution table of microplankton in Section 12-13. (The figures represent specimens tabulated during the counting procedures. Species occurrences observed subsequently, outside the count, are indicated by an "X.")

TABLE 1. MICROPLANKTON SPECIES RECOVERED FROM THE SEYMOUR ISLAND PALEOGENE SECTIONS

Species	3	12-13	15	16	17	18	19
Achomosphaera sp. indet.	X		X				
Alisocysta circumtabulata	X	X	X	X			
Aptea securigera	X						X
Aptea sp. indet.	X						
Arachnodinium antarcticum	X				X		
Aerosphaeridium diktyoplokus	X	X			X	X	X
Batiacasphaera explanata				X	X		X
Botryococcus sp. indet.	X						X
Cerebrocysta bartonensis			X	X			
Cleistosphaeridium ancoriferum	X						
Comasphaeridium cometes	X	X		X			X
Cometodinium cf. *?whitei*	X	X		X		X	X
Cribroperidinium edwardsii							X
Cribroperidinium muderongense	X						
Cribroperidinium sp. indet.	X					X	X
Cyclonephelium distinctum	X				X		
Cyclopsiella elliptica	X						
Cyclopsiella trematophora	X	X	X		X		X
Cymatiosphaera sp. indet.	X		X	X			X
Deflandrea antarctica Type I	X				X	X	X
Deflandrea antarctica Type II	X				X	X	
Deflandrea boloniensis comb. nov.			X	X			
Deflandrea cygniformis	X						X
Deflandrea dartmooria	X		X				
Deflandrea oebisfeldensis sensu Cookson and Cranwell, 1967	X						X
Deflandrea cf. *phosphoritica*	X						
Deflandrea speciosa	X		X				
Deflandrea striata	X						
Deflandrea webbii nov. sp.	X						X
Deflandrea sp. indet.	X	X	X		X	X	X
Diconodinium cristatum	X						X
Diconodinium multispinum	X				X		X
Dinogymnium cf. *curvatum*	X						
Enigmadinium cylindrifloriferum gen. nov. sp. nov	X				X		
Eyrea nebulosa	X	X	X	X	X	X	X
Hystrichosphaeridium cf. *astartes*	X						
Hystrichosphaeridium parvum	X			X	X	X	X
Hystrichosphaeridium salpingophorum			X	X			X
Hystrichosphaeridium truswelliae sp. nov.	X				X	X	
Hystrichosphaeridium tubiferum subsp. *brevispinum*	X		X		X	X	
Hystrichosphaeridium sp. indet.							X
Impagidinium dispertitum	X						
Impagidinium maculatum			X	X			X

Species	3	12-13	15	16	17	18	19
Impagidinium victorianum							X
Impagidinium sp. indet.	X						
Impletosphaeridium clavus sp. nov.	X	X	X	X	X	X	X
Impletosphaeridium ligospinosum	X	X	X		X	X	X
Impletosphaeridium lorum sp. nov.	X	X	X	X	X	X	X
Impletosphaeridium sp. B.	X				X		X
Isabelidinium druggii	X						
Isabelidinium pellucidum	X						X
Kallosphaeridium cf. *capulatum*	X		X	X			
Leiofusa jurassica	X				X		
Lejeunecysta fallax	X						
Lejeunecysta hyalina	X		X				
Lejeunecysta sp. indet.	X		X		X		
Micrhystridium sp. A.	X						X
Octodinium askiniae gen. nov. sp. nov.	X						X
Odontochitina spinosa	X						
Oligosphaeridium complex	X				X		X
Oligosphaeridium pulcherrimum	X						
Oligosphaeridium sp. indet.	X		X				
Operculodinium bergmannii	X		X	X			X
Ophiobolus lapidaris	X		X	X		X	X
Palaeocystodinium australinum			X	X			
Palaeocystodinium glozowense	X	X	X	X			
Palaeocystodinium granulatum	X						
Palaeocystodinium sp indet.	X			X			
Palaeoperidinium pyrophorum	X	X	X	X		X	X
Plamabages Forma B (of Manum and Cookson, 1964)			X				
Palambages morulosa	X	X	X				X
Palambages Type I	X				X		X
Paralecaniella indentata	X	X	X	X	X	X	X
Pareodinia sp. indet.						X	
Phelodinium boldii sp. nov.	X		X		X	X	X
Pterospermella australiensis	X						
Phthanoperidinium echinatum	X	X			X		X
Selenopemphix nephroides	X						
Senegalinium asymmetricum	X					X	X
Spinidinium colemanii sp. nov.	X	X	X	X			
Spinidinium densispinatum	X	X	X	X			X
Spinidinium essoi	X	X			X	X	X
Spinidinium lanterna	X			X			
Spinidinium luciae sp. nov.	X					X	X
Spinidinium macmurdoense	X				X	X	X
Spinidinium sp. indet.	X	X		X	X	X	X
Spiniferites multibrevis	X						X
Spiniferites ramosus		X			X	X	

TABLE 1. MICROPLANKTON SPECIES RECOVERED FROM THE SEYMOUR ISLAND PALEOGENE SECTIONS
(continued)

Species	3	12-13	15	16	17	18	19
Spiniferites ramosus subsp. *granomembranaceus*	X	X					
Spiniferites ramosus subsp. *granosus*	X						
Spiniferites ramosus subsp. *reticulatus*	X	X			X	X	X
Spiniferites sp. indet.	X		X	X			X
Tanyosphaeridium xanthiopyxides			X				
Thalassiphora pelagica	X				X		
Trigonopyxidia ginella			X				
Turbiosphaera filosa	X						
Veryhachium sp. indet.	X					X	
Vozzhennikovia apertura	X	X			X	X	X
Xylochoarion cf. *hacknessense*	X						

Figure 9. Distribution table of microplankton in Sections 15 and 16. (The figures represent specimens tabulated during the counting procedures. Species occurrences observed subsequently, outside the count, are indicated by an "X.")

Epoch	Section	Sample Number	Comasphaeridium cometes	Hystrichosphaeridium parvum	Spinidinium lanterna	Spiniferites ramosus	Spinidinium sp. indet.	Achomosphaera sp. indet.	Alisocysta circumtabulata	Cometodinium cf. ?whitei	Cymatiosphaera sp. indet.	Hystrichosphaeridium salpingophorum	Impletosphaeridium lorum sp. nov.	Kallosphaeridium cf. capulatum	Operculodinium bergmannii	Palaeocystodinium golzowense	Paralecaniella indentata	Spinidinium densispinatum	Spiniferites sp. indet.	Palaeocystodinium sp. indet.	Impagidinium maculatum	Palaeocystodinium australinum	Impletosphaeridium clavus sp. nov.	Ophiobolus lapidaris	Deflandrea boloniensis comb.nov.	Cerebrocysta bartonensis	Eyrea nebulosa	Palaeoperidinium pyrophorum	Spinidinium colemanii sp. nov.	Lejeunecysta hyalina	Palambages morulosa	Trigonopyxidia ginella	Deflandrea dartmooria	Cyclopsiella trematophora	Impletosphaeridium ligospinosum	Deflandrea speciosa	Deflandrea sp. indet.	Hystrichosphaeridium tubiferum subsp. brevispinum	Lejeunecysta sp. indet.	Oligosphaeridium sp. indet.	Phelodinium boldii sp. nov.	Tanyosphaeridium xanthiopyxides	Indeterminate	Total microplankton
EARLY LATE PALEOCENE	16	8499	X	1	238	1	X	1	1	1	1	X	2	2	5	1	5	1	1																								39	300
		8498				1		5								7		1	1		3	1	2	X	3	1	7	2	221	25													3	283
	15	8497			1			16			3		10	X	13	27	2		1	1	8				5	1	3	57	50	1	1	1	X	1			12					93	307
		8496					X	2								1	1				7	2	2	5	1	X	X	39	119	58						5	5						67	312
		8495			1	1	X				3		5		1	3	9		7		2				1	1		9	156	45	2	1	8	5	4	1			X	X	5	1	30	301

Epoch groupings (left margin): "LATE MIDDLE TO LATE EOCENE" (samples 8456, 8455, 8479, 8454); "LATE EARLY EOCENE" (samples 8478 through 8458); "HOLOCENE" (sample 8457). SECTION 3. MICROPLANKTON.

Sample Number	Deflandrea cygniformis	Lejeunecysta fallax	Lejeunecysta hyalina	Selenopemphix nephroides	Octodinium askiniae gen. nov., sp. nov.	Lejeunecysta sp. indet.	Palambages Type I	Deflandrea sp. indet.	Eyrea nebulosa	Impletosphaeridium clavus sp. nov.	Impletosphaeridium sp. B	Operculodinium bergmannii	Ophiobolus lapidaris	Palaeoperidinium pyrophorum	Palambages morulosa	Paralecaniella indentata	Vozzhennikovia apertura	Aptea securigera	Diconodinium cristatum	Diconodinium multispinum	Achomosphaera sp. indet.	Palaeocystodinium sp.	Deflandrea oebisfeldensis sensu Cookson & Cranwell	Cometodinium cf. ?whitei	Pterospermella australiensis	Micrhystridium sp. A	Phelodinium boldii sp. nov.	Spinidinium colemanii sp. nov.	Spinidinium essoi	Spinidinium macmurdoense	Spiniferites sp. indet.	Spinidinium luciae sp. nov.
8456	3	8	X	36	1	X	1	1	14	5	1	9	1	1	2	2	171															
8455	31			5				2	45								7	1	1	1												
8479	9							2	13	109	80	3		1		1	10			4	X	1	1	1	10	3	3	1	1	1	2	5
8454								2	2								6															1
8478	15							12	9			1		1	1	2	19	1	1					X		2						47
8477	4	6	5	19	16			4	13					1		1	145						X		3						1	23
8453								7									29							5								2
8451								4									5	5	1	3												2
8452									1								7	1														15
8450				3					22					1		8	16					1		1		2						
8448									105	1	1					1	1					1		1								
8449						X			145	2		2				5	9					X		6								
8447									94	4		3		1		18	27	1						1		1			13	16	1	
8476									6	14	2				1	3	222				1	X		1	7	1	1	1				
8475									3	40	10		1	3		10	20					1		1		2	1		6	14		
8474									9	3	1		2			13	38						2			1			4	3		
8446									4							27	1															
8444									23							36	1							1								
8445																																
8473									2	1			2		1	17	16										1		3	1		
8472									5	1	1	1			1	23	45	1	X						1	1			4	5		
8443								1	3				1	1	1	15	22					X							4	2		
8469									1							4	3												2	1		
8470			X	X					4			1	1		1	10	35						X						3	3		
8471									1							5	6												1			
8467							X	1	2	1			2			12	11										1		1	9		
8468									1		10		2		1	8	30					X							13	5		
8466			1				1		23				1			253	3															
8465			1						14	12	1		3	1	1	31	62									1	1		3	23		
8464				2				2	17	15		1		49	6	50	11									1	1	8	3	12	1	
8463						X			13	3	2		1	1		7	145												31	2		
8462									17		1		1	X	1	13	36												41			1
8461									3	5			1	1	2		20									1			1			
8459																																
8460						X			104	3			1		5	1	136	8											2	2		
8458						X		3	23	23	2	4	1	1		12	90					X			1	1	2	2			48	3
8457								3	13	39	1	1	1	16	1	8	14								30	1	10		2	2		1

Figure 10 (continued on following two pages). Distribution table of microplankton in Section 3. (The figures represent specimens tabulated during the counting procedures. Species occurrences observed subsequently, outside the count, are indicated by an "X.")

EPOCH	SECTION	Sample Number	Isabelidinium pellucidum	Cyclonephelium distinctum	Hystrichosphaeridium astartes	Phthanoperidinium echinatum	Hystrichosphaeridium parvum	Cyclopsiella trematophora	Aptea sp. indet.	Impletosphaeridium lorum sp. nov.	Manumiella druggii	Deflandrea webbii sp. nov.	Cribroperidinium sp. indet.	Cymatiosphaera sp. indet.	Cleistosphaeridium ancoriferum	Spiniferites ramosus subsp. reticulatus	Oligosphaeridium complex	Oligosphaeridium pulcherrimum	Spinidinium lanterna	Botryococcus sp. indet.	Deflandrea antarctica Type I	Areosphaeridium diktyoplokus	Spinidinium sp. indet.	Comasphaeridium cometes	Veryhachium sp. indet.	Dinogymnium cf. curvatum	Odontochitina spinosa	Leiofusa jurassica	Impletosphaeridium ligospinosum	Deflandrea antarctica Type II	Hystrichosphaeridium truswelliae sp. nov.	Spiniferites ramosus	Kallosphaeridium cf. capulatum	Arachnodinium antarcticum
LATE MIDDLE TO LATE EOCENE		8456																																
		8455																																
		8479																																
		8454																																
		8478	1	1	2	3	2	1																										
		8477	1		X	9	2		1	1	X																							
LATE EARLY EOCENE	3	8453			1										1	X	1																	
		8451									8		2		1	X	2	X	1															
		8452		2												1					1	2												
		8450																			1	8	2											
		8448																			1	3	1											
		8449			1					X											5	3	3	14										
		8447		2																	11	1		1										
		8476		1		1													X	X	1	3				X	X							
		8475		1		3						1	1							1	2	74	5					2	2					
		8474		1		1												X	1		25	185									1			
		8446																			1	4												
		8444				1															1	2												
		8445																		1														
		8473		1																		80									1			
		8472	X		1				X												1	88									1		1	1
		8443			1	1	2	1							1						2	209												X
		8469			X								1									26	1											X
		8470			1	2									1						2	218									1	1		1
		8471																				20												
		8467				4								X			X					238	10											1
		8468		5	1																1	100	1								1	1	1	2
		8466			1																1	11	4			3								
		8465		9	1	1												2			15	77	3							15	10	1	1	X
		8464		2								2	1				X				2	3	33								8	3	1	
		8463																			5	10	18								15	1		
		8462		X		4															9	11	5								3			
		8461		2		1																3												
		8459																				3						1						
		8460				9											X				2	12												
		8458			4	3	1								1		2		1		4	23	2	5							1			X
HOLOCENE		8457				4					2		X		1	X	1				2	12	5	1							8		X	

EPOCH	SECTION	Sample Number	Spiniferites ramosus subsp. granomembranaceus	Xylochoarion cf. hackenssense	Thalassiphora pelagica	Turbiosphaera filosa	Hystrichosphaeridium tubiferum subsp. brevispinium	Deflandrea cf. phosphoritica	Impagidinium sp. indet.	Oligosphaeridium sp. indet.	Alisocysta circumtabulata	Senegalinium asymmetricum	Spinidinium densispinatum	Spiniferites ramosus subsp. granosus	Deflandrea dartmooria	Deflandrea striata	Cribroperidinium muderongense	Spiniferites multibrevis	Cyclopsiella cf. elliptica	Impagidinium dispertitum	Palaeocystodinium granulatum	Deflandrea speciosa	Enigmadinium cylindrifloriferum gen. nov. sp. nov.	Palaeocystodinium golzowense	Indeterminate Microplankton	Total Microplankton	POLLENITES Striatiti
LATE MIDDLE TO LATE EOCENE		8456																							43	300	
		8455																							14	107	5
		8479																							43	301	6
		8454																								11	
		8478																							23	147	2
		8477																							45	300	13
LATE EARLY EOCENE	3	8453																							4	48	
		8451																							22	57	17
		8452																							13	43	12
		8450																							15	80	
		8448																								116	
		8449																							13	207	1
		8447																							47	342	4
		8476																							33	299	3
		8475																							41	245	4
		8474																							21	311	
		8446																							4	41	
		8444																							5	70	
		8445																							1	2	
		8473																							12	138	
		8472																							16	198	
		8443																							20	287	
		8469																							2	41	
		8470	1																						19	305	1
		8471																							4	37	
		8467		X	3	1																			17	316	
		8468		3																					16	202	1
		8466		1																					3	306	
		8465	2	1			1	1	1	X															15	309	1
		8464		X			X					1	8	15	X										47	303	
		8463		1			5										X	X	X	X					50	311	
		8462			1		18	3				6								X	2				38	211	
		8461																							3	43	
		8459																								4	
		8460																					1		12	300	5
		8458	2				X																			300	5
HOLOCENE		8457	X		1		1					2									2	1	87	1	33	272	

bution and frequency of each microplankton species in Sections 12-13, 15, and 16, 3, and 17 through 19, respectively. Figures 15 through 47 show the microplankton recovered during this study.

Age Determinations

The known ranges of the biostratigraphically significant taxa are shown in Table 2. The distribution of these taxa in the Seymour Island sections and the age determinations based on them are discussed below.

Cross Valley Formation. Sections 12-13, 15, and 16 were collected from the Cross Valley Formation (Fig. 6). The palynological results raise questions about what was actually sampled by Section 12-13 (see below). Consequently, Sections 15 and 16 are discussed first.

The biostratigraphically significant taxa in Sections 15 and 16 include *Alisocysta circumtabulata* (late Campanian–late Paleocene), *Deflandrea boloniensis* comb. nov. (Campanian-Paleocene), *D. dartmooria* (Maastrichtian–late early Eocene), *D. speciosa* (late early Paleocene–late early Eocene), *Palaeocystodinium australinium* (late Campanian–late Paleocene), *Palaeoperidinium pyrophorum* (late Campanian–early late Paleocene), and *Spinidinium densispinatum* (late Campanian–late Paleocene). The maximum age of the time interval is set by the first appearance of *Deflandrea speciosa* (late early Paleocene), and the minimum age is set by the extinction of *A. circumtabulata* (late Paleocene) and *P. pyrophorum* (early late Paleocene). The overlapping ranges of these taxa and their co-occurrence in Sections 15 and 16 (Fig. 9) indicate an early late Paleocene age for these sections.

Section 12-13, from the type locality of the Cross Valley Formation in Cross Valley, is not so easily interpreted. The numerous faults in the area, the active mass wasting on the valley slopes, and the lack of detailed geologic mapping at the time of sampling render the biostratigraphic interpretation of the Cross Valley samples tenuous at best.

The stratigraphically significant taxa recovered from Section 12-13 and their age ranges in the literature include *Alisocysta circumtabulata* (late Campanian–late Paleocene), *Areosphaeridium diktyoplokus* (late early Eocene–early Oligocene), *Palaeoperidinium pyrophorum* (late Campanian–early late Paleocene), *Spinidinium densispinatum* (late Campanian–late Paleocene), *S. essoi* (late Cretaceous–early Oligocene), *Spiniferites ramosus* subsp. *granomembranaceus* (late Cretaceous–early Eocene), and *Vozzhennikovia apertura* (Eocene–early Oligocene). The overlapping ranges of *Palaeoperidinium pyrophorum, Spinidinium densispinatum, Alisocysta circumtabulata,* and *S. essoi* suggest an early late Paleocene age for the top three samples (sample numbers 8492 through 8494) of Section 12-13. (The very rare occurrence of *Phthanoperidinium echinatum* in all three samples is considered to be either field or laboratory contamination.) This interpretation is supported by the presence of diagnostic late Paleocene silicoflagellates in the upper laminated beds in Section

12-13 (Harwood, 1985), including *Crassicorbisema disymmetrica disymmetrica, Corbisema naviculoides, C. constriata,* and *C.* cf. *C. cuspis.*

The underlying samples (8480 through 8491) contain very few dinocysts. However, the presence of *Vozzhennikovia apertura, Areosphaeridium diktyoplokus, Phthanoperidinium echinatum,* and *Spiniferites ramosus* subsp. *granomembranaceus* suggest an Eocene age. Sample 8490 is provisionally included with the lower sediments of Eocene age because of the presence of rare specimens of Eocene–early Oligocene dinocysts. However, the lithology (siltstone) and the relative abundance of dinocysts suggest that this sample has more in common with the overlying laminated fine-grained Paleocene sediments.

It is possible that the samples studied were collected, unwittingly, from deposits or fault blocks of differing age that are now recognized in Cross Valley (Trautman, 1976). If this is correct, early late Paleocene and Eocene age determinations are correct for the individual samples indicated.

Alternatively, it is possible that the coarse sand samples 8480 through 8491 are essentially barren, except for reworked taxa and field or laboratory contaminants. If this is correct, none of the age determinations are reliable for Section 12-13. Clearly, reexamination of the stratigraphic interval covered by Section 12-13 is warranted.

La Meseta Formation. The stratigraphic distribution of the microplankton recovered from Section 3 and Sections 17, 18, and 19 are shown in Figures 10 and 11, respectively. The lowermost sample in Section 3, sample 8457, was collected on a modern tidal flat at the base of the section and is Holocene in age. This sample contains a reworked assemblage derived from coastal outcrops that range from Maastrichtian to Eocene in age.

The rest of Section 3 can be divided into a lower interval that extends from 0 to 274 m (Interval A) and an overlying interval (Interval B) from 274 m to the top of the section (at approximately 324 m above the section base at sea level). The base of the known stratigraphic ranges of *Arachnodinium antarcticum* (Eocene), *Areosphaeridium diktyoplokus* (late early Eocene to early Oligocene), *Deflandrea antarctica* (Eocene to early Oligocene), *Spinidinium macmurdoense* (Eocene to early Oligocene), *Vozzhennikovia apertura* (Eocene to early Oligocene), and *Phthanoperidinium echinatum* (late early Eocene to early Oligocene) indicate a maximum age for Interval A of late early Eocene or younger. *Lejeunecysta fallax, Deflandrea cygniformis,* and *Selenopemphix nephroides* are absent from Interval A, but they are present in the overlying Interval B. The absence of these middle Eocene or younger taxa (Table 2) suggests that Interval A is older than the middle Eocene. The combined evidence suggests a late early Eocene age for the samples studied from Interval A.

A recent study of the diatoms from the La Meseta Formation supports an early Eocene age for the lower part of the formation (Harwood, 1985). Diagnostic early Eocene diatoms include *Coscinodiscus bulliens, Pyxilla gracilis,* and *Craspedodiscus molleri.* In addition, the absence of the three diatom genera *Brightwellia, Rylandsia,* and *Asterolampra* that first appear in the

EPOCH	Section	Sample Number	Botryococcus sp. indet.	Octodinium askiniae gen. nov., sp. nov.	Spinidinium densispinatum	Deflandrea webbii sp. nov.	Impagidinium victorianum	Micrhystridium sp. A	Operculodinium bergmannii	Hystrichosphaeridium sp. indet.	Spiniferites sp. indet.	Impagidinium maculatum	Senegalinium asymmetricum	Batiacasphaera explanata	Deflandrea sp. indet.	Hysrichosphaeridium parvum	Paralecaniella indentata	Spinidinium essoi	Cyclopsiella trematophora	Eyrea nebulosa	Impletosphaeridium ligospinosum	Phelodinium boldii sp. nov.	Phthanoperidinium echinatum	Spinidinium sp. indet.	Vozzhennikovia apertura	Cymatiosphaera sp. indet.	Areosphaeridium diktyoplokus	Impletosphaeridium clavus sp. nov.	Impletosphaeridium sp. B	Palambages morulosa	Cometodinium cf. ?whitei	Aptea securigera	Pareodinia sp. indet.	Cribroperidinium sp. indet.	Diconodinium multispinum	Spinidinium macmurdoense
MIDDLE TO LATE EOCENE	19	8512	3	1	5	1	2	1	9	1	x	1	14	1	2	1	1	4	7	20	1	1	113	1	74											
		8511			1	X	14	3				6			1	12	12	2		10	X	1	55	1	104	1	1	25	4							
		8510						1										28		1					2				1	64						
		8509														2		32		12					1					245	1	X	X		1	2
		8508						3								1	10	1					92	3	26	3				65	X					
LATE EARLY EOCENE	18	8507														11	1	6	2				29	1	46					6	5	1			1	
		8506									1	3	1			1	3	15		18	1		2		65					107	2		2		2	23
		8505														2		66					1		2					10						3
		8504											14			3	2	12		8			2		1											88
		8503										1	1		1	128	11	22	4	2			2	7	32	13				2					X	21
	17	8502										1	1			3		18	1	5				4	23					117	7					1
		8501										1	1			1		26	X	4				2	18					131	3				1	1
		8500														1		2	X	X			3	6	8					167	12		1			8

MICROPLANKTON

Figure 11 (continued on next page). Distribution table of microplankton in Sections 17, 18, and 19. (The figures represent specimens tabulated during the counting procedures. Species occurrences observed subsequently, outside the count, are indicated by an "X.")

MICROPLANKTON

EPOCH	Section	Sample Number	Cribroperidinium edwardsii	Deflandrea cygniformis	Deflandrea obisfeldensis sensu Cookson & Cranwell	Diconodinium cristatum	Palaeoperidinium pyrophorum	Spinidinium luciae sp. nov.	Oligosphaeridium complex	Palambages Type I	Spiniferites ramosus subsp. reticulatus	Isabelidinium pellucidum	Comasphaeridium cometes	Hystrichosphaeridium salpingophorum	Spiniferites multibrevis	Impletosphaeridium lorum sp. nov.	Deflandrea antarctica Type I	Ophiobolus lapidaris	Hystrichosphaeridium truswelliae sp. nov.	Hystrichosphaeridium tubiferum subsp. brevispinum	Deflandrea antarctica Type II	Cyclonephelium distinctum	Veryhachium sp. indet.	Enigmadinium cylindrifloriferum gen. nov. sp. nov.	Leiofusa jurassica	Spiniferites ramosus	Arachnodinium antarcticum	Lejeunecysta sp. indet.	Thalassiphora pelagica	Indeterminate Microplankton	Total Microplankton	Striatiti (POLLENITES)
MIDDLE TO LATE EOCENE	19	8512																												32	297	
		8511																												47	300	
		8510																												8	103	2
		8509																												4	300	8
		8508	X	21	1	1	1	6	2	X	1																			42	281	12
LATE EARLY EOCENE	18	8507						4				X																		19	133	4
		8506		X									2	X	1	6	9	1												37	302	
		8505		3													14	1												32	134	
		8504														1	16	1	3	1	37									55	303	
		8503						2								2	3	9	1			X	X							51	315	1
	17	8502				1	1	1									6	10	2	1				65	X					33	301	
		8501						1								1	9	9	1	1				65		1	2	1	1	23	304	
		8500														6	2	4	2					52		1	2	1		23	301	

TABLE 2. THE KNOWN RANGES OF THE BIOSTRATIGRAPHICALLY SIGNIFICANT TAXA RECOVERED FROM THE LOWER TERTIARY DEPOSITS ON SEYMOUR ISLAND*

Taxon	Age Range*
Alisocysta circumtabulata	Late Campanian to late Paleocene
Arachnodinium antarcticum	Eocene
Areosphaeridium diktyoplokus	Late early Eocene to early Oligocene
Deflandrea antarctica	Eocene to early Oligocene
Deflandrea boloniensis	Campanian to Paleocene
Deflandrea cygniformis	Middle to late Eocene
Deflandrea dartmooria	Maastrichtian to late early Eocene
Deflandrea speciosa	Late early Paleocene to late early Eocene
Hystrichosphaeridium salpingophorum	Turonian to early Eocene
Impagidinium victorianum	Eocene to early Oligocene
Isabelidinium pellucidum	Paleocene to early Eocene
Lejeunecysta fallax	Middle Eocene to middle Miocene
Oligosphaeridium complex	Berriasian to middle Eocene
Palaeocystodinium australinum	Late Campanian to late Paleocene
Palaeoperidinium pyrophorum	Late Campanian to early late Paleocene
Phthanoperidinium echinatum	Late early Eocene to early Oligocene
Selenopemphix nephroides	Late middle Eocene to Holocene
Senegalinium asymmetricum	Early Maastrichtian to Eocene
Spinidinium densispinatum	Late Campanian to late Paleocene
Spinidinium essoi	Late Cretaceous to early Oligocene
Spinidinium macmurdoense	Eocene to early Oligocene
Spiniferites ramosus subsp. *granomembranaceus*	Late Cretaceous to early Eocene
Spiniferites multibrevis	Hauterivian to middle Eocene
Vozzhennikovia apertura	Eocene to early Oligocene

*The ranges are based on literature reports that were supported by photomicrographs of specimens of the appropriate species and by our observations.

middle Eocene indicates an early Eocene age for this part of the formation (Harwood, 1985).

The known stratigraphic ranges of *Lejeunecysta fallax* (middle Eocene to middle Miocene) and *Deflandrea cygniformis* (middle to late Eocene) indicates a maximum age of middle Eocene for Interval B. The presence of *Selenopemphix nephroides* (late middle Eocene to Holocene) restricts the maximum age to the late middle Eocene or younger. A minimum age of late Eocene for Interval B is established by the top of the stratigraphic range of *Deflandrea cygniformis* (middle to late Eocene). The combined ranges of these taxa indicate a late middle to late Eocene age for Interval B.

The stratigraphic distribution of taxa in Sections 17, 18, and 19 are shown in Figure 11. As in Section 3, a lower interval (Interval A) and an upper interval (Interval B) can be recognized in the Sections 17, 18, and 19. Interval A consists of samples 8500 through 8507 (0 to 96 m). The base of the stratigraphic ranges of *Arachnodinium antarcticum* (Eocene), *Areosphaeridium diktyoplokus* (late early Eocene to early Oligocene), *Deflandrea antarctica* (Eocene to early Oligocene), *Phthanoperidinium echinatum* (late early Eocene to early Oligocene), *Spinidinium macmurdoense* (Eocene to early Oligocene), and

Vozzhennikovia apertura (Eocene to early Oligocene) indicate Interval A is no older than late early Eocene. The presence of *Isabelidinium pellucidum* (Paleocene–early Eocene) and *Hystrichosphaeridium salpingophorum* (Turonian–early Eocene) indicates an age no younger than early Eocene. The absence of *Deflandrea cygniformis* (middle to late Eocene), which is present in overlying beds, suggests Interval A is older than middle Eocene in age. Interval A is assigned a late early Eocene age.

Interval B consists of the samples 8508 through 8512 (360 to 402 m). The presence of *Arachnodinium antarcticum* (Eocene) and *Deflandrea cygniformis* (middle to late Eocene) and the absence of *Spiniferites multibrevis* (Hauterivian to middle Eocene) indicate a middle to late Eocene age for Interval B.

The age determinations for all lower Tertiary sections are summarized in Figure 12.

The Distribution of Reworked Palynomorphs.

Reworked palynomorphs occur throughout Sections 3, 17, 18, and 19, but are rare or absent in Sections 12-13, 15, and 16. The most abundant reworked palynomorphs are Permo-Triassic Striatiti pollen. These occur in all sections except Sections 15 and

TABLE 3. REWORKED PALYNOMORPHS RECOVERED FROM
SECTION 3

Taxon	Age Range*
Aptea securigera	Early Aptian
Aptea sp. indet.	Berriasian to Santonian
Cleistosphaeridium ancoriferum	Aptian to Coniacian
Cribroperidinium muderongense	Berriasian to Aptian
Cribroperidinium sp. indet.	Portlandian to Campanian
Cyclonephelium distinctum	Kimmeridgian to Maastrichtian
Diconodinium cristatum	Albian to Cenomanian
Diconodinium multispinum	Albian to Santonian
Dinogymnium cf. *curvatum*	Senonian
Manumiella druggii	Maastrichtian to middle Paleocene
Isabelidinium pellucidum	Paleocene to early Eocene
Odontochitina spinosa	Maastrichtian
Oligosphaeridium pulcherrium	Kimmeridgian to Campanian
Ophiobolus lapidaris	?Albian to Paleocene
Palaeocystodinium granulatum	Late Campanian to Maastrichtian
Palaeoperidinium pyrophorum	Late Campanian to early late Paleocene
Spinidinium densispinatum	Late Campanian to late Paleocene
Spinidinium lanterna	Senonian
Spiniferites ramosus subsp. reticulatus	Barremian to Santonian
Striatiti	Permian to Triassic
Xylochoarion cf. *hacknessense*	Callovian

*The age ranges are based on literature reports that were supported by photomicrographs of specimens of the appropriate species.

16. Reworked microplankton range in age from Middle Jurassic (Callovian) to early Eocene.

In Section 3, 15 genera, 17 species, and 1 subspecies of reworked dinocysts were recovered (Table 3). One acritarch species, as well as specimens of Striatiti, was recorded. Fewer than 10 reworked specimens were counted in most samples; however, one sample contained more than 60 specimens.

The reworked assemblages above and below sample 8444 (164.5 m) of Section 3 differ with regard to the species present, relative species abundance, and ages of the dominant species. The reworked microplankton assemblage in the lower half of Section 3 (below 164.5 m) is dominated by the late Senonian–Paleocene species *Palaeoperidinium pyrophorum*, *Spinidinium densispinatum*, and the ?Albian-Paleocene acritarch *Ophiobolus lapidaris*. Reworked accessory species infrequently encountered include *Aptea securigera*, *Aptea* sp. indet., *Cribroperidinium muderongense*, *Cribroperidinium* sp. indet., *Cyclonephelium distinctum*, *Diconodinium cristatum*, *D. multispinum*, *Manumeilla druggii*, *Palaeocystodinium granulatum*, *Spinidinium lanterna*, *Spiniferites ramosus* subsp. *reticulatus*, and *Xylochoarion* cf. *hacknessense*. Specimens of Permo-Triassic Striatiti are common in the two lowermost samples, but they are very rare or absent in the remainder of the samples in the lower half of Section 3.

The reworked palynomorph assemblage above sample 8444

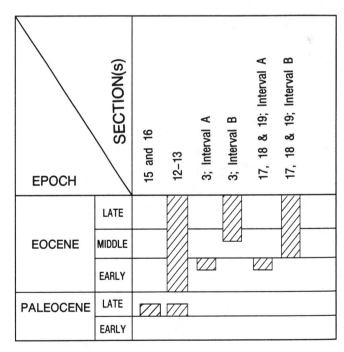

Figure 12. Summary of age determinations for the Seymour Island Paleogene beds.

TABLE 4. REWORKED PALYNOMORPHS RECOVERED FROM
SECTIONS 17, 18, AND 19

Taxon	Age Range*
Aptea securigera	Early Aptian
Cribroperidinium edwardsii	Early Valanginian to late Cenomanian
Cribroperidinium sp. indet.	Portlandian to Campanian
Cyclonephelium distinctum	Kimmeridgian to Maastrichtian
Diconodinium cristatum	Albian to Cenomanian
Diconodinium multispinum	Albian to Santonian
Ophiobolus lapidaris	?Albian to Paleocene
Palaeoperidinium pyrophorum	Late Campanian to early late Paleocene
Pareodinia sp. indet.	Bajocian to Albian
Spinidinium densispinatum	Late Campanian to late Paleocene
Spiniferites ramosus subsp. reticulatus	Berriasian to Santonian
Striatiti	Permian to Triassic

*The age ranges are based on literature reports that were supported by photomicrographs of specimens of the appropriate species.

(164.5 m) is dominated by specimens of Striatiti, *Aptea securigera, Diconodinium multispinium, D. cristatum, Manumiella druggii,* and *Ophiobolus lapidaris.* The ranges of these taxa are pre-Campanian except for *M. druggii* and *O. lapidaris:* these range into the Paleocene. Very rare accessory species include *Aptea* sp. indet., *Cleistosphaeridium ancoriferum, Cribroperidinium* sp. indet., *Cyclonephelium distinctum, Dinogymnium* cf. *curvatum, Isabelidinium pellucidum, Odontochitina spinosa, Oligosphaeridium pulcherrium, Palaeoperidinium pyrophorum, Spinidinium lanterna,* and *Spiniferites ramosus* subsp. *reticulatus.*

In summary, reworked species abundance and diversity are greater in the upper half than in the lower half of Section 3. The ages of the dominant reworked species in the upper half of Section 3 are older than those in the lower half.

Ten genera, 10 species, and 1 subspecies of reworked microplankton and specimens of Striatiti were recorded from Sections 17, 18, and 19 (Table 4). Species diversity increased upsection from Section 17 through Section 19. The most commonly encountered reworked microplankton was *Ophiobolus lapidarus;* this acritarch was most abundant in Section 17 and the basal sample (8503) of Section 18. Only two specimens were observed above sample 8503.

The youngest reworked microplankton, with few exceptions, occur in Sections 17 and 18. These include *Ophiobolus lapidarus* and *Palaeoperidinium pyrophorum,* whose stratigraphic ranges top in the Paleocene and earliest late Paleocene, respectively. Reworked dinocysts in overlying deposits of the upper part of Sections 18 and 19 generally have tops in the Cretaceous. *Spinidinium densispinatum* is the exception; it tops in the Paleocene and was recovered only from the top sample of Section 19. Permo-Triassic Striatiti are most abundant in the beds of Section 19. Overall, the age, abundance, and diversity of the reworked palynomorphs increase up section, as in Section 3.

Relationship to the Regional Geology. In Section 3, the important trends to note in the distribution of reworked palynomorphs are a decrease upsection in the abundance of late Senonian–Paleocene taxa and a concomitant increase in the abundance of pre-Campanian taxa. Considering only the most abundant reworked taxa in Section 3, it is evident that an older reworked palynomorph assemblage overlies a younger reworked assemblage. However, this inverted stratigraphy is partially obscured by the presence in both assemblages of less abundant reworked accessory taxa. These accessory taxa range widely in geologic age and indicate that a simple inverted stratigraphy model cannot explain the reworked palynomorph distribution.

The abundance of reworked late Senonian to Paleocene microplankton in the lower part of Section 3 suggests that they were directly derived from marine deposits of these ages. Most probably, the source beds were relatively close to the Tertiary depocenter. The scarcity of pre-Campanian microplankton indicates either that marine beds of this age were poorly exposed at the time the lower beds of Section 3 were deposited, that the pre-Campanian deposits were far removed from the Paleogene depocenter, or that the concentration of the older taxa had been diluted by an earlier episode of reworking, perhaps during the late Senonian. If there was an earlier reworking episode, it probably occurred during the late Senonian, rather than during the Paleocene. This is suggested by the sparsity of reworked taxa in the Paleocene deposits.

The greater abundance of reworked pre-Campanian microplankton upsection suggests that the rate of erosion of the older marine beds in the source area increased from the early to late Eocene. In addition, the presence of *Manumiella druggii* and *Isabelidinium pellucidum* in the upper half of Section 3 indicates contemporaneous erosion of late Senonian through early Eocene marine beds during the later Eocene.

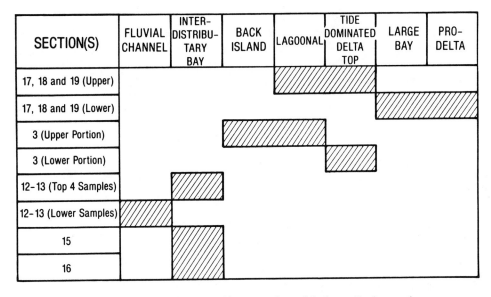

Figure 13. The paleoenvironmental interpretations of the lower Tertiary sections.

The abundance of *Ophiobolus lapidaris* decreases and that of Striatiti increases from the base to the top of Sections 17, 18, and 19. The striking increase in the abundance of Striatiti in Section 19 is analogous to that in Section 3 and indicates that the influx of Striatiti is widespread in the upper La Meseta Formation. These observations confirm those of Askin and Elliot (1982) concerning Striatiti. The marked increase of Striatiti in the upper part of Section 3 was attributed by Askin and Elliot (1982) to a change in provenance, erosional exposure of Striatiti-bearing rocks, or an increased rate of erosion.

The character of the reworked palynofloras is determined by the deposits exposed, the abundance and composition of the palynofloras in those deposits, and the rate at which erosion and redeposition occurred during the early Tertiary. Exactly which deposits are exposed is determined by fluvial processes within the drainage basin, tectonic activity, and sea-level fluctuations.

The palynomorphs available to be reworked are determined by the assemblages in the exposed deposits. Beds containing abundant, diverse, and well-preserved palynomorphs are more likely to be contributors than beds containing sparse, poorly preserved assemblages. Similarly, the abundance and hardiness of previously reworked specimens in the exposed beds influences how well they survive additional reworking.

The abundance of reworked palynomorphs in a deposit may also depend upon the rate of reworking. Rapid sedimentation dilutes palynomorph concentrations in general and reworked palynomorphs, in particular, unless the palynomorphs are exceedingly abundant. The rate of erosion is determined by the climate and climatic fluctuations, local or regional tectonics, and the erodibility of the exposed deposits, as well as sea-level fluctuations.

In summary, recycling may have occurred during the Late Cretaceous and Paleocene but was not substantiated by this investigation. Recycling did occur during the early Eocene and became more active during the middle to late Eocene. This may indicate that mountain building to the west and northwest became more active as the Eocene wore on.

Paleoenvironmental Interpretations

Paleoenvironmental interpretations are based upon palynological, lithological, and total organic carbon (TOC) analyses. The interpretations presented here are summarized from an earlier report (Wrenn, 1982). The depositional environments and energy levels are summarized in Figure 13.

Cross Valley Formation. The inherent biostratigraphic problems of samples from Section 12-13 have been noted previously. These problems do not interfere with palaeoenvironmental interpretations, provided the Paleocene and probable Eocene sediments are considered separately. The fine-grained, laminated Paleocene sediments in Section 12-13 (samples 8492 through 8494) were deposited in an inshore, low-energy, possibly low-salinity environment, such as an estuary or interdistributary bay. This interpretation is indicated by: (1) abundant terrestrial plant debris, (2) common spores and pollen, and (3) low-diversity dinocyst assemblages dominated by *Paralecaniella indentata.* This species apparently preferred a low-salinity, shallow to marginal marine environment (Barss et al., 1979).

Elliot and Trautman (1982) suggested that deposition of the uppermost beds of the Cross Valley Formation occurred in an interdistributary area. This interpretation is compatible with the palynologic evidence.

The fine to coarse sands that characterize the lower samples in Section 12-13 (samples 8480 through ?8490) were deposited in a high-energy environment that was little influenced by marine conditions. This interpretation is indicated by: (1) rare dinocysts, (2) few to moderate amounts of spores and pollen, (3) abundant terrestrial plant debris, and (4) the lithology of the samples (fine to coarse sands, some containing granules).

Elliot and Trautman (1982) considered the lower beds of Section 12-13 to be distributary channel deposits. Palynological data support this interpretation.

Section 15 samples were deposited in a nearshore marine environment subject to moderate wave or current energy. This interpretation is indicated by: (1) the fine-to-medium sand-size of the lithology; (2) the low-diversity dinocyst assemblages; (3) dominance of the dinocyst assemblages by two species (*Palaeoperidinium pyrophorum* and *Spinidinium colemanii* sp. nov. and very low frequencies of the other species; (4) the relative abundance of cavate and proximate dinocysts; and (5) the abundance of spores, pollen, and terrestrial plant debris.

General dinocyst type (i.e., cavate, chorate, etc.) has been used to make paleoenvironmental interpretations. Scott and Kidson (1977) have shown that chorate dinocysts are more commonly associated with low-energy depositional environments, whereas cavate dinocysts are more abundant in high-energy environments. Cavate dinocysts should be more abundant in coarser grained sediments, if sediment grain size is accepted to be a qualitative index of energy in the depositional environment. Conversely, chorate dinocysts should be more common in fine-grained sediments that presumably were deposited in a lower energy depositional environment.

Section 16 samples were deposited in an interdistributary bay into which splay or local channel deposits were introduced. This is suggested by: (1) low-diversity dinocyst assemblage; (2) the dominance of the dinocyst assemblages by one species (*Palaeoperidinium pyrophorum* in sample 8498 and *Spinidinium colemanii* sp. nov. in sample 8499); (3) large carbonized wood fragments; and (4) fine-to-coarse sand-size sediments, suggesting fluctuating energy levels in the environment.

La Meseta Formation. The samples from Section 3 indicate that deposition occurred close to shore in a marine environment that became progressively shallower upsection. This is indicated by: (1) the dominance and abundance of terrestrial plant fragments in all samples; (2) a marked increase in spores upsection; (3) a decrease in the abundance of microplankton, in dinocyst species diversity, and in chorate species diversity upsection; (4) and an increase in the diversity of cavate and proximate dinocysts, in the abundance of reworked palynomorphs, in sediment grain size, and in sediment grain-size variability upsection.

The lower portion of Section 3 was deposited in a moderate-energy marine environment at the front of the delta. However, sample 8459 near the base of the section is from a pebblestone bed that was probably deposited in a high-energy channel. This sample contains abundant terrestrial plant fragments, but very few dinocysts or miospores.

The upper portion of Section 3 was deposited in a lagoonal

or back island paleoenvironment. This low-energy depositional environment was periodically subjected to higher energy storm or splay sedimentation.

Sections 17, 18, and 19 were treated here as a stacked section. The same upward-shoaling trend noted in Section 3 occurs in Sections 17, 18, and 19. This is shown by: (1) an increase in the abundance of terrestrial plant fragments, reworked palynomorphs, spores, and pollen upsection; (2) an overall increase in sediment grain size and variability upsection; (3) an increase in the abundance of cavate and proximate dinocysts upsection; and (4) a decrease in microplankton abundance and chorate dinocyst species diversity upsection.

Comparison of the Seymour Island and southern high-latitude Paleogene floras

Comparison of the dinocyst assemblages recovered from Seymour Island reveal some significant paleogeographic patterns.

The Cross Valley Formation. The early late Paleocene dinocyst species recovered from Sections 15 and 16 (Fig. 9) are cosmopolitan taxa, with perhaps one exception. *Operculodinium bergmannii* has been reported previously only from southern South America (Archangelsky, 1968). Whether *O. bergmannii* is a provincial species is uncertain because reports of this taxa are too few to determine its total geographic distribution. The early late Paleocene dinocyst flora recovered from the upper three samples of Section 12-13 also consists of cosmopolitan taxa. A small cosmopolitan assemblage, interpreted to be Danian in age, has been reported from near the area where Sections 15 and 16 were collected, in the Cape Wiman area (Palamarczuk, 1982).

The cosmopolitan early late Paleocene dinocyst flora indicates that a contemporaneous marine seaway connected the South Atlantic Ocean and the Seymour Island area with Australia, New Zealand, and the Ross Sea area. This assemblage also shows that a distinct southern high latitude dinocyst flora had not yet developed by Paleocene time.

La Meseta Formation. Cosmopolitan dinocyst species constitute the bulk of the Eocene floras in the La Meseta Formation. The remaining dinocysts have a geographic distribution limited entirely or primarily to the southern high latitudes. The existence of this restricted flora was recognized simultaneously by Haskell and Wilson (1975) and by Kemp (1975). This flora (Table 5), referred to as the transantarctic flora (Wrenn and Beckman, 1982), was widely distributed in and around Antarctica during the Eocene (Fig. 14).

Components of the transantarctic flora have been reported in the southern latitudes from Deep Sea Drilling Project (DSDP) Sites 511, 512, and 513A on the Falkland Plateau (Goodman and Ford, 1983), the Rio Turbio Formation of southern South America (Archangelsky, 1968), Seymour Island (Hall, 1977; Wrenn, 1982), the Ross Ice Shelf Project Site J-9 (Wrenn, 1981; Wrenn and Beckman, 1982), the McMurdo Sound area (McIntyre and Wilson, 1966; Wilson, 1967a; Truswell, 1987), the northern Ross Sea (Wilson, 1968), DSDP 270, 274, and 280

TABLE 5. DINOCYST SPECIES COMPOSING THE TRANSANTARCTIC FLORA

Species	Present on Seymour Island
Alterbia? distincta	
Arachnodinium antarcticum	X
Deflandrea antarctica	X
Deflandrea cygniformis	X
Deflandrea granulata	
Deflandrea oebisfeldensis sensu Cookson and Cranwell, 1967	X
Spinidinium macmurdoense	X
Vozzhennikovia apertura	X
Wilsonidium echinosuturatum	

through 283 (Kemp, 1975; Haskell and Wilson, 1975), and the West Ice Shelf area (Kemp, 1972).

Based on these reported occurrences, it is clear that during the Eocene the transantarctic flora was well established in southernmost South America across West Antarctica into the Ross Sea, westward between Australia and Antarctica, and into the West Ice Shelf area (Fig. 14). This distribution supports the existence of a transantarctic seaway connecting the Atlantic and Pacific Oceans prior to the opening of the Drake Passage during the late Oligocene or early Miocene, as suggested by Webb (1978, 1979). Such a seaway would have facilitated the dispersal of microplankton, perhaps by means of a circum-East Antarctic current.

The restricted distribution of the hypothesized transantarctic flora indicates that the endemism characteristic of later Antarctic floras and faunas was becoming established during the Eocene. Whether geographic or oceanographic isolation, or both, initiated this endemism cannot be determined at this time.

Similarly, Zinsmeister (1979) established the Weddelian Province for the late Cretaceous–early Tertiary shallow-water, cool-temperate marine province that extended from southern South America across west Antarctica to east Antarctica and the then closely juxtaposed Australia. New mollusc taxa discovered by Zinsmeister (1979) indicate that the southern circum-Pacific region became a center of molluscan endemism during the early Tertiary. Whatever was initiating this endemism in molluscs was also influencing the development of dinocyst endemism, as noted above.

Occurrences of the Transantarctic Flora Species Outside the Southern High Latitudes

Specimens of some dinocyst species belonging to the hypothesized transantaractic flora occur rarely outside of the southern high latitudes. Griggs (1981) has reported *Arachnodinium antarcticum* (as *"Kolpomacysta"* sp.), *Wilsonidium echinosuturatum*,

and possibly *Deflandrea antarctica* (as *Deflandrea* sp., Griggs, 1981, Plate 10, Fig. 3) from the middle Eocene of the Santos Basin, off the southeast coast of Brazil. The Santos Basin was located at approximately 35°S latitude (Firstbrook et al., 1979) during the middle Eocene.

Displaced Antarctic diatom floras have been noted along the east coast of South America in Pleistocene and Holocene sediments (Burckle and Stanton, 1975; Johnson et al., 1977; Jones and Johnson, 1984). Displaced Antarctic diatoms have also been reported from other areas, including the Pacific Ocean (Booth and Burckle, 1976), the Indian Ocean (Burckle et al., 1974; Burckle, 1981), and the Madagascar Basin (Warren, 1974). These diatoms apparently were transported northward by cold, dense Antarctic bottom water currents. Such cold, high-density currents probably did not exist during preglacial Eocene times. However, relatively cool north-flowing near-surface currents may have existed, which could have carried motile dinoflagellates or their cysts northward from the southern high latitudes into the lower latitudes of Brazil.

Is it possible that the dinoflagellate cyst species found in the Santos Basin did not actually live that far north? Could they, like modern diatoms, have been displaced by currents flowing northward along the east coast of South America? Additional research is needed to answer these questions.

Sparse to abundant specimens of *Vozzhennikovia apertura* (as *V. rotunda*) have been reported from the lower Eocene of Maryland, USA (Goodman, 1975). Goodman has since expressed reservations concerning these identifications (D. G. Goodman, personal communication, 1982). These materials require further study.

More disconcerting has been the discovery of *Spinidinium macmurdoense* in wells on the Alaskan Peninsula (L. S. Satchell, personal communication, 1984). One of us (J.H.W.) has since seen *S. macmurdoense* in additional wells from offshore Alaska. The occurrence of this species in the northern high latitudes raises doubts about the endemic nature of some species included in the hypothesized transantarctic flora. *Spinidinium macmurdoense* apparently had a wider geographic distribution than previously thought, and perhaps it should be excluded from the hypothesized transantarctic flora. The distribution of *Spinidinium macmurdoense* may have been bipolar, because there are no convincing reports of this species from lower latitude deposits. (See discussion of this species in the Systematic Descriptions section.) The absence of *S. macmurdoense* in the intensely studied lower latitude Eocene deposits supports this speculation.

The currently known distribution of the transantarctic flora may be more a reflection of sampling than of its real palaeogeographic distribution. Still, many correlative low-latitude sections in Europe, North America, Australia, on the continental shelves, and in ocean basins outside the southern high latitudes have failed to yield any species of the transantarctic flora. Future research will determine whether part of the hypothesized transantarctic flora is endemic to the southern high latitudes and whether part of it has a bipolar distribution.

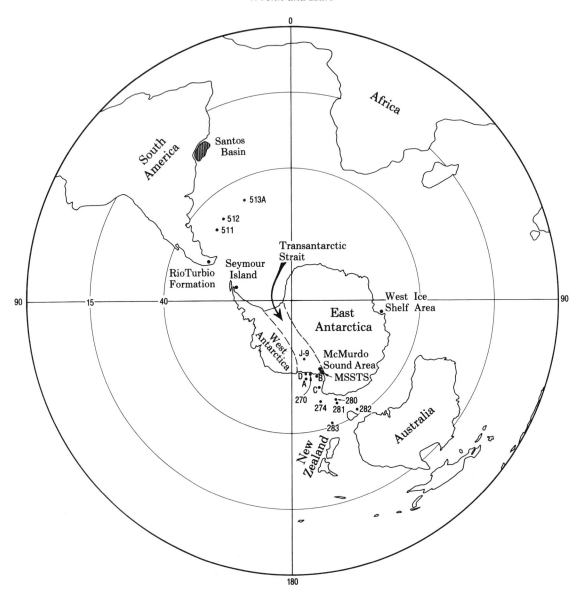

Figure 14. The distribution of the transantarctic flora in the southern high latitudes. The relative positions of the continents are those during the late Eocene. DSDP sites include 270, 274, 280 through 283, and 511 through 513A (Kemp, 1975; Haskell and Wilson, 1975; Goodman and Ford, 1983). The letters A through D represent Wilson's (1968) grab sample stations A452, A459, A461, and A466, respectively. The Santos Basin (Griggs, 1981) is shown by the stipple pattern. The trend of the hypothesized transantarctic strait (Webb, 1979) is shown by the dashed lines. The map was adapted from Firstbrook et al., 1979.

SYSTEMATIC DESCRIPTIONS

The life cycle of many dinoflagellates is characterized by alternating vegetative (motile thecae) and encysted (nonmotile dinocyst) stages (Wall and Dale, 1968). Placing fossil dinocysts into classifications based on living dinoflagellates has proven difficult for several reasons: because the morphologies of the thecae and the dinocyst often are quite different, because some dinoflagellate species apparently produce two or more morphologically different dinocysts, and because many fossil dinocyst species have no living thecate analogs. In addition, palynologists generally work only with the encysted stage, whereas neontologists rarely consider the dinocyst in their classifications. Therefore, in this work dinocysts are treated as form taxa, classified only to the generic and species levels and listed in alphabetic order, for ease of reference.

The synonymy contains only those papers in which the taxonomic status of the species or the subspecies was affected. Taxonomic allocations are generally in accord with those listed in Lentin and Williams (1985).

Division PYRRHOPHYTA Pascher 1914
Class DINOPHYCEA Fritsch 1929
Order PERIDINIALES Haeckel 1894

Genus *Alisocysta* Stover and Evitt 1978
***Alisocysta circumtabulata* (Drugg 1967) Stover and Evitt 1978**
Figure 15.1–3

Eisenackia circumtabulata Drugg 1967, p. 15, Plate 1, Figures 12–13.
Hystrickokolpoma circumtabulatum (Drugg, 1967) Schumacker-Lambry, p. 42.
Alisocysta circumtabulata (Drugg 1967) Stover and Evitt, 1978, Stanford University Publications, Geological Sciences, Vol. 15, p. 16.
Comments. The Seymour Island specimens compare closely with the illustrations of the type material described by Drugg (1967). The dinocysts are subrounded to elliptical and may be compressed in a dorsal-ventral orientation. The punctate endophragm is 2 to 3 μm thick. Membraneous outgrowths of the periphragm are less than 10 μm high and indicate a paratabulation formula of 4′, 6″, 6c, 6‴, 5s, 1P, 1⁗. The presence of preapical paraplates could not be determined because no operculae were observed.

Dimensions. Observed range (two specimens): pericyst length, 46 to 48 μm (mean, 47 μm); pericyst width, 48 to 49 μm; endocyst length, 37 to 39 μm (mean, 38 μm); endocyst width, 41 to 43 μm (mean, 42 μm).

Stratigraphic Occurrence. Cross Valley Formation (Section 12-13, early late Paleocene; Sections 15 and 16, early late Paleocene); La Meseta Formation (Section 3, late early Eocene).

Selected Previous Occurrences. Alisocysta circumtabulata was described from the Danian of California (Drugg, 1967). It has been reported from the early and middle Paleocene of the Scotian Shelf (Williams, 1977; Williams and Bujak, 1977), the Paleocene of the northwest European continental shelf (Caro, 1977), the Thanetian of the Paris Basin (Gruas-Cavagnetto, 1976), and the Danian of Seymour Island (Palamarczuk et al., 1984). Helby et al., (1984) reported a total range for *A. circumtabulata* of late Campanian to middle late Paleocene. Williams and Bujak (1985) reported an early to late Paleocene range for this species, and contended this species does not extend to the top of the Paleocene. Wilson (1984b) reported that this species ranges into the Waipawan stage of Dannevirke Series in New Zealand; he considered the Waipawan Stage to be approximately earliest Eocene (Ypresian Stage) in international terminology. Excluding the New Zealand latest occurrence because of its uncertain relationship to the international stratigraphic divisions, the total known range of *A. circumtabulata* is late Campanian-late Paleocene.

Genus *Aptea* Eisenack (1958)
emend. Davey and Verdier 1974
emend. Dörhöfer and Davies 1980
***Aptea securigera* Davey and Verdier 1974**
Figure 15.7

Aptea securigera Davey and Verdier, 1974, p. 642–643, Plate 91, Figures 2–3.

Comments. Specimens very similar to the illustrations of the type material were recovered from the upper half of the La Meseta Formation. They differ from the type material in possessing more processes in the central middorsal and midventral areas and perforate walls on some

processes. These specimens are reworked from Early Cretaceous (early Aptian) deposits.

Dimensions. (One undamaged specimen): length, 80 μm; width, 92 μm; process length as much as 17 μm; archeopyle width, 68 μm.

Stratigraphic Occurrence. La Meseta Formation (Section 3, Eocene; Section 19, middle to late Eocene).

Selected Previous Occurrences. England (Early Aptian: Batten, 1980). France (Early Aptian: Davey and Verdier, 1974).

Genus *Arachnodinium* Wilson and Clowes 1982
***Arachnodinium antarcticum* Wilson and Clowes 1982**
Figures 15.4–5, 43.1

Aiora fenestrata (Deflandre and Cookson) Cookson and Eisenack, 1960a, sensu Wilson 1967a, p. 69, Figures 2c, 37, 38.
Arachnodinium antarcticum Wilson and Clowes, 1982, p. 98–102, Plate 1, Figures 1–12; Plate 2, Figures 1–10, text-Figure 2.

Comments. The Seymour Island specimens of *Arachnodinium antarcticum* are very similar to the description and illustrations of the original specimens from the McMurdo Sound area in the Ross Sea, Antarctica (Wilson, 1967a; Wilson and Clowes, 1982). Some specimens exhibit a small, hollow, antapical bump, whereas others exhibit a long, distally expanded, hollow process. Ornamentation on the ribbon-like trabeculae may be shagreenate, granulate, denticulate, or striate. One of the specimens recovered had its ?apical operculum in place. Four membranous processes arise from the operculum and join to form an acicular distal termination in the center of the operculum.

Dimensions. Observed range (seven specimens): endocyst length, 45 to 87 μm (mean, 60 μm); endocyst width, 38 to 48 μm (mean, 43 μm); trabeculae length, 66 to 108 μm (mean, 82 μm); trabeculae width, 57 to 110 μm (mean, 87 μm); endophragm thickness <1 μm.

Stratigraphic Occurrence. La Meseta Formation (Section 3, late early Eocene; Section 17, late early Eocene).

Selected Previous Occurrences. Specimens of *Arachnodinium antarcticum*, but attributed to *Aiora fenestrata* (Deflandre and Cookson) Cookson and Eisenack by McIntyre and Wilson (1966) and Wilson (1967a), were recovered from Eocene erratics collected in the McMurdo Sound area of Antarctica. The Eocene age ascribed to these erratics has since been questioned by Stott et al. (1983). Their recovery of Eocene-Pleistocene diatoms from similar McMurdo Sound erratics indicates that the Eocene dinoflagellates are reworked. *A. antarcticum* has been reported from the Eocene Rio Turbio Formation of Argentina (Archangelsky, 1968), the lower Tertiary beds of Tierra del Fuego (Pöthe de Baldis, 1966), and from the middle to late Eocene cores of DSDP 280 and 281 (Haskell and Wilson, 1975). A reexamination of the McMurdo Sound material and specimens from DSDP sites in the south Tasman Sea led Wilson and Clowes (1982) to establish *Arachnodinium antarcticum* for the distinct Eocene dinocysts that had previously been referred to as *Aiora fenestrata*, a Cretaceous species.

Kemp (1975) described a sparse dinocyst assemblage from DSDP 270, including *Arachnodinium antarcticum* (as *Aiora fenestrata*), *Areosphaeridium diktyoplokus*, *Senegalinium asymmetricum*, and *Spinidinium macmurdoense*. The dinocysts were recovered from a calcareous green sand that yielded a potassium-argon age determination of 26 m.y.

(Kemp, 1975). Kemp assigned an Oligocene or late Oligocene age (Fig. 2; Kemp, 1975) to the assemblage based on the radiometric age date and the presence in the assemblage of *Selenopemphix nephroides.* This species was formerly considered to have a restricted stratigraphic range of middle to late Oligocene. *S. nephroides* has since been reported from the late-middle Eocene Barton Beds of southern England (Bujak, 1980). G. L. Williams (personal communication, 1982) considered these beds, and the base of the stratigraphic range of *S. nephroides,* to be as old as late middle Eocene. Thus Kemp's (1975) extensions of the top of the stratigraphic ranges of *Arachnodinium antarcticum, Areosphaeridium diktyoplokus, Senegalinium asymmetricum,* and *Spinidinium macmurdoense* are no longer supported by the stratigraphic range of *Selenopemphix nephroides,* as currently understood. The extensions rest solely on the K-Ar date for the glauconite in the green sand.

Kemp (1975) also reported *Arachnodinium antarcticum* (as *Aiora fenestrata*), *Areosphaeridium diktyoplokus, Vozzhennikovia apertura,* and *Senegalinium asymmetricum* from Unit 4 of DSDP 274. Firm, diatom-based age determinations indicated an Oligocene age for this unit. Lack of evidence indicating reworking, and the excellent preservation of the dinocyst taxa were cited as reasons for extending the stratigraphic ranges of the dinocysts noted above.

Specimens of *Vozzhennikovia apertura* and *Spinidinium macmurdoense* were reported from the overlying Miocene Series, Unit 3. These specimens, however, were considered to be reworked and not indicative of a longer stratigraphic range for *V. apertura* and *S. macmurdoense.* The source of these reworked taxa was thought to be the rich Eocene dinocyst assemblages encountered in Unit 5 (i.e., the unit underlying Unit 4) of DSDP 274. The presence of a rich Eocene dinocyst assemblage below Unit 4, together with a reworked assemblage in Unit 3 above Unit 4, suggests that reworking during the time of deposition of Unit 4 is certainly possible, if not probable.

The presence of reworked, delicate palynomorphs in varying states of preservation is a widespread phenomenon in and around Antarctica (McIntyre and Wilson, 1966; Wilson, 1967a; Kemp, 1972; Wrenn and Beckman, 1982; Truswell, 1983) and elsewhere (Wrenn and Kokinos, 1986). The excellent preservational state of Kemp's (1975) specimens does not necessarily preclude their having been reworked into Unit 4.

These facts suggest reworking cannot be entirely discounted. The evidence against reworking and in favor of extending the stratigraphic ranges is neither compelling nor conclusive.

Hall (1977) assigned an Eocene-Oligocene age to the Seymour Island deposits containing *Arachnodinium antarcticum* (as *Aiora fenestrata*), apparently on the basis of the range extensions proposed by Kemp (1975).

A. antarcticum has also been reported from the late Eocene of DSDP 511 and the middle Eocene of DSDP 512 on the Falkland Plateau in the South Atlantic Ocean (Goodman and Ford, 1983). These dinocyst assemblages were correlated with the middle and late Eocene cores in DSDP 283, in the south Tasman Sea, and the middle and late Eocene Browns Creek Clays of Victoria, Australia.

Griggs (1981) reported *A. antarcticum* (as *"Kolpomacysta"* sp.) in middle Eocene deposits from wells in the Santos Basin, offshore from southeastern Brazil.

Except for the reported Oligocene occurrence of *A. antarcticum* by Kemp (1975) and Hall's (1977) apparent acceptance of her range extensions, the biostratigraphic evidence indicates that this species is restricted to the Eocene. Due to the pervasive nature of reworking in the Ross Sea Basin, we consider these Oligocene occurrences to be the result of reworking. After all, the dinocysts recovered from the glacial erratics and described by Wilson (1967a) were apparently reworked and incorporated into Miocene sediments (Stott et al., 1983). The Miocene deposits were subsequently eroded and transported by post-Miocene glaciers. Whatever the original source of the "Eocene" dinocysts, it is certainly possible that those beds were also being eroded during the Oligocene.

We consider the range of *Arachnodinium antarcticum* to be restricted to the Eocene, based on occurrences outside the Ross Sea.

Genus *Areosphaeridium* Eaton 1971
***Areosphaeridium diktyoplokus* (Klumpp 1953) Eaton 1971**
Figure 15.6

Hystrichosphaeridium diktyoplokus Klumpp, 1953, p. 392, Plate 18, Figures 3–7.
Cordosphaeridium diktyoplokus (Klumpp, 1953) Eisenack, 1963, p. 262.
Areosphaeridium diktyoplokus (Klumpp, 1953) Eaton, 1971, p. 358, Plate 1, Figures 3–8, Plate 2, Figures 1–6.

Comments. This species exhibits a wide range in morphologic variability, including the number, length, and width of the processes. Processes usually consist of a single shaft bearing a distal platform, whereas other shafts bifurcate or trifurcate along their length, and the distal ends may or may not be united by a common platform. Rare specimens bear process complexes. The development of the distal platform varies and they generally have ragged edges. Goodman and Ford (1983) have suggested that Southern Hemisphere specimens attributed to *A. diktyoplokus* are a separate species because of the lack of the entire platform margins characteristic of Northern Hemisphere specimens.

The Southern Hemisphere specimens traditionally assigned to *Areosphaeridium diktyoplokus* not only have ragged platform margins but also irregular fenestrations similar to those on *A. fenestratum.* However, the central bodies of the Southern Hemisphere specimens of *A. diktyoplokus* are generally subspherical to spherical, rather than flattened dorso ventrally as in *A. fenestratum.* However, a detailed comparative study of these two taxa may prove fruitful.

The archeopyle is formed by the loss of a single opercular piece composed of the apical paraplate series. Generally, one intratabular process is centered on each paraplate.

Dimensions. Observed range (10 specimens): overall diameter, 61 to 95 μm (mean, 84 μm); endocyst diameter, 49 to 66 μm (mean, 54 μm); process length, 15 to 29 μm (mean, 22 μm).

Stratigraphic Occurrence. Cross Valley Formation (Section 12-13, Eocene; see discussion of the age determination for this formation). La Meseta Formation (Section 3, late early Eocene; Sections 17, 18, and 19, Eocene).

Selected Previous Occurrences. Worldwide distribution during the late early Eocene–early Oligocene. *Areosphaeridium diktyoplokus* attained a worldwide distribution during the late Eocene (Eaton, 1971). Haskell and Wilson (1975) reported, but did not illustrate, *A. diktyoplokus* from the middle(?) Paleocene interval of DSDP 283. Eaton (1971) and Williams (1977) considered this species to have first appeared during the early Eocene.

A. diktyoplokus has been recovered from widely separated localities, including Eocene deposits of Germany (Morgenroth, 1966a), southern England (Eaton, 1971; Bujak, 1976), the Scotian Shelf and Grand Banks (Williams and Brideaux, 1975), Argentina (Archangelsky, 1968, 1969), southern Chile (Cookson and Cranwell, 1967), southeast Australia (Cookson and Eisenack, 1965a), McMurdo Sound (Cranwell, 1964; McIntyre and Wilson, 1966; Wilson, 1967a), DSDP 270 and 274 (Kemp, 1975), DSDP 270 and 274 (Kemp, 1975), the West Ice Shelf area of Antarctica (Kemp, 1972), and DSDP 280-283 (Haskell and Wilson, 1975).

Goodman and Ford (1983) have argued that Southern Hemisphere specimens referred to *A. diktyoplokus* differ from those of the Northern

Hemisphere in that the distal platforms of the processes lack complete margins. They referred to these Southern Hemisphere forms as *A.* sp. cf. *A. diktyoplokus* and noted their occurrence in the late Eocene–early Oligocene of DSDP 511.

A. *diktyoplokus* has been reported from the lower Oligocene of Europe (Maier, 1959) and the northwestern European continental shelf and adjacent areas (Caro, 1977). Williams and Bujak (1985) listed a late early Eocene to late Eocene age range for *A. diktyoplokus*.

It is assumed for this study that the Northern and Southern Hemisphere dinocysts assigned to *A. diktyoplokus* are one species, although a comparative study of the Northern and Southern Hemisphere forms is in order. The evidence indicates that the biostratigraphic distribution of *A. diktyoplokus* ranges from the late early Eocene to early Oligocene.

Genus *Batiacasphaera* Drugg 1970
Batiacasphaera explanata (Bujak 1980) Islam 1983
Figure 16.4

Chytroeisphaeridia explanata Bujak, 1980, p. 44, Plate 13, Figures 13–14.
Batiacasphaera explanata (Bujak, 1980) Islam, 1983, p. 235.

Comments. These are smooth, thin walled, subspherical dinocysts with an apical archeopyle. These forms compare favorably with the description of *Batiacasphaera explanata,* differing only in their larger size.

Dimensions. Observed range (five specimens): autocyst width, 70 to 107 μm (mean, 85 μm); autocyst height (without operculum), 61 to 114 μm (mean, 88 μm); wall thickness, approximately 0.5 μm.

Stratigraphic Occurrence. La Meseta Formation (Sections 17, 18, and 19, Eocene).

Selected Previous Occurrences. England (Eocene: Bujak et al., 1980; Islam, 1983).

Genus *Cerebrocysta* Bujak 1980
Cerebrocysta bartonensis Bujak 1980
Figure 17.6, 8

Cerebrocysta bartonensis Bujak, 1980, p. 42, Plate 13, Figures 4–11.

Comments. The specimens recovered differ from the type material by possessing coarser sculpture and being slightly larger in diameter. Broken specimens were more common than whole and often were "unrolled," forming a ragged sculptured strip of autophragm. The sculpture patterns are approximately 1.5 μm high and suggest tabulation, as noted by Bujak (1980). It is possible that this is a new species of *Cerebrocysta*. However, insufficient specimens were recovered upon which to establish a new species.

Dimensions. Observed range (two complete specimens recovered): diameter, 29 to 30 μm; wall thickness, approximately 1.5 μm.

Stratigraphic Occurrence. Cross Valley Formation (Sections 15 and 16, early late Paleocene).

Selected Previous Occurrences. Belgium (Ypresian-Lutetian: De Coninck, 1980); England (Bartonian: Bujak, 1973); England (late Eocene: Bujak, 1980).

Genus *Cleistosphaeridium* Davey et al. 1966
Cleistosphaeridium ancoriferum (Cookson and Eisenack 1960) Davey et al. 1966
Figure 17.2

Hystrichosphaeridium ancoriferum Cookson and Eisenack, 1960a, p. 8, Plate 2, Figure 11.
Cleistosphaeridium ancoriferum (Cookson and Eisenack, 1960a) Davey et al., 1966, p. 167–168, Plate 9, Figure 1.

Comments. The rare, usually fragmentary specimens of this species bear hollow processes that constrict distally and then flare into recurved terminations. The process base, when viewed down the process axis, is donut-shaped, emphasizing the hollow structure of the process shaft. Reworked from Cretaceous (Aptian to Cenomanian) deposits.

Dimensions. Observed range (one complete specimen): diameter, 46 μm; processes length, 3.5 to 5 μm.

Stratigraphic Occurrence. La Meseta Formation (Section 3, reworked into late early Eocene deposits).

Previous Occurrence. Worldwide distribution. The stratigraphic range is late Hauterivian to late Campanian.

Genus *Cometodinium*
Cometodinium cf. *?whitei* (Deflandre and Courteville 1939) Stover and Evitt 1978
Figure 17.7

Xanthidium hirsutum Ehrenberg, in Reide, 1839, Plate 9, Figure 3, not Figure 8.
Hystrichosphaeridium whitei Deflandre and Courteville, 1939, p. 103, Plate 3, Figures 5–6.
Baltisphaeridium whitei (Deflandre and Courteville, 1939) Sarjeant, 1959, p. 339.
Impletosphaerdidium whitei (Deflandre and Courteville, 1939) Morgenroth, 1966, p. 37.
Cometodinium ?whitei (Deflandre and Courteville, 1939) Stover and Evitt, 1978, p. 227.

Comments. The central body diameter of the Seymour Island material is approximately half the size of the type material. The solid processes are shorter and may exhibit minute bifurcations distally. No archeopyle was observed.

Dimensions. Observed range (five specimens): central body diameter, 24 to 35 μm (mean, 30 μm); process length, 2 to 11 μm.

Stratigraphic Occurrence. Cross Valley Formation (Section 12–13, Eocene, see discussion on age determinations for this formation; Sections 15 and 16, early late Paleocene); La Meseta Formation (Section 3, Eocene; Section 18, late early Eocene; Section 19, middle to late Eocene).

Selected Previous Occurrences. Worldwide distribution. Cretaceous–late Eocene.

**Genus *Cribroperidinium* Neale and Sarjeant 1962
emend. Davey 1969**
Cribroperidinium edwardsii
(Cookson and Eisenack 1958) Davey 1969
Figure 18.3-4

Gonyaulax edwardsi Cookson and Eisenack, 1958, p. 32–33, Plate 3, Figures 5–6, text-fig. 7.
Gonyaulacysta edwardsi (Cookson and Eisenack, 1958) Clarke and Verdier, 1967, p. 31).
Cribroperidinium edwardsii (Cookson and Eisenack, 1958) Davey, 1969a, p. 128.

Comments. The rare specimens observed compare closely with the illustrations and description of the type material. The stout apical horn interrupts the circular outline of this subspherical dinocyst. The granulate autophragm is approximately 7 μm thick and bears numerous nontabular spines. These short, stout spines are distally capitate to slightly bifid.

Short, stiff spines occur in the apical and antapical regions. Paraplates are delimited by thin, low ridges (<7 μm high). The narrow, slightly laevorotatory paracingulum and the parasulcus are bordered by low, narrow ridges. The precingular archeopyle is formed by the loss of the 3″ paraplate. Reworked from Cretaceous (early Valanginian to late Campanian) deposits.

Dimensions. Body diameter, 130 μm; apical horn length, approximately 28 μm; wall thickness, 7 μm; spine length as much as 6 μm.

Stratigraphic Occurrence. La Meseta Formation (Section 19, reworked into middle to late Eocene deposits).

Previous Occurrence. Worldwide distribution. Cretaceous.

Cribroperidinium muderongense
(Cookson and Eisenack 1958) Davey 1969
Figure 18.1

Gonyaulax muderongensis Cookson and Eisenack, 1958, p. 32, Plate 3, Figure 34.
Gonyaulacysta muderongensis (Cookson and Eisenack, 1958) Sarjeant 1966b, p. 131.
Cribroperidinium muderongense (Cookson and Eisenack, 1958) Davey, 1969a, p. 128.

Comments. The single specimen observed is damaged. The very long, stout apical horn is degraded distally, obscuring the original nature of its tip.

The paraplate areas are delimited by low ridges or by crenulated membranes. Short, squat spines are present on some ridges. Numerous accessory parasutures are present within the paraplate areas. The paracingulum is strongly laevorotatory, very narrow, and bordered by thin, low membranes (approximately 3 μm high). The parasulcus is much broader than the paracingulum. The archeopyle is precingular (3″). This specimen has been reworked from Cretaceous (Barremian-Santonian) deposits.

Dimensions. (one specimen recovered): overall length, 134 μm; width, 85 μm; apical horn length, 34 μm; wall thickness, approximately 2 μm; spines length, <3.5 μm.

Stratigraphic Occurrence. La Meseta Formation (Section 3, reworked into late early Eocene deposits).

Previous Occurrence. Worldwide distribution. Cretaceous (Barremian-Santonian).

**Genus *Cyclonephelium* Deflandre and Cookson 1955
emend. Stover and Evitt 1978**
***Cyclonephelium distinctum* Deflandre and Cookson 1955**
Figure 18.5

Cyclonephelium distinctum Deflandre and Cookson, 1955, p. 285–286, Plate 2, Fiugre 14, text-figs. 47–48.

Comments. The compressed subcircular to elliptical outline of these specimens is the same as that illustrated for the type material. The left antapical horn is well developed on some specimens, but absent on others. The solid processes have acuminate, capitate, bifurcate, or irregular terminations and are less numerous in middorsal and midventral regions than marginally. The archeopyle is apical and the free operculum is usually missing. These specimens are reworked from Cretaceous deposits that may range from Berriasian to early Maastrichtian in age.

Dimensions. One complete specimen measured: Length (without operculum), 107 μm; width, 100 μm; process length, as much as 11 μm.

Stratigraphic Occurrence. La Meseta Formation (Section 3, reworked into Eocene deposits; Section 18, reworked into late early Eocene deposits).

Selected Previous Occurrences. Worldwide distribution during Cretaceous.

Genus *Cyclopsiella* Drugg and Loeblich 1967
***Cyclopsiella* cf. *elliptica* Drugg and Loeblich 1967**

Cyclopsiella elliptica Drugg and Loeblich, 1967, p. 190, Plate 3, Figures 1–6; text-fig. 7.

Comments. The overall shape of these rare, poorly preserved specimens compares closely to the specimens described by Drugg and Loeblich (1967) from the Gulf Coast of North America. However, overall dimensions are slightly larger, and there are no spines on the surface. The paracingulum is indicated by a fold or low ridge. The parasulcus is not evident.

Dimensions. One specimen measured: length: 54 μm; width, 56 μm; pylome diameter, 8 μm.

Stratigraphic Occurrence. La Meseta Formation (Section 3, late early Eocene).

Previous Occurrence. North Atlantic Ocean and contiguous seas during the late Eocene to Miocene.

***Cyclopsiella trematophora* (Cookson and Eisenack 1967)
Lentin and Williams 1977**
Figure 18.2

Leiosphaeridia trematophora Cookson and Eisenack, 1967a, p. 136, Plate 19, Figure 13.
Cyclopsiella trematophora (Cookson and Eisanack, 1967a) Lentin and Williams, 1977b, p. 39.

Comments. The specimens are almost identical to the illustrations of type material from western Tasmania. A granular or smooth rim may be present around the circular aperture.

Dimensions. Observed range (10 specimens): 81 to 108 μm (mean, 90 μm); circular opening, 12 to 17 μm (mean, 15 μm).

Stratigraphic Occurrence. Cross Valley Formation (Section 12-13, Eocene, see discussion of age determinations for this formation; Section 15, early late Paleocene); La Meseta Formation (Section 3, Eocene; Sections 17 and 19, Eocene).

Selected Previous Occurrences. Australia, Victoria (Paleocene: Cookson and Eisenack, 1967b); Australia (late Paleocene: Verdier, 1970); Netherlands (Eocene: De Coninck, 1977); Tasmania (Paleocene: Cookson and Eisenack, 1967a).

Genus *Deflandrea* Eisenack 1938
emend. Williams and Downie
1966 emend. Lentin and Williams 1976

Comments. Lindgren (1984) pointed out that *Ceratiopsis* Vozzhennikova 1963 is a later homonym of *Ceratiopsis* De Wilderman 1896, a genus of fungi. In addition, he considered *Ceratiopsis* Vozzhennikova, 1963, and *Cerodinium* Vozzhennikova, 1967, to be congeneric with *Deflandrea* Eisenack, 1938.

J. K. Lentin and G. L. Williams (1987) have agreed that *Ceratiopsis* and *Cerodinium* are congeneric, but not that they are also congeneric with *Deflandrea*. Rather, they are resurrecting *Cerodinium* to accommodate those species previously assigned to *Ceratiopsis*. We concur with their interpretation and anticipate the transfer to *Cerodinium* of the species formerly included in *Ceratiopsis*. In this chapter, that would include *Deflandrea boloniensis, D. dartmooria, D. speciosa,* and *D. striata.*

Deflandrea antarctica Wilson 1967
Figures 19.1–4, 44.3

Deflandrea antarctica Wilson, 1967a, p. 58, 60, Figures 23–24, 26–27.

Comments. The abundant Seymour Island specimens are nearly identical with the description and illustrations of the McMurdo Sound–type material. Wilson (1967a) noted the presence of atabulate granules, and sometimes tabulate clusters, in the type material. Wide variation in these ornamental features facilitated the recognition of two types (Types I and II) during this study. However, strict categorization was not always possible because transitional forms exist.

The Type I specimens of *D. antarctica* are characterized by a paracingulum delimited by folds and a granulate cyst surface that is generally devoid of paratabulation. Occasional specimens do exhibit very fine parasutural lines similar to those observed on *D. dartmooria* by Stover (1974).

In contrast, the Type II morphology bears paratabular sculpture that varies from low, rounded granules to acicular spines as much as 3 μm long. The paratabulation delimited by these clusters of sculptural elements is 4′, 3a, 7″, 4?C, 5‴, 2⁗, 2-3S. The Type II paracingulum is bordered by clearly defined, but low, ridges capped by a row of granules. The granules are reduced in size, or totally absent, on the adapical paracingular ridge adjacent to the 3″/4″ and 4″/5″ parasutures and on the posterior paracingular ridge over the 2‴/3‴ and the 3‴/4‴ parasutures. Type II specimens are, on average, smaller than Type I specimens.

Transitional forms have been observed in the same sample preparations as Types I and II specimens. For example, specimens with nontabular granules and paracingular ridges capped by granules, as well as specimens covered with uniformly distributed granules and an overprint of paratabulate granules were noted. Consequently, these morphologic types could not be consistently separated.

Extreme morphological variability has been observed in other species of *Deflandrea* by McClean (1971) and Goodman (1975).

Dimensions. Observed range: Type I (10 specimens): pericyst length, 128 to 140 μm (mean, 136 μm); pericyst width, 87 to 101 μm (mean, 94 μm); endocyst length, 67 to 85 μm (mean, 75 μm); endocyst width, 75 to 85 μm (mean, 81 μm); periarcheopyle length, 27 to 43 μm (mean, 36 μm); periarcheopyle width, 44 to 58 μm (mean, 50 μm); endoarcheopyle length, 26 to 29 μm (mean, 27 μm); endoarcheopyle width, 49 to 58 μm (mean, 52 μm).

Type II (10 specimens): pericyst length, 99 to 128 μm (mean, 112 μm); pericyst width, 73 to 92 μm (mean, 84 μm); endocyst length, 51 to 79 μm (mean, 65 μm); endocyst width, 68 to 83 μm (mean, 76 μm); periarcheopyle length, 29 to 36 μm (mean, 32 μm); periarcheopyle width, 43 to 54 μm (mean, 47 μm); endoarcheopyle length, 8 to 24 μm (mean, 17 μm); endoarcheopyle width, 37 to 53 μm (mean, 46 μm).

Stratigraphic Occurrence. La Meseta Formation (Section 3, late early Eocene; Sections 17, 18, and 19, Eocene).

Selected Previous Occurrences. Deflandrea antarctica has been reported from McMurdo Sound erratics (McIntyre and Wilson, 1966, as *D.* aff. *bakeri;* Wilson, 1967a), DSDP 280 and 283 (Haskell and Wilson, 1975), DSDP 511, 512, and 513A (Goodman and Ford, 1983), and from the West Ice Shelf area of Antarctica (Kemp, 1972). Kemp (1975) reported an Eocene-Oligocene age for this species in the Ross Sea. Barss et al. (1979) reported *D. antarctica* from the early Eocene of eastern Canada but did not illustrate any specimens. An Eocene age was reported for every reliable occurrence of *D. antarctica* except by Goodman and Ford (1973); they reported a middle Eocene to early Oligocene range. Current evidence indicates a total stratigraphic range for *D. antarctica* of Eocene to early Oligocene.

Deflandrea boloniensis (Riegel 1974) comb. nov.
Figure 16.3

Deflandrea boloniensis Riegel, 1974, p. 354–355, Plate 1, Figures 6–10.

Ceratiopsis boloniensis (Riegel, 1974) Lentin and Williams, 1977b, p. 20.
Senegalinium boloniense (Riegel, 1974) Stover and Evitt, 1978, p. 123.
Phelodinium boloniense (Riegel, 1974) Riegel and Sarjeant, 1982, p. 296–297.

Comments. The specimens recovered are comparable to illustrations of the type material from southern Spain. Surface sculpture is granular to slightly rugulate and usually aligned longitudinally.

D. boloniensis had been removed from the genus *Ceratiopsis* and placed into the genus *Senegalinium* (Stover and Evitt, 1978). Lentin and Williams (1981), however, rejected this transfer and returned the species *S. boloniensis* to the genus *Ceratiopsis*. Subsequently, Riegel and Sarjeant (1982) also rejected the attribution of *C. boloniensis* to the genus *Senegalinium* and transferred it to the genus *Phelodinium.*

Riegel and Sarjeant (1982) noted that the species *P. boloniensis* is "distinctly cavate (and indeed, as noted, rather more than cornucavate!)"

This indicates that *Phelodinium* may not be the best genus in which to place this species. Therefore, the transfer of *C. boloniensis* to *Phelodinium* is rejected herein because specimens of this species may be bicavate rather than simply cornucavate. (The latter is a generic characteristic of the genus *Phelodinium.*) Bicavate specimens are clearly illustrated in two line drawings (Fig. 3,a–b) of the type material (Riegel, 1974). (We do not consider the specimen illustrated in Fig. 3d and Plate 1, Fig. 6, by Riegel [1979] to be synonymous with the species *D. boloniensis.*) We attribute this species to *Deflandrea*, pending reestablishment of *Cerodinium* by J. K. Lentin and G. L. Williams (1987).

Dimensions. Observed range (seven specimens): pericyst length, 108 to 153 μm (mean, 109 μm); pericyst width, 67 to 83 μm (mean, 75 μm); endocyst length, 47 to 81 μm (mean, 65 μm); endocyst width, 61 to 83 μm (mean, 74 μm); archeopyle length, 18 to 31 μm (mean, 24 μm); archeopyle width, 28 to 41 μm (mean, 40 μm).

Stratigraphic Occurrence. Cross Valley Formation (Sections 15 and 16, early late Paleocene).

Selected Previous Occurrences. Deflandrea boloniensis was described from the Upper Cretaceous Campo de Gibraltar section of southern Spain (Riegel, 1974). Lentin and Williams (1980) reported this species from the Campanian of Texas and Venezuela. The stratigraphic range of *D. boloniensis* is Campanian to Paleocene, according to Williams and Kidson (1975).

Deflandrea cygniformis Pöthe de Baldis 1966
Figures 20.2–5, 21.1–4, 42.3–5, 43.2, 44.4

Deflandrea cygniformis Pöthe de Baldis, 1966, p. 221, Plate 2, Figure C.

Comments. Considerable morphologic variation was evident in the many well-preserved specimens recovered. Some examples were similar to the illustrations of the type specimen, whereas others resembled *D. oebisfeldensis* sensu Cookson and Cranwell (1967). The thin periphragm may be smooth or scabbrate, or bear uniformly distributed granules and/or pits. The thick endophragm is fine to coarsely granular. Some specimens bear small grana on top of the low paracingular ridges; others do not. A flagellar scar is present in the parasulcal area.

The variable length of the paracingular projections determines the shape of the hypopericyst. The periphragm and endophragm are generally appressed laterally in specimens with short paracingular projections. In these specimens the shape of the hypopericyst is similar to the hypoendocyst, except a short posterior extension of the periphragm may be present (Fig. 20.2–4). The posterior edge of this extension is usually flat, but short antapical horns may project from both of its corners.

On specimens bearing large paracingular projections, the hypopericyst is trapezoidal (Fig. 20.5, 21.1–4). The periphragm and the endophragm are not appressed laterally, but may be in contact in the precingular region. The trapezoidal hypopericyst may be truncated (Fig. 20.5) or bear short antapical horns (Fig. 21.1–4).

The shape of the epipericyst is also determined by the length of the paracingular projections. The long apical horn arises from angular epicystal shoulders on specimens with minor paracingular projections (Fig. 20.2–4). In contrast, rounded epicystal shoulders give rise to the apical horn on specimens with well-developed paracingular projections (Fig. 21.1–4).

The epipericyst is as much as three times as long as the hypopericyst, due primarily to the long apical horn. The shape of the apical horn dictates that the apical and intercalary series of paraplates are long and narrow. Paratabulation is indicated by the shape of the archeopyle and

by very faint parasutural lines, usually not visible with the light microscope. The intercalary archeopyle is long and narrow, and the height-to-width ratio (1.0 to 1.3; mean, 1.1 for n=9) indicates it is an attenuated hexa 2a archeopyle. This is at variance with the height-to-width ratio of <1.0 usually ascribed to the genus *Deflandrea* (Lentin and Williams, 1975).

However, the transverse archeopyle index (AI) and the transverse archeopyle ratio (AR) confirm that it is a broad hexa 2a archeopyle (AI = 0.55 to 0.74; mean, 0.64, and AR = 1.3 to 2.8, mean, 1.88; for n = 9). Although the archeopyle is very long or attenuated, its widest portion occurs adjacent to the widest point of the epicystal shoulders. Consequently, a biometric conflict exists between the height to width ratio and the AI and AR indices.

D. cygniformis is similar to *D. antarctica* and may prove to be an evolutionary offshoot of that species.

Dimensions. Observed range (10 specimens): pericyst length, 152 to 186 μm (mean, 170 μm); pericyst width, 99 to 122 μm (mean, 108 μm); endocyst length, 58 to 83 μm (mean, 73 μm); endocyst width, 75 to 99 μm (mean, 83 μm). Archeopyle dimension (9 specimens): periarcheopyle length, 46 to 68 μm (mean, 57 μm); periarcheopyle width, 41 to 58 μm (mean, 50 μm); endoarcheopyle length, 14 to 35 μm (mean, 22 μm); endoarcheopyle width, 46 to 58 μm (mean, 52 μm).

Stratigraphic Occurrence. La Meseta Formation (Section 3, late middle to late Eocene; Section 19, middle to late Eocene.

Selected Previous Occurrences. Deflandrea cygniformis, first described from the lower Tertiary of Tierra del Fuego (Pöthe de Baldis, 1966), was observed in middle to late Eocene samples from DSDP 283 (Haskell and Wilson, 1975). The total known age range for *D. cygniformis* is middle to late Eocene.

Deflandrea dartmooria (Cookson and Eisenack 1965) Lindgren 1984
Figure 16.1–2

Deflandrea dartmooria Cookson and Eisenack, 1965b, p. 133–134, Plate 16, Figures 1–2, text-fig. 1.
Ceratiopsis dartmooria (Cookson and Eisenack, 1965b) Lentin and Williams, 1981, p. 38.
Deflandrea dartmooria Cookson and Eisenack, 1965b, Lindgren, 1984, p. 154.

Comments. Rare specimens comparable to illustrations of the type material were recovered. Parasutural lines noted by Cookson and Eisenack (1965b), and discussed more fully by Stover (1974), are present, although faint.

Dimensions. (One complete specimen recovered): pericyst length, 132 μm; pericyst width, 66 μm; endocyst length, 56 μm; endocyst width, 67 μm; archeopyle height, 23 μm; archeopyle width, 33 μm.

Stratigraphic Occurrence. Cross Valley Formation (Section 15, early late Paleocene); La Meseta Formation (Section 3, late early Eocene).

Selected Previous Occurrences. Deflandrea dartmooria was originally described from the Dartmoor Formation of Australia (Cookson and Eisenack, 1965b). The formation was considered to be Paleocene in age. A subsequent study of type locality material led Stover (1974) to assign an early Eocene age to the Dartmoor Formation. Haskell and Wilson

(1975) recovered specimens similar to *D. dartmooria* (reported as *Deflandrea* cf. *D. dartmooria*) from DSDP 282 core no. 18. A Paleocene age was tentatively accepted for Core 282-18.

D. dartmooria has been reported from early Teurian (early Paleocene: Wilson, 1981a) and Teurian (Paleocene: Wilson, 1981b,c; 1982) deposits of New Zealand.

D. dartmooria has been reported frequently along the east coast of North America. Whitney (1984) reported a Maastrichtian to Danian range for this species in southern Maryland but did not illustrate any specimens. R. J. Witmer (personal communication, 1987) found it in the Paleocene Brightseat(?) and Aquia Formations in southern Maryland. It has also been noted in the Aquia Formation (Paleocene) in northeastern Virginia (Gibson et al., 1980). Benson (1976) noted the occurrence of this species in the Monmouth Formation (Maastrichtian) and the Brightseat Formation (Danian). Farther north, Williams and Bujak (1977) documented a late Paleocene to early Eocene range for *D. dartmooria* in offshore eastern Canada.

Gamerro and Archangelsky (1981) reported *D. dartmooria* from the Maastrichtian-Paleocene interval in two wells in the offshore Colorado Basin of Argentina.

Knox and Harland (1979) report that *D. dartmooria* is restricted to the Thanetian (late Paleocene) beds of Britain. The overall stratigraphic range of *D. dartmooria* is late Paleocene to the lower half of the early Eocene, according to Stover and Williams (1977). Williams and Bujak (1985) cited a total range of late Paleocene to late early Eocene for *D. dartmooria*. We consider the total range to be Maastrichtian to late early Eocene.

Deflandrea oebisfeldensis Alberti sensu Cookson and Cranwell 1967
Figure 20.1

Deflandrea oebisfeldensis Alberti sensu Cookson and Cranwell, 1967, p. 205, Plate 1, Figure 2.
Deflandrea oebisfeldensis Alberti sensu Cookson and Cranwell, 1967; Kemp, 1975, p. 604, Plate 1, Figures 1–3.

Comments. Cookson and Cranwell (1967) assigned specimens of *Deflandrea* recovered from the Lena Dura Formation of southern Chile to *Deflandrea oebisfeldensis* Alberti. They included forms with closely appressed wall layers and others with widely separated wall layers. Kemp (1975) called attention to the closely appressed state of the wall layers and concluded the specimens of Cookson and Cranwell (1967) probably represented a species distinct from *D. oebisfeldensis*. The latter is characterized by separated wall layers.

Kemp (1975) assigned similar specimens of *Deflandrea* recovered from the Ross Sea to *D. oebisfeldensis* Alberti sensu Cookson and Cranwell. The absence of apical papilla and the conically shaped antapical horns distinguished the Ross Sea specimens from *D. antarctica*, although Kemp (1975) did recognize the possibility that some of her specimens were variants of the latter species.

We concur with Kemp's conclusion that *D. oebisfeldensis* Alberti sensu Cookson and Cranwell is distinct from *D. oebisfeldensis* Alberti. However, two of the Lena Dura specimens (Plate 1, Figs. 3, 4) fall within our concept of *D. antarctica*, as does one Ross Sea specimen illustrated by Kemp (1975; Plate 1, Fig. 5). The two Lena Dura specimens are also similar to some of the forms we include in *D. cygniformis*, though they are smaller. This suggests that there is a relationship between *D. antarctica* and *D. cygniformis*. The latter may be an evolutionary offshoot of *D. antarctica*.

Goodman and Ford (1983) equated the specimens identified as *D. oebisfeldensis* by Cookson and Cranwell (1967; p. 205, Plate 1,

Figs. 1–4) and Kemp (1975; p. 604, Plate 1, Figs. 1–6) with *D. antarctica*.

The specimen in Plate 1, Figure 2 of Cookson and Cranwell (1967) is distinctly different from the other illustrated specimens in that the wall layers are completely separated and the paracingulum is less distinct. The degree to which the paracingular projections are developed cannot be ascertained with assurance because the lateral edges of the photograph have been cut off. The Seymour Island specimens compare favorably with what can be seen of this specimen. This form may be an extremely large variant of *D. antarctica*.

Dimensions. Observed range (four specimens): pericyst length, 128 to 168 μm (mean, 149 μm); pericyst width, 91 to 121 μm (mean, 104 μm); endocyst length, 64 to 91 μm (mean, 70 μm); endocyst width, 48 to 90 μm (mean, 74 μm).

Stratigraphic Occurrence. La Meseta Formation (Section 3, Eocene; Section 19, middle to late Eocene).

Selected Previous Occurrences. Chile (Eocene: Cookson and Cranwell, 1967); Ross Sea, DSDP (late Eocene–?early Oligocene: Kemp, 1975).

Deflandrea cf. *phosphoritica* Eisenack 1938
Figure 22.1

Deflandrea phosphoritica Eisenack, 1938, p. 187, Figure 6.

Comments. The morphologically variable La Meseta specimens differ from illustrations of the type material in that the paracingular area protrudes less and the sides of the hypocyst are less posteriorly convergent. The 2a intercalary endoarcheopyle is generally more dorsally situated than often is the case in *D. phosphoritica*. Many of the specimens illustrated in the literature appear to have an apically situated endoarcheopyle, even though it is formed, apparently, by the loss of the 2a intercalary paraplate. The sculpture on the periphragm varies from scabbrate to granular; the latter may be tabular or nontabular. Rare specimens exhibited faint parasutural lines similar to those reported on *C. dartmooria* by Stover (1974) and *D. antarctica* (Wrenn, 1982).

Dimensions. Observed range (10 specimens): pericyst length, 95 to 113 μm (mean, 105 μm); pericyst width, 68 to 85 μm (mean, 76 μm); endocyst length, 51 to 62 μm (mean, 56 μm); endocyst width, 62 to 79 μm (mean, 69 μm); periarcheopyle height, 30 to 38 μm (mean, 30 μm); periarcheopyle width, 37 to 53 μm (mean, 40 μm); endoarcheopyle height, 11 to 24 μm (mean, 16 μm); endoarcheopyle width, 35 to 47 μm (mean, 38 μm).

Stratigraphic Occurrence. La Meseta Formation (Section 3, late early Eocene).

Deflandrea speciosa (Alberti 1959) Lindgren 1984
Figure 17.4

Deflandera speciosa Alberti, 1959, p. 97, Plate 9, Figures 12–13.
Ceratiopsis speciosa (Alberti, 1959) Lentin and Williams, 1977b, p. 21.
Deflandrea speciosa Alberti, 1959, Lindgren, 1984, p. 162.

Comments. Specimens agree closely with holotype illustrations, except some are slightly smaller. The specimens are highly folded, torn, or fragmentary, suggesting fragility and flexibility of the periphragm.

Dimensions. Observed range (three complete specimens): pericyst length, 112 to 160 μm (mean, 128 μm); pericyst width, 75 to 95 μm (mean, 85 μm); endocyst length, 55 to 66 μm (mean, 62 μm); endocyst width, 70 to 90 μm (mean, 80 μm); archeopyle height, 21 to 26 μm (mean, 24 μm); archeopyle width, 37 to 41 μm (mean, 39 μm).

Stratigraphic Occurrence. Cross Valley Formation (Section 15, early late Paleocene); La Meseta Formation (Section 3, reworked into Holocene beach deposits).

Selected Previous Occurrences. Deflandrea speciosa was described from the upper Paleocene deposits of Germany (Alberti, 1959). It has been reported subsequently from the Danian of California (Drugg, 1967), the lower(?) to middle Paleocene of northern Spain (Caro, 1973), the upper Paleocene of the northwestern European continental shelf and adjacent areas (Caro, 1977), and the Paleocene Cannonball Member of South Dakota (Stanley, 1965). Gocht (1969) recovered specimens of this species from the lower Eocene beds of the Meckelfeld well from northwestern Germany. *Deflandrea* aff. *D. speciosa* was reported from DSDP 214 in the Indian Ocean by Harris (1974). Helby et al. (1984) reported a total range for *D. speciosa* of late early Paleocene–early Eocene. Williams and Bujak (1985) cited a total range of late early Paleocene to late early Eocene, and that is the total stratigraphic range of *D. speciosa* recognized herein.

Deflandrea striata (Drugg 1967) Lindgren 1984
Figure 17.5

Deflandrea striata Drugg, 1967, p. 18, Plate 2, Figures 13–14.
Ceratiopsis striata (Drugg, 1967) Lentin and Williams, 1977b, p. 21.
Deflandrea striata Drugg, 1967, Lindgren, 1984, p. 154.

Comments. The rare specimens recovered are very similar to illustrations of the type material. Most of the specimens were damaged.

Dimensions. (One complete specimen, telescopically shortened): pericyst length, 106 μm; pericyst width, 78 μm; endocyst length, 55 μm; endocyst width, 70 μm.

Stratigraphic Occurrence. La Meseta Formation (Section 3, late early Eocene).

Previous Occurrence. Worldwide distribution. Stratigraphic range from the Santonian to Paleocene.

Deflandrea webbii sp. nov.
Figures 22.2–4, 23.4–5, 40.1–4, 41.1–4, 42.2

Deflandrea antarctica Wilson, 1967, pro parte, Goodman and Ford, 1983, p. 864.

Diagnosis. A species of *Deflandrea* bearing well-developed pandasutural striae that delineate paraplate boundaries. Intraparaplate areas are reticulate and bear verracae that may mark the location of trichocyst pores on the theca. A complex flagellar scar lies on the ventral surface at the intersection of the parasulcus and the paracingulum.

Description.
  Shape: The pericyst is peridinioid in outline, whereas the shape of the endocyst varies from subcircular to subpentagonal in outline. The hypocyst bears two short antapical horns. The antapex may be truncated or slightly concave. The length of the apical horn varies.

  Phragma: The thin periphragm (<0.5 μm) is divided into paraplates by pandasutural striae that may be as much as 17 μm wide. The paraplates are irregularly reticulate and bear scattered verrucae that may correspond to trichocyst pores on the original thecae. The endophragm is granular and nontabulate.

  Paratabulation: Pandasutural striae are well developed, and delineate a paratabulation of 4′, ?3a, 7″, 6–7C, 5‴, 2⁗, ?5S.

  Paracingulum: The paracingulum is laevorotatory and offset approximately the width of the paracingulum. The paracingulum is variably incised, bordered by low denticulate ridges and divided into six or seven paraplates by pandasutural striae. The paraplates within the paracingulum are reticulate and bear verrucae.

  Parasulcus: The parasulcal depression widens posteriorly and is bordered by low pandasutural striae. As many as five parasulcal paraplates are delineated by pandasutural striae. A complex flagellar scar is present in the parasulcal area just posterior of the paracingulum. The right parasulcal paraplate projects over the flagellar scar and is convex toward the sulcus; i.e., toward the specimen's left.

  Archeopyle: The broad hexa 2a archeopyle appears to be Type I/I; some specimens, however, suggest a Type I/3I archeopyle. Free, broad hexa 2a (I/I) opercula bearing pandasutural striae along parasutures H2–H6 have been observed. The endoperculum remains attached to the perioperculum. No free 1a or 3a paraplates have been observed, nor have opercula composed of paraplates 1a, 2a, and 3a. This strongly suggests that the archeopyle is Type I/I and not a Type I/3I.

Dimensions. Observed range (six specimens): pericyst length, 81 to 110 μm (mean, 94 μm); pericyst width, 68 to 77 μm (mean, 72 μm); endocyst length, 51 to 80 μm (mean, 63 μm); endocyst width, 50 to 76 μm (mean, 66 μm); endoarcheopyle height, 12 to 22 μm (mean, 15 μm); endoarcheopyle width, 34 to 48 μm (mean, 42 μm). Periarcheopyle (1 measurable specimen): length, 28 μm; width, 46 μm.

Discussion and Comparison with Similar Forms. The distinctive characteristics of *Deflandrea webbii* sp. nov. are the presence of pandasutural striae, reticulate intraparaplate areas bearing verrucae, and a very complex flagellar scar. One of us has observed very faint and scattered pandasutural striae on *D. phosphoritica* (J. H. Wrenn, unpublished data) but never so distinctly or completely developed as that on *D. webbii* sp. nov.

  The flagellar scar is morphologically complex, and the flap that projects over it appears to protect one or more pores. If the pores were the functional site of flagellar insertion, *Deflandrea webbii* sp. nov. is not a cyst but a schizont. This interpretation is supported by the fact that the pandasutural striae are not mere surface features. The striae actually run under the surface paraplate reticulation (Fig. 41.1,3), just as they would if they had been formed by wall growth.

  The overall shape and the flat truncated antapex observed on some specimens is similar to that of *D. antarctica.* However, *D. webbii* sp. nov. differs in having distinctive surface sculpture, a consistently thinner endophragm and by never developing paratabular grana or spines like those on *D. antarctica.*

Holotype. Slide 8511, 117.8 × 7.9 (G23), E/1. Sample 8511. Section 19, La Meseta Formation, middle to late Eocene, Seymour Island, Antarctica.

Stratigraphic Occurrence. La Meseta Formation (Section 3, late early Eocene; Section 19, middle to late Eocene).

Selected Previous Occurrences. Atlantic Ocean, Falkland Plateau, DSDP 511 (middle Eocene–early Oligocene: Goodman and Ford, 1983; as *Deflandrea antarctica,* in part).

Derivation of Name. Deflandrea webbii sp. nov. named after Peter N. Webb in recognition of his pioneering contributions to Antarctic micropaleontology.

Genus *Diconodinium* Eisenack & Cookson, 1960 emend. Morgan 1977
Diconodinium cristatum Cookson & Eisenack 1974 emend. Morgan 1977
Figure 23.1

Diconodinium cristatum Cookson and Eisenack, 1974, p. 77, Plate 24, Figures 3, 5.
Diconodinium cristatum Cookson and Eisenack, 1974, emend. Morgan, 1977, p. 126–127, Plate 1, Figure 3.

Comments. The specimens from the La Meseta Formation exhibit the diagnostic antapical "keel" discussed at length by Morgan (1977). Surface granules range from fine to coarse. The paracingulum is bordered by low ridges capped with granules or minute rounded spines. The broad parasulcus tapers antapically and is bordered by low ridges. The archeopyle is intercalary (2a). Paraplates are suggested by low ridges, but a tabulation formulae could not be determined. The specimens are reworked from Cretaceous (Albian-Cenomanian) deposits.

Dimensions. Observed range (four specimens): length, 72 to 106 μm (mean, 85 μm); width, 41 to 72 μm (mean, 55 μm).

Stratigraphic Occurrence. La Meseta Formation (Section 3, reworked into Eocene deposits; Section 19, reworked into middle to late Eocene deposits).

Selected Prevous Occurrences. Australia (Albian-Cenomanian: Cookson and Eisenack, 1974); Australia (Albian-Cenomanian: Morgan, 1977); Australia, Queensland (Late Albian–Cenomanian: Playford et al., 1975; as *D.* cf. *D. multispinula,* Fig. 5, number 1).

Diconodinium multispinum (Deflandre & Cookson 1955) Eisenack and Cookson 1960 emend. Morgan 1977
Figure 23.3

Palaeohystrichophora multispina Deflandre and Cookson, 1955, p. 257, Plate 1, Figure 5.
Diconodinium multispinum (Deflandre and Cookson, 1955) Eisenack and Cookson, 1960, p. 3.
Diconodinium multispinum (Deflandre and Cookson, 1955) Eisenack and Cookson, 1960, emend. Morgan 1977, p. 127–128, Plate 1, Figures 1, 4, 6, 8.

Comments. The genus *Diconodinium* has been discussed at length by Morgan (1977). Specimens of *D. multispinum* recovered from the La Meseta Formation agree completely with the emended diagnosis of Morgan (1977). Sculpture varies from granulate to spinate, and the spines may be capitate or truncated distally. An intercalary archeopyle with the operculum partially detached was observed on a few specimens. The specimens are reworked from Cretaceous (Albian-Santonian) deposits.

Dimensions. Observed range (seven specimens): length, 69 to 106 μm (mean, 86 μm); width, 37 to 68 μm (mean, 54 μm); spines less than 2 μm.

Stratigraphic Occurrence. La Meseta Formation (Sections 3, 17, 19, reworked into Eocene deposits).

Selected Previous Occurrences. Primarily reported from Australia (Aptian to Santonian). Also recovered in Canada (Albian: Singh, 1971; as *D. pusillum*); France (middle Albian: Verdier, 1975) and the U.S.S.R. (Turonian: Vozzhennikova, 1967).

Genus *Dinogymnium* Evitt et al. 1967
Dinogymnium cf. *D. curvatum* (Vozzhennikova 1967) Lentin and Williams 1973
Figure 23.2

Gymnodium curvatum Vozzhennikova, 1967, p. 43, Plate 1, Figures 10–12; Plate 4, Figures 2–3.
Dinogymnium curvatum (Vozzhennikova, 1967) Lentin and Williams, 1973, p. 48.

Comments. The single specimen recovered shares some morphologic attributes with both *Dinogymnium acuminatum* and *D. curvatum.* The specimen's overall dimensions and the presence of granular surface sculpture across the paracingular area is common to both species. *D. acuminatum* generally lacks granulation at the poles, and its paracingular margins are relatively smooth. In contrast, the Seymour Island specimen is uniformly granular, and its distorted paracingular margin is somewhat scalloped. These are features characteristic of *D. curvatum.* The specimen differs from *D. curvatum* in that the epicyst is slightly less arcuate. The specimen was reworked from Late Cretaceous (?Senonian) deposits.

Dimensions. (One specimen recovered): length, 72 μm; width, 35 μm; wall thickness, 0.5 μm.

Stratigraphic Occurrence. La Meseta Formation (Section 3, reworked into late early Eocene deposits).

Selected Previous Occurrences. Canada, eastern (Maastrichtian: Barss et al., 1979); Gabon (late Senonian: Boltenhagen, 1970; as *D.* aff. *curvatum*); U.S.S.R. (Late Cretaceous: Vozzhennikova, 1967).

Enigmadinium gen. nov.

Type Species. Enigmadinium cylindrifloriferum sp. nov., late early Eocene, La Meseta Formation, Seymour Island, Antarctica.

Diagnosis. The periphragm of these chorate cysts bears numerous hollow, generally flattened nontabular processes. The processes are about the same size and shape. Each process is closed and complexly branched distally. There are no indications of paratabulation. The archeopyle is assumed to be apical.

Enigmadinium differs from *Cordosphaeridium, Hystrichokolpoma, Hystrichosphaeridium* and *Oligosphaeridium* by having undifferentiated nontabular processes; from *Homotryblium* by not having an epicystal archeopyle; from *Adnatosphaeridium* by having two wall layers and by lacking distal trabeculae; from *Areosphaeridium* and *Surculosphaeridium* by having hollow processes; from *Callaiosphaeridium* by lacking indications of paratabulation and by having nontabular processes; and from *Distatodinium* by having a subspherical, rather than an elongate, central body.

Derivation of Name. Latin, *aenigma,* a riddle or mystery—with reference to the unusual processes and their distal terminations.

Enigmadinium cylindrifloriferum gen. nov., sp. nov.
Figures 24.1–3, 45.1–6

Diagnosis. A species of *Enigmadinium* characterized by tapering cylindrical or branched, nontabular processes. The process terminations are complexly branched into as many as eight finger-like projections that are encrusted with submicron sized granules.

Description.
Shape: A subcircular chorate dinocyst bearing numerous hollow, nontabular processes. The central body probably was originally spherical.
Phragma: The endophragm is smooth to finely granular and usually wrinkled. Granules are less than 1 μm in diameter. The periphragm is smooth. Both wall layers are <0.5 μm thick. The periphragm gives rise to as many as 40 nontabular processes. The process shafts may be subconical (tapering distally), cylindrical, or branched midway along their length. Two process shafts also may arise from a common base. The process terminations are closed distally and complexly branched. Each branch may be bifurcate. The more elaborate process terminations form up to eight distal branches, each of which bifurcates into two subbranches whose tips often bifurcate or trifurcate yet again. The processes, and especially their terminations, are heavily encrusted with submicron granules. The more complex terminations are reminiscent of flowers. The process base is broader than the rest of the process except for the distally flared, branching termination. Each process base may form a ring where it contacts the central body. Some processes arise from two or more roots which join to form the shaft.
Paratabulation: None evident.
Paracingulum: None evident.
Parasulcus: None evident.
Archeopyle: Ruptures observed in the central body suggest an apical archeopyle. However, a distinct archeopyle has not been observed.

Dimensions. Observed range (10 specimens): overall length, 39 to 51 μm (mean, 44 μm); overall width, 23 to 40 μm (mean, 33 μm); endocyst length, 29 to 48 μm (mean, 37 μm); endocyst width, 20 to 41 μm (mean, 28 μm); process length, 8 to 10 μm; process width, 0.5 to 5 μm; endophragm thickness, <0.5 μm; periphragm thickness, <0.5 μm.

Holotype. Slide 8500, P72106A01, 107.8 × 19.1. Sample 8500, Section 17, La Meseta Formation, late early Eocene, Seymour Island, Antarctica.

Derivation of Name. Latin, *cylindratus,* in the form of a cylinder, and *florifer,* bearing flowers: in reference to the distinctive processes and terminations—like cylinders bearing flowers at their distal end.

Stratigraphic Occurrence. La Meseta Formation (Section 3, reworked into Holocene beach deposits; Section 17, late early Eocene).

Genus *Eyrea* Cookson and Eisenack 1970
Eyrea nebulosa Cookson and Eisenack 1971
Figure 24.6

Eyrea nebulosa Cookson and Eisenack, 1971, p. 23, Plate 11, Figures 2–6.

Comments. The specimens recovered agree well with illustrations of the type material. The lack of invaginations uniting the inner and outer walls was pointed out by Elsik (1977) as one feature distinguishing *E. nebulosa* Cookson and Eisenack 1971 from *Paralecaniella indentata* (Deflandre and Cookson 1955) Cookson and Eisenack 1970b emend. Elsik 1977. Generally, the outer wall of the Seymour Island specimens is poorly defined.

Dimensions. Observed range (10 specimens): overall length, 42 to 94 μm (mean, 71 μm); overall width, 32 to 72 μm (mean, 66 μm); inner body length, 32 to 72 μm (mean, 46 μm); inner body width, 31 to 42 μm (mean, 38 μm).

Stratigraphic Occurrence. Cross Valley Formation (Section 12-13, Eocene and early late Paleocene; Sections 15 and 16, early late Paleocene); La Meseta Formation (Section 3, Eocene; Sections 17, 18, and 19, Eocene).

Selected Previous Occurrences. Atlantic Ocean, Bay of Biscay, DSDP 400A and 402A (Aptian-Albian: Davey, 1979); Atlantic Ocean, off southwestern Africa (Aptian-Turonian: Davey, 1978); Australia (Albian-Cenomanian: Cookson and Eisenack, 1971).

Genus *Hystrichosphaeridium* Deflandre 1937
emend. Davey and Williams 1966
Hystrichosphaeridium cf. *H. astartes* Sannemann 1955
Figure 24.4–5

Hystrichosphaeridium astartes Sannemann, 1955, p. 325, Plate 4, Figure 1.

Comments. The specimens compare well with illustrations of the type material, but are much smaller. Spherical central body covered with approximately 25 subconical, hollow protuberances, as in *Hystrichosphaeridium astartes* Sannemann 1955. The process bases abut each other. The diameter of the *H. astartes* type specimen is 230 μm, whereas the Seymour Island specimens range from 42 to 50 μm. *H. astartes* Sannemann 1955 was referred to as *?Baltisphaeridium astartes* (Sannemann 1955) without comment or explanation by Eisenack et al. (1973). Although this species undoubtedly does not belong in the Pyrrhophyta, a legitimate transfer to another taxon has never been effected.

Dimensions. (One measurable specimen): central body diameter, 32 μm; overall diameter, 43 μm; process length as much as 10 μm; diameter of process bases as much as 10 μm.

Stratigraphic Occurrence. La Meseta Formation (Section 3, Eocene).

Hystrichosphaeridium parvum Davey 1969

Hystrichosphaeridium parvum Davey, 1969b, p. 5, Plate 1, Figure 8, Plate 2, Figure 1.

Comments. The specimens compare favorably with the illustrations and description of type material from the Cretaceous of South Africa (Davey, 1969b). The endocyst, however, is distinctly granular. The smooth processes bear numerous longitudinal folds that resemble ribbing at first glance.

Dimensions. Observed range (four specimens): overall diameter, 28 to 38 μm (mean, 32 μm); endocyst diameter, 20 to 25 μm (mean, 23 μm); process length, approximately 7 μm.

Stratigraphic Occurrence. La Meseta Formation (Section 3, Eocene; Sections 17, 18, and 19, Eocene).

Selected Previous Occurrences. Canada, offshore eastern (early Eocene: Williams and Brideaux, 1975); South Africa (Campanian-Danian: Davey, 1969b).

Hystrichosphaeridium salpingophorum (Deflandre 1935) emend. Davey and Williams 1966
Figure 25.5–6

Hystrichosphaera salpingophorum Deflandre, 1935, p. 232, Plate 9, Figure 1.
Hystrichosphaeridium salpingophorum (Deflandre, 1935) Deflandre, 1937.
Hystrichosphaeridium salpingophorum (Deflandre, 1935), emend. Davey and Williams, 1969b, p. 61–62, Plate 10, Figure 6.

Comments. Rare specimens agreeing closely with the emended diagnosis occur in Sections 15, 16, and 19. The paratabulation formula, overall shape, number, and structure of processes, and the archeopyle type are the same as the material described by Davey and Williams (1966b). On some specimens these authors noted the presence of a basal ring where the processes join the central body. The basal rings on the Seymour Island specimens are circular, subquadrate, or elliptical.

Dimensions. Observed range (five specimens): endocyst diameter, 33 to 43 μm (mean, 37 μm); process lengths as much as 27 μm.

Stratigraphic Occurrence. Cross Valley Formation (Sections 15 and 16, early late Paleocene); La Meseta Formation (Section 19, late early Eocene).

Selected Previous Occurrences. Worldwide distribution. The reported stratigraphic distribution ranges from Jurassic to Oligocene.

Hystrichosphaeridium truswelliae sp. nov.
Figures 25.1–4, 39.1

Hystrichosphaeridium tubiferum (Ehrenberg) Deflandre 1937 sensu Wilson 1967, Figure 40.

Diagnosis. A species of *Hystrichosphaeridium* characterized by hollow processes with distally open, flared, and fenestrate process terminations. Apical processes shorter than all other processes. Periphragm is smooth, whereas endophragm is granular.

Description.
Shape: Subspherical to spherical skolochorate dinocysts that bear intratabular processes of various lengths. The apical processes are always

the shortest; consequently, the endocyst is asymmetrically placed with respect to the imaginary shell formed by joining the process tips with a line.
Phragma: The finely granular endophragm is < 1.0 μm thick. The smooth periphragm gives rise to infundibular intratabular processes and is closely appressed to the endophragm except beneath each process. A faint ring marks the base of most processes on the central body. The endophragm is continuous beneath the processes, hence there is no communication between the endocoel and the hollow axis of each process. The distally flared process terminations are open and highly fenestrate.
Paratabulation: The paratabulation indicated by the distribution of the processes is 4′, 6″, ?5c 6‴, 1⁗, ?3s.
Paracingulum: The location of the unspiraled paracingulum is indicated by at least five intratabular processes.
Parasulcus: At least three thin infundibular parasulcal processes mark the location of the parasulcus.
Archeopyle: The apical archeopyle is formed by the partial or complete detachment of the simple apical operculum. One infundibular process arises from the center of each opercular paraplate. The operculum often falls through the archeopyle and remains within the central body.

Dimensions. Observed range (five specimens): diameter of endocyst, 39 to 43 μm (mean, 41 μm); overall diameter, 72 to 85 μm (mean, 79 μm); process length, 8 to 15 μm; endophragm thickness, <1 μm; periphragm thickness, <1 μm.

Discussion and Comparison with Similar Species. H. truswelliae is most similar to H. tubiferum (Ehrenberg, 1838) Deflandre, 1937, but differs in that its periphragm is smooth and the endophragm is granular. The reverse is true in H. tubiferum. The fenestrate process terminations of H. truswelliae contrast markedly with the "denticulate to serrate circular margins" (Davey and Williams, 1966b) of H. tubiferum. The fact that the operculum is commonly present, possibly adnate, may be an additional distinguishing characteristic separating H. truswelliae from other species of Hystrichosphaeridium.

The process terminations of H. truswelliae are very similar to the Southern Hemisphere forms of Areosphaeridium diktyoplokus. Both have perforate distal platform or platform-like process terminations with ragged margins. These species are easily separated because H. truswelliae has hollow, distally open processes, whereas those of A. diktyoplokus are solid and distally closed. In addition, the relatively short epicystal processes characteristic of H. trustwelliae are not seen on A. diktyoplokus.

Holotype. Slide 8500W/1, 116.8 x 21.8 (X22). Sample 8500, Section 17, La Meseta Formation, late early Eocene, Seymour Island, Antarctica.

Derivation of Name. Named in recognition of the significant contributions made to the study of Antarctic palynology by Elizabeth M. Truswell.

Stratigraphic Occurrence. La Meseta Formation (Section 3, late early Eocene; Sections 17 and 18, late early Eocene).

Selected Previous Occurrences. Antarctica, Ross Sea (?Eocene: Wilson, 1967a).

356 Wrenn and Hart

Hystrichosphaeridium tubiferum subsp. *brevispinum*
(Davey and Williams 1966) Lentin and Williams 1973
Figure 26.7–9

Hystrichosphaeridium tubiferum var. *brevispinum* Davey and Williams,
1966b, p. 58, Plate 10, Figure 10.
Hystrichosphaeridium tubiferum subsp. *brevispinum* (Davey and Williams, 1966b) Lentin and Williams, 1973, p. 80.

Comments. The Seymour Island specimens differ from the description of the type material in that the endophragm is granular whereas the periphragm is smooth: just the reverse of the type material (Davey and Williams, 1966b). In all other respects they are the same.

Dimensions. Observed range (two specimens): endocyst diameter, 39 μm (mean, 39 μm); process length, 5 to 11 μm. Wall thickness: endophragm, 1 μm; periphragm, <0.5 μm.

Stratigraphic Occurrence. Cross Valley Formation (Section 15, early late Paleocene); La Meseta Formation (Section 3, late early Eocene; Sections 17 and 18, late early Eocene).

Selected Previous Occurrences. England (early Eocene: Davey and Williams, 1966); English Channel (late Ypresian: Auffret and Graus-Cavagnetto, 1975; not illustrated); France (late Turonian: Foucher, 1974); Switzerland (Paleocene: Stuijvenberg et al., 1976); U.S.A., New Jersey (Campanian-Maastrichtian: May, 1980).

Genus *Impagidinium* Stover and Evitt 1978
***Impagidinium dispertitum* (Cookson and Eisenack 1965)**
Stover and Evitt 1978

Leptodinium dispertitum Cookson and Eisenack, 1965a, p. 122–123, Plate 12, Figures 5–7.
Impagidinium dispertitum (Cookson and Eisenack, 1965a) Stover and Evitt, 1978, p. 165.

Comments. Rare specimens were observed, which agree with the illustrations and description of the type material from the upper Eocene Browns Creek Clay of Southwest Victoria, Australia. Differentiation from *I. maculatum* (Cookson and Eisenack 1961b) Stover and Evitt 1978 and *I. victorianum* (Cookson and Eisenack 1961b) Stover and Evitt 1978 was based on contrasting size, sculpture, and tabulation, as indicated by Cookson and Eisenack (1965a).

Dimensions. Observed range (three specimens): diameter, 56 to 65 μm (mean, 63 μm); wall thickness, 2 μm.

Stratigraphic Occurrence. La Meseta Formation (Section 3, late early Eocene).

Previous Occurrence. Worldwide distribution. Late Cretaceous to early Quaternary. (The range and reports of this species need to be reviewed.)

Impagidinium maculatum
(Cookson and Eisenack 1961)
Stover and Evitt 1978
Figure 26.5–6
Leptodinium maculatum Cookson and Eisenack, 1961b, p. 40, Plate 2, Figures 5–6.

Impagidinium maculatum (Cookson and Eisenack, 1961b) Stover and Evitt, 1978, p. 166.

Comments. The specimens from Seymour Island agree with the illustrations and description of the type material. Paratabular crests or ledges on the central body are as much as 4 μm high. The granular endocyst is approximately 3 μm thick. Neither of these dimensions was noted in the original description, so exact comparison is not possible. Most specimens were damaged and incomplete.

Dimensions. Observed range (two complete specimens): diameter, 36 to 42 μm (mean, 39 μm); crest height, approximately 3 μm.

Stratigraphic Occurrence. Cross Valley Formaiton (Sections 15 and 16, early late Paleocene); La Meseta Formation (Section 19, middle-late Eocene).

Previous Occurrence. Worldwide distribution during the Eocene.

***Impagidinium victorianum* (Cookson and Eisenack 1965)**
Stover and Evitt 1978
Figure 26.3–4

Leptodinium victorianum Cookson and Eisenack, 1965a, p. 123, Plate 12, Figures 8–9.
Impagidinium victorianum (Cookson and Eisenack, 1965a) Stover and Evitt, 1978, p. 166.

Comments. Very rare, poorly preserved specimens were recovered from the Seymour Island material.

Dimensions. (One specimen measurable): diameter, 77 μm; wall thickness, approximately 3 μm.

Stratigraphic Occurrence. La Meseta Formation (Section 19, middle to late Eocene).

Selected Previous Occurrences. I. victorianum (Cookson and Eisenack) Stover and Evitt 1978 was described from the late Eocene of Australia (Cookson and Eisenack, 1965a). It has subsequently been reported from the late Eocene of Australia (Verdier, 1970), the late middle–late Eocene of offshore eastern Canada (Williams and Bujak, 1977), the Eocene of eastern Canada (Barss et al., 1979) and California (Damassa, 1979), the Eocene-Oligocene of the Tasman Sea (DSDP 280–283; Haskell and Wilson, 1975), and the early Oligocene of the Falkland Plateau (DSDP 511; Goodman and Ford, 1983). The total stratigraphic range recognized here is Eocene to early Oligocene.

Genus *Impletosphaeridium* Morgenroth 1966
***Impletosphaeridium clavus* sp. nov.**
Figure 27.10–11, 13

Diagnosis. A species of *Impletosphaeridium* characterized by its thin solid, nail-like processes. The processes taper slightly to pad-like terminations that may appear to be bifid, trifid, or multifurcate.

Description.
 Shape: Subrounded to ellipsoidal chorate dinocyst bearing approximately 100 nontabular nail-like processes.

Phragma: Autophragm or closely appressed periphragm and endophragm. The central body is finely granular, whereas the processes are smooth. The difference in surface sculpture noted above suggests that there are two layers. The abundant nontabular processes look like nails driven into a ball. They are solid, thin, and slightly tapered distally and possess pad-like terminations that appear to be bifid, trifid, or multifurcate. The terminations are commonly recurved.

Paratabulation: None evident.
Paracingulum: None evident.
Parasulcus: None evident.
Archeopyle: Indeterminate.

Dimensions. Observed range (10 specimens): central body length, 21 to 47 μm (mean, 28 μm); central body width, 18 to 37 μm (mean, 23 μm); length of processes, 8 to 12 μm; width of processes, approximately 0.5 μm.

Comments. The process termination of *Impletosphaeridium clavus* sp. nov. distinguishes it from all other species of this genus. This species occurs in most of the Seymour Island sections and is quite common in some samples.

Holotype. Slide 8500, W/3, 112.0 x 7.3 (S17). Sample 8500, Section 17, La Meseta Formation, late early Eocene, Seymour Island, Antarctica.

Derivation of Name. Latin, *clavus,* nail, with reference to the nail-like appearance of the processes.

Stratigraphic Occurrence. Cross Valley Formation (Sections 15 and 16, early late Paleocene); La Meseta Formation (Section 3, Eocene; Sections 17, 18 and 19, Eocene).

Impletosphaeridium ligospinosum (De Coninck 1969) Islam 1983
Figure 27.8

Baltisphaeridium ligospinosum De Coninck, 1969, p. 50, Plate 15, Figures 9–19.
Impletosphaeridium ligospinosum (De Coninck, 1969) Islam, 1983, p. 240.

Comments. The numerous solid processes, overall size, and process dimensions noted on the Seymour Island specimens agree closely with illustrations of the type material from the Kallo borehole (De Coninck, 1969). The process terminations are asymmetrically bifurcate, as in type material; however, acuminate tips are also present. No openings were observed in any of the specimens.

Dimensions. Observed range (10 specimens): main body diameter, 17 to 28 μm (mean, 24 μm); process length, 7 to 10 μm; process diameter, 0.5 μm.

Stratigraphic Occurrence. Cross Valley Formation (Section 12-13, Eocene, see discussion of the age determination for this section; Section 15, early late Paleocene). La Meseta Formation (Section 3, late early Eocene and reworked into Holocene beach deposits; Sections 17, 18, and 19, Eocene).

Selected Previous Occurrences. Belgium (late Eocene: De Coninck, 1969); England (Paleocene-Oligocene: Gruas-Cavagnetto, 1976); England (Eocene: Islam, 1983); English Channel, eastern (Ypresian: Auffret

and Gruas-Cavagnetto, 1975); Sweden (late Danian–middle Paleocene: De Coninck, 1975).

Impletosphaeridium lorum sp. nov.
Figure 26.1–2

Diagnosis. A species of *Impletosphaeridium* characterized by acuminate, whip-like solid, but highly flexible, processes.

Description.
Shape: The central body is subspherical to ellipsoidal.
Phragma: The central body is less than 1 μm thick and is composed of an autophragm or closely appressed periphragm and endophragm. The surface texture of the central body is shagreenate to granular. The 50 to 100 nontabular processes are solid, whip-like, and distally acuminate. The processes appear to be highly flexible.
Paratabulation: None evident.
Paracingulum: None evident.
Parasulcus: None evident.
Archeopyle: None evident.

Dimensions. Observed range (four specimens): length, 21 to 33 μm (mean, 27 μm); width, 21 to 25 μm (mean, 22 μm); processes, 10 to 15 μm; autophragm, less than 1 μm thick.

Discussion and Comparison with Other Species. Although *I. lorum* sp. nov. superficially resembles *I. ligospinosum* (De Coninck) Islam, 1983, the former is characterized by whip-like acuminate processes. The processes of *I. ligospinosum,* on the other hand, are quite stiff, and their terminations are bifid or capitate. *I. lorum* sp. nov. has fewer, longer processes (as much as 15 μm long) than *Micrhystridium* sp. A. The latter has a dense covering of processes less than 8 μm long.

Holotype. Slide 8496, W/8, 114.2 x 20.7 (W19). Sample 8496, Section 16, Cross Valley Formation, early late Paleocene, Seymous Island, Antarctica.

Derivation of Name. Latin, *lorum,* whip or scourge, with reference to the whip-like processes.

Stratigraphic Occurrence. Cross Valley Formation (Section 12-13, Paleocene; Sections 15 and 16, early late Paleocene); La Meseta Formation (Section 3, Eocene; Sections 17, 18, and 19, Eocene).

Impletosphaeridium sp. B
Figure 27.1–3, 9

Diagnosis. A species of *Impletosphaeridium* characterized by processes that are triangular in cross section, distally solid, and furcate.

Description.
Shape: Subspherical skolochorate dinocyst.
Phragma: The endophragm is smooth to granular and usually wrinkled. The periphragm is smooth, often wrinkled, and gives rise to 30 or more nontabular processes. The proximally hollow processes are isolated from the endocoel by the endophragm. The distal half of each process is solid, closed, and terminated by what appears to be a bifurcate, trifurcate, or multifurcate tip. The processes are triangular in cross section, with corner ribs extending the length of the process shaft. These

lineations may continue onto the surface of the central body, forming an irregular reticulation by merging with ribs from adjacent processes.

Paratabulation: None evident.

Paracingulum: None evident.

Parasulcus: None evident.

Archeopyle: A distinct archeopyle has not been observed on any specimen. However, an angular opening in one specimen suggests that the archeopyle is apical.

Dimensions. Observed range (two measurable specimens): diameter, 20 to 24 μm (mean, 22 μm); process length, 7 to 10 μm; periphragm thickness, <0.5 μm; endophragm thickness, <0.5 μm.

Discussion and Comparison with Similar Species. This is *Impletosphaeridium* sp. B of Wrenn (1982). The distally solid processes and the uncertainty as to the archeopyle type are the basis for assigning these specimens to the genus *Impletosphaeridium*. The distinctive structure of the processes and the irregular network on the surface of the endocyst distinguish *Impletosphaeridium* sp. B from all other species of *Impletosphaeridium*.

Stratigraphic Occurrence. La Meseta Formation (Section 3, Eocene; Sections 127, 18, and 19, Eocene).

Genus *Isabelidinium* Lentin and Williams 1977
Isabelidinium pellucidum (Deflandre and Cookson 1955) Lentin and Williams 1977
Figure 27.4

Deflandrea bakeri forma *pellucida* Deflandre and Cookson, 1955, p. 251, Plate 4, Figure 3.
Deflandrea pellucida (Deflandre and Cookson, 1955) Cookson and Eisenack, 1958, p. 27, Plate 4, Figure 9.
Isabelidinium pellucidum (Deflandre and Cookson, 1955) Lentin and Williams, 1977a, p. 168.
Alterbia pellucida (Deflandre and Cookson, 1955) Yun, 1981, p. 64.
Isabelidinium pellucidum (Deflandre and Cookson, 1955) Lentin and Williams, 1985, p. 201.

Comments. The Seymour Island specimens agree closely with the description of the type specimen, except with regard to surface sculpture. Although the periphragm of some specimens is smooth or finely granular, like that of the type material, others have a coarsely granular or even rugulate surface. The location of the paracingulum was indicated on one specimen by two parallel rows of low grana.

The transfer of this species to *Alterbia* by Yun (1981) was rejected by Lentin and Williams, 1985. We concur with their retention of this species in *Isabelidinium*.

Dimensions. Observed range (three specimens): pericyst length, 84 to 118 μm (mean, 105 μm); pericyst width, 70 to 87 μm (mean, 78 μm); endocyst length, 54 to 72 μm (mean, 61 μm); endocyst width, 68 to 85 μm (mean, 76 μm); wall thickness, approximately 1 μm.

Stratigraphic Occurrence. La Meseta Formation (Section 3, late middle-late Eocene; Section 19, late early Eocene).

Selected Previous Occurrences. Argentina (Campanian-Eocene: Gamerro and Archangelsky, 1981; Australia (Campanian–lower Maastrichtian: Cookson and Eisenack, 1958; Paleocene–early Eocene: Deflandre and Cookson, 1955); U.S.A., California (Danian: Drugg, 1967).

Genus *Kallosphaeridium* De Coninck 1969
Kallosphaeridium cf. *K. capulatum* Stover 1977
Figure 27.5–7

Kallosphaeridium capulatum Stover, 1977, p. 74, Plate 1, Figures 11–13.

Comments. Specimens comparable to *Kallosphaeridium capulatum* vary somewhat from the original material described by Stover (1977). Some of the Seymour Island specimens have a smaller overall diameter (30 to 33 μm) and a thicker autophragm (1 μm) than the type material (46 to 50 and 0.5 μm, respectively). The low granular sculpture is uniformly distributed. The apical archeopyle remains attached to the autocyst by a narrow parasulcal tab. Accessory archeopyle sutures are generally present.

Dimensions. Observed range (five specimens): length, 29 to 36 μm (mean, 33 μm); width, 29 to 34 μm (mean, 31 μm); wall thickness, 2 μm.

Stratigraphic Occurrence. Cross Valley Formation (Sections 15 and 16, early late Paleocene); La Meseta Formation (Section 3, late early Eocene).

Selected Previous Occurrences. Atlantic Ocean, Falkland Plateau, DSDP 511 (early Oligocene: Goodman and Ford, 1983): Blake Plateau, Atlantic Ocean (middle-late Oligocene: Stover, 1977).

Genus *Lejeunecysta* Artzner and Dörhöfer 1978
emend. Bujak 1980
Lejeunecysta fallax (Morgenroth 1966)
Artzner and Dörhöfer 1978
Figure 28.3

Lejeunia fallax Morgenroth, 1966b, p. 2-3, Plate 1, Figures 6–7.
Lejeunecysta fallax (Morgenroth, 1966b) Artzner and Dörhöfer, 1978, p. 1381–1382.
Lejeunecysta fallax (Morgenroth, 1966b), Artzner and Dörhöfer, 1978, emend. Biffi and Grignani, 1983, p. 132.

Comments. The specimens recovered are less angular than the type material. The short apical and antapical horns are usually solid, subrounded to sharp, and darker than the rest of the autocyst. The autophragm is smooth and without sculpture. The paracingulum, if present, is indicated by parallel folds. The parasulcus is usually not evident. The intercalary archeopyle is not always discernible. When it is visible, the operculum of the standard hexa 2a archeopyle is usually attached along parasuture H4.

Dimensions. Observed range (10 specimens): length, 48 to 92 μm (mean, 70 μm); width, 60 to 81 μm (mean, 67 μm); archeopyle dimensions (6 measurable specimens): height, 14 to 45 μm (mean, 26 μm); width, 22 to 50 μm (mean, 31 μm); wall, less than 1 μm thick.

Stratigraphic Occurrence. La Meseta Formation (Section 3, late middle to late Eocene).

Selected Previous Occurrences. Lejeunecysta fallax was originally described from the middle Oligocene of Europe (Morgenroth, 1966b). Williams and Bujak (1977) recovered *L. fallax* from the Oligocene-middle Miocene of the Scotian Shelf and the Grand Banks. Williams (1978) considered specimens recovered from DSDP 370 to range from middle Eocene to middle Oligocene in age. Griggs (1981) identified specimens of *L. fallax* in middle Miocene samples from the Santos Basin, Brazil.

Additional occurrences include: Argentina (Oligocene-Miocene: Gamerro and Archangelsky, 1981); Belgium (Eocene: De Coninck, 1975); Canada, eastern (Oligocene–middle Miocene: Barss et al., 1979); Canada, offshore Atlantic (middle Oligocene–middle Miocene: Williams, 1975); Canada, offshore Atlantic (middle Miocene: Gradstein et al., 1976); Canada, offshore Atlantic (early Oligocene–middle Miocene: Williams and Bujak, 1977); France (late Oligocene: Charollais et al., 1975); Germany (early-middle Oligocene: Benedek and Müller, 1976); Germany (middle-late Oligocene: Benedek, 1972); Norwegian-Greenland Sea (middle Oligocene–early Miocene: Manum, 1976).

The evidence indicates the stratigraphic distribution *L. fallax* ranges from the middle Eocene to middle Miocene.

Lejeunecysta hyalina (Gerlach 1961)
Artzner and Dörhöfer 1978
Figure 28.5

Lejeunia hyalina Gerlach, 1961, p. 169–171, Plate 26, Figures 10–11.
Lejeunia hyalina (Gerlach, 1961) emend. Kjellström, 1972, p. 469.
Lejeunecysta hyalina (Gerlach, 1961 emend. Kjellström, 1972) Artzner and Dörhöfer, 1978, p. 1381.
Lejeunecysta hyalina (Gerlach, 1961, emend. Kjellström, 1972) emend. Artzner and Dörhöfer, 1978; emend. Sarjeant, 1984.

Comments. The Seymour Island specimens are comparable to the description and illustrations of the type material, although their overall dimensions vary. The autophragm is relatively flexible and transparent, but appears dark wherever it is folded over upon itself. The short apical and antapical horns are solid and generally sharp distally. The operculum is usually attached posteriorly along the H4 parasuture of the standard 2a archeopyle.

Dimensions. Observed range (seven specimens): length, 59 to 121 μm (mean, 92 μm); width, 47 to 105 μm (mean, 84 μm); archeopyle dimensions (one measurable): height, 34 μm; width, 41 μm.

Stratigraphic Occurrence. Cross Valley Formation (Section 15, early late Paleocene); La Meseta Formation (Section 3, late middle to late Eocene).

Selected Previous Occurrences. Argentina (Maastrichtian-Miocene: Gamerro and Archangelsky, 1981); Belgium (early Landinian: Schumacker-Lambry and Châteauneuf, 1976; Eocene: De Coninck, 1975; Ypresian: De Coninck, 1968, 1980); Canada, eastern (Eocene-early Oligocene: Barss et al., 1979); England (early Eocene–middle Eocene: Eaton, 1976; early Oligocene: Liengjarern et al., 1980; Eocene: Bujak et al., 1980; Paleocene-Oligocene: Gruas-Cavagnetto, 1976); France (early Oligocene: Carollais et al., 1980; late Eocene–early Oligocene: Châteauneuf, 1980; Germany (early Eocene: Gocht, 1969; late Oligocene: Eisenack, 1961; Sarjeant, 1984; Gerlach, 1961; middle Oligocene: Benedek, 1972); Korea (Miocene: questionable attribution; Yun, 1981); Netherlands (Eocene: De Coninck, 1977); Norwegian-Greenland Sea (middle Oligocene-early Miocene: Manum, 1976); Sweden (Maastrichtian: questionable attribution, Kjellstrom, 1972, 1973); Tasman Sea DSDP 280-281 (middle Eocene–Oligocene: Haskell and Wilson, 1975).

Genus *Manumiella* Bujak and Davies 1983
Manumiella druggii (Stover 1974)
Bujak and Davies 1983
Figures 28.7–8, 39.3

Deflandrea druggii Stover, 1974, p. 171, Plate 1, Figures 3 and 4.
Isabelia druggii (Stover, 1974) Lentin and Williams, 1976, p. 58.
Isabelidinium druggii (Stover, 1974) Lentin and Williams, 1977a, p. 167.
Manumiella druggii (Stover, 1974) Bujak and Davies, 1983, p. 161.

Comments. Wilson (1978) compared *Manumiella druggii* (as *Isabelidinium druggii*) with the closely related species *M. seelandica* (as *I. seelandica*) and concluded they differ only in the latter possessing a pronounced apical horn. Both species were reported from the late Maastrichtian-Danian of New Zealand (Wilson, 1978). Only *M. druggii* has been observed in the Seymour Island material.

Dimensions. Observed range (seven specimens); pericyst length, 77 to 99 μm (mean, 90 μm); pericyst width, 52 to 70 μm (mean, 60 μm); endocyst length, 60 to 106 μm (mean, 78 μm); endocyst width, 49 to 79 μm (mean, 65 μm).

Stratigraphic Occurrence. La Meseta Formation (Section 3, reworked into Eocene deposits).

Selected previous Occurrences. Southern Hemisphere and North America during the early Campanian to middle Paleocene.

Octodinium gen. nov.

Type Species. Octodinium askiniae sp. nov., early Eocene, La Meseta Formation, Seymour Island, Antarctica.

Diagnosis. Subelliptical to subtriangular cornucavate peridinioid dinocysts characterized by a combination archeopyle, which when fully developed, is of the IPa type. The 2a paraplate may be partially or completely detached; the 4″ paraplate is adnate adcingularly, though the 3″/4″ and 4″/5″ parasutures may be open. The intercalary 2a paraplate is eight-sided.

Comparison. Octodinium gen. nov. differs from *Phelodinium* by having convex sides between the apical and antapical horns, a straight to concave side between the widely separated antapical horns, and a combination archeopyle with an eight-sided 2a paraplate.

Octodinium gen. nov. differs from *Broomea* and *Pareodinia* by having a combination archeopyle, an octagonal 2a intercalary paraplate, two wall layers, and widely separated antapical horns.

Octodinium gen. nov. differs from *Palaeocystodinium, Andalusiella* and *Deflandrea* by having a combination archeopyle and an octagonal 2a intercalary paraplate.

Derivation of Name. Octodinium. Latin, *octo,* eight, referring to the eight-sided 2a intercalary paraplate involved in archeopyle formation.

Octodinium askiniae gen. nov. sp. nov.
Figures 28.1–2, 4; 29.1–7

Diagnosis. A species of *Octodinium* characterized by long tapering apical and antapical horns and periphragm ornamentation consisting of grana or rugulae.

Description.

Shape: The cornucavate cysts vary in outline from subelliptical to subtriangular. Subelliptical cysts are widest in the paracingular area, whereas subtriangular specimens are as wide or wider antapically than they are in the paracingular region. The cysts are convex between the apical and antapical horns, and straight to concave between the two antapical horns. Relatively long apical and antapical horns arise from the three corners of the central body.

Phragma: The periphragm is smooth, granular, or rugulate. The more elaborate sculpture is formed by low, discontinuous features of the periphragm. The periphragm gives rise to long tapering horns that are usually hollow, wrinkled to rugulate, and often darker than the rest of the cyst. Some horns appear solid in the most distal three-quarters of their length. The base of the horns are often striate. The apical horn may exceed the length of the central body, whereas the antapical horns are somewhat shorter, although they too are long. The endophragm generally does not extend into the pericoel of the horns, but always forms a barrier between the endocoel of the central body and pericoel of the horns. This basal barrier is usually quite dark, perhaps due to thickening of the endophragm in this area.

Paratabulation: The only indication of paratabulation is the outline of the combination archeopyle (IPa).

Paracingulum: The paracingulum is not evident on most specimens, although it may be indicated by an unspiraled but distinct fold or by two low ridges.

Parasulcus: Indications of a parasulcus are generally absent, but a shallow depression in the parasulcal area may be present.

Archeopyle: The 2a intercalary and the 4″ paraplates are delineated on specimens with a fully developed archeopyle margin. Accessory parasutures around the archeopyle margin may indicate the position of additional paraplates. Specimens with a completely developed or clearly displayed archeopyle are rare. Four types of archeopyles were observed on various specimens: Ia, I, (IP)a and IPa. These are thought to be part of a developmental continuum, culminating in an IPa type archeopyle, with a free 2a paraplate and an adcingular adnate 4″ paraplate.

Discussion and Comparison with Similar Dinocysts. The morphologic variability of this species is quite pronounced, particularly with regard to the shape of the main body and the archeopyle. Development of the archeopyle varies from none being evident to a completely developed IPa combination archeopyle. Most specimens bore no indication of the archeopyle or only one or two open parasutures along the archeopyle margin. Developed archeopyles of the following types were observed: Ia, I, (IP)a, and IPa (Fig. 46). These different archeopyle types form a developmental continuum resulting from arrested or interrupted development of the IPa type archeopyle.

The specimens recovered were usually broken and exhibited low resistance to oxidation during processing. The abundant, dark specimens observed in unoxidized preparations of sample 8470 were invariably bleached out and corroded in the final palynologic slides. This was true in spite of the minimal oxidative treatment used during sample preparation. In addition, the horns are particularly vulnerable to physical damage and often are sheared off at or near their contact with the central body.

The "primitive" aspect of this species, especially its superficial similarity to *Broomea* and *Pareodinia,* suggests that the specimens recovered have been reworked from older beds into the Eocene La Meseta Formation. Specimens of this species have also been recovered from Eocene beds in Alaska and were considered to be reworked (V. D. Wiggins, personal communication, 1984). Indeed, Askin (this volume) has recovered *O. askiniae* from Maastrichtian deposits on Seymour Island.

Octodinium askiniae differs from species of *Broomea* by having a combination archeopyle with an eight-sided 2a paraplate, two wall layers, and a wide range of sculpturing. The structure of its antapical horns also differs from those of *Broomea.*

O. askiniae differs from species of *Pareodinia* by having two wall layers, two antapical horns, and an eight-sided 2a intercalary paraplate.

Dimensions. Observed range (10 specimens): endocyst length, 44 to 57 µm (mean, 51 µm); pericyst width, 21 to 39 µm (mean, 27 µm); apical horn length, as much as 54 µm; antapical horn length, as much as 36 µm; periphragm thickness, <0.5 µm; endophragm thickness, <0.5 µm.

Holotype. Slide 8470, M/1, 108.5 x 15.0 (Q13). Sample 8470, Section 3, La Meseta Formation, late early Eocene, Seymour Island, Antarctica.

Derivation of Name. Named in recognition of the significant contributions made to Antarctic palynology by Rosemary A. Askin.

Stratigraphic Occurrence. La Meseta Formation (Section 3, Eocene; Section 19, middle to late Eocene).

Previous Occurrences. Askin (this volume) recovered *O. askiniae* (as *Phelodinium* sp.) from the late Maastrichtian on Seymour Island. Very similar or identical dinocysts have been observed in Eocene well samples from the Gulf of Alaska and outcrop samples from the Alaska Peninsula (V. D. Wiggins, personal communication, October 1986).

**Genus *Odontochitina* Deflandre 1935
emend. Davey 1970
Odontochitina spinosa Wilson 1984**
Figure 30.3

Odontochitina spinosa Wilson, 1984a, p. 554–556, Figures 22–26.

Comments. The rare specimens encountered compare closely to the original description of *O. spinosa.* The periphragm bears short, solid, generally nontabular spines with acuminate to bifid distal terminations. The paracingulum was clearly defined on only one specimen by two parallel rows of spines. The spines are longest and most densely concentrated on the periphragm adjacent to the central body. Sculpture on the antapical horns is limited to short spines and scattered grana. The Seymour Island specimens differ from the type material in that a nipple-like bulge of the endocyst protrudes into the pericoel of each antapical horn. Endophragm protrusions have been observed in *Odontochitina costata* and *O. operculata* by Norvick (1975) and in *O. costata* by Davey (1970). This is a variable characteristic within the genus. The specimens are reworked from Late Cretaceous deposits, probably of Maastrichtian age.

Dimensions. (One undamaged specimen): overall length (exclusive of the apical operculum), 150 µm; width overall, 91 µm; endocyst length, 81 µm; endocyst width, 90 µm; periphragm thickness, less than 1 µm; endophragm, approximately 2 µm thick; spine length, 1 to 7 µm.

Stratigraphic Occurrence. La Meseta Formation (Section 3, reworked into late early Eocene deposits).

Selected Previous Occurrences. New Zealand (Maastrichtian: Wilson, 1984a).

Genus *Oligosphaeridium* Davey and Williams 1966
Oligosphaeridium complex (White 1842)
Davey and Williams 1966
Figure 30.4

Xanthidium tubiferum complex White, 1842, p. 39, Plate 4, div. 3, Figure 11.
Hystrichosphaeridium complex (White, 1842) Deflandre, 1946.
Oligosphaeridium complex (White, 1842) Davey and Williams, 1966b, p. 71–74, Plate 7, Figure 1.

Comments. Rare, highly distorted specimens attributable to *Oligosphaeridium complex* were recoverd from the La Meseta Formation samples. The specimens are usually broken; however, the absence of paracingular processes and the presence of distinctive secate and aculeate process terminations faciliates identification. Basal rings mark the contact of the hollow processes with the central body. The periphragm is slightly granular. The archeopyle is apical.

Dimensions. (One measurable specimen): central body length, 50 μm; central body width, 40 μm; process length, as much as 27 μm.

Stratigraphic Occurrence. La Meseta Formation (Section 3, late early Eocene and reworked into Holocene beach deposits; Section 17, late early Eocene; Section 19, middle to late Eocene).

Selected Previous Occurrences. The distribution is worldwide. The total reported stratigraphic range is from Late Jurassic to Miocene. The reported stratigraphic occurrence of this taxa could stand a thorough review. Williams and Bujak (1985) reported *O. complex* ranges from Early Cretaceous (late Berriasian) to the top of the middle Eocene. This range is used herein.

Oligosphaeridium pulcherrimum
(Deflandre and Cookson 1955)
Davey and Williams 1966
Figure 30.5

Hystrichosphaeridium pulcherrimum Deflandre and Cookson, 1955, p. 270–271, Plate 1, Figure 8, text-figs. 21–2.
Oligosphaeridium pulcherrimum (Deflandre and Cookson, 1955) Davey and Williams, 1966b, p. 76–76.

Comments. Very rare specimens of *Oligosphaeridium pulcherrimum* were recognized by their distinctive hollow and distally fenestrate processes. The periphragm is smooth to shagreenate; however, since the specimens are poorly preserved, the original surface sculpture may have been removed. The specimens are reworked from Late Jurassic–Late Cretaceous deposits.

Dimensions. Observed range (two measurable specimens recovered): central body length, 54 to 63 μm; central body width, 44 to 52 μm; wall thickness, approximately 1 μm; process length, 28 to 42 μm; process width, 3 to 7 μm; process terminations, as much as 38 μm across.

Stratigraphic Occurrence. La Meseta Formation (Section 3, reworked into Holocene beach deposits).

Selected Previous Occurrences. This species has been reported worldwide, and its stratigraphic range extends from the Late Jurassic (Kimmeridgian) to Late Cretaceous (Turonian) (Williams and Bujak, 1985).

Genus *Operculodinium* Wall 1967
Operculodinium bergmannii (Archangelsky 1969)
Stover and Evitt 1978
Figure 30.1–2

Cleistosphaeridium bergmannii Archangelsky, 1968, p. 414–415, Plate 2, Figures 8 and 11.
Operculodinium bergmannii (Archangelsky, 1968) Stover and Evitt, 1978, p. 178.

Comments. The thick wall (2 to 5 μm) is spongy in cross section. The surface of the central body may be granulate, as reported by Archangelsky (1968), or microfovelolate. The process bases are composed of root-like strands that arise from the central body and coalesce to form the solid process shafts. Each base may give rise to one or more short processes (<10 μm) with a capitate, acuminate, or bifurcate termination. Two or more shafts may be joined together by one or more strands. The archeopyle is precingular (3″). No other indication of paratabulation was observed.

Dimensions. Observed range (10 specimens): endocyst diameter, 37 to 65 μm (mean, 48 μm); process length, 7 to 15 μm (mean, 9 μm); wall thickness, 1 to 4 μm (mean, 3 μm).

Stratigraphic Occurrence. Cross Valley Formation (Sections 15 and 16, early late Paleocene); La Meseta Formation (Section 3, Eocene and reworked into Holocene beach deposits).

Selected Previous Occurrences. Argentina (Eocene: Archangelsky, 1968, 1969).

Genus *Palaeocystodinium* Alberti 1961

Comments: Lindgren (1984) discussed the relationship between *Palaeocystodinium* and *Svalbardella* and contended that *Svalbardella* is the senior synonym of *Palaeocystodinium*. We do not support this conclusion and consider *Palaeocystodinium* to be distinct from *Svalbardella*.

Overall shape and paratabulation are two of the most important and obvious morphologic features of dinocysts. Stover and Evitt (1978) pointed out that *Palaeocystodinium* differs from *Svalbardella* only in lacking indications of paratabulation, other than the archeopyle, and in having slender, pointed horns (rather than broadly rounded horns as in *Svalbardella*). Lindgren (1984) considered these characteristics to be insufficient for separating these genera: we would agree, if they were highly variable features in the species currently assigned to *Palaeocystodinium* and *Svalbardella*. To our knowlege, they are not.

True, horns in various species of *Palaeocystodinium* may be pointed or bluntly round, but they are still thin and taper markedly from the area of the central body to their terminations. They are quite unlike the broad rounded horns in *Svalbardella cooksoniae* Manum 1960. *Svalbardella hampdenensis* Wilson 1977 (which we transfer to *Palaeocystodinium*; see below) has distally rounded and somewhat wider horns than those typical of species in *Palaeocystodinium*. But they are much longer, narrower, and significantly more tapered distally than those of *Svalbardella*. This may be a transitional form that indicates a relationship between these two genera, but not that they are identical.

Clearly and consistently expressed paratabulation, other than the archeopyle, has not been reported on species currently assigned to *Palaeocystodinium*. Cookson (1965), however, did report that "an equatorial girdle and faintly outlined areas are present on the dorsal surface of *P. australinum*." Cookson did not elaborate on what the "areas" were or how they were outlined.

We reject the synonymizing of *Palaeocystodinium* with *Svalbar-*

della by Lindgren (1984) on the grounds that (1) the shapes of the species in the two genera are distinct and consistent enough to differentiate them; and (2) *Palaeocystodinium* does not bear clear and consistent indications of a paracingulum, a parasulcus, or of paratabulation. *Svalbardella* does bear paratabulation, a strongly and consistently expressed paracingulum and parasulcal depression, although the latter is not as clearly expressed as the paracingulum.

Svalbardella hampdenensis Wilson 1977: We here transfer *Svalbardella hampdenensis* Wilson 1977 to *Palaeocystodinium* Alberti 1961 for the following reasons:

    1. The relative proportions of the cyst dimensions are more like those of species in *Palaeocystodinium* than those of *Svalbardella cooksoniae* Manum 1960, the only other species in the genus *Svalbardella*.

    2. The narrow, tapering nature of the horns is more like species currently in *Palaeocystodinium* than *Svalbardella cooksoniae* Manum 1960.

    3. Paratabulation is not expressed on the pericyst.

    4. The presence of distally blunt or rounded horns does not exclude this species from *Paleocystodinium*. (Consider the horn description from the diagnosis for *P. benjaminii* Drugg 1967: "Two horns, one at each end, tapered and either pointed or slightly blunted.")

    5. The presence of a paracingulum does not necessarily include the species in *Svalbardella*, nor exclude it from *Palaeocystodinium*. Wilson (1977) noted that the paracingulum is consistently present (a characteristic unknown in other species of *Paleocystodinium*) but also that it was faintly expressed (which is the case with the paracingulum and parasulcus where these features are developed on specimens belonging to species of *Palaeocystodinium*). A paracingulum may or may not be present on specimens of the various species of *Palaeocystodinium*.

    In summary, we believe that *Svalbardella cooksoniae* Manum 1960 is a distinct morphologic entity at the generic and specific level and that the species *P. hampdenensis* is more similar to species in *Palaeocystodinium* than to *S. cooksoniae* Manum 1960 or the generic characteristics of *Svalbardella*.

Type Species: Palaeocystodinium golzowense Alberti, 1961, p. 20, Plate 7, Figure 12.
Other Species:
Palaeocystodinium australinum (Cookson, 1965b, p. 140, Plate 25, Figs. 1–4), Lentin and Williams, 1976, p. 89.
P. benjaminii Drugg, 1967, p. 31, Plate 3, Figure 1; Plate 9, Figure 3.
P. ?deflandrei Gruas-Cavagnetto, 1968, p. 92–93, Plate 13, Figures 15–19.
P. ?denticulatum Alberti, 1961, p. 20–21, Plate 7, Figure 9.
P. "gabonense" Stover and Evitt, 1978, p. 115 = *Svalbardella australina* auct. non Cookson, 1965b, of Malloy 1972, p. 63, holotype Plate 1, Figure 17, = *Andalusiella gabonense* (Stover and Evitt, 1978) comb. nov. (See discussion and transfer of *A. gabonense* [Stover and Evitt, 1978] comb. nov., under *Palaeocystodinium australinum* [Cookson 1965] Lentin and Williams 1976, this paper.)
P. granulatum (Wilson, 1967b, p. 226–227, Figs. 7–9) Lentin and Williams, 1976, p. 89.
P. hampdenensis (Wilson, 1977, p. 564–566, Figs. 1–8), comb. nov.
P. ?hyperxanthum (Vozzhennikova, 1963, p. 185, Fig. 20) Vozzhennikova, 1967, p. 152–153.
P. lidiae (Gorka, 1963, p. 37, Plate 5, Fig. 6) Davey, 1969b, p. 12–13.
P. reductum May, 1980, p. 84–85, Plate 21, Fig. 20.
P. ?rhomboides (O. Wetzel, 1933a, p. 168, Plate 2, Fig. 17) Lentin and Williams, 1973, p. 103.
P. ?rhomboides subsp. *filosum* (O. Wetzel, 1933a, p. 169, Plate 2, Fig. 20) Lentin and Williams, 1973, p. 103.
P. ?rhomboides subsp. *incertum* (Deflandre, 1936a, p. 29, Plate 10, Fig. 8-9) Lentin and Wiliams, 1973, p. 103.
P. ?rhomboides subsp. *nodosum* (O. Wetzel, 1933a, p. 169, Plate 2, Fig. 19) Lentin and Williams, 1973, p. 104.

P. ?rhomboides subsp. *ovatum* (O. Wetzel, 1933a, p. 168–169, Plate 2, Fig. 18) Lentin and Williams, 1973, p. 104.
P. scabratum Jain et al., 1975, p. 12-13, Plate 6, Figure 63.
P. stockmansii Boltenhagen, 1977, p. 114-115, Plate 23, figures 1–4.

Palaeocystodinium australinum (Cookson 1965) Lentin and Williams 1976
Figures 32.3, 46.3–4

Svalbardella australina Cookson, 1965, p. 140, Plate 25, Figures 1–4.
non-*Svalbardella australina* (Cookson, 1965) emend. Malloy, 1972, p. 63, Plate 1, Figures 17 and 20.
Palaeocystodinium australinum (Cookson 1965) Lentin and Williams, 1976, p. 89.

Emended Diagnosis: Fusiform, dorso-ventrally flattened, cavate dinocysts bearing one long tapering horn at each end of the central body. The apical horn may be distally acuminate, rounded, or blunt. The antapical horn bears a short, posteriorly projecting accessory horn or spike along its length or almost distally (but not proximally). The accessory horn occurs on the dinocyst's right side of the antapical horn. The structurally simple, spindle-shaped endocyst may or may not extend into the pericoels of the horns formed by the pericyst. The endophragm and the periphragm may be closely appressed in the central body area but are separated in the region of the horns. Wall relationship varies from cornucavate to circumcavate. The thin endophragm and periphragm may be smooth to slightly granular. The endophragm may be clear or dark brown in color, whereas the periphragm is clear. The steno-deltaform intercalary archeopyle is of the I(2a) Type and the operculum is free. Paratabulation is absent or poorly developed. Faint suggestions of the location of the parasulcus and paracingulum may be present. The cysts are large and may be more than 250 μm long.

Comments. Malloy (1972) emended *Palaeocystodinium australinum* (referred to as *Svalbardella asutralina* Cookson 1965 by Malloy, 1972) by: (1) redefining the range of cyst dimensions to include much shorter and broader dinocysts than the type material; (2) including cysts with a proximal location for the accessory spur on the antapical horn, rather than being located well along the horn, as originally described by Cookson (1965); (3) including specimens with a much smaller length-to-width ratio for the species than is indicated by the specimens in photographs of the *P. australinum* type material; (4) including a fundamentally different and significantly more complex structure for the apical horns; (5) citing the presence of a transverse fold as an indication of the location of the paracingulum; and (6) including specimens with a well-developed flagellar scar in the parasulcus.

    We reject the emendation of *Palaeocystodinium australinum* (Cookson) Lentin and Williams 1976 by Malloy (1972) for the following reasons:

    1. The size range of the type material from Australia (length, 293 to 302 μm, width, 40 to 61 μm) is significantly different from the Gabon material of Malloy (1972; length, 110 to 175 μm, width, 49 to 74 μm). The Gabon material is much shorter and does not approach the lower end of the dinocyst length range cited for the type material of *P. australinum* by Cookson (1965). The width of the Gabon material extends from the middle to far beyond the upper range of the dinocyst width of the type material.

    2. The relative proportion of the length to width is markedly different, whether based on measurements or simple visual comparison of the Australian and Gabon materials. The proportions exhibited by the Gabon material are more similar to the proportions of the genus *Andalusiella* than to those of *P. australinum*.

    3. The location of the accesory spur is significantly different on the

Gabon material and is, in fact, more reminiscent of species of *Andalusiella* than of *P. australinum* (Cookson) Lentin and Williams 1976.

4. The fundamentally different and more complex apical and antapical horns on the Gabon material are unlike any reported elsewhere for *P. australinum*. Malloy (1972) noted "The structure of the apical and antapical portions of the inner cyst appears complex (text-fig. 2), with an inner projection often filling the horn and tightly enclosed by the periphragm. This interior portion of the horn is attached basally at an area of thickening which forms a "boss" on the inner capsule." (See Fig. 46.1, 4 for comparative structural sketches of the horns on Morphotypes B and C of Malloy (1972) and those of *P. australinum* of Cookson (1965). Such horn structures have been reported by Riegel and Sarjeant (1982) in *Andalusiella mauthei* Riegel. (See Fig. 46.5–10.) The contrasting structure of the horns is enough to separate the Gabon specimens of *P. australinum* sensu Malloy (Morphotype C) from the Australian type material of *P. australinum* of Cookson (1965).

5. The transverse fold commonly observed marking the paracingulum on the Gabon material (Morphotype C of Malloy, 1972) has not been noted on specimens of *P. australinum* elsewhere.

6. The intercalary 2a archeopyle outlined on the apical horn structural diagram by Malloy (text-fig. 2; 1972) is an omegaform hexa archeopyle, rather than a steno-deltaform archeopyle that is characteristic of *Palaeocystodinium* and *Svalbardella*. The archeopyle in Malloy's figure is also unlike any of those shown on specimens in Plate 1 (Malloy, 1972). We consider this to be a drafting error and not an indication that Malloy (1972) believed *Svalbardella* and *Palaeocystodinium* have an omegaform archeopyle. The "inverted" archeopyle on Malloy's diagram—if turned right-side-up—is more similar to the archeopyle of *Andalusiella* than to that of *Palaeocystodinium*.

7. Finally, the specimens referred to as *S. australina* and shown in Plate 1, Figures 17 and 20 of Malloy (1972) appear to bear a comma-shaped flagellar scar in the parasulcal area. Such structures have not been reported elsewhere for *P. australinum*. They are evident on the Morphotype B specimens of Malloy (Plate 1, Figs. 8–16, 21; 1972), and on specimens of *Andalusiella* (see Riegel, 1974, and Riegel and Sarjeant, 1982, for example).

Stover and Evitt (1978) established *Palaeocystodinium gabonense* to accommodate dinocysts similar to those illustrated in Plate 1, Figures 17 and 20 of Malloy (1972). (A sketch of the former specimen, the Holotype of *P. gabonense*, is shown in Plate 46, Figure 2, of this paper.) However, they did not alter the emendation of *P. australinum* proposed by Malloy (1972).

We believe that *P. gabonense* is more closely related to *Andalusiella* than to *Palaeocystodinium* because of its overall aspect, the relative proportion of cyst length to width, the presence of a distinct flagellar scar in the parasulcal area, and the presence of dark wall thickenings at the poles of the broadly elongate endocyst. These morphologic features are not characteristic of *Palaeocystodinium*, but they are of *Andalusiella*. Therefore, we herein transfer *P. gabonense* Stover and Evitt, 1978, to the genus *Andalusiella*.

The specimen illustrated by Malloy (1972) in Plate 1, Figure 20, is an example of *Andalusiella polymorpha* (Malloy, 1972) Lentin and Williams 1977 and is very similar to specimens of that species illustrated in Plate 1, Figures 8 and 16, by Malloy (1972).

The Seymour Island specimens of *Palaeocystodinium australinum* resemble those from the Pebble Point Formation (Cookson, 1965), but are generally shorter and darker. The darker coloration may be a preservational artifact because other taxa in the same samples are similarly colored. Many Seymour Island specimens of *P. australinum* are broken anteriorly of the paracingular area. Such specimens can only be differentiated from broken specimens of *P. golzowense* by the presence of the accessory spur on the antapical horn of the former. The operculum of the 2a archeopyle is often in place, even though open parasutures may completely surround it.

Dimensions. Observed range (10 specimens): pericyst length, 170 to 226 μm (mean, 196 μm); pericyst width, 29 to 48 μm (mean, 38 μm); endocyst length, 77 to 122 μm (mean, 99 μm); endocyst width, 29 to 48 μm (mean, 38 μm).

Stratigraphic Occurrence. Cross Valley Formation (Sections 15 and 16, early late Paleocene).

Selected Previous Occurrences. Palaeocystodinium australinum was described from the Paleocene Pebble Point Formation of southwest Victoria, Australia (Cookson, 1965). Subsequent planktonic foraminiferal studies indicated a middle Paleocene age for the Pebble Point Formation (McGowan, 1965, 1968). In addition to Australian occurrences (Cookson, 1965; Deflandre and Cookson, 1955), *P. australinum* has been reported from the middle-late Paleocene of DSDP 283 (Haskell and Wilson, 1975) and the middle Paleocene of DSDP 214, in the Indian Ocean (Harris, 1974). *P.* aff. *australinum* has been recovered from the Upper Cretaceous on Campbell Island, south of New Zealand (Wilson, 1967b) and from DSDP 275 cores on the Campbell Plateau (Wilson, 1975). The reported occurrences indicate that *P. australinum* was widely distributed in the high southern latitudes during the Maastichtian and Paleocene. Stover and Williams (1977) reported that the stratigraphic range of *P. australinum* is limited to the upper Paleocene, whereas Williams and Kidson (1975) and Williams (1977) considered the stratigraphic range to be lower Maastrichtian to Paleocene. Williams and Bujak (1985) indicated a total range of late Campanian to late Paleocene; this range is used herein.

Palaeocystodinium golzowense Alberti 1961
Figure 32.1

Palaeocystodinium golzowense Alberti, 1961, p. 20, Plate 7, Figures 10–12, Plate 12, Figure 16.

Comments. The Seymour Island specimens are generally smaller than those described from the Golzow borehole by Alberti (1961). The endophragm and periphragm are smooth and devoid of any indication of a paracingulum or a parasulcus. Paratabulation is indicated only by the intercalary archeopyle. The operculum of the 2a intercalary archeopyle is usually in place, even though it may be completely surrounded by apparently open parasutures.

Dimensions. Observed range (10 specimens): pericyst length, 155 to 289 μm (mean, 195 μm); pericyst width, 27 to 43 μm (mean, 38 μm); endocyst length, 52 to 111 μm (mean, 93 μm); endocyst width, 27 to 42 μm (mean, 37 μm).

Stratigraphic Occurrence. Cross Valley Formation (Section 12-13, early late Paleocene and reworked into Eocene deposits, see discussion on the age determinations of this section; Sections 15 and 16, early late Paleocene); La Meseta Formation (Section 3, reworked into Holocene beach deposits).

Selected Previous Occurrences. Worldwide occurrence. The stratigraphic range of *P. golzowense* extends from Campanian to late Miocene.

Palaeocystodinium granulatum (Wilson 1967)
Lentin and Williams 1976
Figure 32.2

Svalbardella granulata Wilson, 1967b, p. 226–227, Figures 7–9.
Palaeocystodinium granulatum (Wilson, 1967b) Lentin and Williams, 1976, p. 89.

Comments. The Seymour Island specimens are similar to the illustrations and description of the type material, except for surface sculpturing. Wilson (1967b) stated that the most characteristic feature of *Palaeocystodinium granulatum* is "the densely granulated surface of the outer cyst." The specimens studied are more spinate than granulate, being covered with very short evexate or bifurcate spines. Some spines are connected distally. Spine length varies between specimens, as well as on an individual. The spines are generally shorter on the horns and longer near the central body of the cyst. Low grana are scattered between the spines and seem to be more abundant on the horns than adjacent to the paracingulum. The presence of short spines, as well as grana, does not seem to be sufficient reason to establish a new species. The Seymour Island specimens are considered to be intraspecific variants of *P. granulatum.* Reworked from Late Cretaceous (late Campanian–Maastrichtian) deposits.

Dimensions. No complete specimens were observed. However, the periphragm of one specimen, apparently broken in half, was 127 μm long and 42 μm wide. An approximate overall length of 260 μm does not seem unreasonable. These dimensions compare favorably with the range reported for *P. granulatum* by Wilson (1967b).

Stratigraphic Occurrence. La Meseta Formation (Section 3, reworked into late early Eocene and Holocene beach deposits).

Selected Previous Occurrences. Wilson (1967b) described *P. granulatum* from the Garden Cove Formation, Campbell Island, and assigned a Paleocene (Teurian) age to the microflora. Subsequent research led Wilson (1972) to assign a Maastrichtian (Haumurian) age to that formation. Additional occurrences have been reported from Germany (late Oligocene: Benedek, 1972; questionable identification); New Zealand, offshore, DSDP 275 (late Campanian–Maastrichtian: Wilson, 1975), and Seymour Island (Campanian: Palamarczuk et al., 1984).

Palaeocystodinium sp. indet.

Comments. A number of smooth epicysts of *Palaeocystodinium* sp. indet. were encountered. Epicysts of *P. golzowense* and *P. australinum* are identical; differentiation of these species is based on the presence of an accessory spur on the antapical horn of the latter. The absence of the hypocysts precluded species identification of the isolated epicysts.

Genus *Palaeoperidinium* Deflandre 1935 emend. Sarjeant 1967
Palaeoperidinium pyrophorum (Ehrenberg 1838) Sarjeant 1967
Figure 32.4

Peridinium pyrophorum Ehrenberg, 1838, Plate 1, Figures 1, 4.
Peridinium basilium Drugg, 1967, p. 13, Plate 1, Figure 9, Figs. 1a–b.
Palaeoperidinium pyrophorum (Ehrenberg, 1838) Sarjeant, 1967, p. 246–247.
Palaeoperidinium deflandrei Lentin and Williams, 1973, p. 105.

Comments. Some of the Seymour Island specimens exhibit a distinct folding or thickening along the anterior margins of the 2a intercalary paraplate and an apparently contracted endocyst. Otherwise, the specimens conform to the emended species diagnosis (Sarjeant, 1967).

Dimensions. Observed range (10 specimens): length, 88 to 124 μm (mean, 104 μm); width, 85 to 99 μm (mean, 91 μm).

Stratigraphgic Occurrence. Cross Valley Formation (Section 12-13, early late Paleocene and reworked into Eocene deposits, see discussion of age determinations for this section; Sections 15 and 16, early late Paleocene); La Meseta Formation (Section 3, reworked into Eocene deposits and Holocene beach deposits).

Selected Previous Occurrences. Palaeoperidinium pyrophorum was described from the Upper Cretaceous flints of Saxony (Ehrenberg, 1838). Subsequently, this species has been reported from Upper Cretaceous to upper Paleocene deposits worldwide. In particular, from deposits of this age in Europe and the northwest European continental shelf (Caro, 1977), the Danian of California (Drugg, 1967; as *P. basilum*), the Paleocene of DSDP 283 (Haskell and Wilson, 1975) in the South Tasman Sea, and the Danian of Alabama (Drugg, 1970; as *P. basilum*). It has been reported previously from Seymour Island (Paleocene: Hall, 1977; early late Paleocene: Wrenn, 1982; Danian: Palamarczuk, 1982, and Palamarczuk et al., 1984).

Helby et al. (1984) reported a total range for *P. pyrophorum* of Albian to early late Paleocene. G. L. Williams (1987, personal communication) considered an Albian base to be unlikely; we concur. Williams and Bujak (1985) cited a total range of late Campanian to late Paleocene; we recognize this range herein.

Genus *Paralecaniella* Cookson and Eisenack 1970 emend. Elsik 1977
Paralecaniella indentata (Deflandre and Cookson 1955)
Cookson and Eisenack 1970 emend. Elsik 1977
Figure 31.5

Epicephalopyxis indentata Deflandre and Cookson, 1955, p. 292–3, Plate 9, Figures 5–7, text-fig. 56.
Paralecaniella indentata (Deflandre and Cookson, 1955) Cookson and Eisenack, 1970), p. 323.
Paralecaniella indentata (Deflandre and Cookson, 1955) Cookson and Eisenack, 1970b, emend. Elsik, 1977, p. 96, Plate 1, Figures 1–15, Plate 2, Figures 1–11.

Comments. Elsik's (1977) definitive paper on *Paralecaniella indentata* described and clearly illustrated the wide range of morphologic variability commonly seen in specimens of this species. The Seymour Island specimens were equally variable, but exhibited one feature, possibly preservational in nature, not noted by Elsik (1977). Broken specimens often appeared to have been fractured or shattered, suggesting their walls were very stiff, if not brittle, in nature. Other specimens, however, did not convey such an impression. No indications of an archeopyle or a parasulcus were observed. A number of specimens did exhibit a paracingulum.

Dimensions. Observed range (10 specimens): pericyst length, 36 to 80 μm (mean, 53 μm); pericyst width, 35 to 68 μm (mean, 48 μm); endocyst length, 32 to 77 μm (mean, 47 μm); endocyst width, 31 to 62 μm (mean, 41 μm).

Stratigraphic Occurrence. Cross Valley Formation (Section 12-13, Paleocene and Eocene; see discussion of the age determinations for this section; Section 15 and 16, early late Paleocene); La Meseta Formation (Section 3, Eocene; Sections 17, 18, and 19, Eocene).

Selected Previous Occurrences. Worldwide distribution during the Late Cretaceous and Tertiary.

Genus *Pareodinia* Deflandre 1947
emend. Gocht 1970; emend. Johnson and Hills 1973; emend. Wiggins 1975; emend. Stover and Evitt 1978
Pareodinia sp. indet.

Comments. One poorly preserved specimen attributable to *Pareodinia* was recovered. The apical horn has been broken off, and there is no indication of paratabulation. The archeopyle appears to be intercalary. The specimen was reworked from middle Jurassic–Early Cretaceous (Bajocian to Albian) deposits.

Dimensions. (One specimen recovered): length, 63 μm; width, 29 μm.

Stratigraphic Occurrence. La Meseta Formation (Section 19, reworked into middle to late Eocene deposits).

Genus *Phelodinium* Stover and Evitt 1978
Phelodinium boldii sp. nov.
Figure 33.1, 4

Diagnosis. A species of *Phelodinium* characterized by large angular, peridinioid cysts that are as wide as they are long. The periphragm surface may be longitudinally granulate to rugulate. Very short apical and antapical horns. Distinct flagellar scar in the midparasulcal area. Intercalary (2a) archeopyle.

Description.
Shape: Most specimens are squat, slightly cornucavate peridinioid dinocysts whose breadth generally exceeds their length. The short antapical horns are approximately the same length. The short apical horn is usually flat or slightly concave distally. None of the horns are pointed or solid. Each of the two epicystal sides are slightly longer than any of the three sides of the hypocyst.
Phragma: The thin (<0.5 μm) endoprahgm is smooth to finely granulate. The periphragm is smooth, scabbrate, or rugulate, and is less than 0.5 μm thick. Rugulate ornamentation is longitudinally distributed.
Paratabulation: Paratabulation is assumed to be peridinioid, as suggested by parasutures delimiting a 2a intercalary archeopyle.
Paracingulum: The paracingulum is narrow, planar, and bordered by low folds in the periphragm.
Parasulcus: The location of the parasulcus is suggested by two low folds on the hypocyst that diverge posteriorly. This feature is not always evident. A flagellar scar is evident in the central parasulcal area.
Archeopyle: The 2a intercalary archeopyle is often difficult to detect because the operculum usually remains in place and the parasutures around the operculum may not gape open. The presence and location of the archeopyle may be indicated only by occasional open parasutures.
Dimensions. Observed range (10 specimens): length, 63 to 130 μm (mean, 88 μm); width, 61 to 129 μm (mean, 88 μm); periphragm and endophragm, each less than 0.5 μm thick.

Discussion and Comparison with Similar Species. Phelodinium boldii sp. nov. resembles *P. pumilum* Liengjaren et al. (1980) in that the antapical horns of both are rather short. *P. boldii* differs from the latter by being much larger, being generally as broad or broader than long, having an angular to subangular outline, and having scabbrate to rugulate surface sculpture.
Phelodinium boldii sp. nov. is superficially similar to *Palaeoperidinium pyrophorum,* but differs from that species by having a 2a intercalary archeopyle, granulate to longitudinally rugulate ornamentation and by lacking pandasutural striae.

Holotype. Slide 8476, W/10, 121.3 x 11.9 (M27). Sample 8476, Section 3, La Meseta Formation, late early Eocene, Seymour Island, Antarctica.

Derivation of Name. Named after Willem van den Bold, in recognition of his significant contributions to micropalaeontology.

Stratigraphic Occurrence. La Meseta Formation (Section 3, Eocene and reworked into Holocene beach deposits; Sections 17, 18, and 19, Eocene); Cross Valley Formation (Section 15, early late Paleocene).

Genus *Phthanoperidinium* Drugg and Loeblich 1967;
emend. Edwards and Bebout 1981; emend. Islam 1983
Phthanoperidinium echinatum Eaton 1976
Figure 31.3

Phthanoperidinium echinatum Eaton, 1976, p. 298–299, Plate 17, Figures 8–9, 12; text-fig. 23B.

Comments. The Seymour Island specimens closely resemble those from the Bracklesham Beds of England (Eaton, 1976). The rounded polygonal outline and blunt apical horn are the same as those noted by Eaton (1976); however, antapical projections also have been observed on some of the Seymour Island specimens. The left antapical horn is always the more prominently developed if antapical horns are present. The position of the nascent right antapical horn is usually indicated only by a tuft of solid spines, each of which is capped distally by a ball or subspherical termination. The distinctive, spinate, penitabular ornamentation clearly delineates a paratabulation of 4′, 3a, 7″, 5‴, 2⁗ on the surface of the dinocyst. The paratabular areas thus formed assist in the identification of *P. echinatum* even in highly folded and distorted specimens.
The Seymour Island specimens were generally much larger (34 x 43 to 50 x 59 μm) than Eaton's (1976) specimens (22 x 26 to 42 x 48 μm). The spine lengths of specimens from both assemblages were 2 μm or less.

Dimensions. Observed range (10 specimens): length, 43 to 59 μm (mean, 49 μm); width, 34 to 50 μm (mean, 43 μm); spine length <2 μm.

Stratigraphic Occurrence. Cross Valley Formation (Section 12-13, early late Paleocene (contamination?), Eocene; see the discussion of the age determinations for this section); La Meseta Formation (Section 3, Eocene; Section 17, late early Eocene; Section 19, middle to late Eocene).

Selected Previous Occurrences. Phthanoperidinium echinatum was described from the Eocene Bracklesham Beds of southern England (Eaton, 1976). *P. echinatum* has been reported from the upper Eocene of DSDP 370 (Williams, 1978), the lower to upper Eocene of the northwestern European continental shelf and adjacent areas (Caro, 1977), and the late Eocene of the Santos Basin, Brazil (Griggs, 1981).
Additional occurrences include: Atlantic Ocean, Rockall Plateau (early Eocene-middle Eocene: Costa and Downie, 1979); Belgium (Ypresian-Lutetian: De Coninck, 1980); Canada (early Oligocene: Williams and Bujak, 1977); Canada, eastern (late Eocene–early Oligocene: Barss et al., 1979); England (Bartonian: Bujak, 1973); England (Eocene: Bujak et al., 1980); Netherlands (Eocene: De Coninck, 1977); Norwegian-Greenland Sea (middle-late Eocene: Manum, 1976).
Williams and Bujak (1985) recognized a late early Eocene to early Oligocene range for *P. echinatum.* We concur with this range.

Genus *Selenopemphix* Benedek 1972 emend. Bujak 1980
***Selenopemphix nephroides* Benedek 1972 emend. Bujak 1980**
Figure 31.2

Selenopemphix nephroides Benedek, 1972, p. 47–48, Plate 11, Figure 13; Plate 16, Figures 1–4.
Selenopemphix nephroides Benedek, 1972 emend. Bujak, 1980, p. 84.
Selenopemphix nephroides Benedek, 1972; emend. Bujak, 1980; emend. Benedek and Sarjeant, 1981, p. 333–334, 336.

Comments. The Seymour Island specimens are comparable to the description and illustration of the type material from the lower Rhine area of Germany described by Benedek (1972).

Dimensions. Observed range (10 specimens): height, 42 to 68 μm (mean, 52 μm); width, 46 to 66 μm (mean, 59 μm); archeopyle dimension (6 specimens): height, 14 to 24 μm (mean, 20 μm); width, 14 to 29 μm (mean, 24 μm); wall, less than 1 μm thick.

Stratigraphic Occurrence. La Meseta Formation (Section 3, late middle to late Eocene).

Selected Previous Occurrences. Selenopemphix nephroides, discussed previously in conjunction with *Arachnodinium antarcticum,* appeared in the late middle Eocene and was generally believed to have a top in the Oligocene. Williams and Bujak (1985) recognized a late middle Eocene to late Miocene range for the species. However, *S. nephroides* has been recovered from late Miocene to late Pleistocene deposits of the Gulf of Mexico (Wrenn and Kokinos, 1986). Furthermore, *S. nephroides* is the name applied to the cysts produced by the extant motile dinoflagellate *Protoperidinium (Protoperidinium) subinerme* (Paulsen) Loeblich III (Harland, 1983). The total stratigraphic range is late middle Eocene to Holocene.
    Additional occurrences include: Antarctica, Ross Sea, DSDP Leg 28 (Eocene-Oligocene: Kemp, 1975); Antarctica, Ross Sea, DSDP 270 (late Oligocene: Kemp, 1975); Atlantic Ocean, DSDP 370 (Pliocene-Pleistocene, considered reworked from Oligocene deposits, Williams, 1978); Belgium (Ypresian: De Coninck, 1980); Canada, eastern (middle Miocene: Barss et al., 1979); England (Bartonian: Bujak, 1973); England (Eocene: Bujak et al., 1980); France (early Oligocene: Charollais et al., 1980); Germany (middle-late Oligocene: Benedek, 1972); Netherlands (Eocene: De Coninck, 1977).

**Genus *Senegalinium* Jain and Millepied 1973
emend. Stover and Evitt 1978**
***Senegalinium ?asymmetricum* (Wilson 1967)
Stover and Evitt 1978**
Figures 34.4, 42.1

Deflandrea asymmetrica Wilson, 1967a, p. 62–63, Figures 17–21.
Alterbia asymmtrica (Wilson, 1967) Lentin and Williams, 1973, p. 48.
Senegalinium ?asymmetricum (Wilson, 1967a) Stover and Evitt, 1978, p. 123.

Comments. The specimens studied compare closely with the McMurdo Sound specimens (Wilson, 1967a) in regard to the shape, size, archeopyle type (2a), size and asymmetric development of the antapical horns, and the intermittent occurrence of fine marginal serrations.
    The Seymour Island specimens differ most significantly from the type specimens in the granular, rather than smooth, surface of the endocyst. Although this is a consistent feature in the study specimens, it does

not seem to be a significant enough difference upon which to establish a new species.
    Whitney (1979) compared *Alterbia acutula* (Wilson) Lentin and Williams with *A. asymmetrica* (Wilson) Lentin and Williams. (These species now reside in *Alterbidinium* and *Senegalinium,* respectively.) She concluded that they were separate species and that the reported North American occurrences of *A. asymmetrica* should be attributed to *A. acutula.* These included occurrences in Maryland (Benson, 1976) and offshore Canada (Williams and Brideaux, 1975).

Dimensions. Observed range (10 specimens): pericyst length, 62 to 92 μm (mean, 82 μm); pericyst width, 56 to 85 μm (mean, 68 μm); endocyst length, 40 to 61 μm (mean, 47 μm); endocyst width, 51 to 70 μm (mean, 58 μm).

Stratigraphic Occurrence. La Meseta Formation (Section 3, late early Eocene; Sections 18 and 19, Eocene).

Selected Previous Occurrences. Senegalinium? asymmetricum was described from glacial erratics in the McMurdo Sound area of Antarctica (Wilson, 1967a), and an Eocene age was considered likely for the recovered microplankton flora. (See comments concerning the age of the glacial erratics in the discussion of *Arachnodinium antarcticum.*) Eocene occurrences of *S. asymmetricum* are common in the southern high latitudes, including the Eocene of Argentina (Archangelsky, 1969), reworked in the West Ice Shelf area of Antarctica (Kemp, 1972) and DSDP 270, 274, 281, and 283 (Kemp, 1975; Haskell and Wilson, 1975). The total biostratigraphic range for *S. asymmetricum* is from early Maastrichtian to Eocene (Williams and Kidson, 1975).
    Additional occurrences include: Antarctica (early Tertiary: Wilson, 1967a); Europe (Maastrichtian: specimens are probably *Alterbidinium acutulum;* Wilson, 1971); Maryland (early Eocene: Goodman, 1979; not illustrated).

**Genus *Spinidinium* Cookson and Eisenack 1962
emend. Lentin and Williams 1976**
***Spinindinium colemanii* sp. nov.**
Figures 36.1–2, 39.2

Deflandrea macmurdoensis Wilson 1967 sensu Kemp, 1975, Plate 2, Figures 1–3.

Diagnosis. A species of *Spinidinium* characterized by its capitate, penitabular spines, small size, unequal hypo- and epicysts, and its posteriorly adnate 2a intercalary archeopyle.

Description.
    Shape: Cornucavate, subpentagonal to subelliptical dinocysts bearing one hundred or more penitabular spines. The paracingulum separates the epicyst from the hypocyst; the former is usually two to three times the length of the hypocyst. The paracingular area may bulge laterally, giving the dinocyst a subpentagonal shape.
    Phragma: Both the endophragm and the periphragm are smooth. The wall layers are appressed except in the basal regions of the horns. The periphragm gives rise to capitate spines, whereas the endophragm lacks projections. The short apical horn is capped by two or more capitate spines. The right antapical horn is rarely evident, whereas the left horn is a long, spike-shaped projection bearing one or more accessory spines.
    Paratabulation: A paratabulation of x', 3a, 7", xc, 5''', ?2'''' is delineated by penitabular capitate spines. The distribution and number of

paraplates is obscured by the abundance of the closely spaced spines and by the presence of accessory rows of spines. The paracingulum and the parasulcus are not divided by rows of spines.

Paracingulum: Low folds capped with short capitate spines delimit the shallow excavation of the paracingulum. Paraplate divisions within the paracingulum were not observed.

Parasulcus: A broad, bare area bordered by rows of spines indicates the location of the parasulcus. Spines are usually absent within the parasulcal area, although isolated spines occur on some specimens.

Archeopyle: The archeopyle is formed by the partial detachment of the 2a paraplate, and occasionally by the partial detachment of the 4″ paraplate. The operculum is adnate along the H4 parasuture. A row of penitabular capitate spines occurs on each side of all archeopyle parasutures, except the H4 parasuture. The rows of spines along the H2-H3 and the H5-H6 parasutures of the 2a intercalary paraplate continue below the base of that paraplate and along the margins of the 4″ paraplate to the anterior margin of the paracingulum.

Dimensions. Observed range (10 specimens): pericyst length, 42 to 57 μm (mean, 50 μm); pericyst width, 36 to 45 μm (mean, 41 μm); endocyst length, 32 to 40 μm (mean, 37 μm); endocyst width, 30 to 41 μm (mean, 34 μm); process length, 3 to 8 μm; left antapical horn, to 10 μm; apical horn, to 8 μm.

Discussion and Comparison with Similar Species. The capitate spines and their distribution on the pericyst are reminiscent of *Spinidinium macmurdoense.* However, the overall shape, much smaller size, more numerous spines, more complete paratabulation, and unequal division of the dinocyst by the paracingulum differentiate *S. colemanii* sp. nov. from *S. macmurdoense.* The dense spinosity of the pericyst of *S. colemanii* sp. nov. is similar to a specimen of *Vozzhennikovia apertura* illustrated by Haskell and Wilson (1975; Plate 1, Fig. 6). However, the spines on the latter are more numerous, smaller, and generally nontabular, rather than penitabular.

Holotype. Slide 8496, W/8, 134.66 x 12.5 (M40). Sample 8496, Section 15, Cross Valley Formation, early late Paleocene, Seymour Island, Antarctica.

Derivation of Name. Named after James M. Coleman, in recognition of his contributions to the study of deltaic processes and deposits.

Stratigraphic Occurrence. Cross Valley Formation (Section 12-13, Eocene, see discussion of age determinations for this section; Section 15 and 16, early late Paleocene); La Meseta Formation (Section 3, Eocene and reworked into Holocene beach deposits). The sparse Eocene occurrences noted above may be due to reworking.

Spinidinium densispinatum Stanley 1965
Figure 33.2-3

Spinidinium densispinatum Stanley, 1965, p. 226-227, Plate 21, Figures 1-5.

Comments. The specimens studied compare closely with the description and illustrations of the type material from the Cannonball Member of the Fort Union Formation (Stanley, 1965). The dinocyst outline varies from subpentagonal to almost oval in its dorsal-ventral orientation. A short, broad apical and two antapical horns are usually present. The left antapical horn is always more developed than the often-vestigial right antapical horn. The parasulcal area is devoid of the nontabular spines that densely

cover the remainder of the dinocyst. Paracingulum bordered by rows of spines. Spines may be cylindrical and distally acuminate or capitate. Archeopyle not clearly seen on any specimen, but is assumed to be formed by the loss of the 2a paraplate.

Dimensions. Observed range (10 specimens): overall length, 36 to 52 μm (mean, 45 μm); overall width, 31 to 46 μm (mean, 36 μm).

Stratigraphic Occurrence. Cross Valley Formation (Section 12-13, early late Paleocene and reworked into Eocene, see discussion concerning the age determinations of this section; Sections 15 and 16, early late Paleocene); La Meseta Formation (Section 3, reworked into late early Eocene and Holocene beach deposits; Section 19, reworked into middle to late Eocene beds).

Previous Occurrence. Spinidinium densispinatum was described from the Paleocene of South Dakota (Stanley, 1965). Williams and Kidson (1975) considered the biostratigraphic range of this species to be late Campanian to late Paleocene. This range is used here. Post-Paleocene reports of this species are considered by us to be questionable.

Other reported occurrences include: Antarctica, Ross Sea (?Eocene: Wilson, 1967a. This age determination has been questioned by Stott et al., 1983; lower Tertiary: Wilson, 1968); Argentina (Oligocene: Gamerro and Archangelsky, 1981); Canada, eastern (Maastrichtian–early Paleocene: Barss et al., 1979); Canada, offshore Atlantic (Maastrichtian–late Paleocene: Williams and Bujak, 1977); Canada, offshore Atlantic (Maastrichtian–early Paleocene: Bujak and Williams, 1978); France (late Oligocene: Charollais et al., 1975); Germany (middle Oligocene: Benedek, 1972); Germany (Rupelian: Benedek and Müller, 1974); Greenland (Danian: Soper et al., 1976); U.S.A., California (Danian: Drugg, 1967); U.S.A., New Jersey (middle-late Maastrichtian: Koch, 1975); U.S.A., Oklahoma (Campanian: Morgan, 1967); U.S.A., South Carolina (early Paleocene: Hazel et al., 1977); U.S.A., South Dakota (Paleocene: Stanley, 1965); U.S.A., Texas (Maastrichtian: Zaitzeff, 1967); U.S.A., Wyoming (Late Cretaceous: Stone, 1973).

Spinidinium essoi Cookson and Eisenack 1967
Figure 34.3

Spinidinium essoi Cookson and Eisenack, 1967a, p. 135, Plate 19, Figures 1-8.

Comments. The Seymour Island specimens compare quite closely with the description of the Western Tasmanian material by Cookson and Eisenack (1967a). The apical horn is concave distally and usually bears two or more small, solid capitate spines on its end. The left antapical horn terminates in a sharp point that may bear one or more capitate spines distally. The location of the reduced right antapical horn is usually only suggested by a slight posterior bulge and/or a clump of short, capitate spines. Paratabulation is suggested by the linear distribution of spines. Penitabular capitate spines outline the anterior end and the two sides of the posteriorly attached 2a operculum.

Dimensions. Observed range (10 specimens): overall length, 49 to 70 μm (mean, 60 μm); overall width, 34 to 54 μm (mean, 46 μm).

Stratigraphic Occurrence. Cross Valley Formation (Section 12-13, early late Paleocene); La Meseta Formation (Section 3, Eocene; Sections 17, 18, and 19, Eocene).

Previous Occurrence. Worldwide distribution. The total reported range is from Late Cretaceous to the early Oligocene.

Spinidinium lanterna Cookson and Eisenack 1970
Figure 34.1–2

Spinidinium lanterna Cookson and Eisenack, 1970a, p. 144–145, Plate 12, Figures 1–3.

Comments. As noted by Cookson and Eisenack (1970a), paratabulation is most evident on the epicyst. Parasutural spines delineate the 2a and six precingular paraplates. The spines are solid and may be truncated, broadly acuminate, capitate, or bifid distally. The parasulcal and paracingular areas are undivided and devoid of parasutural spines. The broad apical horn is concave distally and usually bears two or more capitate spines on each side of the depression. The stout left antapical horn is always more fully developed than that on the right and may be a very sharp spine or spike. The endocyst is often more globular and less appressed to the pericyst than is indicated for the type material (Cookson and Eisenack, 1970a; see their Plate 12 and Figs. 1–3).

Dimensions. Observed range (10 specimens): pericyst length, 53 to 66 μm (mean, 61 μm); pericyst width, 32 to 46 μm (mean, 39 μm); endocyst length, 20 to 36 μm (mean, 33 μm); endocyst width, 27 to 43 μm (mean, 34 μm); archeopyle height, 8 to 26 μm (mean, 17 μm); archeopyle width, 9 to 14 μm (mean, 12 μm).

Stratigraphic Occurrence. Cross Valley Formation (Section 16, early late Paleocene); La Meseta Formation (reworked into late early Eocene and Holocene beach deposits of Section 3).

Previous Occurrence. Australia (Senonian: Cookson and Eisenack, 1970a); South Atlantic, DSDP 361 (Turonian: Davey, 1978); Seymour Island (Danian: Palamarczuk et al., 1984, as *Gen. et sp.* indet. 2).

Spinidinium luciae sp. nov.
Figures 35.1–3, 38.1–5, 39.4

Diagnosis. A species of *Spinidinium* that is characterized by penitabular and parasutural denticulate crests or rows of isolated denticles. The attenuated intercalary 2a archeopyle may have a free or posteriorly adherent operculum. Accessory parasutures may be developed along the lateral margins of the 4″ paraplate (i.e., the 3″/4″ and the 4″/5″ parasutures).

Description.
Shape: The pericyst outline is subpentagonal to subhexagonal, depending on the distal width of the apical horn. The subtriangular epicyst is terminated by a short, truncated apical horn that often is concave distally. Penitabular or parasutural denticulate crests or rows of individual denticles give the dinocyst margin a serrate appearance. Two lateral transapical crests, each bearing a row of denticles, run from the paracingulum over the apex of the apical horn and down to the paracingulum on the other side. The denticles may be distally rounded or pointed. The subrectangular hypocyst is approximately half the length of the epicyst. It is flattened antapically, but short, unequal antapical horns may be present. The left antapical horn is always more fully developed than is the right antapical horn, which may not be developed at all. The horns vary in shape from subrounded to sharp, spike-like protuberances. The pericyst is broadest in the paracingular region. The endocyst outline is subrounded to subpentagonal and often mimics the outline of the pericyst.
Phragma: Cysts are circumcavate to cornuncavate. The periphragm is approximately 0.5 μm thick and may be shagreenate or uniformly granulate. The granules vary in size, but they are always smaller than the 1- to 3-μm-high denticles capping the penitabular or parasutural

crests. The endophragm is shagreenate to scabbrate and approximately 1 μm thick.
Paratabulation: Paratabulation is expressed on the pericyst by denticulate to serrate crests or isolated denticles. Penitabular rows of denticles delimit all paraplates, except perhaps along the margins of the parasulcus. Parasutural denticles cap the ridges bordering the paracingulum. The penitabular denticles are poorly developed in the parasulcal area. Ventral paratabulation similar to ortho epithecal tabulation. The paratabulation formula is 4′, 3a, 7″, 5-7C, 5‴, ?2⁗. The only indication of paratabulation on the endocyst is the 2a intercalary archeopyle.
Paracingulum: The laevorotatory paracingulum is bordered by two discontinuous parallel denticulate to serrate parasutural crests. The ends of the paracingulum are offset less than one full width of the paracingulum. The discontinuous paracingular crests and/or denticles divide the paracingulum into five to seven paraplates.
Parasulcus: The parasulcus is delimited by denticulate to serrate crests that are less well developed and less continuous than those bordering the paracingulum. The parasulcus is broadest posteriorly and tapers anteriorly, where it runs well up onto the epicyst. A single, large denticle in the right, middle area of the parasulcus may reflect the location of the flagellar pore of the theca. The parasulcus is not expressed on the endocyst.
Archeopyle: Archeopyle type I/I. The operculum of the attenuated 2a archeopyle may be free or adherent along the H4 parasuture. The lateral parasutures of the 4″ paraplate (i.e., 3″/4″ and 4″/5″) may be partially open, but the 4″ is adnate along the anterior margin of the paracingulum. The 4″ may or may not be involved in archeopyle formation; when 4″ is involved, it is probably fortuitous.

Dimensions. Observed range (10 specimens): pericyst length, 51 to 87 μm (mean, 64 μm); pericyst width, 46 to 76 μm (mean, 61 μm); endocyst length, 45 to 66 μm (mean, 52 μm); endocyst width, 42 to 56 μm (mean, 46 μm); archeopyle dimensions (6 specimens): length, 16 to 32 μm (mean, 20 μm); width, 16 to 21 μm (mean, 19 μm); periphragm, approximately 0.5 μm thick; endophragm, 0.75 to 1.0 μm thick.

Discussion. The endocyst varies considerably in shape and size. Consequently, the relationship between the periarcheopyle and the epipericoel also varies. The epipericoel is in communication with the exterior through the intercalary periarcheopyle only on specimens bearing a circular endocyst. Communication is reduced or totally blocked by an elongate endocyst that mimics the outline of the pericyst. The denticulate to serrate transapical crests form distinctive crescent-shaped depressions at the apex of the apical horn.

Comparison with Similar Species. The denticulate parallel transapical crests and the variable archeopyle distinguish *Spinidinium luciae* sp. nov. from all other dinocysts. In particular, *S. luciae* sp. nov. differs from: *S. styloniferum* by having fewer, stouter and distally rounded denticles rather than spines; from *S. macmurdoensis* and *S. sverdrupianum* in overall cyst shape and in the type and distribution of ornamentation; and from *Chichaouadinium vestitum* and *C. boydii* by the type and distribution of ornamentation, in overall shape and archeopyle development.

Holotype. Slide 8452, E/1, 118.3 x 11.4 (L23). Sample 8452, Section 3, La Meseta Formation, late early Eocene, Seymour Island, Antarctica.

Derivation of Name. Named in memory of the senior author's sister, Lucielle Jeni Wrenn Bjorgo.

Stratigraphic Occurrence. La Meseta Formation (Section 3, Eocene; Sections 18 and 19, Eocene).

Spinidinium macmurdoense (Wilson 1967)
Lentin and Williams 1976
Figures 36.6, 44.1

Deflandrea macmurdoensis Wilson, 1967a, p. 60, Figures 2a, 11–16, 22.
Spinidinium macmurodense (Wilson 1967a) Lentin and Williams, 1976, p. 64.

Comments. The Seymour Island specimens compare closely with those described by Wilson (1967a), except that the operculum is not always outlined by small spines. The same extreme variation in the overall cyst outline noted on the McMurdo Sound specimens (Wilson, 1967a; Figs. 11–16) was observed in the Seymour Island material.

Dimensions. Observed range (10 specimens): pericyst length, 69 to 94 μm (mean, 85 μm); pericyst width, 53 to 107 μm (mean, 66 μm); endocyst length, 46 to 61 μm (mean, 55 μm); endocyst width, 48 to 61 μm (mean, 55 μm); archeopyle height, 15 to 24 μm (mean, 19 μm); archeopyle width, 17 to 24 μm (mean, 20 μm).

Stratigraphic Occurrence. La Meseta Formation (Section 3, Eocene; Sections 17, 18, and 19, Eocene).

Previous Occurrence. Goodman (1975) reported *S. macmurdoense* (as *Deflandrea macmurdoensis*) from the lower Eocene Woodstock Member of the Nanjemoy Formation in southern Maryland. The specimens he illustrated and described were recovered from his Unit 16 and are superficially similar to the Antarctic type material of Wilson (1967a). However, they differ significantly: (1) by not being dorso-ventraly flattened; (2) by being one-third to one-half as small; (3) by having much shorter, smaller and apparently fewer spines; and (4) by having the periphragm and endophragm much more closely appressed than the Antarctic-type material.

Although the wall layers of the holotype of *S. macmurdoense* (Wilson, 1967a; Figs. 11–13) are fairly close together, there is a small pericoel separating them. The pericoel is more evident in the paratypes (Wilson, 1967a; Figs. 14–16). The paratype material is typical of the thousands of specimens observed in the Seymour Island samples.

D. K. Goodman (personal communication, 1982) has subsequently expressed strong doubts concerning the identification of *S. macmurdoense* from the Nanjemoy Formation. We believe the specimens that Goodman (1975) assigned to *S. macmurdoense* are closely related to an unpublished species, *Spinidinium "bilineatum,"* that he recognized in the overlying unit, Unit 17. This is *Spinidinium* sp. 17 of Goodman, 1979. The range of *S.* sp. 17 does not overlap with *S. macmurdoense* sensu Goodman (1975) in Units 16 and 17 of the Woodstock Member. The specimens in his Unit 16 (i.e., *S. macmurdoense* sensu Goodman, 1975) may be evolutionary predecessors of his morphologically similar new species in Unit 17 (i.e., *Spinidinium* sp. 17).

De Coninck (1975) reported finding one specimen of *Spinidinium macmurdoense* (as *Deflandrea macmurdoensis*) in the Paleocene deposits of the Klogshamn Quarry, near Malmö, Skåne, southern Sweden. The description and illustrations (Fig. 2, K₁–K₂) indicated that his specimen was more closely related to *Spinidinium* sp. 17 of Goodman (1979) than to *S. macmurdoense*.

Spinidinium macmurdoense was described from reputedly Eocene erratics collected in the McMurdo Sound area (McIntyre and Wilson, 1966; Wilson, 1967a). Recently, the Eocene age of the McMurdo Sound erratics (Wilson, 1967a) has been questioned (Stott et al, 1983). It has been reported from the Eocene Rio Turbio Formation of Argentina (Archangelsky and Fasola, 1971), reworked from the West Ice Shelf area, Antarctica (Kemp, 1972), DSDP 280, 282, and 283 (Haskell and

Wilson, 1975) and DSDP 511 (Goodman and Ford, 1983). *S. macmurdoense* also occurs in the Eocene of the Alaska Peninsula (Loretta Satchel, personal communication, 1984). All of the reported age assignments fall within the Eocene, except in DSDP 282, where a questionable assignment to the Oligocene was cited, and DSDP 511, where a late Eocene–early Oligocene range was documented. The total reported range for *S. macmurdoense* is Eocene to early Oligocene.

Other reports include: Antarctica (late Eocene–early Oligocene: Hall, 1977); Antarctica, Ross Sea (?Eocene: Wilson, 1968); Antarctica, Ross Sea, DSDP Leg 28 (Eocene–late Oligocene: Kemp, 1975); Argentina (Danian: Heisecke, 1970); Argentina (Eocene: Archangelsky, 1969); Pacific Ocean, Campbell Plateau (Late Cretaceous: Wilson, 1975).

Spiniferites multibrevis
(Davey and Williams 1966) Below 1982
Figure 35.5

Hystrichosphaera ramosa var. *multibrevis* Davey and Wiliams, 1966a, p. 35–36, Plate 1, Figure 4; Plate 4, Figure 6; text-fig. 9.
Spiniferites ramosus subsp. *multibrevis* (Davey and Williams, 1966a) Lentin and Williams, 1973, p. 130.
Spiniferites multibrevis (Davey and Williams, 1966a) Below, 1982, p. 35.

Comments. The Seymour Island specimens agree closely with the description of the type material. Integonal processes are very rare on all specimens observed.

Dimensions. Observed range (five specimens): overall length, 53 to 66 μm (mean, 61 μm); overall width, 39 to 63 μm (mean, 53 μm); endocyst length, 39 to 49 μm (mean, 46 μm); endocyst width, 37 to 49 μm (mean, 43 μm); length of processes, as much as 11 μm.

Stratigraphic Occurrence. La Meseta Formation (Section 3, late early Eocene; Section 19, late early Eocene).

Selected Previous Occurrence. *Spiniferites multibrevis* was described from the Eocene London Clay of southern England (Davey and Williams, 1966a). The species has been reported from the Hauterivian to the Ypresian in England and the Aptian of Germany (Davey and Williams, 1966a). The total stratigraphic range is from Lower Cretaceous (Hauterivian) to middle Eocene (Williams and Kidson, 1975).

Spiniferites ramosus (Ehrenberg 1838)
Loeblich and Loeblich 1966
Figure 35.4

Xanthidium ramosus Ehrenberg, 1838, Plate 1, Figures 1, 2, 5.
Spiniferites ramosus Mantell, 1854, text-fig. 17, Nos. 4 and 6.
Ovum hispidum (Xanthidium) ramosum Lohmann, 1904, p. 21, 25.
Hystrichosphaera ramosa (Ehrenberg, 1838) Wetzel, 1932, p. 144.
Spiniferites ramosus (Ehrenberg, 1838) Loeblich and Loeblich, 1966, p. 56–57.

Comments. Specimens referable to this species are rare in the Seymour Island material. The processes are usually rather thin, as are the endophragm and the periphragm.

Dimensions. Observed range (two specimens): endocyst length, 35 to 41 μm; endocyst width, 29 to 36 μm; length of processes, as much as 14 μm.

Stratigraphic Occurrence. Cross Valley Formation (Section 16, early late Paleocene); La Meseta Formation (Section 3, late early Eocene; Section 17, late early Eocene).

Previous Occurrence. Worldwide distribution. Reported range from middle Jurassic to Recent.

Spiniferites ramosus subsp. *granomembranaceus* (Davey and Williams 1966) Lentin and Williams 1973 Figure 36.7

Hystrichosphaera ramosa var. *granomembranacea* Davey and Williams, 1966a, p. 37–38, Plate 4, Figure 4.
Spiniferites ramosus subsp. *granomembranaceus* (Davey and Williams, 1966a) Lentin and Williams, 1973, p. 130.

Comments. The Seymour Island material compares closely to illustrations of the type material. However, some specimens are larger, and the granular membranes connecting the process bases are occasionally perforated.

Dimensions. Observed range (three specimens): overall length, 53 to 67 μm (mean, 61 μm); overall width, 51 to 61 μm (mean, 57 μm); endocyst length, 42 to 53 μm (mean, 51 μm); endocyst width, 41 to 47 μm (mean, 46 μm); membrane height to 10 μm.

Stratigraphic Occurrence. Cross Valley Formation (Section 12-13, early late Paleocene and Eocene, see discussion of the age determinations for this section); La Meseta Formation (Section 3, late early Eocene and reworked into Holocene beach deposits.)

Previous Occurrence. S. ramosus subsp. *granomembranaceus* was described from the lower Eocene London Clay deposits that crop out near Sheppey, Kent, England (Davey and Williams, 1966a). This species has been reported from deposits in France that vary greatly in age. Included among there are Albian (Foucher and Taugourdeau, 1975), late Turonian (Foucher, 1974), Coniacian to Campanian (Foucher, 1976), late Campanian to late Maastrichtian (Foucher and Robaszynski, 1977), late Eocene (Jan du Chêne et al., 1975) and middle Eocene to early Oligocene (Chateauneuf, 1979). However, none of the latter reports illustrated *S. ramosus* subsp. *granomembranaceus.* This species has been reported, but not illustrated, from the Maastrichtian-Danian of Maryland (Whitney, 1984) and the Paleocene of the New Madrid test well 1-X in southeastern Missouri (Frederiksen et al., 1982). The only illustrated report of this species is from the lower Eocene London Clay.

Spiniferites ramosus subsp. *granosus* (Davey and Williams 1966) Lentin and Williams 1973 Figures 36.4–5

Hystrichosphaera ramosa var. *granosa* Davey and Williams, 1966a, p. 35, Plate 4, Figure 9.
Spiniferites ramosus subsp. *granosus* (Davey and Williams, 1966a) Lentin and Williams, 1973, p. 130.

Comments. Specimens attributable to this species are well preserved and compare closely with the description of the type material. Intergonal processes have not been observed on any of the specimens. The size range is somewhat greater than that reported by Davey and Williams (1966a).

Dimensions. Observed range (five specimens): overall length, 49 to 77 μm (mean, 62 μm); overall width, 44 to 68 μm (mean, 53 μm); endocyst length, 36 to 52 μm (mean, 44 μm); endocyst width, 32 to 44 μm (mean, 37 μm); length of processes, 9 to 17 μm.

Stratigraphic Occurrence. La Meseta Formation (Section 3, late early Eocene).

Previous Occurrence. Previously reported from the Northern Hemisphere. The stratigraphic range is from Barremian to early Oligocene.

Spiniferites ramosus subsp. *reticulatus* (Davey and Williams 1966) Lentin and Williams 1973 Figure 35.6

Hystrichosphaera ramosa var. *reticulata* Davey and Williams, 1966a, p. 38, Plate 1, Figures 2–3.
Spinferites ramosus subsp. *reticulatus* (Davey and Williams, 1966a) Lentin and Williams, 1973, p. 130.

Comments. The rare specimens encountered are similar to the description of the type material except the endophragm is distinctly reticulate and the periphragm is granular to reticulate. Davey and Williams (1966a) stated the endophragm is smooth and only the periphragm is reticulate on the type material. The specimens are reworked from Cretaceous (Barremian to Santonian) deposits.

Dimensions. (One complete specimen recovered): overall length, 66 μm; overall width, 61 μm; endocyst length, 52 μm; endocyst width, 44 μm; length of processes, as much as 9 μm.

Stratigraphic Occurrence. Cross Valley Formation (Section 12-13, early late Paleocene); La Meseta Formation (Section 3, late early Eocene; Sections 17, 18, and 19, Eocene).

Selected Previous Occurrences. England (Cenomanian: Davey and Williams, 1966a); Sweden (Senonian-?Danian: Tralau, 1972); DSDP Leg 48, Bay of Biscay (Aptian-Albian: Davey, 1979).

Genus *Tanyosphaeridium* Davey and Williams 1966 *Tanyosphaeridium xanthiopyxides* (O. Wetzel 1933 emend. Morgenroth 1968) Stover and Evitt 1978 Figures 37.2–3

Hystrichosphaera xanthiopyxidies O. Wetzel, 1933, p. 44, Plate 4, Figure 25.
Hystrichosphaeridium xanthiopyxides (O. Wetzel, 1933) Deflandre, 1937, p. 77.
Baltisphaeridium xanthiopyxides (O. Wetzel, 1933) Klement, 1960, p. 59.
?Hystrichosphaeridium xanthiopyxides (O. Wetzel, 1933) emend. Morgenroth, 1968, p. 556, Plate 48, Figures 5, 6.
?Prolixosphaeridium xanthiopyxides (O. Wetzel, 1933) Davey et al., 1969, p. 17.
Tanyosphaeridium xanthiopyxides (O. Wetzel, 1933, emend. Morgenroth, 1968) Stover and Evitt, 1978, p. 85.

Comments. The Seymour Island specimens compare closely with the description of the type material, although some have more processes than reported for the type material (20 to 35 processes).

Dimensions. Observed range (two specimens): central body length, 29 μm (measurable only on one specimen); central body width, 21 to 25 μm; length of processes, 13 to 16 μm; wall, approximately 1.5 μm thick.

Stratigraphic Occurrence. Cross Valley Formation (Section 15, early late Paleocene).

Selected Previous Occurrences. Worldwide distribution. Portlandian to Paleocene.

Genus *Thalassiphora* Eisenack and Gocht 1960 emend. 1968
Thalassiphora pelagica (Eisenack 1954) Eisenack and Gocht 1960
Figure 37.1

Pterospermopsis pelagica Eisenack, 1954, p. 71–72, Plate 12, Figures 17–18.
Pterocystidiopsis velata Deflandre and Cookson, 1955, p. 291, Plate 8, Figure 8.
Thalassiphora pelagica (Eisenack, 1954) Eisenack and Gocht, 1960, p. 513–514.
Thalassiphora velata (Deflandre and Cookson, 1955) Eisenack and Gocht, 1960, p. 514–515.
Thalassiphora pelagica (Eisenack, 1954) Eisenack and Gocht, 1960; emend. Benedek and Gocht, 1981, p. 59–61.

Comments: The Seymour Island specimens compare closely, especially in size, with those reported for the type material (Eisenack, 1954) and with those from southern England (Williams, 1963).

Dimensions. Observed range (five specimens): pericyst length, 174 to 213 μm (mean, 189 μm); pericyst width, 168 to 227 μm (mean, 175 μm); endocyst length, 83 to 106 μm (mean, 95 μm); endocyst width, 71 to 87 μm (mean, 80 μm); archeopyle dimensions (two specimens), length, 31 to 48 μm (mean, 40 μm); width, 36 to 39 μm (mean, 38 μm).

Stratigraphic Occurrence. La Meseta Formation (Section 3, late early Eocene; Section 17, late early Eocene).

Previous Occurrence. Worldwide occurrence; Maastrichtian to early Oligocene (Williams and Bujak, 1985).

Genus *Trigonopyxidia* Cookson and Eisenack 1961
Trigonopyxidia ginella (Cookson and Eisenack 1960) Downie and Sarjeant 1965
Figure 36.3

Trigonopyxis ginella Cookson and Eisenack, 1960a, p. 11, Plate 3, Figures 18–20.
Trigonopyxidia ginella (Cookson and Eisenack, 1960a) Downie and Sarjeant, 1965, p. 149.

Comments. Only one specimen was recovered. It differed from the description and illustrations of the type material in that its surface is shagreenate instead of smooth. The archeopyle margin is smooth and

without accessory parasutures. No indication of a paracingulum or parasulcus is evident. The inner body is rolled up within the pericyst.

Dimensions. (One specimen): periphragm width, 55 μm; periphragm length, 50 μm (without operculum); periphragm thickness, 1 μm; endophragm unmeasurable due to folding.

Stratigraphic Occurrence. Cross Valley Formation (Section 15, early late Paleocene).

Previous Occurrence. Worldwide distribution during the late Albian to late Paleocene.

Genus *Turbiosphaera* Archangelsky 1968
Turbiosphaera filosa (Wilson 1967) Archangelsky 1968
Figure 44.2

Cordosphaeridium filosum Wilson, 1967a, p. 66, Figures 2, 31–32, 34.
Turbiosphaera filosa (Wilson, 1967a) Archangelsky, 1968, p. 408–411.

Comments. The specimens from the La Meseta Formation are comparable to the description of the type material from McMurdo Sound, Antarctica (Wilson, 1967a). Two complete and a few fragmentary specimens were recovered.

Dimensions. Observed range (two complete specimens): pericyst length, 95 to 102 μm; pericyst width, 61 to 62 μm; pericyst thickness, less than 1 μm; endocyst length, 64 to 74 μm; endocyst width, 47 to 51 μm; endocyst thickness, 1 to 2.5 μm.

Stratigraphic Occurrence. La Meseta Formation (Section 3, late early Eocene and reworked into Holocene beach deposits).

Previous Occurrence. Worldwide distribution during the Maastrichtian to middle Eocene (Williams and Bujak, 1985). Reports of this taxa in younger deposits (Evitt and Pierce, 1975; Kemp, 1975; Jan du Chêne, 1977) are rare, but suggest that the stratigraphic top of this species is higher than recognized by Williams and Bujak, 1985. These reports of younger occurrences need to be reviewed.

Genus *Vozzhennikovia* Lentin and Williams 1976
Vozzhennikovia apertura (Wilson 1967) Lentin and Williams 1976
Figures 37.6–9, 43.3–4

Spinidinium aperturum Wilson, 1967a, p. 64–65, Figures 3–5, 8.
Spinidinium rotundum Wilson, 1967a, p. 65–66, Figures 6, 7.
Vozzhennikovia rotunda (Wilson, 1967a) Lentin and Williams, 1976, p. 67.
Vozzhennikovia apertura (Wilson, 1967a) Lentin and Wilson, 1976, p. 65.

Comments. This species and *Vozzhennikovia rotunda* (Wilson, 1967a) Lentin and Williams, 1976 are end members of a morphologic continuum. This view is supported by G. J. Wilson (personal communication, 1983). Consequently, *V. rotunda* is here synonymized with *V. apertura.*

Wilson (1967a) felt V. apertura was characterized by "the relatively large size of the archeopyle, transverse girdle (paracingulum) and longitudinal furrow (parasulcal groove)." *V. rotunda* was distinguished from *V. apertura* "by its circular outline, discontinuous girdle (paracingulum), thinner-walled, single-layered shell, smaller finer spines and by its generally larger size." Wilson (1967a) also noted a few intermediate forms in his material.

If only the end members were observed, they could easily be separated and correctly established as separate species, as Wilson (1967a) did. However, the large, variable populations recovered from the Seymour Island material contain an inseparable mix of specimens. They are best handled as one species, *V. apertura*. The archeopyle is generally formed by the loss of the 2a paraplate. However, a combination archeopyle, as previously noted by Goodman and Ford (1983), was observed on some of the Seymour Island specimens.

Dimension. Observed range (10 specimens): overall length, 32 to 58 μm (mean, 42 μm); overall width, 32 to 46 μm (mean, 38 μm).

Stratigraphic Occurrence. Cross Valley Formation (Section 12-13, Eocene; see discussion of the age determination of this section); La Meseta Formation (Section 3, Eocene; Sections 17, 18, and 19, Eocene).

Previous Occurrence. Vozzhennikovia apertura was described from the McMurdo Sound erratics (Wilson, 1967a). Occurrences of *V. apertura* and its synonym, *V. rotunda,* were subsequently reported from the middle to upper Eocene cores of DSDP 280 and 281 (Haskell and Wilson, 1975). Specimens of *V. apertura* observed in DSDP 283 were considered to be Eocene in age (Haskell and Wilson, 1975), as were reworked specimens recovered from the West Ice Shelf area of Antarctica (Kemp, 1972). Goodman and Ford (1983) reported *V. apertura* from late Eocene–early Oligocene samples of DSDP 511 and middle Eocene samples of DSDP 512. Both the Eocene Rio Turbio Formation of Argentina and the Loreta Formation in southern Chile have yielded *V. apertura* (as *V. rotunda;* Archangelsky and Fasola, 1971). The latter formation was considered to be younger than the Rio Turbio Formation, "possibly Oligocene," although the age assignment is uncertain. Other reports include: Antarctica, Seymour Island (late Eocene–early Oligocene: Hall, 1977; this age determination is not supported by our research); Antarctica, Ross Sea (Eocene: Wilson, 1968); Antarctica, Ross Sea, DSDP 274 (late Eocene–?early Oligocene: Kemp, 1975); Argentina (Eocene, as *Spinidinium rotundum;* Archangelsky, 1968); Argentina, Tierra del Fuego (early Tertiary, as *Dioxya* aff. *villosa* Cookson and Eisenack; Pöthe de Baldis, 1966); Canada, eastern (Eocene; not illustrated, questionable occurrence; Barss et al., 1979); Iran (middle to late Ypresian; not illustrated, questionable occurrence; Stampfli et al., 1978).

Consideration of the available evidence indicates that *V. apertura* has a stratigraphic range of Eocene to early Oligocene.

Genus *Xylochoarion* Erkmen and Sarjeant, 1978
***Xylochoarion* cf. *X. hacknessense* Erkmen and Sarjeant 1978**
Figure 37.4–5

Comments. The single Seymour Island specimen recovered is approximately half the size of the specimens described by Erkman and Sarjeant (1978). The hundreds of nontabular solid processes are capitate, symmetrically or asymmetrically bifurcate distally and have a basal thickening on the central body. The specimen is reworked, possibly from the late middle Jurassic (Callovian).

Dimensions. (One specimen recovered, operculum missing): length, 23 μm; width, 20 μm; process length, 5 to 10 μm; wall thickness, approximately 1 μm.

Stratigraphic Occurrence. La Meseta Formation (Section 3, reworked into late early Eocene deposits).

Previous Occurrence. England (Late Callovian: Erkmen and Sarjeant, 1978).

Group *ACRITARCHA* Evitt 1963

Genus *Comasphaeridium* Staplin, Jansonius and Pocock 1965
***Comasphaeridium cometes* (Valensi 1948) De Coninck 1969**
Figure 17.1, 3

Micrhystridium cometes Valensi, 1948, p. 547, Figure 5 (illustration 6). *Comasphaeridium cometes* (Valensi, 1948) De Coninck, 1969, p. 58, Plate 16, Figure 34–41.

Comments. The specimens had a subspherical to elliptical central body bearing hundreds of short, solid, hair-like acicular spines. No openings were observed in the wall of the central body. The Seymour Island specimens compare closely with the illustrations of the type material (Valensi, 1948) and with the specimens described by De Coninck (1969) from the Kallo borehole.

Dimensions. Observed range (five specimens): central body length, 20 to 29 μm (mean, 23 μm); central body width, 15 to 22 μm (mean, 18 μm); process length, 2 to 5 μm; wall thickness, approximately 0.5 μm.

Stratigraphic Occurrence. Cross Valley Formation (Section 12-13, Paleocene; Section 16, early late Paleocene); La Meseta Formation (Section 3, late early Eocene; Section 19, middle to late Eocene).

Previous Occurrence. Northern Hemisphere. Reported from Santonian to Oligocene deposits.

Genus *Leiofusa* Eisenack 1938
***Leiofusa jurassica* Cookson and Eisenack 1958**
Figure 27.12

Leiofusa jurassica Cookson and Eisenack, 1958, p. 51, Plate 10, Figures 3–4.

Comments. A single specimen bearing a strong resemblance to *Leiofusa jurassica* was recovered. It differs from the description of the type material by having a transverse fold (?paracingulum) and solid polar spines. No opening was observed in the test wall. The hollow central body is closed off at the base of both spines. Each spine base has two or three spindle-shaped hollow areas aligned along the axis of the spine. All areas are less than 0.7 μm long. These areas are separated by solid wall material and may be the result of secondary thickening similar to that reported in *Veryhachium domasia* and some species of *Diexallophasis* (Downie, 1979).

Dimensions. (One complete specimen recovered): overall length, 36 μm; test length, 23 μm; test width, 13 μm.

Stratigraphic Occurrence. La Meseta Formation (Section 3, late early Eocene; Section 17, late early Eocene).

Previous Occurrence. Worldwide distribution. Mesozoic–?Cenozoic.

**Genus *Micrhystridium* Deflandre 1937
emend. Downie and Sarjeant 1963**
Micrhystridium sp. A.
Figure 28.6

Comments. Subspherical to ellipsoidal central body bearing numerous solid spines. The spines are acuminate and as long as 8 μm. Most specimens were enrolled around the longitudinal or transverse axis and looked like a ball of processes.

Dimensions. Observed range (five specimens): central body length, 16 to 33 μm (mean, 22 μm); central body width, 12 to 26 μm (mean, 17 μm); processes as much as 8 μm long.

Stratigraphic Occurrence. La Meseta Formation (Section 3, Eocene and Holocene beach deposits; Section 19, middle to late Eocene).

Genus *Ophiobolus* O. Wetzel 1933 emend. Evitt 1968
Ophiobolus lapidaris O. Wetzel 1933
Figure 30.6

Ophiobolus lapidaris O. Wetzel, 1933, p. 176–179, Plate 2, Figures 30–34, text-fig. 5-7.

Comments. Ophiobolus lapidaris O. Wetzel was discussed in minute detail by Evitt (1968). The surface ornamentation ranges from smooth to microgranulate. The number of processes or strands on the ends of the main body varies from one to five on the Seymour Island specimens. The strands are often difficult to see, as Evitt (1968) noted.

Dimensions. Observed range (six specimens): body length, 18 to 30 μm (mean, 23 μm); body width, 15 x 22 μm (mean, 19 μm); wall thickness, approximately 0.5 μm.

Stratigraphic Occurrence. Cross Valley Formation (Sections 15 and 16, early late Paleocene); La Meseta Formation (Section 3, reworked into Holocene beach and Eocene deposits; Sections 17, 18, and 19, reworked into Eocene deposits).

Previous Occurrence. Northern Hemisphere. Reported from deposits that range from Campanian to Danian in age.

**Genus *Veryhachium* (Denuff 1958)
emend. Downie and Sarjeant 1963**
Veryhachium sp. indet.

Comments. Rare triangular forms bearing a hollow horn at each corner were observed. The wall is smooth, thin, and colorless. One side of the triangular body is half as long as the other two sides. No opening was observed in the wall.

Dimension. (One complete specimen recovered): overall length, 55 μm; overall width, 43 μm; test length, 34 μm; test width, 26 μm; horn length, approximately 20 μm; wall thickness, <0.5 μm.

Stratigraphic Occurrence. La Meseta Formation (Section 3, reworked into Holocene beach and late early Eocene deposits; Section 18, late early Eocene).

**Division *PRASINOPHYTA* Round 1971
Order *PTEROSPERMATALES* Schiller 1925
Family *CYMATIOSPHAERACEAE* Mädler 1963**

**Genus *Cymatiosphaera* (O. Wetzel 1933)
Deflandre 1954**
Cymatiosphaera sp. indet.

Very rare specimens of *Cymatiosphaera* sp. indet. were recovered from Sections 3, 15, 16, and 19. McLean (1971) noted that some of his specimens (which were also rare) occurred in deposits believed to have been deposited under subnormal marine paleoenvironmental conditions.

Family *PTEROSPERMELLACEAE* Eisenack

Genus *Pterospermella* Eisenack 1972
Pterospermella australiensis (Deflandre and Cookson 1955)
Eisenack 1972
Figure 31.1

Pterospermopsis australiensis Deflandre and Cookson, 1955, p. 286, Plate 3, Figure 4, text-figs. 52, 53.
Pterospermella australiensis (Deflandre and Cooksonn, 1955) Eisenack, 1972, p. 596–601.

Comments. The rare specimens recovered from Seymour Island compare well with the illustrations of the type material. They differ only in being slightly larger than the Australian specimens.

Dimensions. (One specimen): overall length, 55 μm; overall width, 44 μm; inner body length, 26 μm; inner body width, 18 μm; both wall layers, <0.5 μm thick.

Stratigraphic Occurrence. La Meseta Formation (Section 3, Eocene).

Previous Occurrence. Worldwide distribution during the Neocomian to Eocene.

**Division *CHLOROPHYTA* Pasher 1914
Class *CHLOROPHYCEAE* Kützing 1843
Order *CHLOROCOCCALES* Marchand 1895 orth. mut. Pasher 1915
Family *CHLOROCOCCACEAE* Blackman & Tansley 1902**

Genus *Palambages* O. Wetzel 1961
Palambages morulosa O. Wetzel 1961
Figure 31.4

Palambages morulosa O. Wetzel, 1961, p. 338, Plate 1, Figure 11.

Comments. The specimens included in *P. morulosa* consist of many subcircular bodies, as in the Baltic type material (Wetzel, 1961). The surface of the individual bodies is smooth and nonuniformly foveolate. Pitting is especially evident along interbody walls. Wall thickness is less than 1 μm, as measured on the outside wall of a peripheral body. The wall thickness between abutting bodies within the colony is as much as 2.5 μm thick. Such interbody walls are much darker than the peripheral walls and the rest of the colony. Ragged openings (tears?) were noted in some individual bodies.
   The thick wall surface, the small size of the constituent bodies, and

the thick interbody boundary walls distinguish these forms from *P.* Forma B Manum and Cookson, 1964 and *P.* Type I.

Dimensions. Observed range (five specimens): overall length, 77 to 153 μm (mean, 121 μm); overall width, 68 to 105 μm (mean, 97 μm); individual body dimensions: overall length, 12 to 20 μm (mean, 19 μm); overall width, 12 to 19 μm (mean, 18 μm).

Stratigraphic Occurrence. Cross Valley Formation (Section 12-13, Eocene, see discussion of age determinations for this section; Section 15, early late Paleocene); La Meseta Formation (Section 3, Eocene and reworked into Holocene beach deposits; Section 19, middle to late Eocene). All of these occurrences represent either a range extension for this species or reworking. The latter is most probable.

Previous Occurrence. Worldwide distribution. Stratigraphic range from the Aptian to the middle Paleocene.

Palambages Forma B Manum and Cookson 1964
Figure 31.6

Palambages Forma B Manum and Cookson, 1964, p. 24, Plate 7, Figure 8.

Comments. The eight individual bodies making up the colony are subcircular and coarsely granulate. Granulation occurs in randomly shaped clumps and as individual grains. Body walls are approximately 1 μm thick. The diameter of the individual bodies greatly exceeds those of *P. morulosa* and *P.* Type I.

Large, angular openings were observed in four of the bodies making up the colony. They did not resemble the openings in *P. morulosa* described by Gocht and Wille (1972). This species is not considered to be synonymous with *P. morulosa,* as suggested by Gocht and Wille (1972), because of: (1) the large diameter of the individual bodies relative to that of the colony; (2) the large, angular opening in the individual bodies; and (3) the coarsely granular nature of the ornamentation.

Dimensions. Observed range (two specimens): diameter, 67 to 102 μm; diameter of individual bodies, 28 to 35 μm.

Stratigraphic Occurrence. Cross Valley Formation (Section 12-13, Eocene, see discussion of the age determinations for this section).

Previous Occurrence. Graham Island, Arctic Canada (Cretaceous: Manum and Cookson, 1964).

Palambages Type I
Figure 32.5

Description. Subcircular colonies formed by numerous smaller, subcircular bodies. Each colony appears to have been a hollow sphere prior to compaction. Each small body is composed of a skeletal network of strands, similar in principle to *Evittosphaerula paratabulata* Manum, 1979. (*P.* Type I and *E. paratabulata* are not synonymous, and no paratabulation formulae was evident on the former.) The smooth to rough strands are round to subelliptical in cross section and as much as 1.5 μm wide. A shagreenate, 0.5-μm-thick membrane covers the skeletal framework, forming the body wall. The constituent bodies appear to be united by strands external to the wall membrane. A wall opening similar to those described on *P. morulosa* (Gocht and Wille, 1972) was observed

in one constituent body. It is not clear whether this is a characteristic opening or an artifact.

Comments. The structure of these colonies indicates a relationship to *Palambages.* The structure of the constituent bodies excludes them from inclusion with *P. morulosa.*

Dimensions. Observed range (five specimens): overall length, 90 to 131 μm (mean, 102 μm); overall width, 71 to 111 μm (mean, 87 μm); diameter of the individual bodies, 19 to 34 μm (mean, 25 μm).

Stratigraphic Occurrence. La Meseta Formation (Section 3, Eocene; Sections 17 and 19, Eocene).

Family *BOTRYOCOCCACEAE* N. Wille

Genus *Botryococcus* Kützing 1849
Botryococcus sp. indet.

Comments. Rare colonies of *Botryococcus* were encountered which agree with the detailed descriptions of Blackburn (1936). Colonies generally consist of 10 to 20 cells.

CONCLUSIONS

1. The La Meseta Formation is Eocene in age. Unit A of Section 3 is late early Eocene, whereas the overlying interval (Unit B) is late middle to late Eocene in age. The two intervals recognized in Sections 17, 18, and 19 (Units A and B) are late early Eocene and middle to late Eocene in age, respectively.

The samples collected from the Cross Valley Formation type section are biostratigraphically suspect. The dinocyst flora indicates that the upper laminated samples of Section 12-13 are early late Paleocene in age. The underlying 130 m may be Eocene in age and could have been collected from a younger, unrecognized fault block in Cross Valley. Cross Valley Formation outcrops in the Cape Wiman area are early late Paleocene in age.

2. The ages and distribution of the commonly observed, reworked palynomorphs revealed an inverted stratigraphy in the La Meseta Formation. This inverted stratigraphy is characterized by a post–late Senonian dinocyst flora overlain by a pre-Campanian dinocyst flora. Permo-Triassic Striatiti pollen are locally present in the lower part of the La Meseta Formation (Sections 3, 17, 18) but are increasingly abundant upsection.

3. Combined palynologic, maceral, TOC, and sedimentologic data (Wrenn, 1982) indicate that the Cape Wiman early late Paleocene beds of the Cross Valley Formation are interdistributary bay deposits.

The early late Paleocene Cross Valley Formation samples from Section 12-13 are very nearshore, possibly estuarine, deposits. The underlying deposits of probable Eocene age in Section 12-13 were deposited in a distributary channel.

A shoaling upward sequence is evident through the La Meseta Formation (Section 3 and Sections 17, 18, and 19). The lower portion of these sections comprise moderate-energy delta top deposits, whereas the overlying beds are lagoonal and storm or splay deposits. The coarse sand Eocene beds of the lower part of Section 12-13 may be contemporaneous and associated with these strandline deposits.

4. The dinocyst floras are composed of cosmopolitan and provincial taxa. The early late Paleocene dinocyst species recovered from the Cross Valley Formation are cosmopolitan taxa. These cosmopolitan floras indicate that the Atlantic and Pacific Oceans were connected during the Paleocene Epoch and that an endemic dinocyst flora had not yet developed in the southern high latitudes.

The La Meseta Formation dinocyst floras are characterized by taxa of the provincial transantarctic flora, including *Arachnodinium antarcticum, Deflandrea antarctica, D. cygniformis, D. oebisfeldensis* sensu Cookson and Cranwell, *Spinidinium macmurdoense,* and *Vozzhennikovia apertura.* During the Eocene Epoch, the transantarctic flora extended from southern South America and Seymour Island across West Antarctica through the Ross Sea and westward between East Antarctica and Australia to the West Ice Shelf area. The distribution of this flora supports the existence of a transantarctic seaway between east and west Antarctica, and perhaps a circum-East Antarctic current during late Paleogene time.

Rare, extra-Antarctic occurrences of some species belonging to the transantarctic flora suggest that the total distribution of the transantarctic flora is not yet known. Some of these taxa may have had a bipolar distribution.

5. The Seymour Island Paleogene deposits document early late Paleocene and Eocene deposition within a deltaic complex and marginal marine waters. Deltaic development occurred adjacent to a rising cordilleran area of the present Antarctic Peninsula, which shed increasingly older sediments and reworked dinocysts southeastward into the Seymour Island area. The Seymour Island deltaic complex developed on the West Antarctic margin of a marine seaway that connected the Weddell and Ross Seas during the Paleogene.

ACKNOWLEDGMENTS

We thank Lewis E. Stover for his comments and for facilitating the processing of some of our samples, and Edde Griffin and Paco Ricalo for preparing the palynologic slides at Exxon Production Research Company. Graham L. Williams and Lucy E. Edwards provided valuable advice and comments on the manuscript. Our thanks are extended to Peter N. Webb and the Office of Polar Programs of the National Science Foundation for providing the samples from Seymour Island and partial financial support of this investigation. (Partial funding was provided from National Science Foundation Grant DPP-79-07043 to Peter N. Webb.) The support provided by the Department of Geology (LSU), the Cartographic Section (LSU), and the Amoco Production Company (Research Center, Tulsa, Oklahoma) was a great assistance. Special thanks are extended to Moira A. Hathcock for typing the many drafts of this manuscript and to George A. Massey for taking some of the SEM photomicrographs. Finally, J.H.W. thanks Allison K. Wrenn for her unflagging support and encouragement during the execution of this investigation.

INDEX OF MICROPLANKTON

Species	Figure(s)
1. *Achomosphaera* sp. indet.	—
2. *Alisocysta circumtabulata*	15.1-15.3
3. *Aptea securigera*	15.7
4. *Aptea* sp. indet.	—
5. *Arachnodinium antarcticum*	15.4-15.5, 43.1
6. *Areosphaeridium diktyoplokus*	15.6
7. *Batiacasphaera explanata*	16.4
8. *Botryococcus* sp. indet.	—
9. *Cerebrocysta bartonensis*	17.6, 17.8
10. *Cleistosphaeridium ancoriferum*	17.2
11. *Comasphaeridium cometes*	17.1, 17.3
12. *Cometodinium* cf. *?whitei*	17.7
13. *Cribroperidinium edwardsii*	18.3-18.4
14. *Cribroperidinium muderongense*	18.1
15. *Cribroperidinium* sp. indet.	—
16. *Cyclonephelium distinctum*	18.5
17. *Cyclopsiella elliptica*	—
18. *Cyclopsiella trematophora*	18.2
19. *Cymatiosphaera* sp. indet.	—
20. *Deflandrea antarctica* (Type I and II)	19.1-19.4, 44.3
21. *Deflandrea boloniensis* comb. nov.	16.3
22. *Deflandrea cygniformis*	20.2-20-5, 21.1-21.4, 42.3-42.5, 43.2, 44.4
23. *Deflandrea dartmooria*	16.1-16.2
24. *Deflandrea oebisfeldensis* sensu Cookson & Cranwell, 1967	20.1
25. *Deflandrea* cf. *phosphoritica*	22.1
26. *Deflandrea speciosa*	17.4
27. *Deflandrea striata*	17.5
28. *Deflandrea webbii* sp. nov.	22.2-22.4, 23.4-13.5, 40.1-40.4, 41.1-41.4, 42.2
29. *Deflandrea* sp. indet.	—
30. *Diconodinium cristatum*	23.1
31. *Diconodinium multispinum*	23.3
32. *Dinogymnium* cf. *curvatum*	23.2
33. *Enigmadinium cylindrifloriferum* gen. nov. sp. nov.	24.1-24.3, 45.1-45.6
34. *Eyrea nebulosa*	24.6
35. *Hystrichosphaeridium* cf. *astartes*	24.4-24.5
36. *Hystrichosphaeridium parvum*	—
37. *Hystrichosphaeridium salpingophorum*	25.5-25.6
38. *Hystrichosphaeridium truswelliae* sp. nov.	25.1-25.4, 39.1
39. *Hystrichosphaeridium tubiferum* subsp. *brevispinium*	26.7-26.9
40. *Hystrichosphaeridium* sp. indet.	—
41. *Impagidinium dispertitum*	—
42. *Impagidinium maculatum*	26.5-26.6
43. *Impagidinium victorianum*	26.3-26.4
44. *Impagidinium* sp. indet.	—
45. *Impletosphaeridium clavus* sp. nov.	27.10-17.11, 27.13
46. *Impletosphaeridium ligospinosum*	27.8
47. *Impletosphaeridium lorum* sp. nov.	26.1-26.2
48. *Impletosphaeridium* sp. B.	27.1-27.3, 27.9
49. *Isabelidinium pellucidum*	27.4
50. *Kallosphaeridium* cf. *capulatum*	27.5-27.7
51. *Leiofusa jurassica*	27.12
52. *Lejeunecysta fallax*	28.3
53. *Lejeunecysta hyalina*	28.5
54. *Lejeunecysta* sp. indet.	—
55. *Manumiella druggii*	28.7-28.8, 39.3
56. *Micrhystridium* sp. A.	28.6
57. *Octodinium askiniae* gen. nov. sp. nov.	28.1-28.2, 28.4, 29.1-29.7, 47.1-47.5
58. *Odontochitina spinosa*	30.3
59. *Oligosphaeridium complex*	30.4
60. *Oligosphaeridium pulcherrimum*	30.5
61. *Oligosphaeridium* sp. indet.	—
62. *Operculodinium bergmannii*	30.1-30.2
63. *Ophiobolus lapidaris*	30.6
64. *Palaeocystodinium australinum*	32.3, 46.3-46.4
65. *Palaeocystodinium golzowense*	32.1
66. *Palaeocystodinium granulatum*	32.2
67. *Palaeocystodinium* sp. indet.	—
68. *Palaeoperidinium pyrophorum*	32.4
69. *Palambages* Forma B. of Manum and Cookson, 1964	31.6
70. *Palambages morulosa*	31.4
71. *Palambages* Type I	32.5
72. *Paralecaniella indentata*	31.5
73. *Pareodinia* sp. indet.	—
74. *Phelodinium boldii* sp. nov.	33.1, 33.4
75. *Phthanoperidinium echinatum*	31.3
76. *Pterospermella australiensis*	31.1
77. *Selenopemphix nephroides*	31.2
78. *Senegalinium ?asymmetricum*	34.4, 42.1
79. *Spinidinium colemanii* sp. nov	36.1-36.2, 39.2
80. *Spinidinium densispinatum*	33.2-33.3
81. *Spinidinium essoi*	34.3
82. *Spinidinium lanterna*	34.1-34.2
83. *Spinidinium luciae* sp. nov.	35.1-35.3, 38.1-38.5, 39.4
84. *Spinidinium macmurdoense*	36.6, 44.1
85. *Spinidinium* sp. indet.	—
86. *Spiniferites multibrevis*	35.5
87. *Spiniferites ramosus*	35.4
88. *Spiniferites ramosus* subsp. *granomembranaceus*	36.7
89. *Spiniferites ramosus* subsp. *granosus*	36.4-36.5
90. *Spiniferites ramosus* subsp. *reticulatus*	35.6
91. *Spiniferites* sp. indet.	—
92. *Tanyosphaeridium xanthiopyxides*	37.2-37.3
93. *Thalassiphora pelagica*	37.1
94. *Trigonopyxidia ginella*	36.3
95. *Turbiosphaera filosa*	44.2
96. *Veryhachium* sp. indet.	—
97. *Vozzhennikovia apertura*	37.6-37.9, 43.3-43.4
98. *Xylochoarion* cf. *hacknessense*	37.5-37.6

Figure 15. Mesozoic and Paleogene dinocysts, Seymour Island, Antarctica. The abbreviations used in the figure descriptions: AS, apical surface; AV, apical view; ANS, antapical surface; ANV, antapical view; DS, dorsal surface; DV, dorsal view; EP, epicyst; H, height; HP, hypocyst; L, length; LLS, left lateral surface; LLV, left lateral view; OS, optical section; ODS, oblique dorsal surface; OVS, oblique ventral surface; RLS, right lateral surface; RLV, right lateral view; VS, ventral surface; VV, ventral view; W, width. 1-3, *Alisocysta circumtabulata* (Drugg) Stover and Evitt 1978; 1, the parasulcal area shows the parasulcal notch in the apical archeopyle margin (VV, VS); 2, (VV, OS); 3, the $4'''$ paraplate and adjacent pandasutural areas (VV, DS), $L \times W$: 48×48 μm, 8497 W/7 (116.2×12.4, M21); 4-5, *Arachnodinium antarcticum* Wilson and Clowes 1982, 4, note perforations in the flat ribbon-like trabeculae and the apical operculum that is in place, 5, note the elliptical shape of the central body and the antapical projection, central body $L \times W$: 77×40 μm, process and trabecular network (orientation indeterminate), $L \times W$: 90×102 μm, 8500 W/10 (119.0×6.5, F24); 6, *Areosphaeridium diktyoplokus* (Klumpp) Eaton 1971, (DV, DS), overall $L \times W$: 102×102 μm; central body $L \times W$: 61×54 μm, 8476 W/10 (120.8×17.2, S26); 7, *Aptea securigera* Davey and Verdier 1974 (VV, DS), $L \times W$: 80×111 μm, 8451E/1 (114.2×13.7, P19).

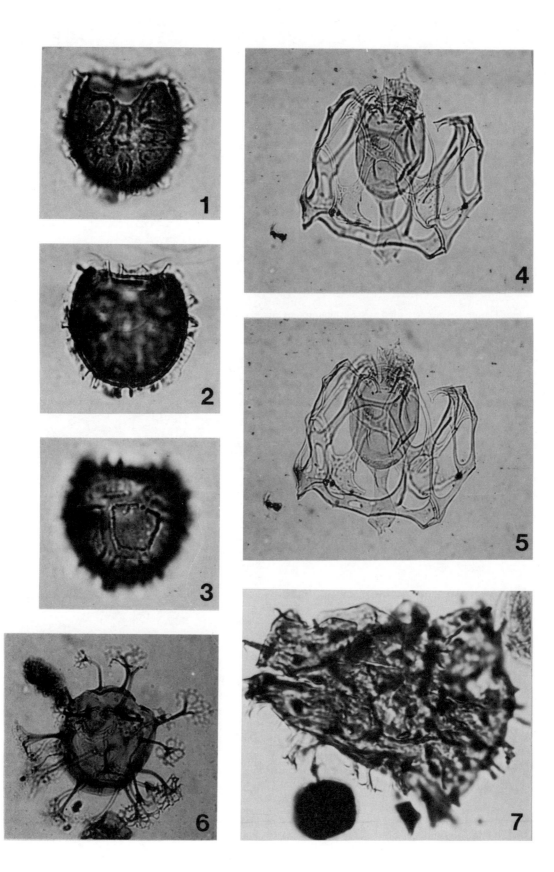

Figure 16. Paleogene dinocysts, Seymour Island, Antarctica. 1-2, *Deflandrea dartmooria* (Cookson and Eisenack) Lindgren 1984, 1, intratabular grana evident below archeopyle (VV, DS), 2, intratabular grana and faint parasultural lines mark ventral paratabulation (VV, VS), L × W: 134 × 70 μm, 8495 W/3 (122.5 × 6.0, F28); 3, *Deflandrea boloniensis* (Riegel) comb. nov., a large hypocoel connects the antapical horns (DV, DS), L × W: 112 × 78 μm, 8498 W/2 (111.1 × 11.8 M16); 4, *Batiacasphaera explanata* (Bujak) Islam 1983 (VV, VS), L × W: 95 × 87 μm, 8503 W/4 (107.8 × 15.3, Q13). (Abbreviations as defined for Fig. 15.)

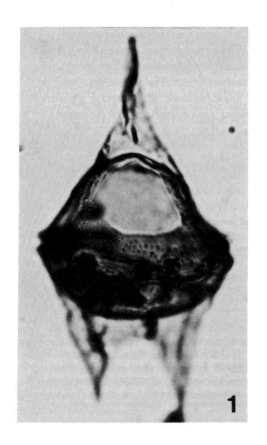

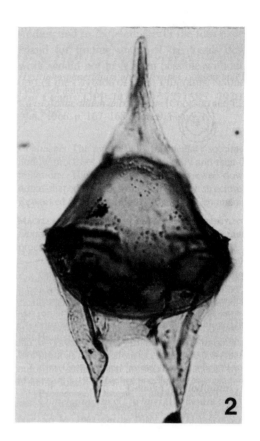

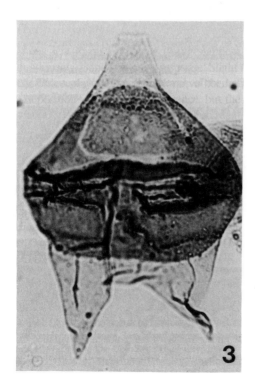

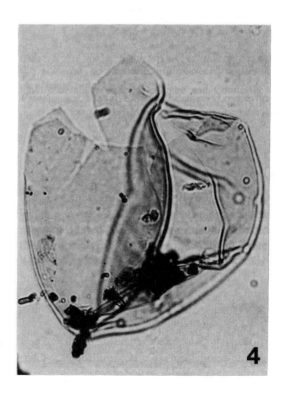

Figure 17. Mesozoic and Paleogene microplankton, Seymour Island, Antarctica. 1, 3, *Comasphaeridium cometes* (Valensi) De Coninck 1969, L × W: 15 × 20 μm, spines 4 μm long, 8499 W/3 (120.7 × 19.3, U26); 2, *Cleistosphaeridium ancoriferum* (Cookson and Eisenack) Davey et al., 1966 (orientation indeterminate), diameter 46 μm, processes 2 to 4 μm, 8451E/1 (115.3 × 14.4, P20); 4, *Deflandrea speciosa* (Alberti) Lindgren 1984 (DV, VS), large archeopyle and intratabular grana are evident, L × W: 155 × 95 μm, 8495 W/2 (123.4 × 12.3, M29); 5, *Deflandrea striata* (Drugg) Lindgren 1984, epicyst partially collapsed into the endocyst (DV, DS), L × W: 116 × 78 μm, 8464 W/5 (106.9 × 17.3, S22); 6, 8, *Cerebrocysta bartonensis* Bujak 1980 (orientation indeterminate), L × W: 31 × 31 μm, 8498 W/2 (104.4 × 6.9, G9); 7, *Cometodinium* cf. ?*whitei* (Deflandre and Courteville) Stover and Evitt 1978, diameter 27 μm, 8509 W/8 (114.6 × 18.8, U20). (Abbreviations as defined for Fig. 15.)

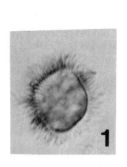

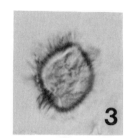

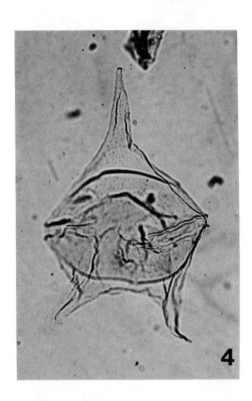

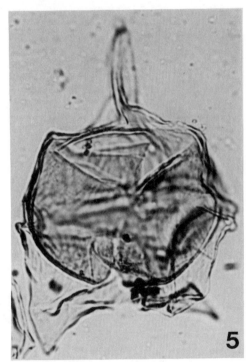

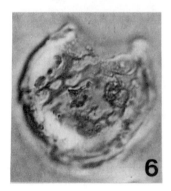

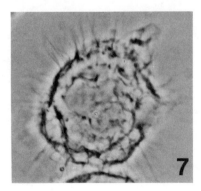

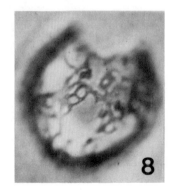

Figure 18. Mesozoic and Paleogene dinocysts, Seymour Island, Antarctica. 1, *Cribroperidinium muder-ongense* (Cookson and Eisenack) Davey 1969 (VV, VS), L × W: 136 × 87 μm, 8463E/1 (131.0 × 2.7, B37); 2, *Cyclopsiella trematophora* (Cookson and Eisenack 1967) Lentin and Williams 1977, L × W: 107 × 80 μm, 8511 W/4 (120.3 × 8.1, H26); 3-4, *Cribroperidinium edwardsii* (Cookson and Eisenack) Davey 1969, 3, (DV, DS) shows large, gapping precingular archeopyle, 4, (DV, VS), L × W: 138 × 136 μm, 8508 E/2 (124.7 × 8.3; H30); 5, *Cyclonephelium distinctum* Deflandre and Cookson 1955 (VV, VS), note offset parasulcal notch, L × W: 104 × 102 μm, 8503 W/5 (119.0 × 11.4, L24). (Abbreviations as defined for Fig. 15.)

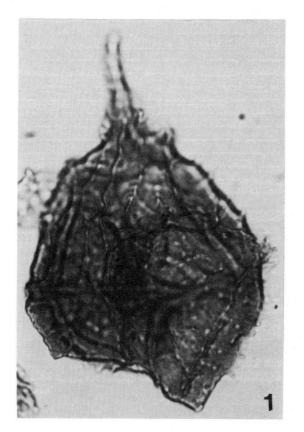

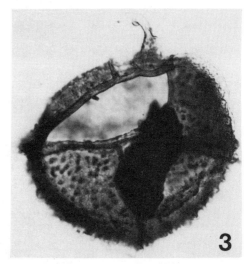

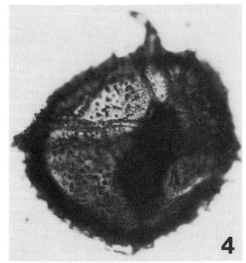

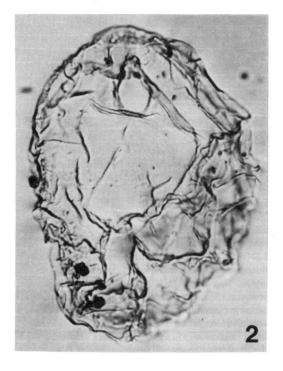

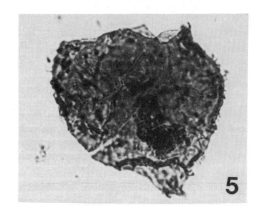

Figure 19. Eocene dinocysts, Seymour Island, Antarctica. 1-2, *Deflandrea antarctica* Wilson 1967, transitional form between *Deflandrea antarctica* Types I and II, granulation well developed, but differentiation into paraplate areas is poorly developed, 1, (DV, DS), 2, (DV, VS), L × W: 120 × 88 μm, 8465 E/1 (132.2 × 18.8, U38); 3-4, *Deflandrea antarctica* Wilson 1967 (Type I), note stout apical horn, broad shoulders on the epipericysts, protruding paracingulum and the posteriorly convergent hypopericyst margins, 3, (DV, VS), 4, (DV, DS), L × W: 127.5 × 90.1 μm, 8506 E/3 (122.7 × 15.9, R28). (Abbreviations as defined for Fig. 15.)

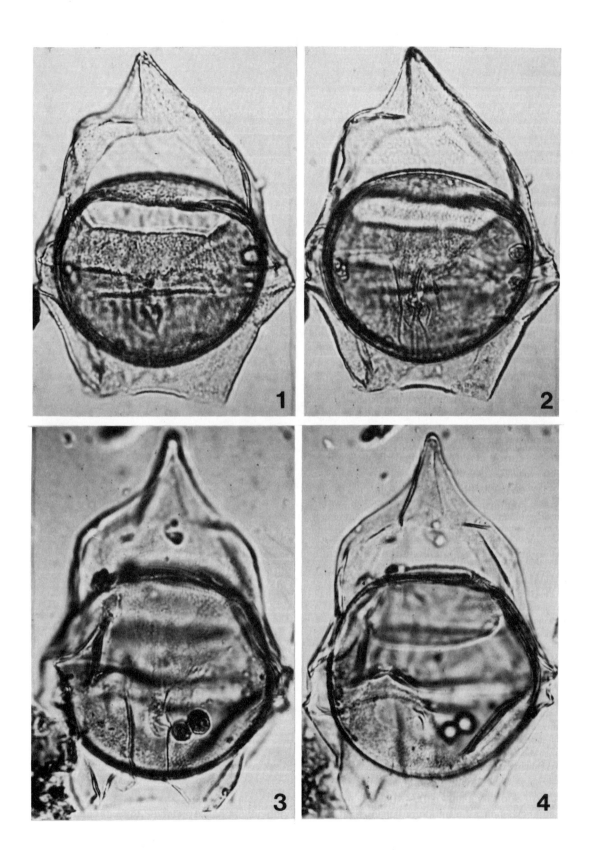

Figure 20. Dinocysts from the Eocene of Seymour Island, Antarctica. 1, *Deflandrea oebisfeldensis* Alberti sensu Cookson and Cranwell, 1967 (VV, VS), L × W: 128 × 94 μm, 8474 W/5, (115.2 × 10.9, K20); 2, *Deflandrea cygniformis* Pöthe de Baldis 1966, note the minor protrusion of the paracingulum and the comma-shaped flagellar sear (DV, DS), L × W: 182 × 163 μm, 8479 E/1 (132.3 × 1.9, A38); 3-4, *Deflandrea cygniformis* Pöthe de Baldis 1966, note that only the ridges project beyond the dinocyst outline in the paracingular area, 3, (VV, DS), 4, (DV, DS), L × W: 204 × 90 μm, 8455 E/3, (105.5 × 19.4, U21); 5, *Deflandrea cygniformis* Pöthe de Baldis 1966, the well-developed perihypocyst exhibits moderately well-developed paracingular projections (DV, DS), L × W: 180 × 92 μm, 8455 E/3 (115.4 × 19.8, V21). (Abbreviations as defined for Fig. 15.)

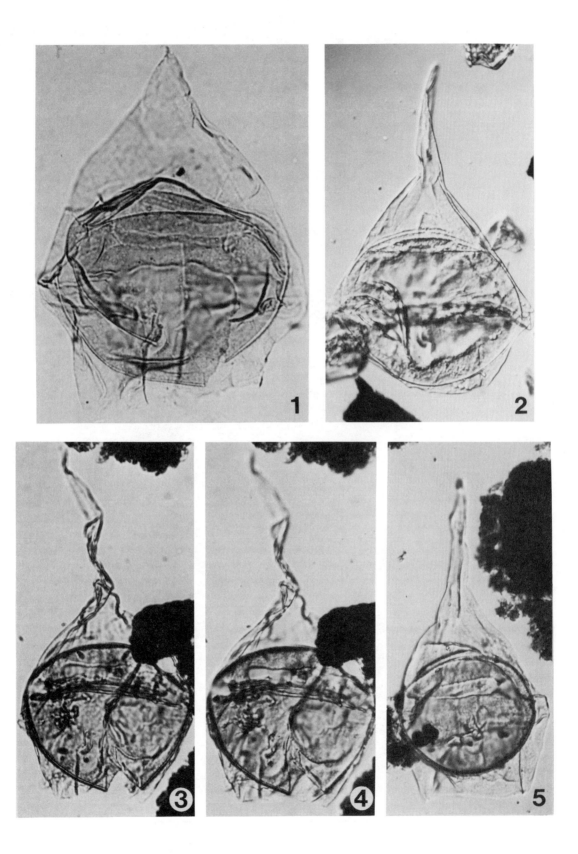

Figure 21. Dinocysts from the Eocene of Seymour Island, Antarctica. 1-2, *Deflandrea cygniformis* Pöthe de Baldis 1966, note scabbrate surface of the periphragm, well-developed perihypocyst, paracingular projections, comma-shaped flagellar scar and large intercalary archeopyle, 1, (VV, DS), 2, (VV, DS), L × W: 182 × 104 µm, 8479 E/3 (121.3 × 10.7, K27); 3-4, *Deflandrea cygniformis* Pöthe de Baldis 1966, note extreme development of the paracingular projections and wide separation of the endohypocyst and the perihypocyst, 3, (VV, VS), 4, (VV, DS), L × W: 196 × 128 µm, 8479 E/1 (128.5 × 19.8, V34). (Abbreviations as defined for Fig. 15.)

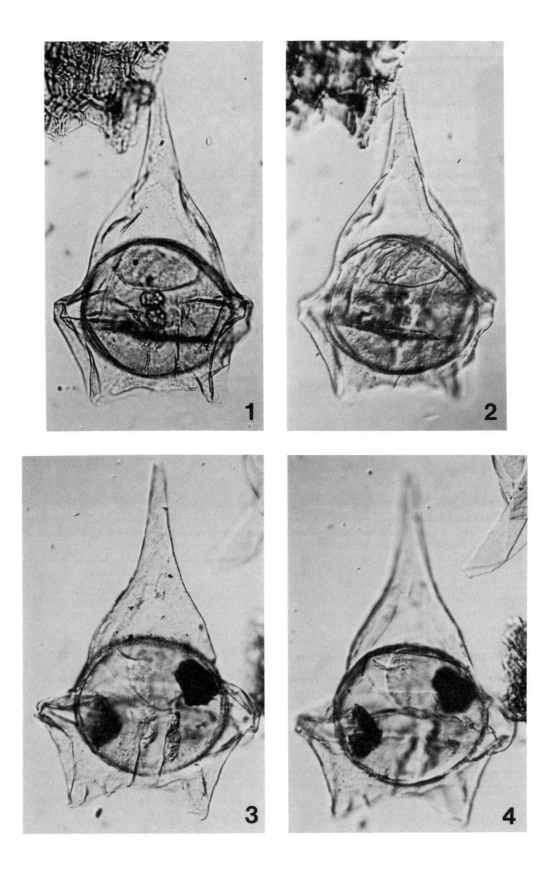

Figure 22. Eocene dinocysts, Seymour Island, Antarctica. 1, *Deflandrea* cf. *phosphoritica* Eisenack 1938, note flagellar scar, large 2a archeopyle and well-developed paracingular projections (VV, VS), L × W: 121 × 88 μm, 8463 E/1 (116.5 × 12.4, M22); 2-4, *Deflandrea webbii* sp. nov. holotype, note pandasutural striae on the pericyst and flagellar scar, 2, (DV, DS), 3, (DV, OS), 4, (DV, VS), L × W: 105 × 77 μm, 8511 E/1 (117.8 × 7.9, G23). (Abbreviations as defined for Fig. 15.)

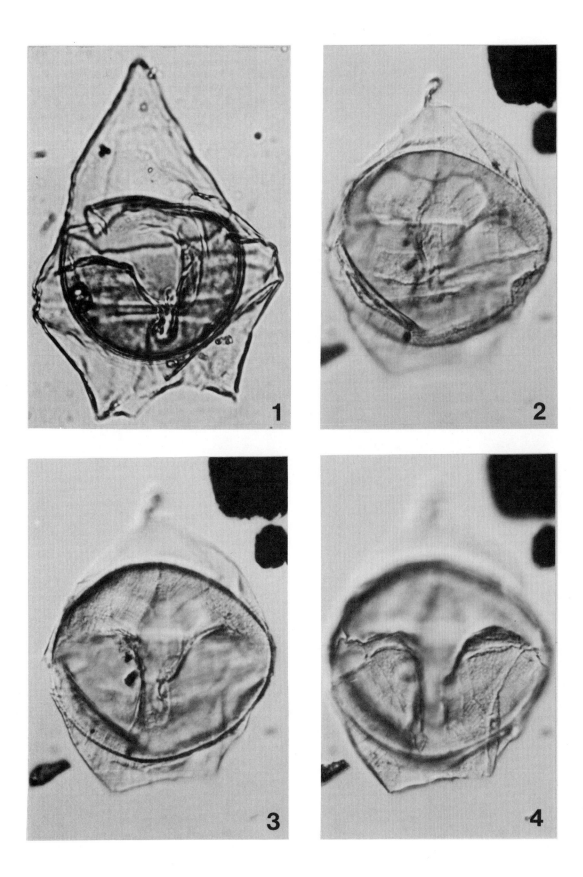

Figure 23. Mesozoic and Eocene dinocysts from Paleogene deposits, Seymour Island, Antarctica. 1, *Diconodinium cristatum* Cookson and Eisenack 1974 emend. Morgan 1977, note distinctive antapical "keel," (DV, DS), L × W: 87 × 60 μm, 8476 E/5 (117.5 × 3.6, C23); 2, *Dinogymnium* cf. *curvatum* (Vozzhennikovia) Lentin and Williams 1973 (DV, VS), L × W: 73 × 36 μm, 8476 W/2 (126.1 × 18.6, U32); 3, *Diconodinium multispinulum* (Deflandre and Cookson) Eisenack and Cookson emend. Morgan 1977 (DV, DS), L × W: 92 × 54 μm, 8443 E/1, (131.5 × 12.3, M37); 4-5, *Deflandrea webbii* sp. nov., 4 (DV, DS), 5, (DV, VS), L × W: 90 × 73 μm, 8511 E/1 (120.9 × 11.8, L26). (Abbreviations as defined for Fig. 15.)

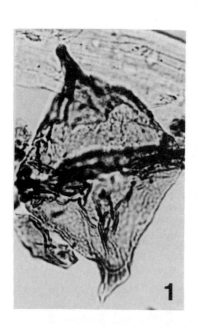

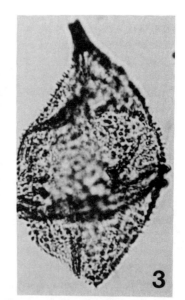
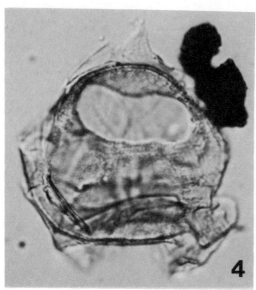
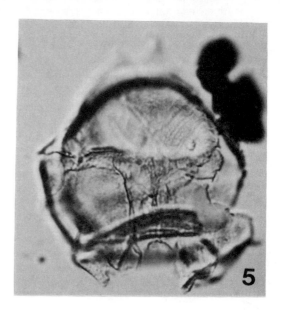

Figure 24. Dinocysts from Eocene deposits on Seymour Island, Antarctica. 1-3, *Enigmadinium cylindro-floriferum* sp. nov., note branching terminations on hollow processes and folds connecting the bases of some processes, 1, holotype (orientation indeterminate), central body L × W: 26 × 22 µm, P72106A01 (107.8 × 19.1, U13), 2, (orientation indeterminate), central body L × W: 24 × 17 µm, P72106A01 (110.3 × 19.2, U15), 3, (orientation indeterminate), central body L × W: 24 × 24 µm, P72106A01 (104.3 × 13.8, P9); 4-5, *Hystrichosphaeridium* cf. *astartes* Sanneman 1955 (note the pattern formed by the hollow process bases on the central body, the outline of the processes, and the open distal pore in each process), 4, (orientation indeterminate), 5, (orientation indeterminate), L × W: 42.5 × 42.5 µm, 8462 E/1 (124.5 × 4.6, D30); 6, *Eyrea nebulosa* Cookson and Eisenack 1971, periphragm diameter 87 µm, endophragm diameter 50 µm, 8512 W/6 (122.9 × 14.4, P28). (Abbreviations as defined for Fig. 15.)

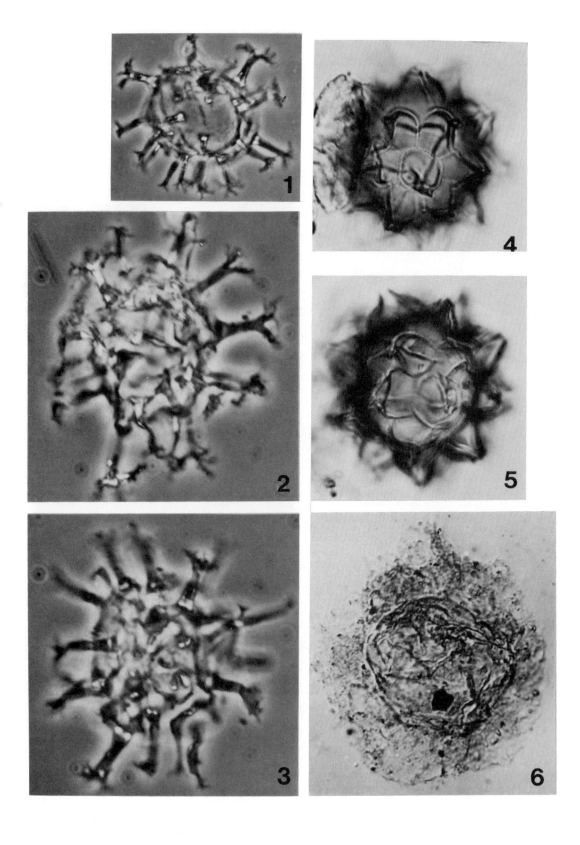

Figure 25. Dinocyst from Paleogene deposits, Seymour Island, Antarctica. 1-3, *Hystrichosphaeridium truswelliae* sp. nov., holotype, note hollow, flexible processes, fenestrate process terminations, the short apical processes, and the much longer hypocystal processes, 1, (LLV, LLS), 2, (LLV, OS), 3, (LLV, RLS), central body diameter: 41 μm, process length <29 μm, 8500 W/1, (116.8 × 21.8, ×22); 4, *Hystrichosphaeridium truswelliae* sp. nov., note operculum inside central body (VV, VS), central body diameter 39 μm, <30 μm, 8500 W/1 (116.8 × 21.8, ×22); 5-6 *Hystrichosphaeridium salpingophorum* Deflandre emend. Davey and Williams 1966, 5, (RLV, RLS), 6 (RLV, OS), central body diameter 36 μm, processes <19 μm, 8495 W/3 (118.5 × 19.3, U24). (Abbreviations as defined for Fig. 15.)

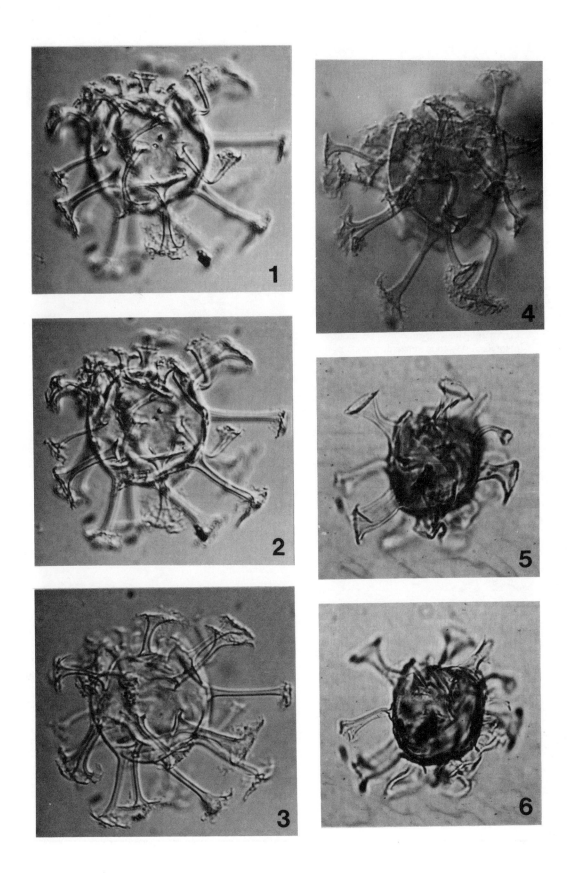

Figure 26. Paleogene dinocysts, Seymour Island, Antarctica. 1-2, *Impletosphaeridium ?lorum* sp. nov., holotype (orientation indeterminate), note flexible, accuminate processes, L × W: 31 × 26 μm, processes < 10μm long, 8496 W/8 (114.2 × 20.7, W19); 3-4, *Impagidinium victorianum* (Cookson and Eisenack) Stover and Evitt 1978, 3, (DV, DS), 4, (DV, DS), L × W: 70 × 77 μm, 8512 W/10 (120.3 × 9.0, J27); 5-6, *Impagidinium maculatum* (Cookson and Eisenack) Stover and Evitt 1978, 5, (LLV, OS), 6, (LLV, DS), L × W: 48 × 43 μm, 8490 W/8 (124.1 × 11.4, L30); 7-9 *Hystrichosphaeridium tubiferum* subsp. *brevispinium* (Davey and Williams) Lentin and Williams 1973, 7, (VV, VS), 8, (VV, OS), 9, (VV, DS), central body diameter 37 μm, processes <11 μm, 8503 W/5 (114.9 × 13.6, N20). (Abbreviations as defined for Fig. 15.)

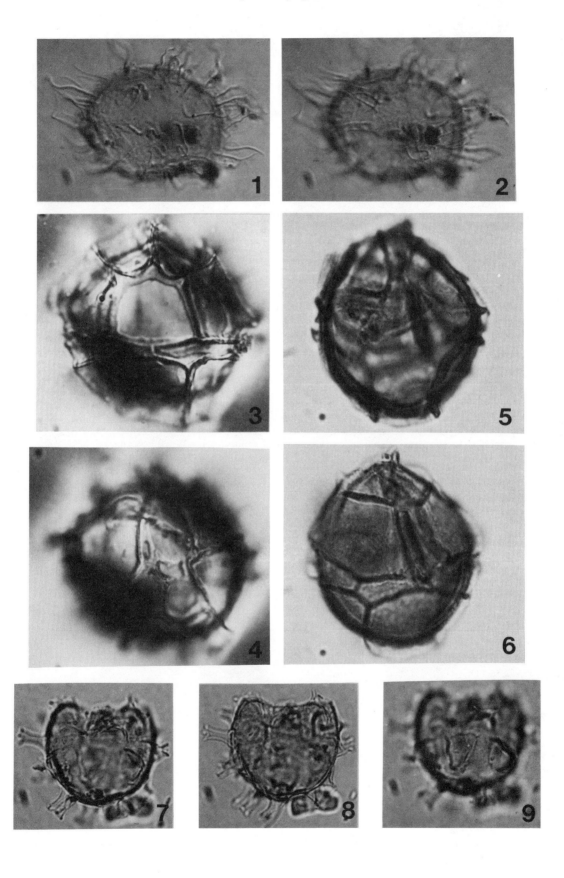

Figure 27. Dinocysts from the Paleogene deposits of Seymour Island, Antarctica. 1-3, *Impletosphaeridium* sp. B., 1, note apical (?) archeopyle (AV?, AS), 2, (AV?, OS), 3, (AV?, ANS), central body diameter 20 μm, processes <6 μm, 8511 E/1 (124.8 × 21.7, ×30); 4, *Isabelidinium pellucidum* (Deflandre and Cookson) Lentin and Williams 1977 (VV, DS), L × W: 116 × 73 μm, 8507 E/2 (126.6 × 5.1, E32); 5-7, *Kallosphaeridium* cf. *capulatum* Stover 1977, note attached operculum inside central body, 5, (DV, VS), 6, (DV, OS), 7, (DV, DS), L × W: 29 × 31 μm, 4898 E/7 (115.7 × 6.9, G21); 8, *Impletosphaeridium ligospinosum* De Coninck 1969 (orientation indeterminate), central body diameter 27 μm, processes <10 μm, P72106A01 (111.3 × 18.8, U16); 9, *Impletosphaeridium* sp. B., holotype, note granular central body (orientation indeterminate), L × W: 20 × 19 μm, spines <6 μm, 8500 W/6 (130.5 × 14.1, P36); 10, *Impletosphaeridium clavus* sp. nov. (orientation indeterminate), central body diameter 29 μm, processes <9 μm, P72106A01 (120.4 × 11.6, L26); 11, 13, *Impletosphaeridium clavus* sp. nov., holotype, (orientation indeterminate), note processes with expanded capitate terminations, L × W: 30 × 24 μm, processes <10 μm, 8500 W/3 (112.0 × 7.3, S17); 12, *Leiofusa jurassica* Cookson and Eisenack 1958, central body L × W: 24 × 12 μm, projections <16 μm, 8502 W/2 (113.3 × 10.0, K18). (Abbreviations as defined for Fig. 15.)

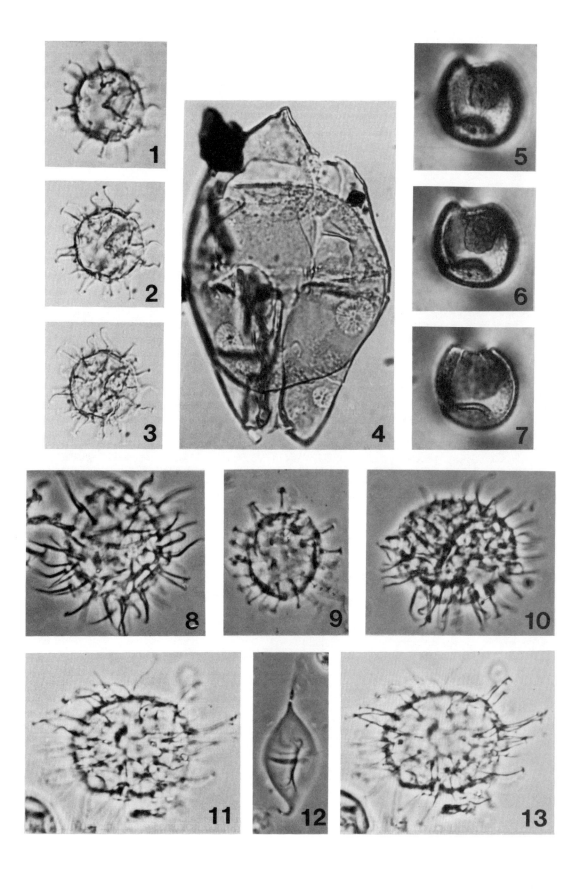

Figure 28. Dinocysts from the Paleogene deposits of Seymour Island, Antarctica. 1-2, *Octodinium askiniae* gen. nov., sp. nov., 1, note archeopyle outline with distinctly sloping portions of the upper right and lower right margin (DV, DS) L × W: 61 × 34 μm, P72105A01 (112.9 × 4.6, D18); 2, note the archeopyle margin and the paracingulum (VV, VS), L × W: 88 × 30 μm, 8470 M/1 (103.4 × 9.2, J8); 3, *Lejeunecysta fallax* (Morgenroth) Artzner and Dörhöfer 1978; note solid antapical horn tips and the archeopyle outline (VV, DS), L × W: 111 × 116 μm, 8477 W/7 (126.2 × 14.2, P32); 4, *Octodinium askiniae* gen. nov., sp. nov.; note paracingulum, thickenings at the horn bases and the widely spaced antapical horns; (DV, DS), L ×n W: 88 × 32, 8470 M/1 (97.4 × 13.0, N2); 5, *Lejeunecysta hyalina* (Gerlach) Artzner and Dörhöfer 1978; note the well-developed paracingular folds and the thin autophragm (VV, VS), L × W: 114 × 104 μm, 8477 W/14 (101.5 × 9.7, K6), 6 *Micrhystridium* sp. A, L × W: 32 × 26 μm, processes <8 μm long, 8457 W/9 (110.5 × 15.2, Q15), 7-8, *Manumiella druggii* (Stover) Bujak and Davies 1983, 7, (DV, DS), L × W: 94 × 94 μm, 8451 E/1 (115.4 × 4.4, D21); 8, (VV, DS) L × W: 94 × 80, 8451 E/1 (115.1 × 4.9, E20). (Abbreviations as defined for Fig. 15.)

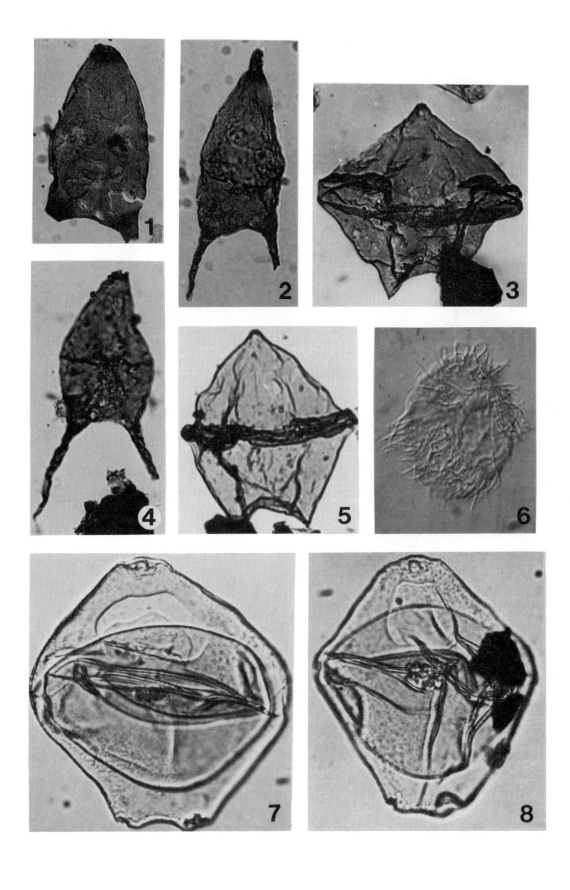

Figure 29. Dinocysts from the Paleogene deposits of Seymour Island, Antarctica. 1-7, *Octodinium askiniae* gen. nov., sp. nov., 1; note separation of the 2a and 4″ paraplates (DV, DS) L × W: 60 × 29 μm, P72105A01 (108.2 × 17.6, T13), 2-3, note outline of operculum and the sloping adcingular parasutures, 2, phase contrast, 3, interference contrast (DV, DS), L × W: 63 × 36 μm, P72105A01 (113.2 × 7.2, G18); 4, (DV, DS), L × W: 77 × 29 μm, 8470M/2 (102.6 × 4.6, D7); 5, holotype, note archeopyle outline (DV, DS), L × W: 121 × 30 μm, 8470 M/1 (108.5 × 15.0, Q13), 6, (VV, VS) L × W: 95 × 26 μm, 8470 M/1 (106.1 × 7.6 μm, G11); 7, (orientation indeterminate), L × W: 111 × 17 μm, P72105A01 (111.6 × 8.3, H17). (Abbreviations as defined for Fig. 15.)

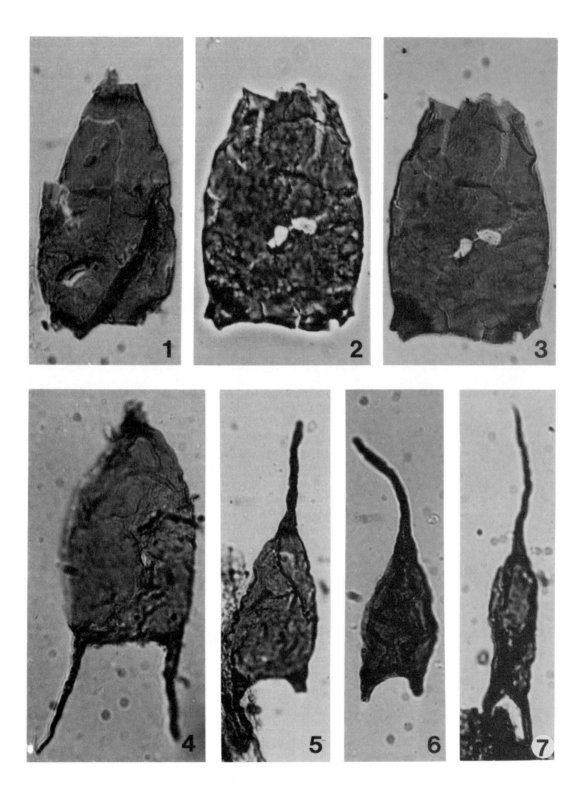

Figure 30. Dinocysts from the Paleogene deposits of Seymour Island, Antarctica. 1-2, *Operculodinium bergmannii* (Archangelsky) Stover and Evitt 1978, 1, (VV, DS), 2 (VV, VS), L × W: 44 × 41 μm, 8412W/10 (131.4 × 14.4, P37); 3, *Odontochitina spinosa* Wilson 1984 (VV, VS), L × W: 145 × 91 μm, 8476 E/5 (122.4 × 11.0, L28); 4, *Oligosphaeridium complex* (White) Davey and Williams 1966 (DV, VS), L × W: 37 × 27 μm, 8495 W/10 (113.5 × 5.3, E18); 5, *Oligosphaeridium pulcherrimum* (Deflandre and Cookson), Davey and Williams, 1966 (DV, DS), central body diameter 44 μm, processes <24 μm, 8479 W/6 (111.3 × 6.3, F16); 6, *Ophiobolus lapidaris* O. Wetzel emend. Evitt 1968, L × W: 31 × 20 μm, 8501 W/2 (114.3 × 10.9, L19). (Abbreviations as defined for Fig. 15.)

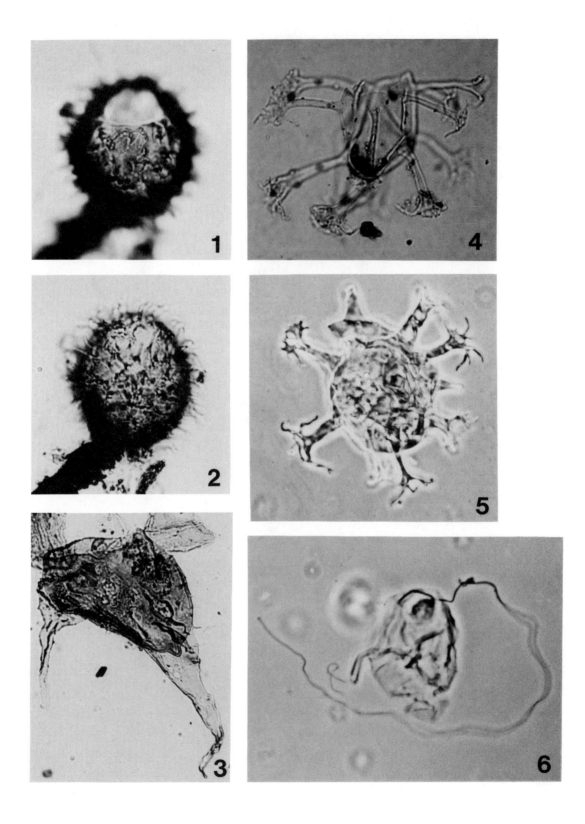

Figure 31. Palynomorphs from the Paleogene deposits of Seymour Island, Antarctica. 1, *Pterospermella australiensis* (Deflandre and Cookson) Eisenack 1972, outer body L × W: 54 × 43 μm; inner body L × W: 27 × 19 μm, 8478 W/4 (123.1 × 5.4, E28); 2, *Selenopemphix nephroides* Benedek emend. Bujak 1980 (polar orientation), H × W: 53 × 63 μm, 8477 W/14 (134.9 × 18.9, U41); 3, *Phthanoperidinium echinatum* Eaton 1976 (LLV, LLS), 49 × 46 μm, 8511 E/1, (119.3 × 4.3 D25); 4, *Palambages morulosa* O. Wetzel 1961, L × W: 162 × 131 μm, 8479 E/3 (135.0 × 4.5, H41); 5, *Paralecaniella indentata* (Deflandre and Cookson) Cookson and Eisenack emend. Elsik 1977, L × W: 51 × 53, 8470 M/2, (114.0 × 8.3 μm, H19); 6, *Palambages* Forma B of Manum and Cookson 1964, colony diameter: 66 μm, 8495 W/10 (129.2 × 8.2, H35). (Abbreviations as defined for Fig. 15.)

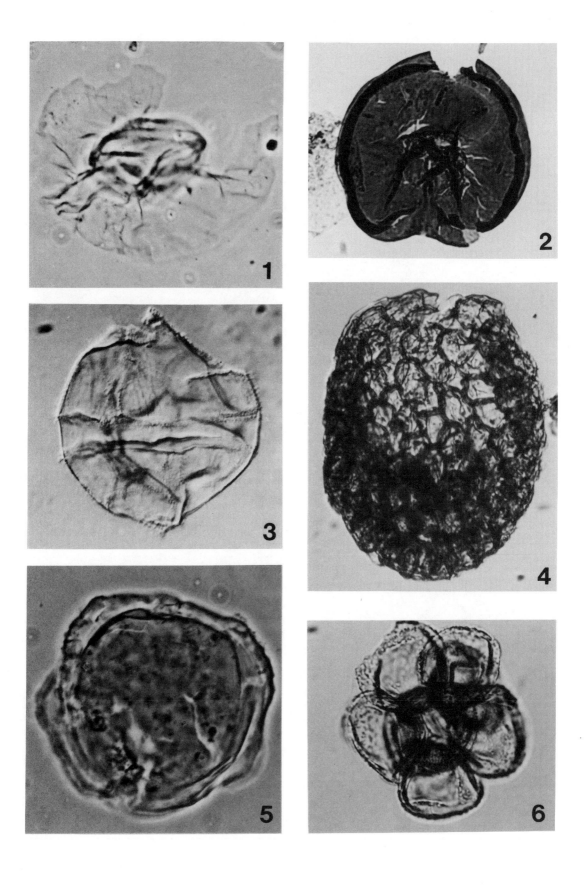

Figure 32. Palynomorphs from the Paleogene deposits of Seymour Island, Antarctica. 1, *Palaeocystodinium golzowense* Alberti 1961 (DV, DS), L × W: 112 × 34 μm, 8495 W/3 (113.8 × 9.7, J119); 2, *Palaeocystodinium granulatum* (Wilson) Lentin and Williams 1976, damaged specimen, note spinate periphragm (VV, DS), L × W: 128 × 43 μm, 8457 W/8 (120.7 × 11.9, M26); 3, *Palaeocystodinium australinum* (Cookson) Lentin and Williams, 1976 (VV, DS), L × W: 204 × 41 μm, 8498 W/3 (132.9 × 12.9, M39); 4, *Palaeoperidinium pyrophorum* (Ehrenberg) Sarjeant 1967 (VV, DS), L × W: 95 × 102 μm, 8498 W/2 (121.5 × 10.7, K27); 5, *Palambages* Type I, L × W: 95 × 77 μm, 8511 E/1 (123.7 × 3.1, B29). (Abbreviations as defined for Fig. 15.)

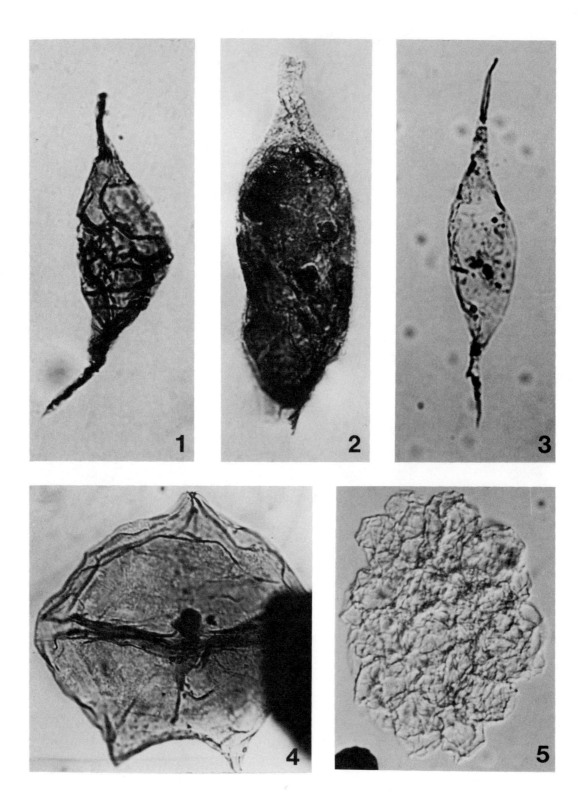

Figure 33. Paleogene dinocysts from Seymour Island, Antarctica. 1, 4, *Phelodinium boldii* sp. nov., holotype, 1, note flat-topped broad epicysts, narrow paracingulum, and restricted pericoel (VV, DS), 4, note granulate surface of the periphragm and the outline of the 2a archeopyle (VV, DS), L × W: 116 × 116 μm, 8476 W/10 (121.3 × 11.9, M27); 2-3, *Spinidinium densispinatum* Stanley 1965, 2, (VV, DS), 3, (VV, VS), L × W: 51 × 46 μm, 8493 W/6 (113.8 × 94.0, J19). (Abbreviations as defined for Fig. 15.)

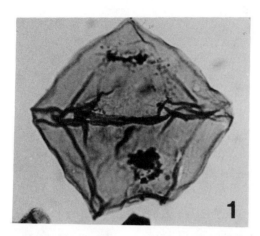

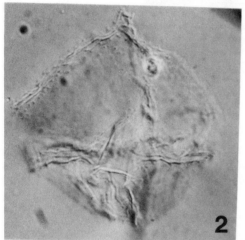

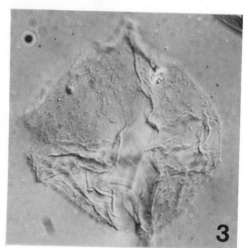

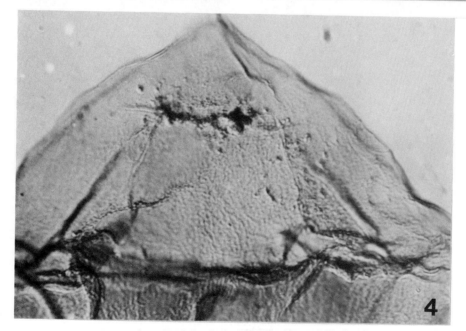

Figure 34. Paleogene dinocysts from Seymour Island, Antarctica. 1-2, *Spinidinium lanterna* Cookson and Eisenack 1970, note posteriorly attached operculum, 1, (DV, OS), 2, (DV, DS), L × W: 70 × 43 μm, 8499 W/3 (112.1 × 15.1, Q17); 3, *Spinidinium essoi* Cookson and Eisenack 1967 (DV, DS), L × W: 63 × 44 μm, 8468 E/1 (120.2 × 17.1, S25); 4, *Senegalinium ?asymmetricum* (Wilson) Stover and Evitt 1978 (DV, DS), L × W: 102 × 78 μm, 8511 E/1 (127.9 × 21.6 μm, X32). (Abbreviations as defined for Fig. 15.)

Paleogene dinoflagellates

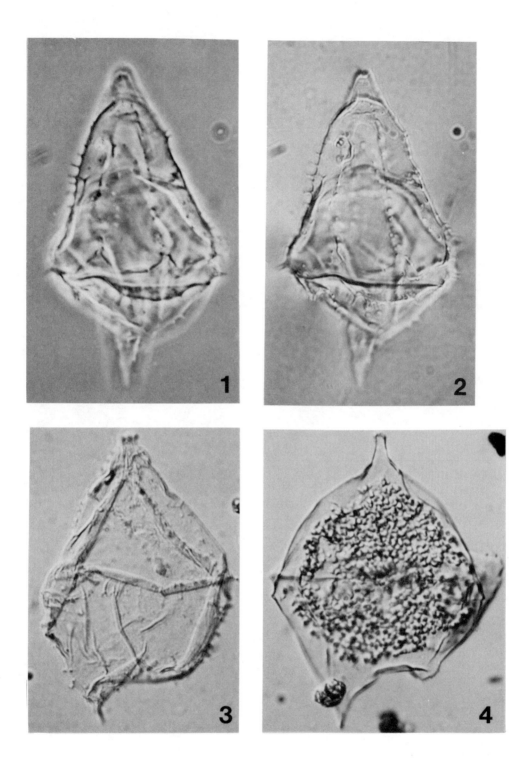

Figure 35. Dinocysts from the Paleogene deposits of Seymour Island, Antarctica. 1-3, *Spinidinium luciae* sp. nov., holotype, 1, (DV, DS), 2, (DV, OS), 3, (DV, VS), L × W: 87 × 76 μm, 8452 E/1 (118.3 × 11.4, L23); 4, *Spiniferites ramosus* (Ehrenberg) Loeblich and Loeblich 1966 (RLV, OS), central body diameter 43 μm, processes <9 μm, 8463 E/1, (132.0 × 13.9, P38); 5, *Spiniferites multibrevis* (Davey and Williams) Below 1982 (RLV, RLS), L × W: 60 × 54 μm, processes <12 μm, 8463 E/1 (129.5 × 12.0, M35); 6, *Spiniferites ramosus* subsp. *reticulatus* (Davey and Williams) Lentin and Williams 1973 (DV, DS), central body 95 × 87 μm, 8451 E/1 (124.5 × 4.1, D30). (Abbreviations as defined for Fig. 15.)

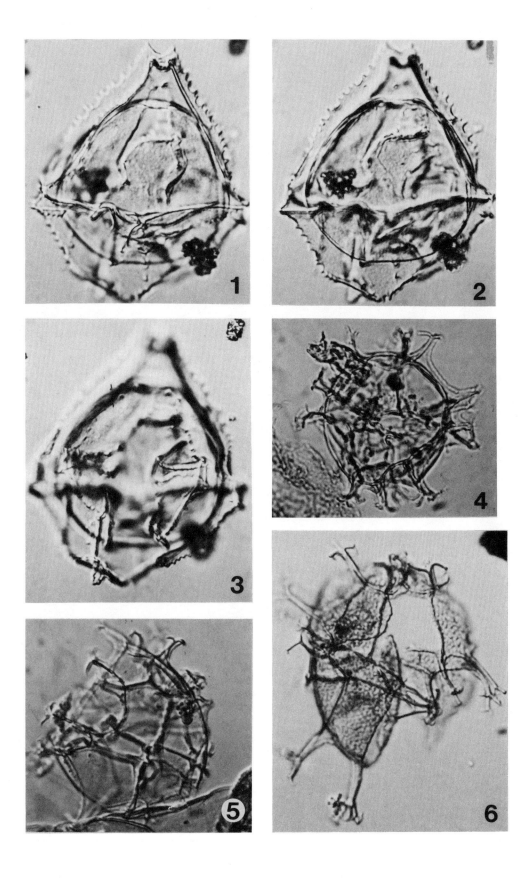

Figure 36. Dinocysts from the Paleogene deposits of Seymour Island, Antarctica. 1-2, *Spinidinium colemanii* sp. nov., holotype, note closely spaced capitate, penitabular and parasutural spines, and vestigial antapical horns, 9, (VV, VS), 10, (VV, OS), L × W: 58 × 46 μm, 8496 W/8 (134.6 × 12.5, M40); 3, *Trigonopyxidia ginella* (Cookson and Eisenack) Downie and Sarjeant 1965, note smooth "apical" archeopyle margin, (orientation indeterminate), L × W: 50 × 55 μm, 8497 W/10, (104.8 × 8.6, H20); 4-5, *Spiniferites ramosus* subsp. *granosus* (Davey and Williams) Lentin and Williams 1973, 4, (LLV, OS), note operculum inside endocyst, 5, (LLV, RLS), L × W: 53 × 44 μm, processes <17 μm, 8464 W/1 (119.9 μm 17.5, T25); 6, *Spiniferites macmurdoense* (Wilson) Lentin and Williams 1976 (DV, DS), note widely spaced capitate spines and the strong dorso-ventral cyst compression, L × W: 105 × 75 μm, 8457 W/9 (121.9 × 10.9, L27); 7, *Spiniferites ramosus* subsp. *granomembranaceus* (Davey and Williams) 1973 (LLV, LLS), L × W: 64.6 × 61.2 μm, 8494 W/1 (127.3 × 10.9, K33). (Abbreviations as defined for Fig. 15.)

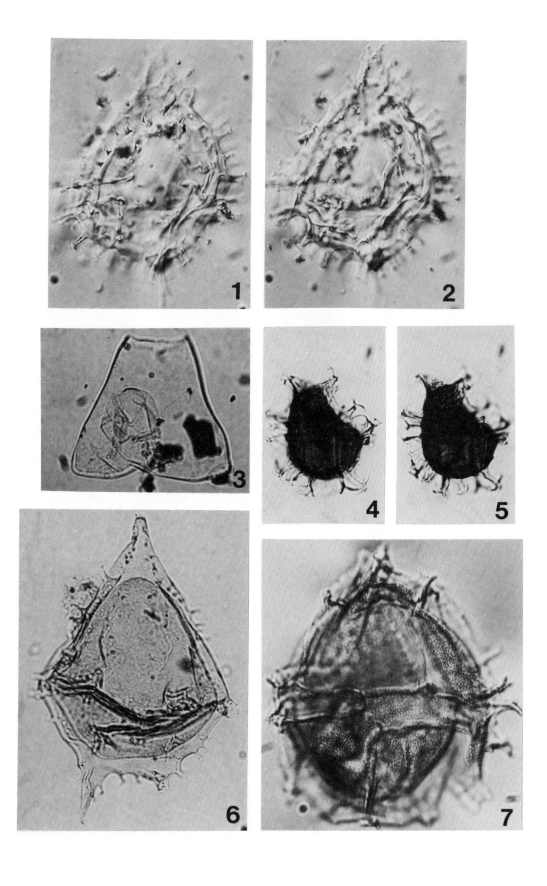

Figure 37. Dinocysts from the Paleogene deposits of Seymour Island, Antarctica. 1, *Thalassiphora pelagica* (DV, VS), endocyst diameter: 92 μm, pericyst diameter 187 μm, 8500 W/10 (107.2 × 5.2, E22), composite photograph; 2-3, *Tanyosphaeridium xanthiopyxides* (O. Wetzel) Stover and Evitt 1978 (orientation indeterminate), central body L × W: 29 × 20 μm, processes <16 μm, 8495 W/3 (122.1 × 13.7, P27); 4-5, *Xylochoarion* cf. *hacknessense* Erkman and Sarjeant 1978 (orientation indeterminate), L × W: 23 × 20 μm, processes <10 μm, 8467 E/1 (132.4 × 12.5, M38); 6-9, *Vozzhennikovia apertura* (Wilson) Lentin and Williams 1976, 6, paracingulum and parasulcus not delineated (VV, VS), L × W: 31 × 43 μm, 8463 W/3 (129.1 × 9.8, J35), 7-9, note clearly defined parasulcus and paracingulum, 7, (VV, DS), 8 (VV, OS), 9, (VV, VS), L × W: 43 × 39 μm, 8463 W/3 (119.1 × 9.8, K35). (Abbreviations as defined for Fig. 15.)

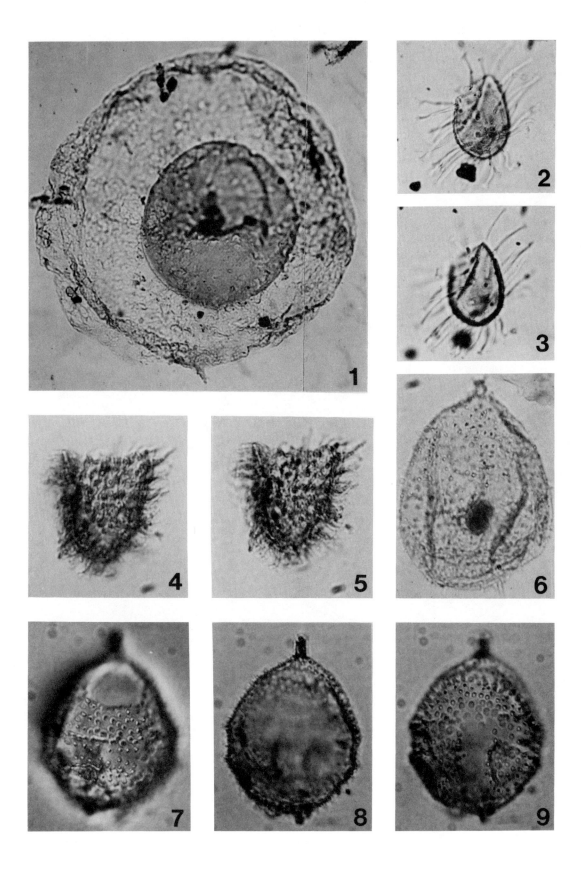

Figure 38. Eocene dinocysts from Seymour Island, Antarctica. All bar scales = 10 μm. 1-5, *Spinidinium luciae* sp. nov., 1-3, SEM stub M-50, specimen 11, ×1,200, 1, (VV, VS), 2, (AV, EP), note periarcheopyle and endoarcheopyle, 3, (AV, OVS); 4-5, SEM stub M50, specimen 12, ×1,200, 4, (RLV, RLS) note paracingular crests curving into the parasulcus on the right side of the photograph, 5, (AV, EP). (Abbreviations as defined for Fig. 15.)

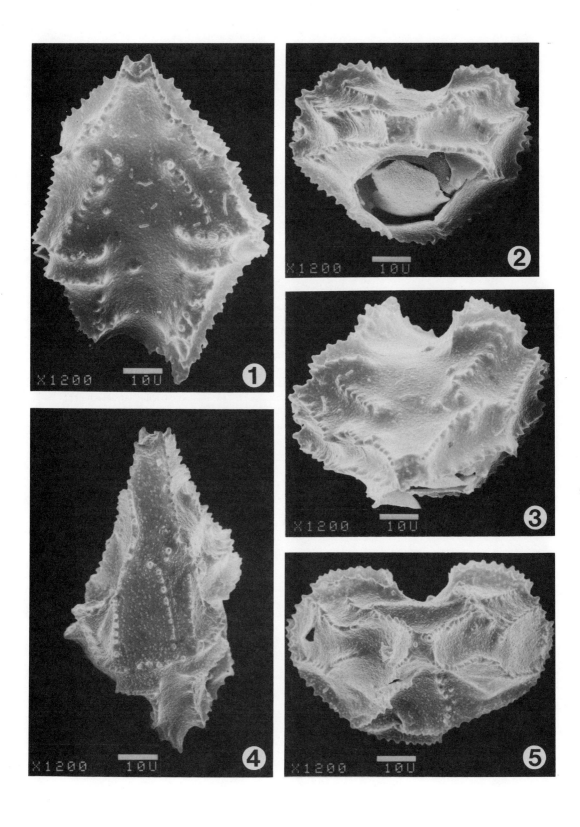

Figure 39. Dinocysts from the Paleogene deposits from Seymour Island, Antarctica. All scale bars = 10 μm. 1, *Hystrichosphaeridium truswelliae* sp. nov., note flexible, hollow processes and the perforate process terminations (orientation indeterminate), ×1,200; SEM stub M13, specimen 13; 2, *Spinidinium colemanii* sp. nov. (DV, DS), note penetabular capitate spines on the epicyst and the hypocyst and parasutural spines on the paracingulum; 3, *Manumiella druggii* (Stover) Bujak and Davies 1983 (DV, DS), ×1,200 SEM stub M19, specimen 27; 4, *Spinidinium luciae* sp. nov., note adcingular detachment of the operculum, probably by tearing (DV, DS), ×1,100, SEM stub M50, specimen 10. (Abbreviations as defined for Fig. 15.)

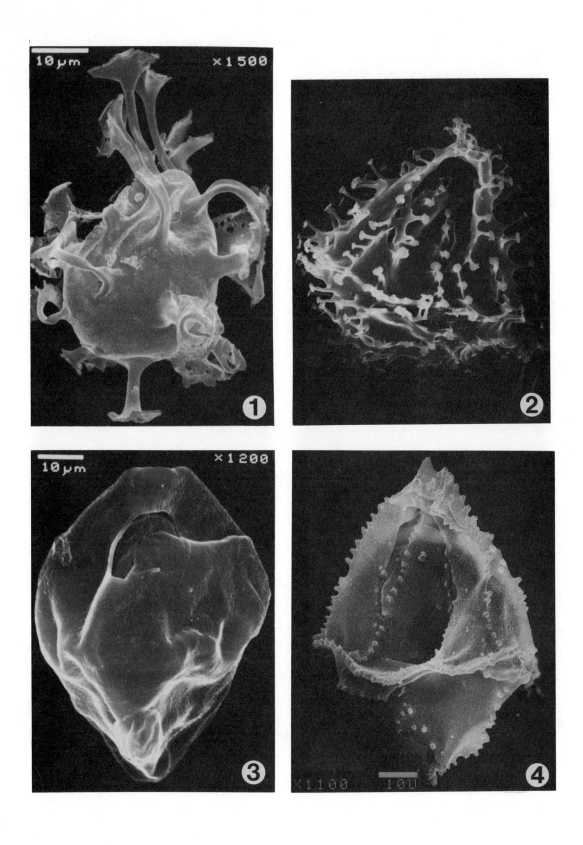

Figure 40. Dinocyst from the Eocene deposits of Seymour Island, Antarctica. All scale bars = 10 µm. 1-4, SEM stub M48, specimen 1, *Deflandrea webbii* sp. nov., 1, (ANV, HP), note parasulcal groove and complex flagellar scar, ×900, 2, (VV, VS), note flagellar scar and pandasutural striae, ×900, 3, note the parasutural paraplate projecting and protecting the area of the flagellar scar, ×3,000, 4, the complex flagellar scar, ×5,000. (Abbreviations as defined for Fig. 15.)

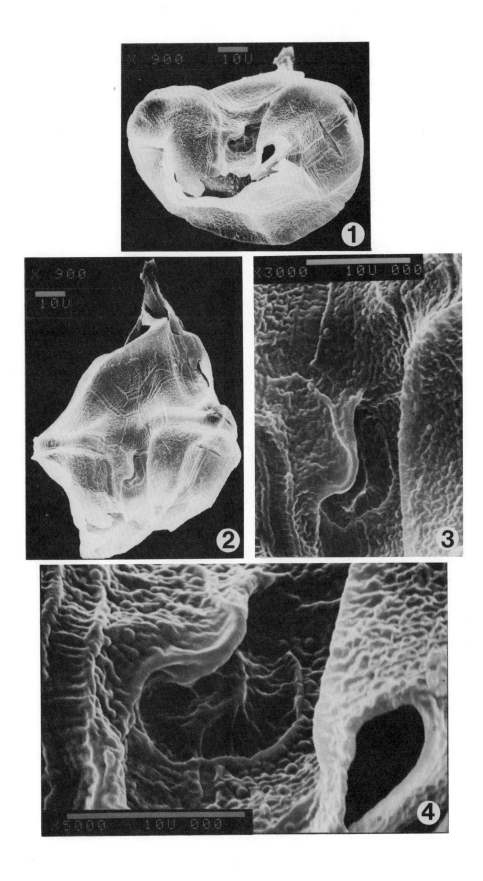

Figure 41. Dinocyst from the Eocene of Seymour Island, Antarctica. All scale bars = 10 μm.) 1-4, *Deflandrea webbii* sp. nov., SEM stub M47, specimen 2, 1, (RLV, RLS), note pandasutural striae, especially that below the surface of the partially removed periphragm (arrow) of the 5″ paraplate, ×1,200, 2, (LLV, LLS), ×1,200, 3, (AV, EP), ×1,200, 4, (ANV, HP), ×1,200. (Abbreviations as defined for Fig. 15.)

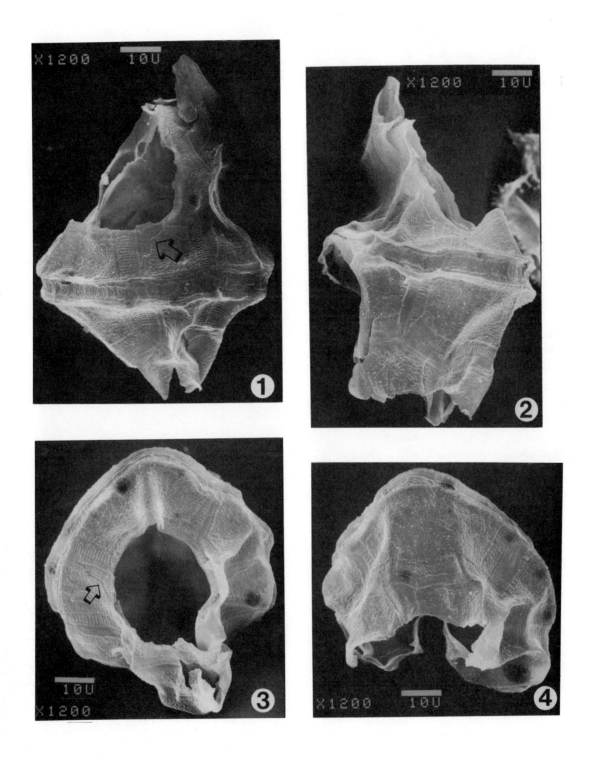

Figure 42. Dinocysts from the Eocene deposits of Seymour Island, Antarctica. 1, *Senegalinium ?asym-metricum* (Wilson) Stover and Evitt 1978 (DV, DS), ×1,000, scale bar = 10 μm, SEM stub M47, specimen 2; 2, *Deflandrea webbii* sp. nov. (RLV, DS), ×1,200, scale bar = 10 μm, SEM stub M47, specimen 1; 3-5, *Deflandrea cygniformis* Pöthe de Baldis 1966, SEM stub M44, specimen 2, scale bars = 100 μm, ×630, 3, (AV, ODS), 4, (ANV, HP), 5 (DV, DS). (Abbreviations as defined for Fig. 15.)

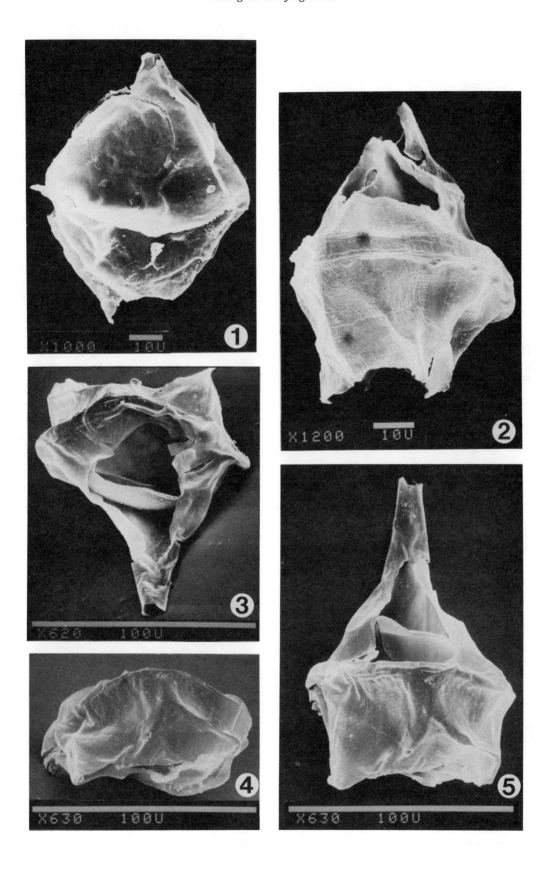

Figure 43. Eocene dinocysts from Seymour Island, Antarctica. 1, *Arachnodinium antarcticum* Wilson and Clowes 1982 (orientation indeterminate), ×1,200, scale bar = 10 μm, SEM stub M13, specimen 19; 2, *Deflandrea cygniformis* Pöthe de Baldis 1966 (RLV, RLS), ×640×, scale bar = 100 μm, SEM stub M44, specimen 2; 3-4, *Vozzhennikovia apertura* (Wilson) Lentin and Williams 1976, 3, (VV, VS), ×1,800, scale bar = 10 μm, SEM stub M47, specimen 14; 4, note 3I intercalary archeopyle (DV, DS), ×1,900, scale bar = 10 μm, SEM stub M47, specimen 12. (Abbreviations as defined for Fig. 15.)

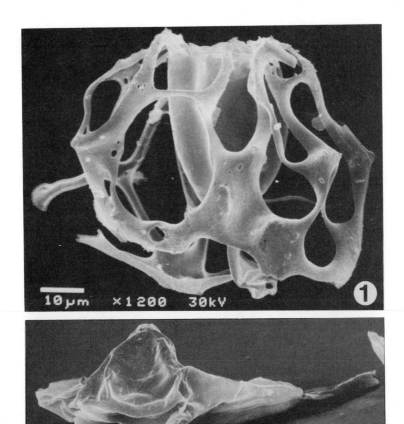

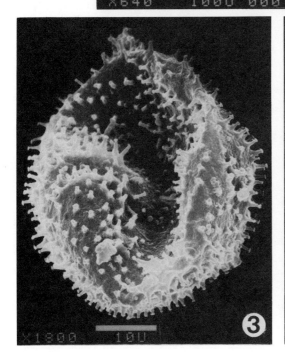

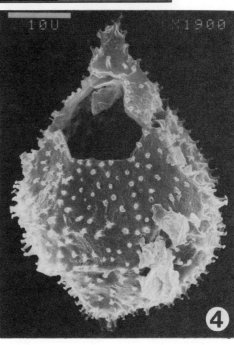

Figure 44. Eocene dinocysts from Seymour Island, Antarctica. 1, *Spinidinium macmurdoense* (Wilson) Lentin and Williams 1976, notice the sparse, widely spaced capitate spines (VV, VS), ×1,200, scale bar = 10 μm, SEM stub M45, specimen 22; 2, *Turbiosphaera filosa* (Wilson) Archangelsky 1969, note ledge-like paracingular processes (LLV, DS), ×1,200, scale bar = 10 μm, SEM stub M11, specimen 4; 3, *Deflandrea antarctica* Wilson 1967, Type II, note fine parasutural lines and paratabular clusters of ornamentation surrounded on three or more sides by pandasutural bands devoid of spines (RLV, RLS), ×1,200, scale bar = 10 μm, SEM stub M13, specimen 29; 4, *Deflandrea cygniformis* Pöthe de Baldis, 1966 (VV, VS), ×700, scale bar = 100 μm, SEM stub M44, specimen 5. (Abbreviations as defined for Fig. 15.)

Figure 45. Dinocysts from the Paleogene deposits of Seymour Island, Antarctica. 1-6, *Enigmadinium cylindrifloriferum* gen. nov., sp. nov., 1-3, SEM stub M45, specimen 15, 1 and 3, details of processes on the specimen in Figure 49.2, ×10,000, scale bar = 1 μm, 2, (orientation indeterminate), ×2,000, scale bar = 10 μm; 4, (orientation indeterminate), ×2,600, scale bar = 10 μm, SEM stub M45, specimen 2; 5-6, SEM stub M45, specimen 17; 5, (orientation indeterminate), ×2,400, scale bar = 10 μm, 6, detail of processes on specimen in Figure 49.5, ×8,000, scale bar = 10 μm. (Abbreviations as defined for Fig. 15.)

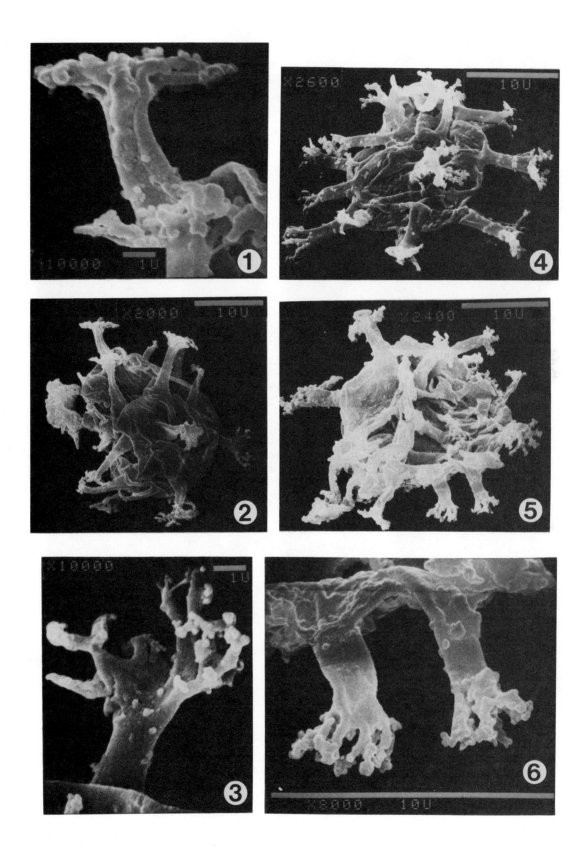

Figure 46. Dinocysts from the Paleogene deposits of Seymour Island or elsewhere. 1, Complex horn structure of dinocysts from Gabon that were assigned to *Svalbardella-Palaeocystodinium* by Malloy (1972). Sketch shows the "outer membrane" (periphragm), "inner capsule" (endocyst), the "apical boss," and the "apical extension" (from Malloy, 1972). 2, *Palaeocystodinium australinum* sensu Malloy (1972). Note the rotund aspect of the cyst and the small length-to-width ratio. Traced from Plate 1, Figure 17 of Malloy (1972). 3, *Palaeocystodinium australinum* Cookson 1965, traced from the illustration of the holotype (Plate 25, Fig. 4, Cookson, 1965). Note the long spindly shape of the cyst and the large length-to-width ratio of the cyst. Dimensions of the specimen are L × W: 270 × 54 μm; 4, simple horn structure of *Palaeocystodinium australinum* Cookson 1965. 5-10, *Andalusiella mauthei* Riegel emend. Riegel and Sarjeant 1982. 5, 8, The structure of the apical horn (from Riegel and Sarjeant, 1982). 6, 9, The structure of the antapical horns (from Riegel and Sarjeant, 1982). 7, 10, Sketch of the type specimen showing faint paratabulation, paracingulum, parsulcal groove, and flagellar scar.

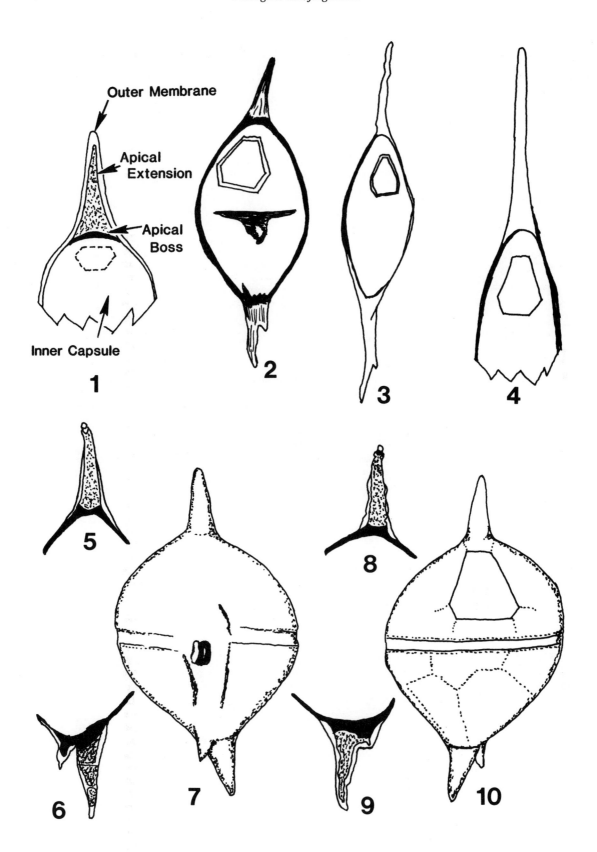

Figure 47. Dinocysts from the Paleogene deposits of Seymour Island, Antarctica. 1-5, *Octodinium askiniae* gen. nov., sp. nov, 1, The tracing of a broken specimen shows the outline of the eight-sided 2a intercalary archeopyle and the attached operculum. Stylized sketch of the 2a opercular piece is shown above the broken specimen. 2, Tracing of an essentially complete specimen of *O. askiniae* gen. nov., sp. nov. showing the outline of the 2a intercalary archeopyle and the free operacular piece. The stylized sketch of the opercular piece is shown above the tracing of the specimen. 3, Tracing of a broken specimen showing the remains of the 2a intercalary paraplate attached to the partially detached 4″ paraplate. The stylized sketch of the opercular pieces is shown above the broken specimen. 4, Tracing of a broken specimen shows the fully developed combination archeopyle with the free 2a intercalary and the adnate 4″ paraplates. Stylized sketch of the opercular pieces is shown above the broken specimen. 5, Composite reconstruction of *O. askiniae* based on light and scanning electron microscopy of hundreds of specimens. Approximate dimensions are 150×30 μm.

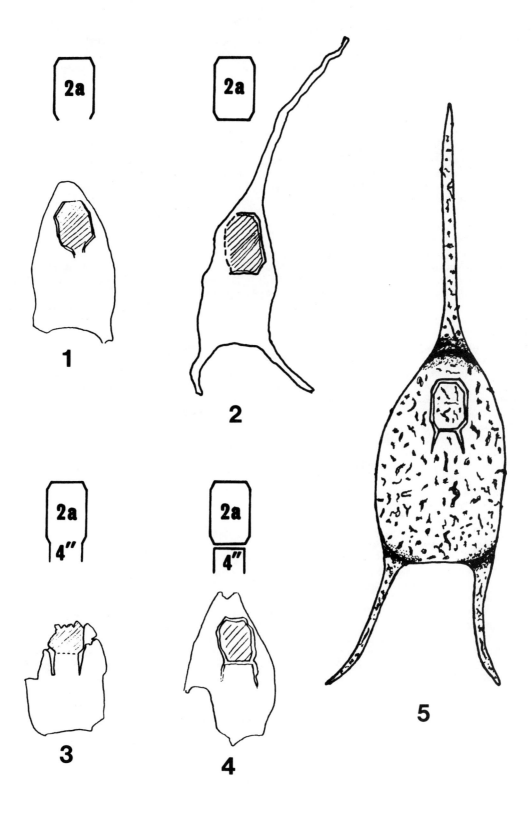

REFERENCES

ADIE, R. J. 1958. Geological investigations in the Falkland Islands Dependencies since 1940. Polar Record, 9(58):3–17.

ALBERTI, G. 1959. Zur Kenntnis der Gattung *Deflandrea* Eisenack (Dinoflag.) in der Kreide und im Alttertiär Nord-und Mittel deutschlands. Mitteilungen aus dem Geologischen Staatsinstitut in Hamburg, 28:93–105.

——. 1961. Zur Kenntnis mesozoischer und alttertiärer Dinoflagellaten und Hystrichosphaerideen von Nord-und Mitteldeutschland sowie einigen anderen europäischen Gebieten. Palaeontographica, Abt. A, 116:1–58.

ANDERSSON, J. G. 1906. On the geology of Graham Land. Uppsala University Geological Institute Bulletin, 7:19–71.

ARCHANGELSKY, S. 1968. Sobre el paleomicroplancton del Terciario inferior de Río Turbio, Provincia de Santa Cruz. Ameghiniana, 5:406–416.

——. 1969. Estudio del paleomicroplancton de la Formacíon Río Turbio (Eoceno), Provincia de Santa Cruz. Ameghiniana, 6:181–218.

——, AND A. FASOLO. 1971. Algunos elementos del paleomicroplancton del Terciario Inferior de Patagonia (Argentina y Chile). Revista del Museo de La Plata, Seccion Paleontologia, Nueva Serie, 6(36):1–18.

ASKIN, R. A., AND D. H. ELLIOT. 1982. Geologic implications of recycled Permian and Triassic palynomorphs in Tertiary rocks of Seymour Island, Antarctic Peninsula. Geology, 10:547–551.

AUFFRET, J. P., AND C. GRAUS—CAVAGNETTO. 1975. Les formations paléogènes sous-marines de le Manche orientale Données palynologiques. Bulletin Societe geologique de France, 7(17):641–655.

BARSS, M. S., AND B. CRILLEY. 1976. A mounting medium for palynological residues. Geological Survey of Canada Paper 76-1B:131–132.

——, AND G. L. WILLAMS. 1973. Palynology and nannofossil processing techniques. Geological Survey of Canada Paper 73-26:25.

——, J. P. BUJAK AND G. L. WILLIAMS. 1979. Palynological zonation and correlation of sixty-seven wells, Eastern Canada. Geological Survey of Canada Paper 78-24:1–118.

BATTEN, D. J. 1980. Aptian and Albian palynomorph assemblages from southern England. IV International Palynological Conference, Lucknow (1976–77), 2:403–408.

BENEDEK, P. N. 1972. Phytoplanktonten aus dem Mittel-und Oberoligozän von Tönisberg (Niederrheingebiet). Palaeontographica, Abt. B, 137:1–71.

——, AND C. MÜLLER. 1974. Nannoplankton–Phytoplankton–Korrelation im Mittle- und Ober-Oligozän von NW-Deutschland. Neus Jahrbuch fur Geologie und Paläontologie: Monatshefte, 7:385–397.

——. 1976. Die Grenze Unter-/Mittel-Oligozän am Dobert bei Bünde/Westfalen. I. Phyto- und Nannoplankton. Neues Jahrbuch fur Geologie und Paläontologie, Monatshefte, 3:129–144.

BENSON, D. G., JR. 1976. Dinoflagellate taxonomy and biostratigraphy at the Cretaceous-Tertiary boundary, Round Bay, Maryland. Tulane Studies in Geology and Paleontology, 12(4):169–233.

BERTRAND, K. J. 1971. Americans in Antarctica 1775–1948. American Geographical Society, Special Publication 39, New York, 554 p.

BLACKBURN, K. B. 1936. A reinvestigation of the alga *Botryococcus Braunii* Kützing. Transactions of the Royal Society of Edinburg, 58:841–854.

BOLTENHAGEN, E. 1970. Microplancton du Crétacé supérieur du Gabon. Cahiers de Paléontologie, Editions du Centre National de la Recherche Scientific, 180 p.

BOOTH, J. D., AND L. H. BURCKLE. 1976. Displaced Antarctic diatoms in the southwestern and central Pacific. Pacific Geology, 11:99–108.

BUCKMAN, S. S. 1908. Antarctic fossil Brachipoda collected by the Swedish South Polar Expedition 1901–03. Wissenschaftliche Ergebnisse der Schwedischen Südpolar-Expedition 1901–1903, Stockholm, 3(7):1–43.

BUJAK, J. P. 1973. Microplankton from the Barton Beds of the Hampshire Basin, England. Unpub. Ph.D. thesis, University of Sheffield, 445 p.

——. 1976. An evolutionary series of Late Eocene dinoflagellate cysts from southern England. Marine Micropaleontology, 1:101–117.

——. 1980. Dinoflagellate cysts and acritarchs from the Eocene Barton Beds of southern England, p. 36–91. *In* J. P. Bujak, C. Downie, G. L. Eaton, and

G. L. Williams, Dinoflagellate cysts and acritarchs from the Eocene of southern England. Palaeontological Association, Special Papers in Palaeontology, vol. 24.

——, AND G. L. WILLIAMS. 1978. Cretaceous palynostratigraphy of offshore southeastern Canada. Geological Survey of Canada Bulletin 297, 19 p.

——, C. DOWNIE, G. L. EATON, AND G. L. WILLIAMS. 1980. Taxonomy of some Eocene dinoflagellate cyst species from southern England, p. 26–86. *In* Bujak, J. P., Downie, C., Eaton, G. L., and Williams, G. L., Dinoflagellate cysts and acritarchs from the Eocene of southern England. Palaeontological Association, Special Papers in Palaeontology, vol. 24.

BURCKLE, L. H. 1981. Displaced Antarctic diatoms in the Amirante Passage. Marine Geology, 39:M39–M43.

——, AND D. STANTON. 1975. Distribution of displaced Antarctic diatoms in the Argentine Basin, p. 283–291. Diatom Symposium Volume, Diatom Conference, Kiel, Federal Republic of Germany.

——, V. KOLLA, AND J. D. BOOTH. 1974. Sediment transport by Antarctic Bottom Water in the Western Indian Ocean (abs.). Annual Meeting of the American Geophysical Union, 55:312.

CARO, Y. 1973. Contribution à la connaissance des dinoflagellés du Paléocène-Eocène inférieur des Pyrenées espagnoles. Revista Española de Micropaleontologia, 5:329–372.

——. 1977. Stratigraphic range charts diagnostic dinocysts: Paleogene. *In* B. Thusu (ed.), Distribution of Biostratigraphically Diagnostic Dinocysts and Miospores from the Northwestern European Continental Shelf and Adjacent Areas. International Palynological Colloquium, September, 1977, Leon, Spain, Charts 1-5.

CHAROLLAIS, J., ET AL. 1980. Les Marnes à Foraminifêres et les Shistes à *Meletta* des chaînes subalpines septentrionales (Haute-Savoie, France). Eclogae géologicae Helvatiae, 73(1):9–69.

——, R. JAN DU CHÊNE, A. LOMBARD, AND J. VAN STUIJVENBERG. 1975. Contribution à l'étude des flyschs des environs de Bonneville (Haute-Savoie, France). Géologie Alpine, 51:25–34.

CHATEAUNEUF, J. J. 1979. Upper Eocene and Oligocene Dinophyceae of the Paris Basin (France). Fourth International Palynological Conference, Proceedings; Birbal Sahni Institute of Palaeobotany, Lucknow, India (1976–77), 2:47–58.

——. 1980. Palynostratigraphie et paléoclimatologie de l'Eocène supérieur et de l'Oligocène du Bassin de Paris. Bureau des Recherches Geologiques et Minières Mémoir, 116:1–360.

COOKSON, I. C. 1965. Microplankton from the Paleocene Pebble Point Formation, southwestern Victoria. Proceedings of the Royal Society of Victoria, 78:137–141.

——, AND L. M. CRANWELL. 1967. Lower Tertiary microplankton, spores and pollen grains from southernmost Chile. Micropaleontology, 13:204–216.

——, AND A. EISENACK. 1958. Microplanton from Australian and New Guinea Upper Mesozoic sediments. Proceedings of the Royal Society of Victoria, 70:19–79.

——. 1960. Microplankton from Australian Cretaceous sediments. Micropaleontology, 6:1–18.

——. 1961a. Upper Cretaceous microplankton from the Belfast No. 4 Bore, southwestern Victoria. Proceedings of the Royal Society of Victoria, 74:69–76.

——. 1961b. Tertiary microplankton from the Rottnest Island Bore, Western Australia. Journal of the Royal Society of Western Australia, 44:39–47.

——. 1962. Additional microplankton from Australian Cretaceous sediments. Micropaleontology, 8:485–507.

——. 1965a. Microplankton from the Browns Creek Clays, SW Victoria. Proceedings of the Royal Society of Victoria, 79:119–131.

——. 1965b. Microplankton from the Dartmoor Formation, SW Victoria. Proceedings of the Royal Society of Victoria, 79:133–137.

——. 1967a. Some Early Tertiary microplankton and pollen grains from a deposit near Strahan, western Tasmania. Proceedings of the Royal Society of

Victoria, 80:131–140.

——. 1967b. Some microplankton from the Paleocene Rivernook Bed, Victoria. Proceedings of the Royal Society of Victoria, 80:247–257

——. 1968. Microplankton from two samples from Gingin Brook No. 4 Borehole, Western Australia. Journal of the Royal Society of Western Australia, 51:110–122.

——. 1969. Some microplankton from two bores at Balcatta, Western Australia. Journal of the Royal Society of Western Australia, 52:3–8.

——. 1970a. Cretaceous microplankton from the Eucla Basin, Western Australia. Proceedings of the Royal Society of Victoria, 83:137–157.

——. 1970b. Die Familie der Lecaniellaceae n. fam.—Fossile Chlorophyta, Volvocales? Neues Jahrbuch für Geologie und Paläontologie, Monatshefte, 1970(6):321–325.

——. 1971. Cretaceous microplankton from Eyre No. 1 Bore Core 20, Western Australia. Proceedings of the Royal Society of Victoria, 84:217–226.

——. 1974. Mikroplankton aus Australischen Mesozoischen und Tertiären Sedimenten. Palaeontographica, Abt. B, 148:44–93.

——, AND N. F. HUGHES. 1964. Microplankton from the Cambaridge Greensand (mid-Cretaceous). Palaeontology, 7:37–59.

COSTA, L. I., AND C. DOWNIE. 1979. The Cenozoic dinocyst stratigraphy of Sites 405 to 406 (Rockall Plateau), IPOD, Leg 48. Initial Reports of the Deep Sea Drilling Project, 48:513–528.

CRANWELL, L. M. 1959. Fossil pollen from Seymour Island, Antarctica. Nature, 184(4701):1782–1785.

——. 1964. Hystrichospheres as an aid to Antarctic dating with special reference to the recovery of *Cordosphaeridium* in erractics at McMurdo Sound. Grana Palynologica, 5(3):397–405.

——. 1969a. Antarctic and Circum-Antarctic palynological contributions. Antarctic Journal of the United States, 4(5):197–198.

——. 1969b. Palynological intimations of some pre-Oligocene Antarctic climates, p. 1–19. In E. M. van Zinderen Bakker, Sr. (ed.), Palaeoecology of Africa. Balkena, Cape Town, South Africa.

CROFT, W. N. 1947. Geological reports for the year ending January 1947, Hope Bay. Falkland Island Dependency Scientific Bureau, No. E89/47, 182.

DAMASSA, S. P. 1979. Eocene dinoflagellates from the coastal belt of the Franciscan Complex Northern California coast ranges. Palynology, 3:290.

DAVEY, R. J. 1969. Some dinoflagellate cysts from the Upper Cretaceous of northern Natal, South Africa. Palaeontologia Africana, 12:1–23.

——. 1970. Noncalcareous microplankton from the Cenomanian of England, northern France and North America, Part II. Bulletin of the British Museum (Natural History) Geology, 18:333–397.

——. 1978. Marine Cretaceous palynology of Site 361, D.S.D.P. Leg 40, off southwestern Africa. Initial Reports of the Deep Sea Drilling Project, 40:883–913.

——. 1979. Marine Apto-Albian palynomorphs from Holes 400A and 402A, IPOD Leg 48, Northern Bay of Biscay. Initial Reports of the Deep Sea Drilling Project, 48:547–577.

DAVEY, R. J., AND J. P. VERDIER. 1974. Dinoflagellates cysts from the Aptian type sections at Gargas and La Bédoule, France. Palaeontology, 17:623–653.

——, AND G. L. WILLIAMS. 1966a. The genera *Hystrichosphaera* and *Achomosphaera*, p. 28–52. In R. J. Davey, C. Downie, W.A.S. Sarjeant, and G. L. Williams, Studies on Mesozoic and Cainozoic Dinoflagellate Cysts. Bulletin of the British Museum (Natural History) Geology, Supplement 3.

——. 1966b. The genus *Hystrichosphaeridium* and its allies, p. 53–106. In R. J. Davey, C. Downie, W.A.S. Sarjeant, and G. L. Williams, Studies on Mesozoic and Cainozoic Dinoflagellate Cysts. Bulletin of the British Museum (Natural History) Geology, Supplement 3.

——, C. DOWNIE, W.A.S. SARJEANT, AND G. L. WILLIAMS. 1966. Studies on Mesozoic and Cainozoic dinoflagellate cysts. Bulletin of the British Museum (Natural History) Geology, Supplement 3:248p.

DE CONINCK, J. 1965. Microfossiles planctoniques du Sable yprésien à Merelbeke. Dinophyceae et Acritarcha. Académie royale du Belgique, Classe des Sciences, Mémoires, Coll. 8(26-2):1–56.

——. 1968. Dinophyceae et Acritarcha de l'Yprésien du sondage de Kallo.

Institut Royal des Sciences Naturelles de Belgique Memoires, 161:1–67.

——. 1969. Dinophyceae et Acritarcha de l'Yprésien du Sondage de Kallo. Institut Royal des Sciences Naturelles de Belgique Mémoire, 161:1–67.

——. 1975. Microfossiles à paroi organique de l'Yprésien du Bassin belge. Service Géologique de Belgique Professional Paper, No. 12:1–151.

——. 1977. Organic walled microfossils from the Eocene of the Woensdrecht borehole, southern Netherlands. Mededelingen Rijks Geologische Dienst, Nieuwe Serie, 28(3):33–64.

——. 1980. Especes indicatrices de microfossiles à paroi organique des depots de l'Yprésien superieur et du Lutetien dans le sondage de Kallo. Bulletin Societe Belge de Geologie, 89(4):309–317.

DEFLANDRE, G. 1937. Microfossiles des silex crétacés. Deuxième partie. Flagellés *incertae sedis* Hystrichosphaeridés. Sarcodinés Organisms divers. Annales de paléontologie, 26:51–103.

——, AND I. C. COOKSON. 1955. Fossil microplankton from Australian Late Mesozoic and Tertiary sediments. Australian Journal of Marine and Freshwater Research, 6:242–313.

DONALD, C. W. 1893. A voyage toward the Antarctic Sea, September 1892 to June 1893. Geographical Journal, 2:433.

DOWNIE, C. 1979. The Acritarchs; short course notes, Louisiana State University, September 10-12, 1979, Baton Rouge, Louisiana, unpaged.

DRUGG, W. S. 1967. Palynology of the Upper Moreno Formation (Late Cretaceous–Paleocene) Escarpado Canyon, California. Palaeontograhpica, Abt. B, 120:1–71.

——. 1970. Some new genera, species, and combinations of phytoplankton from the Lower Tertiary of the Gulf Coast, U.S.A. North American Paleontological Convention, Chicago, 1969, Proc. G, p. 809–843.

——, AND A. R. LOEBLICH, JR. 1967. Some Eocene and Oligocene phytoplankton from the Gulf Coast, U.S.A. Tulane Studies in Geology, 5:181–194.

DUSEN, P. 1908. Uber die Tertiär Flora der Seymour-insel., Wissenschaftliche Ergebnisse der Südpolar-Expedition, 1901-1903, 3(3):1–27.

EATON, G. L. 1971. A morphogenetic series of dinoflagellate cysts from the Bracklesham Beds of the Isle of Wight, Hampshire, England, p. 355–379. In A. Farinacci (ed.), Proceedings of the II Planktonic Conference, Rome 1970, Edizioni Technoscienza, Rome.

——. 1976. Dinoflagellate cysts from the Bracklesham Beds (Eocene) of the Isle of Wight, southern England. Bulletin of the British Museum (Natural History) Geology, 26:227–332.

EHRENBERG, C. G. 1838. Über das Massenverhältniss der jetzt lebenden Kiesel-Infusorien und über ein neues Infusorien-Conglomerat als Polirschiefer von Jastraba in Ungarn. Abhandlungen der Preussischen Akademie der Wissenschaften, 1836, p. 109–135.

EISENACK, A. 1938. Die Phosphoritknollen der Bernsteinformation als Überlieferer tertiären Planktons. Schriften der Physikalischökonomischen Gesellschaft zu Königsberg, 70:181–188.

——. 1954. Mikrofossilien aus Phosphoriten des samländischen Unteroligozäns und über die Einheitlichkeit der Hystrichosphaerideen. Palaeontographica, Abt. A, 105:49–95.

——. 1958. Mikroplankton aus dem norddeutschen Apt. Neues Jahrbuch für Geologie und Paläontologie, Abhandlungen, 106:383–422.

——. 1961. Einige Erörterungen über fossile Dinoflagellaten nebst Übersicht über die zur Zeit Bekannten Gattungen. Neues Jahrbuch für Geologie und Paläontologie, Abhandlungen, 112:281–324.

——, AND I. C. COOKSON. 1960. Microplankton from Australian Lower Cretaceous sediments. Proceedings of the Royal Society of Victoria, 72:1–11.

——, F. H. CRAMER, AND M.D.C.R. DIEZ RODRIGUEZ. 1973. Katalog der fossilen Dinoflagellaten, Hystrichosphären, und verwandten Mikrofossilien Band III Acritarcha, 1. Teil. Stuttgart, E. Schweizerbart'sche Verlagsbuchhandlung, 1,104 p.

ELLIOT, D. H. 1982. Mesozoic evolution of the Antarctic Peninsula, (abs.). Fourth International Symposium on Antarctic Earth Sciences, August, 1982, Australian Academy of Sciences, Adelaide, Australia.

——, AND T. A. TRAUTMAN. 1977. Lower Tertiary strata on Seymour Island, (abs.). Volume of Abstracts, Third Symposium on Antarctic Geology and

Geophysics, 22-27 August 1977, University of Wisconsin, Madison, 49.

——. 1982. Lower Tertiary strata on Seymour Island, Antarctic Peninsula, p. 287–297. *In* C. Craddock (ed.), Antarctic Geoscience. University of Wisconsin Press, Madison.

——, C. RINALDI, W. J. ZINSMEISTER, T. A. TRAUTMAN, W. A. BRYANT, AND R. A. DEL VALLE. 1975. Geological investigations on Seymour Island, Antarctic Peninsula. Antarctic Journal of the United States, 10(4):182–186.

ELSIK, W. C. 1977. *Paralecaniella indentata* (Defl. and Cooks. 1955) Cookson and Eisenack 1970 and allied dinocysts. Palynology, 1:95–102.

ERKMEN, U., AND W.A.S. SARJEANT. 1978. *Xylochoarion,* new genus of dinoflagellate cysts from the Hackness Rock (Middle Jurassic: Callovian) of Yorkshire, England. Neues Jahrbuch für Geologie und Paläontologie, Monatshefte, 1978(7):400–407.

EVITT, W. R. 1968. The Cretaceous microfossil *Ophiobolus lapidaris* O. Wetzel and its flagellum-like filaments. Stanford University Publications in Geological Sciences, 12(3):1–11.

——, AND S. T. PIERCE. 1975. Early Tertiary ages from the coastal belt of the Franciscan complex, northern California. Geology, 3(8):433–436.

FELDMANN, R. M., AND W. J. ZINSMEISTER. 1984. New fossil crabs (Decapoda: Brachyura) from the La Meseta Formation (Eocene) of Antarctica: paleogeographic and biogeographic implications. Journal of Paleontology, 58(4):1046–1061.

FIRSTBROOK, P. L., B. M. FUNNELL, A. M. HURLEY, AND A. G. SMITH. 1979. Paleoceanic Reconstructions, 160–0Ma. National Ocean Sediment Coring Project, La Jolla, CA, 41 p.

FOUCHER, J. J. 1974. Microfossiles des silex du Turonien Supérieur de Ruyaulcourt (Pas-de-Calais). Annales de Paléontologie (Invertébrés), 60(2):113–164.

——. 1976. Microplancton des silex Crétacés du Beauvaisis. Cahiers de Micropaléontologie, 2:28.

——, AND F. ROBASZYNSKI. 1977. Microplancton des silex du Bassin du Mons (Belgique) (Dinoflagellés Crétacés et Daniens). Annales de Paléontologie (Invertébrés), 63(1):19–58.

——, AND P. TAUGOURDEAU. 1975. Microfossiles de l'Albo-Cénomanien de Wissant (Pas-de-Calais). Cahiers de Micropaléontologie, C.N.R.S., Paris, 1975-1:1–30.

FREDERIKSEN, N.O., ET AL. 1982. Biostratigraphy and paleoecology of Lower Paleozoic, Upper Cretaceous, and Lower Tertiary rocks in U.S. Geological Survey, New Madrid test wells, southeastern Missouri. Tulane Studies in Geology and Paleontology, 17(2):23–45.

GAMERRO, J. C., AND S. ARCHANGELSKY. 1981. Palinozonas Neocretácicas y Terciarias de la Plataforma Continental Argentina en la Cuenca del Colorado. Revista Española de Micropaleontologíe, 8(1):119–140.

GERLACH, E. 1961. Mikrofossilien aus dem Oligozän und Miozän Nordwestdeutschlands, unter besonderer Berücksichtigung der Hystrichosphären und Dinoflagellaten. Neues Jahrbuch für Geologie und Paläontologie, Abhandlungen, 112:143–228.

GIBSON, T. G., ET AL. 1980. Biostratigraphy of the Tertiary strata of the core. Geology of the Oak Grove core, p. 14–30. Virginia Division of Mineral Resources Publication, Charlottesville, Vol. 20.

GOCHT, H. 1969. Formengemeinschaften alttertiären Mikroplanktons aus Bohrproben des Erdölfeldes Meckelfeld bei Hamburg. Palaeontographica, Abt. B, 126:1–100.

——, AND W. WILLE. 1972. Untersuchungen an *Palambages morulosa* O. Wetzel (Chlorophyceae inc. sed.). Neues Jahrbuch für Geologie und Paläontologie, Monatshefte, H3:146–161.

GOODMAN, D. K. 1975, Lower Eocene dinoflagellate assemblages from the Maryland coastal plain south of Washington, D.C. Unpubl. M.S. thesis, Virginia Polytechnic Institute and State University, Blacksburg, 1–298.

——. 1979. Dinoflagellate "communities" from the Lower Eocene Nanjemoy Formation of Maryland, U.S.A. Palynology, 3:169–190.

——, AND L. N. FORD, JR. 1983. Preliminary dinoflagellate biostratigraphy for the middle Eocene to lower Oligocene from the southwest Atlantic Ocean. Initial Reports of the Deep Sea Drilling Project, 71:859–877.

GRADSTEIN, F. M., G. L. WILLIAMS, W.A.M. JENKINS, AND P. ASCOLI. 1976. Mesozoic and Cenozoic stratigraphy of the Atlantic continental margin, eastern Canada, p. 103–131. *In* C. J. Yorath, E. R. Parker, and D. J. Glass (eds.), Canada's Continental Margins and Offshore Petroleum Exploration. Canadian Society of Petroleum Geologists Memoir, Vol. 4.

GRIGGS, P. 1981. Tertiary dinoflagellate morphologies, Santos Basin, Brazil. "Hexrose" Conference on Modern and fossil dinoflagellates. Tubingen, West Germany, August 31-September 4, 1981. Poster Session handout, unpaged.

GRUAS-CAVAGNETTO, C. 1976. Les marqueurs stratigraphiques (Dinoflagellés) de L'Éocene du bassin de Paris et de la Manche Orientale. Revue de Micropaléontologie, 18(4)221–228.

HALL, S. A. 1977. Cretaceous and Tertiary dinoflagellates from Seymour Island, Antarctica. Nature, 267:239–241.

HARLAND, R. 1983. Distribution of recent dinoflagellate cysts in bottom sediments from the North Atlantic Ocean and adjacent seas. Palaeontology, 261:(2):321–387.

HARRIS, W. K. 1974. Tertiary nonmarine dinoflagellate cyst assemblages from Australia. Special Publications of the Geological Society of Australia, 4:159–166.

HARWOOD, D. M. 1985. Cretaceous to Eocene Seymour Island siliceous microfossil biostratigraphy (abs.), p. 17–18. Workshop on Cenozoic Geology of the Southern High Latitudes, August 16-17, 1985, Sixth Gondwana Symposium, Ohio State University, Columbus.

HASKELL, T. R., AND G. J. WILSON. 1975. Palynology of sites 280-284, DSDP Leg 29, off southeastern Australia and western New Zealand. Initial Reports of the Deep Sea Drilling Project, 29:723–741.

HAZEL, J. E., L. M. BYBELL, AND R. A. CHRISTOPHER. 1977. Biostratigraphy of the deep corehole (Clubhouse Crossroads Corehole 1) near Charleston, South Carolina. 1977. U.S. Geological Survey Professional Paper 1028, 10(8):1–89.

HEISECKE, A. M. 1970. Microplancton de la Formación Roca de la Provincia de Neuquén. Ameghiniana, 7:225–262.

HELBY, R. J., E. J. KIDSON, L. E. STOVER, AND G. L. WILLIAMS. 1984. Survey of dinoflagellate biostratigraphy: unpublished short course notes. Sixteenth Palynology Short Course, September 17-21, 1984, convened by G. F. Hart, Baton Rouge, Louisiana, unpaged.

ISLAM, M. A. 1983. Dinoflagellate cysts from the Eocene cliff sections of the Isle of Sheppey, southeast England. Revue de Micropaleontologie, 25(4):231–250.

JAIN, K. P., AND P. MILLEPIED. 1973. Cretaceous microplankton from Senegal Basin, N.W. Africa. 1. Some new species and combinations of dinoflagellates. Palaeobotanist, 20:22–32.

JAN DU CHÊNE, R. 1977. Etude palynologique du Miocene supérieur Andalou (Espagne). Revista Española de Micropaleontologia, 9:97–114.

——, J. VAN STUIJVENBERG, J. CHAROLLAIS, AND J. ROSSET. 1975. Sur l'âge du flysch de la nappe inférieure de la Klippe de Sulens (Haute-Savoie, French). Géologie Alpine, 51:79–81.

JOHNSON, D. A., M. LEDBETTER, AND L. H. BURCKLE. 1977. Vema Channel paleo-oceanography: Pleistocene dissolution cycles and episodic bottom water flow. Marine Geology, 23:1–33.

JONES, G. A., AND D. A. JOHNSON. 1984. Displaced Antarctic diatoms in Vema Channel sediments: Late Pleistocene/Holocene fluctuations in AABW flow. Marine Geology, 58:165–186.

KEMP, E. M. 1972. Recycled palynomorphs in continental shelf sediments from Antarctic. Antarctic Journal of the United States, 7(5):190–191.

——. 1975. Palynology of Leg 28 Drill Sites, Deep Sea Drilling Project. *In* A. G. Kaneps (ed.), Initial Reports of the Deep Sea Drilling Project, 28:599–623.

KJELLSTRÖM, G. 1972. Archaeopyle formation in the genus *Lejeunia* Gerlach, 1961 emend. Geologiska Föreningens i Stockholm Förhandlingar, 94:467–469.

——. 1973. Maastrichtian microplankton from the Höllviken Borehole No. 1 in Scania, southern Sweden. Sveriges Geologiska Undersökning, Afhandlingar och Uppsatser, 67:1–59.

KNOX, R. W. O'B., AND R. HARLAND. 1979. Stratigraphical relationships of the early Palaeogene ash-series of NW Europe. Journal of the Geological Society, 136(4):463–470.

KOCH, R. C. 1975. Dinoflagellate biostratigraphy of Maestrichtian formations of the New Jersey coastal plain. Unpubl. Ph.D. thesis, Rutgers University, New Brunswick, N.J., 110 p.

LENTIN, J. K. AND G. L. WILLIAMS. 1975. Fossil dinoflagellates: index to genera and species. Supplement 1. Canadian Journal of Botany, 53:2147–2157.

——. 1976. A monograph of fossil peridinioid dinoflagellate cysts. Bedford Institute Oceanography Report BI-R-75-16, 1–237.

——. 1977. Fossil dinoflagellates: index to genera and species, 1977 ed. Bedford Institute of Oceanography Report Series B1-R-77-8, 209 p.

——. 1980. Dinoflagellate provincialism. American Association of Stratigraphic Palynologists Contributions Series No. 7, 47 p.

——. 1981. Fossil dinoflagellates: index to genera and species, 1981 edition. Bedford Institute of Oceanography Report Series B1-R-81-12, 345 p.

——. 1985. Fossil dinoflagellates: index to genera and species, 1985 edition. Canadian Technical Report of Hydrography and Ocean Sciences No. 60, 451 p.

——. 1987. Status of the fossil dinoflagellate genera *Ceratiopsis* Vozzhennikova 1963 and *Cerodinium* Vozzhennikova 1963 emend. Palynology, 11.

LIENGJARERN, M., L. COSTA, AND C. DOWNIE. 1980. Dinoflagellate cysts from the upper Eocene–lower Oligocene of the Isle of Wight. Palaeontology, 23:475–499.

LINDGREN, S. 1984. Acid resistant peridinioid dinoflagellates from the Maastrichtian of Trelleborg, southern Sweden. Stockholm Contributions in Geology, 39(6):145–201.

MACELLARI, C. E. 1984. Late Cretaceous stratigraphy, sedimentology and macropaleontology of Seymour Island, Antarctic Peninsula. Unpubl. Ph.D. thesis, Ohio State University, Columbus, 599 p.

MAIER, D. 1959. Planktonuntersuchungen in tertiären und quartären marinen Sedimenten. Ein Beitrag zur Systematik, Stratigraphie und Ökologie der Coccolithophorideen, Dinoflagellaten und Hystrichosphaerideen vom Oligozän bis zum Pleistozän. Neues Jahrbuch für Geologie und Paläontologie, Abhandlungen, 107:278–340.

MALLOY, R. E. 1972. An Upper Cretaceous dinoflagellate cyst lineage from Gabon, West Africa. Geoscience and Man, 4:57–65.

MANUM, S. 1960. Some dinoflagellates and hystrichosphaerids from the lower Tertiary of Spitzbergen. Nytt Magasin for Botanikk, 8:17–26.

——. 1976. Dinocysts in Tertiary Norwegian-Greenland Sea sediments (Deep Sea Drilling Project Leg 38), with observations on palynomorphs and palynodebris in relation to environment. Initial Reports of the Deep Sea Drilling Project, 38:897–906.

——. 1979. Two new Tertiary dinocyst genera from the Norwegian Sea: *Lophocysta* and *Evittosphaerula*. Review of Palaeobotany and Palynology, 28:237–248.

——, AND I. C. COOKSON. 1964. Cretaceous microplankton in a sample from Graham Island, Arctic Canada, collected during the second "Fram"-Expedition (1898–1902). With notes on microplankton from the Hassel Formation, Ellef Ringes Island. Schrifter utgitt av Det Norske Videnskaps-Akademi i Oslo, I. Mat-Naturv. Klasse, Ny Series 17, 1–35.

MARPLES, B. J. 1953. Fossil penguins from the mid-Tertiary of Seymour Island. Falkland Islands Dependencies Survey Scientific Reports, 7:26.

MAY, F. E. 1980. Dinoflagellate cysts of the Gymnodiniaceae, Peridiniaceae, and Gonyaulacaceae from the Upper Cretaceous Monmouth Group, Atlantic Highlands, New Jersey. Palaeontographica, Abt. B, 172:10–116.

MCGOWAN, B. 1965. Two Paleocene foraminiferal faunas from the Wangerrip Group, Pebble Point coastal section, Victoria. Proceedings of the Royal Society of Victoria, 79:9–74.

——. 1968. Late Cretaceous and early Tertiary correlations in the Indo-Pacific region. Geological Society of India Memoir, 2:335–360.

MCINTYRE, D. J., AND G. J. WILSON. 1966. Preliminary palynology of some Antarctic Tertiary erratics. New Zealand Journal of Botany, 4(3):315–321.

MCLEAN, D. M. 1971. Transfer of *Baltisphaeridium septatum* Cookson and Eisenack, 1967, from the Acritarcha to the Dinophyceae. Journal of

Paleontology, 45:729–730.

MORGAN, R. 1977. Elucidation of the Cretaceous dinoflagellate *Diconodinium* Eisenack and Cookson, 1960, and related peridinioid species from Australia. Palynology, 1:123–138.

MORGAN, R. A. 1967. Palynology of the Ozan Formation (Cretaceous) McCurtin County, Oklahoma. Unpubl. M.S. thesis, University of Oklahoma, Norman, 120 p.

MORGENROTH, P. 1966a. Mikrofossilien und Konkretionen des nordwesteuropäischen Untereozäns. Palaeontographica, Abt. B, 119:1–53.

——. 1966b. Neue in organischer Substanz erhaltene Mikrofossilien des Oligozäns. Neues Jahrbuch für Geologie und Paläontologie, Abhandlungen, 127:1–12.

NORVICK, M. S. 1975. Mid-Cretaceous microplankton from Bathurst Island, p. 21–113. *In* M. S. Norvick, and D. Burger, Palynology of the Cenomanian of Bathurst Island, Northern Territory, Australia. Australian Bureau of Mineral Resources, Geology and Geophysics, Bulletin 51.

PALAMARCZUK, S. 1982. Dinoflagellate de edad Daniana en La Isla Vicecomodoro Marambio, (Ex Seymour), Antartida Argentina. Ameghiniana, 19(3-4):353–360.

——, G. AMBROSINI, H. VILLAR, F. MEDINA, J. C. MARTINEZ MACCHIAVELLO, AND C. RINALDI. 1984. Las formaciones Lopez de Bertodano y Sobral en la Isla Vice Comodoro Marambio, Antartida. Noveno Congreso Geologico Argentino, S. C. Bariloche, ACTAS, I:399–419.

PLAYFORD, G., D., W. HAIG, AND M. E. DETTMANN. 1975. A mid-Cretaceous microfossil assemblage from the Great Artesian Basin, Northwestern Queensland. Neues Jahrbuch für Geologie und Paläontologie, Abhandlungen, 149(3):333–362.

PÖTHE DE BALDIS, E. D. 1966. Microplancton del Terciario de Tierra del Fuego. Ameghiniana, 4:219–228.

RIEGEL, W. 1974. New forms of organic-walled microplankton from an Upper Cretaceous assemblage in southern Spain. Revista Española de Micropaleontologia, 6:347–366.

——, AND W.A.S. SARJEANT. 1982. Dinoflagellate cysts from the Upper Cretaceous of southern Spain: new morphological and taxonomic observations. Neues Jahrbuch für Geologie und Paläontologie, Abhandlungen, 162:286–303.

RINALDI, C. A. 1977. About the Upper Cretaceous from the James Ross Island Group, (abs.). Volume of Abstracts, Third Symposium on Antartic Geology and Geophysics, 22–27 August 1977, University of Wisconsin, Madison, 127.

——. 1982. The Upper Cretaceous in the James Ross Island Group, p. 281–286. *In* C. Craddock (ed.), Antarctic Geoscience. University of Wisconsin, Madison.

——, A. MASSABIC, J. MORELLI, H. L. ROSEMAN, AND R. DEL VALLE. 1978. Geologia de la Isla Vicecomodoro Marambio. Contribución del Instituto Antarctico Argention, 217:37 p.

ROSS, J. C. 1847. A voyage of discovery and research in the Southern and Antarctic Regions, during the years 1839–43. Murray, London, 335 p.

SANNEMANN, D. 1955. Hystrichosphaerideen aus dem Gotlandium und Unter Devon des Frankenwaldes und ihr Feinbau. Senckenbergiana lethaea, 36:321–346.

SARJEANT, W.A.S. 1967. The genus *Palaeoperidinium* Deflandre (Dinophyceae). Grana Palynologica, 7:243–258.

——. 1984. Restudy of some dinoflagellate cysts from the Oligocene and Miocene of Germany. Journal of Micropalaeontology, 3(2):73–94.

SCHUMACKER-LAMBRY, J., AND J. J. CHATEAUNEUF. 1976. Dinoflagellés et acritarches des marnes Heersiennës de Gelinden (base de Landenien, Paléocène, Belgique). Review of Palaeobotany and Palynology, 21(4):267–294.

SCOTT, R. W., AND E. J. KIDSON. 1977. Lower Cretaceous depositional systems West Texas, p. 169–181. *In* D. G. Bebout and R. G. Loucks (eds.), Cretaceous Carbonates of Texas and Mexico: Applications to Subsurface Exploration. Bureau of Economic Geology, University of Texas, Austin. Report of Investigation 89.

SHARMAN, G., AND E. I. NEWTON. 1894. Notes on some fossils from Seymour Island, in the Antarctic regions, obtained by Dr. Donald. Transactions of the Royal Society of Edinburgh, 37, 3(30):707–709.

——. 1898. Notes on some additional fossils collected at Seymour Island, Graham Land, by Dr. Donald and Captain Larsen. Proceedings of the Royal Society of Edinburgh, 22:58–61.

SIMPSON, G. G. 1971. Review of fossil penguins from Seymour Island. Transactions of the Royal Society of London (Ser. 6), 178:357–387.

SINGH, C. 1971. Lower Cretaceous microfloras of the Peace River area, Northwestern Alberta. Research Council of Alberta, 2:301–542.

SMITH-WOODWARD, A. 1908. On fossil fish-remains from Snow Hill and Seymour Islands. Wissenschaftliche Ergebnisse der Schwedischen Súdpolar-Expediton 1901–1903, Stockholm, 3(4):1–4.

SOPER, N. J., ET AL. 1976. Late Cretaceous–early Tertiary stratigraphy of the Kangerdlugssuaq area, East Greenland, and the age of opening of the north Atlantic. Journal of the Geological Society (London and Edinburgh), 13(1):85–104.

STAMPFLI, G., R. JAN DU CHÊNE, AND R. HERB. 1978. Geologie et micropaleontologie (Nummulites et palynologie) de la formation Eocene de Ziarat, Elbourz Oriental (Iran). Revista Italiana Paleontologia e Stratigrafia, 84(2):383–402.

STANLEY, E. A. 1965. Upper Cretaceous and Paleocene plant microfossils and Paleocene dinoflagellates and hystrichosphaerids from northwestern South Dakota. Bulletin of American Paleontology, 49:179–384.

STONE, J. F. 1973. Palynology of the Almond Formation (Upper Cretaceous), Rock Springs Uplift, Wyoming. Bulletins of American Paleontology, 64(278):1–135.

STOTT, L. D., B. C. MCKELVEY, D. M. HARWOOD, AND P. N. WEBB. 1983. A revision of the age of Cenozoic erratics at Mt. Discovery and Minna Bluff, McMurdo Sound, Antarctica. Antarctic Journal of the United States, 18(5):36–38.

STOVER, L. E. 1974. Palaeocene and Eocene species of *Deflandrea* (Dinophyceae) in Victorian coastal and offshore basins, Australia. Special Publications of the Geological Society of Australia, 4:167–188.

——. 1977. Oligocene and early Miocene dinoflagellates from Atlantic Corehole 5/5B, Blake Plateau. American Association of Stratigraphic Palynologists, Contributions Series, 5A:66–89.

——, AND W. R. EVITT. 1978. Analyses of pre-Pleistocene organic-walled dinoflagellates. Stanford University Publications, Geological Sciences, 15:1–300.

——, AND G. L. WILLIAMS. 1977. Introduction to Tertiary dinoflagellates: unpublished short course notes. Louisiana State University, September 12-16, 1977, convened by G. F. Hart, Baton Rouge, Louisiana, unpaged.

TRALAU, H. 1972. Spores, pollen grains, and planctonic microfossils from Upper Cretaceous Flint Boulders from Halland, South-Western Sweden. Geologiska Föreningens i Stockholm förhandlingar, 94:568–571.

TRAUTMAN, T. 1976. Stratigraphy and petrology of Tertiary clastic sediments, Seymour Island, Antarctica. Unpubl. M.S. thesis, Ohio State University, Columbus, 170 p.

TRUSWELL, E. M., 1983. Recycled Cretaceous and Tertiary pollen and spores in Antarctic marine sediments: a catalogue. Palaeontographica, Abt. B, 186:121–174.

——. 1987. Palynology of MSSTS 1 corehole (in press).

——, AND D. J. DREWRY. 1984. Distribution and provenance of recycled palynomorphs in surficial sediments of the Ross Sea, Antarctica. Marine Geology, 59:187–214.

VALENSI, L. 1948. Sur quelques microorganismes planctoniques des silex du Jurassique moyen du Poitou et de Normandie. Bulletin de la Société géologique de France, 5(18):537–550.

VAN STUIJVENBERG, J., P. MOREL, AND R. JAN DU CHÊNE. 1976. Contribution à l'étude du flysch de la région des Fayaux (Préalpes externes vaudoises). Eclogae géologicae Helvatiae, 69(2):290–326.

VERDIER, J. P. 1970. Addendum au mémoire de G. Deflandre et I. C. Cookson microplankton fossile de sediments du Mésozoique superieur et du Tertiaire d'Australie. Cahiers de Micropaléontologie, Série 2, 4:1–54.

——. 1975. Les kystes de dinoflagellés de la section de Wissant et leur distribution stratigraphique au Crétacé Moyen. Revue de Micropaléontologie, 17(4):191–197.

VOZZHENNIKOVA, T. F. 1963. Pirrofitovye Vodorosli [Phylum Pyrrhophyta]. In Yu. A. Orlov (ed.), Osnovy Paleontologii 14 [Fundamentals of Paleontology], 179–185.

——. 1967. Iskopaemye peridinei Yurskikh, Melovykh i Paleogenovykh otlozhenii SSSR. Akademiya Nauk SSSR, Sibirskoye Otdeleniye, Institut Geologii i Geofiziki, Trudy, Moscow. [Fossilized peridinid algae in the Jurassic, Cretaceous, and Paleogene deposits of the USSR], 347 p.

WALL, D., AND B. DALE. 1968. Modern dinoflagellate cysts and evolution of the Peridiniales. Micropaleontology, 14(3)265–304.

WARREN, B. A. 1974. Deep flow in the Madagascar and Mascarene Basins. Deep Sea Research, 21:1–21.

WEBB, P. N. 1978. Paleogeographic evolution of the Ross Sea and adjacent montane areas during the Cenozoic. Dry Valley Drilling Project (DVDP) Seminar III, National Institute of Polar Research, Tokyo, Bulletin, 8:124.

——. 1979. Paleogeographic evolution of the Ross Sector during the Cenozoic, p. 206–212. In T. Nagata (ed.), Proceedings of the Seminar III on Dry Valley Drilling Project, 1978. National Institute of Polar Research, Tokyo.

WELTON, B. J., AND W. J. ZINSMEISTER. 1980. Eocene Neoselachians from the La Meseta Formation, Seymour Island, Antarctic Peninsula. Natural History Museum of Los Angeles County, Contributions in Science 329:10.

WETZEL, O. 1961. New microfossils from Baltic Cretaceous flintstones. Micropaleontology, 7:337–350.

WHITNEY, B. L. 1979. A population study of *Alterbia acutula* (Wilson) Lentin and Williams from the Maestrichtian (Upper Cretaceous) of Maryland. Palynology, 3:123–128.

——. 1984. Dinoflagellate biostratigraphy of the Maestrichtian-Danian section in southern Maryland, p. 123–136. In N. O. Frederiksen and K. Krafft (eds.), Cretaceous and Tertiary Stratigraphy, Paleontology, and Structure, Southwestern Maryland and Northeastern Virginia. American Association of Stratigraphic Palynologists Field Trip Volume and Guide Book.

WILCKENS, O. 1911. Die Mollusken der antarktischen Tertiär formation. Wissenschaftliche Ergebnisse der Schwedischen Súdpolar-Expedition, 1901–1903, 3:1–62.

WILLIAMS, G. L. 1963. Organic-walled microplankton of the London Clay: a stratigraphic and palaeontological investigation. Unpubl. Ph.D. thesis, University of Sheffield, 423 p.

——. 1975. Dinoflagellate and spore stratigraphy of the Mesozoic–Cenozoic, offshore eastern Canada. Offshore Geology of Eastern Canada, Geological Survey of Canada Paper 30, 2:107–161.

——. 1977. Dinocysts: their classification, biostratigraphy and palaeoecology, p. 1231–1335. In A.T.S. Ramsay (ed.), Oceanic Micropalaeontology. Academic Press, New York.

——. 1978. Palynological biostratigraphy, Deep Sea Drilling Project Sites 367 and 370. Initial Reports of the Deep Sea Drilling Project, 41:783–815.

——, AND W. BRIDEAUX. 1975. Palynologic analyses of Upper Mesozoic and Cenozoic rocks of the Grand Banks, Atlantic continental margin. Geological Survey of Canada Bulletin 236, 163 p.

——, AND J. P. BUJAK. 1977. Cenozoic palynostratigraphy of offshore eastern Canada, p. 14–47. In H. J. Sullivan, W. W. Brideaux, and W. C. Elsik (eds.), Contributions of Stratigraphic Palynology, Vol. 1, Cenozoic Palynology. American Association of Stratigraphic Palynologists Contributions Series 5A.

——. 1985. Mesozoic and Cenozoic dinoflagellates, p. 847–964. In H. M. Bolli, J. B. Saunders, and K. Perch-Nielsen (eds.), Plankton Stratigraphy. Cambridge University Press, Cambridge.

——, AND E. J. KIDSON. 1975. Biostratigraphy and paleoecology of marine Mesozoic-Cenozoic dinoflagellate cysts: unpublished short course notes. Louisiana State University, May 17-21, 1976, Baton Rouge, Louisiana, unpaged.

WILSON, G. J. 1967a. Some new species of Lower Tertiary dinoflagellates from McMurdo Sound, Antarctica; New Zealand Journal of Botany, 5(1):57–83.

——. 1967b. Microplankton from the Garden Cove Formation, Campbell Island. New Zealand Journal of Botany, 5(2):223–240.

——. 1968. On the occurrence of fossil microspores, pollen grains, and microplankton in bottom sediments of the Ross Sea, Antarctica. New Zealand Journal of Marine and Freshwater Research, 2(3):381–389.

——. 1971. Observations on European Lake Cretaceous dinoflagellate cysts. Second Planktonic Conference, Rome, Proceedings, II:1259–1275.

——. 1972. Age of the Garden Cove Formation, Campbell Island. New Zealand Journal of Geology and Geophysics, 15(1):184–185.

——. 1975. Palynology of deep-sea cores from DSDP Site 275, southeast Campbell Plateau. Initial Reports of the Deep Sea Drilling Project, 29:1031–1035.

——. 1977. A new species of *Svalbardella* Manum (Dinophyceae) from the Eocene of New Zealand. New Zealand Journal of Geology and Geophysics, 20(3):563–566.

——. 1978. The dinoflagellate species *Isabelia druggii* (Stover) and *I. seelandica* (Lange): their association in the Teurian of Woodside Creek, Marlborough, New Zealand. New Zealand Journal of Geology and Geophysics, 21(1):75–80.

——. 1981a. Report on Cretaceous-Tertiary dinoflagellates from the south branch of the Waipara River. New Zealand Geological Survey Report, PAL 52:3.

——. 1981b. Dinoflagellate determinations from the Tapuwaeroa Foramtion, Southern Hawkes Bay. New Zealand Geological Survey Report, PAL 52:5.

——. 1981c. Dinoflagellate determinations from an undated section southwest of Glentunnel, Canterbury. New Zealand Geological Survey Report, PAL 52:7.

——. 1982. Early Tertiary biostratigraphy, West branch of the Grey River, Canterbury (M34). New Zealand Geological Survey Report, PAL 52:19.

——. 1984a. Some new dinoflagellate species from the New Zealand Haumurian and Piripauan Stages (Santonian-Maastrichtian, Late Cretaceous). New Zealand Journal of Botany, 22:549–556.

——. 1984b. New Zealand Late Jurassic to Eocene dinoflagellate biostratigraphy: a summary. Newsletter of Stratigraphy, 13(2):104–117.

——, AND C. D. CLOWES. 1982. *Arachnodinium*, a new dinoflagellate genus from the lower Tertiary of Antarctica. Palynology, 6:97–103.

WIMAN, C. 1905. Über die alttertiaaren Vertebraten der Seymourinsel. Wissenschaftliche Ergebnisse de Schwedischen Südpolar-Expedition 1901–1903, Stockholm, 3(1):1–37.

WRENN, J. H. 1981. Preliminary palynology of the RISP Site J-9, Ross Sea, Antarctica. Antarctic Journal of the United States, 16(5):72–74.

——. 1982. Dinocyst biostratigraphy of Seymour Island, Palmer Peninsula, Antarctica. Unpubl. Ph.D. thesis, Louisiana State University, Baton Rouge, 467 p.

——, AND S. W. BECKMAN. 1982. Maceral, total organic carbon, and palynological analyses of Ross Ice Shelf Project Site J-9 cores. Science, 216:187–189.

——, AND J. P. KOKINOS. 1986. Preliminary comments on Miocene through Pleistocene dinoflagellate cysts from De Soto Canyon, Gulf of Mexico, p. 169–229. *In* J. H. Wrenn, S. L. Duffield, and J. A. Stein (eds.), American Association of Stratigraphic Palynologists Contributions Series 17.

YUN, H. S. 1981. Dinoflagellaten aus der Oberkreide (Santon) von Westfalen. Palaeontographica Abt. B, 177:1–89.

ZAITZEFF, J. B. 1967. Taxonomic and stratigraphic significance of dinoflagellates and acritarchs of the Navarro Group (Maestrichtian) from eastcentral and southwest Texas. Unpubl. Ph.D. thesis, Michigan State University, East Lansing, 172 p.

ZINSMEISTER, W. J. 1976a. A new genus and species of the gastropod family Struthioloriidae, *Antarctodarwinella ellioti*, from Seymour Island, Antarctica. Ohio Journal of Science, 76(3):111–114.

——. 1976b. Intertidal region and molluscan fauna of Seymour Island, Antarctic Peninsula. Antarctic Journal of the United States, 11(4):222–225.

——. 1977. Note on a new occurrence of the Southern Hemisphere aporrhaid gastropod *Struthioptera*, Finlay and Marwick on Seymour Island, Antarctica. Journal of Paleontology, 51(2):399–404.

——. 1979. Biogeographic significance of the late Mesozoic and early Tertiary molluscan faunas of Seymour Island (Antarctic Peninsula) to the final breakup of Gondwanaland, p. 349–355. *In* J. Gray and A. J. Boucot (eds.), Historical Biogeography, Plate Tectonics, and the Changing Environment. Oregon State University Press, Corvallis.

——. 1982. Review of the Upper Cretaceous–Lower Tertiary sequence on Seymour Island, Antarctica. Journal of the Geological Society, London, 139:779–785.

——, AND H. H. CAMACHO. 1982. Late Eocene (to possibly earliest Oligocene) molluscan fauna of the La Meseta Formation of Seymour Island, Antarctic Peninsula, p. 299–304. *In* C. Craddock (ed.), Antarctic Geoscience. University of Wisconsin Press, Madison.

MANUSCRIPT ACCEPTED BY THE SOCIETY SEPTEMBER 1, 1987

Printed in U.S.A.

Geological Society of America
Memoir 169
1988

Brachiopoda from the La Meseta Formation (Eocene), Seymour Island, Antarctica

Lawrence A. Wiedman
Department of Geology, Monmouth College, Monmouth, Illinois 61462
Rodney M. Feldmann
Department of Geology, Kent State University, Kent, Ohio 44242
Daphne E. Lee
Department of Geology, University of Otago, Dunedin, New Zealand
William J. Zinsmeister
Department of Geosciences, Purdue University, West Lafayette, Indiana 47907

ABSTRACT

Twelve species of brachiopods have been collected from the La Meseta Formation (Eocene) of Seymour Island, Antarctica, arrayed in one inarticulate and nine articulate genera. None of these brachiopods is known to reside currently in the region near Seymour Island. Modern congeneric descendants of some of these taxa are found in both shallow- and deep-water habitats in a wide variety of Southern Hemisphere localities. These adaptations to a broad spectrum of habitats and the restriction to lower latitudes may have been triggered by changing conditions in the region subsequent to the development of the circum-Antarctic current following deposition of the La Meseta Formation. The articulate brachiopod genera ?*Probolarina, Tegulorhynchia,* ?*Plicirhynchia,* and *Terebratulina* are noted from the La Meseta Formation for the first time, although the latter three genera have been identified previously from the Tertiary beds of nearby Cockburn Island, Antarctica.

INTRODUCTION

Collections of marine macroinvertebrate body fossils found in the La Meseta Formation, an upper Eocene sandstone from Seymour Island, Antarctic Peninsula, Antarctica, have been obtained over the last decade. Analysis of this material reveals a diverse assemblage including bivalves, gastropods, scaphopods, decapod crustaceans, barnacles, echinoids, crinoids, asteroids, ophiuroids, a coral, and articulate and inarticulate brachiopods, as well as wood fragments, chordate bones and teeth, and trace fossils. This chapter analyzes the brachiopods found in the La Meseta Formation and interprets their paleoecological significance. The decapods (Feldmann and Zinsmeister, 1984; Feldmann, 1985; Feldmann and Wilson, this volume), bivalved molluscs (Zinsmeister, 1979), asteroids (Blake and Zinsmeister, 1979), and trace fossils (Wiedman and Feldmann, this volume) have been discussed elsewhere and are not considered herein.

Articulate and inarticulate brachiopods were first noted by early Swedish (Buckman, 1910) and Scottish (Jackson, 1912) south polar exploration teams working in the Seymour Island

region at the turn of the century. Buckman's collections were sufficient to describe most of the brachiopod fauna of the Cretaceous and Tertiary of the region (Owen, 1980). Owen (1980) studied the brachiopods from three islands, including Seymour Island, off the Antarctic Peninsula. Despite this recent, careful study, some of the taxa are reported herein for the first time and represent first occurrences not only for the La Meseta Formation, but for the continent of Antarctica as well.

Environmental conditions at the time of deposition of the La Meseta Formation have been interpreted from previously studied taxa of body fossils (Zinsmeister, 1979; Feldmann and Zinsmeister, 1984), the trace fossils (Wiedman and Feldmann, this volume), and by examination of the sedimentary sequences (Zinsmeister, 1979; Elliot and Trautman, 1982; Sadler, this volume). Each concluded that the unit was deposited in a shallow-water, nearshore, wave- and tide-dominated setting. The studies of the brachiopod taxa are in general agreement with that assessment.

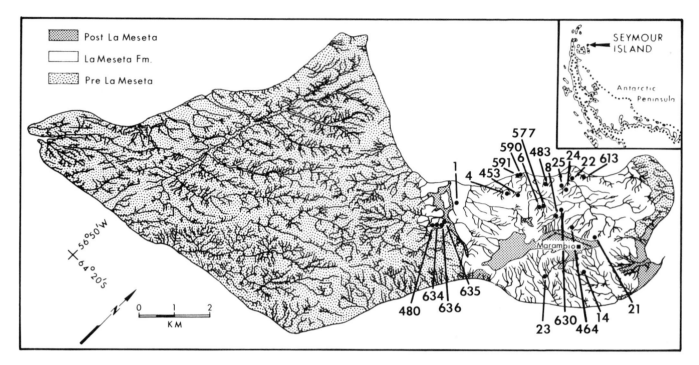

Figure 1. Map showing location of Seymour Island, Antarctica, and outcrops of the Eocene La Meseta Formation where brachiopods were collected. Locality numbers are keyed to occurrences listed in Table 1. The numbers cited are IPS (Institute for Polar Studies, Ohio State University) locality numbers and represent collections made by Zinsmeister, except localities, 1, 4, 6, 8, 14, 21, 22, 24, and 25, which represent sites collected by Feldmann.

ANALYSIS OF THE BRACHIOPOD FAUNA

Sixteen localities from various stratigraphic positions throughout the La Meseta Formation contain brachiopods. The unit is very fossiliferous in some areas, whereas body fossils are rare or absent in others. Concentrations of one or only a few taxa are common, especially among the molluscs. Of the seven subunits of the La Meseta Formation recognized by Sadler (this volume) in his comprehensive lithofacies mapping of the unit, only three have been noted to contain brachiopods. The subunits sampled represent the vast majority of the surficial exposures. Time and logistics prevented the entire extent of the formation from being sampled when the collections were made. Other specimens were collected in float or do not have precise locality information. The subunits recognized by Sadler have been designated Telm 1 through Telm 7, generally based on their stratigraphic position. Brachiopods were collected from subunits Telm 2, 5, and 7. Locality positions are shown in Figure 1, and a faunal list for each locality is shown in Table 1. Few of the brachiopod taxa preserved in the La Meseta have congeneric descendants in recent forms living in the Scotian biogeographical subprovince (Knox, 1960), the region of Seymour Island. However, many of the taxa do have congeneric descendants, which are known from elsewhere in the Southern Hemisphere.

The La Meseta Formation contains 11 articulate and 1 inarticulate brachiopod species, of which 9 can be identified, with certainty, to species level. Some have been included in prior reports of the fauna from Seymour Island (Buckman, 1910; Thomson, 1918; Owen, 1980) or the Antarctic Peninsula area (Foster, 1974), but only within limited stratigraphic ranges. *Liothyrella* cf. *L. lecta* (Guppy) (Fig. 2.13), *Liothyrella anderssoni* Owen, *Notosaria seymourensis* Owen (Fig. 2.10,11), *Magellania antarctica* Buckman (Fig. 2.5–8), *Terebratulina buckmani* Owen (Fig. 2.14,15), *Bouchardia zitteli* von Ihering, and *Bouchardia antarctica* Buckman (Figs. 2.1–4, 3) have each been found in significant numbers within the unit, although *B. zitteli* and *B. antarctica* are by far the most abundant in our collections, with more than 200 complete representatives of the genus. *Tegulorhynchia imbricata* (Buckman) (Fig. 2.9) is found in limited numbers, and *Liothyrella* Thomson is generally represented only by a few pedicle valves, although articulated specimens are known (Fig. 2.13). It is not uncommon for specimens assigned to *Liothyrella* to have the brachial valves detached and destroyed preferentially, as they are quite thin in comparison to the pedicle valves (G. A. Cooper and R. Grant, personal communication, 1984). Specimens of *Notosaria* Cooper have both valves preserved, but only eight specimens have been collected and are referred to as *N. seymourensis.*

TABLE 1. SYSTEMATIC LIST OF BRACHIOPOD TAXA COLLECTED FROM THE LA MESETA FORMATION*

Name	Localities	Units
Class **INARTICULATA** Order **LINGULIDA** Superfamily **LINGULACEA** Family **LINGULIDAE** *Lingula antarctica* Buckman	6, 464, 630	5, 7
Class **ARTICULATA** Order **RHYNCHONELLIDA** Superfamily **RHYNCHONELLACEA** Family **BASILIOLIDAE** ?*Probolarina* sp.	float	
Family **HEMITHYRIDIDAE** *Notosaria seymourensis* Owen	4, 480, 634	2, 5
Plicirhynchia sp.	float	
Tegulorhynchia imbricata (Buckman)	636	2
Order **TEREBRATULIDA** Suborder **TEREBRATULIDINA** Superfamily **TEREBRATULACEA** Family **TEREBRATULIDAE** *Liothyrella anderssoni* Owen	(*in* Owen, 1980)	
Liothyrella cf. *L. lecta* (Guppy)	453	5
Family **CANCELLOTHYRIDIDAE** *Terebratulina buckmani* Owen	635, 636	2
Suborder **TEREBRATELLIDINA** Superfamily **TEREBRATELLACEA** Family **TEREBRATELLIDAE** *Bouchardia antarctica* Buckman	1, 8, 14, 21, 23 24, 25, 453, 590, 591, 613	2, 5, 7
Bouchardia zitteli Ihering	1, 6, 8, 14, 21, 22, 24, 25	2, 5, 7
"Terebratella" crofti Owen	639	7
Magellania antarctica (Buckman)	483, 577, 591, 630	5, 7

*Numbers represent localities and stratigraphic units (Sadler, 1987) from which each was collected. Locality numbers are keyed to Figure 1.

A close relative of *Bouchardia zitteli* and *B. antarctica*, *Bouchardia rosea* (Mawe), is currently found living at depths of 10 to 20 m off the South American coast at Rio de Janeiro and Montevideo (Thomson, 1927) and São Paulo (Tommasi, 1970). Water temperatures at these locations are significantly higher than around Seymour Island today. Other closely related species, of Tertiary age, have been recorded from Chile and Patagonia (Levy, 1964), and a related genus, *Neobouchardia* Thompson, occurs in the Tertiary of New Zealand and Australia.

Numerous specimens of the taxon *Magellania antarctica*, a large, smooth terebratellacean, have been identified from the La Meseta Formation. This genus is absent from New Zealand Tertiary and Recent seas, although it occurs in the late Eocene of Patagonia (Levy, 1964) and is relatively common in modern subantarctic waters (Foster, 1974). Most reports of Recent occurrences have been in water depths ranging from 10 to 1,894 m (Foster, 1974). The temperature at these depths is presumed to be

approximately 4°C. The related *Magellania venosa* (Solander) ranges from the subtidal to 1,900 m off southern South America today. It is most common on the shelf and has temperatures ranging from 3 to 12°C (McCammon, 1973).

Another terebratellacean, *"Terebratella" crofti* Owen (Fig. 2.12), reported by Owen (1980), is similar to an unnamed species from the Chatham Islands (Lee, 1980). Lee has determined both to have most of the characteristic features of *Terebratella* d'Orbigny, but they are unique in the Terebratellidae in possessing strong dental plates.

Liothyrella, a terebratulacean, has been reported from several Recent localities including South America (Foster, 1974), Antarctica (Foster, 1974; Ayala et al., 1975), Australia (Foster, 1974), and New Zealand (Allan, 1932a,b; Foster, 1974). The genus has a wide bathymetric range. The occurrences from Antarctica (Ayala et al., 1975), and New Zealand (Foster, 1974; Allan, 1932a,b) can occur in quite shallow water ranging from 0

to 20 m. South American and Antarctic occurrences are much deeper, at least to 2,310 m (Foster, 1974). Most New Zealand records are from 0 to 100 m.

As mentioned earlier, generally only pedicle valves of this taxon seem to be preserved. However, a few entire specimens have been examined and are referred to *L.* cf. *L. lecta* (Guppy) (Fig. 2.13); Owen (1980) named *L. anderssoni* from the La Meseta Formation. The specimens in this study bear some resemblance to the extant *L. uva* (Broderip), a taxon known only from South America and Antarctica (Foster, 1974). Allan (1949) and Foster (1974) each studied various species of *Liothyrella* from localities throughout the Southern Hemisphere.

The genus *Terebratulina* d'Orbigny, another terebratulacean, is considered by Thomson (1927) to be "the most prolific in species of all Recent and Tertiary genera," and although not previously reported from the unit, is fairly common in the lower La Meseta Formation. Thus, *Terebratulina buckmani* represents one of the few truly cosmopolitan brachiopod genera reported from Seymour Island. It is worth noting that, in spite of its wide distribution in the southern circum-Pacific in the Paleogene, *Terebratulina* is extremely rare in the region today. Foster (1974) listed only three somewhat doubtful records, and only a handful of specimens are known from the New Zealand region, limiting the utility of depth or temperature information.

Notosaria, a rhynchonellide in the family Hemithyrididae, occupies a wide range of depths from the intertidal of South Island, Stewart Island (Percival, 1960; Allan, 1960; Rudwick, 1962; Lee and Wilson, 1979) and the Chatham Islands, New Zealand (Allan, 1932b) to a maximum known depth of about 800 m (Lee, 1978; Lee and Wilson, 1979). The Seymour Island occurrence of *N. seymourensis* is the oldest known for the genus. The dispersal pattern of *Notosaria* is still unclear, although some have previously addressed the issue indirectly (Allan, 1932b, 1949; Lee, 1978).

Tegulorhynchia imbricata (Buckman) (Fig. 2.9) is represented by a few fine-ribbed, spinose specimens whose stratigraphic position in the La Meseta Formation is uncertain. This genus was originally recognized on Cockburn Island by Buckman (1910) and is the first recorded occurrence from Seymour Island. A

Figure 3. Cluster of *Bouchardia antarctica* Buckman from the La Meseta Formation.

closely related New Zealand species, *T. squamosa* (Hutton), is recorded from the late Paleocene–early Eocene to middle Miocene (Lee, 1980). The related *T. doederleini* (Davidson) has a fossil record in Japan extending back to the Pliocene (Cooper, 1959) and an Indo-Pacific distribution in modern seas.

?Plicirhynchia Allan, another rhynchonellide, is noted for the first time from the lower subunits of the La Meseta Formation, although previously recorded from Cockburn Island (Owen, 1980). The assignment must remain questionable until such time as better preserved material is available for study. This genus appears to be restricted to the Eocene of the Antarctic Peninsula and South America (Williams, 1965, p. H624). Although it may serve as an index to the age of the La Meseta Formation, little information is available regarding its paleoecology.

One other rhynchonellide has been collected, from the lower subunits of the La Meseta Formation, that bears some resemblance to *Probolarina* Cooper (Fig. 2.16,17). Until additional specimens can be collected, verification of this occurrence is impossible. The genus has not been reported previously from Antarctica.

A single inarticulate brachiopod species, *Lingula antarctica* Buckman (Fig. 2.18–20), is fairly common at many localities, primarily in subunit Telm 5. This shallow-water burrowing form was originally described from two incomplete specimens by Buckman (1910), and later figured from an entire specimen by Owen (1980). Many specimens were found in association with gastropod and bivalved molluscs, the echinoids *Stigmatopygus* d'Orbigny and *Abatus* Truschel (McKinney et al., this volume), as well as the ichnogenus *Skolithos* Haldemann. Several small blocks contain multiple specimens in random orientation on an irregular bedding surface (Fig. 2.18). Many of these specimens

Figure 2. Representative brachiopods found in the La Meseta Formation. Scale bars equal 1 cm. 2.1-4, *Bouchardia antarctica* Buckman, brachial exterior, pedicle exterior, brachial interior and pedicle interior views. 2.5-8, *Magellania antarctica* (Buckman), brachial exterior, brachial interior with a portion of the posterior part of the pedicle valve attached, brachial interior and brachial exterior. 2.9, *Tegulorhynchia imbricata* (Buckman), brachial exterior. 2.10, 11, *Notosaria seymourensis* Owen, brachial exterior and pedicle exterior. 2.12, *"Terebratella" crofti* Owen, brachial exterior. 2.13, *Liothyrella* cf. *L. lecta* (Guppy), brachial exterior. 2.14, 15, *Terebratulina buckmani* Owen, brachial exterior and pedicle exterior. 2.16, 17, *?Probolarina* sp., pedicle interior and brachial interior. 2.18-20, *Lingula antarctica* Buckman, specimen with several individuals, unwhitened, individual with gastropod bore hole and individual with preserved color pattern, respectively.

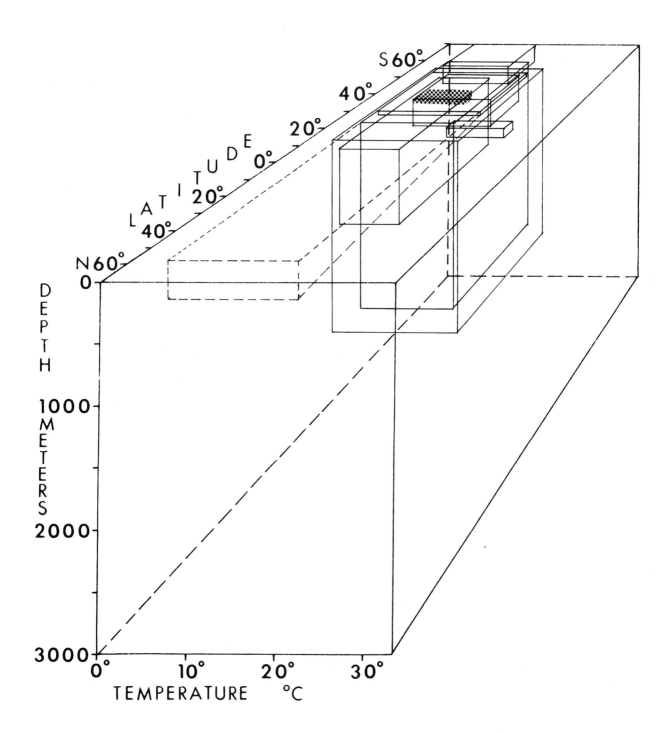

Figure 4. Plots of depth/temperature/latitude ranges of modern descendants for the six extant articulate brachiopod genera from the La Meseta Formation of Seymour Island, Antarctic Peninsula. Above, range distribution for all taxa showing (stippled area) the area of overlap of all taxa, excluding *Bouchardia*. Facing page, range distribution for common overlap (course stipple), the La Meseta Formation (intermediate stipple), and *Bouchardia* (fine stipple).

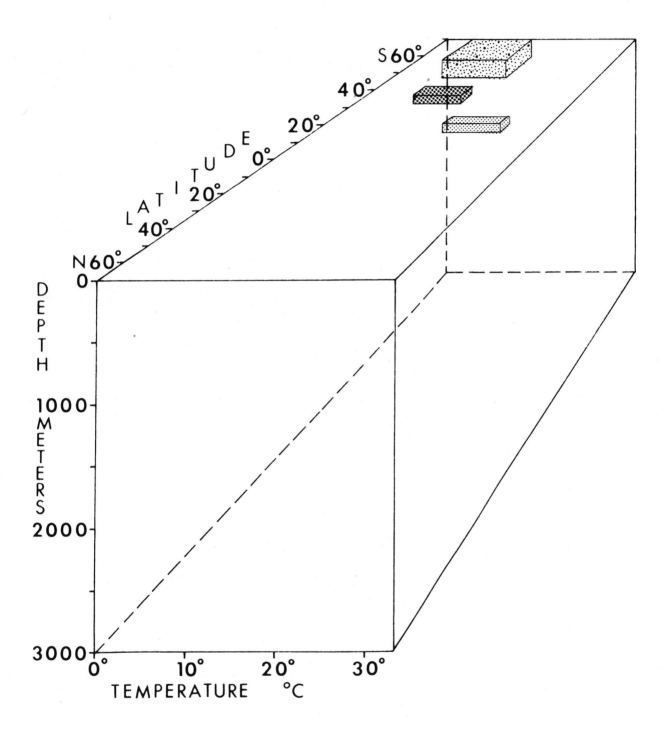

show black and bluish-gray color bands parallel to growth markings (Fig. 2.20). Although these may be interpreted as original color patterns, they do not necessarily represent true original colors. Some, preserved with conjoined valves, are filled with a fine quartzose sand. Two specimens appear to have been bored by gastropods (Fig. 2.19). Associations with other brachiopods, primarily *Bouchardia* Davidson, that can also live partially buried in coarse sand, and with the ichnogenera *Skolithos* and *Diplocraterion* Torell (Wiedman and Feldmann, this volume) indicate a very shallow-water depth (Paine, 1970). The possible low-salinity environment, suggested by the ichnofossil assemblage, would not have been tolerated by the other brachiopod taxa found in the unit. A related species, *Lingula waikatoensis* Penseler, has a long and nearly continuous record in the New Zealand Tertiary from the Paleocene to the Miocene (Lee and Campbell, 1987).

The presence of a large fossil population of *Lingula* Bruguière, whose recent forms are confined to tropical and subtropical seas, suggests that surface-water temperatures during deposition of the La Meseta Formation may have been at least 10 to 15°C warmer than those near Seymour Island today (Buckman, 1910; Thomson, 1927; Owen, 1980). The areas currently occupied by *Bouchardia* and *Tegulorhynchia* Chapman and Crespin, both widespread in the Paleogene southern oceans, are restricted to the regions much closer to the equator. The most southern recent occurrence of *Bouchardia* is approximately 40°S (Thomson, 1927; Owen, 1980). Tommasi (1970) measured sea temperatures ranging from 19 to 26°C off the coast of Rio de Janeiro where *Bouchardia* now flourishes. This warm temperature range is in agreement with other evidence offered by Churchill (1973), who reported well-developed mangrove vegetation in the late Eocene of southwestern Australia and by the warm-water nature of the mid- to late Eocene biotas of New Zealand (Hornibrook, 1978; Fleming, 1979) at latitudes assumed to be from between 40 and 50°S.

When temperature, depth, and latitude ranges are plotted for each of the six extant articulate brachiopod genera with the best documented recent records (*Bouchardia, Magellania, Terebratella, Liothyrella, Terebratulina,* and *Notosaria*), some range overlap can be seen (Fig. 4). All but *Bouchardia,* which is found in lower latitudes, have distributions that overlap near 35 to 40°S latitudes. Although depth information is incomplete in some cases and suspect in others, each appears capable of living in shallow depths from approximately 10 to 100 m. There is no uniform temperature at which each can be found. All except *Bouchardia* can be found living in temperatures ranging from 10 to 14°C, although this is at the extreme cool tolerance level for some. *Bouchardia* is found only in warmer waters from approximately 17 to 26°C (Tommasi, 1970; Foster, 1974). This may suggest that species of *Bouchardia* have become adapted to warmer water conditions subsequent to the Eocene. Certainly this type of comparison to modern congeneric descendents is not conclusive; however, it can be used to deduce general conditions during the depositional history of the La Meseta Formation. The temperature range of 10 to 14°C is in general agreement with late Eocene paleotemperatures derived by Shackleton and Kennett (1975), based on oxygen isotope analysis of planktonic foraminifera in deep-sea cores in the Antarctic Peninsula region. Thus, the study of the brachiopods from the unit suggests that conditions during Eocene time on Seymour Island were similar to those found in shallow-water settings in temperate lower latitudes with temperature ranges much warmer than are encountered in the Antarctic Peninsula region today.

DISCUSSION AND SUMMARY

Although none of the 12 brachiopod taxa from the La Meseta Formation is currently found living in the Seymour Island region, many have closely related congeneric descendants that now occupy habitats different from the temperate shallow-water, sandy environments interpreted to have occurred during deposition of the unit during late Eocene time. These taxa may reflect major dispersal events to lower latitudes with the onset of cooler conditions in the area resulting from the initiation of the circum-Antarctic current in Oligocene time. No uniform trend exists. Some of the taxa moved northward into warmer, shallow-water habitats; some dispersed to colder, deeper water conditions; still others became extinct. As observed earlier, with respect to the molluscs and decapods (Zinsmeister and Feldmann, 1984), the developmental role of high-latitude faunas may not have received the attention warranted as a factor in the evolution of biotas throughout Phanerozoic time. We offer the explanation that these taxa may have dispersed to lower latitudes, and in some cases, to habitats quite different from those found during La Meseta deposition—such as to colder, deep-water conditions. The fossil record of deep-water faunas from the Paleogene of the Southern Hemisphere is virtually nonexistent. It is possible that the taxa discussed herein and those identified earlier (Feldmann and Zinsmeister, 1984; Zinsmeister and Feldmann, 1984) may have been adapted to these deeper and colder habitats by Eocene time, and may have since become restricted to the habitats where their descendants are currently found. Certainly the role of ecological pioneering seems to be recognizable in many taxa from many higher taxonomic groups.

ACKNOWLEDGMENTS

Support for the field work on Seymour Island, for Rodney M. Feldmann, was provided by a National Science Foundation grant (to W.J.Z.). Support for laboratory work was provided by NSF Grant DPP-8411842 (to R.M.F.). Peter Sadler, Mike Woodburne, and William Daily of the Department of Earth Sciences, University of California at Riverside, and one of us (W.J.Z.) provided the specimens used in this study. Peter Sadler provided a manuscript copy of the geologic map of Seymour Island and permitted extracting major unit boundaries from the map for preparation of the location map. Contribution 339, Department of Geology, Kent State University, Kent, Ohio 44242.

REFERENCES

ALLAN, R. S. 1932a. The genus *Liothyrella* (brachiopoda) in New Zealand. Transactions of the New Zealand Institute, 63:1–10.

——. 1932b. Tertiary brachiopods from the Chatham Islands, New Zealand. Transactions of the New Zealand Institute, 63:11–23.

——. 1949. Notes on a comparison of Tertiary and Recent Brachiopoda of New Zealand and South America. Royal Society of New Zealand Transactions, 77:288–289.

——. 1960. The succession of Tertiary brachiopod faunas in New Zealand. Records of the Canterbury Museum, 7:233–268.

AYALA, F. J., J. W. VALENTINE, T. E. DELACA, AND G. S. ZUMWALT. 1975. Genetic variability of the Antarctic brachiopod *Liothyrella notorcadensis* and its bearing on mass extinction hypotheses. Journal of Paleontology, 49:1–9.

BLAKE, D. B., AND W. J. ZINSMEISTER. 1979. Two Early Cenozoic sea stars (Class Asteroidea) from Seymour Island, Antarctic Peninsula. Journal of Paleontology, 53:1145–1154.

BUCKMAN, S. S. 1910. Antarctic fossil brachiopoda collected by the Swedish south polar expedition. Ergebn. schwed. Sudpolarexped., Stockholm, 3(7):1–40.

CHURCHILL, D. M. 1973. The ecological significance of tropical mangroves in the Early Tertiary of Southern Australia. Special Publications of the Geological Society of Australia, 4:79–86.

COOPER, G. A. 1959. Genera of Tertiary and Recent rhynchonelloid brachiopods. Smithsonian Miscellaneous Collections 139, p. 90.

ELLIOT, D. H., AND W. J. TRAUTMAN. 1982. Lower Tertiary strata on Seymour Island, Antarctica Peninsula, p. 287–297. *In* C. Craddock (ed.), Antarctic Geosciences. University of Wisconsin Press, Madison.

FELDMANN, R. M. 1985. Eocene decapod crustaceans from Antarctica: ecological and biogeographical pioneers. New Zealand Geological Survey, Hornibrook Symposium, Extended Abstracts:44–46.

——, AND W. J. ZINSMEISTER. 1984. New fossil crabs (Decapoda; Brachyura) from the La Meseta Formation (Eocene) of Antarctica: paleogeographic and biogeographic implications. Journal of Paleontology, 58:1046–1061.

FLEMING, C. A. 1979. The geological history of New Zealand and its life. Auckland University Press, Auckland, 141 p.

FOSTER, M. W. 1974. Recent Antarctic and subantarctic brachiopods. Antarctic Research Series, American Geophysical Union, 21:1–189.

HORNIBROOK, N. DE B. 1978. Tertiary Paleontology, p. 407–443. *In* R. P. Suggate et al. (eds.), The Geology of New Zealand. Government Printing Office, Wellington.

JACKSON, J. W. 1912. The brachiopods of the Scottish national Antarctic expedition (1902 to 1904). Transactions of the Royal Society of Edinburgh, 48(19):367–390.

KNOX, G. A. 1960. Littoral ecology and biogeography of the southern oceans. Proceedings of the Royal Society of London, p. 577–624.

LEE, D. E. 1978. Aspects of the ecology and paleoecology of the brachiopod *Notosaria nigricans* (Sowerby). Journal of the Royal Society of New Zealand, 8:395–417.

——. 1980. Cenozoic and Recent Rhynchonellid brachiopods of New Zealand: systematics and variation in the genus *Tegulorhynchia*. Journal of the Royal Society of New Zealand, 10:223–245.

——, AND J. D. CAMPBELL. 1987. Cenozoic records of the genus *Lingula* (Brachiopoda: Inarticulata) in New Zealand. Journal of the Royal Society of New Zealand, 17 (in press).

——, AND J. B. WILSON. 1979. Cenozoic and Recent rhynchonellid brachiopods of New Zealand: systematics and variation in the genus *Notosaria*. Journal of the Royal Society of New Zealand, 9:437–463.

LEVY, R. 1964. Acerca de los Generos *Bouchardiella* y *Bouchardia* (Braquiopodes) en el Terciario de Patagonia (Argentina). Ameghiniana, 3:212–220.

McCAMMON, H. M. 1973. The ecology of *Magellania venosa*, an articulate brachiopod. Journal of Paleontology, 47:266–278.

OWEN, E. F. 1980. Tertiary and Cretaceous brachiopods from Seymour, Cockburn, and James Ross Islands, Antarctica. Bulletin of the British Museum (Natural History), Geology, 33:123–145.

PAINE, R. T. 1970. The sediment occupied by Recent lingulid brachiopods and some paleoecological implications. Paleaeogeography, Palaeoclimatology, Palaeoecology, 7:21–31.

PERCIVAL, E. 1960. A contribution to the life-history of the brachiopod *Tegulorhynchia nigricans*. Quarterly Journal of Microscopical Science, 101:439–457.

RUDWICK, M.J.S. 1962. Notes on the brachiopods of New Zealand, Transactions of the Royal Society of New Zealand (Zoology), 1:327–335.

SHACKLETON, N. J., AND J. P. KENNETT. 1975. Paleotemperature history of the Cenozoic and the initiation of Antarctic glaciation; oxygen and carbon isotope analysis in DSDP 277, 279, and 281, Vol. 29, p. 743–756. *In* J. P. Kennett et al., Initial Reports of the Deep Sea Drilling Project, Washington, DC.

THOMSON, J. A. 1918. Brachiopoda. Australasian Antarctic expedition. Science Reports, Series C, Zoology and Botany, 4:5–76.

——. 1927. Brachiopod morphology and genera (Recent and Tertiary). New Zealand Board of Science, Manual 7, 338 p.

TOMMASI, L. R. 1970. Sobre o braquiopode *Bouchardia rosea* (Mawe, 1828). Bolm Institute de Oceanography, São Paulo, 19:33–42.

WILLIAMS, A. 1965. Part H, Brachiopoda, p. H1–H927. *In* R. C. Moore (ed.), Treatise on Invertebrate Paleontology. University of Kansas Press and Geological Society of America, Lawrence.

ZINSMEISTER, W. J. 1979. The Struthiolaridae (Gastropoda) fauna from Seymour Island, Antarctic Peninsula. Congreso Geologico Argentino, Actas Buenes Aires, 1:609–618.

——, AND R. M. FELDMANN. 1984. Cenozoic high latitude heterochroneity of southern hemisphere marine faunas. Science, 224:281–283.

MANUSCRIPT ACCEPTED BY THE SOCIETY SEPTEMBER 1, 1987

Printed in U.S.A.

Geological Society of America
Memoir 169
1988

Balanomorph Cirripedia from the Eocene La Meseta Formation, Seymour Island, Antarctica

Victor A. Zullo
Department of Earth Sciences, University of North Carolina at Wilmington, Wilmington, North Carolina 28403
Rodney M. Feldmann
Department of Geology, Kent State University, Kent, Ohio 44242
Lawrence A. Wiedman
Department of Geology, Monmouth College, Monmouth, Illinois 61462

ABSTRACT

Two extinct species of balanomorph barnacles are identified from the La Meseta Formation (Eocene) of Seymour Island, Antarctic Peninsula. The report of *Austrobalanus macdonaldensis* Buckeridge in subunits Telm 5 and 7 represents the first noted occurrence of the species outside of South Island, New Zealand, and the first record of the genus *Austrobalanus* Pilsbry outside of Australasia. The notice of *Solidobalanus* sp. in subunits Telm 2, 5, and 7 represents the earliest known occurrence of the genus in the Southern Hemisphere. Records of the genus from middle Eocene deposits in Western Europe and North America predate those from the La Meseta Formation. Each taxon is typically associated with subtidal and inner shelf faunal assemblages in moderate- to high-energy environments.

INTRODUCTION

The La Meseta Formation, which crops out over much of the northeastern portion of Seymour Island (Fig. 1), is a sequence of marine sandstone and siltstone approximately 300 m thick. The sedimentologic and stratigraphic framework of the unit have been described by Elliot and Trautman (1982) and Sadler (1987). Based on enclosed mollusks, the unit has been assigned a late Eocene age (Zinsmeister, 1982). The primary attention focused on the unit has resulted from the observation that an abundant and diverse invertebrate fauna occurs in the La Meseta Formation. This chapter describes the balanomorph barnacles collected from the unit, and represents the first record of barnacles from Paleogene rocks in Antarctica.

The balanomorph barnacles represent only a small portion of the total assemblage found within the La Meseta Formation, but they are fairly abundant locally. The dominant bivalved and gastropod mollusks (Zinsmeister, 1979, 1984), as well as the decapod crustaceans (Feldmann and Zinsmeister, 1984; Feldmann and Wilson, this volume), brachiopods (Owen, 1980;

Wiedman et al., this volume), asteroids (Blake and Zinsmeister, 1979, this volume), echinoids (McKinney et al., this volume), and several trace fossils (Wiedman and Feldmann, this volume), collected from the unit have been discussed elsewhere.

Two species of balanomorph barnacles have been obtained from the Eocene La Meseta Formation of Seymour Island. The tetraclitid genus *Austrobalanus* Pilsbry is represented by *A. macdonaldensis* Buckeridge, previously known from the upper Eocene and lower Oligocene (Runungan-Whaingaroan) of South Island, New Zealand. The La Meseta specimens are abundant and well preserved, with opercular plates in growth position. The scuta differ slightly from those of the New Zealand examples in that they possess crests in the lateral depressor muscle pit. Possession of the crests is a feature characteristic of *Austrobalanus*, but apparently they are absent in typical *A. macdonaldensis*. Specimens of *A. macdonaldensis* were collected from four localities (5, 14, 21, and 602) (Fig. 1) and from loose float material within subunits Telm 5 and 7.

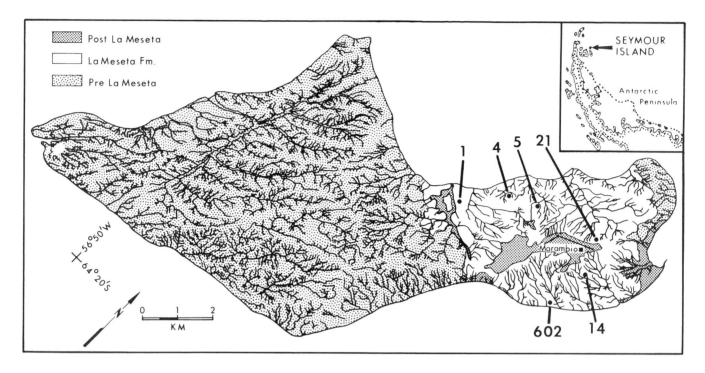

Figure 1. Location map showing the position of sites from which barnacles have been collected on Seymour Island, Antarctica.

The second La Meseta species is an archaeobalanid that does not occur typically in association with *A. macdonaldensis*. There is only one co-occurrence in subunit Telm 7 at locality 14. The high conical to cyindrical shell is smooth and rather thick, and bears well-developed radii. Poorly preserved opercular plates are often present in growth position within the indurated matrix that fills in the body chambers. Specimens of *Solidobalanus* were found in subunits, Telm 2 at locality 1, Telm 5 at locality 4, and Telm 7 at locality 14.

Archaeobalanids are also found in association with *Austrobalancus macdonaldensis* in the New Zealand Paleogene: *Paleobalanus lornensis* Buckeridge in the upper Eocene Waiareka Volcanic Formation, and *Solidobalanus everetti* Buckeridge in the lower to middle Oligocene MacDonald Limestone. The broad radii and the absence of a discernable tripartite division in the rostral plate of the La Meseta specimens indicate identification of this barnacle with *Solidobalanus* Hoek, in the broad sense, rather than with *Paleobalanus* Buckeridge.

Although the La Meseta barnacles provide us with only a glimpse of the Antarctic Paleogene cirriped fauna, they do permit comparison with the better known contemporary faunas in Australasia and the Northern Hemisphere, and begin to fill a major gap between the previously described cirriped faunas of Antarctic Cretaceous and upper Cenozoic strata.

SYSTEMATIC PALEONTOLOGY

Subclass CIRRIPEDIA Burmeister, 1834
Order THORACICA Darwin, 1854
Suborder BALANOMORPHA Pilsbry, 1916
Superfamily PACHYLASMATOIDEA Utinomi (Buckeridge, 1983)
Family TETRACLITIDAE Gruvel, 1903
Subfamily AUSTROBALANINAE Newman and Ross, 1976

Genus *AUSTROBALANUS* Pilsbry, 1916
AUSTROBALANUS MACDONALDENSIS Buckeridge, 1983
Figure 2.1–3, 6

Austrobalanus macdonaldensis BUCKERIDGE, 1983, p. 74, Figure 56a–f.

Holotype. Shell without opercular plates, no. OAR106, Department of Geology, University of Otago, New Zealand.

Type locality. MacDonald Limestone, Everetts quarry, South Island, New Zealand.

Geologic and geographic range. Late Eocene through early Oligocene (Runangan-Whaingaroan), South Island, New Zealand; late Eocene, Seymour Island, Antarctica.

Disposition of specimens. Figured hypotypes USNM 413819 and 413820 and unfigured hypotypes USNM 413821 through 413826 are

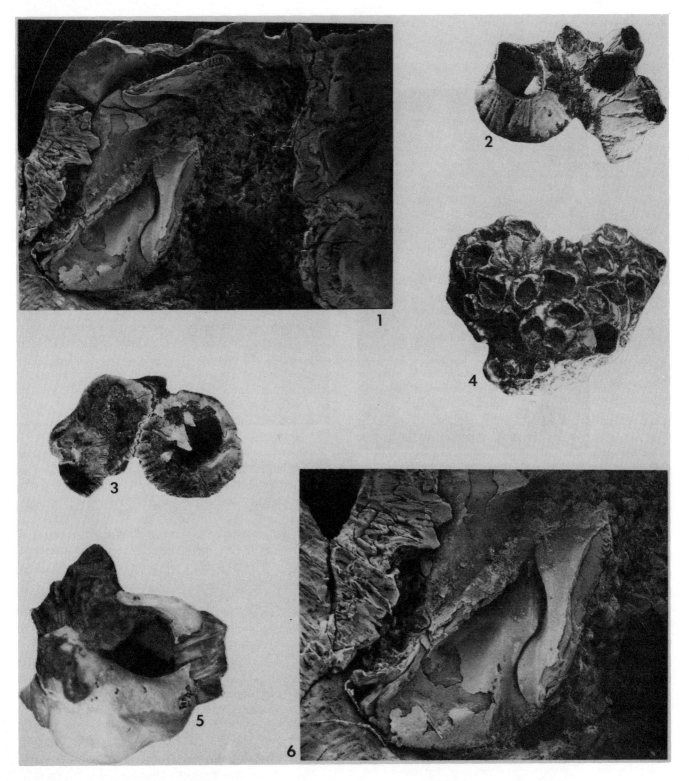

Figure 2. 1-3, 6, *Austrobalanus macdonaldensis* Buckeridge. 1, Scanning electron micrograph (SEM) of basal view of interior of shell containing articulated opercular pyramid, hypotype USNM 413819, collected from float, ×12. 2, 3, Orifice and basal views of shells, USNM hypotype 413820, collected from float, ×2.5. 6, SEM of one pair of opercular plates shown in figure 2.1, ×20. Note crests in lateral depressor muscle pit of scutum. 4, 5, *Solidobalanus* sp. 4, Shell cluster, hypotype USNM 413827, locality 14, ×2.5. 5, Large, cylindrical specimens on gastropod, hypotype USNM 413828, locality 1, ×1.5.

deposited in the collections of the Department of Paleobiology, National Museum of Natural History, Washington, D.C.

Remarks. The La Meseta specimens attributed to this species are small (as much as 6 mm in height and 11 mm in basal diameter), thin-shelled, high conical, and regularly plicate, with intervening sulci forming internal ribs. The interior of the basal shell margin bears irregular short ribs. The radii are narrow to moderately broad, with thin, smooth, sutural edges. The scutum agrees well with Buckeridge's (1983) description and illustrations, except that faint crests are present in the lateral depressor muscle pit (Fig. 2.1,6). The first author does not believe that this slight difference is sufficient grounds for specific or subspecific distinction. The tergum is narrow and arcuate, and has a short, basally rounded, tergal spur that is barely separated from the basiscutal angle. The basis is membranous. The La Meseta specimens occur in small, partially crowded clumps. Although none of the specimens is attached to a substratum, the smooth, regularly convex basal margins of the clumps suggest that they grew on smooth pebbles or the shells of gastropods.

Superfamily BALANOIDEA Darwin (Newman and Ross, 1976)
Family ARCHAEOBALANIDAE Newman and Ross, 1976
Subfamily ARCHAEOBALANINAE Newman and Ross, 1976

Genus *SOLIDOBALANUS* Hoek, 1913 (*sensu lato*)
SOLIDOBALANUS sp.
Figure 2.4,5

Disposition of Specimens. Figured specimens USNM 413827 and 413828 and all additional specimens, USNM 413829 through 413832, are deposited in the collections of the Department of Paleobiology, National Museum of Natural History, Washington, D.C.

Remarks. The specimens of *Solidobalanus* sp. are distinguished from those of *Austrobalanus* by their usually thicker shells, smooth parietes, broad radii with septate sutural edges, and thick, solid calcareous bases. *Solidobalanus* sp. also tends to be larger, attaining a maximum height of 20 mm and a basal diameter of 13 mm. The largest La Meseta specimens are crowded, cylindrical individuals on gastropod shells.

PALEOECOLOGY

Paleogene species of *Austrobalanus* and *Solidobalanus* are typically associated with subtidal and inner shelf faunal assemblages in moderate- to high-energy environments. The La Meseta barnacles are often still attached to substrata of pebbles, cobbles, and gastropod shells, and are preserved with their opercular plates in near-growth postition. This type of preservation is accomplished through rapid burial of living specimens and is characteristic of barnacle preservation in all but the latest high-stand deposits of coastal onlap sequences (Harris and Zullo, 1984). In contrast, barnacle assemblages of late progradational high-stand deposits are preserved as disarticulated, often worn, compartmental and opercular plate hashes, and represent deposition immediately prior to rapid sea-level fall.

Austrobalanus is represented by a single extant species restricted to the warm, temperate southeast coast of Australia. Extant *Solidobalanus* achieves its greatest diversity in the tropical and subtropical Indo-West Pacific region, but species are also known from warmer regions of the Atlantic Ocean. According to

Buckeridge (1983), Eocene and Oligocene faunal associates of *Austrobalanus macdonaldensis* are indicative of warm to subtropical hydroclimates at depths ranging from the immediate subtidal to 200 m.

GEOLOGIC HISTORY OF THE ANTARCTIC CIRRIPED FAUNA

The balanomorph barnacles of the La Meseta Formation help to fill an important gap in the geologic record of the Antarctic cirriped fauna. Although five taxa have been described from Cretaceous rocks and two from Pleistocene deposits, prior to this discovery no cirripeds were known from the Paleogene of Antarctica (Table 1). Even though only one of the two La Meseta barnacles is identifiable to species level, both are sufficiently diagnostic to permit some conjecture on the geologic history of the Antarctic cirriped fauna as compared to that of better known faunas in Australasia, North America, and Europe.

Buckeridge (1979, 1983) discussed the strong Tethyan influence on the Australasian Cretaceous cirriped fauna. A similar Tethyan influence is indicated for the Antarctic Cretaceous fauna. There are no endemic Antarctic genera, and although all the species-level taxa are endemic, two of the five differ only slightly from western European Cretaceous species. Clearly, the Antarctic Cretaceous cirriped fauna is part of the cosmopolitan fauna previously documented for western Europe, eastern North America, and Australasia.

Buckeridge's (1983) monographic study of Australasian fossil cirripeds provides the foundation for interpretation of the Eocene La Meseta fauna. Three major features distinguish Cretaceous cirriped faunas of the world from those of the Paleogene. The cosmopolitan nature of the shallow-water Cretaceous fauna gave way to increasing endemism, several Cretaceous genera became extinct at or near the Cretaceous-Tertiary boundary, and the lepadomorph-dominated Cretaceous fauna was replaced by a mixed lepadomorph-balanomorph fauna. These changes are readily observed in the development of the Australasian fauna as documented by Buckeridge (1983). As reconstructions of continental positions during the Eocene indicate that both New Zealand and Australia were in contact with, or in close proximity to, Antarctica (e.g., Smith et al., 1981), and that prevailing westerly circum-Antarctic currents would have provided an excellent pathway for the dispersal of cirriped species between Australasia and Antarctica, we can presume that the evolution of the Antarctic Paleogene fauna was similar to that of Australasia. Thus, the occurrences of *Austrobalanus macdonaldensis* in the upper Eocene of both New Zealand and the Antarctic Peninsula is readily acceptable on paleobiogeographic grounds. The specimens of *Solidobalanus* from the La Meseta fauna represent the earliest known occurrence of the genus in the Southern Hemisphere, as it does not appear in the Australiasian record until the early Oligocene. However, records of *Solidobalanus* from deposits of middle Eocene age (late Lutetian and Bartonian) in western Europe and North America predate those of the Southern Hemisphere, sug-

TABLE 1. FOSSIL CIRRIPEDIA OF ANTARCTICA

Cretaceous

Order **ACROTHORACICA**
†*Brachyzapfes elliptica gigantea* Taylor — belemnite rostra, upper Aptian, Alexander Island, Antarctic Peninsula (Taylor, 1965)

Order **THORACICA**
Suborder **LEPADOMORPHA**
†*Zeugmatolepas georgiensis* Withers — upper Aptian, Annenkov Island, South Georgia (Wilckens, 1947)
†*Cretiscalpellum aptiensis antarcticum* Taylor — upper Aptian, Alexander Island, Antarctic Peninsula (Taylor, 1965)
†*Euscalpellum antarcticum* Withers — middle Campanian, Antarctic Peninsula (Withers, 1951)

Suborder **BRACHYLEPADOMORPHA**
†*Pycnolepas articulata* Collins — (?)lower Aptian, Alexander Island, Antarctic Peninsula (Collins, 1980)

Paleogene

Order **THORACICA**
Suborder **BALANOMORPHA**
†*Austrobalanus macdonaldensis* Buckeridge — upper Eocene, Seymour Island, Antarctic Peninsula (reported herein)
†*Solidobalanus* sp. — upper Eocene, Seymour Island, Antarctic Peninsula (reported herein)

Pleistocene

Order **THORACICA**
Suborder **BALANOMORPHA**
Bathylasma corrolliforme (Hoek) — Scallop Hill Formation, McMurdo Sound (Speden, 1962; Newman and Ross, 1971)
†*Fosterella hennigi* (Newman) — Cockburn Island, Antarctic Peninsula (Hennig, 1911; Newman, 1979; Buckeridge, 1983)

† Dagger symbol indicates extinct species.

gesting an origin of the genus outside the Australasian-Antarctic region. Some caution must be taken with this conclusion, however, as most if not all, of the Paleogene Northern Hemisphere species identified with *Solidobalanus* are not assignable to this extant genus. The La Meseta fauna, however, scanty, is typical of inshore Eocene faunas; this suggests that the development of the Antarctic Paleogene cirriped fauna did indeed parallel that of Australasia after the demise of the cosmopolitan Cretaceous fauna.

The fate of the Antarctic Paleogene fauna was probably similar to that proposed by Buckeridge (1979, 1983) for the New Zealand fauna. As an example, *Austrobalanus* is found in the Oligocene and lower Miocene of New Zealand, but is absent in younger Cenozoic deposits (Buckeridge, 1983). Modern *Austrobalanus* is restricted to the southeastern Australian coast, where it also occurs in the Pleistocene. Buckeridge (1979) concluded that the disappearance of *Austrobalanus* from the New Zealand fauna was the result of late Cenozoic cooling trends and the develop-

ment of migration barriers between Australia and New Zealand. The demise of the Antarctic Paleogene fauna was undoubtedly related to climatic cooling, but could well have been initiated during the global cooling event of the latest Eocene and Oligocene.

Our knowledge of the upper Cenozoic fauna of Antarctica is limited to two occurrences. The extant Antarctic species *Bathylasma corolliforme* (Hoek) (=*Hexelasma antarctica* Borradaile) occurs in the Pleistocene Scallop Hill Formation of McMurdo Sound (Speden, 1962; Newman and Ross, 1971). *Fosterella hennigi* (Newman), first reported by Hennig (1911) as *Balanus* sp., occurs in a presumed Pleistocene deposit on Cockburn Island, off Graham Land. Presently, the only balanomorph known from the region south of the Antarctic convergence is *Bathylasma corolliforme*. Species of the extinct genus *Fosterella* Buckeridge are otherwise known from subantarctic regions, including the lower Pliocene through lower Pleistocene of Chatham Island, the Pleistocene of the Argentine shelf off Tierra del Fuego, and the lower

Pliocene through lower Pleistocene of New Zealand. According to Newman (1979), *Fosterella* was present in subantarctic regions until at least 35,000 yr ago, and thus did not become extinct until the last major glacial advance. It is likely that populations of *Fosterella* were established several times in Antarctica during interglacial periods in the late Pliocene and Pleistocene through recruitment from more permanent, northerly, subantarctic populations.

ACKNOWLEDGMENTS

Support for the field work on Seymour Island, for Rodney M. Feldmann, was provided by a National Science Foundation grant to William J. Zinsmeister. Support for laboratory work was provided by NSF Grant DPP-8411842 to Feldmann. Acknowlegment is made to the donors of the Petroleum Research Fund, administered by the American Chemical Society, for research support (to V.A.Z.). Peter Sadler, Department of Earth Sciences, University of California at Riverside, provided specimens and a manuscript copy of the geologic map of Seymour Island, and permitted extracting major unit boundaries from the map for preparation of the location map. Thoughtful reviews of the manuscript were provided by William A. Newman, Scripps Institution of Oceanography, and John Buckeridge, University of Auckland, New Zealand. Contribution 340, Department of Geology, Kent State University, Kent, Ohio 44242.

REFERENCES

BLAKE, D. B., AND W. J. ZINSMEISTER. 1979. Two early Cenozoic sea stars (Class Asteroidea) from Seymour Island, Antarctic Peninsula. Journal of Paleontology, 53:1145–1154.

BUCKERIDGE, J. S. 1979. Aspects of Australasian biogeography. Cirripedia: Thoracica. *In* Proceedings of the International Symposium on Marine Biogeography and Evolution in the Southern Hemisphere, Auckland, New Zealand, 1978. New Zealand Department of Scientific and Industrial Research Information Series 137(2):485–490.

——. 1983. Fossil barnacles (Cirripedia: Thoracica) of New Zealand and Australia. New Zealand Geological Survey Palaeontological Bulletin, 50:1–151.

COLLINS, J.S.H. 1980. A new *Pycnolepas* (Cirripedia) from the (?)lower Aptian of Alexander Island. British Antarctic Survey Bulletin, 50:21–26.

ELLIOT, D. H., AND W. J. TRAUTMAN. 1982. Lower Tertiary strata on Seymour Island, Antarctic Peninsula, p. 287–297. *In* C. Craddock (ed.), Antarctic Geosciences. University of Wisconsin Press, Madison.

FELDMANN, R. M., AND W. J. ZINSMEISTER. 1984. New fossil crabs (Decapoda; Brachyura) from the La Meseta Formation (Eocene) of Antarctica: paleogeographic and biogeographic implications. Journal of Paleontology, 58:1041–1061.

HARRIS, W. B., AND V. A. ZULLO. 1984. Recognition of coastal onlap sequences on basin margins using barnacle assemblages. Geological Society of America, Abstracts with Programs, 16(6):530.

HENNIG, A. 1911. Le conglomerate Pleistocene a Pecten de l'ile Cockburn. Schwedischen Sudpolar Expedition 1901–1903, vol. III. Geologie and Paleontology, 72 p.

NEWMAN, W. A. 1979. On the biogeography of balanomorph barnacles of the Southern Ocean including new balanid taxa: a subfamily, two genera and three species. Proceedings of the International Symposium on Marine Biogeography and Evolution in the Southern Hemisphere, Auckland, New Zealand, July 1978. New Zealand Department of Scientific and Industrial Research Information Series, 137(1):279–306.

——, AND A. ROSS. 1971. Antarctic Cirripedia. Antarctic Research Series, American Geophysical Union, 14:1–257.

OWEN, E. F. 1980. Tertiary and Cretaceous brachiopods from Seymour, Cockburn, and James Ross Islands, Antarctica. Bulletin of the British Museum (Natural History), Geology, 33:123–145.

SMITH, A. G., A. M. HURLY, AND L. C. BRIDEN. 1981. Phanerozoic paleocontinental world maps. Cambridge Earth Science Series, Cambridge University Press, Cambridge, 102 p.

SPEDEN, I. G. 1962. Fossiliferous Quaternary marine deposits in the McMurdo Sound Region, Antarctica. New Zealand Journal of Geology and Geophysics, 5(5):746–777.

TAYLOR, B. J. 1965. Aptian cirripedes from Alexander Island. British Antarctic Survey Bulletin, 7:37–42.

WILCKENS, O. 1947. Palaontologische und geologische Ergebnisse der Reise von Kohl-Larson (1928–29) nach Sud-Georgien. Abhandllung senckenbergiana naturforsch. Gesselschaft, 474:1–66.

WITHERS, T. H. 1951. Cretaceous and Eocene peduncles of the cirripede *Euscalpellum*. Bulletin of the British Museum (Natural History), A. Geology, 1(5):149–162.

ZINSMEISTER, W. J. 1979. The Struthiolaridae (Gastropoda) fauna from Seymour Island, Antarctic Peninsula. Congreso Geologico Argentino, Actas Buenes Aires, 1:609–618.

——. 1982. Review of the Upper Cretaceous–Lower Tertiary sequence on Seymour Island, Antarctica. Journal of the Geological Society of London, 139:776–786.

——. 1984. Late Eocene bivalves (Mollusca) from the La Meseta Formation, collected during the 1974–1975 joint Argentine-American expedition to Seymour Island, Antarctic Peninsula. Journal of Paleontology, 58:1497–1527.

MANUSCRIPT ACCEPTED BY THE SOCIETY SEPTEMBER 1, 1987

Geological Society of America
Memoir 169
1988

Eocene decapod crustaceans from Antarctica

Rodney M. Feldmann and Margaret T. Wilson
Department of Geology, Kent State University, Kent, Ohio 44242

ABSTRACT

Two species of anomuran decapods and six species of brachyuran decapods were identified from 14 localities in the Eocene La Meseta Formation on Seymour Island, Antarctica. Of these, six have not previously been identified from Antarctica, and four—*Munidopsis scabrosa, Homolodromia chaneyi, Calappa zinsmeisteri,* and ?*Micromithrax minisculus*—are new species. All of the records, with the exception of *Protocallianassa* cf. *P. faujasi,* represent the oldest occurrences of the genera in the fossil record. Three extant genera, *Munidopsis, Homolodromia,* and *Chasmocarcinus,* are known from the fossil record only in the La Meseta Formation. The fauna was preserved in sediments deposited in a cool temperate, nearshore, shallow-water habitat. Modern descendants of three of the genera—*Munidopsis, Homolodromia,* and *Lyreidus*—are known primarily from offshore, deep-water habitats.

INTRODUCTION

Eight species of decapod crustaceans, arrayed in the Anomura and Brachyura, have been identified (Table 1) from the La Meseta Formation that crops out on Seymour Island, a part of the Antarctic Peninsula, Antarctica (Fig. 1). Two of these species have been described (Feldmann and Zinsmeister, 1984b). The remaining six taxa not previously described are described herein. Additionally, paleoecological interpretations, primarily derived from analysis of the ecological requirements of recent congeners, are presented, which may be useful in explaining the general paleoecological setting of the La Meseta Formation.

Previous studies of the La Meseta Formation have established that the rocks were deposited in a variety of nearshore, shallow-water, wave and tidally dominated habitats interpreted to represent a deltaic complex (Elliot et al., 1982). Sadler (1986, this volume) suggested that the sequence of mappable sedimentary facies was deposited in a northwest-southeast–trending trough, approximately 6 km wide as exposed on Seymour Island, and that the sequence was more complex than had previously been thought. These facies were identified by a numerical sequence, Telm1 through 7, along with appropriate modifiers, representing general stratigraphic position. None of the units could be demonstrated to be continuous across the area. Rather, the lowermost unit, Telm1, was deposited in isolated sites along the margins of the trough and was superseded by Telm2, which seems to have been deposited along the walls of the trough, with beds dipping as much as 15°, generally toward the axis. Overlying

units, typically exhibiting dips to the southeast in the direction of plunge of the trough, have progessively more discontinuous distributions and represent a complex of shallowing-upward deposits in which textures and structures were controlled by local conditions.

Based in part on the paleoceanographic setting resulting from the Eocene configuration of continents, cool temperate water conditions, exhibiting high seasonal temperature fluctuation, were postulated by Zinsmeister and Feldmann (1984). The Eocene, or possibly earliest Oligocene, age was established by Zinsmeister (1982a), primarily on the basis of the abundant molluscan fauna. Subsequent studies have yielded few fossils that serve as good age indices. Wiedman et al. (this volume) have identified the brachiopod *Plicirhynchia* Allan, which is known only from late Eocene deposits in South America. The evidence for an Eocene age, based on study of the decapods, is not conclusive.

Feldmann and Zinsmeister (1984b) described the first two decapod taxa collected from the formation, *Lyreidus antarcticus* and *Chasmocarcinus seymourensis.* Subsequent collecting has yielded a few hundred specimens and has increased the decapod faunal diversity to eight taxa (Table 1). They represent the entire sampling of known Eocene decapods from Antarctica.

The arthropod fauna is significant in several ways. Four of the taxa—*Munidopsis scabrosa* n. sp., *Homolodromia chaneyi* n. sp., *Chasmocarcinus seymourensis* Feldmann and Zinsmeister,

**TABLE 1. SYSTEMATIC LIST OF
ANOMURAN AND BRACHYURAN DECAPOD CRUSTACEANS
COLLECTED FROM THE
EOCENE AGE LA MESETA FORMATION, SEYMOUR ISLAND, ANTARCTICA**

Infraorder **ANOMURA** H. Milne Edwards, 1832
Superfamily **THALASSINOIDEA** Latreille, 1831
Family **CALLIANASSIDAE** Dana, 1852
Protocallianassa cf. *P. faujasi*

Superfamily **GALATHEOIDEA** Samouelle, 1819
Family **GALATHEIDAE** Samouelle, 1819
Munidopsis scabrosa n. sp.[1,2]

Infraorder **BRACHYURA** Latreille, 1803
Section **PODOTREMATA** Guinot, 1977
Subsection **DROMIACEA** de Haan, 1833
Superfamily **HOMOLODROMIOIDEA** Alcock, 1899
Family **HOMOLODROMIIDAE** Alcock, 1899
Homolodromia chaneyi n. sp.[1,2]

Subsection **ARCHAEOBRACHYURA** Guinot, 1977
Superfamily **RANINOIDEA** de Haan, 1839
Family **RANINIDAE** de Haan, 1839
Lyreidus antarcticus Feldmann and Zinsmeister, 1984[3]

Section **HETEROTREMATA** Guinot, 1977
Superfamily **CALAPPOIDEA** de Haan, 1833
Family **CALAPPIDEA** de Haan, 1833
Calappa zinsmeisteri n. sp.[3]

Superfamily **PORTUNOIDEA** Rafinesque, 1815
Family **PORTUNIDAE** Rafinesque, 1815
?Callinectes sp.[3]

Superfamily **XANTHOIDEA** McLeay, 1838
Family **GONEPLACIDAE** McLeay, 1838
Chasmocarcinus seymourensis Feldmann and Zinsmeister, 1984[1,2]

Superfamily **MAJOIDEA** Samouelle, 1819
Family **MAJIDAE** Samouelle, 1819
?Micromithrax minisculus n. sp.[1]

[1]Absolute oldest occurrence in the fossil record of the genus.
[2]Sole occurrence of the genus in the fossil record.
[3]The first occurrence of the genus is Eocene.

and *?Micromithrax minisculus* n. sp.—represent the oldest known occurrence of the respective genera in the fossil record, and an additional three—*Lyreidus antarcticus* Feldmann and Zinsmeister, *Calappa zinsmeisteri* n. sp., and *?Callinectes* sp.—represent genera that have known geologic ranges of Eocene to Recent. Only one genus, *Protocallianassa,* represents a taxon that ranges back beyond the Eocene. Thus, the sample would seem to more closely represent a pioneer population than a relict one in the sense that several generic-level taxa are first noted in the fossil record of the La Meseta Formation. Furthermore, three of the taxa, *Munidopsis scabrosa, Homolodromia chaneyi,* and *Lyreidus antarcticus,* are precursors of modern species characteristic of deeper water. The remaining genera, with the exception of *Protocallianassa,* contain at least some species adapted to deep-water habitats in modern oceans.

Although corroborative work has not been completed, there seems to be some relationship between this Antarctic assemblage and the Eocene decapod fauna of New Zealand. *Lyreidus* spp. are dominant elements in New Zealand decapod faunas (Glaessner, 1960, 1980) as in the La Meseta assemblage, and majids and portunids are also present. The New Zealand Eocene assemblage may also be a pioneer population, strong in components ancestral to quiet, deeper water organisms. These similarities, although neither strong nor conclusive, corroborate the biotic relationship described by Zinsmeister (1982b) for the molluscs. He coined the term Weddellian Province to reflect the unit of this southern circum-Pacific assemblage (Zinsmeister, 1979).

The interpretation of these observations is that the Eocene circum-Antarctic ocean was probably characterized by highly seasonal, cool temperate water conditions which served as a site

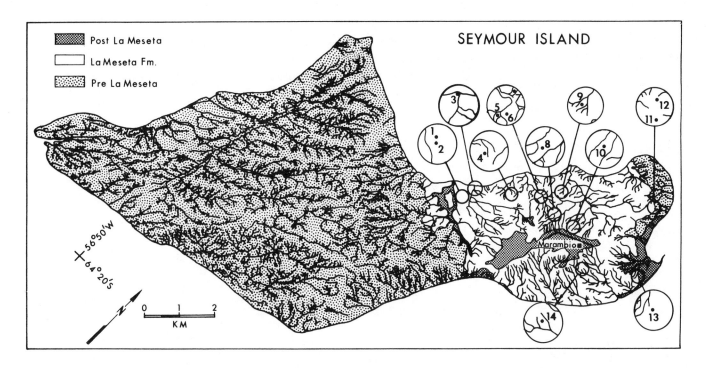

Figure 1. Map of Seymour Island showing the localities from which decapod crustaceans have been collected in the La Meseta Formation.

for the origin of at least some galatheid, homolodromiid, raninid, calappid, portunid, goneplacid, and majid genera that are now more widely dispersed geographically and ecologically.

LOCALITIES

Decapod crustaceans have been collected from 14 localities in the La Meseta Formation, which crops out on the eastern end of Seymour Island. Specific descriptions of the localities are given below. The relative position of the locations is illustrated in Figure 1.

Locality 1. This locality is a low, rounded knoll, at an elevation of about 15 m, overlooking the main drainage of the Cross Valley. *Lyreidus antarcticus,* barnacles, gastropods, bivalves, brachiopods, shark teeth, vertebrate bones, and trace fossils have been collected from the lower part of the La Meseta Formation, unit I of Elliot et al. (1982) and Telm2 of Sadler (1986, this volume).

Locality 2. This site is about 100 m south of Locality 1, and is another domal knob at an elevation of approximately 15 m, overlooking the Cross Valley. The lower part of the La Meseta Formation, unit I of Elliot et al. (1982) and Telm2 of Sadler (1986, this volume), is exposed, allowing collection of *Lyreidus antarcticus, Protocallianassa* cf. *P. faujasi,* gastropods, and bivalves.

Locality 3. This is a coastal section located on the western side of the mouth of a small valley. The lower part of the La Meseta Formation, unit I of Elliot et al. (1982) and Telm2 of Sadler (1986, this volume), is exposed, and *Protocallianassa* cf. *P. Protocallianassa faujasi, Munidopsis scabrosa, ?Micromithrax minisculus,* and several echinoderms have been collected.

Locality 4. A north-south–trending ridge crest, at an elevation of approximately 70 m, exposes the middle part of the La Meseta Formation, unit II of Elliot et al. (1982) and Telm4 of Sadler (1986, this volume). *Homolodromia chaneyi,* crinoids, brachiopods, gastropods, and bivalves have been collected from this locality.

Locality 5. *Lyreidus antarcticus,* barnacles, asteroids, brachiopods, gastropods, bivalves, vertebrate bones, teeth of polydolophid marsupials, and numerous trace fossils have been collected from this site, referred to as IPS (Institute of Polar Studies) locality 445 and as the "Rocket Site." This is the type locality for *L. antarcticus* Feldmann and Zinsmeister (1984). It is a rounded hill at the base of the meseta, at an elevation of about 45 m, in the middle part of the La Meseta Formation, unit II of Elliot et al. (1982) and Telm5 of Sadler (1986, 1987).

Locality 6. About 100 m northeast of Locality 4, at an elevation of about 50 m, *Lyreidus antarcticus* has been collected from the middle part of the La Meseta Formation, unit II of Elliot et al. (1982) and Telm5 of Sadler (1986, 1987), on the nose of a northwest-southeast–trending ridge.

Locality 7. This locality is a sloping divide between two small drainages dissecting the middle part of the La Meseta Formation, unit II of Elliot et al. (1982) and Telm5 of Sadler (1986, 1987), at an elevation of approximately 40 m. At this site, *Lyreidus antarcticus* is extremely abundant, and is associated with a fauna including relatively few brittle stars, echinoids, gastropods, bivalves, bryozoans, and trace fossils. Small, unidentifiable plant fragments are common.

Locality 8. This site is a steep, north-facing slope, at an elevation of approximately 150 m, exposing the upper part of the La Meseta Formation, unit III of Elliot et al. (1982) and Telm7 of Sadler (1986, this volume). *Homolodromia chaneyi*, large numbers of brittle stars, echinoids, gastropods, bivalves, brachiopods, bryozoans, teredid bored wood, shark teeth, vertebrate bones, and several trace fossils have been collected from this locality.

Locality 9. This locality is a steep, north-facing slope, at an elevation of approximately 30 m, overlooking the primary drainage on the eastern end of Seymour Island. *Lyreidus antarcticus* has been collected in association with numerous small gastropods and bivalves in the middle part of the La Meseta Formation, unit II of Elliot et al. (1982) and Telm5 of Sadler (1986, 1987).

Locality 10. *Homolodromia chaneyi* has been collected in association with crinoids, brittle stars, gastropods, bivalves and lingulide brachiopods at this site near the crest of the meseta, just below the memorial cross at Marambio. The locality, at an elevation of about 170 m, is in the upper part of the La Meseta Formation, unit III of Elliot et al. and Telm7 of Sadler (1986, this volume).

Locality 11. *Lyreidus antarcticus* was collected in the middle part of the La Meseta Formation, unit Telm3 of Sadler (1986, this volume) near the base of a steep, east-facing slope at an elevation of approximately 20 m.

Locality 12. The type locality of *Chasmocarcinus seymourensis* Feldmann and Zinsmeister (1984) is located at the top of a small hill, at an elevation of approximately 160 m. This is IPS locality 14. The middle part of the La Meseta Formation, unit II of Elliot et al. (1982) and Telm3 of Sadler (1986, this volume), is exposed at this site.

Locality 13. *Lyreidus antarcticus* was collected at this site, at an elevation of approximately 15 m, by Dan Chaney, U.S. National Museum of Natural History. The locality is in the middle part of the La Meseta Formation, unit II of Elliot et al. (1982) and Telm3 of Sadler (1986, this volume).

Locality 14. Specimens were collected on a divide between two small drainages at an elevation of approximately 70 m. *Homolodromia chaneyi, Calappa zinsmeisteri, ?Callinectes* n. sp., ascothoracican barnacles, ophiuroids, crinoids, gastropods, bivalves, brachiopods, serpulids, wood bored by teredid bivalves, and vertebrate bone fragments have been collected at this locality in the upper part of the La Meseta Formation, unit III of Elliot et al. (1982) and Telm7 of Sadler (1986, this volume).

SYSTEMATIC PALEONTOLOGY

Superclass CRUSTACEA, Pennant, 1777
Class MALACOSTRACA Latreille, 1806
Order DECAPODA Latreille, 1803
Infraorder ANOMURA H. Milne Edwards, 1832
Superfamily THALASSINOIDEA Latreille, 1831
Family CALLIANASSIDAE Dana, 1852
Subfamily PROTOCALLIANASSINAE Beurlen, 1930

Genus *Protocallianassa* Beurlen, 1980
Protocallianassa cf. *P. faujasi* (Desmarest)
Figures 2.1–3, 3

Description. Material basis for taxon limited to crushed remains of major and minor claws, arm of major claw, and aureole of decompositional products of cephalothorax of one specimen and interior surfaces of right major cheliped of a second specimen.

Major claw moderate size for genus, quadrate, fingers short, stout; no distinct denticles developed. Hand tapers distally from maximum height near carpus-propodus. Maximum length of manus developed along upper surface. Proximal termination, carpus-propodus joint, intercepts base of manus at angle of about 120°. Fixed finger terminating distally in sharp, upturned point. Occlusal surface broadly undulose with no apparent denticles. Dactylus tapering to downturned, pointed termination crossing over inner surface of fixed finger. Occlusal surface undulose, edentate. Surface ornamentation unknown.

Carpus crushed, fragmented, but appears relatively short. Merus ovoid in lateral aspect, about 5.3 mm long and 3.6 mm high, with greatest height near midlength. Ischium poorly preserved but appears to have flabellate enlargement distally and upwardly curved, narrow proximal termination. Surfaces of these elements appears smooth.

Minor claw smaller, more elongate, more delicate than major claw. Manus subquadrate, slightly higher proximally than distally, upper and lower surfaces smoothly convex. Fixed finger slender, slightly downturned, occlusal surface with single undulation near proximal end. Dactylus nearly circular in cross section, slender, no denticles or undulations evident. Surfaces smooth.

Additional remains consist of numerous small fragments of integument and an ill-defined area of stained, weakly fluorescent material located in the position of the cephalothorax.

Measurements. Measurements, in millimeters, are given in Table 2.

Studied material. The two specimens, USNM 404849a and b and 404850, are deposited in the U.S. National Museum of Natural History, Washington, D.C.

Localities and stratigraphic position. The specimens were collected near the base of the late Eocene La Meseta Formation at localities 2 (USNM 404849) and 3 (USNM 404850) (Fig. 1), Seymour Island, Antarctica.

Remarks. Only two specimens, referable to the Callianassidae, have been collected from the La Meseta Formation. This would appear to be anomalous in that burrows referable to the ichnogenus *Ophiomorpha* and attributed to the work of callianassids (Weimer and Hoyt, 1964) are abundant throughout the unit. However, the frequency of *Ophiomorpha* burrows is low in the stratigraphic position of the callianassids but tends to be very high in the position from which another decapod, *Lyreidus,* is found. This suggests that *Lyreidus* might be the producer of *Ophiomorpha* in this instance. The callianassids in the La Meseta were burrowers, but the absence of evidence of their burrowing activity must be attributed to taphonomic processes, rather than to a change in their lifestyle.

Figure 2. *Protocallianassa* cf. *P. faujasi.* 1, USNM 404850, showing portions of right and left pereiopods. 2 and 3, USNM 404849a and b, showing interiors of a major cheliped. Note the geopetal structure preserved as a calcite coated mass of sand in the lower part of the hand. Scale bars = 1 cm.

One of the specimens upon which the above description was based (USNM 404849) consists of the inner surface of a right cheliped preserved in a small ovoid concretion. The chela was apparently separated from the carpus prior to burial. The lower part of the hand is filled with sediment similar to that in the surrounding concretion and constitutes a geopetal structure. In addition to the sediment fill, a thin lining of calcite obscures interior detail of the claw. Thus, the only clues to identity of the organism, based upon this specimen, lie in the general outline and relative proportion of elements of the propodus and dactylus.

The second specimen (USNM 404850), which is proportionally smaller than the first, preserves more elements of the arm of the major cheliped as well as the minor claw. This fortunate circumstance demonstrates differences in size and proportions of the claws, characters important in the classification of the group. In addition to the chelipeds, an aureole of tiny fragments and a dark stain on the rock surface seem to define the position of the cephalothoracic region. Examination of this surface in plain and ultraviolet light, however, does not reveal enough detail to allow adequate description of the structure.

Although confusion surrounds the distinction between *Callianassa* and *Protocallianassa*, these specimens most closely conform to the description of the chelipeds of *Protocallianassa* as discussed by Mertin (1941, p. 199). He outlined criteria useful in distinguishing the two genera, two of which can be applied to the identification of this material. The reentrant, along the distal margin of the manus and just below the base of the dactylus, is relatively shallow. In *Callianassa* this feature is often quite deep. More important, the articulation between the carpus and propodus lies at an angle of about 120° to the long axis of the propodus. By contrast, the axis of rotation of this joint on the overwhelming majority of specimens referred to *Callianassa* is nearly 90°. If this evidence is correctly interpreted, it documents an upward range extension for *Protocallianassa* from the Paleocene (Glaessner, 1969, p. R478) into the Eocene. Most specimens of *Protocallianassa* are Late Cretaceous in age.

Several callianassids have been identified from Antarctica. Ball (1960) described *C. meridionalis* from Upper Cretaceous rocks of James

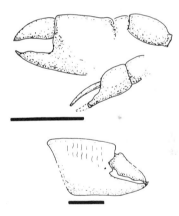

Figure 3. Line drawings of chelipeds of *Protocallianassa* cf. *P. faujasi* based on USNM 404850 (upper) and 404849 (lower). Bar scales = 1 cm.

Ross Island. Those specimens tend to be smaller than ours, and the carpus-propodus joint is clearly at right angles to the base of the hand. That same morphologic condition distinguishes the present specimen from *C. symmetrica* Feldmann and Zinsmeister (1984a). Specimens referable to *Protocallianassa* have been recognized in lower Aptian rocks from Alexander Island by Taylor (1979), but neither of the taxa that he identified closely approximates the specimens from the La Meseta. *Protocallianassa antarctica* Taylor has a relatively slender hand and fingers that greatly exceed the length of the manus, a condition generally not known in other representatives of the genus. *Protocallianassa* sp. (Taylor, 1979, p. 22) is easily distinguished from the specimens from the La

Specimen	L-manus	H-manus	W-manus	L-finger	H-finger	L-dactyl	H-dactylus
404849-major	15.2	12.0	10.5	8.3	3.6	ca. 8	5.2
404850-major	14.1	——	8.2	5.2	3.0	6.1	2.9
404850-minor	ca. 4.4	3.5	2.9	>4.3	1.4	5.8	1.2

Note: All measurements in millimeters.

Meseta in that, in the former, the height of the hand exceeds the length. Additionally, the longest part of the hand on the specimen from Alexander Island lies along the lower surface rather than along the upper surface. No described fossil callianassids from New Zealand bear close comparison with the La Meseta forms.

The species to which the La Meseta specimens can be compared most closely is *Protocallianassa faujasi* (Desmarest). They are similar in relative proportions (Mertin, 1941, p. 201) and in general outline. *Protocallianassa faujasi* has been described from a number of Cretaceous sites in Germany. In the absence of additional bases for comparison, however, it would seem inappropriate to apply that name, with certainty, to this Eocene form. For the same reason it seems unwise to designate a new name.

Superfamily GALATHEOIDEA Samouelle, 1819
Family GALATHEIDAE Samouelle, 1819
Subfamily MUNIDOPSINAE Ortmann, 1898

Genus *Munidopsis* Whiteaves, 1874

Remarks. Fossils previously assigned to the Galatheidae possess a carapace that is longer than wide, which bears transverse ornamentation over some or all of the surface, and a well-developed triangular rostrum. As with other members of the Galatheoidea, the epistome is not fused with the carapace. These features are all demonstrable on these specimens of *Munidopsis,* rendering placement in the family a certainty. Genera within the family are distinguished from one another on the basis of details of the rostrum, frontal margin, definition of regions, and development of transverse sculpture.

Ambler (1980) provided a diagnosis of *Munidopsis,* which is represented by over 140 living species, and until this report, no fossil forms. Important points of comparison relative to material from Seymour Island are evident on the carapace. The rostrum tends to be keeled and typically is smooth or only slightly serrated on the lateral margins. The frontal region may lack spines or possess small antennal spines; large supraorbital spines, such as those seen in *Munida,* are not present. The lateral margins may be variously arrayed with spines, or may lack spines. The gastric, cardiac, and branchial regions are well defined. The dorsal surface may be transversely rugose, nodose, squamose, or nearly smooth. On living forms, the eyes are not pigmented and lack facets.

Until recently (Via Boada, 1981, 1982), most fossil galatheids had been referred to the Galatheinae. Yet no described genera within this subfamily can accomodate the Antarctic specimens. The rostrum in all galatheins is serrated or spined, except that of *Munida* Leach and *Protomunida* Beurlen. In both of these genera, however, spines are developed at the base of the rostrum. The transverse ornamentation is not as strongly developed as it is in *Galathea* Fabricius, *Munida, Paleomunida*

Lorenthey, and *Protomunida. Rugafarius* Bishop (1985) was defined on the basis of possession of three transverse grooves, none of which cross the midline. Instead, the midline in this genus is defined by a ridge that is prominent in the thoracic region, termed the "scapular arch" by Bishop (1985, p. 615), and subtle in the cephalic region.

Within the Munidopsinae, transverse sculpture is well developed in *Paragalathea* Patrulius and *Eomunidopsis* Via Boada; both have dentate rostra (Via Boada, 1982). Only *Munidopsis* may have weakly developed transverse ornamentation and a smooth, keeled rostrum. Transverse ornamentation in the *Munidopsis* described herein is limited to rows of nodes and scales best developed on the branchial region, and to a lesser extent, on the axial regions. The rostrum is strongly keeled and devoid of spines or serrations.

As summarized by Via Boada (1981, p. 249), the concept of generic and subgeneric units in the Galatheidae has been variously interpreted by recent workers. Even the attempts to refine the classification of species within the single genus *Munidopsis* (Milne Edwards and Bouvier, 1894) have been proven to be inadequate (Chace, 1942). Clearly, the generic descriptors employing details of spinosity and ornamentation of the rostrum and lateral margins may be viewed as of considerable significance. Perhaps it may be more reasonable to consider the nature of the groove patterns—as they define major regions on the carapace—and gross aspects of the sculpture as being more significant generic descriptors.

The fossils from Seymour Island may be referred to *Munidopsis,* in the Munidopsinae, with confidence and appear to be quite different from taxa in the Galatheinae. Thus, this notice represents the first record of *Munidopsis* in the fossil record.

Munidopsis scabrosa n. sp.
Figure 4.1–3, 5

Description. Carapace small size for family; outline quadrate, slightly longer than wide, weakly convex longitudinally, strongly vaulted transversely; rostrum prominent.

Frontal region about one-third the width of carapace, strongly depressed below level of gastric region; defined posteriorly by narrow transverse sulcus interrupted mesially by axial ridge extending from mesogastric region anteriad onto rostrum. Rostrum elongate triangular, about one-third total length of remainder of carapace, margin with narrow, well-defined smooth rim, prominently keeled axially at least in posterior one-half the length; remainder of surface slightly arched, finely pustulose. Orbits, when viewed from above, expressed as shallow concavities at base of rostrum. Anterolateral corner with two nodose protuberances, which may be spine bases, separated by broad, shallow reentrant. Lateral margins subparallel, slightly convex, greatest width of cephalothorax near midlength of carapace where cervical groove crosses midline. Posterior margin weakly concave with smooth border and narrow, well-defined rim anterior to border.

Figure 4. *Munidopsis scabrosa* n. sp. 1, Holotype, USNM 404851, dorsal aspect of cephalothorax. 2, Paratype, USNM 404860, dorsal aspect of cephalothorax. 3, Paratype, USNM 404859, sternum and portions of two pereiopods. Scale bar = 1 cm.

Carapace regions and grooves well defined as pustulose or scabrose, domed regions and broad, shallow, smooth depressions, respectively. Gastric region with two pairs of subtle, ovoid elevations identifiable as epigastric and protogastric regions, former smaller than latter; mesogastric region narrow anteriorly and broadening abruptly posteriorly. Gastric regions with transverse rows of scabrous ornamentation, steeply sloping anteriorly and gently sloping posteriorly. Hepatic regions reflected as circular domed areas ornamented by numerous fine nodes. Cervical groove lyrate, deeply and broadly impressed in the axial region, becoming broad and poorly defined in mesolateral areas, and narrow and well defined near lateral terminations. Cardiac region broad, about three-fourths the total width of carapace, well defined by smooth sulci, narrowest axially and widest near adaxial terminations; surface with pair of large nodes adjacent to midline and numerous smaller nodes laterally. Intestinal region large, triangular, irregularly domed; greatest width, at anterior, about one-half the total carapace width; posteriormost expression as an axial triangular swelling. Epibranchial region triangular, ornamented by numerous nodes; separated from remainder of branchial region by subtle depression. Remainder of branchial region smoothly arched, ornamented by rows of nodes and forward-directed scabrose ornamentation.

Sternal region with at least five discrete pairs of elements, each separated from others by narrow, deeply incised depressions; anteriormost elements, fused sternites of MXP1-3, small, forming a triangular unit; subsequent elements widening uniformly from P1–P3, those of P4 apparently slightly narrower than P3; length of somites, measured along midline, decreases posteriorly, P2 only slightly shorter than P1, P3, and P4 much shorter.

Abdomen unknown.

Appendages known only from separate, long, slender, finely spinose articles, largest 13 mm long and 1.5 mm wide.

Measurements. Measurements on the specimens referred to this species are given in Table 3. Specimens representing part and counterpart of the

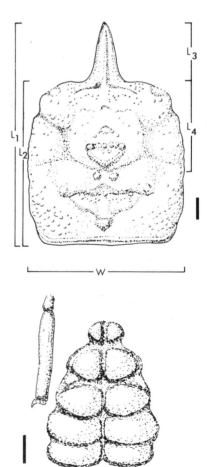

Figure 5. Line drawings of the cephalothorax (upper) and sternum (lower) of *Munidopsis scabrosa,* showing positions and orientations of measurements. Bar scales = 1 mm.

TABLE 3. MEASUREMENTS TAKEN ON SPECIMENS OF
MUNIDOPSIS SCABROSA FROM THE LA MESETA FORMATION

Specimen	L-1	L-2	L-3	L-4	W
404851	11.7	8.6	3.1	4.4	7.7
404852-mold	——	4.0	——	2.1	4.4
404853	——	ca. 7.2	——	3.9	ca. 5.2
404854	9.1	6.9	2.2	3.5	5.9
404855-mold	——	7.2	——	4.9	ca. 6.0
404856	5.2	3.7	1.5	1.8	3.6
404857-mold	14.2	11.5	2.7	5.4	12.4
404858-mold	——	——	1.2	1.8	ca. 3.4
404859	——	8.6	——	4.8	9.0
404860	8.3	6.4	1.9	3.4	6.2
404861-mold	——	5.8	——	ca. 3.0	5.4
404862-mold	11.7	9.2	2.5	5.1	8.7
404863-mold	——	6.2	——	3.5	5.4
404864-mold	5.2	3.9	1.3	2.0	ca. 7.0
404865-mold	——	4.3	——	2.3	——
404866	13.4	10.4	3.0	5.5	9.1
404867	——	7.2	——	3.8	ca. 7.2

Note: All measurements in millimeters.

same individual are measured and recorded as separate specimens because the nature of preservation results in slightly different measurements on each, and part and counterpart are difficult to relate.

Types. The holotype USNM 404851, and 18 paratypes, USNM 404852 through 404869, are deposited in the U.S. National Museum of Natural History, Washington, D.C.

Etymology. The trivial name was derived from the Latin word *scab(e)r* = rough, reflecting the nature of the ornamentation, considered to be a distinguishing characteristic of this species.

Locality and stratigraphic position. All specimens referred to this species were collected from a single exposure near the base of the late Eocene La Meseta Formation at Locality 3 (Fig. 1), Seymour Island, Antarctica.

Remarks. Most fossil galatheids have been distinguished mainly on morphology of the cephalothorax. Because of the fine quality of preservation of the La Meseta specimens, comparison has been made with adequate detail to assure the uniqueness of *Munidopsis scabrosa.*

The sternal region (USNM 404869), which is preserved in association with carapace material, may represent only the second such part known from the fossil record; for this reason, it offers little basis for comparison, except with extant forms. Via Boada (1982, Plate 3, Figs. 1, 2) illustrated two sterna he referred to two different, undetermined species. It is not clear that the two specimens differ from one another, and both may be referrable to *Galathea.*

Examination of sternal elements of a variety of living representatives of the Galatheoidea suggests that three general forms can be identified, based upon the relationship of the MXP1-3 element and that of P1. In one group, represented *Munidopsis* and *Munida,* MXP1-3 is clearly separated from the posterior elements by a narrow constriction. By contrast, that region in *Galathea, Sadayoshia,* and *Eumunida* shows a distinct separation of the two elements, but the constriction is less

pronounced or is reduced to a narrow depression. Finally, in the chirostylid genus *Uroptychus* the anterior elements are almost entirely fused, and their positions are marked only by marginal reentrants. The sternum of *Munidopsis scabrosa* appears to most closely resemble those of living *Munidopsis* and *Munida,* in which the sternal elements supporting attachments for the maxillipeds are discrete, large, and obviously separated from those of the pereiopods. This line of evidence tends to confirm the generic placement, but until much more comparative work is conducted on sternal elements in the family, the evidence must be taken as suggestive.

The ecological settings in which living galatheids have been collected are varied. They range through most of the oceanic regions of the world and throughout nearly the entire spectrum of temperatures and depths, except in the Antarctic. Baba (1979), for example, reported 32 species of galatheids from the region of the Moluccas, in the East Indian Archipelago. Seventeen of these species were dominantly deep-water forms, and 14 were confined to shallow-water, reef habitats. Similar variations in ecological requirements have been noted in many other regions. *Munidopsis,* however, is almost exclusively restricted to deepwater, aphotic settings (Austin Williams, personal communication), including at least one species that is adapted to the thermal vents of the Galapagos Rift (Corliss and Ballard, 1977). One of the few records of a shallow-water occurrence of a species of *Munidopsis* is that of *M. polymorpha* Koelbel, which is known only from subterranean habitats in the Canary Islands (Miyake and Baba, 1970).

Although it is tempting to generalize on morphological variations between taxa inhabiting shallow-water realms and those living in deeper water, no such contrasts are apparent in the Galatheoidea. Shallow-water species may have shorter, stouter walking legs and prominent ornamentation, as for example, in *Galathea inflata* Potts, 1915 (fide, Baba, 1979); deep-water forms may have more delicate ornamentation and longer, more slender appendages, as in the various species of the chirolistid genus *Uroptychus,* as described for example, by Baba (1981). On the other hand, *Munidopsis gibbosa* Baba, 1978 was taken from a depth of 520 to 560 fathoms in the South China Sea and has stout appendages and coarse ornamentation. Based solely upon observations of the morphology of this species, it would probably be considered a shallow-water species. The scabrous ornamentation and general outline of the carapace may lead to the conclusion that the form had burrowing capability, backing into the burrow (Savazzi, 1986).

For the above-stated reason, therefore, little can be said about the ecological requirements of *Munidopsis scabrosa* based on functional morphology. Its preservation in sediments interpreted as having been deposited in shallow-water, nearshore habitats may be taken as the best available interpretation of its living site. Thus, the adaptation of living representatives of *Munidopsis* to bathyal and abyssal regions represents a habitat preference that is significantly different from that of *M. scabrosa.*

Section PODOTREMATA Guinot, 1977
Subsection DROMIACEA de Haan, 1833
Superfamily HOMOLODROMIOIDEA Alcock, 1899
Family HOMOLODROMIIDAE Alcock, 1899

Genus *Homolodromia* A. Milne Edwards, 1880

Remarks. Placement of *Homolodromia* within the Brachyura has been the subject of much controversy. Although it has been generally agreed that the genus represents one of the more generalized brachyurans, suprageneric assignment differs widely. Rathbun (1937) placed the genus in the Homolodromiidae, which was included, along with the Dromiidae and Dynomenidae, in the superfamily Dromiidea. The essential characters, most useful in paleontological studies, uniting these groups were possession of a common orbito-antennulary pit, a triangular epistome, and an abdomen with seven segments. This position has been followed, essentially, by numerous subsequent zoologists, including Balss (1957) and Sakai (1976).

Glaessner (1969) placed *Homolodromia* in the subfamily Homolodromiinae within the Prosopidae. This family also embraces two subfamilies of Mesozoic dromiaceans, the Prosopinae and the Pithonotinae (Glaessner, 1969, p. R484–R486). Together with the Eocarcinidae, Dromiidae, and Dynomenidae, the Prosopidae were united in the superfamily Dromioidea. Members of the superfamily were characterized by absence of dorsal lineae and dorsal position of at least the fifth pereiopod. The Prosopidae were distinguished from other families in the superfamily by strong development of both cervical and branchiocardiac grooves.

Subsequent to these works, Guinot (1977, 1978) presented a new classification of brachyurans in which sections were defined on the basis of placement of genital openings. She recognized that the homolodromiians were substantially different than the Dromiidae and the Dynomenidae and placed the latter two families in a separate superfamily, the Dromioidea. *Homolodromia* and another genus, *Dicranodromia* A. Milne Edwards, were considered the sole representatives of the Homolodromiidae in the superfamily Homolodromioidea. The essential distinguishing descriptors of the family, of primary use to paleontologists, are possession of a weakly calcified carapace with poorly defined, subvertical lateral margins, an absence of lineae, and reduced, subchelate fourth and fifth pereiopods that are carried in a dorsal position. Guinot noted (1978, p. 226) that the terminal pereiopods were not utilized, however, as grasping devices to hold camouflaging organisms. Finally, she (1978, p. 228) supported the observations of several paleontologists in considering the Prosopidae as the progenitors of the Homolodromiidae. The Homolodromiidae are deep-water organisms.

Although it is not the expressed intent of this work to comment on suprageneric classification of brachyurans, it is necessary to consider the range of placements of homolodromians to properly assess the significance of the fossils from Seymour Island. In comparing both Recent and fossil material to the Antarctic fossils, numerous points of comparison can be defined to demonstrate that *Homolodromia,* previously known only from Recent records, has a long ancestry. Further, the generic descriptors defined (Rathbun, 1937, p. 58; Glaessner, 1969, p. R486) and alluded to (Guinot, 1978, p. 226–229) serve to separate the genus from previously defined fossil, as well as Recent, brachyurans.

Homolodromia is characterized by having a quadrate carapace that is longer than wide, with subvertical and poorly sclerotized lateral margins that are not separated from the dorsal surface by a rim. The frontal region is attenuated into two dominant spines separated by a depressed axial region. The dorsal surface of the carapace is crossed by distinct cervical and branchiocardiac grooves and there are no dorsal lineae. The abdominal region is composed of seven segments, of which three are visible in dorsal aspect. Although the pereiopods tend to be elongate and slender, the last two are reduced in size, and at least in Recent species, subchelate. The buccal region is quadrate, the epistome distinct.

There is a tendency to consider fossil forms as representative of distinct genera, especially when separated from Recent counterparts by a considerable hiatus. However, the Seymour Island fossils so closely conform to the above description that it would seem more prudent to refer them to *Homolodromia.* Few morphological characters, none of apparent generic significance, distinguish them from their Recent descendents.

Homolodromia chaneyi n. sp.
Figures 6.1–10, 7

Description. Carapace moderate size; outline pentagonal, longer than wide, widest near posterolateral corner, weakly convex longitudinally and transversely; sides well defined, vertical or slightly inturned.

Front narrow, about one-third the total width; produced anteriorly, sulcate dorsally and apparently terminated by downturned sulcate surface; anteriormost extensions form small, blunt spines. Orbits nearly vertical, slightly concave, extending obliquely from front posterolaterally to well-defined blunt projection on anterolateral corner, with single supraorbital projection at midlength; lateral margins of orbits defined by sharp, pustulose ridges; ventral margin undefined. Lateral margins long, about two-thirds the total length, nearly straight, slightly converging anteriorly, slightly constricted at level of cervical groove. Sides well defined along entire length, bounded dorsally in branchial region by pustulose ridge; height of sides greatest anteriorly, tapering to termination at posterolateral corner. Posterolateral corner slightly produced as posteriorly directed flabellate extension. Posterior margin straight or slightly sinuous, bordered by well-defined marginal ridge and furrow.

Regions and carapace grooves well defined. Mesogastric region about one-third the total width posteriorly, tapering abruptly to narrow termination anteriorly; surface finely pustolose, bordered by shallow, smooth depressions. Gastric and hepatic regions distinguishable only as two broadly domed areas; surfaces pustulose mesially, becoming smooth laterally. Cervical groove well defined, gently arcuate, broad, smooth, extending from midline anteriorly and laterally to side and onto side to terminate at swollen region. Thoracic region with distinguishable gastric, cardiac, and intestinal regions axially and epibranchial, mesobranchial, and metabranchial regions laterally. Gastric region broad, swollen, weakly defined, concave anteriorly; surface pustulose. Cardiac region narrower, less swollen, pustulose, defined laterally by deep, narrow, well-developed branchiocardiac groove which originates near posterolateral corner of gastric region at epimeral muscle scar, extends posteromesially to anterior corner of cardiac region, then posteriorly, and then anterolaterally, paralleling cervical groove and separating epibranchial and mesobranchial regions from metabranchial region; cardiac region tapers posteriorly and terminates at low domed region. Intestinal region triangular, poorly defined, smoother than other axial regions. Epibranchial region relatively small, swollen, with prominent broad node situated posterior to cervical groove; mesobranchial region subtly distinguishable from epibranchial region, surface pustulose mesially, nearly smooth laterally. Metabranchial region broadly arched, pustulose throughout.

Figure 6. *Homolodromia chaneyi* n. sp. 1-3, Holotype, USNM 404870, dorsal aspect of cephalothorax and proximal segments of abdominal terga, left lateral, and ventral views. 4 and 5, Paratype, USNM 404872, left lateral and dorsal views. 6, 10, Paratype, USNM 404871, dorsal and right lateral views. 7, Paratype, USNM 404873, dorsal aspect of cephalothorax and proximal abdominal terga. 8 and 9, Paratype, USNM 404874, ventral view of abdomen with basal portions of pereiopods 1 through 3 and left cheliped, and dorsal view of left meri and carpi of pereiopods 1 through 3. Bar scales = 1 cm.

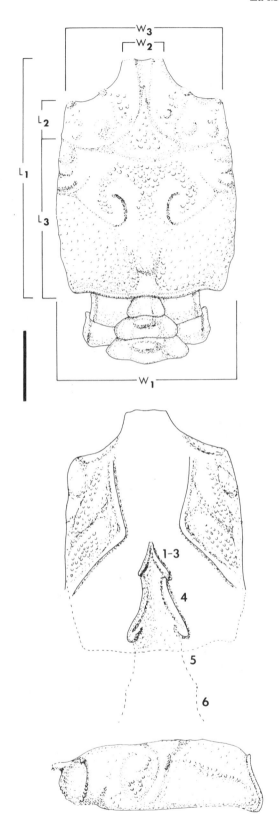

Figure 7. Line drawings of dorsal (upper), ventral (middle), and left lateral (lower) surfaces of the cephalothorax of *Homolodromia chaneyi* showing the positions and orientations of measurements. Bar scale = 1 cm.

First three abdominal somites exposed dorsally, remainder folded ventrally. Width of first three somites increases posteriorly from about one-fifth to nearly one-half the carapace width; width apparently decreases slightly on somites 3 through 6. Length of abdominal somites increases to somite 3 and decreases thereafter. Surfaces of somites pustulose; axial region elevated; pleural regions depressed, margins apparently smoothly rounded; prominent transverse swelling on anterior part of axial regions.

Sternum poorly known; narrow anteriorly, widening posteriorly to approximately maximum width posterior to first pereiopod; depressed axially.

Pterygostomial region large, well defined, triangular surface broadly corrugated, pustulose. Buccal cavity generally quadrate, widens anteriorly.

Thoracic appendages strong. Third maxilliped poorly known, apparently slender. Cheliped wider, thicker, and shorter than second and third pereiopods. Fourth and fifth pereiopods smaller, thinner, dorsal in position. Articles of appendages generally wider than high, ornamented by fine spines, more or less arranged in longitudinal rows, and by a row of somewhat longer spines on leading edges of articles. Meri and carpi with prominent distal ridges and sulci. Cheliped moderately straight, long, slender. Hand apparently ovoid in cross section with nodose ridge on upper surface. Fixed finger with longitudinal sulcus near lower edge. Dacylus with sulcus near upper edge. Denticles unknown. Termination of remaining appendages unknown.

Types. The holotype, USNM 404870, and five paratypes, USNM 404871 through 404875, are deposited in the U.S. National Museum of Natural History, Washington, D.C.

Etymology. The trivial name honors Dan Chaney, U.S. National Museum of Natural History, who collected one of the key specimens referable to this species.

Measurements. Measurements, in millimeters, are given in Table 4.

Geographical and stratigraphical position. Specimens referable to this taxon were collected from Localities 4, 8, 10, and 14 (Fig. 1), in the upper part of the La Meseta Formation on Seymour Island, Antarctica.

Remarks. Five specimens of *Homolodromia chaneyi* were collected from three separate localities, making this the second most abundant decapod in the La Meseta Formation. The lengths of the carapaces range from a minimum of about 11 mm to a maximum of nearly 38 mm, but the relative proportions and general aspect remain similar. Both dorsal and ventral aspects of the cephalothorax are preserved, along with the proximal elements of the abdomen and parts of several pereiopods. Thus, it is possible to make a detailed comparison of this species with Recent forms.

Homolodromia chaneyi seems to be most like *H. paradoxa* A. Milne Edwards, 1880, type species of the genus (Rathbun, 1937, p. 59). The cephalothorax of the Recent species is somewhat more vaulted and less coarsely ornamented than that of *H. chaneyi.* Additionally, the lateral margins on the former tend to be less well calcified and less well defined than on the fossils. A pustulose ridge gives a suggestion of a demarcation between the dorsal and lateral parts of the branchial region.

The pereiopods are smoother and tend to have a more nearly circular cross section on the modern form than on the fossils. The relative proportions of appendages are similar, although those of *H. chaneyi* are somewhat stouter throughout. Unfortunately, dactyli of pereiopods 2 through 5 are not available for comparison.

The form of the abdomen of the two species appears similar, to the

TABLE 4. MEASUREMENTS TAKEN ON SPECIMENS OF
HOMOLODROMIA CHANEYI FROM THE LA MESETA FORMATION*

Specimen	L1	L2	L3	W1	W2	W3	LA4	WA1	LA2	WA2	LA3	WA3	LA4	WA4	LA5
404870	37.6	6.1	23.9	29.6	7.4	25.8	3.4	6.8	3.8	10.0	6.0	12.9	4.8*	12.8	4.8
404871	20.4*	3.7	13.8	17.4		15.5									
404872	20*	3.6	12.5	16.0	4.9	13.8									
404873	10.9			7.3	8.8										

Specimen†	Merus		Carpus		Propodus		Dactylus	
	L	W	L	W	L	W	L	W
404872-1			5.3	3.6	12.1	3.9		
404874-1	16.3	7.2	11.3	6.3	23.3	7.8	10.7	3.0
404874-2	22.3	5.3	10.7	4.4	10.7*			
404874-3	25.5	4.6	11.0	4.3	6*			
404874-4			6.3	2.6				

Note: All measurements in millimeters. Position and orientation of measurements taken on the cephalothorax are illustrated in Figure 7.
*Indicates an approximated measurement.
†Refers to the USNM catalogue number to which the number of the pereiopod has been appended.

extent that comparison can be made. The tergal regions on both species increase to a maximum width at somite 3, and the pleura are smooth or finely pustulose, gently rounded, and reduced. The abdomen is carried in such a fashion that somites 1, 2, and 3 are visible in dorsal aspect and project above the bases of pereiopods 4 and 5.

On the ventral surface of the cephalothorax, the general quadrate form of the buccal region on both forms results in similar appearance of the anteroventral area. Additionally, the anterior halves of the sternal regions on the two species are comparable. On *H. chaneyi,* the sternum originates as a small triangular anterior element that widens posteriorly to the approximate position of the insertion of the first pereiopod, at which point the sternum narrows and then widens progressively to the point of insertion of the second pereiopod. Although not investigated in detail, the position of insertion of pereiopods on *H. paradoxa* would suggest a similar sternal outline.

In 1983, Birkenmajer et al. published an illustration of a crab fossil collected from the early Miocene glaciomarine Cape Melville Formation on King George Island, approximately the northernmost island on the Antarctic Peninsula. They referred (p. 58) to the crab as a "crab of the section Dromiacea de Haan, 1833." Subsequently, the material formed the basis for description of a new genus and species, *Antarctidromia inflata* Förster, Gazdzicki and Wrona, 1985, of the Homolodromioidea. The branchial regions on the King George Island material are more inflated transversely, the thoracic region is much broader than the cephalic area, and the frontal region is more attenuated than on *H. chaneyi.* Although there are some differences in the relative proportions of regions, as defined by well-developed grooves, the groove patterns are similar. Therefore, it would appear that the Miocene specimens represent a different, but possibly closely related, species.

The only other family of organisms reported from the fossil record that contains species comparable to *H. chaneyi* is the Torynommidae Glaessner, 1980. This family was erected to embrace five genera of Cretaceous crabs, two of which—*Torynomma* and *Eodorippe*—are austral forms and a third—*Dioratiopus*—is cosmopolitan. These genera are characterized (Glaessner, 1980, p. 181) by having a square carapace with

a spatulate frontal region, and distinctly defined regions with broad, well-defined grooves, including a branchiocardiac groove. The chelipeds on representatives of the Torynommidae are subequal, and pereiopods 4 and 5, or just pereiopod 5, are reduced and dorsal in position. Applying these, and the remainder of the familial descriptors, to *H. chaneyi* would seem to suggest the possibility of its placement in this taxon with as much confidence as offered by placement in the Homolodromiidae.

Examination of species within the Torynommidae, however, leads to the conclusion that most differ in significant ways from *H. chaneyi.* Most have very broad and deeply impressed carapace grooves or groove patterns that differ significantly from those of the Seymour Island material. The genus with members that most closely resemble *H. chaneyi* is *Dioratiopus* Woods, 1953 (= *Glaessneria* Wright and Collins, 1972; *non Glaessneria* Takeda and Miyake, 1964; = *Glaessnerella* Wright and Collins, 1975). However, careful comparison of trivial characters of the seven species discussed by Wright and Collins (1972) and the two discussed by Glaessner (1980) reveals substantial differences that separate each from *H. chaneyi.* Most representatives of the genus have a more drawn out frontal and anterolateral form, often with a downturned rostrum. In all cases, details of the groove patterns can be considered significant points of distinction. Because most of the species are apparently known only from remains of the cephalothoracic region, no adequate comparison of the abdomen or the walking legs can be made.

Thus, it would appear that the homolodromiid material from Seymour Island may represent the earliest record of the genus *Homolodromia* in the fossil record, extending the range of that genus into the late Eocene. Furthermore, considering *Antarctidromia inflata* as the first fossil record of the Homolodromiidae, *H. chaneyi* represents only the second, and the earliest, occurrence of the family. It is also possible that the Homolodromiidae arose from the Torynommidae, possibly through *Dioratiopus.* A much more detailed examination of the Torynommidae must be made before this suggestion can be confirmed.

Records of the sites from which living representatives of the genus *Homolodromia* have been collected (Rathbun, 1937, p. 58) range from 356 to 472 fathoms in the regions of the West Indies and east Africa.

Guinot (1977, p. 229) noted that the entire range of the superfamily Homolodromioidea (comprising a half-dozen species in two genera, *Homolodromia* and *Dicranodromia*) was in deep-water settings. Specimens have been collected at few, widely separated sites in the east and west Atlantic, the Indian Ocean, and Japan. Therefore, the occurrence of *Homolodromia chaneyi* in shallow-water sediments in Eocene rocks of Antarctica documents the first occurrence of a representative of the superfamily in shallow-water habitats and documents the inshore environments as the site of origin of taxa now known to exist only in bathyal environments.

Subsection ARCHEOBRACHYURA Guinot, 1978
Superfamily RANINOIDEA De Haan, 1839
Family RANINIDAE De Haan, 1839
Subfamily RANININAE Serene and Umali, 1972

Genus *Lyreidus,* De Haan, 1841
***Lyreidus antarcticus* Feldmann and Zinsmeister, 1984**
Figures 8, 9.1–11, 10.1–8

Lyreidus antarcticus FELDMANN and ZINSMEISTER, 1984, p. 1048–1056, Figs. 3A–K, 4A–I, 5, 6B, 7.

Emended Description. With the recent collection of 190 additional specimens of *Lyreidus antarcticus,* some refinements of the original description of the species may be made. These emendations follow in the order of the original description.

Fronto-orbital margin approximately two-fifths the maximum carapace width; frontal width relatively greater in smaller, presumably younger, individuals and more narrow in larger, or mature, specimens. Rostrum subacute, relatively less acute than that of some other species of *Lyreidus;* rostrum 1.2 times as long as wide, postorbital spines slightly divergent and just longer than rostrum. In addition to setal pits, originally described as only feature of ornamentation, carapace possesses prominent cardiac grooves located in posterior one-third, centered on dorsal surface; as in Recent species, grooves present as longitudinal, paired, curved depressions; gastric apodeme pits not found, although typically extremely small and not readily preserved in fossil material.

Sternal elements 1 through 3 fused, apex sharp; anterolateral portion of element 4 acute and upturned, unlike Recent species of *Lyreidus;* sternum attains greatest width across element 5 and narrows abruptly across element 6, gradually tapering posteriorly. Abdomen articulated with sternum between elements 5 and 6, with female specimens exhibiting a small, raised, outwardly curved projection of sternum.

Thoracic appendages, excluding first pereiopod, not commonly preserved completely, previously undescribed. Second pereiopod, described and measured from two complete specimens, much smaller than first; basis rounded; ischium about 0.80 times as long as wide; merus nearly 2.5 times as long as wide, upper and lower surfaces thin and sharp, sides ornamented with minute spines, nodes, or setal pits; carpus nearly twice as long as wide, widens distally; propodus nearly square, flat; dactylus extremely long, nearly 2.5 times as long as wide, blade shaped. Third pereiopod described and measured from three incomplete specimens, longest walking appendage; merus uniformly about 3.5 times as long as wide, merus proportionally longer than merus of second pereiopod, upper surface sharp, lower surface rounded; carpus much stouter than merus, about twice as long as wide, thinning proximally, upper surface sharp; propodus approximately 1.5 times as long as wide, tapers distally, upper and lower surfaces thin and sharp; dactylus blade shaped, stouter than dactylus of second pereiopod, only slightly more than 1.5 times as long as wide, upper and lower surfaces extremely thin and sharp. Fourth pereiopod described and measured from three specimens. Appendage smaller than third pereiopod; merus about twice as long as wide, slightly

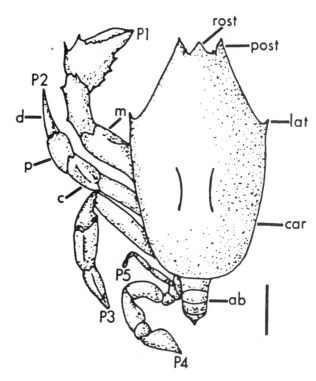

Figure 8. Reconstruction of *Lyreidus antarcticus* with major morphological features labeled. Key: ab = abdomen; car = carapace; c = carpus; d = dactylus; lat = lateral carapace spine; m = merus; Pl = pereiopod 1; P2 = pereiopod 2; P3 = pereiopod 3; P4 = pereiopod 4; P5 = pereiopod 5; post = postorbital spine; p = propodus; rost = rostrum. Bar scale = 1 cm.

wider distally, upper surface rounded, surface thin and sharp; carpus stout, just slightly longer than wide; propodus extremely stout, approximately twice as wide as long; dactylus twice as long as wide, very thin in cross section, and elongate chordate shaped. Fifth pereiopod described and measured from three incomplete specimens. Appendage radically reduced, lies in dorsal plane of body, aligned with first abdominal somite; males exhibit greater proportional length of basis, which extends nearly to posterior portion of second abdominal somite; females exhibit comparatively reduced basis, extending posteriorly only to about one-half the length of second abdominal somite; ischium and merus extremely reduced; carpus, propodus, and dactylus unknown.

Types. The holotype, USNM 365441, paratypes, USNM 365442–365450 and 365454, and hypotypes USNM 404881–404922 are deposited in the U.S. National Museum of Natural History, Washington, D.C. Additional specimens KSU 5038–5048 are deposited at Kent State University, Kent, Ohio.

Measurements. Measurements, in millimeters, of previously undescribed thoracic appendages (pereiopods 2, 3, 4) are given in Table 5. Measurements were derived from maximum lengths and breadths of the individual appendage segments. Measurements for the fifth pereiopod are not presented, as only incomplete ischia and meri are marginally preserved on two specimens and only approximations could be given.

**TABLE 5. MEASUREMENTS TAKEN ON THORACIC APPENDAGES
OF SPECIMENS OF *LYREIDUS ANTARCTICUS***

Specimen	Mer-2 L	Mer-2 W	Carp-2 L	Carp-2 W	Prop-2 L	Prop-2 W	Dactyl-2 L	Dactyl-2 W
404882	12.2	5.0	9.5	5.0	9.2	8.1	10.9	5.0
404895	10.0	4.7	8.5	4.5	7.6	—	10.3	4.4

Specimen	Mer-3 L	Mer-3 W	Carp-3 L	Carp-3 W	Prop-3 L	Prop-3 W	Dactyl-3 L	Dactyl-3 W
404883	12.9	4.7	8.8	5.0	—	—	—	—
404886	12.6	3.4	8.5	4.7	11.0	6.8	10.4	6.3
404888	14.9	5.2	—	—	—	—	—	—

Specimen	Mer-4 L	Mer-4 W	Carp-4 L	Carp-4 W	Prop-4 L	Prop-4 W	Dactyl-4 L	Dactyl-4 W
404883	8.8	4.1	7.5	6.5	3.3	6.3	—	—
404886	—	—	—	—	—	—	11.1	5.0
404888	8.9	5.0	8.2	6.4	4.2	7.5	—	—

Note: Measurements in millimeters. Length and width measurements represent the maxima for each segment of each measured appendage. Mer = merus, Carp = carpus, Prop = propodus, Dactyl = dactylus. Numbers beside these abbreviations (2-4) represent the corresponding numbers of each pereiopod.

Remarks. Differentiation of the sexes of *Lyreidus* and some other raninids is very difficult, especially in fossil material. Prior to this study, only Sakai (1937) had described any external feature as part of sexual dimorphism in *Lyreidus*. As there are no obvious external signs of sexual dimorphism, such as claw or carapace size differences, definitive dimorphic characters must be based on analogy with Recent specimens. Sex is easily determined in such specimens by the position and number of pleopods; the male possesses one pair of modified pleopods, the female, two. Sex may also be determined by position of the female genital pore, which is located on the coxa of the third pereiopod; the male genital opening is positioned on the coxa of the fifth pereiopod. Another subtle, but sexually dimorphic feature, is found on the sternum of both sexes. Bourne (1922) noted that *Lyreidus* has the capability to tightly "lock" its abdomen to its sternum, a feature not found in any other raninid. Bourne (1922) referred to these raised, curved features found on the sternum between the fifth and sixth sternal elements as "pterygoid processes." However, he failed to notice that these raised portions of the sternum are sexually dimorphic. The males, in mature Recent species of *Lyreidus* and *Lysirude*, possess straight, bladed, raised flanges; whereas the female possesses notably reduced, outwardly curved projections. Bliss (1982, p. 109) noted that "male crabs have a 'locking device,' consisting of small tubercules on the fifth thoracic segment that secure the triangular or T-shaped abdomen in a depression on the ventral side of the thorax." Bliss (1982) did not describe a locking device in female crabs, but noted that females tightly hold their abdomina within a depression in the sternum. The smaller curved areas in the female sterna of *Lyreidus* are probably lateral extensions of this depression. Another dimorphic aspect was described by Sakai (1937), who noted that the male *L. tridentatus* has a proportionally longer fifth abdominal somite, and the female a proportionally wider one. This dimorphic character is apparent in mature specimens of *L. brevifrons* and *L. stenops*. This relationship is not yet established for *L. antarcticus*.

In addition to the two primary types which have well-preserved abdomina, there are four additional specimens. All six exhibit a relatively broad fifth abdominal somite, indicating that all are female. One last feature may be used to determine the sex of Recent specimens. The basis

Figure 9. *Lyreidus antarcticus* Feldmann and Zinsmeister, 1 through 11 figured hypotypes. 1, USNM 404881, ventral aspect of cephalothorax, including third maxilliped, well-preserved lower sternal elements, and right and left basal segments and left meri of the first three pereiopods. 2, USNM 404882, complete elements of right and left second pereiopod. 3, USNM 404883, carpus, propodus, and dactylus of right cheliped. 4, USNM 404884, propodus and dactylus of left cheliped. 5, USNM 404883, lateral view of cephalothorax, including elements of pereiopods 1 through 4. 6, USNM 404885, complete upper sternal plate, lower elements (5, 6, 7) lacking. 7, USNM 404886, pereiopods 3 and 4, fourth pereiopod on left shows complete dactylus (arrow), third pereiopod on right shows complete merus, carpus, and propodus. 8, USNM 404889, posterior view of carapace edge showing detached abdomen, abdominal somites 1 through 4, right and left ischia and meri of fifth pereiopod. 9, USNM 404881, posterior view of carapace showing attached abdomen, abdominal somites 1 through 4, ischium and merus of left fifth pereiopod. 10, USNM 404887, dorsal view of cephalothorax, with complete rostrum, postorbital spines, lateral carapace spines, and with merus of left cheliped, integument attached to posterior portion of carapace. 11, USNM 404888, ventral aspect of cephalothorax with complete right and left third maxilli, right and left basal segments, and left meri of pereiopods 1 through 4, and abdominal somites 4 through 7. Bar scales = 1 cm.

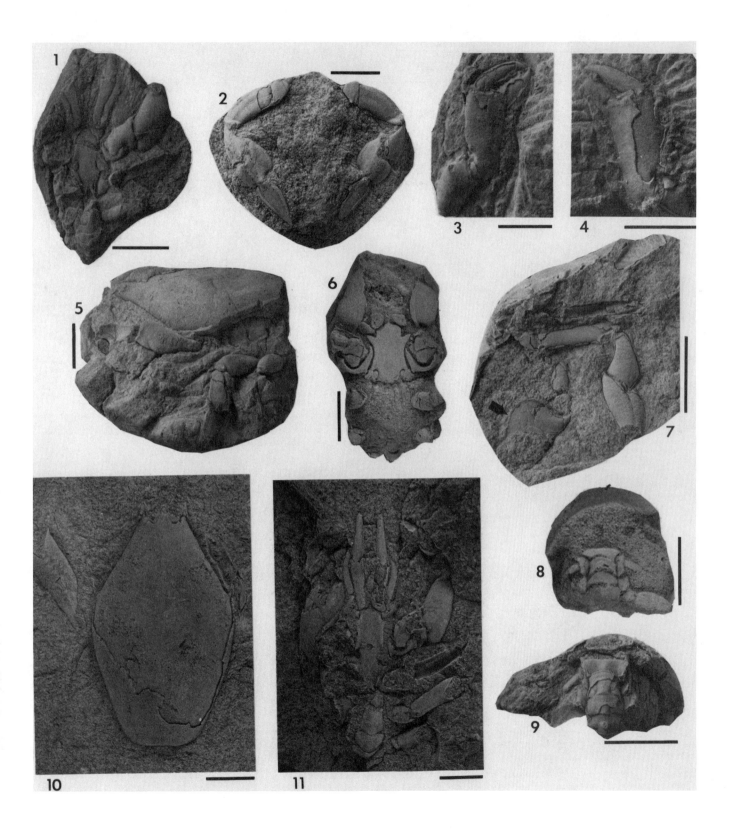

of the fifth pereiopod is proportionally longer in the male of each species than that of the female. The basis extends to the posterior portion of the second abdominal somite in the male, and extends only midway to the second abdominal somite of the female. The sexual determinations by Feldmann and Zinsmeister (1984b) were not supported by these criteria, and are possibly in error. However, definite sex assignments may not be made at this time.

The hard-part preservation of individual specimens of *Lyreidus antarcticus* from the La Meseta Formation is variable; the extent and detail of preservation of some specimens are exceptional. Preparation and study of specimens have revealed the morphology and ornamentation of the carapace, rostrum, thoracic and cephalic appendages, sternum, and abdomen. One specimen (Fig. 10.1,2) shows a remarkably well-preserved cast of the soft-bodied portion of the cephalothorax. Of the 201 specimens examined, including the type specimens, most are preserved within calcite cemented, fine-grained sandstone concretions. Many of these concretions were exposed at the surface, and were collected as part and counterpart. Typically, part and counterpart display the dorsal aspect of the carapace with replaced integument broken and covering portions of both the internal and external molds. These specimens, generally, do not appear to be crushed, inflated, or otherwise distorted. However, one (Fig. 10.7) shows unusual breakage along the right anterolateral margin.

Lyreidus, like other raninids, is a burrowing crab (Bourne, 1922; Glaessner, 1969). The specimen shown in Figure 10.4 consists of a carapace of *L. antarcticus* within a very fine-grained structure which is sinuous and tubular in shape. This fossil might represent a *Lyreidus* fossil preserved in its burrow. *Ophiomorpha,* a trace fossil which has been attributed to burrowing activity of callianassid shrimp, is found in the La Meseta Formation in close association with *Lyreidus antarcticus.* Although fossil callianassids are found in the La Meseta, they typically do not occur in close association with *Ophiomorpha.* Wiedman and Feldmann (this volume) have suggested that the *Ophiomorpha* from the La Meseta Formation might have been produced by the burrowing activity of *Lyriedus antarcticus.*

The right or left merus of the first pereiopod is present and exposed

Figure 10. *Lyreidus antarcticus* Feldmann and Zinsmeister, 1 through 7 figured hypotypes; 8 and 9, scanning electron micrographs. 1 and 2, USNM 404890, internal mold of cephalothorax, dorsal and ventral, dorsal view with rostrum, muscle attachment areas; ventral view with oral and sternal areas, coxae of pereiopods 1 through 4. 3, USNM 404891, oblique anterior view, rostrum missing, showing typical juxtaposition of chelae. 4, USNM 404893, dorsal view of central portion of carapace embedded in fine-grained, sinuous-tubular matrix, possible burrow structure. 5 and 6, USNM 404892, dorsal view of cephalothorax with right and left meri of chelae and external view of concretion with weathered paired dactyli of pereiopods 2, 3, and 4 emerging at surface. 7, USNM 404894, dorsal view of cephalothorax with complete rostrum, right and left meri of chelae, showing crushed right anterolateral margin. 8 and 9, Scanning electron micrographs of integument taken from carapace of *L. antarcticus* (8) and from carapace of Recent *L. nitidus* (=*L. bairdii*). Bar scales for 1 through 7 = 1 cm, bar scales for 8 and 9 = 0.1 mm.

on nearly every specimen, often with some replaced integument still attached (Fig. 9.10). The carpus of the first pereiopod is preserved less often. Where intact, the claws are commonly held in parallel, with one claw in front of the other, just in front of the rostrum (Fig. 10.3). There is no preferred orientation to this arrangement of the claws; that is, the left claw is not more commonly in front of the right claw. The remaining thoracic appendages are typically held beneath the body, subperpendicular to the plane of the carapace, and are not exposed on the same surface with the carapace (Figs. 9.5,10,10.5). Paired distal tips of the second through fifth thoracic appendages are frequently exposed along the outside, weathered surface of the concretion (Fig. 10.5,6). Only four prepared specimens (Fig. 9.5,7) show specimens of the second through fifth pereiopods, including the dactyli. These are extremely fragile elements and were probably easily scattered upon the death of the animal.

In several specimens, the abdomen is preserved attached, intact, to the carapace (Fig. 9.8). However, most specimens with preserved abdomina are found in Salter's position. They exhibit the abdomen split from the carapace and projected downward with respect to the body. Although it is commonly assumed that when brachyuran crabs are preserved in Salter's position they represent molts, this manner of preservation may result from other processes (Schäfer, 1951). When found in the fossil record, Salter's position has been interpreted as occurring through molting, decomposition of the animal, or from the action of water or wind currents (Schäfer, 1951). The specimens that possess preserved abdomina, either attached to or detached from the cephalothorax, typically possess equally well-preserved thoracic appendages and ventral aspects. It is possible that sedimentary conditions were such that these animals were buried and preserved so quickly that true molts and nonmolts were equally well preserved.

The integument is replaced with calcite in nearly every specimen and is best preserved on the carapace. In the less well-calcified areas, such as the individual elements and joints of the appendages, the fragile integument is usually lost, in some cases due to exposure to weathering. The finest specimens are black in color, and possess very detailed preservation of the carapace and appendage ornamentation such as the fine setal pits, minute spines, and large cardiac groove. One black specimen, encased in an extremely fine-grained and well-cemented concretion, possesses some "geodized" appendages (Fig. 9.11).

Samples of integument, taken from the carapace and appendages of *Lyreidus antarcticus* and a Recent form, *Lysirude nitidus,* were examined by scanning electron microscopy (Fig. 10.8,9). The samples revealed few morphologic differences. Each exhibits the layered integument, typical of decapods, with thin exoskeletal and endoskeletal layers, and an intermediate, thick, prismatic layer. Samples of the integument of *L. antarcticus* even show the morphology of setal pits. The similar structures and thicknesses of these samples suggest that, in this case, the fossil is not a molt.

Of the 190 specimens of *L. antarcticus* collected during the austral summer of 1983–1984, 180 were derived from nine major localities; the remainder were from various miscellaneous or undetermined localities. All of these localities are found stratigraphically in Telm2, 3, and 5 (Sadler, this volume), which correspond to Unit I and Unit II of Elliot et al. (1982). No specimens were collected from the lowermost or uppermost sections of the La Meseta Formation. The localities are listed in Table 1, associated decapods are given therein, and their locations are illustrated on Figure 1. The majority of these specimens were collected from Telm3 and 5, the middle of the La Meseta Formation. *Lyreidus* fossils do not appear to occur commonly in the deeper water facies of the unit, the lowermost portion stratigraphically, nor in the shallow-water facies of the unit, the uppermost portion stratigraphically. *Lyreidus antarcticus* was found in association with a wide variety of fossil animals and plants.

The occurrence of the concretions at Locality 7 is particularly interesting. Most are spherical to ovoid in cross section; occasionally, two ovoid masses coalesce to form a dumbbell shape. The size of the concretions vary from a diameter of about 2 cm to a maximum of 25 cm; the typical size range is 7 to 10 cm. Nearly all the concretions were broken open in the field, perhaps by processes of freezing and thawing. Specimens that had been exposed were subject to abrasion by wind-blown sediment, and the surfaces had the appearance of desert varnish. In 78 cases, both halves of the concretions were found lying within a few centimeters of one another and could be positively identified as representing all the fragments of a single unit. All but two of the concretions were single units; two were doubles. Of the 78 concretions examined on the exposure, only 2 appeared to be totally devoid of organic matter. The overwhelming majority, 64, contained plant debris, primarily in the form of carbonized stem material. Of the 64 concretions containing plant fragments, 11 also contained other fossil material, bryozoans, small bivalves, small gastropods, or burrows. Ten of the concretions contained *Lyreidus antarcticus* as the nucleus. The remaining two concretions were cored—by a bivalve and a gastropod in one case, and by a burrow structure in the other. The megafauna of these concretions was delicately preserved, suggesting that little, if any, transportation had occurred prior to emtombment. Therefore, based upon these observations, it can be concluded that *Lyreidus antarcticus* was a significant element in the benthic fauna at this site and that other inhabitants were small bivalves and gastropods along with bryozoans. Furthermore, formation of the concretions was induced by the presence of the organic remains; conversely, preservation of the assemblage was assured by the protective encasement of the concretion. From these occurrences and associations, it is clear that *L. antarcticus* was a part of an extremely diverse, shallow-water, variable-energy marine environment.

Recent species of *Lyreidus* are deep water dwellers. Griffin (1970) recognized and summarized the distribution of five Recent species of *Lyreidus. Lyreidus tridentatus,* the type species of the genus, has by far the broadest geographic range and the longest geologic range. This species is found throughout the western and central Pacific in depths ranging from 27 to 425 m; this species has a geologic range of middle Oligocene to Recent (Jenkins, 1972). *Lyreidus brevifrons* is restricted to the Indo-west Pacific and is commonly found in depths ranging from 188 to 440 m. *Lyreidus stenops,* a western Pacific form, is commonly found in depths of 55 to 160 m. As understood by Griffin (1970), the genus *Lyreidus* also included *L. channeri* and *L. nitidus.* However, Goeke (1985) recently reassigned these two species to a closely related new genus, *Lysirude. Lysirude channeri* is restricted to the northern Indian and western Pacific oceans and is found in depths of 366 to 820 m (Griffin, 1970). Examination of preserved biological specimens of *Lyreidus* and *Lysirude* in the National Museum of Natural History provided further depth information exceeding Griffin's maxima: *L. tridentatus,* 669 m; *L. brevifrons,* 776 m; *L. stenops,* 503 m; and *Lysirude channeri,* 1,455 m. *Lysirude nitidus* has a geographic range from the coast of Maine to the Gulf of Mexico and a bathymetric range of 119 to 475 m (Griffin, 1970). Powers (1977) listed the geographic range of this species as Massachusetts to Puerto Rico in depths of 119 to 823 meters. Powers (1977) referred to the habitat of *L. nitidus* (= *L. bairdii*) as "soft mud substrates."

Workers including Bourne (1922) and Sakai (1937) have studied the probable life habits of these animals. Such studies have been based on observations regarding functional morphology and on the nature of the sediment dredged up with the crab. Direct observations of these animals in their natural setting have not been documented. However, because we do know the approximate geologic, geographic, and bathymetric limits on each species, we can say that these are, today, deep-dwelling crabs.

Section HETEROTREMATA Guinot, 1977
Superfamily PORTUNOIDEA Rafinesque, 1815
Family PORTUNIDAE Rafinesque, 1815
Subfamily PORTUNINAE Rafinesque, 1815

Genus *Callinectes* Stimpson, 1860
?Callinectes sp.
Figures 11.1–3, 12

Description. Specimen moderately large, strong, spined, consisting of partial well-preserved merus, carpus, propodus, and dactylus of right cheliped and mold of exterior of carapace fragments.

Carapace fragment appears to be right posterolateral corner with well-defined, smooth border and coarsely punctate surface.

Merus greater than 34 mm long and 8 mm wide. Upper margin angular, defined by row of closely spaced spines with diameters at base of about 1 mm. Outer surface convex with nodes, smaller than spines on upper margin, arranged in poorly defined longitudinal rows. Inner surface nodose near upper margin.

Carpus about 14.5 mm long, 7.5 mm wide. Upper surface triangular, with greatest width at midlength, flattened, ornamented by coarse spines and interspersed pustules. Distal margin finely serrate.

Propodus, including all but distal portion of fixed finger, greater than 37.5 mm long, 16.8 mm high, 9.1 mm wide. Hand generally smoothly convex; outer surface finely granular, with shallow sulcus just below upper margin extending from carpus-propodus joint distally one-half the length of hand; upper surface slightly convex in profile, narrow, ornamented by moderate-sized nodes; inner surface smooth, convex; lower margin slightly sinuous in profile. Fixed finger tapers distally; sulcus with numerous setal pits, near lower margin of outer surface; ovoid cross section. Denticles on occlusal surface low, broad domes.

Dactylus broad proximally, tapering in width and height distally; outer surface with longitudinal, pitted sulcus just above midline; denticles as on fixed finger. Maximum height, 7.0 mm; width, 4.7 mm; length, greater than 10 mm.

Type. The sole specimen, USNM 404880, is deposited in the U.S. National Museum of Natural History, Washington, D.C.

Locality and stratigraphic position. This specimen was collected from the upper part of the late Eocene La Meseta Formation at Locality 14 (Fig. 1).

Remarks. The right arm and claw described above is unquestionably distinct from other decapods collected from the La Meseta Formation. The elements are large, robust, and have relative proportions unlike any of the other species. The most distinctive morphologic characteristics are the great length of the merus and the strength of the ornamentation on the various segments. In most crab genera, the cheliped is carried close to the front in a defensive posture. The merus tends to be relatively short, and the ornamentation becomes more pronounced on the distal elements, the propodus and dactylus. Just the opposite condition is observed on this specimen. The merus is nearly as long as the propodus, and the ornamentation on the merus is composed of coarse tubercles, or possibly spines, whereas ornamentation on the propodus is limited to fine granulations.

Using the general proportions of the various arm elements as the primary basis for identification, the specimen would seem to represent a species within the Portunidae. Many of these swimming crabs have chelipeds with elongate, strongly ornamented meri. More specific identification is difficult, however, owing to the absence of carapace material and the unique combination of characters expressed on the arm. The

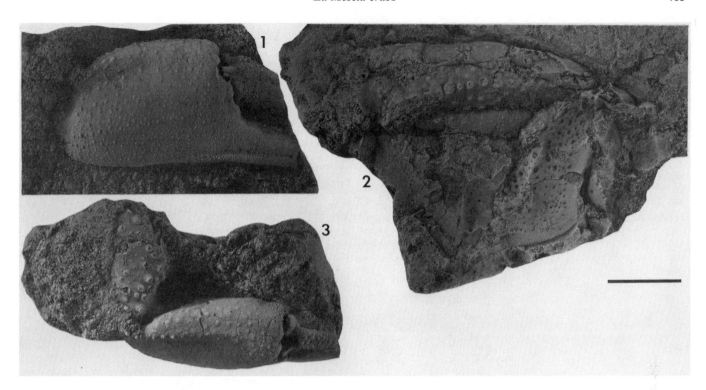

Figure 11. ?*Callinectes* sp., USNM 404880. 1, Outer surface of propodus and dactylus of right cheliped. 2, Upper surface of merus and mold of fragments of carapace. 3, upper surface of carpus, propodus, and dactylus of right cheliped. Bar scale = 1 cm.

specimen is tentatively referred to *Callinectes,* based on the absence of a mesiodistal spine on the carpus (Williams, 1984, p. 355) and coupled with the elongate, well-ornamented merus. However, the propodus on typical *Callinectes* is longitudinally keeled, which is not the case with this specimen. Taken alone, the propodus is more like that seen on cancrids than on portunids, but the merus is typically short and relatively smooth on the former. Given this uncertainty, it seems most prudent to questionably refer the specimen to *Callinectes* sp.

Superfamily CALAPPOIDEA de Haan, 1833
Family CALAPPIDAE de Haan, 1833
Subfamily CALAPPINAE de Haan, 1833

Genus *Calappa* Weber, 1795
***Calappa zinsmeisteri* n. sp.**
Figures 13.1,2, 14

Description. Taxon represented by one nearly complete right claw and partial hands of one left and one right cheliped. Right and left claws similar size; strong, thick, with short, stout fingers.

Propodus generally triangular, narrow at articulation with carpus, highest at point of articulation with dactylus. Carpus-propodus joint steeply inclined to long axis of carpus. Upper margin smoothly convex, ornamented with row of more than five prominent spines. Lower margin, including base of fixed finger, straight or slightly downturned when viewed from side. Fixed finger short, about one-fourth the total length of propodus, narrows uniformly distally, occlusal surface intercepts lower

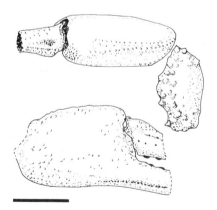

Figure 12. Line drawings showing the interpreted morphology of the carpus, propodus, and dactylus of ?*Callinectes* sp. Bar scale = 1 cm.

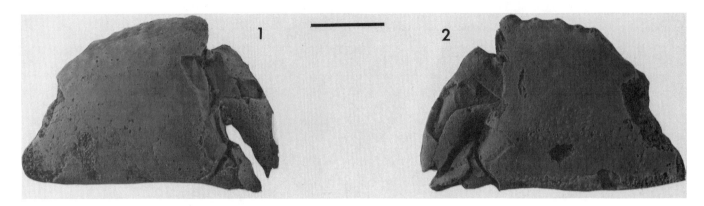

Figure 13. *Calappa zinsmeisteri* n. sp., USNM 404877. 1, Outer surface of propodus and dactylus of right cheliped. 2, Inner surface of propodus and dactylus of right cheliped. Bar scale = 1 cm.

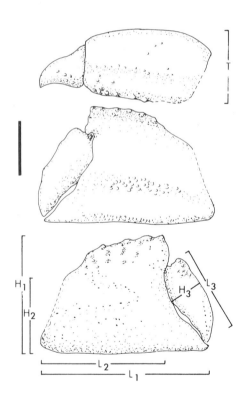

Figure 14. Line drawings showing interpreted morphology of upper, inner, and outer surfaces of the right cheliped of *Calappa zinsmeisteri* and position of measurements. Bar scale = 1 cm.

margin at about 40° angle. Outer surface pustulose; pustules small, some arranged in approximately longitudinal rows from base of hand to approximate level of articulation with dactylus; pustules become larger, less numerous, less systematically arranged above. Inner surface a sinuous curve with hand generally concave near upper and lower margins and thickest and convex mesially. Surface finely pustulose mesially and otherwise smooth. Claws are convex anteriorly when viewed from above.

Dactylus short, stout, pustulose only on proximal portion of upper surface, convex anteriorly when viewed from above. Upper surface gently arched. Occlusal surface of propodus and dactylus apparently with broad, domed denticles. Long axis of dactylus, when closed, makes approximate angle of 60° with base of propodus.

Measurements. Measurements, in millimeters, are given in Table 6. All measurements represent minimum dimensions, as none reflects estimates of the amount of material broken or exfoliated.

Etymology. The trivial name honors Dr. William J. Zinsmeister, Department of Geosciences, Purdue University, who made it possible to make the collections upon which this study is based.

Types. The holotype, USNM 404877, and two paratypes, USNM 404878 and 404879, are deposited in the U.S. National Museum of Natural History, Washington, D.C.

Locality and stratigraphic position. Specimens referred to *Calappa zinsmeisteri* were collected from the upper part of the late Eocene La Meseta Formation, at Locality 14 (Fig. 1), Seymour Island, Antarctica.

Remarks. Although *Calappa zinsmeisteri* is represented only by cheliped material, placement in the genus can be made with a high degree of certainty. The very strong, triangular hand is smooth and generally concave on the inner surface to conform to the front of the carapace. On its outer surface, the hand is more heavily ornamented on the upper half, and the ornamentation is in the form of rows of nodes. The upper margin projects well above the point of articulation of the dactylus, and it is ornamented by strong spines; the lower margin is smooth. The dactylus rotates downward against the short, stout fixed finger at a steep angle. The base of the fixed finger is not downturned.

This combination of characters is typical of most species of *Calappa*

TABLE 6. MEASUREMENTS TAKEN ON SPECIMENS OF *CALAPPA ZINSMEISTERI* FROM THE LA MESETA FORMATION†

Specimen	L1 (mm)	L2 (mm)	L3 (mm)	H1 (mm)	H2 (mm)	H3 (mm)	T (mm)
404877 Right	35.2	28.0	18.3	23.3	7.6	8.6	12.0
404878 Right				23.7	8.4		12.0
404879 Left				24.9	7.9		12.3

†Position and orientation of measurements are illustrated in Figure 14.

Figure 15. *Chasmocarcinus seymourensis* Feldmann and Zinsmeister. Holotype, USNM 365455. Dorsal aspect of carapace and pereiopods. Bar scale = 1 cm.

and serves to distinguish the species from other members of the subfamily (Rathbun, 1937; Sakai, 1976). Species of *Matuta* typically have a longer fixed finger, and the occlusal surface is not as oblique to the long axis of the propodus. In species of *Mursia* and *Cycloes* the fixed finger tends to be elongate and downturned. The lower margin is spinose in many species of *Matuta, Paracyclois* and some *Mursia* and is serrate in species of *Acanthocarpus* and *Cycloes.* that surface is smooth or beaded in species of *Calappa,* but not typically spinose.

Prior to this notice, there have been only two references to *Calappa* in Eocene rocks. Bittner (1875) reported *Calappa* sp. from Eocene rocks in the Vicentia region of Italy (Glaessner, 1929, p. 72), and Toniolo (1909) noted *Calappa* sp. in rocks on Mount Staraj, in the Istria. Glaessner (1929, p. 72) concluded that this occurrence was Lutetian in age; however, the reference to age in Toniolo (1909, p. 250) is equivocal. An apparent typographical error in setting the table summarizing the stratigraphic distribution of fossils resulted in no record of occurrence of *Calappa* sp., even though the species is listed in the table. Glaessner (1969, p. R494) questioned Eocene occurrences of the genus and indicated the certain range of the genus to be Oligocene to Recent. Whether or not the Eocene records of *Calappa* cited by Bittner (1875) and Tonioli (1909) are authentic, the occurrence of the genus on Seymour Island represents the southernmost collection site and one of the oldest occurrences. It documents the genus in the Eocene with certainty.

Glaessner (1969, p. R494) reported the geographic range of fossil forms of *Calappa* to include North America, Europe, Central America, Egypt, Burma, and the East Indies. Most of the records are from the Northern Hemisphere and range from Oligocene through Pleistocene. Many, if not most, of the records are derived from identification of chelipeds. Carapace material is more scarce; however, as indicated above, chelipeds are fairly diagnostic and can be identified with confidence.

Modern representatives of *Calappa* have been described from a variety of shelf habitats and are distributed through warm temperate and tropical regions worldwide. Rathbun (1937) recorded collecting sites in North and South America on rocky and sandy substrates varying in depth from sea level to 125 fathoms. Sakai (1979) reported specimens from the western Pacific, referable to the genus, on substrates ranging from hard beaches and coral reefs to soft sand in depths varying from 10 to 150 m. A somewhat wider range of habitats was reported for African specimens (Manning and Holthuis, 1981), owing in large part to records of *C. granulata* (Linnaeus) to depths of 400 to 700 m (p. 52). Nonetheless, the bulk of collections of individuals referred to *Calappa* are from warm, shallow-water, firm substrate habitats. These conditions would seem to define the preferred habitat of *Calappa.*

Superfamily XANTHOIDEA McLeay, 1838
Family GONEPLACIDAE McLeay, 1838
Subfamily CHASMOCARCININAE Serene, 1964

Genus *Chasmocarcinus* Rathbun, 1898
Chasmocarcinus seymourensis Feldmann and Zinsmeister, 1984
Figure 15

Chasmocarcinus seymourensis FELDMANN and ZINSMEISTER, 1984, p. 1056–1058, Figs. 8, 9.
Chasmocarcinus. ZINSMEISTER and FELDMANN, 1984, p. 282; FELDMANN, 1984, p. 16, unnumbered figure; FELDMANN, 1984, cover photograph; FELDMANN and ZINSMEISTER, 1984, p. 507.

Type. The holotype and sole specimen, USNM 365455, is deposited in the U.S. National Museum of Natural History, Washington, D.C.

Locality and stratigraphic position. This species was collected from unit II of Elliot et al. (1982) and Unit Telm3 of Sadler (1986, this volume) of the La Meseta Formation at Locality 12 (Fig. 1), Seymour Island, Antarctica.

Remarks. A single specimen of this species was collected by W. J. Zinsmeister during the 1981–1982 expedition. In the 1983–1984 season, and again in 1986–1987, Feldmann visited the type locality and found no additional specimens. If more specimens are present, they would have to be considered extremely rare. Therefore, no new information is available to expand on the description or the statements regarding the ecological setting of this little crab.

Superfamily MAJOIDEA Samouelle, 1819
Family MAJIDAE Samouelle, 1819

Genus ?*Micromithrax* Noetling, 1881

Remarks. Assignment of the sole specimen referable to this taxon is extremely difficult, owing to partial preservation of the material. Preserved as a mold of the interior of the cephalothorax, the regions are distinctly defined and can be interpreted readily. However, the anterolateral and fronto-orbital margins are partially missing and difficult to interpret. It is difficult to determine whether the anterolateral margin is denticulate or generally smooth. We interpret it to be smooth, in the absence of evidence to the contrary. It appears that either a short, bifid rostrum is present, or that the rostral region is attenuated, sulcate, and broken. Orbits are either poorly developed or lacking.

Placement in *Micromithrax*, rather than in one of the other closely related genera within the Majidae, was based on the conformation and relative development of the carapace regions and on the general outline. *Micromithrax* tends to have a more circular outline than many other majids that are more strongly attenuated in the frontal region.

The tentative nature of the placement must be emphasized. It is possible, but unlikely, that the specimen belongs to one of the Eocene genera of the Portunidae. For example, *Liocarcinus* Stimpson-Pourtalis, 1870, has a groove pattern and conformation of carapace regions very similar to that of the specimen in question, particularly *L. rakosensis* (Lorenthey, *in* Lorenthey and Beurlen), 1979 (fide, Müller, 1984, p. 83, Plate 69, Figs. 2 through 6). This species, however, has a strongly denticulate anterolateral margin which is not likely to be present on the sole specimen from the La Meseta Formation. Furthermore, *L. rakosensis* has a reentrant at the posterolateral corner, as is the case in many portunids. That corner is not modified in *Micromithrax* and is not typically modified in any of the Majidae.

On the basis of the above observations, it seems prudent to refer the specimen questionably to *Micromithrax* until such time as more and better material is available for study.

?*Micromithrax minisculus* n. sp.
Figure 16

Description. Carapace small, width 5.1 mm, length including rostral region, 5.0 mm, uniformly vaulted transversely and longitudinally, outline hexagonal.

Frontal region attenuated into short, bifid rostral area with axial depression. Fronto-orbital margin nearly straight with no orbital depressions evident; anterolateral and posterolateral margins approximately equally long, straight segments joined at point of greatest carapace width just posterior to midlength; posterior margin about 45 percent total width.

Regions well defined as swollen areas separated by shallow, subtle sulci. Mesogastric region long and slender, merging posteriorly with broad, nearly circular metagastric region. Gastric regions circular with diameter about 25 percent total carapace width. Hepatic regions swollen, triangular, separated from adjacent regions by broad, smooth depressions. Urogastric region three times as wide as long, well defined. Cardiac region circular, vaulted, 30 percent of carapace width. Intestinal region not differentiated. Branchial regions not well differentiated, broadest near cervical groove, narrowing posteriorly.

Surface ornamented by fine pustules, best developed on gastric and hepatic regions, more subtle elsewhere.

Abdomen and appendages unknown.

Type. The holotype and sole specimen of this species USNM 404876, is deposited in the U.S. National Museum of Natural History, Washington, D.C.

Etymology. The trivial name is taken from the Latin word, *minisculus* = rather small, in reference to the diminutive size of the specimen.

Figure 16. ?*Micromithrax minisculus* n. sp. Holotype, USNM 404876. Dorsal view of exfoliated carapace. Bar scale = 1 cm.

Location and stratigraphic position. The specimen was collected from unit Telm2 of Sadler (1986) in the Eocene La Meseta Formation at Locality 3, Seymour Island, Antarctica.

Remarks. Little can be added to what has been said above. Comparison of the La Meseta material with fossil and Recent majids confirms that it cannot be referred to a previously described species, and it certainly differs significantly from any of the other decapods from Seymour Island. However, confirmation of the generic placement and elaboration of the morphologic details must await discovery of better specimens. Studies dealing with living majids—for example, Rathbun (1925), Sakai (1976), Manning and Holthius (1981), Williams (1984), and McLay (1985)—confirm that most of these organisms have an attenuated frontal region. The poor preservation of the La Meseta specimen simply does not permit thorough interpretation of that region, and therefore, the placement must remain enigmatic.

ACKNOWLEDGMENTS

During the preparation of this manuscript, Raymond Manning and Austin Williams at the National Museum of Natural History provided access to the library and the collections of that institution, as well as facilitating loans of comparative material. Portions of the manuscript were read by Manning, as well as by Richard Jenkins, University of Adelaide, Australia; Barry Miller, Kent State University; and Loren Babcock, University of Kansas. Field work for collection of specimens was provided to R.M.F. by a grant from the National Science Foundation to William J. Zinsmeister, and laboratory work was supported by NSF Grant DPP-8411842 (to R.M.F.). Critical reviews of the paper were provided by Austin Williams and Frederick Schram. Contribution 341 of the Department of Geology, Kent State University, Kent, Ohio.

REFERENCES CITED

AMBLER, J. W., 1980. Species of *Munidopsis* (Crustacea, Galatheidae) occurring off Oregon and in adjacent waters. Fishery Bulletin, 78(1):13–34.

BABA, K. 1979. Expedition Rumphius II (1975) Crustaces parasites, Comensaux, etc. (Th. Monod et R. Serène, ed.). VII. Galatheid Crustaceans (Decapoda, Anomura). Bulletin Musée national Histoire naturelle, Paris, 4th Series, 1, Section A, 3:643–657.

——, 1981. Deep-sea galatheidean Crustacea (Decapoda, Anomura) taken by the R/V *Soyo-maru* in Japanese waters. Bulletin of the National Science Museum, Tokyo, Series A (Zoology) 7(3):111–134.

BALL, H. W. 1960. Upper Cretaceous Decapoda and Serpulidae from James Ross Island, Graham Land, Volume 24. Falkland Islands Dependencies Survey Scientific Reports, 30 p.

BALSS, H. 1957. Decapoda. VIII. Systematik. Bronns Klassen und Ordnungen des Tierreichs. Band 5, Abteilung I, Buch 7, Lieferung 12:1505–1672.

BIRKENMAJER, J., A. GZDZICKI, and R. WRONA. 1983. Cretaceous and Tertiary fossils in glacio-marine strata at Cape Melville, Antarctica. Nature, 303:56–59.

BISHOP, G. A. 1985. Fossil decapod crustaceans from the Gammon Ferruginous Member, Pierre Shale (Early Campanian), Black Hills, South Dakota. Journal of Paleontology, 59:605–624.

BITTNER, A. 1975. Die Brachyuren des Vicentinischen tertiargebirges. Denkschriften der Kais. Akademie der Wissenschaften in Wien, 34:63–106.

BLISS, D. 1982. Shrimps, Crabs, and Lobsters: Their Fascinating Life Story. New Century Publishers, Piscataway, NJ, 242 p.

BOURNE, G. C. 1922. The Raninidae: a study in carcinology. Journal of the Linnaean Society of London (Zoology), 35:25–79.

CHACE, F. A. JR. 1942. Reports on the scientific results of the Atlantic expeditions to the West Indies, under the joint auspices of the University of Havana and Harvard University. The anomuran Crustacea. I. Galatheidea. Torreia, 11:1–106.

CORLISS, J. B., AND R. D. BALLARD. 1977. Oases of life in the cold abyss. National Geographic Magazine, 152:441–451.

ELLIOT, D. H., C. A. RINALDI, W. J. ZINSMEISTER, AND W. J. TRAUTMAN, 1982. Lower Tertiary strata on Seymour Island, Antarctic Peninsula, p. 287–297. *In* C. Craddock (ed.), Antarctic Geosciences. University of Wisconsin Press, Madison.

FELDMANN, R. M., AND W. J. ZINSMEISTER. 1984a. First occurrence of fossil decapod crustaceans (Callianassidae) from the McMurdo Sound region, Antarctica. Journal of Paleontology, 58(4):1041–1045.

——. 1984b. Fossil crabs (Decapoda: Brachyura) from the La Meseta Formation (Eocene) of Antarctica: paleoecologic and biogeographic implications. Journal of Paleontology, 58(4):1046–1061.

FÖRSTER, R., A. GAZDZICKI, AND R. WRONA. 1985. First record of a homolodromiid crab from a Lower Miocene glacio-marine sequence of West Antarctica. Neues Jarbuch für Geologie und Palaontologie Mh., 6:340–348.

GLAESSNER, M. F. 1929. Fossilium Catalogus; I: Animalia, Editus a J. F. Pompeckj, pars 41: Crustacea Decapoda. W. Junk, Berlin, 464 p.

——. 1960. The fossil decapod Crustacea of New Zealand and the evolution of the order Decapoda. New Zealand Geological Survey Paleontological Bulletin, 31:1–78.

——. 1969. Decapoda, p. R399–R628. *In* R. C. Moore (ed.), Treatise on Invertebrate Paleontology, Part R, Arthropoda 4, volume 2. Geological Society of America and University of Kansas Press, Lawrence.

——. 1980. New Cretaceous and Tertiary crabs (Crustacea: Brachyura) from Australia and New Zealand. Transactions of the Royal Society of South Australia, 104(6):171–192.

GOEKE, G. D. 1985. Decapod Crustacea: Raninidae. Mémoire Museum nationale Histoire naturelle, série A, Zoologie, 133:205–228.

GRIFFIN, D.J.G. 1970. A revision of the Recent Indo-west Pacific species of the genus *Lyreidus* de Haan (Crustacea, Decapoda, Raninidae). Transactions of the Royal Society of New Zealand, Biological Series, 12(10):89–112.

GUINOT, D. 1977. Propositions pour une nouvelle classification des Crustacés Décapodes Brachyoures. Compte Rendu Academie Science Paris, Série D, 285:1049–1052.

——. 1978. Principles d'une classification évolutive des Crustacés Décapodes Brachyoures. Bulletin Biologique de la France et de la Belgique, 112(3):211–292.

JENKINS, R. 1972. Australian fossil decapod Crustacea: faunal and environmental changes, Volume 1. Unpubl. Ph.D. thesis, University of Adelaide, South Australia, 392 p.

MANNING, R. B., AND L. B. HOLTHUIS. 1981. West African brachyuran crabs (Crustacea: Decapoda). Smithsonian Contributions to Zoology, 306:1–379.

MCLAY, C. L. 1985. Brachyura and crab-like Anomura of northern New Zealand. Leigh Laboratory Bulletin, 19:1–387.

MERTIN, H. 1941. Decapode Krebse aus dem Subhercynen und Braunschweiger Emscher und Undersenon, sowie Bemerkungen über einige verwandte Formen in der Oberkreide. Nova Acta Leopoldina, n.f., 10(68):152–257.

MILNE EDWARDS, A., AND E.-L. BOUVIER. 1894. Considerations generales sur la Famille des Galatheides. Annales Science Naturelle, Zoologie, Series 7, 16:191–327.

MIYAKI, S., AND K. BABA. 1970. The Crustacea Galatheidae from the tropical-subtropical region of West Africa, with a list of the known species. Scientific Results of the Danish Expedition to the Coasts of Tropical West Africa, 1945–1946, Atlantide Report No. 11:61–97.

MÜLLER, P. 1984. Decapod Crustacea of the Badenian. Institutum Geologicum Hungaricum, Geologica Hungarica, Series Palaeontologica, Fasciculus 42, 317 p.

POWERS, L. W. 1977. A Catalogue and Bibliography to the Crabs (Brachyura) of the Gulf of Mexico. Port Aransas Marine Laboratory, University of Texas Marine Science Institute, 190 p.

RATHBUN, M. J. 1925. The spider crabs of America. United States National Museum Bulletin, 129:1–461.

——. 1937. The oxystomatous and allied crabs of America. United States National Museum Bulletin, 166:1–278.

SADLER, P. M. 1986. Sequence and lenticular form of Paleogene units on Seymour Island, Antarctic Peninsula. Geological Society of America Abstracts with Programs, 18(4):322.

SAKAI, T. 1937. Studies on the crabs of Japan. II. Oxystomata. Science Report of the Tokyo Bunrika Daigaku (B), 2:155–177.

——. 1976. Crabs of Japan and the Adjacent Seas, 2 volumes. Kodansha Ltd., Tokyo, 773 p.

SAVAZZI, E. 1986. Burrowing sculptures and life habits in Paleozoic lingulacean brachiopods. Paleobiology, 12:46–63.

SCHÄFER, W. 1951. Fossilisations bedingungen brachyurer krebse. Adhandlungen der Senckenbergischen Naturforschender Gesellschaft, 485:221–238.

TAYLOR, B. J. 1979. Macrurous Decapoda from the Lower Cretaceous of southeastern Alexander Island. British Antarctic Survey Scientific Reports, 81:1–39.

TONIOLO, A. R. 1909. L'Eocene dei dintorni di Rozzo in Estriane las sua fauna. Palaeontographia Italica, 15:237–295.

VIA BOADA, L., 1981. Les crustacés décapodes du Cenomanien de Navarra (Espagne): premiers resultats de l'etude des Galatheidae. Geobios, 14(2):247–251.

——. 1982. Les Galatheidae du Cenomanien de Navarra (Espagne). Annales de Paleontologie, 68(2):107–128.

WEIMER, R. J., AND J. H. HOYT. 1964. Burrows of *Callianassa major* Say, geologic indicators of littoral and shallow neritic environments. Journal of Paleontology, 38:761–767.

WILLIAMS, A. B. 1984. Shrimps, Lobsters, and Crabs of the Atlantic Coast of the Eastern United States, Maine to Florida. Smithsonian Institution Press, Washington, D.C., 550 p.

WRIGHT, C. W., AND J.S.H. COLLINS. 1972. British Cretaceous crabs. Palaeontographical Society Monographs, 126:1–114.

ZINSMEISTER, W. J. 1979. Biogeographic significance of the Late Mesozoic and

Early Tertiary molluscan faunas of Seymour Island (Antarctic Peninsula) to the final breakup of Gondwanaland, p. 349–355. *In* J. Gray and A. Boucot (eds.), Historical Biogeography, Plate Tectonics and the Changing Environment. Proceedings of the 37th Annual Biology Colloquium and Selected Papers, Oregon State University Press, Corvallis, Oregon.

——. 1982a. Review of the Upper Cretaceous–Lower Tertiary sequence on Seymour Island, Antarctica. Journal of the Geological Society of London, 139:779–785.

——. 1982b. Late Cretaceous–Early Tertiary molluscan biogeography of the southern circum-Pacific. Journal of Paleontology, 56:84–102.

——, AND R. M. FELDMANN. 1984. Cenozoic high latitude heterochroneity of Southern Hemisphere marine faunas. Science, 224:281–283.

MANUSCRIPT ACCEPTED BY THE SOCIETY SEPTEMBER 1, 1987

Geological Society of America
Memoir 169
1988

Eocene asteroids (Echinodermata) from Seymour Island, Antarctic Peninsula

Daniel B. Blake
Department of Geology, University of Illinois, Urbana, Illinois 61801
William J. Zinsmeister
Department of Geosciences, Purdue University, West Lafayette, Indiana 47907

ABSTRACT

This chapter augments the work of Blake and Zinsmeister (1979) on asteroids of the upper Eocene La Meseta Formation, Seymour Island, Antarctic Peninsula. *Buterminster elegans* n. gen. n. sp. (Goniasteridae) is described, and small *Zoroaster* aff. *Z. fulgens* (Zoroasteridae), a four-armed *Ctenophoraster downeyea* (Astropectinidae), and an undetermined species of *Sclerasterias*(?) (Asteriidae) are reported and evaluated.

Asteroids are rare in most fossil faunas but common in the La Meseta Formation; the poor record of asteroids is attributed to body construction and habits rather than to a geologically recent diversification. Asteroids, especially members of the Asteriidae, are important in determining structure of many modern communities. The presence of an asteriid species in the La Meseta Formation fauna suggests a community structure parallel to certain modern examples. Elsewhere, the La Meseta Formation has been inferred to have been deposited in moderately high-energy, shallow water; in contrast, modern *Sclerasterias* (in Antarctica), *Zoroaster,* and *Ctenophoraster* are known only from relatively deeper waters. Three small *Zoroaster* aff. *Z. fulgens* are preserved with their arms extended above the disc, apparently buried while suspension-feeding. This posture is rare among asteroids and has not been reported among modern members of the Zoroasteridae. Morphologic differences between La Meseta Formation asteroids and their closest modern biologic allies are relatively minor, suggesting slow evolution. Modern species closely related to the fossil species are known from southern oceans; no major biogeographic changes are evident.

INTRODUCTION

Asteroids are rare or absent from most fossiliferous strata, and many of those that have been collected are incomplete or poorly preserved. The La Meseta Formation, with its relatively abundant asteroids, is important for what it can reveal about asteroid history. In spite of a poor overall fossil record, these fossils also offer some insights into problems of the geological and biological evolution of the Southern Hemisphere.

Important works on modern asteroids of Antarctica include Koehler (1920), Fisher (1940), A. M. Clark (1962), H.E.S. Clark (1963), and Bernasconi (1970). A few fossils have been described by Blake and Zinsmeister (1979) and Medina and del Valle (1983).

GEOLOGIC SETTING AND METHODS

Specimens were collected individually by a number of workers (not including D.B.B.) working on a variety of problems; no bulk samples were collected for asteroids. The localities are illustrated in Figure 1. The asteroids were then turned over (to D.B.B.) for study. No preparation was required beyond simple

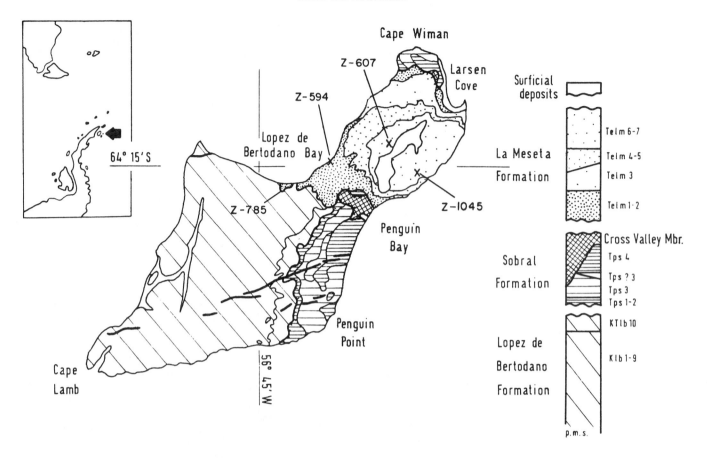

Figure 1. Four localities for sea stars discussed in text, Seymour Island, Antarctic Peninsula.

water and brush cleaning. Specimens were painted with alizarin red S dye to obtain an even background tone, and then coated with ammonium chloride for photography. Illustrated specimens were reposited in the National Museum of Natural History (USNM).

ASTEROID PRESERVATION: GENERAL NATURE AND PALEOECOLOGICAL IMPLICATIONS

The fossil record of asteroids is sparse, yet many modern faunas are taxonomically diverse and species are commonly represented by many individuals. On Seymour Island, as of the time of this writing, four species, two of them represented by numerous specimens, have been collected. The implications of this unusually abundant fossil occurrence are important to the interpretation of the history of asteroids and for their biological interactions.

The generally poor fossil record of asteroids reflects construction and typical life habits rather than a geologically recent diversification. Asteroids have a relatively large coelomic cavity and a near-surface skeleton of relatively small, unfused ossicles; with

death and tissue decay, the skeleton usually is quickly dispersed. Although some asteroids (e.g., Porcellanasteridae) spend at least part of their lives at shallow depths beneath the surface, most are epifaunal and therefore must be quickly buried in order to be preserved. Those few fossil asteroid faunas that are relatively large and diverse accumulated under ideal conditions; the La Meseta Formation, characterized by rapid deposition and limited physical and biological reworking, and the Cretaceous chalk faunas of northwest Europe are examples. Rosenkranz (1971) and Goldring and Stephenson (1972) described other unusually well-preserved echinoderm faunas.

Differential preservation of *Zoroaster* aff. *Z. fulgens, Ctenophoraster downeyae,* and perhaps *Sclerasterias*(?) reflect differences in body construction. *Z.* aff. *Z. fulgens* and *C. downeyae* attained broadly comparable sizes, with disc radii of 1 or 2 cm and arm lengths probably between 10 and 20 cm. *Ctenophoraster* appears to have been the more fragile of the two because its arms are flat, and the two sides of the arm (including the marginal series) in life are linked only by the somewhat delicate paxillae and their associated tissues, and the ambulacral dentition and tissues. *Zoroaster* arms are columnar with heavy tissues and ossicles, forming a continuous, stout, and closely fitted frame; the cross-furrow ambulacral articulation also is more closely fitted than in *Ctenophoraster* (Fig. 2).

It is not possible to determine living abundance ratios between the two species, but fossil fragments are superficially similar to one another in appearance; there should have been no significant collecting bias. About 83 percent of the 72 *Ctenophoraster* arm fragments were broken near the arm axial plane and consist of only one side of the arm, whereas only 26 percent of the 120 *Zoroaster* specimens were broken in this manner. The disc reinforces the arms; 29 percent of 14 *Ctenophoraster* and 19 percent of 21 *Zoroaster* specimens that included portions of the disc were preserved with only one side of arms. Many of the more incomplete fragments include only one interbrachium from the disc. Size of fragments also appears to reflect preservability; the 141 *Zoroaster* specimens have a mean weight of 3.5 gm, that of the 86 *Ctenophoraster* is 2.2 gm. Weight provides an accurate and convenient measurement of size because specimens of both species are similar in skeletal material and preservation of ossicles, and they are relatively free of matrix material. Finally, *Sclerasterias*(?) sp. has an open, rather delicate construction, and perhaps as a result, only two arm fragments were collected. Again it is important that there is no information available on original abundance.

The fossil record of marine invertebrates is dominated by relatively passive and well-armored small-particle feeders, whereas the mobile, less heavily armored predators are generally uncommon. Many asteroids are predators (Sloan, 1980; Jangoux, 1982); their abundance in the La Meseta Formation example suggests that these animals in general were more abundant in the past than their record suggests, but that they were preserved in large numbers only under ideal conditions.

Modern predatory asteroids, and especially asteriids such as *Sclerasterias,* have been shown to significantly affect community structure and species abundance of many temperate and colder environments (Menge, 1982). It is not possible to determine the dynamics of biotic interactions among La Meseta Formation species; however, the presence of asteriids and other asteroids in a molluscan-rich fossil suite does strongly suggest this ancient community was structurally and functionally parallel to certain modern communities.

OTHER ASPECTS OF PALEOECOLOGY

The most striking aspect of the paleoecology of Seymour Island is the presence of numerous molluscan, crustacean, and asteroid genera known from shallow-water Eocene sediments on that island, but found today only in deeper waters. Because the Seymour Island occurrences represent the first appearance in the fossil record of many of these taxa, it has been suggested that they evolved in a southern Weddellian Province and then migrated into deeper waters with the breakup of Gondwanaland and declining Cenozoic climates (Zinsmeister, 1982; Zinsmeister and Feld-

Figure 2. Cross sections of arms of *Zoroaster* (above) and *Astropecten* (below). The sturdier arrangement of *Zoroaster* is less prone to taphonomic breakage. Approx. ×2.

mann, 1984). As described earlier (Blake and Zinsmeister, 1979; Zinsmeister and Feldmann, 1984), *Zoroaster* and *Ctenophoraster* fit this pattern, with modern occurrences of *Ctenophoraster* ranging from 48 to 216 m (Marsh, 1974) and those of the Zoroasteridae at depths greater than 200 m (Downey, 1970). These genera occur in both the upper and lower part of the La Meseta Formation, suggested by Trautman and Elliot (1976) to have been deposited in deeper water than that inferred for the lower interval. Specimens of *Sclerasterias* and *Buterminaster* have been recovered from the lower, apparently shallower water portion of the La Meseta Formation. Modern *Sclerasterias* has been reported from the Bellingshausen Sea, west of the Antarctic Peninsula, at depths between 450 and 560 m, again reflecting the depth relationship noted for other groups. *Sclerasterias* is widely distributed today and occurs off California in waters as shallow as 180 m (Fisher, 1930), still considerably below the apparent depth of the La Meseta Formation occurrence. *Buterminaster* is extinct, although *Pentagonaster,* which is broadly similar, occurs today in depths from 0 to 160 m in Australia (Marsh, 1976).

The La Meseta Formation provides the only known fossil

occurrences of all four asteroid genera, and therefore the time of origin and migrational histories are unknown, although biologic controls appear important in asteroid distributions. The distribution of the Asteriidae (which includes *Sclerasterias*) apparently is restricted today relative to that of the past because the family is now virtually unknown from warm, tropical seas, yet occurred in such settings in the Jurassic of Europe. Menge (1982) and Blake (1983) independently suggested that exclusion of certain asteroids from warm, shallow waters might result from predation pressure. The Zoroasteridae does not appear to have been excluded from shallower waters just by declining climates, because it is essentially worldwide in deeper waters today and presumably could reinvade shallower, warmer water if not constrained by some unidentified factor.

A specimen of *Ctenophoraster downeyae* (Fig. 3.1) is unusual but by no means unique among asteroids in that it has only four symmetrically disposed arms. A number of authors have discussed occurrences of four-armed asteroids, including Hotchkiss (1979), who summarized the literature. Based on many dissections, Hotchkiss concluded that most four- and six-armed individuals in typically five-armed species result from regeneration of injuries rather than originating at metamorphosis. This is true even of specimens that show no obvious signs of damage. Modern asteroids commonly show signs of damage by predators, such as abruptly truncated and partially regenerated arms. Although many fragments of *Zoroaster* and *Ctenophoraster* are available, no other predator damage was recognized.

The only three available small specimens of *Zoroaster* aff. *Z. fulgens* are unusual and important to an understanding asteroid biology in that all are preserved with arms extended aboral to the disc. Arm axes of hypotype 406171 are parallel, with tips separated; arm tips of 406170 are appressed tightly together in an arrangement suggestive of the bud of a flower (Fig. 3.7–10).

We consider the uplifted arms to be a natural living posture, probably reflecting a suspension-feeding habit. The orientation does not appear to be related to death because of the alignment of arms in specimen 406171, and because of the lack of significant distortion. Further, very few asteroid specimens in the U.S. National Museum collections, either dried or in preservatives, are in a comparable posture; these appear to have been distorted during preservation. Dorsal musculature of small Recent zoroasterids could not be investigated because such specimens are rare and none was available for dissection. Asteroids in general have limited muscle tissue developed in the body wall; muscles were found to be present but not obviously enlarged in a dissected larger specimen (R > 130 mm) of *Zoroaster*. The anatomy of the modern forms needs to be more carefully investigated.

Modern *Hymenaster caelatus* (Pterasteridae, Velatida) orient their arms in a similar manner (Sladen, 1889, Plate 85). Seilacher (1983) based the Carboniferous trace fossil species *Asteriacites gugelhopf* on specimens he interpreted as representing infaunal asteroids living with arms raised, and Chamberlain (1971) and Branson (1960) described similar occurrences. Radwanski (1970, Fig. 3) suggested similar behavior for fossil *Astro-*

pecten, although representatives of this genus apparently are not capable of such a posture (Christensen, 1970).

It is not known whether the immature fossil *Zoroaster* were infaunal or epifaunal. Infaunal habits have not been described among the Forcipulatida, to which the zoroasterids belong, although a number of species dig pits in pursuit of prey, and they can be at least partially covered by shifting sediments. Young *Zoroaster* aff. *Z. fulgens* could have been epifaunal, clinging to the substrate by means of the tube feet around the mouth and extending arms into the water column. Presumably, the zoroasterids did not relax their upright orientation because of abrupt burial. Alternatively, *Zoroaster* could have lived infaunally or semi-infaunally, as suggested for the sea star that formed *Asteriacites gugelhopf.*

Comparisons of diet between young and adult fossil and modern zoroasterids are difficult. Collected modern small zoroasterids are few; we are aware of none in the bud posture, but presumably they could have relaxed during collection. Reports of gut contents of modern zoroasterids are few, but include gastropods, spatangoids, crustaceans, and sediment particles (Sloan, 1980; Jangoux, 1982). Neither shells nor sediments are apparent in the fossilized discs of larger individuals of *Z.* aff. *Z. fulgens,* although shells occur in *Ctenophoraster* specimens. None of the adult *Z.* aff. *Z. fulgens* has its arms upraised. If adult *Z.* aff. *Z. fulgens* were predatory, then an ontogenetic dietary shift is implied; shifts are known among modern species (summarized by Sloan, 1980), but they generally involve changes in prey size rather than feeding type.

Adult and small fossil zoroasterids have not been found together. The different source rocks represent local subenvironments in a larger depositional setting. Modern occurrences of physical separation of physical separation of smaller and larger individuals resulting from feeding and protection requirements have been reported (e.g., Birkeland et al., 1971; Yamaguchi, 1975; Shivji et al., 1983), and therefore, physical separation presents no problems of interpretation.

BIOGEOGRAPHY

The Eocene asteroid genera are poorly represented in contemporary Antarctic seas. *Sclerasterias* has been reported from the Bellinghausen Sea, west of the Antarctic Peninsula, from waters between 450 and 560 m; other asteriids occur at shallow depths (A. M. Clark, 1962). All zoroasterids are rare in the Antarctic today; one occurrence has been reported between Gough and Bouvet Islands, northeast of Seymour, and a species has been reported off southern Australia and New Zealand (A. M. Clark, 1962). *Ctenophoraster* is known from three modern species, one from 48 to 93 m in the Marquesas Islands, one from about 100 m in Hawaii, and the third from 216 m, near southern Luzon, Philippines (Marsh, 1974). *Buterminaster* is extinct, although the broadly similar genus *Pentagonaster* is known from western and southern Australia (Marsh, 1976).

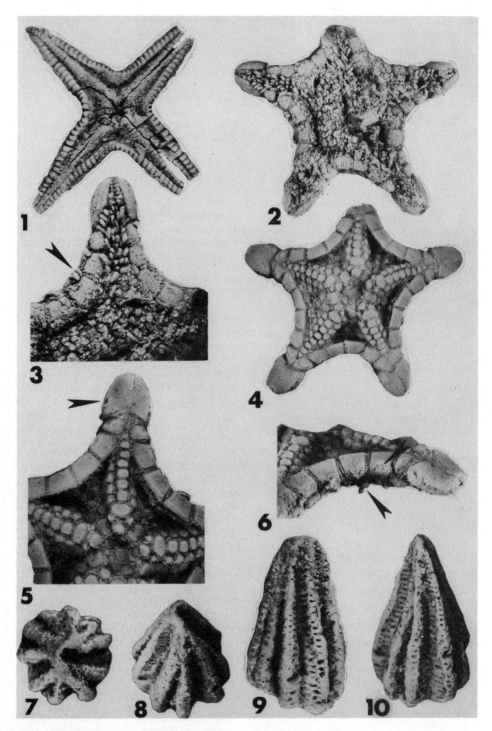

Figure 3. 1, Specimen of *Ctenophoraster downeyae* with four arms, USNM 406169, X1. 2-6, *Buterminaster elegans* n. gen. n. sp., USNM holotype 406170; 2, 4, X2; 3, 5, 6, X3; 2, 3, Actinal views showing disrupted actinal and furrow ossicles inside of marginal frame, part of marginal fringe of spinelets (arrow), furrow spinelets, granules along edges of marginals, numerous adambulacrals preserved in terminal furrow groove; 4-6, Abactinal views showing much enlarged terminal, shape and arrangement of abactinals, marginals, and granules, and marginal and terminal spines and spine bases (arrows). 7-10, *Zoroaster* aff. *Z. fulgens,* USNM 406171, X2 1/2, some arms lifted in bud-like, suspension-feeding(?) posture; closely fitted arms, especially at tips, suggests a life orientation.

A. M. Clark (1962, p. 6) noted an "extraordinary" tendency of modern sea stars in the seas around Antarctica to converge on a single form, one with curved interbrachial angles, small marginal ossicles, and an abactinal surface rather uniformly covered with spinelets. Modern Antarctic sea stars thus seem typified by an overall morphology not shared by the known Eocene representatives; this single morphology among a number of families might suggest convergence comparable to that identified from fossil materials (e.g., Zinsmeister and Feldmann, 1984).

EVOLUTION

Ctenophoraster downyaea, Zoroaster aff. *Z. fulgens,* and—insofar as can be determined—*Sclerasterias*(?) sp. are all similar to extant forms. *Buterminaster* is extinct; however, its overall form is not significantly different from that of certain modern goniasterids. Neither the approximately 40 m.y. of evolution since deposition of the La Meseta Formation, nor the exclusion of *Ctenophoraster* and *Zoroaster* from shallower water settings has had major effects on the morphology of the asteroids.

First recorded occurrences of many surviving asteroid families are in Jurassic rocks, and many fossil taxa from Jurassic and younger strata are remarkably modern in appearance (Wright, 1967; Blake, 1982). The La Meseta Formation asteroids are consistent with this record.

SYSTEMATIC PALEONTOLOGY

Class ASTEROIDEA de Blainville, 1830
Order PAXILLOSIDA Perrier, 1884
Family ASTROPECTINIDAE Gray, 1840

Genus *Ctenophoraster* Fisher, 1905
Ctenophoraster downeyae Blake and Zinsmeister, 1979
Figure 3.1

Discussion. Numerous additional specimens of this species have been collected since Blake and Zinsmeister (1979). Although most are fragmentary, one is relatively complete and unusual in that only four arms are present; the implications are discussed in the paleoecology section.

Repository. U.S. National Museum of Natural History Department of Paleobiology, hypotype 406169.

Occurrence. Locality Z-1045. The specimen was collected from the east side of Seymour Island in a dark brown, weathering sandy siltstone with rounded calcareous concretions. In contrast to other facies in the La Meseta Formation, invertebrates in this facies other than sea stars are virtually absent.

Order VALVATIDA Perrier, 1884
Family GONIASTERIDAE Forbes, 1841

Buterminaster Blake, n. gen.
Figure 3.2–6

Diagnosis. A goniasterid with few tabulate, partially bare abactinals, much enlarged terminals forming the entire arm tip and encompassing distal ambulacral column ossicles, a lateral fringe of spinelets on the inferomarginals and terminals, and spines on the superomarginals and terminals.

Type species. Buterminaster elegans, n. sp.

Description. Rays five, body flat with interbrachial arcs broadly rounded, arm distal intervals prominent, marginal frame defining body outline. Madreporite polygonal, small, placed above and immediately distal to oral frame, surrounded by three larger ossicles. Abactinals tabulate, flattened abactinally, edges beveled. Central enlarged, clearly defined, surrounded by ring of five large ossicles at head of arm columns, and one to as many as four (in the area of the anus) smaller ossicles in each interbrachial area. Carinal series differentiated, not reaching terminal, ossicles nine in number, second distinctly enlarged relative to adjacent ossicles, remainder only weakly enlarged. First interbrachial much enlarged, remainder in rows paralleling carinals, alternate rows offset, second row with about seven ossicles, third with five, remaining area filled with about 12 ossicles that diminish in size toward the marginal frame. Superomarginals and inferomarginals paired and aligned, with three pair between adjacent arm tips; distalmost pair of superomarginals separated proximally by abactinals, abut distally at arm midline. Distal marginals subtrigonal in outline, remainder are subrectangular with long axis parallel to disc edge. Superomarginal cross-sectional outline with fairly distinct shoulder, diameter of one circular spine base approximately 1.0 mm, medially placed on shoulder. Inferomarginal surfaces rather poorly preserved but similar in general form to superomarginals. Terminal proportionately much enlarged, length 4.0 to 4.5 mm, breadth about 5.0 mm, outline subelliptical but truncated proximally by concave juncture for marginals; abactinal surface slightly concave, actinal surface with large groove for ambulacral column, eight adambulacrals preserved in one terminal, with space for about two more.

Ossicles of actinal surface disrupted, poorly exposed, but appear generally characteristic of goniasterids; actinals apparently stout, tabular, surface finely irregular suggesting presence of encrusting ossicles, adambulacrals in terminals are short, stout, block-like.

Granules line marginal, abactinal boundaries, granule positions where not preserved suggested by small pits on ossicular surfaces, granules possibly more or less covered laterally directed faces of superomarginals but absent or few in number over the remainder of surfaces of marginals and abactinals. Inferomarginals, terminals bore columnar spines; one incomplete spine about 1.6 mm long.

Comparisons. Buterminaster is unique among known adult asteroids in the presence of stout, much enlarged terminals that form the entire arm tip. Enlarged distal marginals are uncommon but known among both fossil (e.g., *Trichasteropsis*) and modern (e.g., *Pentagonaster, Sphaeriodiscus*) genera. Terminals are also enlarged in the porcellanasterid *Benthogenia cribellosa,* illustrated by Fisher (1919, Plates 1:1, 8:2c). In this species, and unlike *Buterminaster elegans,* the ossicle is extended proximally on the aboral surface of the attenuated arm tip, and the ambulacral column is not enclosed. These ossicles appear to provide some protection to the tips of the arms and their sensory tube feet; presumably the enlarged terminal would as well. Flat, somewhat rounded abactinals occur in a number of goniasteroid genera; presence of marginal spines, an absence of abactinal spines and incompletely granulose abactinals serve to separate *Buterminaster* from such similar genera as *Goniaster, Calliaster, Pentagonaster,* and *Tosia.*

Cladaster has been reported by Medina and del Valle (1983) from the Cretaceous of Seymour Island (called Isla Vicecomodoro Marambio by the Argentines), although preservation suggests the La Meseta Formation. This genus is similar in general form and distribution of granules to *Buterminaster,* and it has a somewhat enlarged terminals. It lacks marginals spines and has relatively more marginals.

Buterminaster elegans Blake, n. sp.
Figure 3.2-6

Diagnosis and description. As for the genus.

Material. Buterminaster elegans is known from a single specimen preserved free of significant matrix, with the distal tip of one arm lost. The specimen is rather small, and might not have approached maximum size for the species before burial. The abactinal surface is complete and uncompressed; the ambulacral columns have maintained their arches. The interbrachial abactinal surfaces collapsed into the disc but they were not ruptured. Positions of abactinal ossicles were otherwise preserved. The marginal frame also is intact but the actinal surface within the frame is largely disrupted and difficult to interpret. Ossicles are limonitic but detail of abactinal surface ossicles is well preserved.

Dimensions. Primary radii of four arms between 18 and 20 mm, interbrachial radii between 8 and 10 mm.

Repository. National Museum of Natural History Department Department of Paleobiology, holotype USNM 406170.

Etymology. Buterminaster elegans is the elegant sea star with large terminals.

Occurrence. Locality Z-785. The specimen was collected from the discontinuous gravel at base of La Meseta Formation, which consists of ferruginous weathering cobbles and shell debris deposited in a moderately high-energy environment during Late Eocene transgression. The gravel varies in thickness from a few centimeters to a meter, reflecting the uneven surface of the underlying Late Cretaceous rocks on which the sediments were deposited. The fauna of the gravel is dominated by brachiopods, bryozoans with hemispheric and fenestrate colony forms, and crinoids. Only a few molluscs are present with oysters and coarsely ribbed pectens being most common. This low molluscan diversity contrasts sharply with the abundant and diverse molluscan assemblages that characterize most of the La Meseta Formation. Most invertebrates from the basal conglomerate are rare or absent from higher in the La Meseta Formation.

Order FORCIPULATIDA Perrier, 1884
Suborder ZOROCALLINA Downey, 1970
Family ZOROASTERIDAE Thomson, 1873

Genus *Zoroaster* Thomson, 1873
Zoroster aff. *Z. fulgens* Thomson, 1873
Figures 3.7-10, 4.1-4

Discussion. Many more specimens of this species, almost all of them fragmentary, have been collected since Blake and Zinsmeister's work

(1979); most have contributed little new information about the species. The only three known small individuals, however, are unusual in that all are preserved with arms extended aboral to the disc; these are discussed further in the section on paleoecology.

Material. Three small specimens of *Z.* aff. *Z. fulgens* (USNM 460171-460173) are available. Preservation is quite poor because of the presence of many sand grains, which are quite large relative to ossicular size; further, postpreservational surface abrasion has partially obscured ossicular detail. Nonetheless, identification is clear, based on arrangement of the ambulacral column (four rows of tube feet proximally; alternate adambulacrals strongly carinate), arrangement of the abactinal skeleton (prominent, keel-like carinal series; stout, closely fitted marginals and actinals) and the presence of elongate, triangular, and pointed arms.

Repository. National Museum of Natural History, Department of Paleobiology, hypotypes USNM 406171-406173.

Occurrence. Localities Z-607 (USNM 406171), Z-598 (USNM 406172-406173): specimens collected from the crinoid horizon near the top of Unit III, La Meseta Formation (see Zinsmeister, 1984). The crinoids occur in a fine-grained, greenish weathering glauconitic horizon. This glauconite sand contains an assemblage of invertebrates (crinoids, ophiuroids, and molluscs) unique to this horizon.

Suborder ASTERIIDINA
Family ASTERIIDAE
Sclerasterias (?) sp.
Figure 4.5-9

Description. Carinal series clearly defined, ossicles of series of single type, gradually becoming smaller distally, ossicles cruciform with lateral flanges approximately perpendicular to proximal-distal axis, length and breadth of medial ossicle about 5 mm, central region quite small with single medial spine base; spines very poorly preserved, might have been short, about 1 to 2 mm in length. Positions of ossicles between carinals and superomarginals (i.e., laterals and secondaries) somewhat distorted by preservation; however, ossicular arrangement might have appeared primarily transverse proximally, more longitudinal medially with a relatively large gap between carinals and marginals proximally on arm, becoming relatively smaller medially. One lateral row present on each side of carinal series; secondaries linked laterals to superomarginals proximally, but might have been absent medially, distally on arm; no secondaries recognized between laterals, carinals. Laterals rather flat, outlines somewhat irregular, some at least with medial spine base. Secondaries simple, rod-like. Ossicles of two marginal series paired, quite strongly overlapping proximally. Superomarginals rather low, cruciform, with large central area, at least one spine base; medial ossicle approximately 4 mm long, with actinal flange about 2½ mm long, abactinal flange about two-thirds of length of actinal, lateral flanges offset to proximal-distal ossicular axis and axis of abactinal flange slightly offset proximally relative to that of actinal flange; ossicles strongly overlapping proximally and distally, without distinct proximal and distal flanges. Inferomarginals similar in size to superomarginals but thicker, with more irregular central area; medial abactinal flange length about 3 mm, other flanges probably weakly defined; with one or perhaps two spine bases; spines on medial ossicles about 3 mm in length, 4 mm on proximal ossicle. Central area of disc probably lightly constructed, interbrachial ossicles stout. One row of actinals present medially on arm; ossicles rather bulbous with medial spine base, small lateral flanges. Ambulacral column strongly compressed as in typical asteriid pattern, ossicular form poorly preserved; a

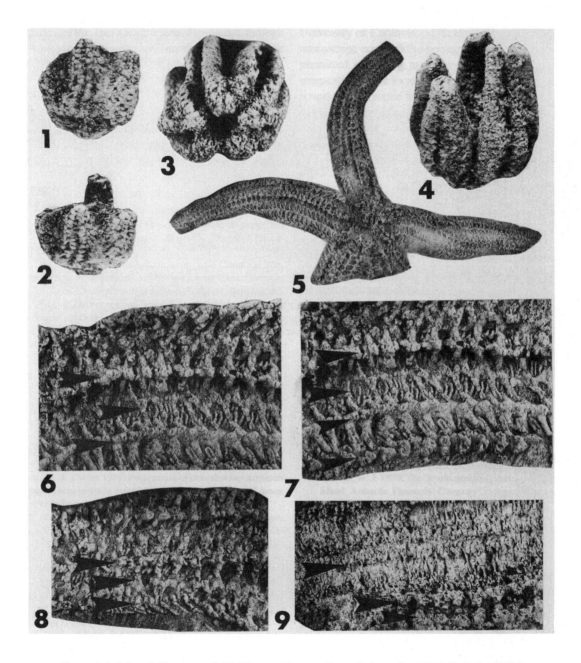

Figure 4. 1-4, Small *Zoroaster* aff. *Z. fulgens* with arms directed above disc, X2 1/2; 1, 2, USNM 406172, distal portions of arms lost, surfaces weathered; 3, 4, USNM 406173, 3 shows oral view of disc 4, arms are raised but not drawn together as in USNM 406171, Figure 3:7-10. 5-9, *Sclerasterias*(?) sp., USNM 406174, artificial cast of abactinal (5-8) and actinal (9) surfaces; 5, complete specimen, arm to left is illustrated in 6-8, X 1/2, others X2; 6, right medial part of arm, distal left, arrows (top to bottom) show carinals, laterals and ambulacrals, superomarginals; 7, left proximal part of arm, distal left, arrows show carinals, laterals and ambulacrals, superomarginals, inferomarginals; 8, distal arm interval, distal left, arrows showing carinals, narrowed lateral column, marginals, and spines; 9, actinal surface showing ambulacrals, actinals, inferomarginals.

few adambulacrals with one well-developed spine base near furrow edge, relative positions of adjacent ossicles suggest abradial ends of adambulacral may have been covered by dermal and ossicular material, as in *Pisaster.* Other ossicles not clearly preserved.

Comparisons. Among post-Paleozoic asteroids, fossil members of the Asteriidae probably are the easiest to recognize at the familial level and the most difficult to assign to genus. Characteristic familial characters include a much compressed ambulacral column with four rows of tube feet; a small disc and columnar, and, as generally preserved, typically somewhat sinuously flexed arms; and a generally more or less openly reticulated skeleton of rather small, cruciform ossicles. Asteriid generic concepts (summarized by Fisher, 1930, and A. M. Clark, 1962) make much use of features typically distorted and displaced in fossils, including numbers of spines on the adambulacrals, nature of the pedicellariae and their arrangement, arrangement of abactinals and of soft organs (i.e., gonads and papulae) and number of rows of actinal and abactinal ossicles. These ossicular rows can only be identified relative to the marginals, yet marginals can be difficult to identify because they tend to be similar in size and form to ossicles of adjacent series.

Recognition of marginals in USNM 406174 is based on linear and transverse alignment of ossicles of the two series and on the relative size of the ossicles. Interpretation of arrangement of actinal ossicles is based on one small interval of one arm (Fig. 4.9). Comparison of the two surfaces suggests the inferomarginal row in this interval is exposed on the actinal surface, and only a single actinal row is present between the inferomarginals and the adambulacrals.

Assignment to *Sclerasterias* is based on very strong similarities of ossicular form and arrangement between the fossil and modern species. Because modern taxon concepts depend so strongly on spine and pedicellariae development, however, generic identification is considered uncertain.

Material. Parts of five specimens are available; three are relatively complete, and two are small arm fragments. The ossicles of the relatively complete specimens are largely collapsed and partially leached so that, although body form and the compressed nature of the ambulacra can be recognized, most ossicles are poorly preserved. A part of the largest specimen (USNM hypotype 406174) is quite well preserved, however. This specimen was flattened and it is preserved on the two sides of a single slab broken along a fracture plane. The two slabs were collected in separate field seasons. The actinal surface was collected first and retains the leached ossicles; relatively little can be seen. The abactinal surface was exposed an additional year, and the ossicles were more completely leached; an artificial cast shows arrangement of much of the skeleton. Because the matrix is calcareous, the actinal surface cannot be artificially leached.

USNM 406147 consists of the disc, most of three arms, and the base of the other two. The abactinal surface has arm radii of 120, 115, 100, 32 and 28 mm, and disc radii between 24 and 20 mm. The arms of the actinal surface have radii of 125, 95, 85, 25 and 23 mm, and disc radii of 24 to 12 mm.

The other two relatively complete specimens are smaller, and both are preserved with abactinal surfaces exposed. Arm radii of USNM 412358 are between 29 and 53 mm, with disc radii of about 10 mm, whereas those of USNM 412359 are between 33 and 72 mm, with disc radii of about 18 mm. Arm fragments in the UC Riverside collections are of lengths of 28 and 30 mm. Unlike the more complete specimens, ossicles of the arm fragments are not leached. To the limited extent determinable, the five specimens are similar and likely to belong to a single species, but this is uncertain. The description is based on the single large specimen.

Repository. National Museum of Natural History, Department of Paleobiology, hypotypes USNM 406174, 412358, 412359.

Occurrence. Locality Z-594: The specimens were collected from a calcareous concretionary horizon at the base of a low sea cliff about 2 km from the mouth of Cross Valley. Concretions form discontinuous lenses as much as 1 m in thickness. Associated sediments are flaggy weathering, well-bedded, fine-grained silty sandstones. Plant debris is abundant in concentrations, and molluscs and echinoids are locally numerous on some bedding planes.

ACKNOWLEDGMENTS

Field work on Seymour Island was supported by the Division of Polar Programs, National Science Foundation Grants DPP-8020096 and DPP-8213585. One of us was partially supported by NSF Grant BSR-8106922 (to D.B.B.).

REFERENCES

BERNASCONI, I., 1970. Equinodermos Antarticos, II Asteroideos, 3 Asteroideos de la Peninsula Antarctica. Revista del Museo Argentino de Ciencias Naturales, 9(10):211–282.

BIRKELAND, C., F.-S. CHIA, AND R. R. STRATHMANN. Development, substratum, selection, delay of metamorphosis, and growth in the seastar *Mediaster aequalis* Stimpson. Biological Bulletin, 141:99–108.

BLAKE, D. B. 1982. Recognition of higher taxa and phylogeny of the Asteroidea, p. 105–108. *In* J. M. Lawrence (ed.), Echinoderms, Proceedings of the International Conference, Tampa Bay. A. A. Balkema, Rotterdam.

——. 1983. Some biological controls on the distribution of shallow water sea stars (Asteroidea; Echinodermata). Bulletin of Marine Science, 33(3):703–712.

——, AND W. J. ZINSMEISTER 1979. Two early Cenozoic sea stars (class Asteroidea) from Seymour Island, Antarctic Peninsula. Journal of Paleontology, 53:1145–1154.

BRANSON, C. C. 1960. *Conostichus.* Oklahoma Geology Notes, 20:195–207.

CHAMBERLAIN, C. K. 1971. Morphology and ethology of trace fossils from the Quachita Mountains, southeast Oklahoma. Journal of Paleontology, 45:212–246.

CHRISTENSEN, A. M. 1970. Feeding biology of the sea-star *Astropecten irregularis* Pennant. Ophelia, 8:1–134.

CLARK, A. M. 1962. Asteroidea. Reports B.A.N.Z.A.R.E. Antarctic Research Expedition, B9:1–104.

CLARK, H.E.S. 1963. The fauna of the Ross Sea, Part 3, Asteroidea. New Zealand Oceanographic Institute Memoir 21, 84 p.

DOWNEY, M. E. 1970. Zorocallida, new order, and *Doraster constellatus,* new genus and species, with notes on the Zoroasteridae (Echinodermata: Asteroidea). Smithsonian Contributions to Zoology 64, 18 p.

FISHER, W. K. 1919. Starfishes of the Philippine Seas and adjacent waters. United States National Museum Bulletin, 100:3, 712 p.

——. 1930. Asteroidea of the North Pacific and adjacent waters, Part 3, Forcipulata (concluded). United States National Museum Bulletin 76, 356 p.

——— . 1940. Asteroidea. Discovery Reports, 20:69–306.

GOLDRING, R., AND D. G. STEPHENSON. 1972. The depositional environment of three starfish beds. Neues Jahrbuch für Geologie und Paläontologie, Monatshefte, 1972:611–624.

HOTCHKISS, F. H. 1979. Case studies in the teratology of starfish. Proceedings of the Academy of Natural Sciences of Philadelphia, 131:139–157.

JANGOUX, M. 1982. Food and feeding mechanisms: Asteroidea, p. 117–160. *In* M. Jangoux and J. M. Lawrence, eds. Echinoderm Nutrition. A. A. Balkema, Rotterdam.

KOEHLER, R. 1920. Echinodermata: Asteroidea. Scientific Reports of the Australasian Antarctic Expedition, c8(1):1–308.

MARSH, L. M. 1974. Shallow water asterozoans of southeastern Polynesia. 1. Asteroidea. Micronesica, 10:65–104.

——— . 1976. Western Australian Asteroidea since H. L. Clark. Thalassia Jugoslavica, 12(1):213–226.

MEDINA, F. A., AND R. A. DEL VALLE. 1983. Un nuevo asteroideo Cretacico de la Isla Vicecomodoro Marambio. Direccion Nacional del Antartico, Instituto Antartico Argentino Contribucion 286, 7 p.

MENGE, B. A. 1982. Effects of feeding on the environment: Asteroidea, p. 521–552. *In* M. Jangoux and J. M. Lawrence (eds.), Echinoderm Nutrition. A. A. Balkema, Rotterdam.

RADWANSKI, A. 1970. Dependence of rock-borers and burrowers on the environmental conditions within the Tortonian littoral zone of Southern Poland, p. 371–390. *In* T. P. Crimes and J. C. Harper (eds.), Trace Fossils. Geological Journal Special Issue 3.

ROSENKRANZ, D. 1971. Zur Sedmentologie und Okologie von Echinodermen-Lagerstätten. Neues Jahrbuch für Geologie und Paläontologie, Abhandlungen, 138:221–238.

SEILACHER, A. 1983. Upper Paleozoic trace fossils from the Gilf Kebir–Abu Ras area in southwestern Egypt. Journal of African Earth Sciences, 1:21–34.

SHIVJI, M., et al. 1983. Feeding and distribution study of the sunflower sea star *Pycnopodia helianthoides* (Brandt, 1935). Pacific Science, 37:133–140.

SLADEN, W. P. 1889. Asteroidea. Reports of the scientific results of the voyage of the Challenger. Zoology, 30:1–935.

SLOAN, N. A. 1980. Aspects of the feeding biology of asteroids, p.57–124. *In* M. Barnes (ed.), Oceanography and Marine Biology, v. 18, Aberdeen University Press.

TRAUTMAN, T. A., AND D. H. ELLIOT. 1976. Sedimentology and petrology of lower Tertiary deltaic sediments of Seymour Island, Antarctic Peninsula. Geological Society of America Abstracts with Programs, 8(6):1144–1145.

WRIGHT, C. W. 1967. Evolution and classification of Asterozoa, p. 159–162. *In* N. Millott (ed.), Echinoderm Biology. Academic Press, New York.

YAMAGUCHI, M. 1975. Coral-reef asteroids of Guam. Biotropica, 7:12–23.

ZINSMEISTER, W. J. 1982. Late Cretaceous–Early Tertiary molluscan biogeography of the southern circum-Pacific. Journal of Paleontology, 56(1):84–102.

——— . 1984. Late Eocene bivalves (Mollusca) from the La Meseta Formation, collected during the 1974–1975 joint Argentine-American expedition to Seymour Island, Antarctic Peninsula. Journal of Paleontology, 58:1497–1527.

——— , AND R. M. FELDMANN. 1984. Cenozoic high latitude heterochroneity of southern hemisphere marine faunas. Science, 224:281–283.

MANUSCRIPT ACCEPTED BY THE SOCIETY SEPTEMBER 1, 1987

Geological Society of America
Memoir 169
1988

Echinoids from the La Meseta Formation (Eocene), Seymour Island, Antarctica

Michael L. McKinney
Department of Geological Sciences, University of Tennessee, Knoxville, Tennessee 37916
Kenneth J. McNamara
Department of Paleontology, Western Australian Museum, Perth, Australia 6000
Lawrence A. Wiedman
Department of Geology, Monmouth College, Monmouth, Illinois 61462

ABSTRACT

Two species of echinoids are identified from the La Meseta Formation (Eocene) of Seymour Island, Antarctic Peninsula, Antarctica. The occurrence of *Stigmatopygus* d'Orbigny in subunits Telm 5 and 7 extends its range upward from the Upper Cretaceous to the upper Eocene and represents the first recorded occurrence of the genus in Antarctica. The presence of *Abatus* Troschel in subunits 2, 5, and 7 are its first fossil records. The genus is currently found living off the coast of Antarctica. Both genera are known to be shallow-water dwellers. *Abatus kieri* n. sp. is described herein.

ECHINOID FAUNA

More than 200 specimens of echinoids have been collected from the La Meseta Formation of Seymour Island, Antarctica, over the past decade (Fig. 1). The La Meseta Formation crops out over much of the northeastern portion of the island and is believed to be late Eocene in age (Zinsmeister, 1982). The stratigraphy of this part of the island (Sadler, this volume) and the geological setting for the unit (Zinsmeister, 1982, 1984) have been addressed elsewhere and are not repeated here.

The echinoids represent only a small portion of the total assemblage found within the La Meseta Formation, but they are fairly abundant locally. The dominant bivalved and gastropod molluscs (Zinsmeister, 1979, 1984), the decapod crustaceans (Feldmann and Zinsmeister, 1984; Feldmann and Wilson, this volume), the brachiopods (Owen, 1980; Wiedman et al., this volume), the asteroids (Blake and Zinsmeister, 1979, this volume), the barnacles (Zullo et al., this volume), and several trace fossils (Wiedman and Feldmann, this volume) collected from the unit have been discussed elsewhere.

Two species of echinoids are identified. Both species significantly extend the stratigraphic ranges of each genus. Both genera are known to be shallow-water inhabitants (Mortensen, 1951; Kier, 1962), a fact that agrees with information derived from other taxa collected from the formation.

Stigmatopygus is a cassiduloid known previously only from the Upper Cretaceous (Cenomanian to Senonian) of Africa and Europe. We assign the Eocene fossils to this taxon because of the characteristic deep transverse groove below the supramarginal periproct and the shape of the petals. The best known species, *S. lamberti* Besairie, is found in coarse Upper Cretaceous sandstones of Madagascar (Kier, 1962). The Antarctic form differs mainly in having a significantly lower test and is considered to represent a different species (Fig. 2).

Stigmatopygus is less abundant than *Abatus* in the La Meseta Formation, with fewer than 30 specimens collected, only a few of which are nearly complete. The specimens used in this study were collected by R. M. Feldmann, D. S. Chaney, and Peter Sadler at localities 6, 8, 17, and 31 (Fig. 1), in subunits 5 and 7 (Sadler, this volume). The paucity of specimens may be an artifact of collecting procedures, and more collecting needs to be done to increase the number and quality of the specimens available. This genus is often found associated with the starfish *Zoroaster* Blake and Zinsmeister, the echinoid *Abatus,* an indeterminate ophiuroid, the articulate brachiopod *Bouchardia antarctica* (Buckman), and the inarticulate brachiopod *Lingula antarctica* Buckman, as well as the ichnofossils *Skolithos* Haldemann and *Palaeophycus* Hall.

Abatus is a burrowing spantangoid previously identified only from the Recent of Antarctica and South America (Mortenson,

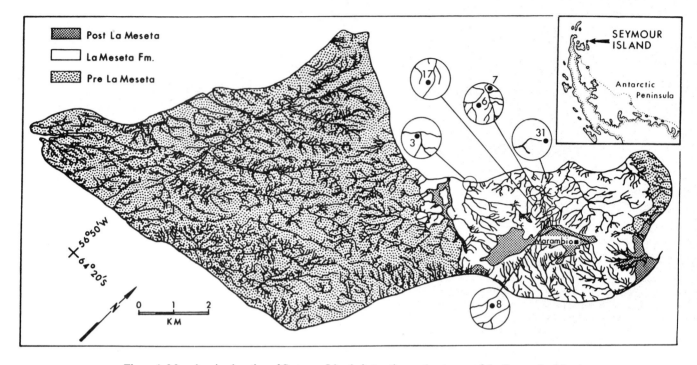

Figure 1. Map showing location of Seymour Island, Antarctica, and outcrops of the Eocene La Meseta Formation where the echinoids used in this study were collected. Locality 30 has specimens included from it, but the exact geographic position is uncertain.

Figure 2. *Stigmatopygus andersoni* Lambert 1910. 1, Apical view. 2, Lateral view. Bar = 1 cm.

Figure 3. *Abatus kieri* n. sp. 1, 2, Apical views. Note presence of three gonopores and exceptional spine preservation. Bar = 1 cm.

1951). The Eocene form differs slightly in its broad anterior ambulacrum and notch and posterior apical system, and represents a different species (Fig. 3). Importantly, McKinney and McNamara consider that a species of echinoid from the Paleocene of Madagascar—previously designated as *Tripylus pseudoviviparus* (Lambert, 1933)—is probably congeneric and should be referred to *Abatus*.

Abatus is primarily a shallow-water, burrowing form, dwelling in littoral to somewhat deeper waters (Mortensen, 1951). It has a very wide peripetalous fasciole that is unusual in forms inhabiting coarse-grained sediments, because the fasciolar area tends to increase in finer grained sediments (Mortensen, 1951). The presence of such a wide fasciole suggests that *Abatus* may be a very deep burrower. This notion is supported by the excellent preservation of spines on one of the specimens studied, indicating probable in situ fossilization in its burrow. Alternatively, the large fasciole may be related to the fact that *Abatus* is a marsupiate echinoid. Perhaps the primary function of this large fasciole is to maintain an effective lining to the burrow, preventing sediment from falling onto the brooded young that nestle in the petals.

The specimens of *Abatus* used in this study were collected by Feldmann, Chaney, and Sadler from seven localities (3, 6–8,

17, 30, 31) (Fig. 1) and in float. The localities are from three different subunits, Telm 2, 5, and 7. Specimens are locally abundant; at some sites they are the dominant taxon present. Most of the more than 200 specimens used in the study are crushed or broken. As reported above, however, portions of the broken specimens show exceptional preservation. The genus is found in assemblages similar to those containing *Stigmatopygus*.

SYSTEMATIC PALEONTOLOGY

Order CASSIDULOIDA Claus, 1880
Family FAUJASIIDAE Lambert, 1905

Genus *Stigmatopygus* d'Orbigny, 1856
***Stigmatopygus andersoni* (Lambert), 1910**
Figure 2.1,2

Cassidulus andersoni LAMBERT, 1910, p. 8; Plate 1, Figs. 34–37.

Description. Test moderately large; gently inflated, orally concave; reaching 5.0 cm in length; width variable 83 to 92 percent test length (TL), widest slightly posterior of center; height 35 to 45 percent TL, maximum height slightly posterior of apical system. Apical system situated 45

percent TL from anterior ambitus; tetrabasal, with four gonopores. Anterior and posterior petals of equal length, each occupying 30 percent TL in length; maximum width, at about midlength, 12 percent TL; petals closed distally; interporal area nearly twice as wide as each row of pore pairs; as many as 40 pore pairs in each row. Periproct supramarginal, longitudinal, forming narrow slits adapically, widening ventrally into large, broad, sunken, pear-shaped depressions, on either side of which are shallow transverse grooves; length of periproct as much as 25 percent TL; width of transverse depressed area, 30 percent TL. Peristome anterior of center, 45 percent TL from anterior ambitus; pentagonal, with strongly developed bourrelets. Spines on adoral surface up to 6 mm in length.

Materials. Fifteen specimens, USNM 416180, positively identified, several possible fragments.

Occurrence. Localities 6, 8, 17, and 31 (Fig. 1) of subunits Telm 5 and 7 of the La Meseta Formation (Sadler, this volume).

Remarks. Lambert's (1910) figured specimen of *Cassidulus andersoni* is virtually identical to our specimens; there can be no doubt that the two are conspecific. The species does not belong in *Cassidulus,* a genus restricted by Kier (1962) to forms with small tests; monobasal apical systems; and straight, open petals. The species clearly belongs to *Stigmatopygus,* a genus uniquely characterized by the possession of a deep subanal transverse groove. *Stigmatopygus andersoni* is similar to the type species *S. lamberti* from the Campanian of Madagascar, differing mainly in its lower test height (35 to 45 percent TL vs. 44 to 62 percent TL for *S. lamberti*). The Seymour Island species can also be distinguished by its slightly shorter petals, smaller peristome but more strongly developed bourrelets, and slightly narrower transverse subanal groove.

Order SPATANGOIDA Claus, 1876
Family SCHIZASTERIDAE Lambert, 1905

Genus *Abatus* Troschel, 1851
Abatus kieri n. sp.
Figure 3.1,2

?*Schizaster anatarcticus* LAMBERT, 1910, p. 11, Plate 1, Figs. 47–49.

Diagnosis. Species of *Abatus* with apical system posterior of center; broad ambulacrum III and relatively shallow petals.

Description. Test to 5.0 cm in length; width, 92 to 95 percent TL; height about 50 to 55 percent TL; maximum height slightly posterior of apical system. Apical system ethmolytic, with three gonopores; situated about 55 percent TL from anterior ambitus. Anterior petals 42 to 45 percent TL in length, diverging anteriorly at about 105°. Posterior petals 20 percent TL in length. Petal depth variable, deep to shallow. Ambulacrum III broad, width being up to 16 percent TL. Anterior notch moderately deep and broad. Peripetalous fasciole wide; very thin lateroanal fasciole observed in one specimen. Peristome about 15 percent TL in width; situated 18 percent TL from anterior ambitus. Labrum projects strongly anteriorly and has a raised rim. Plastron width 40 percent TL. Spines on aboral surface as much as 5 mm in length; petals and ambulacrum III covered by dense canopy of spines, each 2 to 3 mm in length. On lateral interambulacrum, aborally, spines as much as 6 mm in length. Spines on plastron with flattened distal terminations. Subanally spines drawn out into two bundles, dense array of spines surrounds periproct.

Materials. Fifty-nine crushed or fragmented specimens were examined, though preservation of detail, both internal and external, is excellent. Holotype, USNM 416181; paratypes, USNM 416182.

Occurrence. Localities 3, 6–8, 17, 30, and 31 (Fig. 1) from subunits Telm 2, 5, and 7, and in float of the La Meseta Formation (Sadler, this volume).

Etymology. In honor of Porter M. Kier, in recognition of his major contributions to echinoid research.

Remarks. This species is similar to the living species of *Abatus* from Antarctica (Mortensen, 1951), particularly in its possession of three gonopores and a wide peripetalous fasciole, with the lateroanal fasciole being weakly developed. It differs mainly in its broader, more open ambulacrum III; deeper anterior notch; more posterior apical system; and relatively shorter posterior petals. This is the first reported fossil occurrence of the genus. It bears some similarity to species of *Tripylaster,* but differs in its broader peripetalus fasciole and broader petals.

Lambert (1910) described *Schizaster antarcticus* from Seymour Island. His drawings (Lambert, 1910, Plate 1, Figs. 47, 48) suggest the presence of a lateroanal fasciole. In his description he considered the fasciole to be "peu distinct." In the material of *Abatus kieri* that we have examined, there is evidence of a lateral fasciole on only one specimen. Mortensen (1951) noted that it is rudimentary in adults. If Lambert's specimen is conspecific with the form that we describe as *Abatus kieri,* his name *S. antarcticus* is invalid, being a junior homonym of *S. antarcticus* Doederlein, 1906. Because of some uncertainty of the conspecificity of the two forms, we designate one of the specimens that we have studied to be the holotype. As noted above, *Tripylus pseudoviviparus* Lambert, 1933 is probably congeneric with *A. kieri. Abatus kieri* can be distinguished by its longer anterior petals and more posterior apical system.

Small specimens of *A. kieri* show differences from larger specimens. The smallest specimen, a juvenile with unopened gonopores, 16 mm TL, has a more posterior apical system than the larger specimens; shorter posterior petals(?); broader petals, the anterior pair of which are more sinuous; broader ambulacrum III; and less anterior projecting labrum. Intermediate-sized specimens suggest that during the ontogeny of the species the apical system migrates anteriorly; consequently, the anterior petals become less sinuous. Furthermore, the posterior petals undergo a relatively greater increase in length than the anterior pair, ambulacrum III narrows slightly, and the labrum lengthens anteriorly.

There is some intraspecific variation between adults. Distinguishing this variation from postmortem distortion is difficult; however, some specimens seem to have much deeper petals than others. It is possible that some of those with the deepest petals may represent marsupiate females.

The persistence of the subanal spines of two discrete bundles suggests that this species constructed two sanitary tunnels. Nichols (1959) has observed how species inhabiting either coarse sands or mud construct two such tunnels. *Abatus kieri* inhabited coarse sand. The presence of paddle-shaped plastronal spines and a dense concentration of long, lateral, interambulacral aboral spines provide further evidence that this species burrowed quite deeply into the sediment.

ACKNOWLEDGMENTS

Support for the field work on Seymour Island, for Rodney M. Feldmann, was provided by a National Science Foundation grant to William J. Zinsmeister. Support for laboratory work was provided by NSF Grant DPP-8411842 to Feldmann, Kent State University. Feldmann and Dan S. Chaney, Smithsonian Institution, graciously provided some of the specimens used in this study. Peter Sadler, Department of Earth Sciences, University of California, Riverside, provided specimens and a manuscript copy of the geologic map of Seymour Island and permitted use of the map for preparation of the location map.

REFERENCES

BLAKE, D. B., AND W. J. ZINSMEISTER. 1979. Two Early Cenozoic sea stars (Class Asteroidea) from Seymour Island, Antarctic Peninsula. Journal of Paleontology, 53:1145–1154.

FELDMANN, R. M., AND W. J. ZINSMEISTER. 1984. New fossil crabs (Decapoda; Brachyura) from the La Meseta Formation (Eocene) of Antarctica: paleogeographic and biogeographic implications. Journal of Paleontology, 58:1041–1061.

KIER, P. M. 1962. Revision of the cassiduloid echinoids. Smithsonian Miscellaneous Collection, 144:262 p.

LAMBERT, J. 1910. Les echinoides fossiles des iles Snow-Hill et Seymour. Wiss. Ergebr. Schwed. Sudpolar Expedition. 1900–1903. 3:11:15 p.

——. 1933. Echinides de Madagascar communiquès par M. H. Besairie. Annales Geologiques du Service des Mines, Madagascar, 3:1–49.

MORTENSEN, T. 1951. A monograph of the echinoidea 5(2), Spangoidea II. Reitzel, Copenhagen, 593 p.

NICHOLS, D. 1959. Changes in the chalk heart-urchin *Micraster* interpreted in relation to living forms. Philosophical Transactions of the Royal Society of London, Series B(693), 242:347–437.

OWEN, E. F. 1980. Tertiary and Cretaceous brachiopods from Seymour, Cockburn, and James Ross Islands, Antarctica. Bulletin of the British Museum (Natural History), Geology, 33:123–145.

ZINSMEISTER, W. J. 1979. The Struthiolaridae (Gastropoda) Fauna from Seymour Island, Antarctic Peninsula. Congreso Geologico Argentino, Actas Buenes Aires: 1:609–618.

——. 1982. Review of the Upper Cretaceous–Lower Tertiary sequence on Seymour Island, Antarctica. Journal of the Geological Society of London, 139;776–786.

——. 1984. Late Eocene bivalves (Mollusca) from the La Meseta Formation, collected during the 1974–1975 joint Argentine–American expedition to Seymour Island, Antarctic Peninsula. Journal of Paleontology, 58:1497–1527.

MANUSCRIPT ACCEPTED BY THE SOCIETY SEPTEMBER 1, 1987

Geological Society of America
Memoir 169
1988

A new genus of polydolopid marsupial from Antarctica

Judd A. Case and Michael O. Woodburne
Department of Earth Sciences, University of California at Riverside, Riverside, California 92521
Dan S. Chaney
Department of Paleobiology, U.S. National Museum, Smithsonian Institution, Washington, D.C. 20560

ABSTRACT

Eurydolops seymourensis is a new genus and species of polydolopid marsupial found in strata of late Eocene age on Seymour Island, Antarctic Peninsula. It may be the sister taxon of *Antarctodolops dailyi,* also recovered from the same deposits. The new species, known only from an isolated upper premolar, is much smaller and differently specialized than *A. dailyi.* Both of these Antarctic species differ from South American forms and suggest that regional differentiation had taken place with respect to the marsupial faunas of South America and peninsular Antarctica by late Eocene time. The time of greatest potential dispersal by land mammals between South America and Antarctica in the Tertiary Period may have been about late Paleocene, ca. 60 Ma, after which the Seymour Island stocks underwent endemic evolution, with later, and apparently terminal, representatives being preserved in the La Meseta Formation at about 40 Ma. During the Late Cretaceous–late Eocene interval it is likely that only marsupials composed the land mammal population of the Antarctic region, which, by reason of their adaptation to an arboreal habitus, were favored over placental mammals for dispersal (especially in the Late Cretaceous) between South America and Australia through forests dominated by *Nothofagus.*

INTRODUCTION

This chapter records the presence of a second polydolopid marsupial in strata of the La Meseta Formation (late Eocene in age) of Seymour Island, Antarctic Peninsula (Fig. 1), and assesses its phyletic and zoogeographic significance. *Antarctodolops dailyi* was recovered in 1982 from the same deposits (Woodburne and Zinsmeister, 1982, 1984), and demonstrated a faunal link between peninsular Antarctica and South America in the Eocene or possibly late Paleocene. The new material described here is of an animal that was much smaller than, was differently specialized from, and may be the sister taxon of, *A. dailyi.* As such, *Eurydolops seymourensis* n. gen., n. sp. indicates that the land mammal fauna of the Antarctic Peninsula during the late Eocene was significantly greater than appreciated heretofore. Inasmuch as both Antarctic species appear to have their closest relative in or near *Polydolops thomasi,* which lived in South America during the early Eocene (Casamayoran Mammal Age [Fig. 2]; Marshall, 1982), the common ancestor of the Antarctic species may have entered that region well in advance of the late Eocene, at which

time the currently known Antarctic polydolopids became entombed in the La Meseta Formation.

On the one hand, the new material strengthens the faunal affinity between South America and the Antarctic Peninsula during the Eocene; on the other, it shows that peninsular Antarctica had undergone faunal differentiation relative to South America. Whether the barrier that promoted the development of this differentiation was sufficient to exclude placental land mammals from Antarctica remains to be evaluated. Based on botanical evidence reviewed here, it appears likely that, from the Late Cretaceous through the Eocene, the Antarctic Peninsula, mainland Antarctica, New Zealand, Australia, and southern South America were clothed by dense forests composed primarily of *Nothofagus,* the Southern Beech. The arboreal habitus of small marsupials, in contrast to the presumed terrestrial adaptation of contemporaneous placental mammals and medium- to large-sized marsupials (e.g., borhyaenids), suggests that small marsupials were selectively favored over other land mammals for

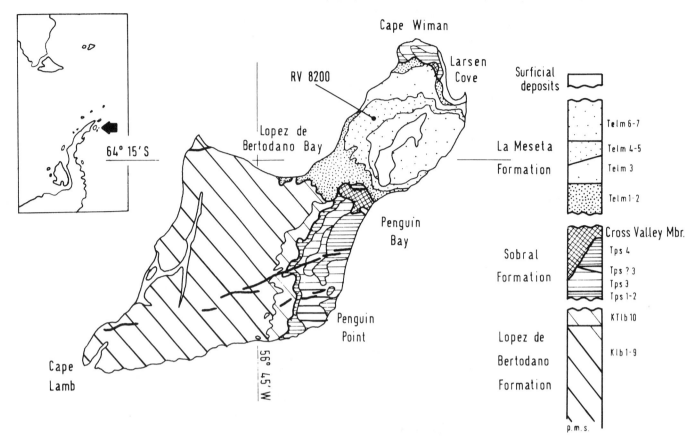

Figure 1. Map showing Seymour Island and the Antarctic Peninsula. RV-8200 is the locality that produced the remains of the polydolopids discussed herein. Ages of rock units: López de Bertodano, Late Cretaceous; Sobral, Late Cretaceous–Paleocene; Cross Valley Member of the Sobral; La Meseta, Eocene; Basaltic dikes (heavy lines on map), ?Late Tertiary.

dispersal between South America and Australia, through forests of *Nothofagus.*

CONVENTIONS

All dimensions are metric. Radioisotopic age assignments have been recalculated, when necessary, to the IUGS constants as presented in Steiger and Jaeger (1977). The classification used here follows Clemens and Marshall (1976). Dental terminology follows Marshall (1982).

The following abbreviations are used:

UCR: Specimen number of the fossil mammal collections of the Department of Earth Sciences, University of California at Riverside

RV-8200: Fossil mammal locality of the Department of Earth Sciences, University of California at Riverside

AMNH: Department of Vertebrate Paleontology, American Museum of Natural History, New York

MACN: Museo Argentino de Ciencias Naturales "Bernardino Rivadavia," Buenos Aires, Argentina

MLP: Museo de La Plata, La Plata, Argentina

MNRJ: Museo Nacional e Universidade Federale do Rio de Janeiro, Rio de Janeiro, Brazil.

UCMP: University of California Museum of Paleontology, Berkeley.

ARW: Width of anterior root of P3/.

PRW: Width of posterior root of P3/.

L/W ratio: Ratio between length and width of P3/.

Ma: Megannum, or a million years in the radioisotopic time scale. The relevant part of the Paleogene time scale is shown in Figure 2.

METHODS

The specimens of *Antarctodolops dailyi* were obtained by surface prospecting at RV-8200 in northern Seymour Island. In 1982, dry screening of the productive unit yielded concentrate that was inspected for fossil mammals. Only an edentulous dentary of a small mammal (as yet unidentified) was recovered. In 1984, a technique of obtaining concentrate by wet-screening

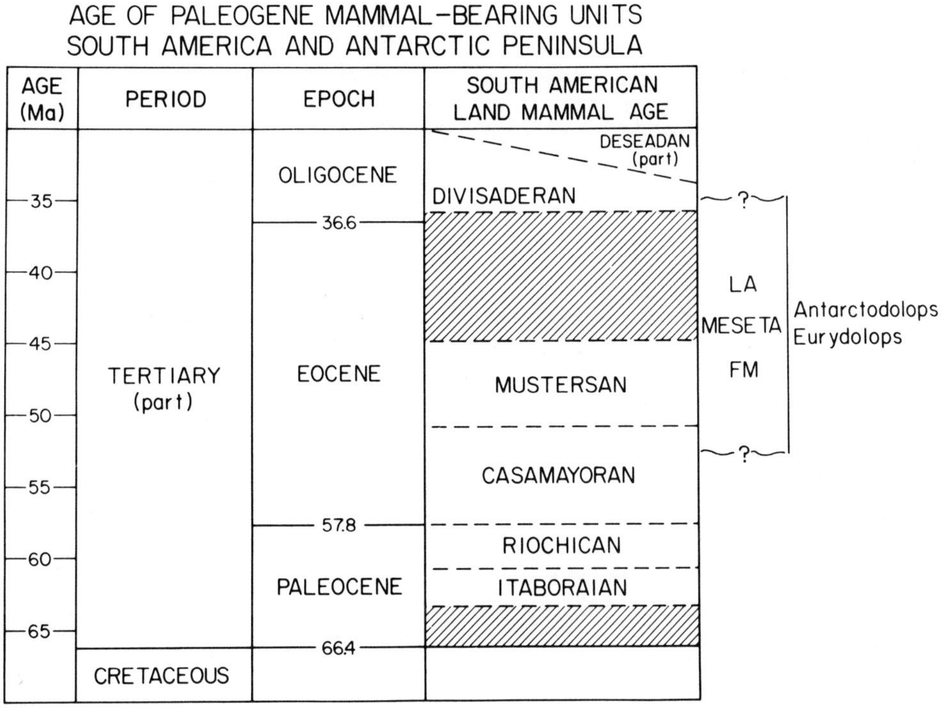

AGE OF PALEOGENE MAMMAL−BEARING UNITS
SOUTH AMERICA AND ANTARCTIC PENINSULA

Figure 2. Paleogene time scale pertinent to discussions in the text. Modified after Marshall, 1985.

methods was utilized. Such methods have long been used by vertebrate paleontologists (e.g., Hibbard, 1949), with some significant changes being developed in materials utilized (Grady, 1979). In 1984 it was not known whether the La Meseta sediments on Seymour Island would be susceptible to the wet-screening process, inasmuch as the low ambient temperatures associated with Antarctic conditions suggested against the use of that technique.

In 1982 and 1984, however, it was realized that despite the near-freezing ambient temperatures, the dark-colored sediments that had been blown onto the local snow and ice cover absorbed sufficient solar radiation, even on cloudy days, to produce enough meltwater to facilitate wet-screening techniques. Hand-sewn bags of mosquito netting (75 × 25 cm), a shallow box 25 × 25 cm and floored with 5-mm hardware cloth, oval galvanized steel tubs 1 m long, rubberized gloves, burlap bags, and muslin cloths were used in this procedure.

The matrix at RV-8200 is an unconsolidated sand with a few large sandstone concretions and large angular fragments of mollusc shells. The concretions and shell fragments made it necessary to process the dry matrix through hardware cloth to remove these large clasts, which otherwise would pierce or abrade holes in the mosquito-net bags. The matrix was placed in the burlap bags for transport to the wash site.

The sediment was transferred to mosquito-net bags, which were then knotted at the top and placed into water-filled tubs. After the sediment had become saturated, the bags were gently agitated, resulting in the fine sediment fraction passing out of the bags into the tubs. When the volume of concentate no longer decreased with additional agitation, the bags were removed from the tub and spun in the air to remove excess water. The concentrate was then dumped from the bags onto muslin sheets for drying by evaporation. Muslin was found to be superior to black plastic sheeting for drying the concentrate, inasmuch as it absorbed moisture even at near-freezing temperatures, whereas the plastic sheeting retained moisture on its surface.

The screen-washing process resulted in the reduction of several tons of matrix to several hundred pounds of concentrate. The reduction of material was important in facilitating the handling and shipping of the potentially fossiliferous matrix and in shortening the time required to sort the material for fossils. In addition, the mosquito-net bags proved to be easier to handle than conventional wooden-framed screen boxes, both in the field and in transport to and from the United States.

MATERIALS

Specimens studied in the course of this study were *Antarctodolops dailyi*, UCR 20912, LM2/ and associated alveolus and roots for P3/; *Polydolops thomasi*, MACN 10338, RP3/-M2/ (type; cast), AMNH 28440, LP3/-M3/ (cast); *Amphidolops serrula*, AMNH 28929, left maxilla fragment with roots for P3/, M1/-M2/, roots for M3/; and *Epidolops ameghinoi*, MNRJ 1407-V, RP3/-M2/ (cast), as well as specimens illustrated and measured in Marshall (1982).

SYSTEMATIC PALEONTOLOGY

Order MARSUPIALIA Illiger, 1811
Superfamily POLYDOLOPOIDEA
(Ameghino, 1897)
Family POLYDOLOPIDAE Ameghino, 1897
Subfamily POLYDOLOPINAE (Ameghino, 1897)

Eurydolops n. gen.

Etymology. Eury-, Greek for wide or broad, in reference to the much greater posterior width compared to the length of P3/; *-dolops,* the suffix commonly employed in generic names of other members of the Polydolopidae. Gender is masculine.

Type species. Eurydolops seymourensis n. sp.

Diagnosis. Eurydolops differs from all other polydolopids in the following combination of characters: in small size [of P3/ only *Polydolops clavulus* (Casamayoran) probably was smaller]; in having a length/width ratio in P3/ (0.80) that is the lowest for any polydolopid; in P3/ being higher crowned relative to its length than in any other polydolopid; in having a P3/ in which the anteroposteriorly flattened posterior root is nearly twice as wide as the triangularly shaped anterior root; in the presence of P2/; and in P2/ and P3/ probably being oriented parallel to the molar row (in contrast to species of *Epidolops*).

Eurydolops seymourensis n. sp.
Figures 3, 4; Table 1

Holotype. UCR 22355, an isolated left P3/, complete, but worn to the extent that dentine is exposed at the apex of the main cusp. Tooth identified as P3/ on basis of comparison with other polydolopids; on presence of interdental wear facets on anterior and posterior crown surfaces; and presence of cuspules on anterior, rather than posterior, half of shearing blade.

Type locality. UCR RV-8200, approximately 550 m above base of La Meseta Formation, unit Telm 5 of Sadler (this volume), Breakwind Canyon, Seymour Island, Antarctic Peninsula; 64°13′S latitude, 56°40′W longitude (Fig. 1).

Age and distribution. Late Eocene, based on associated marine invertebrates, approximately equivalent to both Runangan Stage (marine) of New Zealand and Bartonian Stage (marine) of Europe. Genus and species known only from Seymour Island at this time.

Diagnosis. The diagnosis is that of the genus until other species are described.

Etymology. In reference to the fact that this species is now known only from Seymour Island, Antarctic Peninsula.
Description of the holotype. The following description is made with P3/ being oriented as though in occlusal position, i.e., the roots being dorsally, crown ventrally, situated. The holotype (UCR 22355) is a left upper third premolar (P3/) of a small-sized polydolopid marsupial (Figs. 3, 4). The tooth is bilaterally subtriangular in occlusal outline, and the posterior width is essentially twice that of the anterior portion of the tooth (Figs. 3B, 4B). The posterior width of the tooth is also greater than its length; consequently, the length-to-width ratio is less than 1

(L/W = 0.80; Table 1). The tooth is double-rooted (Figs. 3C,E; 4A,E), with the posterior root (width, 1.9 mm; Figs. 3A,D; 4A,C) almost twice as wide as the anterior root (1.05 mm; Figs. 3D; 4A,E). The ratio of posterior root to width to anterior root is thus 1.81 (Table 1), a ratio greater than that for any other polydolopid for which P3/ is known. The anterior root is triangular in cross section; that of the posterior root is greatly flattened anteroposteriorly (Fig. 4A); thus this root is very short, with its length about one-third its width.

Despite its worn condition and dimunitive size, the crown of the tooth is nearly as tall (3.3 mm) as the unworn P3/ of *P. thomasi* (3.8 mm, cast of MACN 10338). The lengths of P3/ in *P. thomasi* (3.40 mm; MACN 10338; Fig. 5A–C) and *Pseudolops princeps* (5.9 mm; Fig. 5D–F) are greater than that of *Eurydolops seymourensis* (2.25 mm), so that P3/ in the Seymour Island form is relatively more hypsodont than in the other two species. The ratio of P3/ height to length in *E. seymourensis* is 1.46 vs. 1.17 in *P. thomasi,* and 1.02 in *P. princeps.* The shortness of P3/ in the Seymour Island form is not due to interdental wear, although interdental wear facets are present on both the anterior and posterior surfaces of the tooth. The short coronal length of P3/ in *E. seymourensis* is mirrored by the closely appressed, vertically directed, roots of that tooth, and by the fact that coronal features, such as the V-shaped pair of lingual ribs, also are more closely appressed than are comparable features in *P. thomasi.* P3/ roots in *P. thomasi* and *A. dailyi* appear to diverge distally in contrast to their vertical arrangement in the new Seymour Island form (see other polydolopid species; Marshall, 1982). Although more evidence is needed, the attenuated length of P3/ in *E. seymourensis,* as shown by features of its crown and roots, suggests that the snout of this Seymour Island species was actually and relatively shorter than in any other polydolopid.

The cutting edge of P3/ in *Eurydolops seymourensis* is straight and is divided into two approximately equal lengths on either side of the single, central, major cusp (Figs. 3B, 4B; note that, in these figures, the crown of the tooth is oriented with its anterior end directed toward the bottom of the page). The cutting blade is serrated along its occlusal surface. Three small cuspules are aligned along the anterior portion of the blade. At the stage of wear shown here, no cuspules can be identified on the posterior half of the blade. In labial view (Figs. 3E, 4F), the anterior half of the shearing crest of the tooth extends to its apex at an angle that is shallower than that of the more steeply ascending posterior half. A single, narrow, yet distinct labial, rib extends vertically from the central cusp toward the base of the crown, at which point the rib thickens and curves posteriorly at an angle of about 30° relative to the vertical axis of the tooth, and terminates below the anterior base of the posterior root. A less distinct, narrowly V-shaped pair of lingual ribs rises vertically from the apex of the major cusp and terminates below the base of the posterior root (Figs. 3C, 4D). These features are generally similar to those seen in P3/ of *P. thomasi* (Fig. 5C), but the anteroposterior attentuation of P3/ in the Seymour Island form results in the lingual pair of ribs being more closely appressed than in *P. thomasi.*

Anterior to the labial and lingual ribs, both the labial and lingual surfaces of the tooth are convex relative to the midsagittal plane of the tooth (Figs. 3B, 4B). In contrast, the labial and lingual surfaces of the tooth posterior to the ribs are concave with respect to the midsagittal plane of the tooth and flare out near the base of the crown. This results in P3/ of *Eurydolops seymourensis* being bilaterally subtriangular in occlusal view (Figs. 3B, 4B), and reflects the fact that the tooth is much wider posteriorly than anteriorly at the base of the crown. P3/ of *E. seymourensis* is in this manner generally similar to *P. thomasi,* and differs from P3/ of *Pseudolops princeps* in being labially convex rather than nearly flat (Fig. 5E; see also Marshall, 1982).

The anterior (Figs. 3D, 4E) and posterior (Figs. 3A, 4C,F) faces of P3/ bear triangular interdental wear facets that begin immediately above the cutting edge of the blade and widen toward the base of the crown. These facets represent, respectively, contact points with a P2/ anteriorly and a M1/ posteriorly. Although it is difficult to predict tooth size from

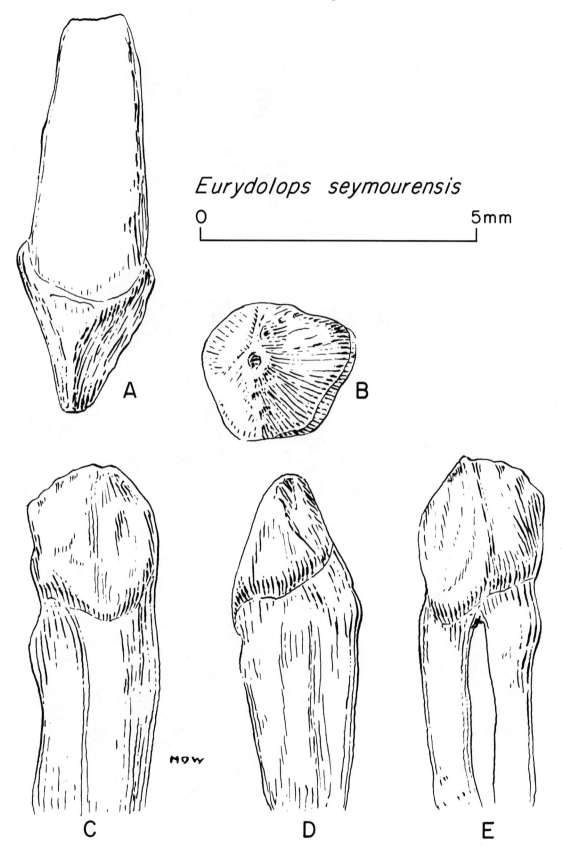

Figure 3. *Eurydolops seymourensis*, n. sp., from the La Meseta Formation, Antarctic Peninsula. Holotype, UCR 22355, LP3/. A, Posterior; B, occlusal; C, lingual; D, anterior; E, labial views. Bar = 5 mm.

Figure 4. Scanning electron microphotographs of *Eurydolops seymourensis.* A, Ventral view of roots; B, stereopair of crown; C, posterior view; D, lingual view; E, anterior view; F, labial view. Bar = 1 mm.

TABLE 1. P3/ MEASUREMENTS AND RATIOS OF
EURYDOLOPS SEYMOURENSIS AND OTHER POLYDOLOPIDS

Species	P3/ Length	Width	P3/ L/W	P3/ Root Width Anterior	Posterior	P3/ PRW/ARW [a]
Eurydolops seymourensis						
UCR 22355	2.25	2.82	0.80	1.05	1.90	1.81
Antarctodolops dailyi [b, c]						
UCR 20912	2.38	2.50	0.95	1.55	2.50	1.61
Polydolops thomasi [d, e]						
MACN 10338	3.40	3.80	0.89	2.20	3.42	1.55
AMNH 28440	3.00	3.00	1.00	1.65	2.30	1.39
MLP 66-V-4-26	3.20	3.10	1.03	——	——	——
Amphidolops serrula [c]						
AMNH 28929	5.05	4.00	1.26	2.35	4.00	1.70
Pseudolops princeps [d]						
MACN 10332	5.90	4.30	1.37	——	——	——
Eudolps tetragonus [d]						
MLP 77-VI-14-5	3.50	3.00	1.17	——	——	——
Epidolops ameghinoi [f]						
from (n = 21)	4.30	3.40	0.92	——	——	——
to (n = 21)	5.80	5.30	1.29	——	——	——
Mean values	4.97	4.40	——	——	——	——
S.D.	0.55	0.57	——	——	——	——

[a] Ratio between width of posterior root/anterior root.
[b] From Woodburne and Zinsmeister (1984).
[c] Root dimensions only.
[d] From Marshall (1982).
[e] Measurements made by us on a cast of the type specimen.
[f] Calculated from data in Marshall (1982, Appendix 12).

interdental wear facets, it can be said that these indicate at least the minimal width of the next adjacent tooth. Thus, P2/ of *E. seymourensis* was at least as wide as the wear facet developed on the anterior surface of P3/ (1.0 mm). Similarly, M1/ of *Eurydolops seymourensis* was at least as wide as the posterior wear facet on P3/ (2.0 mm). Although not conclusive, this would be consistent with the condition in polydolopine polydolopids in general (in which P2/ is nearly as wide as or wider than P3/, and M1/ is nearly as wide as or slightly wider than P3/), but would be in contrast to the condition seen in the epidolopine polydolopid, *Epidolops ameghinoi*, in which P2/ is minute (e.g., Marshall, 1982, Figs. 67, 68). Regarding this, P3/ is actually present only in *Polydolops thomasi* (Fig. 5B) and *Eudolops tetragonus*; Marshall, 1982, Figs. 58, 59). P3/ is represented by its roots in *Amphidolops serrula* and *Antarctodolops dailyi* and, to judge from root dimensions, P3/ in these species would have been as large relative to M1/ and P2/ as seen in other polydolopines.

The anterior and posterior interdental wear facets of P3/ in *Eurydolops seymourensis* are perpendicular with respect to the midsagittal axis of the tooth, suggesting that the adjacent teeth were aligned parallel to P3/, as well. Although other orientations of P3/ are conceivable, that suggested here is similar to other polydolopine polydolopids where this condition is known (*Polydolops thomasi, Eudolops tetragonus*), or inferred from root orientations (*Amphidolops serrula, Antarctodolops dai-*

lyi). The linear arrangement of this part of the dentition contrasts with that found in the epidolopine polydolopid, *Epidolops ameghinoi*, in which P3 is labially deflected relative to the molar row at an angle of about 30°, and P2/ is lingually inflected at an angle of about 30° relative to P3/ (e.g., Marshall, 1982).

Comparisons. P3/ of *Eurydolops seymourensis* resembles that of all polydolopids, except *Pseudolops princeps*, in having a bilaterally subtriangular occlusal outline. P3/ of *P. princeps* is gently convex labially (Fig. 5E). The new Seymour Island form additionally differs from *P. princeps* in being smaller, in having fewer serrations on the anterior and posterior cutting edges of P3/, in having a much shorter length/width ratio (0.80 vs. 1.37 in *P. princeps*; Table 1), in having more distinct labial and lingual ribs, and in being proportionately higher crowned.

P3/ of *Eurydolops seymourensis* differs from that of species of the largest polydolopid, *Eudolops*, in having more serrations on the cutting blade, in having a shorter length/width ratio (0.80 vs. 1.17 in *E. tetragonus*; Table 1), and in having a posterior root that is nearly twice as wide as the anterior root. By contrast, in *E. tetratgonus* the difference in widths of the anterior and posterior root are not nearly so great. P3/ of *Eurydolops seymourensis* also is relatively and actually higher crowned than *E. tetragonus*. P3/ in *E. tetragonus* is 3.6 mm long and 2.8 mm high

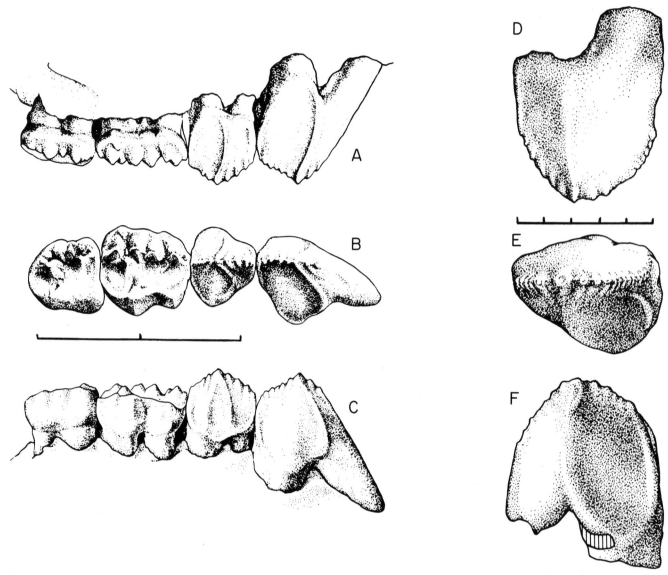

Figure 5. Upper dentition of *Polydolops thomasi* (after Marshall, 1982, Fig. 32; A-C), and *Pseudolops princeps* (D-F), after Marshall, 1982, Figure 41. A, D, Labial; B, E, occlusal; C, F, lingual views. Bar = 5 mm.

(based on illustrations in Marshall, 1982), so that the length is greater than the height, in contrast to the Seymour Island form. The Seymour Island form also differs from *E. tetragonus* in having a pair of lingual ribs, rather than a single lingual rib on P3/.

P3/ of *Eurydolops seymourensis* differs from that of species of *Amphidolops* in its smaller size, in having what appears to have been a shorter length/width ratio (0.80 based on the tooth crown for *E. seymourensis*, vs. a ratio of 1.26 based on roots in *A. serrula*; Table 1), in having a posterior root that is nearly twice as wide as the anterior root (a ratio of 1.81 vs. 1.70) for *A. serrula* (PRW = 4.00; ARW = 2.35), in the vertically oriented vs. the apparently distally divergent roots of P3/ in *A. serrula*, and in having a P2/ (absent in *A. serrula*).

P3/ of *Eurydolops seymourensis* differs from that of *Antarctodolops dailyi*, known only from roots, in being smaller, in having a lower length/width ratio (0.80 vs. 0.95 for *A. dailyi*; Table 1), in the posterior

root being nearly twice as wide as the anterior root (the ratio being 1.81 vs. 1.61 for *A. dailyi*; Table 1), and in the vertically oriented rather than apparently distally divergent P3/ roots of *A. dailyi*. The anterior root of P3/ in *A. dailyi* is set labial to the midline of the tooth, whereas that root in *E. seymourensis* is aligned with the posterior root along the midaxis of the tooth.

P3/ of *Eurydolops seymourensis* differs from that of species *Polydolops*, known only in *P. thomasi*, in being smaller, in having a smaller length/width ratio (0.80 vs. 0.89 to 1.03 in *P. thomasi*; Table 1), in the posterior root being nearly twice as wide as the anterior (ratio of 1.81 vs. 1.55 to 1.39 in two specimens of *P. thomasi*, taken at the base of the crown). *Polydolops clavulus* (Casamayoran, nominally early Eocene of Patagonia) is the only species of *Polydolops* that (based on its lower dentition) would have a P3/ estimated to be close in size to that of *Eurydolops seymourensis* (Fig. 6). This is discussed further below.

P3/ of *Eurydolops seymourensis* differs from that of the epidolopine polydolopid, *Epidolops ameghinoi* in being smaller, in having a more centrally located main cusp, in having more cuspules anterior, and, apparently, fewer posterior to the main cusp, in having a much shorter length/width ratio (0.80 vs. 1.13 in *E. ameghinoi*; Table 1), in apparently being aligned with M1/ and P2/ rather than having been deflected labially as in *E. ameghinoi*, and probably in having a much larger P2/ (minute in *E. ameghinoi*). P3/ of the Seymour Island form also differs from *E. ameghinoi* in having a pair of ribs on the midlingual surface of the tooth (absent in the epidolopine species).

DISCUSSION

Eurydolops seymourensis is morphologically most similar to *Polydolops thomasi*, the only polydolopine polydolopid for which P3/ is known other than *Pseudolops princeps* and *Eudolops tetragonus*. According to Marshall (1982) P3/ of *P. thomasi* is represented only by three specimens MACN 10338, AMNH 28440, and MLP 66-V-4-26. The coronal morphology of these teeth is similar to that of *Eurydolops seymourensis*, but differs in having a greater length/width ratio, in being relatively lower crowned, and in being actually, as well as relatively, much longer. Due to the fact that P3/ is not represented in any other species of *Polydolops*, it is important to ascertain whether P3/ of *E. seymourensis* could pertain to a species of *Polydolops* in which P3/ is yet unknown.

According to Marshall (1982, Appendix 5), three specimens of *Polydolops thomasi* show that the length of P3/ ranges from 3.0 to 3.4 mm, the mean is 3.20, and the standard deviation is 0.20; comparable figures for the width of P3/ are 3.0 to 3.8 mm, 3.30, and 0.44.

The only larger sample of P3/ for a polydolopid species is that for the epidolopine, *Epidolops ameghinoi*, of Itaboraían age (Fig. 3; probably medial Paleocene, ca. 61.0 to 63.6 Ma; Marshall, 1985). All specimens were recovered from fissure fillings in the limestone quarry at São José de Itaboraí, Brazil (Marshall, 1982, p. 82) and should constitute a reasonably homogeneous sample for statistical analysis. According to Marshall (1982), the length of P3/ in *E. ameghinoi* ranges from 4.1 to 5.8 mm. We calculate that the mean dimension for P3/ length in *E. ameghinoi* (N = 22) is 4.97, the coefficient of variation is 0.55. The width of P3/ in *E. ameghinoi* ranges from 3.9 to 5.3 mm, according to Marshall (1982, Appendix 10). From these data the calculated mean for the width of P3/ (N = 21) is 4.40, and the coefficent of variation is 0.57. The range shown by these parameters falls within two (±) standard deviations of the mean.

The analyses presented above suggest that the length and width of P3/ of *Eurydolops seymourensis* could have been included within the lower end of the range of P3/ dimensions of *P. thomasi*, and within the range of P3/ dimensions of smaller sized species of *Polydolops* for which that tooth is yet unknown. These analyses presume that the single specimen parameters given for *Eurydolops seymourensis* represent the mean for the population being analyzed. Regardless of the ultimate evaluation (with more specimens) of this question, P3/ of *E. seymourensis* is interpreted

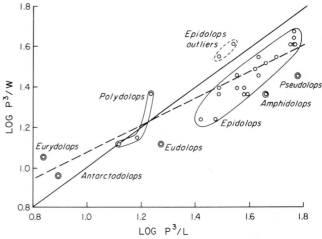

Figure 6. Log-log plot, length P3/vs. width P3/of polydolopid marsupials. The best-fit regression line (dashed; $Y = 0.655 \times + 1.046$) satisfactorially accounts for about 78 percent of the variance for polydolopine species and clearly separates these from epidolopines. The solid line represents a slope of 1.0 for P3/length/width ratios. Circled outliers (of *Epidolops ameghinoi*) were found to be significantly different from other plotted specimens of *E. ameghinoi* at the 0.1 level of significance, utilizing the Dixon text for outliers (Sokal and Rolfe, 1981). As indicated in the text, these specimens may pertain to another species of *Epidolops*. Of the various polydolopid species, only individuals of *Polydolops*, *Antarctodolops*, and *Eurydolops* plot near or above the solid line, having length/width ratios ≤ 1.00 (see Table 1 for details). Note that the plot for *Eurydolops seymourensis* falls farthest above the solid line, consistent with the fact that this species has the lowest length/width ratio of P3/ for any polydolopid yet known.

to have been morphologically (and possibly functionally) different from that tooth in other polydolopines in which tooth and crown proportions are known (*P. thomasi*) or presumed (*A. dailyi*). This is shown by the strongly appressed and vertically oriented P3/ roots and the apparently correlated short crown length and relatively taller crown height in the new Seymour Island form.

The remaining character that is possibly distinctive of *Eurydolops seymourensis* relative to species of *Polydolops* is the great difference in width between the anterior and posterior roots of P3/ (Fig. 4A). It may be instructive that the dimensional differences between *Antarctodolops dailyi* and *Polydolops thomasi* are less than those between *Eurydolops seymourensis* and the other two species. *A. dailyi* differs from species of *Polydolops* in a number of other characters, in addition to tooth proportions (Woodburne and Zinsmeister, 1984). Were other teeth known for *E. seymourensis*, they presumably could differ as well from those of *Polydolops* species.

Another way to evaluate the distinctiveness of *Eurydolops seymourensis* is to assess the overally stability of P3/ dimensions among polydolopids, and then to compare P3/ dimensions with those for P/3 with which P3/ occludes. If the dimensions of P/3 can be used to predict those for P3/, it should be possible to estimate a potential degree of relationship between taxa for which only one or the other tooth is known.

The plot (Fig. 6) of P3/ length vs. P3/ width for polydolopid genera suggests a strong linear relationship of the two parameters within the family. The best-fit regression line ($Y = 0.649 \times 1.098$) explains more than 85 percent of the variance ($\mathbf{P} \ll 0.0001$) for the scatter plot. The circled outliers (*Epidolops*) are excluded at the 0.1 level of probability, utilizing the Dixon test for outliers (Sokal and Rohlf, 1981). This suggests that these specimens of *Epidolops ameghinoi* (DGM 203-M and DGM 200-M; Marshall, 1982, Appendix 9) may not pertain to the same population as other members of that species. All specimens assigned to *E. ameghinoi* are, nevertheless, significantly different at the subfamilial rank from the other specimens plotted in Figure 6 (see discussion herein, and Marshall, 1982).

Eurydolops seymourensis and *Antarctodolops dailyi* fall at the lowermost end of the best-fit regression line in Figure 6. Of the various polydolopid species, only *P. thomasi* and the two Antarctic species have a P3/ length/width ratio of less than 1.0. Note, however, that some specimens of *P. thomasi* have length/ width ratios of P3/ that are equal to or greater than 1.0, and some specimens of *Epidolops ameghinoi* have length/width ratios of P3/ that are less than 1.0. This raises the possibility that P3/ dimensions (as yet unknown) in other species of *Polydolops* could yield ratios similar to that for *Eurydolops seymourensis* and that this parameter alone is not distinctive of all *Polydolops* species.

In an earlier test of the general situation, Woodburne and Zinsmeister (1984, Fig. 7) showed that either the length or width of the lower premolar (P/3) was a reasonably good predictor of the other dimension of that tooth. The regression line computed on the basis of known dimensions for P/3 for species of *Polydolops* explains more than 75 percent of the variance of the dimensions observed. The apparent stability of the width and length of P/3 in *Polydolops* species thus suggests that P/3 dimensions are reasonably stable parameters by which to estimate comparable dimensions for P3/.

Figure 7 shows a plot of P3/ length vs. P3/ width for species of *Polydolops, A. dailyi* and *E. seymourensis*. In this plot, only the points for *P. thomasi* and *E. seymourensis* represent actual measurements of the crown of P3/. All other data points have been determined from estimates calculated from P/3 dimensions, except for *A. dailyi* where P3/ dimensions were calculated from the dimensions of P/3 as well as from the roots of P3/. Thus, using *P. thomasi* as a standard in which both P3/ and P/3 are known, it appears that the ratio in length between P3/ and P/3 is 0.61; that for the width between P3/ and P/3 is 0.75. The best-fit regression line plotted in Figure 7 explains 81 percent of the variance ($P = 0.001$) and thus seems to be about as good a predictor of P3/ dimensions from those of P/3 as is the length or width of P/3 for the other dimension of that tooth (Woodburne and Zinsmeister, 1984, Fig. 7). In the present example, the esti-

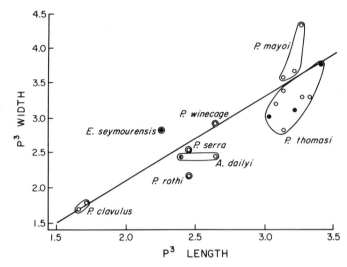

Figure 7. Bivariate plot of P3/length vs. P3/width for species of *Polydolops* and the Antarctic taxa. The determination of the loci for estimated P3/dimensions of certain species is explained in the text. Open circles are estimated parameters; solid circles are actual dimensions. The half-open, half-solid circle indicates that the predicted values of P3/length and width for *Antarctodolops dailyi* coincide with the dimensions taken from the actual roots of that tooth in UCR 20912.

mated length and width of P3/ for *A. dailyi* fall almost exactly at the same point on the regression line as determined from actual measurements of its roots (Fig. 7).

In this context, P3/ dimensions of *E. seymourensis* seem to differ from polydolopids of generally similar size. For its width, P3/ of the Seymour Island form should be about as long as *P. winecage* (Riochican); for its length, P3/ of the Seymour Island form should be about as wide as *P. serra* (Cassamayoran) or *A. dailyi* (ca. 40 Ma; Fig. 2), although the range in predicted values (*P. mayoi*; Mustersan) and known and predicted values (*P. thomasi*; Casamayoran) suggest caution when evaluating the significance of single teeth.

The plots of P3/ length and width (Figs. 6, 7) show that *E. seymourensis* has the lowest length/width ratio (0.80) of any polydolopid (Table 1). This species also has a much greater posterior crown width than would be predicted from its crown length. Coincidental with that greater width is the ratio of posterior root width compared to that of the anterior root. The ratio between these dimensions (1.81; Table 1) is the largest value for that ratio among those members of the Polydolopidae in which that parameter may be determined, keeping in mind that only a few specimens are available for measurement.

Other factors support the recognition of *Eurydolops seymourensis* as a species distinct from those embraced within *Polydolops*. Based on the length of P3/, *E. seymourensis* is one of the smallest polydolopids (but apparently had the most hypsodont

P3/) yet known. Only *Polydolops clavulus* (Casamayoran, nominally early Eocene of Argentina; Fig. 2) is smaller (Fig. 7). One of the major trends shown by species of *Polydolops* is an increase in size from the Itaboraían and Riochican (medial and late Paleocene; Marshall, 1985) to the Deseadan, approximately Oligocene (Marshall, 1982, Fig. 40). If *E. seymourensis*, which most closely resembles *P. thomasi*, were considered a late Eocene member of the genus *Polydolops*, then it would represent a dramatic reversal in the morphocline for size increase exhibited by other species of that genus. In view of this, one of two possibilities must be more likely. One is that *E. seymourensis* is a member of *Polydolops* that has persisted unchanged as to size (but modified P3/ proportions and relative crown height) since having been derived about 20 m.y. ago from an ancestral condition (generally similar to *P. thomasi*) that had been present in the late Riochican or early Casamayoran of South America. The other possibility is that *E. seymourensis* represents a new genus of polydolopid that probably was derived from *Polydolops thomasi* but was differently specialized than was *Antarctodolops dailyi*, also considered to have been possibly derived from or near *P. thomasi* (Woodburne and Zinsmeister, 1984).

The following analysis suggests that *Eurydolops seymourensis* and *Antarctodolops dailyi* could be sister taxa derived from *Polydolops thomasi*, and that each deserves recognition at the generic rank. Based on Woodburne and Zinsmeister (1984), characters that supprt affinity between *Antarctodolops dailyi* and *Eurydolops seymourensis* (and suggest ultimate ancestry with *Polydolops*) include P2/? and P3/ being relatively broad, P3/ thus probably (*A. dailyi*) and actually (*E. seymourensis*) biconvex posteriorly (contra *Pseudolops princeps*), and otherwise similar to the morphology of that tooth in *Polydolops thomasi*. The single labial and pair of lingual vertical ribs on P3/ in *E. seymourensis* also suggest affinity with species of *Polydolops* in contrast with any other polydolopid (e.g., Marshall, 1982, Appendix 13). Just as *A. dailyi* appears nevertheless to have diverged from an ancestry near *P. thomasi* (or *P. serra;* P3/ not known; Woodburne and Zinsmeister, 1984), *Eurydolops seymourensis* is also considered divergent from *P. thomasi* in modification of tooth proportions and concomitant attenuation in length of P3/, reduced spacing between the lingual V-shaped pair of vertical ribs, and the development of a proportionately increased height of crown of that tooth. *Eurydolops seymourensis* is not considered to be a species of *Antarctodolops*, nor a species of *Polydolops*. *Eurydolops seymourensis* most likely had an ancestry near that of, but became differently specialized than, *A. dailyi*. *Eurydolops seymourensis* therefore deserves recognition as a distinct generic-rank taxon.

INTRAFAMILIAL GROUPS OF POLYDOLOPIDS

Marshall (1982) has summarized evidence in favor of two subfamilies of Polydolopidae being recognized: the Epidolopinae, represented by *Epidolops* of Itaboraían and Riochican age, and the Polydolopinae, which contains species of *Amphidolops* (Riochican and Casamayoran), *Eudolops* (Casamayoran), *Pseudolops*

(Casamayoran), and *Polydolops* (Riochican to Deseadan). Together with Woodburne and Zinsmeister (1984), we add the late Eocene Antarctic species *Antarctodolops dailyi* and *Eurydolops seymourensis* to the Polydolopinae. Marshall (1982; Fig. 74) also suggested that *Polydolops* was closer to *Amphidolops* than either of these were to *Eudolops*. Although the coronal morphology of P3/ and P/ is unknown in *Amphidolops*, evidence we review and present here suggests to us that *Amphidolops* is more likely to have had a P3/ morphology that was more like that tooth in *Eudolops* and *Pseudolops*, and that these taxa differ strongly from a clade that included *Polydolops, Antarctodolops*, and *Eurydolops*.

In support of affinity of *Polydolops* and the two Antarctic species, we cite evidence of the coronal morphology of the available dentition of these species cited here and in Woodburne and Zinsmeister (1984). We now add the evidence of size and proportions of P3/ as set out in Table 2. This table shows the means of P3/ dimensions and ratios for all polydolopids, and suggests that they can be arrayed on this basis into four groups (see also Fig. 6). Group I consists of *Polydolops* and the two Antarctic species; it is separated from all other groups on the distinctly smaller dimensions of P3/. Group I taxa also differ from all but members of Group IV in the length/width ratio of P3/ being less than 1.00, and in P3/ having a more complex morphology as indicated by at least the presence of a lingual pair of ribs on its crown.

Polydolopines of Group II (including dimensions of P3/ of *Amphidolops* having been estimated from its roots; Table 1) differ from Group I taxa in P3/ being much larger and in having a length/width ratio distinctly greater than 1.00. Although smaller in size, Group II forms differ from those of Groups III and IV in the large length/width ratio of P3/. Note that P3/ in at least *Eudolops* and *Pseudolops* is more simply constructed (single lingual rib) than that tooth in those Group I forms where the coronal morphology is known (e.g., Marshall, 1982; this chapter).

Groups III and IV consist of epidolopine specimens that have been referred by Marshall (1982) to *Epidolops ameghinoi*. We suggest here (Table 2; Fig. 6) that specimens in Group IV fall outside the population limits for P3/ dimensions shown by the other specimens referred to this species (Group III) and that these Group IV "outliers" (Fig. 6) seem to pertain to another species of *Epidolops*.

Based on the above, we propose that, relative to epidolopines, polydolopine species overall reduced the size of P3/. In addition, Group II polydolopines are characterized by having a P3/ that is relatively the longest of any other polydolopids, and is simply ornamented (single lingual rib). On the other hand, Group I taxa further reduced the overall size and especially the length of P3/, while at the same time having increased the complexity of its coronal morphology by the addition of a pair of lingual ribs on the tooth. Whereas the separation of *Eudolops* and *Amphidolops* still can be defended on the data adduced by Marshall (1982), we suggest that the clade here proposed as consisting of *Polydolops* and the two Antarctic taxa is of no particular affinity to any of the other polydolopine forms, at least based on the morphology and dimensions of P3/. We suspect, but cannot yet demonstrate, that

TABLE 2. MEAN VALUES FOR P3/ IN POLYDOLOPID GROUPS

Group	P3/ Length	P3/ Width	P3/ L/W Ratio
I (n = 5) *Antarctodolops* *Eurydolops* *Polydolops*	2.85 (II+, III*, IV+)	3.04 (III*, IV+)	0.94 (.009) (II+, III*)
II (n = 3) *Amphidolops* *Eudolops* *Pseudolops*	4.88 (I+)	3.70 (III#, IV#)	1.31 (.014) (I+, III+, IV+)
III (n = 19) *Epidolops*	5.04 (I*)	4.35 (I*, II#)	1.16 (.003) (I*, II+, IV*)
IV (n = 2) *Eipdolops* outliers	4.50 (I*)	4.85 (I+, II#)	0.93 (.0001) (II+, III*)

Explanation: The means of P3/ dimensions and the length/width (L/W) ratio are shown for the various polydolopid genera, assembled into groups (Roman numerals) as discussed in the text. The means for each group are accompanied by symbols that indicate the level of significant difference between them, shown by + (0.01 confidence level), * (0.001 confidence level), or # (0.05 confidence level). For example, Group I is statistically different in mean length of P3/ from Group II at the 0.01 confidence level (+), from Group III at the 0.001 confidence level (*), and from Group IV at the 0.01 confidence level (+). Means of the length/width ratios are based on calculations for each specimen in the group. The variance (s^2) (in parentheses) about the mean is shown to the right of the value for the mean in each group.

the clade formed by *Polydolops* and the Antarctic forms represents an early divergence from the basal polydolopine stock.

EOCENE SPECIES DIVERSITY

Woodburne and Zinsmeister (1984) suggested that the presence of *Antarctodolops dailyi* in rocks of the La Meseta Formation of Seymour Island could be compatible with the hypothesis that the stock most directly ancestral to that species entered the Antarctic Peninsula as early as the late Paleocene (ca. 60 Ma). This is much older than the correlated age for the fossil mammals themselves, which occur in rocks approximately 40 Ma old. The synchronous presence of a differently specialized polydolopid genera in the La Meseta Formation lends credence to that interpretation.

The presence of these two taxa at the same locality in the middle part of the La Meseta Formation (RV-8200; Woodburne and Zinsmeister, 1984), along with the supposition that both have had a separate ancestry from or near a species that lived about 15 to 20 Ma earlier, suggests that a colonizing species or species-group entered the Antarctic Peninsula from South America about 55 or 60 Ma, and that vicariant isolation of the Antarctic Peninsula resulted in the evolution of the two endemic Antarctic species recognized in this study. If one were to speculate as to the time of greatest potential for dispersal of land mammals from

South America to the Antarctic Peninsula (or vice versa), as suggested by the data and inferences cited here, that time would appear to have been during the late Paleocene, or about 60 Ma, roughly equivalent to the time of final separation of Australia from mainland Antarctica (ca. 56 Ma; e.g., Woodburne and Zinsmeister, 1984).

SOUTHERN HEMISPHERE LAND MAMMAL DISPERSAL

One of the major questions that faces students of marsupial biogeography is how to account for the diversity of an effectively endemic suite of those animals in Australia, which, until the appearance of Asian rodents about 4.5 Ma, had evolved on that continent in isolation from placental mammals (e.g., Woodburne et al., 1985). Although the fossil record of Australian marsupials extends to only about 23 Ma, the oldest member of each family-level taxon is considered to be as distinct from contemporaneous members of other families as are their living representatives (Woodburne et al., 1985). This suggests that the common ancestor of the Australian marsupials had evolved considerably prior to the known fossil record of the group. Under any of the several mobilist concepts of the past distribution of marsupials (summarized by Marshall, 1980), the basic Australian marsupial stock or stocks must have been present in that continent by the time it finally separated from Antarctica, about 56 Ma (e.g., Woodburne and Zinsmeister, 1984; Fig. 8, this chapter).

Growing evidence (Muizon et al., 1984; Marshall et al., 1985) indicates that both marsupial and placental mammals inhabited South America in the Late Cretaceous. The presence of both groups in Cenozoic strata of that continent has long been known (e.g., Simpson, 1948, 1967, 1978). Whether marsupials originated in Australia (e.g., Kirsch, 1977) or in the Americas (e.g., Tedford, 1974), biogeographers are faced with the problem of either (in the first example) keeping placental mammals out of an area already populated by marsupials or (in the second instance) allowing marsupials, but not placentals, to enter Australia from South America. This dilemma is explored further below.

Evidence from the La Meseta Formation

Current evidence from the La Meseta Formation suggests that the land mammal population of the Antarctic Peninsula was composed solely of polydolopoid marsupials. This seems incredible for the following reasons:

1. Forest Habitat. R. A. Askin (personal communication, 1987) suggests on the basis of palynofloral assemblages, that the upper Maastrichtian terrestrial flora of Seymour Island was composed primarily of podocarpaceous conifers; some angiosperms (such as *Nothofagus*) and ferns were present, but less abundantly. This association is characterized by R. A. Askin (personal communication, 1987) as a humid coniferous forest. By contrast, the land flora of the late Eocene (La Meseta Formation) was a cooler adapted mixed beech (*Nothofagus*)–conifer association. Elsewhere in this volume, Case has recorded the presence of a flora of Paleocene age (Cross Valley member of the Sobral Formation), as well as a sample of nominally medial Eocene age, from sites lower in the La Meseta Formation than previously sampled. These appear to be dominated by megafossil specimens of *Nothofagus*. Cranwell et al. (1960) indicated that mainland Antarctica supported a forest dominated by *Nothofagus* in the Late Cretaceous.

If these ancient *Nothofagus* and *Nothofagus*–conifer forests were as densely structured as those now seen, for instance, in southern South America (Pearson and Pearson, 1982), and on the South Island of New Zealand (J. A. Case and M. O. Woodburne, personal observation), and given the well-known observation that such dense forests allow the passage through them of only small-sized animals (Pearson and Pearson, 1982), it is possible that land mammals (marsupial or placental) larger in size than a mouse or a rat would not have been able to disperse through the Antarctic Peninsula either to or from South America and Australia (Fig. 8). According to Pearson and Pearson (1982), medium- to large-size mammals are absent in the modern *Nothofagus* forests of southern Argentina, even though those forests show levels of species diversity that are comparable to those seen in other types of forests worldwide.

The Argentinian forests described by Pearson and Pearson (1982) range from dense, humid, *Nothofagus*-dominated associations that live under an annual rainfall regime of 3,000 mm, through somewhat more open ("intermediate") structures (the absence of a sapling layer) still dominated by species of *Nothofagus*, with a rainfall regime of about 2,000 mm/yr, to a mixed deciduous *Nothofagus*–bamboo forest with a rainfall of about 1,000 mm/yr. Even though these examples record a progression, from first to last, of decreasing density and progressively more open areas, with ground cover of herbs and grasses in the most open situation, *no mammal larger than a rat inhabited these forests,* and the only truly arboreal form, the insect-eating, microbiotheriid ameridelphian marsupial *Dromiciops australis* was limited to the dense and transitional forests. It was not found in the most open example.

The humid, temperate to cool, moist climates suggested above for the Late Cretaceous to late Eocene of Seymour Island, and the evidence from the fossil plant associations there, suggest that, whereas a flora similar in structure to the "intermediate" floral example (above) may have been present in the Antarctic Peninsula during the times in question, a dense *Nothofagus* forest apparently was present on mainland Antarctica from the Late Cretaceous (Cranwell et al., 1960). As seems likely from their presence in the Tertiary to Recent in New Zealand, Australia (Van Steenis, 1972), and South America, this same factor would also have precluded animals larger than a mouse or rat from dispersing across mainland Antarctica in the early Cenozoic.

2. Marsupials. Small-sized mammals (defined as mouse- or rat-sized) that are known from deposits in South America of Late Cretaceous to Eocene age include the marsupial groups Pediomyidae, Didelphidae, Caroloameghiniidae, Microbiotheriidae, and Polydolopidae (Case and Woodburne, 1986; Marshall et al., 1985). Any or all of these groups could be expected to have inhabited the Antarctic Peninsula during the times under discussion (80 to 40 Ma; above), although arboreal or ground-dwelling animals greater in weight than 200 g probably would not be found (Pearson and Pearson, 1982).

3. Placentals. Furthermore, small-sized placental mammals that pertain to the groups Proteutheria, Notoungulata, and, possibly (as to size) Edentata, also are known from some or all of the deposits that represent the Late Cretaceous to Eocene interval in South America (Simpson, 1948; McKenna, 1981; Marshall et al., 1985). Some or all of these mammals could be expected to have been present on the Antarctic Peninsula and mainland Antarctica, as well.

Thus, based on considerations of size alone, both marsupial and placental mammals can be considered as possibly having inhabited greater Antarctica (including its peninsula) in the Late Cretaceous to the Eocene, when climatic conditions began to deteriorate significantly (e.g., Woodburne and Zinsmeister, 1984; Fig. 9, this chapter). Considerations developed below, however, suggest that placental mammals may not have been as likely as marsupials to have been present in the Antarctic region.

MIDDLE CRETACEOUS PALEOGEOGRAPHY ANTARCTIC REGION

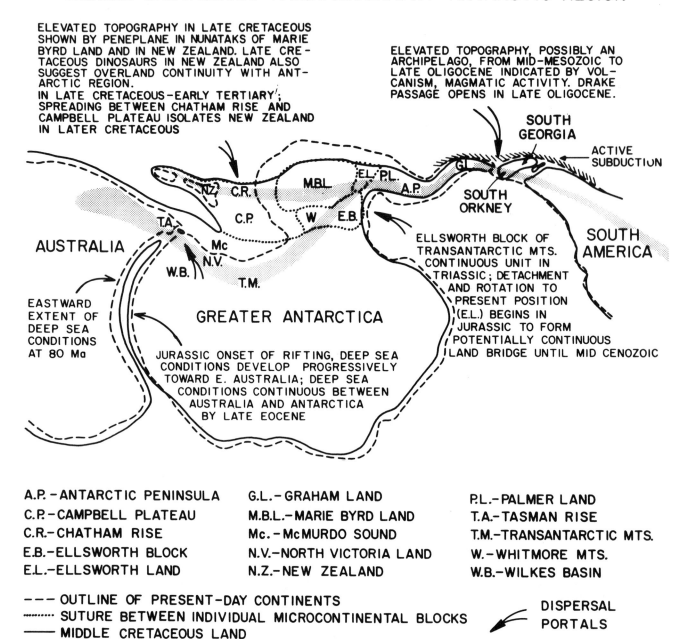

ELEVATED TOPOGRAPHY IN LATE CRETACEOUS SHOWN BY PENEPLANE IN NUNATAKS OF MARIE BYRD LAND AND IN NEW ZEALAND. LATE CRE-TACEOUS DINOSAURS IN NEW ZEALAND ALSO SUGGEST OVERLAND CONTINUITY WITH ANT-ARCTIC REGION.
IN LATE CRETACEOUS-EARLY TERTIARY'; SPREADING BETWEEN CHATHAM RISE AND CAMPBELL PLATEAU ISOLATES NEW ZEALAND IN LATER CRETACEOUS

ELEVATED TOPOGRAPHY, POSSIBLY AN ARCHIPELAGO, FROM MID-MESOZOIC TO LATE OLIGOCENE INDICATED BY VOL-CANISM, MAGMATIC ACTIVITY. DRAKE PASSAGE OPENS IN LATE OLIGOCENE.

SOUTH GEORGIA

ACTIVE SUBDUCTION

SOUTH ORKNEY

SOUTH AMERICA

ELLSWORTH BLOCK OF TRANSANTARCTIC MTS. CONTINUOUS UNIT IN TRIASSIC; DETACHMENT AND ROTATION TO PRESENT POSITION (E.L.) BEGINS IN JURASSIC TO FORM POTENTIALLY CONTINUOUS LAND BRIDGE UNTIL MID CENOZOIC

AUSTRALIA

EASTWARD EXTENT OF DEEP SEA CONDITIONS AT 80 Ma

GREATER ANTARCTICA

JURASSIC ONSET OF RIFTING, DEEP SEA CONDITIONS DEVELOP PROGRESSIVELY TOWARD E. AUSTRALIA; DEEP SEA CONDITIONS CONTINUOUS BETWEEN AUSTRALIA AND ANTARCTICA BY LATE EOCENE

A.P.—ANTARCTIC PENINSULA
C.P.—CAMPBELL PLATEAU
C.R.—CHATHAM RISE
E.B.—ELLSWORTH BLOCK
E.L.—ELLSWORTH LAND

G.L.—GRAHAM LAND
M.B.L.—MARIE BYRD LAND
Mc.—McMURDO SOUND
N.V.—NORTH VICTORIA LAND
N.Z.—NEW ZEALAND

P.L.—PALMER LAND
T.A.—TASMAN RISE
T.M.—TRANSANTARCTIC MTS.
W.—WHITMORE MTS.
W.B.—WILKES BASIN

--- OUTLINE OF PRESENT-DAY CONTINENTS
.... SUTURE BETWEEN INDIVIDUAL MICROCONTINENTAL BLOCKS
——— MIDDLE CRETACEOUS LAND

DISPERSAL PORTALS

Figure 8. Reconstruction of middle Cretaceous paleogeography of the Antarctic region, after Wood-burne and Zinsmeister (1984; Fig. 12). The stippled pattern indicates possible dispersal routes for land animals. By the end of the Cretaceous, New Zealand was isolated, but the pathway between Australia and South America probably persisted into the early Eocene, when Australia separated from Antarctica.

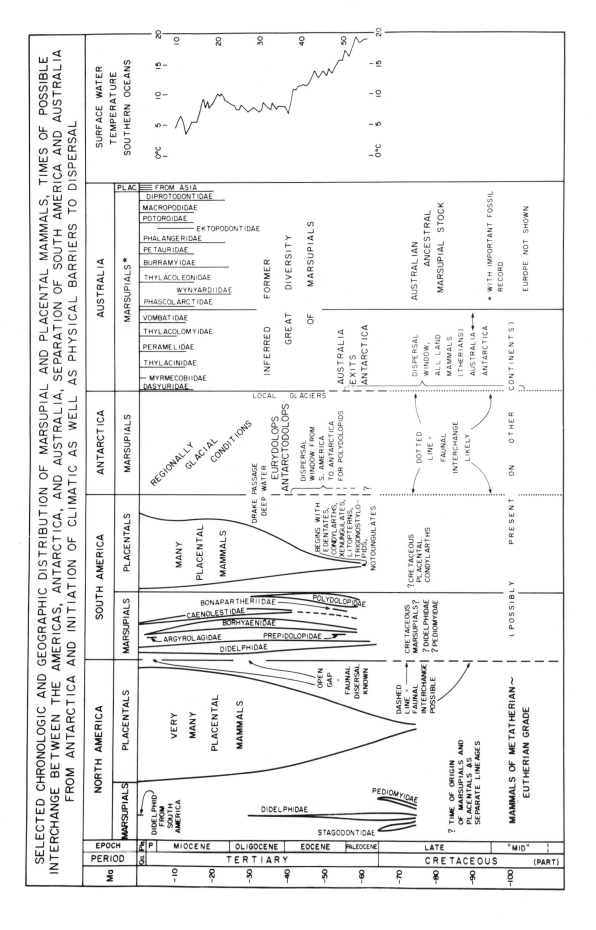

Figure 9. Distribution of therian mammals in the Americas, Antarctica, and Australia, and selected oceanographic and dispersal events (modified from Woodburne and Zinsmeister, 1984, Fig. 13).

Arboreality

In recent years (Kielan-Jaworowska et al., 1979), it has been suggested that a major difference between placental mammals and marsupials was that the original adaptive zone was effectively terrestrial for placentals and arboreal for marsupials. Most living didelphids have an arboreal habit, as shown by their opposite hallux (thumb) and prehensile tail. Living Australian marsupials include a large variety of forms primarily adapted to an arboreal and volant habitus (phascolarctids, phalangerids, pseudocheirids, petaurids, burramyids, acrobatids, tarsipedids). *All* (fossil and Recent) Australian marsupials differ from *all other* marsupials, for which the skeleton of the hind foot is known, in having an articulation between the astragalus (ankle bone) and calcaneus (heel bone) that is a single, continuous surface. Marsupials with this tarsal morphology have been grouped as Australidelphia by Szalay (1982a,b), and it is important to note that the South American microbiotheriid *Dromiciops australis* belongs in this clade. All other marsupials have a two-faceted surface in this part of the hind foot (Szalay, 1982a,b). Szalay (1982a) suggested, and we concur, that the condition in Australian marsupials was derived to enhance the ability of the pollex (big toe) to be opposable to the other digits so as to increase the capability of the hind foot to grasp when climbing (as is already possible for the front feet), and that nonarboreal Australian marsupials (which nevertheless retain this ankle joint construction) have secondarily diverged from an originally arboreal habitus.

Furthermore, we suggest here that not only was the continuous ankle joint facet described above for Australian marsupials developed in the common ancestor for those animals (the basic Australidelphian), but that either by selection in the context of the issue at question (dispersal through *Nothofagus* forests) or by having been preadaptively derived in South America, this ankle joint morphology gave a dispersal advantage to the ancestor of Australia's marsupials as it passed through, rather than under, the presumed dense Antarctic *Nothofagus* forests. Pearson and Pearson (1982) indicated that the only truly arboreal mammal in the dense *Nothofagus* forests of southern Argentina is the insect-eating microbiotheriid marsupial *Dromiciops australis*. The scansorial placental rodent *Irenomys tarsalis* also lives above ground in these *Nothofagus* forests. Importantly, none of the placental mammals that lived in southern South America in the Late Cretaceous to the Eocene are thought to have been arboreal. The other five or six mammal species that successfully inhabit dense to intermediate *Nothofagus* forests of southern Argentina are rodents (Pearson and Pearson, 1982), and it has been suggested (e.g., Marshall, 1982, p. 88) that polydolopids were adaptively similar to rodents or primates, both of which groups were to have their earliest records in deposits of nominally Oligocene age in South America. If polydolopid marsupials also were adaptive vicars of rodents or primates, their presence in the *Nothofagus* forests of Seymour Island is additionally not surprising. The data reviewed here suggest that the animals most likely to have inhabited the *Nothofagus* forests of South America, Antarctica (and its

peninsula), and Australia in the Late Cretaceous to Eocene were small-sized arboreal marsupials (especially Australidelphian ones), or if terrestrial, small-sized marsupials or placentals of rodent-like adaptation. To the extent that the Seymour Island polydolopids are distinctly larger than *Dromiciops australis* (but not much bigger than a large rat), they may have been rodent-like, but not primarily of arboreal habitus.

The scenario outlined here suggests reasons for which placental mammals were not present in the archaic land mammal fauna of Australia, and why one might not expect to find them in deposits of Late Cretaceous to Eocene age in the Antarctic region. It also suggests why one might not expect to find larger sized marsupials (stagodontids, borhyaenids) or placentals of presumed terrestrial (but not rodent-like) habitus in the Seymour Island or other Antarctic deposits.

SUMMARY

1. We describe a new genus and species, *Eurydolops seymourensis* of polydolopid marsupial, and the techniques by which it was collected, from the late Eocene La Meseta Formation of Seymour Island.

2. It appears that *E. seymourensis* had a South American ancestry near *Polydolopos thomasi* or *P. rothi* similar to that proposed for the contemporaneous La Meseta species *Antarctodolops dailyi*. This suggests a Paleocene age for the origin of the Seymour Island polydolopids, and that the most likely time of dispersal from South America to the Antarctic Peninsula was about 55 to 60 Ma.

3. If the above reasoning is sound, it follows that vicariant isolation of the land mammal fauna of the Antarctic Peninsula resulted in *A. dailyi* and *E. seymourensis* being differently specialized from each other, as well as from their South American counterparts, and especially divergent from the evolutionary trends shown by South American polydolopids during the Paleocene through the Eocene.

4. The land surface in southern South America, Antarctica (and its peninsula), New Zealand, and Australia apparently was covered by a dense to intermediate density forest of *Nothofagus*. It appears that marsupials (especially the Australidelphian marsupials), being arboreal in contrast to contemporaneous placental mammals, would have been selectively advantaged over placentals for dispersal through such forests.

5. Polydolopids apparently filled the rodent or primate adaptive niches in South America and the Antarctic Peninsula, prior to the introduction of those placental groups in South America in the Oligocene, and based on size, a rodent-like—but secondarily terrestrial—habitus (as opposed to primarily arboreal habitus) is here proposed for the Seymour Island polydolopids.

ACKNOWLEDGMENTS

This work was supported by National Science Foundation Grant DPP-8213985 to W. J. Zinsmeister (1982) and by Grants

DPP-8215493 and 8521368 to Woodburne (1983, 1986). This support is gratefully acknowledged. We also have benefited during these years from collaborative efforts in the field or post season discussions and collaborations with W. R. Daily and P. M. Sadler (University of California, Riverside), W. J. Zinsmeister (Purdue University), Rosemary Askin (formerly at the Colorado School of Mines, now at U.C., Riverside), Carlos Macellari (University of South Carolina), Brian Huber (Ohio State University); from logistical support given by personnel of International Telephone and Telegraph, by the officers and crews of the U.S. Coast Guard Icebreakers *Glacier* and *Westwind* and of the Research Vessel *Polar Duke*. Illustrations used herein were prepared by Linda Bobbitt, stereophotographs and scanning electron micrographs (from Philips model 515) by Marcia Kooda-Cisco (both U.C., Riverside), and line drawings by M. O. Woodburne. We also thank Richard L. Cifelli and Larry G. Marshall for reviewing the manuscript on which this chapter is based, and for providing many helpful comments.

REFERENCES CITED

CASE, J. A., AND M. O. WOODBURNE. 1986. South American marsupials: a successful crossing of the Cretaceous-Tertiary boundary. Palaios, 2(3):413–416.

CLEMENS, W. A., AND L. G. MARSHALL. 1976. Fossilium Catalogus 1, Animalia: American and European Marsupialia, Pars 123. W. Junk, The Hague, 114 p.

CRANWELL, L. M., H. J. HARRINGTON, AND I. G. SPEDEN. 1960. Lower Tertiary microfossils from McMurdo Sound, Antarctica. Nature, 186:700–702.

GRADY, F. B. 1979. Some new approaches to microvertebrate collecting and processing. Newsletter, Geological Curators Group, 2(7):439–442.

HIBBARD, C. W. 1949. Techniques of collecting microvertebrate fossils. Contributions of the Museum of Paleontology, University of Michigan, 7:91–105.

KIELAN-JAWOROWSKA, Z., T. M. BOWN, AND J. A. LILLEGRAVEN. 1979. Eutheria, p. 221–258. *In* J. A. Lillegraven, Z. Kielan-Jaworowska, and W. A. Clemens (eds.), Mesozoic Mammals: The First Two-Thirds of Mammalian History. University of California Press, Berkeley.

KIRSCH, J.A.W., 1977. The comparative serology of Marsupialia, and a classification of Marsupials. Australian Journal of Zoology, supplementary series, 52:152 p.

MARSHALL, L. G., 1980. Marsupial biogeography, p. 345–386. *In* L. L. Jacobs (ed.), Aspects of Vertebrate History. Museum of Northern Arizona Press, Flagstaff.

——. 1982. Systematics of the extinct South American marsupial family Polydolopidae. Fieldiana (Geology), new series, no. 12, 107 p.

——. 1985. Geochronology and land-mammal biochronology of the transamerican faunal interchange, p. 49–85. *In* F. G. Stehli and S. D. Webb (eds.), The Great American Biotic Interchange. Plenum Press, New York.

——, C. DE MUIZON, M. GAYET, A. LAVENU, AND B. SIGÉ. 1985. The "Rosetta Stone" for mammalian evolution in South America. National Geographic Research, p. 273–288.

MCKENNA, M. C., 1981. Early history and biogeography of South America's extinct land animals, p. 43–77. *In* R. L. Chiochon and A. B. Chiarelli (eds.), Evolutionary Biology of the New World Monkeys and Continental Drift. Plenum Press, New York.

MUIZON, C. DE, L. G. MARSHALL, AND B. SIGÉ. 1984. The mammalian fauna from the El Molino Formation (Late Cretaceous, Maestrichtian) at Tiupampa, southcentral Bolivia. Bulletin du Muséum National d'Histoire Naturalle,

Paris, Section C, 6(4):327–351.

PEARSON, O. P., AND A. K. PEARSON. 1982. Ecology and biogeography of the southern rainforests of Argentina, p. 129–142. *In* M. A. Mares and H. H. Genoways (eds.), Mammalian Biology in South America. Pymatuning Laboratory of Ecology, Special Publication 6.

SIMPSON, G. G. 1948. The beginning of the Age of Mammals in South America, Part 1. Bulletin of the American Museum of Natural History 91:1–232.

——. 1967. The beginning of the Age of Mammals in South America, Part 2. Bulletin of the American Museum of Natural History, 137:1–260.

——. 1978. Early mammals of South America: fact, controversy, and mystery. Proceedings of the American Philosophical Society, 122:318–328.

SOKAL, R. R., AND F. J. ROLF. 1981. Biometry. W. H. Freeman, San Francisco, 859 p.

STEIGER, R. H., AND E. JAEGER. 1977. Subcommission on geochronology: convention on the use of decay constants in geo- and cosmochronology. Earth and Planetary Science Letters, 36:359–362.

SZALAY, F. S. 1982a. A new appraisal of marsupial phylogeny and classification, p. 621–640. *In* M. Archer (ed.), Carnivorous Marsupials. Royal Zoological Society of New South Wales.

——. 1982b. Phylogenetic relationships of the marsupials. Géobios, Mémoire Spécial, 6:177–190.

TEDFORD, R. H. 1974. Marsupials and the new paleogeography, p. 109–126. *In* C. A. Ross (ed.), Paleogeographic Provinces and Provinciality. Society of Economic Paleontologists and Mineralogists Special Publication 21.

VAN STEENIS, C.G.G.S., 1972. *Nothofagus,* key genus to plant geography, p. 275–288. *In* D. H. Valentine (ed.), Taxonomy, Phytogeography and Evolution. Academic Press, London.

WOODBURNE, M. O., AND W. J. ZINSMEISTER. 1982. Fossil land mammal from Antarctica. Science, 218:284–286.

——. 1984. The first land mammal from Antarctic and its biogeographic implications. Journal of Paleontology, 58:913–948.

WOODBURNE, M. O., R. H. TEDFORD, M. ARCHER, W. D. TURNBULL, M. D. PLANE, AND E. L. LUNEDIUS, 1985. Biochronology of the continental mammal record of Australia and New Guinea. South Australia Department of Mines and Energy Special Publication, 5:347–363.

MANUSCRIPT ACCEPTED BY THE SOCIETY SEPTEMBER 1, 1987

Geological Society of America
Memoir 169
1988

Paleogene floras from Seymour Island, Antarctic Peninsula

Judd A. Case
Department of Earth Sciences, University of California, Riverside, Riverside, California 92521

ABSTRACT

Numerous carbon films and leaf impressions have been recovered from a Paleocene locality of the Cross Valley Member of the Sobral Formation and from two horizons of the middle to late Eocene La Meseta Formation, Seymour Island, Antarctic Peninsula. These fossil megafloral assemblages are dominated by specimens of the genus *Nothofagus,* the Southern Beech, with representatives of araucarian and podocarp conifers and several species of ferns. The fossil material of *Nothofagus* exhibits a strong similarity to species of that genus from southern South America and indicates that the paleoclimate of the Antarctic Peninsula was cool temperate during the early Tertiary. The association of this plant material with fossil marsupials of the family Polydolopidae and the dependence of *Nothofagus* on overland seed dispersal strengthen the contention that the hypothesized Weddellian Province reconstruction of Gondwanaland during late Cretaceous and early Tertiary time was a continuous coastal, cool, temperate biogeographical province.

INTRODUCTION

Well-preserved specimens of leaf impressions and carbon films of the Southern Beech, *Nothofagus,* araucarian and podocarp conifers, and several species of ferns were collected during the 1983–1984 and 1985 field seasons on Seymour Island, Antarctic Peninsula. The field expedition combined geologic, stratigraphic, palynologic, and invertebrate and vertebrate paleontologic studies of the Upper Cretaceous through Upper Eocene strata on Seymour Island (Fig. 1). The leaf fossils were recovered from three localities ranging from Paleocene to late Eocene in age. One flora (RV-8265) was collected from the Cross Valley Member of the Sobral Formation (Paleocene), and two floras (RV-8200 and RV-8425) were collected from the La Meseta Formation (middle to late Eocene). The uppermost stratigraphic site, RV-8200 (University of California, Riverside locality number designated for the mammal locality on Seymour Island), has yielded specimens of the first land mammals discovered from Antarctica, the late Eocene polydolopids, *Antarctodolops dailyi* (Woodburne and Zinsmeister, 1982, 1984) and *Eurydolops seymourensis* (Case et al., this volume). The three fossil floras represent the terrestrial indicators of the paleoclimate of the Antarctic Peninsula from the Paleocene through the late Eocene to be compared with hypotheses of paleoclimate based on marine invertebrates. The plant megafossils and fossil marsupials represent the only terrestrial components from the three localities (RV-8200, RV-8265, and RV-8425). The implications of this

association of plant and marsupial material are important in correlating biogeographical and paleoecological relationships of Antarctica, southern South America, and pertinent land masses of the Australasian region during the early Tertiary and possibly as early as the late Cretaceous.

The genus *Nothofagus,* the predominant angiosperm taxon in each of the three floras, has a Gondwana distribution, with the living species present in southern South America, southeastern Australia, Tasmania, New Guinea, New Caledonia, and New Zealand. The fossil *Nothofagus* pollen (=*Nothofagidites*) is known from all of the above locations with the exception of New Caledonia (Hanks and Fairbrothers, 1976; Van Steenis, 1971), and with the addition of Antarctica (Cranwell, 1959; Cranwell et al., 1960; Askin and Fleming, 1982). In addition, palynomorphs of *Nothofagus* have been recovered from Deep Sea Drilling Project (DSDP) sites in the areas of the Campbell plateau (DSDP site 275, South Pacific, southeast of New Zealand), the Falklands plateau (DSDP sites 327, 328 and 511, South Atlantic), and the Ninetyeast Ridge (DSDP sites 214 and 254, Indian Ocean) (Tanai, 1984). Megafossils of *Nothofagus* are known from New Zealand, mainland Australia, Tasmania, southern South America, and Antarctica (Van Steenis, 1971; Tanai, 1984).

Previously collected fossil material of the Southern Beech from Antarctica was described by Dusen (1908). However, the descriptions of the material were of such a superficial nature that

523

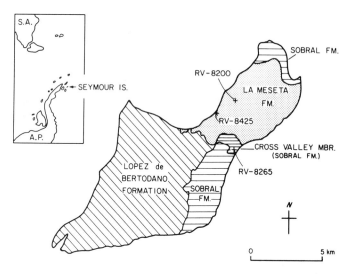

Figure 1. Seymour Island (right) with the three paleofloral localities: RV 8265 (Paleocene) within the Cross Valley Member of the Sobral Formation, RV-8425 (middle Eocene) and RV-8200 (late Eocene) within the La Meseta Formation. *Inset,* The position of Seymour Island relative to the Antarctic Peninsula and South America. Redrawn from Sadler (this volume).

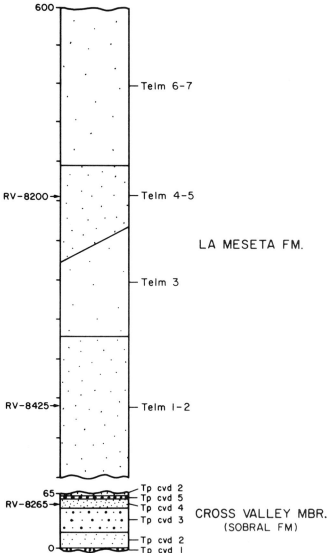

Figure 2. Placement of the three paleofloral localities with the stratigraphic columns (scale in meters) of the Cross Valley Member of the Sobral Formation (Paleocene) and the La Meseta Formation (middle to late Eocene). Units of the Cross Valley Member are Tp cvd 1 through 5. Units of the La Meseta Formation are Telm 1 through Telm 7. For the lithology of these units, refer to Sadler (this volume).

they could be applied to a number of different taxa. In addition, Dusen (1908) attributed some material incorrectly to the genus *Fagus* (Van Steenis, 1971). The overall quality of preservation of the material that was described has been judged to be quite poor (Cranwell, 1963). This material was equivalent to the Paleocene plant material collected from the Cross Valley locality, RV-8265.

Fossil wood of *Nothofagus* from Seymour Island was described by Gothan (1908) and can be found in both the Cross Valley Member of the Sobral Formation and in the La Meseta Formation. Wood is most abundant in Units Telm 2 through Telm 5 of the La Meseta Formation (see Fig. 2).

GEOLOGIC SETTING

The Paleocene megafloral specimens are from Unit Tpcvd 4 (Fig. 2) of the Cross Valley Member of the Sobral Formation (all units presented here are according to Sadler, this volume) a coarse, resistant layer of gray sands. The leaf specimens are present as carbon films in fine-grained, calcium carbonate–cemented, blocky sandstone concretions with medium- to fine-grained, yellow clay inclusions that are frequently encountered within this unit at this locality (RV-8265).

The middle Eocene locality (RV-8425) is represented by a large lens in a coastal cliff of Unit Telm 2 of the La Meseta Formation (Fig. 2). This unit consists of alternating laminated beds of loosely consolidated mudstones and sandstones. The lens is a massive, fine-grained, calcium carbonate–cemented, gray sandstone block cutting down into the laminated bedding. The

megafloral specimens are carbon films in distinct horizons within the massive sandstone lens. Associated with the leaf specimens within these horizons are small, coalified bits of wood, abundant molluscan shell fragments, and starfish remains.

The megafossil leaf material of the late Eocene locality, RV-8200, is from Unit Telm 5 of the La Meseta Formation (Fig. 2). The mammal and plant fossils at RV-8200 were recovered from a thin-bedded arkosic sandstone, overlain by a thicker, fine-grained sandstone, and superimposed on a series of laminated, fine-grained sand and silty clay. From the sedimentology of the units and the remains of crabs, *Lyreidus antarcticus,* the environment has been interpreted as being a shallow-water, high-energy facies, possibly a beach deposit (Feldmann and Zinsmeister, 1984). The leaf impressions and carbon films are preserved in a gray, calcareous siltstone that is darkened internally due to an abundance of fine particles of organic material. The outer surface of the siltstone is light gray due to surface weathering.

PALEOFLORAS

The Paleocene Cross Valley flora (RV-8265; Fig. 3) is dominated by ferns, with at least three different species present based on frond pinna morphology. The three pinna morphs are long and slender (Fig. 3A), short and stout (Fig. 3B), and pinnately compound (Fig. 3C). Angiosperm specimens are primarily *Nothofagus* with a microphyllous leaf size, plus an unidentified large-leaf taxon. Specimens of podocarp conifers round out the flora. Cranwell (1959) recorded many fern spores, plus podocarp and araucarian conifer, and *Nothofagus* pollen from the rocks containing the fossil leaves studied by Dusen (1908). The palynomorph data come from samples that are equivalent to the specimens discussed here from the Cross Valley Member of the Sobral Formation, and are in agreement with composition of the flora indicated by the megafossil data.

The middle Eocene La Meseta flora (RV-8425; Fig. 4) is dominated by a large-leafed (= notophyllous, 7.6 to 12.7 cm [3 to 5 in] in length; Webb, 1959) species of *Nothofagus* (Fig. 4,A–C) of a significantly greater size (based on means of leaf widths) than the species of *Nothofagus* that occur at the other two localities. The ferns are represented by two species exhibiting the short and stout pinna morphology. The two species differ by having either an alternate (Fig. 4E) or opposite (Fig. 4F) arrangement of pinna along the frond. Also represented in the flora is an araucarian conifer (Fig. 4D).

A combination of 18 leaf impressions and carbon films was recovered from the late Eocene La Meseta Formation locality RV-8200 (Fig. 5). Morphological analysis and terminology in the following description are according to Hickey (1973). The megafloral fossils represent the remains of small to medium-small leaves pinnately veined, ovate leaves with acute apecies, and obtuse bases. The leaf margins are crenate. The secondary veination is craspedodromous and has a divergence angle of approximately 50° for the basal and medial secondary veins, with the angle becoming more acute in the apical region of the leaf. The craspedodromous secondary veins extending directly out from the midvein to the leaf margin, the divergence angle of the secondary veins from the midvein, the leaf size and shape, plus the biogeographic locality all indicate that the specimens belong to the genus *Nothofagus* (Bentham and Hooker, 1880; Langdon, 1947). In specimens that exhibit the appropriate detail, it can be noted that the areolation is irregular, the areoles are four- or five-sided, and the tertiary veins are embedded within the leaf reticulation. The above features have been used by Tanai (1984) to characterize the genus *Nothofagus.*

Analysis of differences in the intercostal distances between secondary veins, along the midvein of the leaf impressions that were suitable for such an analysis, suggests the presence of two different species of *Nothofagus* on Seymour Island. Mean values for the two leaf morphologies were calculated from measurements of the intercostal distances between the secondary veins along the midvein. A statistical analysis—utilizing a t-test to distinguish if a difference between the two group means exists (for a small sample size)—indicates two distinct morphologies in regard to intercostal distance. The resultant *t*-value (8.319) indicates that there is a significant difference between the mean values of each of the two leaf morphs (d.f. = 7; $p = 0.0001$) (Table 1). This agrees well with fossil pollen data (Cranwell, 1963; Askin and Fleming, 1982), which indicate the presence of more than one species of *Nothofagus* on Seymour Island, as each of the three *Nothofagus* pollen groups have been collected and described (Hanks and Fairbrothers, 1976). The leaf specimens from Seymour Island most closely resemble some of the South American deciduous species (Leathart, 1977; Poole, 1958; Cockayne, 1926). Specimens with sufficient detail exhibit no apparent fimbrial vein along the leaf margin, and the areoles appear to be demarcated by thin fourth- or fifth-order veins. The absence of a fibrial vein and the presence of areoles with thin veins are character states that would place these Southern Beech specimens into *Nothofagus* species groups I or II of Tanai (1984). Species groups I and II are composed almost exclusively of South American species of Southern Beech (*N. gunnii* of New Zealand is the sole exception).

PALEOCLIMATE

Nothofagus is considered to be of critical importance as an indicator of paleoclimate. The Southern Beech, with the exception of the species in New Caledonia (New Guinea species are found only in the cool montane zones), is an inhabitant of cool to cold temperate zones with abundant rainfall (Moore, 1972; Van Steenis, 1971; Ward, 1965; Du Rietz, 1960; Couper, 1960).

The plant megafossils recovered from the three Paleogene localities on Seymour Island suggest that the terrestrial environment that both the land flora and fauna inhabited was a cool, temperate, rain forest. The idea that the paleoenvironment from the late Cretaceous through the late Eocene for the Antarctic Peninsula was under a cool, temperate climatic regime had been proposed earlier (Zinsmeister, 1982; Woodburne and Zinsmeister, 1984). The interpretations presented by these authors was based on the presence of particular taxa of marine molluscs and gastropods. Support for the cool, temperate paleoclimate is now provided by two biological indicators, the marine fauna and the terrestrial flora.

The larger leaf size of the middle Eocene flora may indicate a situation of ameliorating climatic conditions during this time period. This may be especially true if the middle Eocene flora is

Figure 3. Suite of megafloral specimens representing the Cross Valley Paleoflora (RV-8265) of Paleocene age. Specimens A through C are the three fern morphs present in the flora (A, ×1.25, B, ×1.65, C, ×2.05. Specimens D and E are two species of *Nothofagus*. D, ×1.75, E, ×0.90 Specimen F is a podocarp conifer (×1.25).

Figure 4. Suite of megafloral specimens representing the middle Eocene, La Meseta paleoflora (RV-8425). Specimens A through C are the large-leafed species of *Nothofagus* typical of this locality. A, ×0.60, B, ×0.70, C, ×0.90. Specimen D is an araucarian conifer (×0.90). Specimens E and F are the two fern morphs based on the arrangement of pinna along the frond (E, alternate, F, opposite) from the locality. E, ×1.20, F, ×1.45.

Figure 5. Specimens A through D are four representative *Nothofagus* leaf specimens from the late Eocene La Meseta paleoflora (RV-8200). The two types of morphologies in regard to the intercostal distance between the secondary veins are shown, with specimens A and B exhibiting the narrow intercostal morph and specimen C exhibiting the wide intercostal morph. Specimen D contains an example of each intercostal morph. All specimens are ×1.65. Note that only specimens of *Nothofagus* are represented in this flora.

TABLE 1. TEST OF SIGNIFICANCE BETWEEN GROUP MEANS OF INTERCOSTAL DISTANCE*

Leaf Morph	$\overline{X}$ (mm)	S.D.	N
A	5.40	0.708	4
B	2.60	0.253	5

*$t = 8.319$, for d.f. = 7, so group means are significantly different at $p = .0001$.
Note: Analysis indicates that two species of *Nothofagus* are present in the flora from the late Eocene locality RV-8200.

compared to either the Paleocene flora with its predominance of smaller leafed specimens or the late Eocene flora with the smaller sized leaves of the two species of *Nothofagus* that compose that flora.

BIOGEOGRAPHY

Nothofagus has been considered a "key" genus in plant biogeography of the Southern Hemisphere (Van Steenis, 1971). The association of marsupials and cool temperate, *Nothofagus*-dominated rain forests in Antarctica is similar to the contemporary association of ancient lineages of marsupials, such as microbiotheriids and caenolestids, in such floral settings in southern Chile and Patagonia (Marshall, 1980, 1982).

Species of *Nothofagus* are currently distributed in southern South America, in New Zealand, and in Australia and associated islands (e.g., New Guinea and New Caledonia). The position of Antarctica, relative to the other southern continents, is critical to the dispersal of this genus when considering the current geographic range of *Nothofagus*. The dispersal pattern for *Nothofagus* becomes even more intriguing when the continental arrangement suggested by the hypothesis of the Weddellian Province is considered (Fig. 6). The Weddellian Province is hypothesized as a shallow, cool, temperate marine region that extended from southern South America, along western Antarctica, to New Zealand, Tasmania, and southeastern Australia. This province existed from the later portion of the Cretaceous through the Eocene when Australia, Antarctica, and southernmost South America were in proximity (Zinsmeister, 1979, 1982; Woodburne and Zinsmeister, 1984).

A reconstruction of the Weddellian Province for the late Cretaceous and early Tertiary, prior to the separation of New Zealand and Australia from Antarctica, brings together within a single geographic region those areas (i.e., South America, Antarctica, and Australia) that have for the most part two unique features: marsupials and species of *Nothofagus* (other genera of plants also exhibit this pattern of distribution). Schuster (1976) coined the term "marsupial route" for the dispersal track followed by the distributions of both marsupials and species of *Nothofagus* from southern South America, through Antarctica, to Australia.

Terrestrially, the Weddellian Province must represent a long, continuous (i.e., not an island chain, although very narrow water gaps [<1 km] could exist) coastal region bordered by shallow seas, which would facilitate the dispersal of forms, both plant and animal, that are incapable of long-range dispersal. Dispersal is defined here as the distance of seed dispersal from the parent plant or the distance migrated from the place of birth during the life time of an animal.

Modern species of *Nothofagus* are known to have a very restricted overland seed dispersal (less than 1 km; Preest, 1963). The seeds are incapable of oceanic dispersal, as they sink in salt water (Hollaway, 1954). *Nothofagus* is thus limited in its dispersal capabilities and in its dispersal vectors. Its wingless seeds are

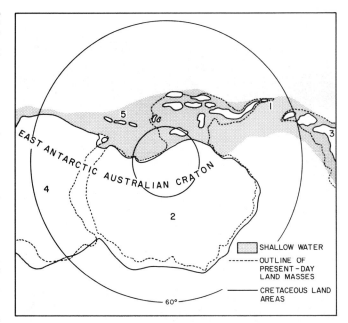

Figure 6. Reconstruction of the Weddellian Zoogeographical Province (stippled area) hypothesized by Zinsmeister (1979, 1982) during the late Cretaceous and early Tertiary. The pertinent land masses are numbered as follows: 1, Seymour Island; 2, Antarctic craton; 3, South America; 4, Australia; 5, New Zealand. Note: A closer proximity between southern South America and the Antarctic Peninsula than is indicated here has been proposed by W. J. Zinsmeister (personal communication). With species of *Nothofagus* and marsupials limited to overland dispersal, it is hypothesized here that the stippled area should represent a continuous, coastal environment bordered by shallow seas.

not adequate for dispersal by wind, and the seeds will not germinate after having passed through the alimentary canal of an animal (Preest, 1963; Van Steenis, 1971). Thus, the Southern Beech is restricted to short-range terrestrial dispersal of its seeds (or possibly over the shortest of seaways).

The dependence of *Nothofagus* on overland seed dispersal supports the Weddellian Zoogeographical Province hypothesis of an arrangement of the southern continents that placed them in juxtaposition, facilitating the terrestrial dispersal of both *Nothofagus* and marsupials via a continuous (or nearly so) coastal land connection from South America through Antarctica to the Australasian land masses. This equals the "marsupial route" of Schuster (1976). Thus, the concept of a shallow-water marine zoogeographic province proposed by Zinsmeister (1979, 1982) must be expanded to a concept of a distinct terrestrial and marine paleobiogeographic province (from the late Cretaceous to the end of the Eocene) in order to accommodate the distribution of the terrestrial biota of this region.

SUMMARY

The recovery of plant megafossils of *Nothofagus,* the Southern Beech, from the three early Tertiary localities on Seymour

Island reveals the association of both marsupials and *Nothofagus* in Antarctica, resembling the associations of these two taxa in both southern South America and southeastern Australia. There are striking similarities between the biogeography and paleoecology of the species of this plant genus and that of primitive marsupials, as both the plant and animal species appear to have the same requirements for habitat and for the dispersal of their respective offspring.

As both the marsupial species and species of *Nothofagus* require overland dispersal routes, the likelihood of a continuous, cool, temperate coastal land mass, stretching from southern South America through Antarctica to Australia (i.e., the Weddellian Paleobiogeographical Province) would appear to be quite strong. Thus the proposal for this Southern Hemisphere biogeographic region based on marine invertebrates is supported by the terrestrial paleofloras and paleofauna from the three Paleogene localities from Seymour Island.

ACKNOWLEDGMENTS

I am grateful to M. O. Woodburne and W. J. Zinsmeister for the opportunity to be a member of the 1983–1984 and 1985 field parties. Two most eventful voyages were provided by the personnel of the U.S. Coast Guard Cutter *Westwind* (1983–1984) and the research ship *Polar Duke* (1985). I thank Jack Wolfe (U.S. Geological Survey) and Rudolph Schuster (University of Massachusetts) for their helpful comments on the manuscript. Photographic work was provided by Bob Hicks, Department of Earth Sciences, and Marcia Kooda-Cisco, Department of Biology, University of California, Riverside. Illustrations were prepared by Linda Bobbitt, Department of Earth Sciences, University of California, Riverside. This work was supported by National Science Foundation Grants 8215493 and DPP-8521368 to M. O. Woodburne.

REFERENCES

ASKIN, R. A., AND R. F. FLEMING. 1982. Palynological investigations of Campanian to lower Oligocene sediments on Seymour Island, Antarctic Peninsula. Antarctic Journal of the United States, 17:70–71.

BENTHAM, A., AND J. D. HOOKER.. 1880. *Fagus,* section *Nothofagus.* Genera Plantarum, 3:410.

COCKAYNE, L. 1926. Monograph of New Zealand beech forests: Part I, The ecology of the forests and taxonomy of the beeches. New Zealand State Forestry Bulletin, 4:1–67.

COUPER, R. A. 1960. Southern Hemisphere Mesozoic and Tertiary Podocarpaceae and Fagaceae and their paleogeographic significance. Proceedings of the Royal Society of London, Series B, 152:491–500.

CRANWELL, L. M. 1959. Fossil pollen from Seymour Island, Antarctica. Nature, 184:1782–1785.

——. 1963. *Nothofagus:* living and fossil, p. 387–400. *In* J. K. Gressitt (ed.), Pacific Basin Biogepgraphy. Bishop Museum Press, Honolulu, Hawaii.

——, H. J. HARRINGTON, AND I. G. SPEDEN. 1960. Lower Tertiary microfossils from McMurdo Sound, Antarctica. Nature, 186:700–702.

DU RIETZ, G. E. 1960. Remarks on the botany of the southern cold temperate zone. Proceedings of the Royal Society of London, Series B, 152:500–507.

DUSEN, P. 1908. Uber die Tertiare flora der Seymour-Insel. Wissenschaftliche Ergebnesse de Schwedischen Sudpolar-Expedition 1901–1903, Stockhold, 3(3):1–27.

FELDMANN, R. M., AND W. J. ZINSMEISTER. 1984. New fossil crabs (Decapoda: Brachyura) from the La Meseta Formation (Eocene) of Antarctica: paleogeographic and biogeographic implications. Journal of Paleotology, 58(4):1046–1061.

GOTHAN, W. 1908. Die fossilen holzer von der Seymour und Snow Hill Insel. Wissenschaftliche Ergebnisse de Schwedischen Sudpolar-Expedition 1901–1903, Stockholm, 3(8):1–33.

HANKS, S. L., AND D. E. FAIRBROTHERS. 1976. Palynotaxonomic investigations of *Fagus* L. and *Nothofagus* Bl.: light microscopy, scanning electron microscopy, and computer analysis, p. 1–142. *In* V. H. Heywood (ed.), Botanical Systematics, Academic Press, London.

HICKEY, L. J. 1973. Classification of the architecture of dicotyledonous leaves. American Journal of Botany, 60:17–33.

HOLLAWAY, J. T. 1954. Forest climates in the South Island of New Zealand. Transactions of the Royal Society of New Zealand, 82:329–410.

LANGDON, L. M. 1947. The comparative morphology of the Fagaceae. The genus *Nothofagus.* Botanical Gazette, 108:350–401.

LEATHART, S. 1977. Trees of the World. A & W Publishers, New York, p. 112.

MARSHALL, L. G. 1980. Systematics of the South American marsupial family Caenolestidae. Fieldiana, Geology, n.s., no. 5:1–145.

——. 1982. Systematics of the South American marsupial family Microbiotheriidae. Fieldiana, Geology, n.s., no. 8:1–75.

MOORE, D. M. 1972. Connections between cool temperate floras, with particular reference to southern South America, p. 115–138. *In* D. H. Valentine (ed.), Taxonomy, Phytogeography and Evolution. Academic Press, London.

POOLE, A. L. 1958. Studies of New Zealand *Nothofagus* species: Part III, The entire-leaved species. Transactions of the Royal Society of New Zealand, 85:551–564.

PREEST, D. S. 1963. A note on the dispersal characteristics of seeds of the New Zealand podocarps and beeches and their biogeographic significance, p. 415–424. *In* J. K. Gressitt (ed.), Pacific Basin Biogeography. Bishop Museum Press, Honolulu, Hawaii.

SCHUSTER, R. M. 1976. Plate tectonics and its bearing on the geographical origin and dispersal of angiosperms, p. 48–138. *In* C. B. Beck (ed.), Origin and Early Evolution of Angiosperms. Columbia University Press, New York.

TANAI, T. 1984. Phytogeographic and phylogenetic history of the genus *Nothofagus* BL. (Fagaceae) in the Southern Hemisphere. Journal of the Faculty of Science, Hokkaido University, Series IV, 21(4):505–582.

VAN STEENIS, C.G.G.J. 1971. *Nothofagus,* key genus of plant geography, in time and space, living and fossil, ecology and phylogeny. Blumea, 19:65–98.

WARD, R. T. 1965. Beech (*Nothofagus*) forests in the Andes of southwestern Argentina. American Midland Naturalist, 74:50–56.

WEBB, L. J. 1959. A physiognomic classification of Australian rain forests. Journal of Ecology, 47:551–570.

WOODBURNE, M. O., AND W. J. ZINSMEISTER. 1982. Fossil land mammal from Antarctica. Science, 218:284–286.

——. 1984. The first land mammal from Antarctica and its biogeographic implications. Journal of Paleontology, 58(4):913–948.

ZINSMEISTER, W. J. 1979. Biogeographic significance of the late Mesozoic and early Tertiary molluscan faunas of Seymour Island (Antarctic Peninsula) to the final breakup of Gondwanaland, p. 349–355. *In* J. Gray and A. Bouot (eds.), Historical Biogeography, Plate Tectonics and the Changing Environment, Proceedings, 37th Annual Biological Colloquium and Selected Papers. Oregon State University Press, Corvallis.

——. 1982. Late Cretaceous–Early Tertiary molluscan biogeography of southern circum-Pacific. Journal of Paleontology, 56:84–102.

MANUSCRIPT ACCEPTED BY THE SOCIETY SEPTEMBER 1, 1987

Printed in U.S.A.

Geological Society of America
Memoir 169
1988

Ichnofossils, tubiform body fossils, and depositional environment of the La Meseta Formation (Eocene) of Antarctica

Lawrence A. Wiedman and Rodney M. Feldmann
Department of Geology, Kent State University, Kent, Ohio 44242

ABSTRACT

Characteristics of a diverse assemblage of ichnofossils and tubiform-body fossils collected from the La Meseta Formation (Eocene age) of Seymour Island, Antarctica, indicate deposition in a shallow, nearshore environment. Specimens were collected from three of seven subunits within the La Meseta Formation. These units comprise the bulk of the exposures of the formation. The La Meseta Formation is composed of quartzose sandstone that is fine to coarse and subangular to subrounded, ranging in color from dusky yellow-green to light olive-gray, cemented primarily by sparry calcite. Moderate to steep cross-bedding is common in the medial and upper units. Body fossils, in a wide range of sizes, are common at many locations. More than 14 ichnogenera and tubiform-body fossils were collected and described from the unit, including *Ophiomorpha, Muensteria, Diplocraterion, Skolithos, Helminthopsis, Spirorbis, Serpula, Oichnus,* several "rind" and "halo" burrows, and fossilized wood with abundant *Teredolites.* The previously defined autecology of the less cosmopolitan ichnogenera, along with the sediment analysis, indicates a nearshore environment of deposition, generally above wave base, for the sampled units of the La Meseta Formation. These conclusions corroborate and strengthen those derived from similar studies using taxa of bivalve and gastropod molluscs and malacostracan arthropods.

INTRODUCTION

During 1983–1984, a collection of ichnofossils was made by one of us (R.M.F.) from exposures throughout the Eocene La Meseta Formation of Seymour Island, Antarctic Peninsula, Antarctica (Fig. 1). A diverse assemblage of arthropods, molluscs, brachiopods, bryozoans, corals, echinoderms, wood fragments, and chordate bones and teeth was collected from a total of 30 localities within the unit. Widely distributed trace fossils were collected at nine localities (Fig. 1). Many of the traces, which include borings, burrows, tracks, trails, and predation marks, are abundant (Table 1). Epizoans were also found.

This chapter records the trace fossils and utilizes them as paleoenvironmental indicators. Environmental conditions have been interpreted from studies of other taxa collected from the unit (Zinsmeister, 1979; Feldmann and Zinsmeister, 1984) and from examination of the sedimentary sequence (Zinsmeister, 1979; Elliot and Trautman, 1982). Feldmann and Zinsmeister (1984), using malacostracan arthropods, and Zinsmeister (1979), using bivalves and gastropods, both concluded that the sediments making up the La Meseta were deposited in a shallow-water, near-shore environment. This assessment agrees with conclusions, presented herein, that have been derived from analysis of the trace fossil assemblage. Although many of the traces are cosmopolitan and do not indicate a particular set of depositional circumstances, some are more restrictive with respect to the environments in which they are found. We also note first-reported occurrences in the Eocene of the Antarctic of several of these traces and tubiform fossils, including *Diplocraterion, Serpula,* and *Muensteria.*

The trace fossils are housed in the National Museum of Natural History (USNM), in Washington, D.C. Additional specimens are deposited in the Department of Geology, Kent State University (KSU), in Kent, Ohio.

The grains composing the La Meseta Formation range from fine to predominantly coarse sand. Quartz grains constitute more than 60 percent of the total particles; generally, they range from angular to subrounded, with grains from some localities having uniform high sphericity. The quartz grains tend to have fresh, fractured surfaces, in most cases with almost no etching or frost-

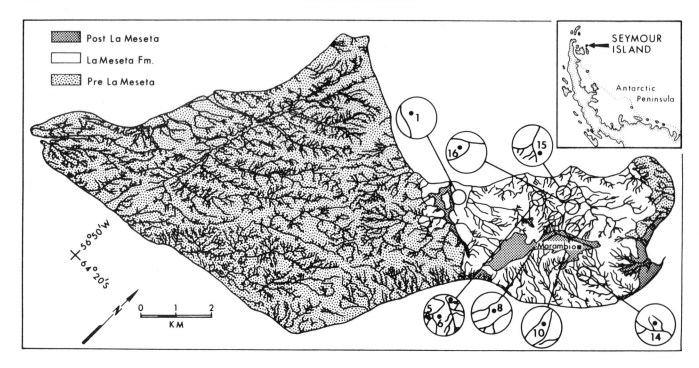

Figure 1. Map showing location of Seymour Island, Antarctica, outcrops of the Eocene La Meseta Formation, and sites where ichnofossils and tubiform fossils used in this study were collected.

ing. Dark green grains of glauconite, uniformly 50 to 100 percent larger than the quartz grains, make up most of the rest of the rock. As a rule, grains of glauconite show a much greater degree of rounding than do the quartz grains. Mica and organic grains are present; at some locations these compose as much as 5 percent of the sediment composition, especially near burrow linings.

The cementing agent is primarily sparry calcite, with locally abundant iron staining and iron cement. The dusky yellow-green (5Y 7/4) to light olive-gray (5Y 6/1) color is derived primarily from the colors of the cement and the glauconitic material.

The grains of the La Meseta seem to indicate a nearby origin; little reworking is suggested by their lack of textural maturity. Multi-directional, moderate to high-angle cross-bedding is common throughout the medial and upper units of the formation; however, the lower units, Telm1 and 2 (described below), typically show only planar bedding. Body fossils of a wide range of sizes are common at numerous localities. In the middle portion of the La Meseta, the most common occurrence of fossils is as molluscan shell beds. Trace fossils and burrows are locally abundant.

The difficulty with making generalizations about the lithologies results from the observation that there are numerous vertical lithologic variations on the scale of fractions of a meter to a few meters. These variations seem to reflect variations in wave energy level such that fine-grained, laminated sands with numerous vertical burrows appear to have been deposited in relatively protected, quiet-water reaches. Coarser grained, cross-bedded sands

were the product of deposition in a higher energy regime. Both of these rock types are replaced by lenticular masses of moderately coarse sandstone containing numerous large, thick-shelled bivalved molluscs, with valves articulated in some accumulations and disarticulated in others. Alternations in lithologies of these types and on this scale typify deposition in the shallow-water, nearshore habitats comparable to barrier island and lagoonal complexes. Although a detailed facies analysis would be necessary to document this environmental setting conclusively, that is not the focus of this work. Sadler (this volume) has summarized the depositional setting of each of the units within the La Meseta Formation.

The salinity conditions prevailing in this marginal marine habitat were probably in the range of high hyposaline to normal marine. Echinoderms, one group of invertebrates generally not abundant in areas of low salinity, occur in large numbers at a variety of localities throughout the La Meseta. The presence of a large volume of wood fragments suggests both nearshore conditions and proximity to rivers or streams to transport the fallen timbers into the marine habitat. These observations, coupled with the recognition that the region of Seymour Island lay at a high latitude during the time of deposition, strongly reduce the likelihood of hypersaline conditions in the lagoonal habitats.

Within the La Meseta, seven lithofacies have been defined and mapped as a result of field work by Peter M. Sadler, Department of Earth Sciences, University of California at Riverside. Changes in characteristics of grain size, sorting, and bedding, as

TABLE 1. LIST OF ICHNOFOSSILS AND TUBIFORM BODY FOSSILS, ARRANGED BY LOCALITIES, COLLECTED FROM THE LA MESETA FORMATION (EOCENE) SEYMOUR ISLAND, ANTARCTICA

Locality 1 Telm 2	**Locality 4** Telm 5	**Locality 5** Telm 5
"Halo" burrows	*Diplocraterion*	*Helminthopsis* *Muensteria* *Ophiomorpha* *Oichnus*
Locality 6 Telm 5	**Locality 7** Telm 5	**Locality 8** Telm 5
Serpula Lobed rod *Skolithos*	"Halo" burrows *Euromyliid* *Ophiomorpha*	*Teredolites*
Locality 14 Telm 7	**Locality 15** Telm 5	**Locality 16** Telm 5
Serpula *Spirorbis* *Diplocraterion* *Teredolites*	*Skolithos* *Diplocraterion*	*Ophiomorpha* *Oichnus*

well as variations in the distribution and dominant organisms in prominent molluscan shell beds, formed the primary bases for distinguishing units within the formation. These units are designated Telm1 through Telm7 (Sadler, this volume), and a more comprehensive lithofacies description is presented.

SYSTEMATIC PALEONTOLOGY

Diplocraterion Torell, 1870
Figure 2.5

Material studied. Three specimens.

Description of material. Planar U-shaped tubular traces, perpendicular to bedding with subparallel vertical elements. Those with high height/width (>2) ratios have side tubes along lower portion of U, connected by a series of subparallel, nonuniform arcs or semicircles with relief of 0.2 to 0.7 cm, in same plane as tubes and in same orientation, open end up, as bottom of trace. Those with low height/width ratios (<1) have vertical elements connected throughout. Exterior boundary of trace greatly accentuated by weathering. Spreite range from well-defined to obscure. Distance between arcs, 1 to 3 cm.

Trace height range 6 to 40 cm; width, 6 to 12 cm; tube diameter range, 0.6 to 1.8 cm, same as trace breadth; radii at bottom of U, 3 to 6 cm.

Fill uniform medium sand; uniform lithic components, light olive-gray, slightly more indurated than matrix.

Matrix same as fill with coarse crossbeds of 5 to 25° in several directions.

"Halo" burrows *sensu* Chamberlain, 1975
Figure 3.2

Material. One specimen.

Description of material. Lined, straight, cylindrical, tubular trace; 1.7 cm in diameter, 9 to 12 cm in length. Fill different material than lining; similar to matrix, light gray, uniform size, fine sand, uniform lithic constituents, some small mollusc fragments, no preferred orientation, no differentiation. Lining wall uniformly circular, 0.4 cm thick; distinct inner boundary, gradational outer margin. Dark brown organic flakes found throughout concretion are concentrated within and just outside lining.

Found in sandy, egg-shaped concretion; 12 cm long, 6 cm wide; trace penetrates only one end; lining slightly more weathered than fill or matrix on exposed terminous. No bedding or banding apparent in concretion; only one trace collected, centered in concretion.

Helminthopsis Heer, 1877
Figure 2.3

Material studied. One specimen.

Description of material. Continuous, convex hyporelief, meandering planar trace; uniformly 1.4 to 1.6 cm wide, 1.0 to 1.6 cm in relief; as much as 65 cm long, parallel to bedding. Slightly undulatory surface smooth, grooved, or slightly lobate; smoothest at curves generally; grooves 0.1 to 1.0 cm long, 0.1 to 0.3 cm wide, as much as 0.6 cm deep, primarily transverse, often crescentic to irregularly shaped, along the

Figure 3. Examples of diagnostic ichnofossils and tubiform body fossils found in unit 7 of the La Meseta Formation, not illustrated in Figure 2. 1, *Serpula;* 2, "halo" burrow; 3 through 5, examples of fossilized wood heavily infested by *Teredolites,* the borings of teredid bivalves. All figures X0.5.

Figure 2. 1 through 6, Examples of the diagnostic trace fossils found in unit 5 of the La Meseta Formation. 1, *Ophiomorpha,* as found in field. 2, *Ophiomorpha.* Numerous burrowing crabs, *Lyreidus antarcticus,* have been collected from this site. 3, *Helminthopsis.* 4, *Muensteria.* 5, *Diplocraterion,* shown in place. 6, *Skolithos.* Laboratory photographs, each X0.5. Specimens 2 through 4 were coated with ammonium chloride before photography. 7, Enigmatic lobed rod found at Locality 6. Actual size; coated with ammonium chloride.

entire trace surface. Where weathered, trace has a cone-in-cone appearance. Occasional unidirectional recumbent lobes overlap and grade into main trace.

Meanders seldom converge or cross; three apparent exceptions produced by trace segment similar to, but not continuous with, main trace. Meander loops range 4 to 11 cm long, eight present on a 65-cm length; often reverse trace direction, curve radii 1 to 2 cm, longest semi-straight section, 14 cm.

Fill uniform medium sand, uniform lithic components, same as matrix, trace/matrix boundary indistinguishable on basis of lithology. Matrix poorly bedded. Entire trace area 17 × 21 cm. Abundant hyporelief nodes on matrix and trace surface, uniform size 0.1 to 0.7 cm long, 0.1 to 0.7 cm wide, as much as 0.3 cm of relief, very common, no preferred orientation.

Muensteria von Sternberg, 1833
Figure 2.4

Material studied. One specimen.

Description of material. Continuous trace, series of uniformly sized and directed crescentic markings of low positive relief on an undulatory bedding surface. Surficial crescents normally unidirectional, a few exceptions as much as 90° askew; grayish brown to light brown color, uniformly 0.1 to 1.5 cm wide, 0.8 to 1.1 cm long, arc radii 0.3 to 0.4 cm. Spaces between crescents range 0.1 to 1.0 cm wide. Spacing of crescents regular when crescent direction is most uniform, irregular when not; appears uniform in transverse profile as well as on surface. In transverse profile trace, a uniform convex ridge 0.1 to 0.3 cm above bedding surface, flanks dip gently and grade into matrix. Crescents continue into matrix as semi-cylinders at steep angle, nearly 70°, to the 3.5-cm maximum thickness of the sample.

Oichnus Bromley, 1981

Material studied. More than 25 specimens.

Description of material. Chamfered, circular hole, completely penetrating shells of bivalves and gastropods; greatest diameter always on exterior of shell, range 0.35 to 0.40 cm; interior 0.25 to 0.30 cm. Hole walls smooth to tangentially ridged, with very slight relief markings that may result from weathering of shell layers. On bivalves, *Oichnus* is usually found slightly dorsal of center near widest part of shell; on gastropods always on last whorl near point of involution. On thick shells, holes taper only slightly and take on more cylindrical form.

Ophiomorpha Lundgren, 1881
Figures 2.1, 2

Material studied. Five specimens.

Description of material. Narrow, straight to slightly arcuate, cylindrical rod-like traces, segments of which are as much as 13 cm long and 1.8 cm in diameter. Fill of medium to fine sand. Surface nodose throughout except at the termini, which are either planar or bluntly rounded. Nodes relatively uniform-size hemispheres, 0.25 to 0.50 cm in diameter, arranged in poorly defined longitudinal rows.

Trace fragments show no indication of bifurcation. Bulbous protuberances locally expand normal diameter by 25 to 75 percent, enlarging one side of tube and extending longitudinally 1.5 to 3.5 cm. None collected show more than one bulbous area; most have one.

Internal fill uniform size and composition; grains range from medium to fine sand.

Serpula Linnaeus, 1768
Figure 3.1

Material studied. More than 25 specimens.

Description of material. Clustering, calcareous, randomly oriented tubes, hollow or with consolidated to lithified sediment fill. Tube walls generally parallel within the clusters of two to nine tubes; most often in continuous contact with one another. Wall ranges 0.05 to 0.2 cm thick, depending on proximity to next tube lining; thinnest if walls shared with another tube, thickest when lining fills gaps between adjacent tubes; white to gray color with no observable color pattern. Wall composed of concentric laminae in transverse cross section. Traces often anastomosing; clusters often change course; in extreme cases, cluster doubles back, 90° shifts in direction of growth common.

Tube diameter ranges 0.2 to 0.5 cm, longest seen 5.2 cm; few termini apparent, each end open; some terminate attached to mollusc or brachiopod shell fragments; if so, attached side flairs at attachment, forming greater surface area there; interior diameter remains constant. Interior generally smooth with some slight exceptions, which are transverse, very low-relief undulations. Exterior well defined, but highly variable and much rougher in appearance, often with matrix sand pressed into lining or pits apparent where matrix has become dislodged. Where preservation exceptional, exterior surface has continuous low-relief, wavy to scallop-shaped ridges around and along trace.

Fill and matrix similar in composition, matrix uniformly slightly coarser sand, light grayish orange to pale yellowish orange to pale yellowish brown.

Tubes act as host to other tubiform fossils including *Spirorbis* and several encrusting bryozoan taxa.

Skolithos Haldeman, 1840
Figure 2.6

Material studied. Twenty-four specimens.

Description of material. Straight, lined, vertical to slightly inclined solitary traces, few sharply inclined; preserved as hollow or sediment-filled ovoid tubes in a massive matrix. Oval openings with long axis 0.1 to 0.2 cm, short axis 0.25 to 0.35 cm, as much as 6 cm long. Lining uniformly about 1.0 mm thick with very low relief, transverse markings closely spaced along entire interior surface; similar in color and composition to matrix, except for fine, darker interior lining; outer limit of lining accentuated by surficial weathering.

Spacing of traces irregular, none closer than 1 cm center to center, over 30 traces in one 10-cm sample.

Matrix fine sand to coarse silt, light olive-gray, no bedding, some mollusc fragments.

Spirorbis Daudin, 1800

Material studied. Seventeen whole or nearly whole specimens.

Description of material. Small, coiled, planispiral evolute to trocoid, unchambered, calcareous tubes attached primarily to other larger calcareous tube hosts. Interior wall smooth; exterior wall transverse low-relief lines around traces. Trace size ranges from 0.1 to 0.2 cm; greatest number of complete whorls is three, each successive whorl is larger than the preceding. Trace primarily found in recessed areas of host, density as great as 15 in an area 1.0 × 1.5 cm, some exteriors in contact.

Teredolites Leymerie, 1842
Figures 3.3–5

Material studied. Hundreds, on several pieces of lithified wood.

Description of material. Thin-walled, undulose, solitary, irregularly ovoid to circular calcareous tubes preserved individually as lined hollow tubes with sand or calcite fillings; penetrating lithified wood. Tubes oriented generally parallel to each other when preserved in multiples, both perpendicular and parallel to wood growth rings, only rarely oblique. Traces not branching, often in contact, but never crossing one another. External tangential rings of very low relief seen, some closely spaced often at irregular intervals; some longitudinal lines visible. Both ornaments may be related to matrix growth rings.

When sediment filled, uniform-size, light gray silt; when calcite filled, druzzy concentric lining.

Possibly two separate trace types present. One smaller, consistently about 0.2 cm in diameter, mainly perpendicular to growth rings, found primarily on a single large sample with hundreds of examples; many have very thin calcite crack fills in walls. The larger tubes of the bimodal distribution range primarily from 0.4 to 0.8 cm in diameter, found both perpendicular and parallel to grain. Several samples of lithified wood with many traces each found, very few traces seen of diameter between two peaks in bimodal distribution.

Zapfella De Saint-Seine, 1955

Material studied. Three specimens.

Description of material. Very small, n-shaped excavations; sides cylindrical to ovate in cross section perpendicular to surface of the host; uprights parallel, 0.3 to 0.5 cm apart, <1.0 cm deep. The connecting arc at top and visible on surface is about 30 to 75° angle to sides, seen as semicircular surficial indentation. The median wall top is nearly flush with surface of the host; slightly longer than wide. Three examples found, all within 0.25 cm, aligned semi-parallel, but not in contact with each other, on calcareous tube trace host. Little interior structure apparent, sides seem smooth.

TRACES IN THE LA MESETA FORMATION

Of the 14 trace and tubiform types collected and described, eight can be assigned to preexisting ichnogenera. Six of these trace types, including *Diplocraterion, Ophiomorpha, Skolithos,* plus the lithified wood, have been used previously as indicators of littoral to very shallow sublittoral environments of deposition (Frey, 1975; Frey and Pemberton, 1984). Some of the tubiform fossils that are epibionts on body fossils may have been transported into the depositional framework of the La Meseta Formation from elsewhere. Some of the body fossils do show indications of slight abrasion and sorting of valves, suggesting the possibility of transport, although there is no evidence that they have moved a great distance. The possibility of transportation, however, limits the use of these specimens as environmental indicators. Although ichnofossils were frequently encountered, only two of the seven units recognized by Sadler, units Telm5 and Telm7, provided sufficient specimens for paleoenvironmental analysis. The first, unit Telm5, is characterized by *Helminthopsis, Muensteria, Ophiomorpha, Skolithos, Diplocraterion, Spirorbis, Oichnus,* possible examples of *Euromyliid, Serpula, Teredolites,* and an enigmatic lobed rod collected from six localities (Fig. 1, localities 5, 6, 7, 10, 15, and 16).

Helminthopsis Heer (Fig. 2.3) has been associated with deep-water deposits, especially flysch deposits by Crimes (*in* Frey, 1975). Crimes and Crossley (1968) reported the occurrence of this trace in lower Paleozoic flysch deposits in Ireland, and Crimes and Anderson (1985) noted the ichnogenus in early Cambrian deep- and shallow-water deposits in Newfoundland. Other reports have been made of this trace from Europe, Asia, Alaska, and Antarctica (Häntzschel, 1975). The known range of *Helminthopsis* extends from the Cambrian to the Tertiary (Häntzschel, 1975). In regard to the interpreted shallow-water environment of deposition in which this trace was found during this study, it is entirely possible that, through time, the tracemaker may have altered its position with respect to water column depth, or that an entirely different tracemaker made a trace in the Eocene similar to that in the lower Paleozoic rocks.

Muensteria von Sternberg (Fig. 2.4) is composed of a series of uniform-size, directed crescentic markings of low positive relief on an undulatory surface (Frey and Howard, 1984). Only one example of this trace and *Helminthopsis* was found, both in float at locality 5. However, it is concluded later that transportation subsequent to exhumation was minimal, and that these traces were likely to have been collected very near their original position.

Ophiomorpha Lundgren (Fig. 2.2) is common in many Recent deposits as well as ancient ones. It has had a continuous record since the Permian and has been noted from diverse localities worldwide (Häntzschel, 1975; Kennedy and Sellwood, 1970; Asgaard and Bromley, 1974; Frey et al., 1978). The trace is believed to be the result of preservation of an arthropod dwelling tube (Simpson, 1975). Often in the past this trace has been associated with the ghost shrimp, *Callianassa* (Häntzschel, 1952;

Weimer and Hoyt, 1964; Kennedy and MacDougall, 1969; Chamberlain, 1975), which would indicate extremely shallow-water conditions and/or an intertidal setting. The direct association of callianassids with *Ophiomorpha* has been demonstrated several times (Waage, 1968; Beikirch and Feldmann, 1980; Pemberton and Frey, 1984). This association may not always be valid (Kennedy and Sellwood, 1970; Asgaard and Bromley, 1974), as there are deeper water occurrences (Crimes, 1977; Frey et al., 1978; Crimes et al., 1981). Frey et al. (1978) also indicated that the potential relationship may be considerably more complex than previously thought. Callianassid shrimps, *Protocallianassa* cf. *P. faujasi,* have been found in unit Telm2 in the La Meseta Formation (Feldmann, 1984; Feldmann and Wilson, this volume). However, *Ophiomorpha* is rarely encountered in unit Telm2.

Another decapod, *Lyreidus antarcticus* Feldmann and Zinsmeister, 1984, does occur in close association with *Ophiomorpha* (Fig. 2.1). *Lyreidus* is the appropriate size to have produced these burrows and is adapted to a burrowing lifestyle in Recent habitats. However, *Lyreidus* is currently found only in deep-water environments. Discussions of the possibility that *Lyreidus* has adapted to greater depths in the water column since Eocene time appeared in Feldmann and Zinsmeister (1984) and Feldmann and Wilson (this volume). Most workers consider *Ophiomorpha* to be a primary indicator of littoral and sublittoral conditions (Frey, 1975), a position with which we concur.

Diplocraterion Torell (Fig. 2.5) is common at three localities within unit 5. The tracemaker is uncertain (Fürsich, 1974; Osgood, 1975). *Diplocraterion* ranges from the Cambrian through the Mesozoic (Häntzschel, 1975) and has been found on several continents. This report, however, is believed to be the first for this taxon in Antarctica. Richter (1927), Goldring (1962), Osgood (1970), Frey (1975), and Simpson (1975) each considered *Diplocraterion* to be a primary index to littoral and very shallow sublittoral conditions.

Skolithos Haldeman, 1840 (Fig 2.6) is attributed to worms or phoronids (Fenton and Fenton, 1934, Häntzschel, 1975). *Skolithos* ranges from the Cambrian to Recent and has been found worldwide. Both Seilacher (1967) and Frey (1975) used *Skolithos* as an index trace and as the name bearer to the zone indicating littoral to shallow sublittoral water depth and relatively high-energy conditions. Pemberton and Frey (1984) have shown *Skolithos* to be found in a wide range of depths. Skolithos was found abundantly at locality 14 and noted at other localities.

An enigmatic rod-like trace (Fig. 2.7) was found at locality 6. This cylindrical trace is characterized by being straight to very slightly arcuate, 5.5 cm long, and 0.8 cm in diameter, with a series of four imbricate lobes of approximately equal size that grade into one another on one side of the trace. Only one specimen was collected. This specimen had one broken terminus; the blunt ends of the lobes were directed toward the broken end. The trace was preserved as a fill that was slightly coarser than the surrounding matrix.

Other traces and tubiform fossils found in this unit include

Spirorbis Daudin, the ubiquitous, small, flat, coiled worm tube known from Devonian to Recent sediments (Häntzschel, 1975); *Oichnus* Bromley, gastropod predation borings found on bivalved and gastropod molluscs of various taxa; many examples of *Teredolites* Leymerie (Bromley et al., 1984); and *Serpula* Linnaeus, 1768. The implications of the occurrence of both *Teredolites* and *Serpula* are discussed below.

The environment of deposition for this lithofacies of the La Meseta Formation is inferred to have been a littoral to very shallow sublittoral clastic marine terrain, above storm wave base and most likely above normal wave base as well. This interpretation agrees with, and refines, the conclusions of Zinsmeister (1979) and Feldmann and Zinsmeister (1984), determined by work on the bivalved molluscs and decapod arthropods of the unit.

The second unit in which numerous ichnofossils were collected was Sadler's unit Telm7. Collections from both localities in this unit of the La Meseta Formation yielded fossil traces (Fig. 1, localities 8 and 14, Fig. 4). Present at site 14 were specimens of *Serpula,* many examples of well-preserved fossil wood with abundant *Teredolites,* and several specimens of "rind" or "halo" burrows (Chamberlain, 1975). At site 8, additional samples of *Teredolites* and several specimens of *Ophiomorpha* were found.

Serpula Linnaeus (Fig. 3.1) are produced by members of the family Serpulidae Burmeister, 1837. This cosmopolitan trace type is known from a variety of ichnofacies and from rocks ranging from the Silurian to Recent (Häntzschel, 1975).

Several examples of fossilized wood were found within this unit. In each case, *Teredolites* were common. There probably are two separate types of borings present. The traces are bimodally distributed with respect to size. One is uniformly small, consistently about 0.2 cm in diameter (Fig. 3.3). This smaller boring is primarily oriented perpendicular to growth rings. This trace has been identified from a single large sample with hundreds of examples, many of which have very thin calcite crack fillings in the walls of the tube. The larger tubes range primarily from 0.4 to 0.8 cm in diameter, and are oriented perpendicular and parallel to the wood grain (Figs. 3.3 to 3.5). Several samples of lithified wood, each with many traces, were collected. In the large sample of bored wood, very few traces were seen of diameter between the two peaks in this bimodal distribution. Bromley et al. (1984) suggested the bimodal distribution may indicate separate tracemakers, but in this case, we feel the evidence is inconclusive, since only one specimen showed this distribution clearly. Although wood can float a great distance from its source, the abundance of bored wood may indicate that the source was relatively close.

"Halo" burrows are lined (Fig. 3.2), and are generally either too nondescript or too cosmopolitan to use as an indicator of either water depth or energy levels at the time of deposition. Chamberlain (1975) described "halo" and "rind" burrows from DSDP cores, and thus, in a deep-water environment. Others have described the same types of trace from much shallower water conditions (Frey, 1975). The tracemakers are probably numerous and varied. This trace term is used as a catchall for traces of this

description that do not fit current ichnogeneric descriptions, although some of these specimens may be additional *Skolithos* or examples of *Bergaueria,* a sea anemone trace (S. G. Pemberton, personal communication).

The remaining units of the La Meseta Formation were not sufficiently sampled for trace fossils for any environmental interpretation to be made. However, additional examples of "halo" and "rind" burrows were found in unit 2. *Zapfella* De Saint-Seine, attributed to acrothoracican barnacles, were found throughout the entire formation, especially on bivalved mollusc shells.

There were some difficulties in working with the specimens used in this study. Some, where indicated, were not found in situ. Float material was used in the study if it could be determined that only minimal transportation had occurred subsequent to exhumation. This was not considered a particular problem in the case of the La Meseta traces because there is little running water on Seymour Island and the major source of erosion and subsequent transportation is by wind and gravity. Matching local lithologies with the matrix material of transported traces, distances of postexposure transport were determined. Additionally, some of the traces could only be described from field observations. This was due to the fact that some traces were too large or too delicate to transport.

SUMMARY

Utilizing the models of paleobathymetry and energy conditions derived by Seilacher (1967), Frey (1975), Frey and Pemberton (1984), and Howard and Frey (1984), the exposures of the La Meseta Formation on Seymour Island, in unit Telm5, can be interpreted as having been deposited in a littoral to shallow sublittoral setting. Several of the trace types found in this subunit are indicative of the *Skolithos* zone. This includes *Skolithos, Diplocraterion,* and *Ophiomorpha. Helminthopsis* has normally been associated with a deeper water environment, but has been seen in a littoral to shallow sublittoral environment by Crimes and Anderson (1985). *Muensteria* has been associated with littoral, as well as terrestrial, deposits (Frey, 1975).

The other unit containing abundant trace fossils, unit Telm7, is not so easily characterized or assigned to an ichnofaunal zone. Far fewer trace types are known from this unit than from unit Telm5, and those that are present are far less restricted in their ranges with respect to water depth and energy conditions. Both *Serpula* and the "halo" and "rind" burrows are cosmopolitan. It is only through sediment analysis and the large volumes of teredid bivalve bored fossilized wood (Bromley et al., 1984) that this unit was determined to be of shallow-water origin.

We agree with Elliot and Trautman (1982) that their units, the upper Units II and III, probably represent a tidally dominated environment. Many similarities in trace fossils and depositional patterns were seen between the barrier bar system described by Heinberg and Birkelund (1984) and the La Meseta sequence. Elliot and Trautman favored a deltaic model. The features mentioned by Elliot and Trautman to support deltaic model such as ripple-in-phase and ripple-in-drift laminations, climbing ripples, flaser structures, and wavy, steep cross-bedding are also consistent with a barrier bar model. The La Meseta sequence does not represent a simple shallowing upward sequence. The additional information derived from the trace analysis and that of the body fossils (Zinsmeister, 1979; Feldmann and Zinsmeister, 1984) indicates a variety of very shallow, nearshore environments varying with respect to wave energy and organic content. More paleontological and sedimentological data need to be obtained to more finely resolve this interpretation.

ACKNOWLEDGMENTS

Support for the field work on Seymour Island (for R.M.F.) was provided by a National Science Foundation grant to William J. Zinsmeister. Support for laboratory work was provided by NSF Grant DPP-8411842 (to R.M.F.). Peter Sadler, Department of Earth Sciences, University of California at Riverside, provided a manuscript copy of the geologic map of Seymour Island and permitted extracting major unit boundaries from the map for preparation of the location map. The manuscript was critically read by Alan Coogan, Department of Geology, Kent State University. Careful reviews were provided by Peter Crimes, Molly Miller, Robert Frey, and S. George Pemberton, and we gratefully acknowledge their assistance. Contribution 302, Department of Geology, Kent State University, Kent, Ohio 44242.

REFERENCES

ASGAARD, U., AND R. G. BROMLEY. 1974. Sporfossiler fra den mellem Miocaene transgression; Soby–Fasterholt Omradet, Dansk Geologisk Forening Arsskrift: 1973:11–19 (in Danish with English summary).

BEIKIRCH, D. W., AND R. M. FELDMANN. 1980. Decapod crustaceans from the Pflugerville Member, Austin Formation (Late Cretaceous: Campanian) of Texas. Journal of Paleontology, 54:309–324.

BROMLEY, R. G., S. G. PEMBERTON, AND R. A. RAHMANI. 1984. A Cretaceous woodground: the *Teredolites* ichnofacies. Journal of Paleontology, 58:488–498.

CHAMBERLAIN, C. K. 1975. Trace fossils in DSDP cores of the Pacific. Journal of Paleontology, 49:1074–1096.

CRIMES, T. P. 1977. Trace fossils of an Eocene deep-sea sand, northern Spain, p. 71–90. *In* T. P. Crimes and J. C. Harper (eds.), Trace Fossils II. Geological Journal, Special issue, Volume 9.

——, AND M. M. ANDERSON. 1985. Trace fossils from late Precambrian–early Cambrian strata of southeastern Newfoundland (Canada): temporal and environmental implications. Journal of Paleontology, 59:310–343.

——, AND J. D. CROSSLEY. 1968. The stratigraphy, sedimentology, ichnology and structure of the lower Paleozoic rocks of northeastern County Wexford. Royal Irish Academy, Section B, Proceedings, 67:185–215.

——, R. GOLDRING, P. HOMEWOOD, J. VAN STUIJVENBERG, AND W. WINKLER. 1981. Trace fossil assemblages of deep-sea fan deposits, Gurnigel and Shlieren flysch (Cretaceous-Eocene, Switzerland). Ecologae Geologicae Helvatiae, 74:953–995.

ELLIOT, D. H., AND W. J. TRAUTMAN. 1982. Lower Tertiary strata on Seymour Island, Antarctic Peninsula, p. 287–297. *In* C. Craddock (ed.), Antarctic Geosciences. University of Wisconsin Press, Madison.

FELDMANN, R. M. 1984. Seymour Island yields a rich fossil harvest. Geotimes, 29(2):16–18.

——, AND W. J. ZINSMEISTER. 1984. New fossil crabs (Decapoda; Brachyura) from the La Meseta Formation (Eocene) of Antarctica: paleogeographic and biogeographic implications. Journal of Paleontology, 58:1041–1061.

FENTON, C. L., AND M. A. FENTON, 1934. *Scolithus* as a fossil phoronoid. Pan American Geologist, 61:341–348.

FREY, R. W. 1975. The realm of ichnology, its strengths and limitations, p. 13–38. *In* R. W. Frey (ed.), The Study of Trace Fossils. Springer-Verlag, New York.

——, AND J. HOWARD, 1984. Trace fossils from the Panther Member, Star Point Formation (Upper Cretaceous), Coal Creek Canyon, Utah. Journal of Paleontology, 59:370–404.

——, AND PEMBERTON, S. G., 1984. Facies Models. *In* R. G. Walker (ed.), Facies models. Geoscience Canada, Reprint Series 1, 189–207.

——, J. D. HOWARD, AND W. A. PRYOR, 1978. *Ophiomorpha:* its morphologic, taxonomic, and environmental significance. Palaeogeography, Palaeoclimatology, Palaeoecology, 23:199–229.

FÜRSICH, F. T. 1974. On *Diplocraterion* Torell 1870 and the significance of morphological features in vertical, spreiten-bearing, U-shaped trace fossils. Journal of Paleontology, 48:952–962.

GOLDRING, R. 1962. The trace fossils of the Baggy Beds (Upper Devonian) of North Devon, England. Palaontologische Zeitschrift, 36:232–251.

HÄNTZSCHEL, W. 1952. Die lebensspur *Ophiomorpha* Lundgren im Miozan bei Hamburg, ihre weltweite verbreitung und synonymie. Geological Staatsinstitute, Hamburg, Mitteil., 19:77–84.

——. 1975. Trace fossils and problematica. *In* Curt Teichert (ed.), Treatise on Invertebrate Paleontology, Pt. W, Miscellanea Supplement 1. Geological Society of America and University of Kansas Press, Lawrence, 269 p.

HEINBERG, C., AND T. BIRKELUND. 1984. Trace-fossil assemblages and basin evolution of the Vardekloft Formation (Middle Jurassic), central east Greenland. Journal of Paleontology, 58:362–397.

HOWARD, J., AND R. W. FREY. 1984. Characteristic trace fossils in nearshore to offshore sequences, Upper Cretaceous of east-central Utah. Canadian Journal of Earth Science, 21:200–219.

KENNEDY, W. J., AND J.D.S. MACDOUGALL. 1969. Crustacean burrows in the Weald Clay (Lower Cretaceous) of south-eastern England and their environmental significance. Palaeontology, 12:459–471.

——, AND B. W. SELLWOOD. 1970. *Ophiomorpha nodosa* Lundgren, a marine indicator from the Sparnacian of south-east England. Geologists' Association, Proceedings. 81:99–110.

OSGOOD, R. G. 1970. Trace fossils of the Cincinnati area. Palaeontographica Americana, 6(41):281–444.

——. 1975. The history of invertebrate ichnology, p. 3–12. *In* R. W. Frey (ed.), The Study of Trace Fossils. Springer-Verlag, New York.

PEMBERTON, S. G., AND R. W. FREY. 1984. Quantitative methods in ichnology: spacial distribution among populations. Lethaia, 17:33–49.

RICHTER, R. 1927. Die fossilen Fahrten und Bauten der Wurmer, ein Uberblick uber ihre biologischen Grundformen und deren geologische Bedeutung. Palaeontologische Zeitschrift, 9:193–240.

SEILACHER, A. 1967. Bathymetry of trace fossils. Marine Geology, 5:413–428.

SIMPSON, S. 1975. Classification of trace fossils, p. 39–54. *In* R. W. Frey (ed.), The Study of Trace Fossils. Springer-Verlag, New York.

WAAGE, K. M. 1968. The type Fox Hills Formation, Cretaceous (Maastrichtian), South Dakota. Peabody Museum of Natural History Bulletin, 25:1–175.

WEIMER, R. J., AND J. H. HOYT. 1964. Burrows of *Callianassa major* Say: geologic indicators of littoral and shallow neritic environments. Journal of Paleontology, 38:761–767.

ZINSMEISTER, W. J. 1979. The Struthiolaridae (Gastropoda) fauna from Seymour Island, Antarctic Peninsula. Congreso Geologico Argentino. Actas Buenos Aires, 1(6):609–618.

MANUSCRIPT ACCEPTED BY THE SOCIETY SEPTEMBER 1, 1987

Geological Society of America
Memoir 169
1988

Tectonic setting and evolution of the James Ross Basin, northern Antarctic Peninsula

David H. Elliot
Byrd Polar Research Center and Department of Geology and Mineralogy, The Ohio State University, Columbus, Ohio 43210

ABSTRACT

The upper Mesozoic to lower Cenozoic sequence in the region of James Ross Island is the only exposed marine succession of that age in Antarctica. The sequence makes up part of the fill of the James Ross Basin and includes: (1) an upper Jurassic mudstone-tuff sequence, the Nordenskjöld Formation; (2) a Lower to Upper Cretaceous conglomerate-sandstone-mudstone-tuff assemblage, the Gustav Group and equivalents; (3) an Upper Cretaceous to Paleocene poorly consolidated sand, silt, mud and tuff sequence, the Marambio Group and equivalents; and (4) an Eocene sequence of weakly consolidated, nonvolcanic fine sands and silts—the La Meseta Formation. Sedimentary facies include proximal submarine fans, shelf settings, and deltaic environments.

Sea-floor anomaly data from the Pacific Ocean suggest that development of Upper Mesozoic to Cenozoic fore-arc, magmatic arc, and back-arc terrains of the Peninsula resulted from the subduction of the Phoenix Plate until the early Tertiary, and, after reorganization of spreading centers in Late Cretaceous time, subduction of the Aluk Plate. Strata in the James Ross Island region constitute the sedimentary and volcanic fill of an ensialic back-arc basin developed on the Weddell Sea flank of the Antarctic Peninsula. Broad correlations can be made between the strata and evolution of the James Ross Basin, the tectonic and magmatic history of the peninsula, and plate subduction.

INTRODUCTION

Exploration by the Swedish South Polar Expedition at the beginning of this century led to the discovery of fossiliferous rocks of Cretaceous and Tertiary age on James Ross Island and adjacent islands (Figs. 1, 2), and to the establishment of the basic geological framework for the northern Antarctic Peninsula (Andersson, 1906; Nordenskjöld, 1905). The two structural zones recognized were a cordilleran belt of folded strata, which is unconformably overlain by undeformed sedimentary and volcanic beds of Mesozoic age and intruded by plutons; and a platform to the east, which consists of only slightly deformed upper Mesozoic and Cenozoic strata overlain by upper Tertiary basalts.

Systematic survey of the Antarctic Peninsula began with the establishment of the Falkland Islands Dependencies Survey, now the British Antarctic Survey. Knowledge of the geology increased rapidly (Adie, 1964, 1972). The first plate tectonic interpretation of the geologic evolution of the peninsula was presented by Dalziel and Elliot (1971, 1973), and since then the simple model of Pacific crust subduction beneath the western continental margin

has been amplified and refined. Suarez (1976) interpreted the Upper Jurassic and Cretaceous rocks of the southern Antarctic Peninsula in terms of three tectono-stratigraphic units that represent fore-arc, magmatic arc, and back-arc terrains associated with Pacific crust subduction. Cande and others (1982) presented sea-floor magnetic data for the southeastern Pacific Ocean. Barker (1982) described the ridge crest-trench collision that occurred along the western margin of the Peninsula and also discussed the Mesozoic plate tectonic history. The evolution of the fore-arc, magmatic arc, and back-arc terrains of late Mesozoic to early Cenozoic age in the peninsula region was described in a previous paper (Elliot, 1983).

This chapter reviews the stratigraphy, tectonic setting, and evolution of the back-arc basin in the James Ross Island region, and relates it to the subduction history of the Antarctic Peninsula.

REGIONAL SETTING

An extensive sequence of largely undeformed marine rocks, ranging in age from Late Jurassic to early Tertiary, crops out

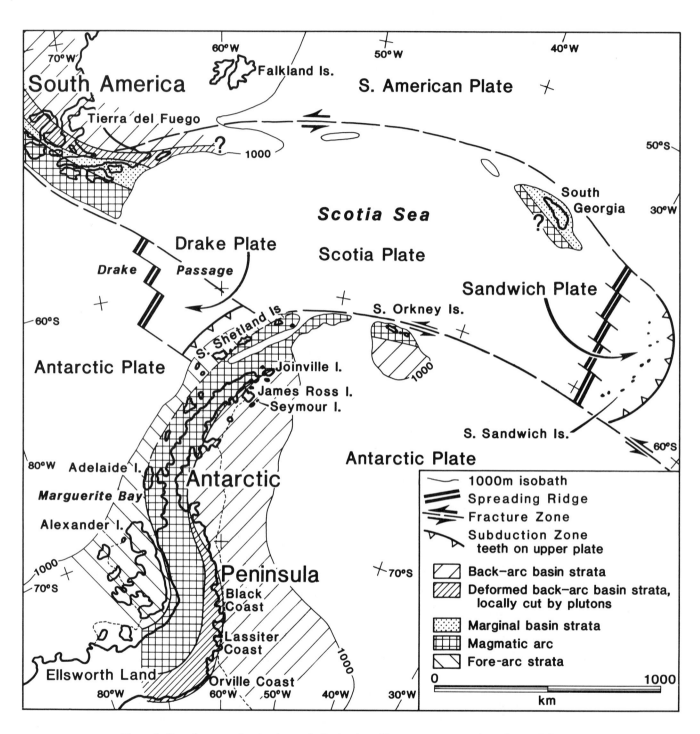

Figure 1. Location map for the Antarctic Peninsula, with present geotectonic setting and Andean lithotectonic units superimposed (data for South America and South Georgia from Dalziel et al., 1974, and Storey et al., 1977).

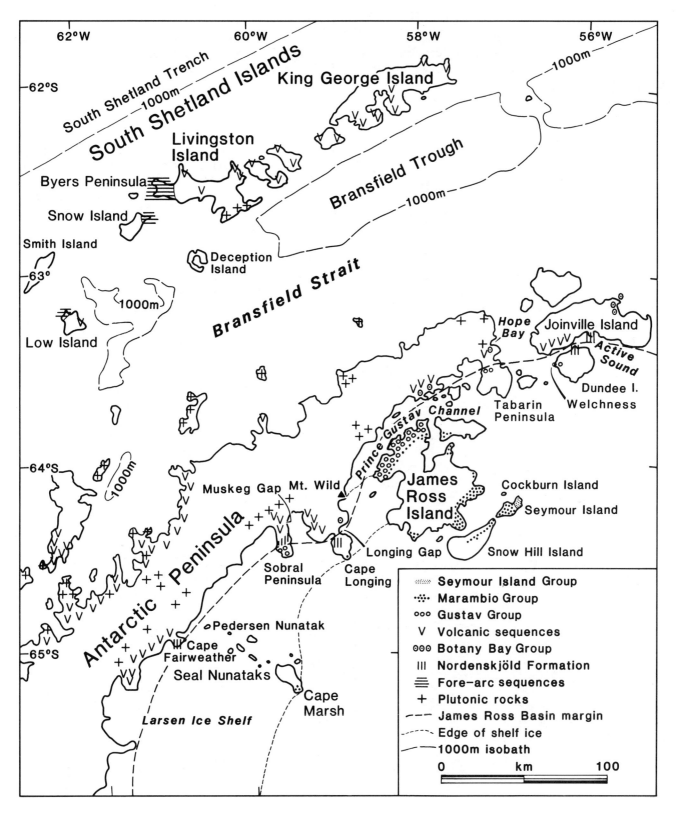

Figure 2. Location and simplified geologic map for the northern Antarctic Peninsula.

from Dundee Island and Active Sound in the northeast to Cape Fairweather and Cape Marsh in the southwest (Fig. 2). The sequence, which includes the Upper Cretaceous and Paleogene strata exposed on Seymour Island, constitutes the sedimentary fill of the James Ross Basin and records an important part of the geologic history of the Antarctic Peninsula.

East- to southeast-directed subduction beneath the Antarctic Peninsula is strongly suggested by the asymmetry displayed in the compositional trends of the upper Mesozoic to Cenozoic magmatic arc rocks (Saunders et al., 1982; Weaver et al., 1982), and supports the inferences that can be drawn from Cenozoic sea-floor magnetic data for the southeastern Pacific Ocean (Barker, 1982; Cande et al., 1982; Fig. 3). Thus, the peninsula region is regarded as an active plate margin with plate convergence located along its western boundary. Three major lithotectonic units are recognized in the Antarctic Peninsula and can be related in a general manner to plate interaction.

Along the length of the peninsula, substantial thicknesses of volcanic rocks with sparse sedimentary intercalations record the onset of magmatism. Volcanism began in the mid-Jurassic in eastern Ellsworth Land (Quilty, 1977), extended to the whole of the peninsula by latest Jurassic time, and continued to the mid-Tertiary. Emplacement of plutonic bodies accompanied the extensive volcanic activity. Magmatic activity shifted first to the southeast in Palmer Land during the early Cretaceous, but the subsequent trend was a gradual migration to the north and west across the peninsula, and by the mid-Tertiary was confined to the northwestern flank of the peninsula and the South Shetland Islands (Elliot, 1983; Smellie et al., 1984).

Early stages in magmatic arc development in the northern Antarctic Peninsula are marked by volcanic rocks interbedded with, or overlying, sequences of alluvial fan deposits and associated plant-bearing beds—the Botany Bay Group of Farquharson (1984). Contemporaneous volcanism is recorded at Hope Bay (Elliot and Gracanin, 1983), the Longing Gap area, and Camp Hill (Farquharson, 1982b), but volcanic rocks are conspicuously absent from the conglomerate clasts in the basal parts of these sequences and the sequences on Joinville Island and the South Orkney Islands (Elliot and Wells, 1982; Elliot et al., 1977). Volcanic strata are not known to occur between the pre-Jurassic basement and the basal conglomerate beds. The age of this well-defined lithotectonic unit is uncertain, although it is probably in the range of latest Jurassic to Early Cretaceous (Farquharson, 1984). Evidence for a magmatic arc along the Antarctic Peninsula is also found in the adjacent sedimentary basins.

Fore-arc basin deposits are preserved on the western side of the peninsula, principally on Alexander Island and the South Shetland Islands. The shallow marine sedimentary rocks on Alexander Island, consisting of a thick sequence of conglomerates, sandstones, and mudstones, range in age from Late Jurassic to middle Cretaceous (Aptian-Albian). Lavas and pyroclastic rocks are interbedded in the Jurassic part (Elliott, 1974), and airfall detritus is abundant in the Cretaceous beds (Horne and Thomson, 1972). Most of the fore-arc sediments are marine;

however, the presence of fossil tree trunks in growth position at a number of horizons (Jefferson, 1982) confirms nonmarine deposition for the upper part of the sequence. Although unequivocal evidence for the provenance of the sequence is lacking (Thomson, 1982a), the bulk of the sediments were probably derived from an active volcanic arc to the east along the axis of the southern Antarctic Peninsula (Elliot, 1974).

In the South Shetland Islands (Fig. 2), the deposits are only locally preserved. On Livingston Island, Upper Jurassic and lowermost Cretaceous marine shales and sandstones with a detrital volcanic component pass up into subaerial volcanic rocks (Smellie et al., 1980, 1984), possibly as young as Barremian (Askin, 1983). A thin nonmarine sequence that includes tuffs and lavas is present on an adjacent headland, President Head, on Snow Island (D. H. Elliot, unpublished data; Smellie et al., 1980, 1984). The sequence is Valanginian to Barremian in age (Askin, 1983). Upper Jurassic volcaniclastic rocks are also present on Low Island (Smellie, 1979) and have been assigned an Oxfordian and possibly younger age (Thomson, 1982b).

Two rather different successions deposited in back-arc environments occur on the east side of the magmatic arc. A volcanogenic sequence of conglomerates, sandstones, and shales crops out in an arc from eastern Ellsworth Land to the Black Coast (Singleton, 1980; Laudon et al., 1983). This marine sequence is interbedded with and overlain by intermediate to silicic extrusive rocks: it is clearly related to an active magmatic arc. Invertebrate faunas indicate a Middle to Late Jurassic age for the sedimentary succession (Quilty, 1977, 1983; Rowley and Williams, 1982; Thomson, 1983). These volcanogenic sequences show considerable deformation and are cut by plutonic complexes. In eastern Ellsworth Land and the Black Coast, the deformation was latest Jurassic to earliest Cretaceous in age (Quilty, 1977; Singleton, 1980), whereas on the Lassiter Coast it was pre–middle Cretaceous (Rowley and Williams, 1982).

The other succession, located at the northern end of the Peninsula, constitutes the exposed strata of the James Ross Basin. This basin contains a thick sequence of Upper Jurassic to lower Tertiary clastic sediments and tuff. In fact it is just a small part of an extensive region of late Mesozoic and Cenozoic sedimentation located on the continental shelf east of the peninsula. Details of the James Ross succession are described below.

Although Mesozoic sea-floor anomalies are absent from the southeastern Pacific Ocean, reconstructions strongly suggest that the Late Jurassic and Cretaceous magmatic and tectonic history of the peninsula is related to subduction of oceanic crust, the Phoenix Plate, which formed at the same time as M anomalies in the Western Pacific (Larson, 1977; Barker, 1982). Reorganization of spreading centers occurred toward the end of the Cretaceous (Fig. 4). Spreading between New Zealand and West Antarctica was initiated at about 84 Ma (anomaly 34), but the developing Antarctic and Bellingshausen Plates (Stock and Molnar, 1987) extended no farther northeastward (Larson et al., 1979) than the Tharp Fracture Zone of the Eltanin Fracture Zone system. To the northeast, spreading on the preexisting Pacific-Phoenix Ridge con-

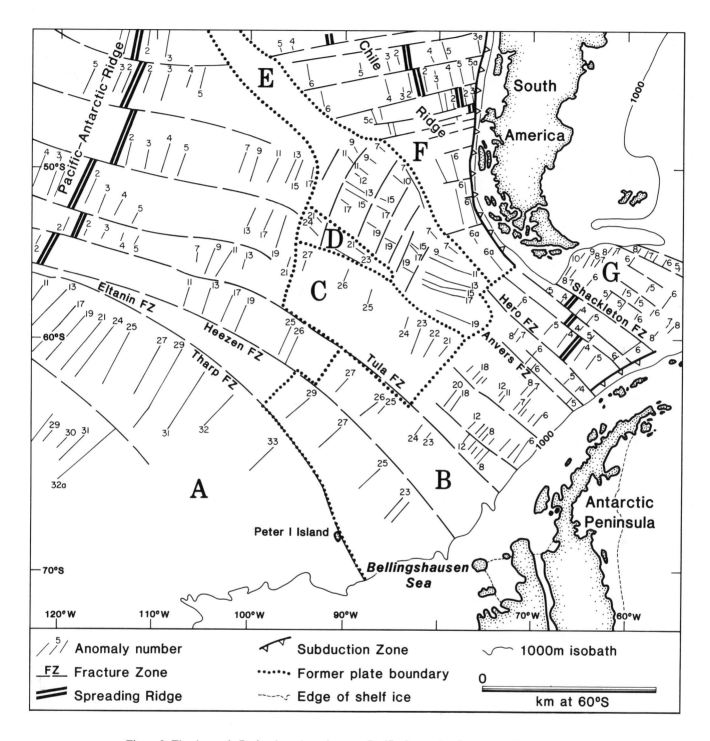

Figure 3. The Antarctic Peninsula and southeastern Pacific Ocean. Sea-floor magnetic anomaly data from Barker (1982), Cande et al. (1982), and Cande (personal communication, 1987). Sea floor divided into regions based on spreading ridge at which oceanic crust was generated (from Cande et al., 1982; Stock and Molnar, 1987): A = Pacific-Antarctic/Bellingshausen; B = Aluk-Antarctic/Bellingshausen; C = Pacific-Phoenix (Aluk); D = Pacific-Farallon; E = Antarctic-Farallon; F = Antarctic-Nazca; G = Spreading on a now-extinct ridge east of the Shackleton Fracture Zone.

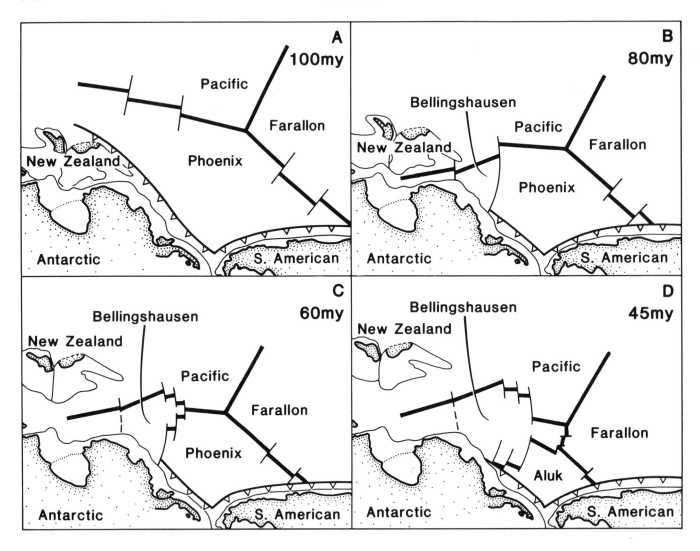

Figure 4. Schematic diagrams of the evolution of the sea floor of the southeastern Pacific Ocean with lithospheric plates identified by name, with the exception that New Zealand is used only in a geographic sense (adapted from Barker, 1982). The sea-floor data for the Pacific Ocean suggest that thousands of kilometers of oceanic crust were subducted beneath the Antarctic margin. Convergence rate and orientation can be determined only for the Cenozoic. Spreading between New Zealand and West Antarctica is defined by anomalies generated at the Pacific-Antarctic/Bellingshausen ridge system. The migration of triple junctions in the southeastern Pacific has left a complicated anomaly pattern (see Fig. 3). The Aluk Plate was generated by spreading on ridges between it and the Bellingshausen, Antarctic, and Farallon Plates.

tinued. A new ridge system, the Bellingshausen-Aluk Ridge (referred to as the Antarctic-Aluk Ridge prior to recognition of the Bellingshausen Plate), was initiated between the Tharp and Heezen Fracture Zones by 66 Ma (anomaly 29). The Bellingshausen-Pacific and Bellingshausen-Aluk Ridges propagated northeastward, replacing the Pacific-Phoenix Ridge. The Bellingshausen Plate was locked onto the Antarctic Plate at about 43 Ma (anomaly 18). The Aluk Plate, formed by spreading on the Bellingshausen-

Aluk Ridge, and after 43 Ma, the Antarctic-Aluk Ridge, was subducted beneath the Antarctic Peninsula.

Subduction, however, was shut off as the Bellingshausen/Antarctic-Aluk Ridge collided with the trench located along the western margin of the Antarctic Peninsula (Barker, 1982). In a northeasterly direction, progressively younger sea floor lies adjacent to the peninsula and suggests that ridge crest–trench collision, in segments defined by fracture zones, started at about 50

Ma and continued to about 4 Ma. A remnant of the Aluk Plate, the Drake Plate of Barker (1982), forms the sea floor off the South Shetland Islands (Fig. 1). Subduction in that sector slowed markedly about 4 m.y. ago, and shortly thereafter, opening of the Bransfield Trough—a young marginal basin—was initiated.

The Phoenix Plate was subducted during the Mesozoic; the rate of convergence was high, possibly >10 cm/yr, but the orientation of convergence is unknown. The late Cretaceous and Cenozoic history is more complicated because of the development of the Pacific-Antarctic, Pacific-Bellingshausen (anomaly 32 to anomaly 18 time), and Bellingshausen-Aluk (anomaly 29 to anomaly 18 time) and Antarctic-Aluk (<43 Ma) spreading centers. A 15- to 20-m.y. hiatus in subduction in the Palmer Land sector, between the Tharp and Heezen Fracture Zones, was followed by renewed subduction until ridge-trench collision at about 50 Ma. To the northeast, subduction continued, first of the Phoenix Plate, and then, with development and propagation of the Bellingshausen/Antarctic–Aluk Ridge, of Aluk Plate crust. The direction of convergence since 66 Ma (anomaly 29) is constrained by the Tharp, Heezen, and Tula Fracture Zones. The rates of convergence were high, >10 and possibly >15 cm/yr, for those segments associated with Pacific-Phoenix spreading. After about 50 Ma, the rates of convergence dropped to less than about 5 cm/yr (Barker, 1982).

The eastern margin of the Antarctic Peninsula continental crust abuts the Weddell Sea floor (Fig. 5); unfortunately, details of the continent-ocean boundary are poorly known because of perennial pack ice and icebergs. Published bathymetric charts (e.g., LaBrecque, 1986) suggest the presence of a thick sedimentary prism at the foot of the continental slope. Sea-floor magnetic anomaly data indicate that at least part of the eastern Weddell Sea floor is Mesozoic in age, but exactly how old remains uncertain. The age of the oldest anomaly is 155 to 160 m.y. (M25 or M29, LaBrecque [1986] and British Antarctic Survey [1985], respectively). Even older crust is present between the oldest anomaly and the Explora-Andenes Escarpment (Kristoffersen and Haugland, 1986). The depth of the Weddell Sea floor is consistent with a Mesozoic age. It seems likely that the Antarctic Peninsula margin is of long standing. The margin may be a passive margin whose origin lies in the early stages in the dispersal of Gondwanaland and in the initial rifting between Africa–South America and Antarctica–India–Australia. On the other hand, the margin might be a transform boundary now relatively deeply buried by sediment derived from the magmatic arc, sited on the Peninsula, that was active for much of Jurassic to late Tertiary time. The lack of a pronounced magnetic signature on the few tracks crossing the margin (La Brecque, 1986) adds to the uncertainties.

STRATIGRAPHY OF THE JAMES ROSS BASIN

The principal exposures of strata forming the sequence that fills the basin occur on James Ross Island or in its vicinity (Fig. 2; Table 1). Scattered outcrops of Upper Jurassic and Lower Cre-

taceous strata to the northeast and southwest indicate the probable minimum extent of the basin, but the relations to the principal exposures are uncertain. Seismic data suggest its northeastward extension is to be found on the platform south of the South Orkney Islands (Farquharson et al., 1984; Harrington et al., 1972). Aeromagnetic data indicate thick sedimentary sequences are present under the Larsen Ice Shelf as far south as the base of the peninsula. Estimates of stratigraphic thickness have been made only for the James Ross Island region; del Valle et al. (1983) suggest 6.4 km, and Ineson et al. (1986) suggest 4.8 km, excluding the Tertiary Cross Valley and La Meseta Formations on Seymour Island. Although the Nordenskjöld Formation (Farquharson, 1982a), the Ameghino Formation of Medina and Ramos (1980), does not represent deposition in a deepening basin, it is included here because it probably forms the basal sequence above the basement that lies east of the peninsula. The Upper Jurassic (Kimmeridgian-Tithonian) Nordenskjöld Formation crops out at isolated localities between Cape Fairweather and Joinville Island (Fig. 2). It consists of a succession of alternating thin-bedded, radiolarian-rich mudstones and airfall tuffs.

Lower Cretaceous rocks identified at Sobral Peninsula consist of a thick sequence of coarse clastic marine strata with a large volcanic component. They have been assigned a late Hauterivian-Barremian age (Farquharson, 1982a). At Pedersen Nunatak, to the southwest, the marine conglomeratic sequence lacks an airfall volcanic component. The age of these beds is uncertain. According to Medina et al. (1981), it is Early Cretaceous, but Farquharson (1983a) assigned the beds a Late Cretaceous (Turonian-Maastrichtian) age. The strata at Sobral Peninsula and Pedersen Nunatak have been placed in the Pedersen Formation (see del Valle and Fourcade, 1986). The Cape Sobral beds are lateral equivalents of the lower part of the sequence on James Ross Island, although possibly extending further back in time, and are part of the same tectono-stratigraphic unit. An isolated outcrop of conglomerate, sandstone, shale, and tuff on Tabarin Peninsula (Scasso et al., 1986) also belongs to this unit.

The stratigraphic scheme proposed by Ineson et al. (1986) for the Cretaceous sequence on James Ross Island area is followed here. The Gustav Group, which is restricted to the northwest part of the island, forms the lower part. It is predominantly a conglomerate and sandstone succession, though finer grained beds are important locally. The oldest rocks, the Lagrelius Point Formation, are a sequence of pebble and cobble conglomerates with minor interbedded sandstones. They are in contact only with the Upper Cenozoic James Ross Island Volcanic Group (Nelson, 1975). The Kotick Point Formation consists of thin-bedded, fine-grained sandstones and silty mudstones or clays, with interbedded breccia and conglomerate intervals that include clasts up to several meters across. Slide blocks of Nordenskjöld Formation lithology are an important component of the formation. The succeeding Whiskey Bay Formation shows great lateral facies variation; locally, however, the succession has been broken out into members. Pebble and boulder conglomerates, breccias, pebbly sandstones, sandstones, and mudstones are all present. The

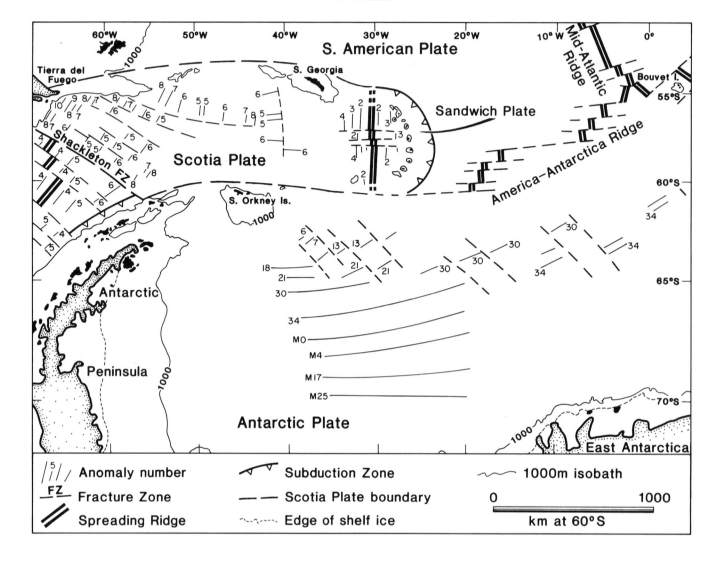

Figure 5. The Antarctic Peninsula, Scotia and Weddell Seas. Sea-floor magnetic data from LaBrecque (1986). Younging of Weddell Sea anomalies away from Antarctica suggests that the spreading ridge at which the oceanic crust was formed and the complementary crust on the other flank of the ridge have been subducted beneath the South American Plate, and, in later time (less than 30 m.y.), beneath what is now the Scotia Sea region.

uppermost unit, the Hidden Lake Formation, is a sequence of coarse to fine sandstones, with interbedded conglomerates in the lower part and siltstones in the upper part. Volcanic detritus and airfall debris are common throughout the Gustav Group.

The beds at Cape Sobral are lithologic equivalents of those of the Gustav Group, but the lack of volcanic detritus in the Pedersen Nunatak sequence tends to support the younger age assigned to them by Farquharson (1983a). Other Gustav Group equivalents include the Aptian-Albian sandstones and volcaniclastic beds at Welchness on Dundee Island (Ramos et al., 1978; Crame, 1979) and the beds at Cape Longing (Crame, 1981). At the latter locality, airfall tuffs are interbedded with sandstones and siltstones (Farquharson, 1982a); the age of the sequence is early to middle Turonian (Crame, 1983).

The Marambio Group (as defined by Rinaldi et al., 1978) and its equivalents crop out over the central and southeastern parts of James Ross Island, on the southern part of Vega Island, on Snow Hill, Seymour, and Cockburn Islands, and at Cape Marsh (del Valle and Medina, 1985). It consists of fine-grained clastic strata that include, from the base upward, the Santa Marta Formation (Olivero et al., 1986), beds as yet unnamed (except as part of Bibby's [1966] Snow Hill Island Series), the López de Bertodano Formation, and the Sobral Formation, which is now known to be of Paleocene age. The Santa Marta Formation is

TABLE 1. STRATIGRAPHY OF THE JAMES ROSS BASIN REGION

Group	Formation	Thickness (m)	Age
Seymour Island	La Meseta	750	L. Eocene
	Cross Valley*	105	? Paleocene
Marambio	Sobral	255	Paleocene
	López de Bertodano	1,190	L. Campanian-? E. Paleocene
	Unnamed strata	?1,000	? Campanian
	Santa Marta	1,000	Santonian-Campanian
Gustav	Hidden Lake	400	? Coniacian-Santonian
	Whiskey Bay	950	Albian-Coniacian
	Kotick Point	1,000	Aptian-Albian
	Lagrelius Point	500	? Barremian-Aptian
	Nordenskjöld	550	Kimmeridgian-Tithonian

Note: The Gustav Group has an estimated maximum thickness of only 2,300 m because of lateral changes in formation thicknesses (Ineson et al., 1986). Thickness for the Santa Marta Formation from Olivero et al. (1986), the unnamed strata from the text, the López de Bertodano and Sobral Formations from Macellari (this volume), Cross Valley Formation from Elliot and Trautman (1982), and La Meseta Formation from Zinsmeister and DeVries (1982).

*Recent field work suggests the assignment of the Cross Valley Formation to the Seymour Island Group may need revision (see text).

---

made up of silty and muddy sandstones with intercalated conglomerates and pelitic beds, calcite-cemented sandstones and marls, and concretionary beds containing invertebrates (Olivero et al., 1986). The fossils suggest an age of ?Santonian to Campanian.

Detailed stratigraphical and sedimentological studies have not been carried out on the unnamed beds; however, Bibby (1966) established that they consist largely of sands with interbedded sandy clays and clays. Abundant calcareous concretions, many of which contain well-preserved ammonites and other invertebrates, occur at a number of horizons. Ammonite faunas suggest a Campanian age (Spath, 1953; Howarth, 1958, 1966). The thickness of this interval is uncertain. Estimates for the complete sequence below the López de Bertodano range from 1,300 to 1,900 m (Ineson et al., 1986, and del Valle and Fourcade, 1986, respectively), but this interval includes the Santa Marta Formation. The unnamed beds may be as much as 1,000 m in thickness.

The López de Bertodano Formation, described by Macellari (1987), consists of 1,190 m of very fine sands and sandy siltstones together with sparse carbonate-cemented sandstones. Calcareous concretions, many containing well-preserved invertebrates, are abundant in parts of the section. The upper part of the formation contains volcanic debris and probable tuff beds. Glauconitic beds are also prominent. The age of the formation is late Campanian

to Paleocene, based on studies of the ammonites (Macellari, 1986), foraminifera (Huber, this volume), palynomorphs (Askin, this volume), and siliceous microfossils (Harwood, this volume). The Cretaceous-Tertiary boundary occurs in the uppermost part of the formation. It has yet to be precisely located in the southern half of the outcrop belt; in the northern half it is close to the boundary between units 9 and 10 of the López de Bertodano Formation. At least the upper part of unit 10 is Danian.

The upper part of the sequence cropping out on the northern promontory of Snow Hill Island is equivalent to the lowest López de Bertodano beds on Seymour Island; successively older beds are present in a northwesterly direction. Preliminary studies suggest that at least some of the unnamed strata in the region of The Naze and Cape Lamb may be equivalent to beds exposed on Snow Hill Island (Huber, this volume). The strata at Cape Marsh, to which del Valle and Medina (1985) gave a Senonian (Campanian) age, are also lateral equivalents of the Marambio Group.

The disconformably overlying Sobral Formation is a 255-m-thick sequence of silts, sands, and subordinate sandstones (Macellari, this volume). The beds are commonly clay-rich in the lower part and increasingly coarse grained up-section. Tuff beds are present in the lower part of the section, and glauconite is a common constituent of the sands above the basal 50 to 60 m of the formation.

Younger beds on Seymour Island, comprising the Cross

Valley and La Meseta Formations, were placed in the Seymour Island Group by Elliot and Trautman (1982). The Cross Valley Formation, as originally defined, consists of 82 m of weakly stratified coarse sands and conglomerates overlain by 23 m of finer grained sands and sandstones. The lower beds in particular have a significant volcanic component. Field work in the 1986–87 season (W. J. Zinsmeister and D. H. Elliot, unpublished data, 1987) casts some doubt on previous age assignments and correlations. The youngest formation, the La Meseta Formation, rests unconformably on Sobral and Cross Valley beds. Uncemented silts and sands with sparse carbonate-cemented sandstones form a sequence about 750 m thick. Fossils are abundant in many beds, and locally form conspicuous shell banks. The age of the formation is late Eocene, according to studies on the invertebrate faunas (Zinsmeister, 1984; Weidman et al., this volume); however, dinoflagellate cysts suggest the lower part is middle Eocene (Wrenn and Hart, this volume). The Tertiary beds on Cockburn Island have some lithologic similarities to the La Meseta beds (Elliot and Rieske, in press). The fauna recovered is too poorly preserved for age determination; thus, the relations to Seymour Island Tertiary strata remain unclear.

DEPOSITIONAL SETTINGS

The interlayered radiolarian-rich mudstones and tuffs of the Nordenskjöld Formation were deposited in a euxinic environment (Farquharson, 1982a) that extended along the eastern flank of a low-relief volcanic arc situated on the Antarctic Peninsula. This formation and the equivalent successions on the South Shetland Islands were the first widespread sedimentary units to be deposited subsequent to the Gondwanian tectonism that is recorded in the pre-Jurassic basement; they signal the development of a magmatic arc and a period of subsidence in Middle to Late Jurassic time throughout the northern peninsula region.

The two lower stages in the Neocomian have yet to be recognized, and therefore a significant gap exists in the record. The arc region was probably uplifted during the Neocomian, thus providing a suitable source terrain for much of the coarse clastic debris in the Gustav Group and its equivalents. The western margin of the basin was fault-controlled (Farquharson et al., 1984), and uplift of the arc region was accompanied by subsidence of the basin. Proximal deposits, represented by the coarse-grained components of the Gustav Group and the beds at Cape Sobral, were emplaced by sediment gravity flows and turbidity currents. Much penecontemporaneous volcanic material, both reworked and airfall, was also deposited (Farquharson et al., 1984). Slide blocks of the Nordenskjöld Formation that are up to 200 m thick and 800 m in outcrop length testify to the instability of the basin margin. Submarine fan deposition is inferred for much of the coarser debris. The finer grained traction-current deposits, together with the mudstones and clays, presumably represent environments away from the main fan channels or times of tectonic quiescence. The influence of fault control on the basin

margin diminished during Hidden Lake and Santa Marta time. Beds constituting those formations represent the transition to a more stable continental shelf that persisted during deposition of much of the Marambio Group. Nevertheless, if the Late Cretaceous age inferred for the strata at Pedersen Nunatak is correct, parts of the basin margin remained as sites for accumulation of coarse proximal deposits. The finer grain size of the remaining exposed Campanian and Maastrichtian strata must be due in part to a more distal setting.

The rather monotonous sandy, muddy silts, and fine sands of the López de Bertodano Formation represent a somewhat quieter depositional environment than that of the coarser grained strata lower in the succession. Macellari (this volume) postulated shelf environnments for the López de Bertodano beds. The presence of a diverse agglutinated foraminiferal assemblage in the lower part (Units 1 through 6) has led Huber (this volume) to infer a nearshore, shallow, restricted environment, and the sparsity of macrofauna and the dominance within it of rotularids support this interpretation. The siliceous microfossils, on the other hand, indicate open marine conditions (Harwood, this volume). The presence of bioturbation points to an active bottom fauna and may explain the absence of sedimentary structures that might be expected in a shallow environment. Water depths greater than 75 m are suggested by the absence of benthic diatoms, although shallower depths as a result of turbid conditions are possible and are supported by the occurrence of pyritized siliceous microfossils (Harwood, this volume). On balance, a shelf environment below storm wave base and with turbid waters seems most likely.

Units 7, 8, and 9 carry an abundant invertebrate fauna. A middle to outer shelf restricted depositional environment is inferred by Macellari (this volume). Glauconite increases upsection and is particularly prominent in the lower part of Unit 10. An abrupt decrease in macrofauna at the base of Unit 10 is accompanied by changes in the microfossil assemblages; foraminiferal abundance decreases (Huber, this volume), and the nature of the preserved fauna is interpreted by Huber (this volume) to indicate a dissolution interval. Monospecific dinocyst assemblages (Askin, this volume) suggest a restricted, possibly nearshore, environment. The sediments themselves do not change markedly, although clay decreases upsection in Unit 10.

A disconformity at the top of the López de Bertodano is demonstrated by the presence of concretions showing signs of boring (W. J. Zinsmeister, personal communication), by broad channelling accompanied by changes in thickness in the lowermost Sobral, and by a lag conglomerate. Marked shallowing to near sea level due to either uplift or regression is indicated.

Deposition of the lower Paleocene, thin-bedded, lowermost Sobral marks a return to a shallow shelf setting below wave base. Periods of slow deposition in the overlying Sobral strata are suggested by abundant glauconite. At the same time, there is a progressive increase in grain size, and in the upper third of the formation, cross-bedding is preserved, showing that the region was again above wave base. Overall, the Sobral represents a regressive phase that Macellari (this volume) interpreted as a sequence of environments from delta front through coastal barrier sands to delta top.

The regressive phase and uplift must have persisted long enough for the cutting of a broad valley or canyon, in which, following subsidence, the Cross Valley Formation was deposited. The contacts between the Cross Valley and Sobral strata are steep and irregular. The crude stratification and poor sorting of the coarse sands of the Cross Valley suggest rapid deposition, and the rare invertebrates indicate the setting was marine, and probably therefore a submarine canyon.

Another regressive phase is represented in La Meseta sediments. The sequence has been interpreted as delta-front and tide-dominated delta-top depositional environments (Elliot and Trautman, 1982). Large-scale lenticularity has been documented by Sadler (this volume).

GEOLOGIC HISTORY

Storey and Garrett (1985) argued that subduction and related processes were virtually continuous from late Paleozoic time onward. Certainly, pre–Middle to Late Jurassic fore-arc deposits are widespread from Alexander Island (LeMay Group) to the northern Antarctic Peninsula (Trinity Peninsula Group) and the South Orkney Islands (Greywacke-Shale Formation), and locally in Alexander Island are as young as Early Jurassic (Thomson and Tranter, 1986). Nevertheless, the tectonic events that culminated in the deformation and metamorphism of those rocks that now form part of the pre–Middle to Late Jurassic metasedimentary, metamorphic, and igneous basement of the Antarctic Peninsula (Dalziel, 1983, 1984) mark a distinct point in the geological history of the region. Sometime thereafter, deposition began along the whole of the eastern flank of the peninsula, and the James Ross Basin was initiated. Throughout the Mesozoic and early Cenozoic history of the James Ross Basin, plate subduction occurred along the western margin of the Antarctic Peninsula. The convergence rate was probably controlled by the spreading rate and migration of the Pacific-Phoenix Ridge, but at least part of the Phoenix Plate subducted beneath the northern peninsula was probably generated at the Farallon-Phoenix spreading center.

In Late Jurassic time a volcanic arc of low relief was constructed on the basement terrain, although subduction of the Phoenix Plate must have been occurring for some millions of years prior to the onset of recorded igneous activity. Interbedded mudstones and tuffs, the Nordenskjöld Formation, were deposited in an anoxic environment in the back-arc terrain, and somewhat similar beds form part of the sequence in the fore-arc region (Farquharson, 1983b).

The magmatic arc was uplifted during the Early Cretacous (Fig. 6). Only rare evidence exists for accompanying volcanism in the South Shetland Islands (Smellie et al., 1980, 1984). On the east flank of the arc, none is known to be older than the 130-m.y.-old rocks at Longing Gap (Pankhurst, 1982). Elsewhere on the northern peninsula, the volcanic rocks overlie, or are interbedded

with alluvial fan sequences of the Botany Bay Group. Although the age and relationships of the Botany Bay Group to beds in the James Ross Basin are unclear, the conglomerates may be in part contemporaneous with the Gustav Group. The Botany Bay Group clearly indicates a major period of uplift, fault block tectonism, and alluvial fan deposition, prior to local onset of volcanism. Fluvial systems originating in these fault block valleys probably drained out into the developing James Ross Basin and contributed sediment to the basin fill.

Uplift and erosion of the arc was accompanied by subsidence of the adjacent back-arc basin. The margin of the basin was fault-controlled and led to the proximal deposition of coarse clastics on submarine fan complexes, the Gustav Group. The submarine fans extended to the southeast, and doubtless, deposition of finer grained sediment occurred farther out and now forms the lower part of the fill in the central and southeastern parts of the basin. It is likely that subsidence was initiated before the Barremian and that a longer record is preserved away from the basin margin.

Volcanic activity expressed in the Gustav Group was accompanied by a major phase of pluton emplacement (Pankhurst, 1982). This is coincident with an increase in Pacific-Phoenix spreading rate at 131 Ma, anomaly M 11 (Larson, 1977), and with the worldwide fast spreading during the Cretaceous quiet interval from about 120 to 85 Ma (Larson and Pitman, 1972). Fast spreading on the Pacific-Phoenix Ridge continued through the end of the Cretaceous and into the Tertiary. Although plutonism decreased in Late Cretaceous time, volcanic activity is well represented in the fore-arc terrain in the South Shetlands and in the sediments of the upper part of the Marambio Group, reaching a peak in the latest Maastrichtian and Paleocene, with deposition of airfall debris in the Seymour region.

Rates of uplift in the arc region diminished in the late Cretaceous, and most of the succession deposited in the proximal part of the basin—the Hidden Lake and Santa Marta Formations and the unnamed beds of the Marambio Group—lacks coarse conglomerate and is dominated by sand. In the more distal parts of the basin, in the region of Snow Hill and Seymour islands, the sediments are fine sands and sandy siltstones deposited on a broad shelf, largely below wave base. The Danian was a time of regression or uplift, bringing the Seymour Island region close to sea level. Transgression or subsidence followed, and fine clastic sediments of the Sobral Formation accumulated above the basal lag just formed. The depositional environment differed little from that of the López de Bertodano, but with time, it changed to a wave-dominated environment of a shallow shelf, once again reflecting regression.

In the late Paleocene, the Seymour Island region was tilted, and the Maramio Group eroded to yield a canyon in which volcaniclastic sands and conglomerates of the Cross Valley Formation were dumped. This was a time of intense silicic volcanism, although in contrast, in the fore-arc region of the South Shetland Islands the activity was basaltic. Further uplift, tilting, and erosion produced a surface on which deposition was renewed

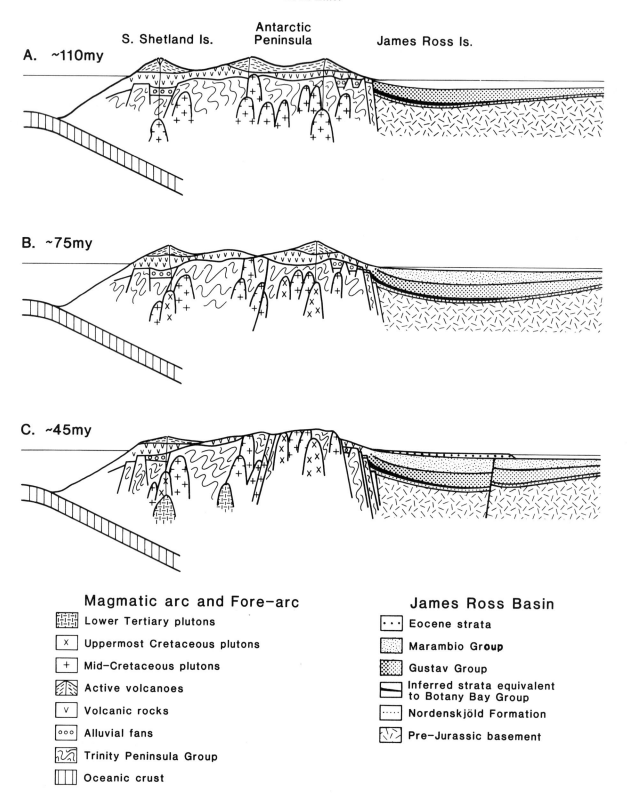

Figure 6. Schematic cross sections of the northern Antarctic Peninsula. A, Mid-Cretaceous deposition of Gustav Group beds. B, Late Cretaceous deposition of Marambio Group beds. C, Late Eocene deposition of Seymour Island Group beds. No attempt has been made to illustrate the compressional tectonism that is evident at Pedersen Nunatak and Cape Sobral (see text).

during the Eocene, the sediments lapping onto now-exposed Paleocene and uppermost Cretaceous strata.

The Gustav Group and equivalent beds at Dundee Island and Cape Sobral, and the strata at Pedersen Nunatak, all exhibit structural deformation associated with evolution of the arc and basin. The Gustav Group beds adjacent to the Prince Gustav Channel are tilted steeply, but dips diminish rapidly toward the southeast. The Dundee Island beds also dip toward the southeast. The simplest explanation is that the deformation is the result of differential movements along the basin margin. Tilting may be partly syn-depositional, but later movements in post-Gustav and Marambio Group time cannot be excluded. Late Cenozoic uplift has brought these beds to their present level of exposure. The beds at Pedersen Nunatak are displaced by a high-angle reverse fault (Elliot, 1967; Del Valle and Nuñez, 1986), which is post-Hauterivian (<124 Ma) or post-Maastrichtian (66 Ma), depending on the age assigned to the rocks. The conglomeratic beds at Cape Sobral are cut by an arc-directed thrust, and the Nordenskjöld (Ameghino) Formation beds that crop out immediately to the north are strongly folded. The deformation can be dated only as younger than late Hauterivian–Barremian (<119 Ma). The deformation observed in the volcanic strata in the region of Muskeg Gap, an open anticline and steeply dipping sequences, may be related to the same episode of tectonism. Thrusting in the Trinity Peninsula Group at Mt. Wild (Aitkenhead, 1975) has been correlated with the structures cutting the Cretaceous beds (Del Valle and Nuñez, 1986).

The relations of these structures to the tectonic evolution of the arc and adjacent basin are unclear. Arc-directed structures are not common in back-arc or foreland fold and thrust-belt deformation. These localities, in fact, contain the only known evidence for late Mesozoic or early Cenozoic compressional tectonism, whether arc-directed or basin-directed, in the back-arc region for the whole of the peninsula north of 68°S. Structural thickening of the crust is commonly associated with compressional tectonism and is likely to be accompanied or followed by increasing rates of erosion and coarsening of detritus delivered to adjacent basins. On these grounds the deformation was more likely to have been mid-Cretaceous or late Paleocene. In the latter case it would have been approximately contemporaneous with the tilting of the Marambio Group strata on Snow Hill and Seymour Islands. The role of compressional tectonism in the evolution of the northern peninsula remains uncertain. On the other hand, subsidence of the basin and uplift of the adjacent arc terrain can be inferred from stratigraphic and sedimentologic data for much of the Cretaceous, and intra-basin uplift and inferred faulting are documented for the early Tertiary.

Deltaic sediments, built up through Eocene time on the tilted and eroded upper Cretaceous and Paleocene strata, mark another regressive phase in the history of the basin. These deltaic sediments conspicuously lack a detrital or airfall volcanic component. However, uplift and erosion of the arc terrain must have proceeded, because at this time, granite clasts first appear; they provide convincing evidence of the unroofing of the plutonic complexes of the magmatic arc. It is at this stage also that subduction rates diminished, although basaltic volcanism continued to dominate the western flank of the arc in the South Shetland region. The subsequent history of the basin doubtless included further subsidence and deposition out on the shelf.

In late Cenozoic time the James Ross Island region was the site of alkaline basaltic volcanism and uplift. Although both of these have been related to tectonic processes associated with the virtual cessation of spreading on the Antarctic-Aluk Ridge and the opening of the Bransfield Strait marginal basin, the alkaline volcanism predates the change in spreading rate by at least 2 m.y. Nevertheless, an obvious spatial relation exists, and the exposures of the James Ross Basin strata owe their existence to those tectonic and magmatic events.

CONCLUSIONS

The James Ross Isla d region evolved as an ensialic back-arc basin during late Mesozoic and Cenozoic time. Subduction of Phoenix and Aluk Plate crust along the western margin of the peninsula was directly related to development of a magmatic arc and cordillera that formed the source for the sedimentary and volcanic fill of the James Ross Basin. Increasing rates of convergence in Early Cretaceous time were reflected in major uplift of the magmatic arc and subsidence of the back-arc region with accompanying deposition of coarse clastic sequences. The place of compressional tectonism in the evolution of the back-arc region is unclear. The James Ross Basin succession records peaks in volcanic activity in the mid-Cretaceous, and in latest Cretaceous to Paleocene time. The plutonic complexes in the arc region were unroofed by the late Eocene. Shorelines migrated basinward and lay in the general vicinity of Seymour Island in the Paleogene. Further shifts in shorelines doubtless occurred in post-Eocene time. The present setting of the east coast of the Antarctic Peninsula represents a major transgression, except in the immediate region of James Ross Island, which has been affected by late Cenozoic volcanism and tectonism.

ACKNOWLEDGMENTS

Field work conducted in the general region of the northern Antarctic Peninsula has been supported by grants from the Division of Polar Programs, National Science Foundation. Preparation of this review has been supported by NSF Grant DPP-8213985. Contribution No. 589 of the Byrd Polar Research Center (formerly, the Institute of Polar Studies).

REFERENCES

ADIE, R. J. 1964. Geological history, p. 118–162. *In* R. Priestley, R. J. Adie, and G. DeQ. Robin (eds.), Antarctic Research. Butterworths, London.

——. 1972. Recent advances in the geology of the Antarctic Peninsula, p. 121–124. *In* R. J. Adie (ed.), Antarctic Geology and Geophysics. Universitetsforlaget, Oslo.

AITKENHEAD, N. 1976. The geology of the Duse Bay–Larsen Inlet area, north-east Graham Land (with particular reference to the Trinity Peninsula Series). British Antarctic Survey, Scientific Reports, 51:1–62.

ANDERSSON, J. G. 1906. On the geology of Graham Land. Geological Institute, University of Uppsala, Bulletin, 7:19–71.

ASKIN, R. A. 1983. Tithonian (uppermost Jurassic)–Barremian (Lower Cretaceous) spores, pollen, and microplankton from the South Shetland Islands, Antarctica, p. 295–297. *In* R. L. Oliver, P. R. James, and J. B. Jago (eds.), Antarctic Earth Science. Australian Academy of Sciences, Canberra.

BARKER, P. F. 1982. The Cenozoic subduction history of the Pacific margin of the Antarctic Peninsula: ridge crest–trench interactions. Journal of the Geological Society, London, 139:787–802.

BIBBY, J. S. 1966. The stratigraphy of part of north-east Graham Land and the James Ross Island Group. British Antarctic Survey, Scientific Reports, 53:1–37.

BRITISH ANTARCTIC SURVEY. 1985. Tectonic map of the Scotia Arc. Sheet BAS (Misc.) 3, scale 1:3,000,000.

CANDE, S. C., E. M. HERRON, AND B. R. HALL. 1982. The early Cenozoic tectonic history of the southeast Pacific. Earth and Planetary Science Letters, 57:63–74.

CRAME, J. A. 1979. The occurrence of the bivalve *Inoceramus concentricus* on Dundee Island, Joinville Island Group. Bulletin, British Antarctic Survey, 49:283–286.

——. 1981. Upper Cretaceous inoceramids (Bivalvia) from the James Ross Island Group and their stratigraphic significance. Bulletin, British Antarctic Survey, 53:29–56.

——. 1983. Cretaceous inoceramid bivalves from Antarctica, p. 298–302. *In* R. L. Oliver, P. R. James, and J. B. Jago (eds.), Antarctic Earth Science. Australian Academy of Science, Canberra.

DALZIEL, I.W.D. 1983. The evolution of the Scotia Arc: a review, p. 283–288. *In* R. L. Oliver, P. R. James, and J. B. Jago (eds.), Antarctic Earth Science, Australian Academy of Science, Canberra.

——. 1984. Tectonic evolution of a fore-arc terrane, Southern Scotia Ridge, Antarctica. Geological Society of America Special Paper 200, 32 p.

——, AND D. H. ELLIOT. 1971. Evolution of the Scotia arc. Nature, 233:246–252.

——. 1973. The Scotia Arc and Antarctic Margin, p. 171–246. *In* F. G. Stehli and A.E.M. Nairn, (eds.), The Ocean Basins and Margins: 1. The South Atlantic. Plenum, New York.

——, M. J. DEWIT, AND K. R. PALMER. 1974. Fossil marginal basin in the Southern Andes. Nature, 250:291–294.

DEL VALLE, R. A., AND N. H. FOURCADE. 1986. La cuenca sedimentaria pos-Triasica del extremo nororiental de la Peninsula Antárctica. Contribuciones Instituto Antártico Argentino, 323:1–22.

——, AND F. MEDINA. 1985. Geologia de Cabo Marsh, Isla Robertson, Antartida. Contribuciones Instituto Antártico Argentino, 309:1–27.

——, AND H. J. NUÑEZ. 1986. Estructuras tectonicas en el borde oriental de la Peninsula Antartica y su posible relacion con el cinturon de compresion del arco volcanico mesozoico. Geoacta, 13:313–324.

——, N. H. FOURCADE, AND F. A. MEDINA. 1983. Geología del extremo norte del borde oriental de la Península Antártica e islas adyacentes entre los 63° 25′ y 63° 15′ de latitud sur. Contribuciones Instituto Antártico Argentino, 276:1–19.

ELLIOT, D. H. 1967. Geology of the Nordenskjöld Coast and a comparison with north-west Trinity Peninsula. Bulletin, British Antarctic Survey, 10:1–43.

——. 1983. The mid-Mesozoic to mid-Cenozoic active plate margin of the Antarctic Peninsula, p. 347–351. *In* R. L. Oliver, P. R. James, and J. B. Jago

(eds.), Antarctic Earth Science. Australian Academy of Science, Canberra.

——, AND D. E. RIESKE. 1987. Field investigations of the Tertiary strata on Seymour and Cockburn Islands. Antarctic Journal of the United States, 22 (in press).

——, AND T. A. TRAUTMAN. 1982. Lower Tertiary strata on Seymour Island, Antarctic Peninsula, p. 287–297. *In* C. Craddock (ed.), Antarctic Geoscience. University of Wisconsin Press, Madison.

——, AND N. A. WELLS. 1982. Mesozoic alluvial fans of the South Orkney Islands, p. 235–244. *In* C. Craddock (ed.), Antarctic Geoscience. University of Wisconsin Press, Madison.

——, R. WATTS, R. B. ALLEY, AND T. M. GRACANIN. 1978. Geologic studies in the northern Antarctic Peninsula: R/V *Hero* Cruise 78-1B. Antarctic Journal of the United States, 13(4):12–13.

ELLIOTT, M. H. 1974. Stratigraphy and sedimentary petrology of the Ablation Point area, Alexander Island. Bulletin, British Antarctic Survey, 39:87–113.

FARQUHARSON, G. W. 1982a. Late Mesozoic sedimentation in the northern Antarctic Peninsula and its relationship to the southern Andes. Journal of the Geological Society, London, 139:721–727.

——. 1982b. Lacustrine deltas in a Mesozoic alluvial sequence from Camp Hill, Antarctica. Sedimentology, 29:717–725.

——. 1983a. Evolution of Late Mesozoic sedimentary basins in the northern Antarctic Peninsula, p. 323–327. *In* R. L. Oliver, P. R. James, and J. B. Jago (eds.), Antarctic Earth Science. Australian Academy of Science, Canberra.

——. 1983b. The Nordenskjöld Formation of the northern Antarctic Peninsula: an Upper Jurassic radiolarian mudstone and tuff sequence. Bulletin, British Antarctic Survey, 60:1–22.

——. 1984. Late Mesozoic, non-marine conglomerate sequences of northern Antarctic Peninsula (the Botany Bay Group). Bulletin, British Antarctic Survey, 65:1–32.

——, R. D. HAMER, AND J. R. INESON. 1984. Proximal volcaniclastic sedimentation in a Cretaceous back-arc basin, northern Antarctic Peninsula, p. 219–229. *In* B. P. Kokelaar and M. F. Howell (eds.), Marginal Basin Geology. Geological Society of London, Special Publication 16.

HALL, S. A. 1977. Cretaceous and Tertiary dinoflagellates from Seymour Island, Antarctica. Nature, 267:239–241.

HARRINGTON, P. K., P. F. BARKER, AND D. H. GRIFFITHS. 1972. Crustal structure of the South Orkney Islands area from seismic refraction and magnetic measurements, p. 27–32. *In* R. J. Adie (ed.), Antarctic Geology and Geophysics. Universitetsforlaget, Oslo.

HORNE, R. R., AND M.R.A. THOMSON. 1972. Airborne and detrital volcanic material in the Lower Cretaceous sediments of south-eastern Alexander Island. Bulletin, British Antarctic Survey, 29:103–111.

HOWARTH, M. K. 1958. Upper Jurassic and Cretaceous ammonite faunas of Alexander Island and Graham Land. Falkland Islands Dependencies Survey, Scientific Reports, 21:1–16.

——. 1966. Ammonites from the Upper Cretaceous of the James Ross Island Group. Bulletin, British Antarctic Survey, 10:55–69.

INESON, J. R., J. R. CRAME, AND M.R.A. THOMSON. 1986. Lithostratigraphy of the Cretaceous strata of west James Ross Island. Cretaceous Research, 7:141–159.

JEFFERSON, T. H. 1982. The Early Cretaceous fossil forests of Alexander Island, Antarctica. Palaeontology, 25:681–708.

KRISTOFFERSEN, Y., AND K. HAUGLAND. 1986. Geophysical evidence for the East Antarctic plate boundary in the Weddell Sea. Nature, 322:538–541.

LABRECQUE, J. L. 1986. South Atlantic Ocean and adjacent Antarctic continental margin. Ocean Margin Drilling Program, Regional Data Synthesis Series, Atlas 13. Marine Science International, Woods Hole.

——, AND P. BARKER. 1981. The age of the Weddell Basin. Nature, 290:489–492.

LARSON, R. L. 1977. Late Jurassic and early Cretaceous evolution of the western central Pacific Ocean. Journal of Geomagnetism and Geoelectricity, 28:219–236.

——, AND W. C. PITMAN. 1972. World-wide correlation of Mesozoic magnetic anomalies, and its implications. Bulletin, Geological Society of America, 83:3645–3662.

——, S. C. CANDE, J. H. BODINE, AND A. B. WATTS. 1979. The origin of the Eltanin Fracture Zone. EOS, 60:957.

LAUDON, T. S., M.R.A. THOMSON, P. L. WILLIAMS, K. L. MILLIKEN, P. D. ROWLEY, AND J. M. BOYLES. 1983. The Jurassic Latady Formation, southern Antarctic Peninsula, p. 308–314. *In* R. L. Oliver, P. R. James, and J. B. Jago (eds.), Antarctic Earth Science. Australian Academy of Science, Canberra.

MACELLARI, C. E. 1984. Revision of the serpulids of the genus *Rotularia* (Annelida) at Seymour Island (Antarctic Peninsula) and their value in stratigraphy. Journal of Paleontology, 58:1098–1116.

——. 1986. Late Campanian-Maastrichtian ammonites from Seymour Island, Antarctic Peninsula. Journal of Paleontology, Supplement, 60(2):1–55. Paleontological Society Memoir 18.

MEDINA, F. A., AND A. RAMOS. 1980. Geología del refugio Ameghino y alrededores, Antártida. Contribuciones Instituto Antártico Argentino, 229:1–22.

——, E. B. OLIVERO, AND C. A. RINALDI. 1981. Estratigrafía del Jurásico y Cretácico del Arco de Scotia y Peninsula Antártica, p. 157–179. *In* Cuencas sedimentarias del Jurásico y Cretácico de America del Sur, Vol. 1. Comite Sudamericano del Jurásico y Cretácico.

NELSON, P.H.H. 1975. The James Ross Island Volcanic Group of northeast Graham Land. British Antarctic Survey, Scientific Reports, 54:1–62.

NORDENSKJÖLD, O. 1905, Petrographische untersuchungen aus dem Westantarktischen Gebiete. Geological Institute, Uppsala University, Bulletin, 6(12):234–246.

OLIVERO, E., R. A. SCASSO, AND C. A. RINALDI. 1986. Revision of the Marambio Group, James Ross Island, Antarctica. Contribuciones Instituto Antártico Argentino, 351:1–29.

PANKHURST, R. J. 1982. Rb-Sr geochronology of Graham Land, Antarctica. Journal of the Geological Society, London, 139:701–712.

QUILTY, P. G. 1977. Late Jurassic bivalves from Ellsworth Land, Antarctica: their systematics and paleogeographic implications. New Zealand Journal of Geology and Geophysics, 20:1033–1080.

——. 1983. Bajocian bivalves from Ellsworth Land, Antarctica. New Zealand Journal of Geology and Geophysics, 26:395–418.

RAMOS, A. M., F. M. MEDINA, J.C.A. MARTINEZ MACCHIAVELLO, AND R. A. DEL VALLE. 1978. Informe preliminar sobre las sedimentitas del Cretácico medio de Cabo Welchness, Isla Dundee, Antártida. Contribuciones Instituto Antártico Argentino, 249:3–10.

RINALDI, C. A., A. MASSABIE, J. MORELLI, L. H. ROSENMAN, AND R. A. DEL VALLE. 1978. Geología de la isla Vicecomodoro Marambio, Antártida. Contribuciones Instituto Antártico Argentino, 217:1–37.

ROWLEY, P. D., AND P. L. WILLIAMS. 1982. Geology of the northern Lassiter Coast and southern Black Coast, Antarctic Peninsula, p. 339–348. *In* C. Craddock (ed.), Antarctic Geoscience. University of Wisconsin Press, Madison.

SAUNDERS, A. D., S. D. WEAVER, AND J. TARNEY. 1982. The pattern of Antarctic Peninsula plutonism, p. 305–314. *In* C. Craddock (ed.), Antarctic Geoscience. University of Wisconsin Press, Madison.

SCASSO, R. A., R. A. DEL VALLE, AND G. AMBROSINI. 1986. Caracterización litologica y paleoambiental de las sedimentitas cretacicas del nunatak Troilo, peninsula Tabarin, Antártida. Contribuciones Instituto Antártico Argentino, 328:1–15.

SINGLETON, D. C. 1980. The geology of the central Black Coast, Palmer Land. British Antarctic Survey, Scientific Reports, 102:1–50.

SMELLIE, J. L. 1979. The geology of Low Island, South Shetland Islands, and Austin Rocks. Bulletin, British Antarctic Survey, 49:239–257.

——, R.E.S. DAVIES, AND M.R.A. THOMSON. 1980. The geology of a Mesozoic intra-arc sequence on Byers Peninsula, Livingston Island, South Shetland Islands. Bulletin, British Antarctic Survey, 50:55–76.

——, R. J. PANKHURST, M.R.A. THOMSON, AND R.E.S. DAVIES. 1984. The geology of the South Shetland Islands: VI. Stratigraphy, geochemistry, and evolution. British Antarctic Survey, Scientific Reports, 87:1–85.

SPATH, L. F. 1953. The Upper Cretaceous cephalopod fauna from Graham Land. Falkland Islands Dependencies Survey, Scientific Reports, 3:1–66.

STOCK, J., AND P. MOLNAR. 1987. Revised history of early Tertiary plate motion in the south-west Pacific. Nature, 325:495–499.

STOREY, B. C., AND S. W. GARRETT. 1985. Crustal growth of the Antarctic Peninsula by accretion, magmatism, and extension. Geological Magazine, 122:5–14.

——, B. F. MAIR, AND C. M. BELL. 1977. The occurrence of Mesozoic oceanic floor and ancient continental crust on South Georgia. Geological Magazine 114:203–208.

SUAREZ, M. 1976. Plate tectonic model for southern Antarctic Peninsula and its relation to southern Andes. Geology, 4:211–214.

THOMSON, M.R.A. 1982a. Mesozoic paleogeography of West Antarctica, p. 331–338. *In* C. Craddock (ed.), Antarctic Geoscience, University of Wisconsin Press, Madison.

——. 1982b. Late Jurassic fossils from Low Island, South Shetland Islands. Bulletin, British Antarctic Survey, 56:25–36.

——. 1983. Late Jurassic ammonites from the Orville Coast, Antarctica, p. 315–319. *In* R. L. Oliver, P. R. James, and J. B. Jago (eds.), Antarctic Earth Science. Australian Academy of Science, Canberra.

——, AND T. H. TRANTER. 1986. Early Jurassic fossils from central Alexander Island and their geological setting. Bulletin, British Antarctic Survey, 70:23–39.

WEAVER, S. D., A. D. SAUNDERS, AND J. TARNEY. 1982. Mesozoic-Cenozoic volcanism on the South Shetland Islands and the Antarctic Peninsula: geochemical nature and plate tectonic significance, p. 263–273. *In* C. Craddock (ed.), Antarctic Geoscience. University of Wisconsin Press, Madison.

ZINSMEISTER, W. J. 1984. Late Eocene bivalves (Mollusca) from the La Meseta Formation collected during the 1974–75 Joint Argentine-American Expedition to Seymour Island, Antarctic Peninsula. Journal of Paleontology, 58:1497–1527.

——, AND T. H. DEVRIES. 1982. Observations on the stratigraphy of the Lower Tertiary Seymour Island Group, Seymour Island, Antarctic Peninsula. Antarctic Journal of the United States, 17(5):71–72.

MANUSCRIPT ACCEPTED BY THE SOCIETY SEPTEMBER 1, 1987

Index

[Italic page numbers indicate major references]

Typeset by WESType Publishers Services, Inc., Boulder, Colorado
Printed in U.S.A. by Malloy Lithographing, Inc., Ann Arbor, Michigan